"GREENBOOK"

Standard Specifications for Public Works Construction

2012 Edition

Written And Promulgated By

Public Works Standards, Inc.

Published By

BNi Building News

Division of BNi Publications, Inc.

990 Park Center Drive, Suite E Vista, CA 92081

1-760-734-1113

BNi Building News

BNi PUBLICATIONS, INC.
990 Park Center Drive, Suite E
Vista, CA 92081

VISTA
990 Park Center Dr. Ste E.
Vista, CA 92081

ANAHEIM
1612 S. Clementine St.
Anaheim, CA 92802

NEW ENGLAND
PO Box 14527
East Providence, RI 02914

ISBN 978-1-55701-741-3

..... *By Way Of Explanation*

The Standard Specifications for Public Works Construction, popularly known as the "Greenbook", was originally published in 1967. The 2012 Edition is the 16th edition of this book, which is updated and republished every three years.

The original edition of the Greenbook was the product of almost four years of intensive work by a 400-person task force which then comprised the Joint Cooperative Committee. The City and County of Los Angeles were major contributors to this effort and these two agencies were the first to adopt the Greenbook. To date, more than 200 other cities, counties, and agencies in the area have also adopted it as their standard for public works construction specifications. Interest in the Greenbook has been worldwide, with copies being acquired and studied by public works officials throughout the United States, Canada, and many countries overseas.

Immediately after publication of the Greenbook, a permanent Joint Committee, comprised of representatives from cities, counties, utility companies, and contractors throughout Southern California, was established to carry on the work. The function of this committee is to study and update the provisions of these specifications to reflect the constantly changing technology and advanced thinking of the construction industry.

The first 10 editions of the Greenbook were the product of the Joint Cooperative Committee of the Southern California Chapter of the American Public Works Association, and the Southern California Districts of the Associated General Contractors of California.

In December 1995, the Joint Committee formally ceased operations. In January 1996, Public Works Standards, Inc., a mutual benefit corporation, began producing the Greenbook. The Corporation's Board of Directors is comprised of nine members – five representing the American Public Works Association, and the other four from the Associated General Contractors of California, the Engineering Contractors Association, the Southern California Contractors Association, and BNi Publications, Inc. The Board appoints a 25-member Greenbook Committee which carries on the tradition and function of the original Joint Committee. This committee convenes regularly each month to consider new changes.

In each of the two years between publication of a new Greenbook edition, the changes which have been researched and approved by the committee during the preceding year are published in pamphlet form as amendments to the current edition. This program maintains a "living" document in public works specifications. Stripes in the margin of each new edition point out significant changes in the text adopted since the preceding edition.

The original edition of the Greenbook consisted of three parts: General Provisions, Construction Materials, and Construction Methods.

In 1970, 1991, and 1994, Parts 4, 5 and 6 were added to provide specifications for alternate aggregate materials, pipeline system rehabilitation, and modified asphalt products, respectively. In 2012, Part 6 was incorporated into Parts 2 and 3.

The Greenbook is designed to aid in furthering uniformity of plans and specifications accepted and used by those involved in public works construction and to take such other steps as are designed to promote more competitive bidding by private contractors. The Greenbook provides specifications that have general applicability to public works projects. The Greenbook does NOT test or approve products. It is the function of public agencies and private project owners that utilize the Greenbook to determine whether products proposed by a contractor satisfy the Greenbook specifications or the Special Provisions.

Interested parties who wish to suggest changes or amendments to this book may communicate with Public Works Standards, Inc., c/o Associated General Contractors of California, 1906 W. Garvey Avenue South, Suite 100, West Covina, CA 91790. The change form and further information about the Greenbook Committee are available at www.greenbookspecs.org

In 1984, a companion to the Greenbook entitled the Standard Plans for Public Works Construction was adopted by the joint committee. Both books are available from BNi Publications, Inc., or other technical bookstores.

TABLE OF CONTENTS

PART 1
GENERAL PROVISIONS

SECTION 1 - TERMS, DEFINITIONS, ABBREVIATIONS, UNITS OF MEASURE, AND SYMBOLS

1-1 GENERAL – Unless otherwise stated, the words *directed, required, permitted, ordered, instructed, designated, considered necessary, prescribed, approved, acceptable, satisfactory*, or words of like meaning, refer to actions, expressions, and prerogatives of the Engineer.

1-2 TERMS AND DEFINITIONS.

Addendum – Written or graphic instrument issued prior to the opening of Bids which clarifies, corrects, or changes the Contract Documents. The term "Addendum" shall include bulletins and all other types of written notices issued to potential bidders prior to opening of the Bids.

Agency – The legal entity for which the Work is being performed.

Agreed Price – The cost for new or unforeseen work, or for adjustments in the Contract Unit Price for changes in the character of the work as specified in 3-2.4, established by mutual agreement between the Contractor and the Agency.

Agreement – See Contract.

Assessment Act Contract – A Contract financed by special assessments authorized under a State Act or procedural ordinance of a City or County.

Base – A layer of specified material of planned thickness placed immediately below the pavement or surfacing.

Bid – The offer or proposal of the Bidder submitted on the prescribed form setting forth the prices for the Work.

Bidder – Any individual, firm, partnership, corporation, or combination thereof, submitting a Bid for the Work, acting directly or through a duly authorized representative.

Board – The officer or body constituting the awarding authority of the Agency.

Bond – Bid, performance, or payment bond or other instrument of security.

Caltrans – The State of California Department of Transportation.

Cash Contract – A Contract financed by means other than special assessments.

Certificate of Compliance – A written document signed and submitted by a supplier or manufacturer that certifies that the material or assembled material supplied to the Work site conforms to the requirements of the Contract Documents.

Change Order – A written order to the Contractor signed by the Agency directing an addition, deletion, or revision in the Work, or an adjustment in the Contract Price or the Contract time issued after the effective date of the Contract. A Change Order may or may not also be signed by the Contractor.

Code – The terms *Government Code*, *Labor Code*, etc., refer to codes of the State of California.

Contract – The written agreement between the Agency and the Contractor covering the Work.

Contract Documents – The Contract, Addenda, notice inviting bids, instructions to bidders; Bid (including documentation accompanying the Bid and any post-bid documentation submitted prior to the Notice of Award) when attached as an exhibit to the Contract, the Bonds, permits from jurisdictional regulatory agencies, Special Provisions, Plans, Standard Plans, Standard Specifications, Reference Specifications, Change Orders, and Supplemental Agreements.

Contractor – The individual, partnership, corporation, joint venture, or other legal entity having a Contract with the Agency to perform the Work. In the case of work being done under a permit issued by the Agency, the permittee shall be construed to be the Contractor. The term "prime contractor" shall mean the Contractor.

Contract Price – The total amount of money for which the Contract is awarded.

Contract Unit Price – The amount stated in the Bid for a single unit of an item of work.

County Sealer – The Sealer of Weights and Measures of the county in which the Contract is awarded.

Days – Days shall mean consecutive calendar days unless otherwise specified in the Special Provisions.

Disputed Work – Work in which the Agency and the Contractor are in disagreement.

Electrolier – Street light assembly complete, including foundation, standard, mast arm, luminaire, etc.

Extra Work – New or unforeseen work not covered by a Contract Unit Price or Stipulated Unit Price.

Engineer – The Chief Engineer of the Agency, Director of Public Works, or other person designated by the Board, acting either directly or through authorized agents.

House Connection Sewer—A sewer, within a public street or right-of-way, proposed to connect any parcel, lot, or part of a lot with a mainline sewer.

Luminaire – The lamp housing including the optical and socket assemblies (and ballast if so specified).

Mast Arm – The structural member, or bracket, which, when mounted on a Standard, supports the luminaire.

Modification – Includes Change Orders and Supplemental Agreements. A Modification may only be issued after the effective date of the Contract.

Notice of Award – The written notice by the Agency to the successful Bidder stating that upon its compliance with the required conditions, the Agency will execute the Contract.

Notice to Proceed – A written notice given by the Agency to the Contractor fixing the date on which the Contract time will start.

Person – Any individual, firm, association, partnership, corporation, trust, joint venture, or other legal entity.

Plans – The drawings, profiles, cross sections, Standard Plans, working drawings, and shop drawings, or reproductions thereof, approved by the Engineer, which show the location, character, dimensions, or details of the Work.

Private Contract – Work subject to Agency inspection, control, and approval, involving private funds, not administered by the Agency.

Proposal – See Bid.

Reference Specifications – Those bulletins, standards, rules, methods of analysis or testing, codes, and specifications of other agencies, engineering societies, or industrial associations referred to in the Contract Documents. These shall refer to the latest edition, including amendments in effect and published at the time of advertising the Contract or issuing the permit, unless specifically referred to by edition, volume, or date.

Roadway – The portion of a street reserved for vehicular use.

Service Connection – All or any portion of the conduit cable or duct including meter, between a utility distribution line and an individual consumer.

Service Lateral Connection – The interface of the House Connection Sewer with the host pipe.

Sewer – Any conduit intended for the reception and transfer of sewage and fluid industrial waste.

Shop Drawings – Drawings showing details of manufactured or assembled products proposed to be incorporated into the Work.

Special Provisions – Additions and revisions to the Standard Specifications setting forth conditions and requirements peculiar to the Work.

Specifications – Standard Specifications, Reference Specifications, Special Provisions, and specifications in Change Orders or Supplemental Agreements between the Contractor and the Board.

Standard – The shaft or pole used to support street lighting luminaires, traffic signal heads, mast arms, etc.

Standard Plans – Details of standard structures, devices, or instructions referred to on the Plans or in the Specifications by title or number.

Standard Specifications – The Standard Specifications for Public Works Construction (SSPWC), the "Greenbook".

State – State of California.

Stipulated Unit Price – Unit prices established by the Agency in the Contract Documents.

Storm Drain – Any conduit and appurtenances intended for the reception and transfer of storm water.

Street – Any road, highway, parkway, freeway, alley, walk, or way.

Subbase – A layer of specified material of planned thickness between the base and the subgrade.

Subcontractor – An individual, firm, or corporation having a direct contract with the Contractor or with any other Subcontractor for the performance of a part of the Work.

Subgrade – For roadways, that portion on which pavement, surfacing, base, subbase, or a layer of other material is placed. For structures, the soil prepared to support a structure.

Supervision – Supervision, where used to indicate supervision by the Engineer, shall mean the performance of obligations, and the exercise of rights, specifically imposed upon and granted to the Agency in becoming a party to the Contract. Except as specifically stated herein, supervision by the Agency shall not mean active and direct superintendence of details of the Work.

Supplemental Agreement – A written amendment of the Contract Documents signed by the Agency and the Contractor.

Surety – Any individual, firm, or corporation, bound with and for the Contractor for the acceptable performance, execution, and completion of the Work, and for the satisfaction of all obligations incurred.

Utility – Tracks, overhead or underground wires, pipeline, conduits, ducts, structures, sewers, or storm drains owned, operated, or maintained in or across a public right of way or private easement.

Work – That which is proposed to be constructed or done under the Contract or permit, including the furnishing of all labor, materials, equipment, and services.

Working Day – Any day within the period between the date of the start of the Contract time as specified in 6-1 and the date of field acceptance of the Work by the Engineer, other than:

a) Saturday,
b) Sunday,
c) any day designated as a holiday by the Agency,
d) any other day designated as a holiday in a Master Labor Agreement entered into by the Contractor or on behalf of the Contractor as an eligible member of a contractor association,
e) any day the Contractor is prevented from working at the beginning of the workday for cause as specified in 6-6.1, or
f) any day the Contractor is prevented from working during the first 5 hours with at least 60 percent of the normal work force for cause as specified in 6-6.1.

Working Drawings – Drawings showing details not shown on the Plans which are required to be designed by the Contractor.

1-3 ABBREVIATIONS.

1-3.1 General. The abbreviations are applicable to these Standard Specifications and the Special Provisions. Additional abbreviations shall be as specified on the Plans or in the Special Provisions.

1-3.2 Common Usage.

Abbreviation	Word or Words
ABS	Acrylonitrile Butadiene Styrene
ADA	Americans with Disabilities Act of 1990 (Public Law 101-336, 104 Sat. 1990, 42 USC 12101-12213 (as amended))
APC	Air Placed Concrete
ARHM	Asphalt Rubber Hot Mix
ARAM	Asphalt Rubber and Aggregate Membrane
BMPs	Best Management Practices
CAB	Crushed Aggregate Base
CAPA	Corrugated Aluminum Pipe Arch
CAP	Corrugated Aluminum Pipe
CBR	California Bearing Ratio
CCFRPM	Centrifugally Cast Fiberglass Reinforced Plastic Mortar
CCR	California Code of Regulations
CCTV	Closed Circuit TV
CHDPE	Corrugated High Density Polyethylene
CIP	Cast Iron Pipe
CIPCP	Cast-In-Place Concrete Pipe
CIPP	Cured-In-Place Pipe
CLSM	Controlled Low Strength Material
CMB	Crushed Miscellaneous Base
CRM	Crumb Rubber Modifier
CRUMAC	Crumb Rubber Modified Asphalt Concrete

CRUMAC-GG .... Crumb Rubber Modified Asphalt Concrete Gap Graded
CQS .... Cationic Quick-Setting
CRS .... Cationic Rapid-Setting
CSP .... Corrugated Steel Pipe
CSEP .... Confined Space Entry Plan
CSPA .... Corrugated Steel Pipe Arch
CSP .... Corrugated Steel Pipe
CSS .... Cationic Slow-Setting
CT .... California Test
DIP .... Ductile Iron Pipe
GG .... Gap-Graded
HC .... House Connection
HDPE .... High Density Polyethylene
HRWRA .... High Range Water Reducing Admixture
LS .... Lump Sum
MC .... Medium Curing
Min .... Minimum
MSDS .... Material Safety Data Sheet
MUTCD .... Manual on Uniform Traffic Control Devices
N/A .... Not Applicable
OD .... Outside Diameter
PAV .... Pressure Aging Vessel
PCC .... Portland Cement Concrete
PE .... Polyethylene
PG .... Performance Graded
PLI .... Pounds Per Linear Inch
PMB .... Processed Miscellaneous Base
PRCB .... Precast Reinforced Concrete Box
PVC .... Polyvinyl Chloride
RC .... Rapid Curing
R .... Resistance Value
RA .... Reclaimed Aggregates
RAC .... Recycled Asphalt Concrete
RAP .... Reclaimed Asphalt Pavement
RCP .... Reinforced Concrete Pipe
RPPCC .... Reclaimed Plastic Portland Cement Concrete
RTFO .... Rolling Thin Film Oven
RW .... Reclaimed Water
S .... Hveem Stability
SAPPA .... Structural Aluminum Plate Pipe Arch
SAPP .... Structural Aluminum Plate Pipe
SC .... Slow Curing
SCMs .... Supplementary Cementitious Materials

SDR........................Standard thermoplastic pipe dimension ratio(ratio of pipe O.D. to minimum wall thickness)
SE ....Sand Equivalent
SG........................Specific Gravity
SI..........................International System of Units (Metric)
| SLC ....Service Lateral Connection
SS.........................Slow-Setting
SSPPA ....Structural Steel Plate Pipe Arch
SSPP ....Structural Steel Plate Pipe
SWPPP ....Storm Water Pollution Prevention Plan
TCP......................Traffic Control Plan
TRMAC..............Tire Rubber Modified Asphalt Concrete
TTC......................Temporary Traffic Control
U.S.C. ....United States Code
U.S. ....United States
VCP ....Vitrified clay pipe
VTCSH................Vehicle Traffic Controls Signal Heads
| WATCH ....Work Area Traffic Control Handbook
WTAT..................Wet Track Abrasion Test
X ....By

1-3.3 Institutions.

Abbreviation	Word or Words
AASHTO	American Association of State Highway and Transportation Officials
AISC	American Institute of Steel Construction
ANSI	American National Standards Institute
AREA	American Railway Engineering Association
ASTM	American Society for Testing and Materials
AWPA	American Wood Preservers Association
AWS	American Welding Society
AWWA	American Water Works Association
EIA	Electronic Industries Alliance
EPA	Environmental Protection Agency
ETL	Electrical Testing Laboratories
FCC	Federal Communications Commission
GRI	Geosynthetic Research Institute
IEEE	Institute of Electrical and Electronics Engineers
IMSA	International Municipal Signal Association
ITE	Institute of Transportation Engineers
NEMA	National Electrical Manufacturers Assn.
RUS	Rural Utilities Service
UL	Underwriters' Laboratories Inc.

1-4 UNITS OF MEASURE.

1-4.1 General. The U.S. Standard Measures, also referred to as the U.S. Customary System, is the principal measurement system in these Specifications and shall be used for construction, unless otherwise specified in the Special Provisions. The International System of Units, also referred to as SI or the metric system, is included in parenthesis. U.S. Standard Measures units may or may not be exactly equivalent to the SI units in parenthesis. If SI is specified for use in the Contract Documents, then all values used for construction shall be the SI units shown in parenthesis. Certain material specifications and test requirements contained herein use SI units specifically and U.S. Standard Measures have not been included in those circumstances.

Reference is also made to ASTM E380 for definitions of various units of the SI system and a more extensive set of conversion factors.

1-4.2 Units of Measure and Their Abbreviations.

U.S. Customary Unit (Abbreviations)	Equal To	SI Unit (Abbreviations)
1 mil (= 0.001 inch)		25.4 micrometer (μm)
1 inch		25.4 millimeter (mm)
1 inch		2.54 centimeter (cm)
1 foot (ft)		0.3048 meter (m)
1 yard (yd)		0.9144 meter (m)
1 mile (mi)		1.6093 kilometer (km)
1 square foot (ft^2)		0.0929 square meter (m^2)
1 square yard (yd^2)		0.8361 square meter (m^2)
1 cubic foot (ft^3)		0.0283 cubic meter (m^3)
1 cubic yard (yd^3)		0.7646 cubic meter (m^3)
1 acre		0.4047 hectare (ha)
1 U.S. gallon (gal)		3.7854 Liter (L)
1 fluid ounce (fl. oz)		29.5735 milliliter (mL)
1 pound mass (lb) (avoirdupois)		0.4536 kilogram (kg)
1 ounce mass (oz)		0.02835 kilogram (kg)
1 Ton (= 2000 lb avoirdupois)		0.9072 Tonne (= 907 kg)
1 Poise		0.1 pascal · second (Pa · s)
1 centistoke (cs)		1 square millimeters per second (mm^2/s)
1 pound force (lbf)		4.4482 Newton (N)
1 pounds per square inch (psi)		6.8948 Kilopascal (kPa)
1 pound force per foot (lbf/ft)		1.4594 Newton per meter (N/m)
1 foot-pound force (ft-lbf)		1.3558 Joules (J)
1 foot-pound force per second (ft-lbf)		1.3558 Watt (W)
1 part per million (ppm)		1 milligram/liter (mg/L)

Temperature Units and Abbreviations

Degree Fahrenheit (°F):
°F = (1.8 x °C) + 32

Degree Celsius (°C):
°C = (°F – 32)/1.8

SI Units (abbreviation) Commonly Used in Both Systems

1 Ampere (A)
1 Volt (V)
1 Candela (cd)
1 Lumen (lm)
1 second (s)

Common Metric Prefixes

kilo (k) .... 10^3
centi (c) .... 10^{-2}
milli (m) .... 10^{-3}
micro (μ) .... 10^{-6}
nano (n) .... 10^{-9}
pico (p) .... 10^{-12}

1-5 SYMBOLS.

% .... Percent
' .... Feet or minutes
" .... Inches or seconds
.... Number
/ .... per or (between words)
° .... Degree
x .... Times

SECTION 2 - SCOPE AND CONTROL OF THE WORK

2-1 AWARD AND EXECUTION OF THE CONTRACT. Award and execution of the Contract shall be as provided for in the Special Provisions, Instructions to Bidders, or Notice Inviting Bids.

2-2 ASSIGNMENT. No Contract or portion thereof may be assigned without consent of the Board, except that the Contractor may assign money due or which will accrue to it under the Contract. If given written notice, such assignment will be recognized by the Board to the extent permitted by law. Any assignment of money shall be subject to all proper withholdings in favor of the Agency and to all deductions provided for in the Contract. All money withheld, whether assigned or not, shall be subject to being used by the Agency for completion of the Work, should the Contractor be in default.

2-3 SUBCONTRACTS.

2-3.1 General. Each Bidder shall comply with Chapter 4 of the Public Contract Code including Sections 4100 through 4113.

The Bidder shall set forth in the Bid, pursuant to 4104:

"a) The name and location of the place of business of each subcontractor who will perform work or labor or render service to the prime contractor in or about the construction of the work or improvement, or a subcontractor licensed by the State of California who, under subcontract to the prime contractor, specially fabricates and installs a portion of the work or improvement according to detailed drawings contained in the plans and specifications, in an amount in excess of one-half of 1 percent of the prime contractor's total bid, or, in the case of bids or offers for the construction of streets or highways, including bridges, in excess of one-half of 1 percent of the prime contractor's total bid or ten thousand dollars ($10,000), whichever is greater."

"b) The portion of the work which will be done by each such subcontractor under this act. The prime contractor shall list only one subcontractor for each such portion as is defined by the prime contractor in his or her bid."

If the Contractor fails to specify a Subcontractor, or specifies more than one Subcontractor for the same portion of the Work to be performed under the Contract (in excess of one-half of 1 percent of the Contractor's total Bid), the Contractor shall be qualified to perform that portion itself, and shall perform that portion itself, except as otherwise provided in the Code.

Pursuant to Section 4107, no Contractor whose Bid is accepted shall substitute any person as a Subcontractor in place of a Subcontractor listed in the Bid, except for the causes and by the procedures established in Section 4107.5. This section provides procedures to correct a clerical error in the listing of a Subcontractor.

Section 4110 provides that a Contractor violating any of the provisions of Chapter 4 violates the Contract and the Board may exercise the option either to cancel the Contract or assess the Contractor a penalty in an amount of not more than 10 percent of the subcontract involved, after a public hearing.

2-3.2 Self Performance. The Contractor shall perform, with its own organization, Contract work amounting to at least 50 percent of the Contract Price except that any designated "Specialty Items" may be performed by subcontract and the amount of any such "Specialty Items" so performed will be deducted from the Contract Price before computing the amount required to be performed by the Contractor with its own organization. "Specialty Items" will be identified by the Agency in the Bid or in the Special Provisions. Where an entire item is subcontracted, the value of work subcontracted will be based on the Contract Unit Price. When a portion of an item is subcontracted, the value of work subcontracted will be based on the estimated percentage of the Contract Unit Price. This will be determined from information submitted by the Contractor, and subject to approval by the Engineer.

2-3.3 Status of Subcontractors. The Contractor shall give personal attention to the fulfillment of the Contract. The Contractor shall keep the Work under its control.

Subcontractors shall be considered employees of the Contractor, and the Contractor shall be responsible for their work.

In addition to the requirements of 2-3.1, before the work of any Subcontractor is started, the Contractor shall submit to the Engineer for approval a written statement listing the name, contractor license number, and business address of each Subcontractor and a description and value of each portion of the Work to be so subcontracted.

2-4 CONTRACT BONDS. Before execution of the Contract, the Bidder shall file surety bonds with the Agency to be approved by the Board in the amounts and for the purposes noted below. Bonds issued by a surety who is listed in the latest version of U.S. Department of Treasury Circular 570, who is authorized to issue bonds in California, and whose bonding limitation shown in said circular is sufficient to provide bonds in the amount required by the Contract shall be deemed to be approved unless specifically rejected by the Agency. Bonds from all other sureties shall be accompanied by all of the documents enumerated in the Code of Civil Procedure, Section 995.660 a). The Bidder shall pay all bond premiums, costs, and incidentals.

Each bond shall incorporate, by reference, the Contract and be signed by both the Bidder and the Surety. The signature of the authorized agent of the Surety shall be notarized.

The Bidder shall provide 2 good and sufficient surety bonds. The "Payment Bond" (Material and Labor Bond) shall be for not less than 100 percent of the Contract Price, to satisfy claims of material suppliers and mechanics and laborers employed by it on the Work. The bond shall be maintained by the Contractor in full force and effect until the performance of the Contract is accepted by the Agency and until all claims for materials and labor are paid, and shall otherwise comply with the Civil Code.

The "Performance Bond" shall be for 100 percent of the Contract Price to guaranty faithful performance of all work, within the time prescribed, in a manner satisfactory to the Agency, and that all materials and workmanship will be free from original or developed defects. The bond must remain in effect until the end of all warranty periods set forth in the Contract Documents.

Should any bond become insufficient, the Contractor shall renew the bond within 10 days after receiving notice from the Agency.

Should any surety at any time be unsatisfactory to the Board, notice to that effect will be given to the Contractor. No further payments shall be deemed due or will be made under the Contract until a new surety qualifies and is accepted by the Board.

Changes in the Work or extensions of time, made pursuant to the Contract, shall in no way release the Contractor or the Surety from its obligations. Notice of such changes or extensions shall be waived by the Surety.

2-5 PLANS AND SPECIFICATIONS.

2-5.1 General. The Contractor shall keep at the Work site a copy of the Plans and Specifications, to which the Engineer shall have access at all times.

The Plans, Specifications, and other Contract Documents shall govern the Work. The Contract Documents are intended to be complementary and cooperative. Anything specified in the Specifications and not shown on the Plans, or shown on the Plans and not specified in the Specifications, shall be as though shown on or specified in both.

The Plans shall be supplemented by such working drawings and shop drawings as are necessary to adequately control the Work.

The Contractor shall, upon discovering any error or omission in the Plans or Specifications, immediately call it to the attention of the Engineer.

2-5.2 Precedence of the Contract Documents. If there is a conflict between any of the Contract Documents, the document highest in the order of precedence shall control. The order of precedence, from highest to lowest, shall be as follows:

a) Permits issued by jurisdictional regulatory agencies.
b) Change Orders and Supplemental Agreements; whichever occurs last.
c) Contract/Agreement.
d) Addenda.
e) Bid/Proposal.
f) Special Provisions.
g) Plans.
h) Standard Plans.
i) Standard Specifications.
j) Reference Specifications.

Detail drawings shall take precedence over general drawings. The precedence of the notice inviting bids and instructions to bidders shall be as specified in the Special Provisions.

2-5.3 Submittals.

2-5.3.1 General. Submittals shall be provided, at the Contractor's expense, as specified in 2-5.3.2, 2-5.3.3 and 2-5.3.4, when required by the Plans or Special Provisions, or when requested by the Engineer.

Materials shall neither be furnished nor fabricated, nor shall any work for which submittals are required be performed, before the required submittals have been reviewed and accepted by the Engineer. Neither review nor acceptance of submittals by the Engineer shall relieve the Contractor from responsibility for errors, omissions, or deviations from the Contract Documents, unless such deviations were specifically called to the attention of the Engineer in the letter of transmittal. The Contractor shall be responsible for the correctness of the submittals.

The Contractor shall allow a minimum of 20 working days for review of submittals unless otherwise specified in the Special Provisions. Each submittal shall be accompanied by a letter of transmittal.

2-5.3.2 Working Drawings. Working Drawings shall be of a size and scale to clearly show all necessary details.

Six copies and one reproducible shall be submitted. If no revisions are required, 3 of the copies will be returned to the Contractor. If revisions are required, the Engineer will return one copy along with the reproducible for resubmission. Upon acceptance, the Engineer will return 2 of the copies to the Contractor and retain the remaining copies and the reproducible.

Working Drawings are required in the subsections shown in Table 2-5.3.2 (A).

TABLE 2-5.3.2 (A)

Item	Subsection No.	Title	Subject
1	7-8.5.2	Sanitary Sewers	Sewage Bypass and Pumping
2	7-8.6.3	Water Pollution Control	Storm Water Pollution Prevention Plan
3	7-8.6.4	Water Pollution Control	Dewatering Plan
4	7-10.2.2	Work Area Traffic Control	Traffic Control Plan
5	7-10.4.2.2	Safety	Shoring Plan
6	300-3.2	Cofferdams	Structure Excavation & Backfill
7	303-1.6.1	General	Falsework
8	303-1.7.1	General	Placing Reinforcement
9	303-3.1	General	Prestressed Concrete Construction
10	304-1.1.2	Falsework Plans	Structural Steel
11	306-2.1	General	Jacking Operations
12	306-3.1	General	Tunneling Operations
13	306-3.4	Tunnel Supports	Tunneling Operations
14	306-6	Remodeling Existing Sewer Facilities	Polyethylene Liner Installation
15	306-8	Microtunneling	Microtunneling Operations

Working Drawings listed above as Items 4, 5, 6, 7, 9, 10, 12, and 14 shall be prepared by a Civil or Structural Engineer registered by the State of California.

2-5.3.3 Shop Drawings. Shop Drawings are required in the subsections shown in Table 2-5.3.3 (A).

TABLE 2-5.3.3 (A)

Item	Subsection No.	Title	Subject
1	207-2.5	Joints	Reinforced Concrete Pipe
2	207-8.4	Joints	Vitrified Clay Pipe
3	207-10.2.1	Joints	Vitrified Clay Pipe
4	304-1.1.1	Shop Drawings	Structural Steel
5	304-1.1.2	Falsework Plans	Structural Steel
6	304-2.1	General	Metal Hand Railings
7	307-1	General	Shop Drawings

2-5.3.4 Supporting Information. Supporting information is information required by the Specifications for the purposes of administration of the Contract, analysis for verification of conformance with the Specifications, the operation and maintenance of a manufactured product or system to be constructed as part of the Work, and other information as may be required by the Engineer. Six copies of the supporting information shall be submitted to the Engineer prior to the start of the Work unless otherwise specified in the Special Provisions or directed by the Engineer. Supporting information for systems shall be bound together and include all manufactured items for the system. If resubmittal is not required, 3 copies will be returned to the Contractor. Supporting information shall consist of the following and is required unless otherwise specified in the Special Provisions:

1) List of subcontractors per 2-3.2.
2) List of materials per 4-1.4.
3) Certificates of Compliance per 4-1.5.
4) Construction schedule per 6-1.
5) Spill Prevention and Emergency Response Plan per 7-8.5.3.
6) Confined Space Entry Program per 7-10.4.5.1.
7) Lean concrete base mix designs per 200-4.
8) Concrete mix designs per 201-1.1.
9) Asphalt concrete mix designs per 203-6.1.
10) Pipeline layout diagrams per 207-2.1.
11) Equipment and materials list per 307-1.
12) Controller cabinet wiring diagrams per 307-17.2.2.
13) Data, including, but not limited to, catalog sheets, manufacturer's brochures, technical bulletins, specifications, diagrams, product samples, and other information necessary to describe a system, product or item. This information is required for irrigation systems, street lighting systems, and traffic signals, and may also be required for any product, manufactured item, or system.

2-6 WORK TO BE DONE. The Contractor shall perform all work necessary to complete the Contract in a satisfactory manner. Unless otherwise provided, the Contractor shall furnish all materials, equipment, tools, labor, and incidentals necessary to complete the Work.

2-7 SUBSURFACE DATA. All soil and test hole data, groundwater elevations, and soil analyses shown on the Plans or included in the Special Provisions apply only at the location of the test holes and to the depths indicated. Soil test reports for test holes which have been drilled are available for inspection at the office of the Engineer. Additional subsurface exploration may be performed by Bidders or the Contractor at their own expense.

The indicated groundwater elevation is that which existed on the date specified in the data. It is the Contractor's responsibility to determine and allow for the groundwater elevation on the date the Work is performed. A difference in groundwater elevation between what is shown in soil boring logs and what is actually encountered during construction will not be considered as a basis for Extra Work per 3-3.

2-8 RIGHT-OF-WAY. Rights-of-way, easements, or rights-of-entry for the Work will be provided by the Agency. Unless otherwise specified in the Special Provisions, the Contractor shall make arrangements, pay for, and assume all responsibility for acquiring, using, and disposing of additional work areas and facilities temporarily required. The Contractor shall indemnify and hold the Agency harmless from all claims for damages caused by such actions.

2-9 SURVEYING.

2-9.1 Permanent Survey Markers. The Contractor shall notify the Engineer, or the owner on a Private Contract, at least 7 days before starting work to allow for the preservation of survey monuments, lot stakes (tagged), and benchmarks. The Engineer, or the owner on a Private Contract, will, at its cost, file a Corner Record Form referencing survey monuments subject to disturbance in the Office of the County Surveyor prior to the start of construction and also prior to the completion of construction for the replacement of survey monuments. The Contractor shall not disturb survey monuments, lot stakes (tagged), or benchmarks without the consent of the Engineer or the owner on a Private Contract. The Contractor shall bear the expense of replacing any that may be disturbed without permission. Replacement shall be done only under the direction of the Engineer by a Registered Land Surveyor or a Registered Civil Engineer authorized to practice land surveying within the State of California.

When a change is made in the finished elevation of the pavement of any roadway in which a permanent survey monument is located, the Contractor shall adjust the monument cover to the new grade within 7 days of finished paving unless otherwise specified in the Special Provisions.

2-9.2 Survey Service. Except for Private Contracts, the Engineer will perform and be responsible for the accuracy of surveying adequate for construction. The Contractor shall preserve construction survey stakes and marks for the duration of their usefulness. If any construction survey stakes are lost or disturbed and need to be replaced, such replacement will be performed by the Engineer at the expense of the Contractor.

The Contractor shall notify the Engineer in writing at least 2 working days before survey services will be required in connection with the laying out of any portion of the Work. The Contractor shall dig all holes necessary for line and grade stakes.

Unless otherwise specified in the Special Provisions, stakes will be set and stationed by the Engineer for curbs, headers, sewers, storm drains, structures, and rough grade. A corresponding cut or fill to finished grade (or flowline) will be indicated on a grade sheet.

2-9.3 Private Engineers. Surveying by private engineers on the Work shall conform to the quality and practice required by the Engineer.

2-9.4 Line and Grade. All work shall conform to the lines, elevations, and grades shown on the Plans.

Three consecutive points set on the same slope shall be used together so that any variation from a straight grade can be detected. Any such variation shall be reported to the Engineer. In the absence of such report, the Contractor shall be responsible for any error in the grade of the Work.

Grades for underground conduits will be set at the surface of the ground. The Contractor shall transfer them to the bottom of the trench.

2-10 AUTHORITY OF THE BOARD AND THE ENGINEER. The Board has the final authority in all matters affecting the Work. Within the scope of the Contract, the Engineer has the authority to enforce compliance with the Plans and Specifications. The Contractor shall promptly comply with instructions from the Engineer.

The decision of the Engineer is final and binding on all questions relating to: quantities; acceptability of material, equipment, or work; execution, progress or sequence of work; and interpretation of the Plans, Specifications, or other Contract Documents. This shall be precedent to any payment under the Contract, unless otherwise ordered by the Board.

2-11 INSPECTION. The Work is subject to inspection and approval by the Engineer. The Contractor shall notify the Engineer before noon of the working day before inspection is required. Work shall be done only in the presence of the Engineer, unless otherwise approved. Any work done without proper inspection will be subject to rejection. The Engineer and any authorized representatives shall at all times have access to the Work during its construction at shops and yards and while in storage, as well as to the Work site. The Contractor shall provide every reasonable facility for ascertaining that the materials and workmanship are in accordance with these Specifications. Inspection of the Work shall not relieve the Contractor of the obligation to fulfill all conditions of the Contract.

2-12 SPECIAL NOTICES. When specified in the Specifications or directed by the Engineer, any notice required to be served in accordance with this subsection shall be in writing, dated, and signed by the Contractor or the Engineer. Such notices shall be served by any of the following methods:

a) Personal delivery with proof of delivery which may be made by declaration under penalty of perjury by any person over the age of 18 years. The proof of delivery shall show that delivery was performed in accordance with these provisions. Service shall be effective on the date of delivery. Notices given to the Contractor by personal delivery may be made to the Contractor's authorized representative at the Work site; or

b) Certified mail addressed to the mailing address of the recipient postage prepaid; return receipt requested. Service shall be effective on the date of the receipt of the mailing.

Simultaneously, the Agency may send the same notice by regular mail. If a notice that is sent by certified mail is returned unsigned, then delivery shall be effective pursuant to regular mail, provided the notice that was sent by regular mail is not returned.

SECTION 3 - CHANGES IN WORK

3-1 CHANGES REQUESTED BY THE CONTRACTOR.

3-1.1 General. Changes in the Plans and Specifications, requested in writing by the Contractor, which do not materially affect the Work and which are not detrimental to the Work or to the interests of the Agency, may be granted by the Engineer. Nothing herein shall be construed as granting a right to the Contractor to demand acceptance of such changes.

3-1.2 Payment for Changes Requested by the Contractor. If such changes are granted, they shall be made at a reduction in cost or no additional cost to the Agency.

3-2 CHANGES INITIATED BY THE AGENCY.

3-2.1 General. The Agency may change the Plans, Specifications, character of the work, or quantity of work provided the total arithmetic dollar value of all such changes, both additive and deductive, does not exceed 25 percent of the Contract Price. Should it become necessary to exceed this limitation, the change shall be by written Supplemental Agreement between the Contractor and Agency, unless both parties agree to proceed with the change by Change Order.

Change Orders shall be in writing and state the dollar value of the change or establish the method of payment, any adjustment in the Contract time of completion, and when negotiated prices are involved, shall provide for the Contractor's signature indicating acceptance.

3-2.2 Contract Unit Prices.

3-2.2.1 General. If a change is ordered in an item of work covered by a Contract Unit Price, and such change does not involve a substantial change in character of the work from that shown on the Plans or specified in the Specifications, then an adjustment in payment will be made. This adjustment will be based upon the increase or decrease in quantity and the Contract Unit Price.

If the actual quantity of an item of work covered by a Contract Unit Price and constructed in conformance with the Plans and Specifications varies from the Bid quantity by 25 percent or less, payment will be made at the Contract Unit Price. If the actual quantity of said item of work varies from the Bid quantity by more than 25 percent, payment will be made per 3-2.2.2 or 3-2.2.3 as appropriate.

If a change is ordered in an item of work covered by a Contract Unit Price, and such change does involve a substantial change in the character of the work from that shown on the Plans or specified in the Specifications, an adjustment in payment will be made per 3-2.4.

3-2.2.2 Increases of More Than 25 Percent. Should the actual quantity of an item of work covered by a Contract Unit Price and constructed in conformance with the Plans and Specifications, exceed the Bid quantity by more than 25 percent, payment for the quantity in excess of 125 percent of the Bid quantity will be made on the basis of an adjustment in the Contract Unit Price mutually agreed to by the Contractor and the Agency, or at the option of the Engineer, on the basis of Extra Work per 3-3.

The Extra Work per 3-3, basis of payment, shall not include fixed costs. Fixed costs shall be deemed to have been recovered by the Contractor through payment for 125 percent of the Bid quantity at the Contract Unit Price.

3-2.2.3 Decreases of More Than 25 Percent. Should the actual quantity of an item of work covered by a Contract Unit Price, and constructed in conformance with the Plans and Specifications, be less than 75 percent of the Bid quantity, an adjustment in payment will not be made unless so requested in writing by the Contractor. If the Contractor so requests, payment will be made on the basis of an adjustment in the Contract Unit Price mutually agreed to by the Contractor and the Agency, or at the option of the Engineer, on the basis of Extra Work per 3-3; however, in no case will payment be less than would be made for the actual quantity at the Contract Unit Price nor more than would be made for 75 percent of the Bid quantity at the Contract Unit Price.

3-2.3 Stipulated Unit Prices. Stipulated Unit Prices may be used for the adjustment of Contract changes when so specified in the Special Provisions.

3-2.4 Agreed Prices. If mutual agreement can not be reached, the Engineer may direct the Contractor to proceed on the basis of Extra Work per 3-3, except as otherwise specified in 3-2.2.2 and 3-2.2.3.

3-2.5 Eliminated Items. Should any Bid item be eliminated in its entirety, payment will be made to the Contractor for its actual costs incurred in connection with the eliminated item prior to notification in writing from the Engineer so stating its elimination.

If material conforming to the Plans and Specifications is ordered by the Contractor for use in the eliminated item prior to the date of notification of elimination by the Engineer, and if the order for that material can not be canceled, payment will be made to the Contractor for the actual cost of the material.

In this case, the material shall become the property of the Agency. Payment will be made to the Contractor for its actual costs for any further handling. If the material is returnable, the material shall be returned and payment will be made to the Contractor for the actual cost of charges made by the supplier for returning the material and for handling by the Contractor.

Actual costs, as used herein, shall be computed on the basis of Extra Work per 3-3.

3-3 EXTRA WORK.

3-3.1 General. New or unforeseen work will be classified as Extra Work when the Engineer determines that it is not covered by Contract Unit Prices or Stipulated Unit Prices.

3-3.2 Payment.

3-3.2.1 General. When the cost for Extra Work cannot be agreed upon, the Agency will pay for Extra Work based on the accumulation of costs as provided herein.

3-3.2.2 Basis for Establishing Costs.

3-3.2.2.1 Labor. The cost of labor shall be the actual cost for wages of workers performing the Extra Work at the time the Extra Work is done, plus employer payments of payroll taxes, workers compensation insurance, liability insurance, health and welfare, pension, vacation, apprenticeship funds, and other direct costs, resulting from Federal, State, or local laws, as well as assessments or benefits required by lawful collective bargaining agreements.

The use of a labor classification which would increase the Extra Work cost will not be permitted unless the Contractor establishes the necessity for such additional costs. Labor costs for equipment operators and helpers shall be reported only when such costs are not included in the invoice for equipment rental. The labor cost for foremen shall be proportioned to all of their assigned work and only that applicable to the Extra Work will be paid

Nondirect labor costs, including superintendence, shall be considered part of the markup specified in 3-3.2.3.1.

3-3.2.2.2 Materials. The cost of materials reported shall be at invoice or lowest current price at which such materials are locally available and delivered to the Work site in the quantities involved, plus sales tax, freight, and delivery.

The Agency reserves the right to approve materials and sources of supply, or to supply materials to the Contractor if necessary for the progress of the Work. No markup shall be applied to any material provided by the Agency.

3-3.2.2.3 Tool and Equipment Rental. No payment will be made for the use of tools which have a replacement value of $200 or less.

Regardless of ownership, the rates to be used in determining equipment rental costs shall not exceed listed rates prevailing locally at equipment rental agencies, or distributors, at the time the Extra Work is performed.

The rental rates paid shall include the cost of fuel, oil, lubrication, supplies, small tools, necessary attachments, repairs and maintenance of any kind, depreciation, storage, insurance, and all incidentals. Necessary loading and transportation costs for equipment used on the Extra Work shall be included.

If equipment is used intermittently and, when not in use, could be returned to its rental source at less expense to the Agency than holding it at the Work site, it shall be returned, unless the Contractor elects to keep it at the Work site, at no expense to the Agency.

All equipment shall be acceptable to the Engineer, in good working condition, and suitable for the purpose for which it is to be used.

The reported rental time for equipment already at the Work site shall be the duration of its use on the Extra Work. This time shall begin when the equipment is first used on the Extra Work, plus the time required to move it from its previous site and back, or to a closer site.

3-3.2.2.4 Other Items. The Agency may authorize other items which may be required on the Extra Work, including labor, services, material, and equipment. These items must be different in their nature from those required for the Work, and be of a type not ordinarily available from the Contractor or Subcontractors.

3-3.2.2.5 Invoices. Vendors' invoices for material, equipment rental and other expenditures shall be submitted with the daily report per 3-3.3. If the daily report is not substantiated by invoices or other documentation, the Agency may establish the cost of the item involved at the lowest price which was current at the time of the report.

3-3.2.3 Markup.

3-3.2.3.1 Work by Contractor. Unless otherwise specified in the Special Provisions, a reasonable allowance for overhead and profit shall be added to the Contractor's costs as determined under 3-3.2.2 and shall constitute the markup for all overhead and profit on Extra Work done by the Contractor. The Contractor shall also be compensated as specified in the Special Provisions for the actual increase in the Contractor's bond premium caused by the extra work.

3-3.2.3.2 Work by Subcontractor. When any of the Extra Work is performed by a Subcontractor, the markup established in 3-3.2.3.1 shall be applied to the Subcontractor's costs as determined under 3-3.2.2. Unless otherwise specified in the Special Provisions, a reasonable allowance for the Contractor's overhead and profit shall be added to the sum of the Subcontractor's costs and markup and shall constitute the markup for all overhead and profit for the Contractor on Extra Work done by the Subcontractor.

3-3.3 Daily Reports by Contractor. When the cost for the Extra Work cannot be agreed upon, the Contractor shall submit a daily report to the Engineer on forms approved by the Agency. Applicable delivery tickets, listing all labor, materials, and equipment involved for that day, and other services and expenditures when authorized shall be included. Failure to submit the daily report by the close of the next working day may waive any rights for that day. An attempt shall be made to reconcile the report daily, and it shall be signed by the Engineer and the Contractor. In the event of a disagreement, pertinent notes shall be entered by each party to explain points which cannot be resolved immediately. Each party shall retain a signed copy of the report. Reports by Subcontractors or others shall be submitted through the Contractor.

The report shall:

a) List the names of workers, classifications, and hours worked.

b) Describe and list quantities of materials used.

c) List the type of equipment, size, identification number, and hours of operation, including loading and transportation, if applicable.

d) Describe other services and expenditures in such detail as the Agency may require.

3-4 CHANGED CONDITIONS. The Contractor shall notify the Engineer of the following Work site conditions (hereinafter called "changed conditions"), in writing, upon their discovery and before they are disturbed:

a) Subsurface or latent physical conditions differing materially from those represented in the Contract Documents;

b) Unknown physical conditions of an unusual nature differing materially from those ordinarily encountered and generally recognized as inherent in work of the character being performed; and

c) Material differing from that represented in the Contract Documents which the Contractor believes may be hazardous waste, as defined in Section 25117 of the Health and Safety Code, that is required to be removed to a Class I, Class II, or Class III disposal site in accordance with the provisions of existing law.

The Engineer will investigate conditions which appear to be changed conditions. If the Engineer determines that the conditions are changed conditions and will materially affect costs, a Change Order will be issued adjusting the compensation for such portion of the Work in accordance with 3-2 or 3-3. If the Engineer determines that the conditions are changed conditions and will materially affect performance time, the Contractor, upon submitting a written request, will be granted an extension of time in accordance with 6-6.

If the Engineer determines that the conditions do not justify an adjustment in compensation, the Contractor will be so notified in writing. This notice will also advise the Contractor of its obligation to notify the Engineer in writing if the Contractor disagrees.

If the Contractor disagrees with the decision, it may submit a written notice of potential claim to the Engineer before commencing the disputed work. In the event of such a dispute, the Contractor shall not be excused from any scheduled completion date provided by the Contract and shall proceed with all work to be performed under the Contract. However, the Contractor shall retain any and all rights provided by either the Contract or by law which pertain to the resolution of disputes and protests between the contracting parties. The Contractor shall proceed in accordance with 3-5.

The Contractor's failure to give notice of changed conditions promptly upon their discovery and before they are disturbed shall constitute a waiver of all claims in connection therewith.

3-5 DISPUTED WORK. If the Contractor and the Agency are unable to reach agreement, the Agency may direct the Contractor to proceed with the Disputed Work. Payment shall be as later determined by 3-2, 3-3, mediation or arbitration, if the Agency and Contractor agree thereto, or as fixed in a court of law.

Although not to be construed as proceeding under 3-3, the Contractor shall keep and furnish records of Disputed Work to the Engineer in accordance with 3-3.

SECTION 4 - CONTROL OF MATERIALS

4-1 MATERIALS AND WORKMANSHIP.

4-1.1 General. All materials, parts, and equipment furnished by the Contractor in the Work shall be new, high grade, and free from defects. Used or secondhand materials, parts, and equipment may be used only if so specified in the Special Provisions.

The quality of materials and workmanship shall be subject to approval by the Engineer. Materials and workmanship of quality not conforming to the requirements of the Specifications shall be considered defective and will be subject to rejection. Defective work or material, whether in place or not, shall be removed immediately from the Work site by the Contractor, at its expense, when so directed by the Engineer.

If the Contractor fails to replace any defective or damaged work or material after reasonable notice, the Engineer may cause such work or materials to be replaced. The replacement expense will be deducted from the amount to be paid to the Contractor.

4-1.2 Protection of Work and Materials. The Contractor shall provide and maintain storage facilities and employ such measures as will preserve the specified quality of materials to be used in the Work. Stored materials shall be reasonably accessible for inspection. The Contractor shall also adequately protect new and existing work and all items of equipment for the duration of the Contract.

The Contractor shall not, without the Agency's consent, assign, sell, mortgage, hypothecate, or remove equipment or materials which have been installed or delivered and which may be necessary for the completion of the Work.

4-1.3 Inspection Requirements.

4-1.3.1 General. Unless otherwise specified in the Special Provisions, inspection is required at the source for asphalt concrete pavement mixtures, structural concrete, metal fabrication, metal casting, welding, concrete pipe manufacture, protective coating application, and similar shop or plant operations. Additional materials and fabricated items which require inspection at the source shall be as specified in the Special Provisions.

Steel pipe in sizes less than 18 inches (450mm) and vitrified clay and cast iron pipe in all sizes are acceptable upon certification as to compliance with the Specifications, subject to sampling and testing by the Agency. Standard items of equipment such as electric motors, conveyors, elevators, plumbing fixtures, etc., are subject to inspection at the Work site only. Special items of equipment such as designed electrical panel boards, large pumps, sewage plant equipment, etc., are subject to inspection at the source for performance testing only. Inspection at the source for other items shall be as specified in the Special Provisions.

4-1.3.2 Inspection by the Agency. The Agency will provide inspection and testing laboratory services within the continental United States within a 50-mile radius of the geographical limits of the Agency. Inspection and testing laboratory services beyond this radius or outside the continental United States shall be provided by the Contractor and approved by the Engineer.

4-1.3.3 Inspection of Items Not Locally Produced. When the Contractor intends to purchase materials, fabricated products, or equipment from sources located more than 50 miles (80km) outside the geographical limits of the Agency or outside the United States, an inspector or accredited testing laboratory approved by the Engineer, shall be engaged by the Contractor at its expense, to inspect the materials, equipment or process. This approval shall be obtained before producing any material or equipment. The inspector or representative of the testing laboratory shall evaluate the materials for conformance with the requirements of the Plans and Specifications. The Contractor shall forward reports required by the Engineer. No materials or equipment shall be shipped nor shall any processing, fabrication or treatment of such materials be done without proper inspection by the approved agent. Approval by said agent shall not relieve the Contractor of responsibility for complying with the requirements of the Contract Documents.

4-1.4 Test of Materials. Before incorporation into the Work, the Contractor shall submit samples of materials, as the Engineer may require, at no cost to the Agency. The Contractor, at its expense, shall deliver the materials for testing to the place and at the time designated by the Engineer. Unless otherwise specified in the Special Provisions, all initial testing and a reasonable amount of retesting will be performed under the direction of the Engineer, and at no expense to the Contractor. If the Contractor is to provide and pay for testing, it will be so specified in the Special Provisions.

The Contractor shall notify the Engineer in writing, at least 15 days in advance, of its intention to use materials for which tests are specified, to allow sufficient time to perform the tests. The notice shall name the proposed supplier and source of material.

If the notice of intent to use is sent before the materials are available for testing or inspection, or is sent so far in advance that the materials on hand at the time will not last but will be replaced by a new lot prior to use on the Work, it will be the Contractor's responsibility to re-notify the Engineer when samples which are representative may be obtained.

Third party independent testing and quality control testing shall be performed within the United States.

4-1.5 Certificate of Compliance. A Certificate of Compliance shall be furnished to the Engineer prior to the use of any material or assembled material for which these Specifications so require or if so required by the Engineer.

The Engineer may waive the materials testing requirements of the Specifications and accept a Certificate of Compliance. Material test data may be required by the Engineer to be included with the submittal.

Materials used on the basis of a Certificate of Compliance may be sampled and tested at any time. The submission of a Certificate of Compliance shall not relieve the Contractor of responsibility for incorporating material into the Work which conforms to the requirements of the Contract Documents, and any material not conforming to the requirements will be subject to rejection by the Engineer whether in place or not.

4-1.6 Trade Names or Equals. The Contractor may supply any of the materials specified or offer an equivalent. The Engineer will determine whether the material offered is equivalent to that specified. Adequate time shall be allowed for the Engineer to make this determination.

A listing of materials is not intended to be comprehensive, or in order of preference. The Contractor may offer any material, process, or equipment considered to be equivalent to that indicated. The substantiation of offers shall be submitted as provided in the Contract Documents.

The Contractor shall, at its expense, furnish data concerning items offered by it as equivalent to those specified. The Contractor shall have the material tested as required by the Engineer to determine that the quality, strength, physical, chemical, or other characteristics, including durability, finish, efficiency, dimensions, service, and suitability are such that the item will fulfill its intended function.

Test methods shall be subject to the approval of the Engineer. Test results shall be reported promptly to the Engineer, who will evaluate the results and determine if the substitute item is equivalent. The Engineer's findings shall be final. Installation and use of a substitute item shall not be made until approved by the Engineer.

If a substitute offered by the Contractor is not found to be equal to the specified material, the Contractor shall furnish and install the specified material.

The specified Contract completion time shall not be affected by any circumstance developing from the provisions of this subsection.

4-1.7 Weighing and Metering Equipment. Scales and metering equipment used for proportioning materials shall be inspected for accuracy and certified within the past 12 months by the State of California Bureau of Weights and Measures, by the County Sealer, or by a scale mechanic registered with or licensed by the County.

The accuracy of the work of a scale service agency, except as stated herein, shall meet the standards of the Business and Professions Code and the Code of Regulations pertaining to weighing devices. A Certificate of Compliance shall be presented, prior to operation, to the Engineer for approval and shall be renewed whenever required by the Engineer at no cost to the Agency.

Scales shall be arranged so they may be read easily from the operator's platform or area. They shall indicate the true net weight without the application of any factor. The figures of the scales shall be clearly legible. Scales shall be accurate to within 1 percent when tested with the plant shut down.

Weighing equipment shall be so insulated against vibration or moving of other operating equipment in the plant area that the error in weighing with the entire plant running will not exceed 2 percent for any setting nor 1.5 percent for any batch.

4-1.8 Calibration of Testing Equipment. Testing equipment, such as, but not limited to pressure gages, metering devices, hydraulic systems, force (load) measuring instruments, and strain-measuring devices shall be calibrated by a testing agency acceptable to the Engineer at intervals not to exceed 12 months and following repairs, modification, or relocation of the equipment. Calibration certificates shall be provided when requested by the Engineer.

4-1.9 Construction Materials Dispute Resolution (Soils, Rock Materials, Concrete, Mortar and Related Materials, Masonry Materials, Asphalt Materials, Rock Products, and Modified Asphalts). In the interest of safety and public value, whenever credible evidence arises to contradict the test values of materials, the Agency and the Contractor will initiate an immediate and cooperative investigation. Test values of materials are results of the materials' tests, as defined by these Specifications or by the Special Provisions, required to accept the Work. Credible evidence is process observations or test values gathered using industry accepted practices. A contradiction exists whenever test values or process observations of the same or similar materials are diverse enough such that the Work acceptance or performance becomes suspect. The investigation shall allow access to all test results, procedures, and facilities relevant to the Disputed Work and consider all available information and, when necessary, gather new and additional information in an attempt to determine the validity, the cause, and if necessary, the remedy to the contradiction. If the cooperative investigation reaches any resolution mechanism acceptable to both the Agency and the Contractor, the contradiction shall be considered resolved and the cooperative investigation concluded.

Whenever the cooperative investigation is unable to reach resolution, the investigation may then either conclude without resolution or continue by written notification of one party to the other requesting the implementation of a resolution process by committee. The continuance of the investigation shall be contingent upon recipient's agreement and acknowledged in writing within 3 calendar days after receiving a request. Without acknowledgement, the investigation shall conclude without resolution. The committee shall consist of 3 State of California Registered Civil Engineers. Within 7 calendar days after the written request notification, the Agency and the Contractor will each select 1 engineer. Within 14 calendar days of the written request notification, the 2 selected engineers will select a third engineer. The goal in selection of the third member is to complement the professional experience of the first 2 engineers. Should the 2 engineers fail to select the third engineer, the Agency and the Contractor shall each propose 2 engineers to be the third member within 21 calendar days after the written request notification. The first 2 engineers previously selected shall then select 1 of the 4 proposed engineers in a blind draw.

The committee shall be a continuance of the cooperative investigation and will re-consider all available information and if necessary gather new and additional information to determine the validity, the cause, and if necessary, the remedy to the contradiction. The committee will focus upon the performance adequacy of the material(s) using standard engineering principles and practices and to ensure public value, the committee may provide engineering recommendations as necessary. Unless otherwise agreed, the committee will have 30 calendar days from its formation to complete their review and submit their findings. The final resolution of the committee shall be by majority opinion, in writing, stamped and signed. Should the final resolution not be unanimous, the dissenter may attach a written, stamped, and signed minority opinion.

Once started, the resolution process by committee shall continue to full conclusion unless:

a) Within 7 days of the formation of the committee, the Agency and the Contractor reach an acceptable resolution mechanism; or

b) Within 14 days of the formation of the committee, the initiating party withdraws its written notification and agrees to bear all investigative related costs thus far incurred; or

c) At any point by the mutual agreement of the Agency and the Contractor.

Unless otherwise agreed, the Contractor shall bear and maintain a record for all the investigative costs until resolution. Should the investigation discover assignable causes for the contradiction, the assignable party, the Agency or the Contractor, shall bear all costs associated with the investigation. Should assignable causes for the contradiction extended to both parties, the investigation will assign costs cooperatively with each party or when necessary, equally. Should the investigation substantiate a contradiction without assignable cause, the investigation will assign costs cooperatively with each party or when necessary, equally. Should the investigation be unable to substantiate a contradiction, the initiator of the investigation shall bear all investigative costs. All claim notification requirements of the contract pertaining to the contradiction shall be suspended until the investigation is concluded.

SECTION 5 - UTILITIES

5-1 LOCATION. The permittee (in the case of Private Contracts) or the Agency (in the case of Cash or Assessment Act Contracts), will search known substructure records and furnish the Contractor with copies of documents which describe the location of utility substructures, or will indicate on the Plans those substructures (except for service connections) which may affect the Work. Information regarding removal, relocation, abandonment, or installation of new utilities will be furnished to prospective Bidders.

Where underground main distribution conduits such as water, gas, sewer, electric power, telephone, or cable television are shown on the Plans, the Contractor shall assume that every property parcel will be served by a service connection for each type of utility.

Pursuant to Section 4216 of the Government Code, at least 2 working days prior to commencing any excavation, the Contractor shall contact the regional notification center (Underground Service Alert of Southern California) and obtain an inquiry identification number.

Caltrans is not required by Section 4216 to become a member of the regional notification center. The Contractor shall contact Caltrans for the location of its subsurface installations.

The Contractor shall determine the location and depth of all utilities, including service connections, which have been marked by the respective owners and which may affect or be affected by its operations. Unless otherwise specified in the Special Provisions, payment shall be considered as included in the prices in the Bid for other items of work.

5-2 PROTECTION. The Contractor shall not interrupt the service function or disturb the support of any utility without authority from the utility owner or direction from the Engineer. Valves, switches, vaults, and meters shall be maintained readily accessible for emergency shutoff.

Where protection is required to ensure support of utilities located as shown on the Plans or in accordance with 5-1, the Contractor shall, unless otherwise specified on the Plans or in the Special Provisions, furnish and place the necessary protection at its expense.

Upon learning of the existence and location of any utility omitted from or shown incorrectly on the Plans, the Contractor shall immediately notify the Engineer in writing. When authorized by the Engineer, support or protection of the utility will be paid for as provided in 3-2 or 3-3.

The Contractor shall immediately notify the Engineer and the utility owner if any utility is disturbed or damaged. The Contractor shall bear the costs of repair or replacement of any utility damaged if located in accordance with 5-1.

When placing concrete around or contiguous to any non-metallic utility installation, the Contractor shall at its expense:

a) Furnish and install a 2-inch (50mm) cushion of expansion joint material or other similar resilient material; or

b) Provide a sleeve or other opening which will result in a 2-inch (50mm) minimum-clear annular space between the concrete and the utility; or

c) Provide other acceptable means to prevent embedment in or bonding to the concrete.

Where concrete is used for backfill or for structures which would result in embedment, or partial embedment, of a metallic utility installation; or where the coating, bedding or other cathodic protection system is exposed or damaged by the Contractor's operations, the Contractor shall notify the Engineer and arrange to secure the advice of the affected utility owner regarding the procedures required to maintain or restore the integrity of the system.

5-3 REMOVAL. Unless otherwise specified in the Special Provisions, the Contractor shall remove all interfering portions of utilities shown on the Plans as "abandoned" or "to be abandoned in place". Before starting removal operations, the Contractor shall ascertain from the Agency whether the abandonment is complete, and the costs involved in the removal and disposal shall be included in the Bid for the items of work necessitating such removals.

5-4 RELOCATION. When feasible, the owners responsible for utilities within the area affected by the Work will complete their necessary installations, relocations, repairs, or replacements before commencement of the Work by the Contractor. When the Plans or Special Provisions indicate that a utility installation is to be relocated, altered, or constructed by others, the Agency will conduct all negotiations with the owners and utility work will be done at no cost to the Contractor, except as otherwise specified in 301-1.6. Utilities which are relocated in order to avoid interference shall be protected in their position and the cost of such protection shall be included in the Bid for the items of work necessitating such relocation.

After award of the Contract, portions of utilities which are found to interfere with the Work will be relocated, altered or reconstructed by the utility owners, or the Engineer may order changes in the Work to avoid interference. Such changes will be paid for in accordance with 3-2 or 3-3.

When the Plans or Special Provisions provide for the Contractor to alter, relocate, or reconstruct a utility, all costs for such work shall be included in the Bid for the items of work necessitating such work. Temporary or permanent relocation or alteration of utilities requested by the Contractor for its convenience shall be its responsibility and it shall make all arrangements and bear all costs.

When the Plans or Specifications provide for the Contractor to alter, relocate, or reconstruct a utility, all costs for such work shall be included in the Bid for the items of work necessitating such work. Temporary or permanent relocation or alteration of utilities requested by the Contractor for its convenience shall be its responsibility and it shall make all arrangements and bear all costs.

The utility owner will relocate service connections as necessary within the limits of the Work or within temporary construction or slope easements. When directed by the Engineer, the Contractor shall arrange for the relocation of service connections as necessary between the meter and property line, or between a meter and the limits of temporary construction or slope easements. Payment for the relocation of such service connections shall be in accordance with 3-2 or 3-3. Payment will include the restoration of all existing improvements which may be affected thereby. The Contractor may agree with the owner of any utility to disconnect and reconnect interfering service connections. The Agency will not be involved in any such agreement.

5-5 DELAYS. The Contractor shall notify the Engineer of its construction schedule insofar as it affects the protection, removal, or relocation of utilities. Said notification shall be included as a part of the construction schedule in accordance with 6-1. The Contractor shall notify the Engineer in writing of any subsequent changes in the construction schedule which will affect the time available for protection, removal, or relocation of utilities.

The Contractor will not be entitled to damages or additional payment for delays attributable to utility relocations or alterations if correctly located, noted, and completed in accordance with 5-1.

The Contractor may be given an extension of time for unforeseen delays attributable to unreasonably protracted interference by utilities in performing work correctly shown on the Plans.

The Agency will assume responsibility for the timely removal, relocation, or protection of existing main or trunkline utility facilities within the area affected by the Work if such utilities are not identified in the Contract Documents. The Contractor will not be assessed liquidated damages for any delay caused by failure of the Agency to provide for the timely removal, relocation, or protection of such existing facilities.

If the Contractor sustains loss due to delays attributable to interferences, relocations, or alterations not covered by 5-1, which could not have been avoided by the judicious handling of forces, equipment, or plant, there shall be paid to the Contractor such amount as the Engineer may find to be fair and reasonable compensation for such part of the Contractor's actual loss as was unavoidable and the Contractor may be granted an extension of time.

5-6 COOPERATION. When necessary, the Contractor shall so conduct its operations as to permit access to the Work site and provide time for utility work to be accomplished during the progress of the Work.

SECTION 6 - PROSECUTION, PROGRESS, AND ACCEPTANCE OF THE WORK

6-1 CONSTRUCTION SCHEDULE AND COMMENCEMENT OF THE WORK.

6-1.1 Construction Schedule. After notification of award of the Contract and prior to start of any work, the Contractor shall submit its proposed construction schedule to the Engineer for approval. The construction schedule shall be in the form of a tabulation, chart, or graph and shall be in sufficient detail to show chronological relationship of all activities of the Work. These include, but are not limited to: estimated starting and completion dates of various activities, submission of submittals per 2-5.3, procurement of materials and scheduling of equipment. The construction schedule shall incorporate the requirements of 5-5 and reflect completion of the Work within the specified Contract time and in conformance with the Contract Documents.

If the Contractor desires to make a major change in the method of operations after commencing construction, or if the schedule fails to reflect the actual progress, the Contractor shall submit to the Engineer a revised construction schedule in advance of beginning revised operations.

The Engineer may waive these requirements for work constructed under a permit.

6-1.2 Commencement of the Work. Unless otherwise specified in the Special Provisions, the Contract time shall commence upon the date of issuance of the Notice to Proceed. The Work shall start within 15 days thereafter, and be diligently prosecuted to completion within the Contract time.

6-2 PROSECUTION OF THE WORK. To minimize public inconvenience and possible hazard and to restore street and other work areas to their original condition and state of usefulness as soon as practicable, the Contractor shall diligently prosecute the Work to completion. If the Engineer determines that the Contractor is failing to prosecute the Work to the proper extent, the Contractor shall, upon orders from the Engineer, immediately take steps to remedy the situation. All costs of prosecuting the Work as specified herein shall be included in the Contract Price. Should the Contractor fail to take the necessary steps to fully accomplish said purposes, after orders of the Engineer, the Engineer may suspend the Work in whole or part, until the Contractor takes said steps.

If the Work is suspended through no fault of the Agency, all expenses and losses incurred by the Contractor during such suspensions shall be borne by the Contractor. If the Contractor fails to properly provide for public safety, traffic, and protection of the Work during periods of suspension, the Agency may elect to do so, and deduct the cost thereof from monies due the Contractor. Such actions will not relieve the Contractor from liability.

6-3 SUSPENSION OF THE WORK.

6-3.1 General. The Work may be suspended in whole or in part when determined by the Engineer that the suspension is necessary in the interest of the Agency. The Contractor shall comply immediately with any written order of the Engineer. Such suspension shall be without liability to the Contractor on the part of the Agency except as otherwise specified in 6-6.3.

6-3.2 Archaeological and Paleontological Discoveries. If discovery is made of items of archaeological or paleontological interest, the Contractor shall immediately cease excavation in the area of discovery and shall not continue until ordered by the Engineer. When resumed, excavation operations within the area of discovery shall be as directed by the Engineer.

Discoveries which may be encountered may include, but not be limited to, dwelling sites, stone implements or other artifacts, animal bones, human bones, and fossils.

The Contractor shall be entitled to an extension of time and compensation in accordance with 6-6.

6-4 TERMINATION OF THE CONTRACT FOR DEFAULT.

6.4.1 General. If, prior to the acceptance of the Work, the Contractor:

a) becomes insolvent, assigns its assets for the benefit of its creditors, is unable to pay its debts as they become due, or is otherwise financially unable to complete the Work,

b) abandons the Work by failing to report to the Work site and diligently prosecute the Work to completion,

c) disregards written instructions from the Engineer or materially violates provisions of the Contract Documents,

d) fails to prosecute the Work according to the schedule approved by the Engineer,

e) disregards laws or regulations of any public body having jurisdiction, or

f) commits continuous or repeated violations of regulatory or statutory safety requirements, then the Agency will consider the Contractor in default of the Contract.

Notices, and other written communications regarding default between the Contractor, the Agency, and the Surety shall be transmitted in accordance with 2-12.

6-4.2 Notice to Cure. The Agency will issue a written notice to cure the default to the Contractor and its Surety. The Contractor shall commence satisfactory corrective actions within 5 Working Days after receipt.

6-4.3 Notice of Termination for Default. If the Contractor fails to commence satisfactory corrective action within 5 Working Days after receipt of the notice to cure, or to diligently continue satisfactory and timely correction of the default thereafter, then the Agency will recommend to the Board that the Contractor be found in default of the Contract and upon such finding by the Board:

a) will terminate the Contractor's right to perform under the Contract by issuing a written notice of termination for default to the Contractor and its Surety,

b) may use any materials, equipment, tools or other facilities furnished by the Contractor to secure and maintain the Work site, and

c) may furnish labor, equipment, and materials the Agency deems necessary to secure and maintain the Work site.

The provisions of this subsection shall be in addition to all other legal rights and remedies available to the Agency.

6-4.4 Responsibilities of the Surety. Upon receipt of the written notice of termination for default, the Surety shall immediately assume all rights, obligations and liabilities of the Contractor under the Contract. If the Surety fails to protect and maintain the Work site, the Agency may do so, and may recover all costs incurred. The Surety shall notify the Agency that it is assuming all rights, obligations and liabilities of the Contractor under the Contract and all money that is due, or would become due, to the Contractor shall be payable to the Surety as the Work progresses, subject to the terms of the Contract.

Within 15 Working Days of receipt of the written notice of termination for default, the Surety shall submit to the Agency a written plan detailing the course of action it intends to take to remedy the default. The Agency will review the plan and notify the Surety if the plan is satisfactory. If the Surety fails to submit a satisfactory plan, or if the Surety fails to maintain progress according to the plan accepted by the Agency, the Agency may, upon 48 hours written notice, exclude the Surety from the premises, take possession of all material and equipment, and complete the Work in any way the Agency deems to be expedient. The cost of completing the Work by the Agency shall be charged against the Surety and may be deducted from any monies due, or which would become due, the Surety. If the amounts due under the Contract are insufficient for completion, the Surety shall pay to the Agency, within 30 days after the Agency submits an invoice, all costs in excess of the remaining Contract Price.

6-4.5 Payment. The Surety will be paid for completion of the Work in accordance with 9-3 less the value of damages caused to the Agency by acts of the Contractor.

6-5 TERMINATION OF THE CONTRACT FOR CONVENIENCE. The Board may terminate the Contract if it becomes impossible or impracticable to proceed, or because of conditions or events beyond the control of the Agency.

The Agency will issue a written notice of termination for convenience in accordance with 2-12. Upon receipt, the Contractor shall immediately cease work, except work the Contractor is directed to complete by the Engineer or required to complete for public safety and convenience. The Contractor shall immediately notify Subcontractors and suppliers to immediately cease their work.

The Contractor will be paid without duplication for:

a) work completed in accordance with the Contract Documents prior to the effective date of termination for convenience;

b) reasonable costs incurred in settlement of terminated contracts with Subcontractors, suppliers and others; and

c) reasonable expenses directly attributable to termination.

The Contractor shall submit a final termination settlement proposal to the Agency no later than 90 days from the effective date of termination, unless extended, in writing, by the Agency upon written request by the Contractor.

If the Contractor fails to submit a proposal, the Agency may determine the amount, if any, due the Contractor as a result of the termination. The Agency will pay the Contractor the amount it determines to be reasonable. If the Contractor disagrees with the amount determined by the Agency as being reasonable, the Contractor shall provide notice to the Agency within 30 days of receipt of payment. Any amount due shall be as later determined by arbitration, if the Agency and the Contractor agree thereto, or as fixed in a court of law.

6-6 DELAYS AND EXTENSIONS OF TIME.

6-6.1 General. If delays are caused by unforeseen events beyond the control of the Contractor, such delays will entitle the Contractor to an extension of the Contract time as provided herein, but the Contractor will not be entitled to damages or additional payment due to such delays, except as otherwise specified in 6-6.3. Such unforeseen events may include: war, government regulations, labor disputes, strikes, fires, floods, adverse weather or elements necessitating cessation of work, inability to obtain materials, labor or equipment, required Extra Work, or other specific events as may be further described in the Special Provisions.

No extension of time will be granted for a delay caused by the Contractor's inability to obtain materials unless the Contractor furnishes to the Engineer documentary proof. The proof must be provided in a timely manner in accordance with the sequence of the Contractor's operations and the approved construction schedule.

If delays beyond the Contractor's control are caused by events other than those mentioned above, the Engineer may deem an extension of time to be in the best interests of the Agency. The Contractor will not be entitled to damages or additional payment due to such delays, except as otherwise specified in 6-6.3.

If delays beyond the Contractor's control are caused solely by action or inaction by the Agency, such delays will entitle the Contractor to an extension of time per 6-6.2.

6-6.2 Extensions of Time. Extensions of time, when granted, will be based upon the effect of delays to the Work. They will not be granted for non-controlling delays to minor portions of the Work unless it can be shown that such delays did or will delay the progress of the Work.

6-6.3 Payment for Delays. Pursuant to Section 7102 of the Public Contract Code, the Contractor will be compensated for damages incurred due to delays for which the Agency is responsible. Such actual costs will be determined by the Engineer. The Agency will not be liable for damages which the Contractor could have avoided by any reasonable means, such as judicious handling of forces, equipment, or plant. The determination of what damages the Contractor could have avoided will be made by the Engineer.

6-6.4 Written Notice and Report. If the Contractor desires payment for a delay as specified in 6-6.3 or an extension of time, it shall file with the Engineer a written request and report of cause within 30 days after the beginning of the delay. The request for payment or extension must be made at least 15 days before the specified completion date. Failure by the Contractor to file these items within the times specified will be considered grounds for refusal by the Agency to consider such request.

6-7 TIME OF COMPLETION.

6-7.1 General. The Contractor shall complete the Work within the time specified in the Contract or the Special Provisions. The Contractor shall complete each portion of the Work within the time specified as in the Contract or the Special Provisions for such portion. Unless otherwise specified in the Contract or the Special Provisions, the time of completion of the Contract shall be expressed in working days.

6-7.2 Not Used.

6-7.3 Contract Time Accounting. The Engineer will make a daily determination of each working day to be charged against the Contract time. These determinations will be discussed and the Contractor will be furnished a periodic statement showing the allowable number of working days of Contract time, as adjusted, at the beginning of the reporting period. The statement will also indicate the number of working days charged during the reporting period and the number of working days of Contract time remaining. If the Contractor does not agree with the statement, it shall file a written protest within 15 days after receipt, setting forth the facts of the protest. Otherwise, the statement will be deemed to have been accepted.

6-8 COMPLETION, ACCEPTANCE, AND WARRANTY.

6-8.1 Completion. The Contractor shall submit a written assertion that the Work has been completed. If, in the Engineer's judgment, the Work has been completed in accordance with the Contract Documents, the Engineer will set forth in writing the date the Work was completed. This will be the date when the Contractor is relieved from responsibility to protect and maintain the Work and to which liquidated damages will be computed.

6-8.2 Acceptance. Acceptance will occur after all of the requirements contained in the Contract Documents have been fulfilled. If, in the Engineer's judgment, the Contractor has fully performed the Contract, the Engineer will recommend to the Board that the Contractor's performance of the Contract be accepted.

6-8.3 Warranty. The Work shall be warranted by the Contractor against defective materials and workmanship for a period of 1 year. The warranty period shall start on the date the Work was completed as determined by the Engineer.

The warranty period for specific items covered under manufacturers' or suppliers' warranties shall commence on the date they are placed into service at the direction of or as approved by the Engineer in writing.

All warranties, express or implied, from subcontractors, manufacturers, or suppliers, of any tier, for the materials furnished and work performed shall be assigned, in writing, to the Agency, and such warranties shall be delivered to the Engineer prior to acceptance of the Contractor's performance of the Contract.

The Contractor shall replace or repair defective materials and workmanship in a manner satisfactory to the Engineer, after notice to do so from the Engineer, and within the time specified in the notice. If the Contractor fails to make such replacement or repairs within the time specified in the notice, the Agency may perform the replacement or repairs at the Contractor's expense. If the Contractor fails to reimburse the Agency for the actual costs, the Contractor's Surety shall be liable for the cost thereof.

6-9 LIQUIDATED DAMAGES. Failure of the Contractor to complete the Work within the time allowed will result in damages being sustained by the Agency. Such damages are, and will continue to be, impracticable and extremely difficult to determine. For each consecutive calendar day in excess of the time specified for completion of the Work, as adjusted in accordance with 6-6, the Contractor shall pay to the Agency, or have withheld from monies due it, the sum of $250, unless otherwise specified in the Special Provisions.

Execution of the Contract shall constitute agreement by the Agency and Contractor that $250 per day is the minimum value of the costs and actual damage caused by the failure of the Contractor to complete the Work within the allotted time. Such sum is liquidated damages and shall not be construed as a penalty, and may be deducted from payments due the Contractor if such delay occurs.

6-10 USE OF IMPROVEMENT DURING CONSTRUCTION. The Agency reserves the right to take over and utilize all or part of any completed facility or appurtenance. The Contractor will be notified in writing in advance of such action. Such action by the Agency will relieve the Contractor of responsibility for injury or damage to said completed portions of the improvement resulting from use by public traffic or from the action of the elements or from any other cause, except Contractor operations or negligence. The Contractor will not be required to reclean such portions of the improvement before field acceptance, except for cleanup made necessary by its operations. Nothing in this subsection shall be construed as relieving the Contractor from full responsibility for correcting defective work or materials.

In the event the Agency exercises its right to place into service and utilize all or part of any completed facility or appurtenance, the Agency will assume the responsibility and liability for injury to persons or property resulting from the utilization of the facility or appurtenance so placed into service, except for any such injury to persons or property caused by any willful or negligent act or omission by the Contractor, Subcontractor, their officers, employees, or agents.

SECTION 7 - RESPONSIBILITIES OF THE CONTRACTOR

7-1 THE CONTRACTOR'S EQUIPMENT AND FACILITIES.

7-1.1 General. The Contractor shall furnish and maintain in good condition all equipment and facilities as required for the proper execution and inspection of the Work

The Contractor shall provide and maintain enclosed toilets for the use of employees engaged in the Work. These accommodations shall be maintained in a neat and sanitary condition, and regularly pumped out.

7-1.2 Temporary Utility Services. The Contractor shall, at its own expense, make all arrangements necessary for the provision of temporary utility services necessary for its own use during performance of the Work.

The Contractor shall not draw water from any fire hydrant (except to extinguish a fire), without obtaining permission from the water utility owner.

7-1.3 Crushing and Screening Operations. Unless otherwise specified in the Special Provisions, the establishment and operation of portable screens and crushers will not be allowed on or adjacent to the Work site.

7-2 LABOR.

7-2.1 General. The Contractor, its agents, and employees shall be bound by and comply with applicable provisions of the Labor Code and Federal, State, and local laws related to labor.

Any worker found by the Engineer to be incompetent, intemperate, troublesome, disorderly, or otherwise objectionable, or who fails to perform the Work properly and acceptably, shall be immediately removed from the Work site by the Contractor and shall not be reemployed in the performance of the Work.

7-2.2 Prevailing Wages. Pursuant to Section 1773.2 of the Labor Code, the current prevailing rate of per diem wages at the time of the Bid as determined by the Director of the Department of Industrial Relations (DIR) are on file at the office of the Engineer. The Contractor shall post a copy of these rates at the Work site. Pursuant to Section 1774 of the Labor Code, the Contractor and any Subcontractors shall pay not less than the specified prevailing rates of wages to workers employed on the Contract. If the Contract is Federally-funded, the Contractor and any Subcontractors shall not pay less than the higher of these rates or the rates determined by the United States Department of Labor. Pursuant to Section 1775 of the Labor Code, the Contractor and any Subcontractors, shall, as a penalty to the Agency, forfeit the prescribed amounts per calendar day, or portion thereof, for each worker paid less than the prevailing wage rates.

7-2.3 Payroll Records. Pursuant to Section 1776 of the Labor Code the Contractor shall keep, make available, and submit to the Engineer upon request, certified payroll records.

7-2.4 Hours of Labor. Pursuant to Section 1810 of the Labor Code, 8 hours of labor shall constitute a legal day's work. Pursuant to Section 1813 of the Labor Code, the Contractor and any Subcontractors, shall, as a penalty to the Agency, forfeit the prescribed amount per calendar day for each worker required or permitted to work more than 8 hours in any 1 calendar day and 40 hours in any 1 calendar week without being compensated in accordance with Section 1815.

7-3 LIABILITY INSURANCE. The Contractor shall furnish the Agency a policy or certificate of liability insurance in which the Agency is the named insured or is named as an additional insured with the Contractor. The policy furnished by the Contractor shall be issued by an insurance company authorized by the Insurance Commissioner to transact business in the State of California. The insurance company shall have a policy holder rating of "A" or higher and a Financial Class VII or higher as established by A.M. Best, or equivalent or greater rating established by another recognized rating entity such as Moody's. Notwithstanding any inconsistent statement in the policy or any subsequent endorsement, the Agency shall be the insured or named as an additional insured covering the Work, whether liability is attributable to the Contractor or the Agency. The policy shall insure the Agency, its officers, employees, and agents, while acting within the scope of their duties on the Work, against all claims arising out of or in connection with the Work, except as otherwise specified in 6-10.

The Contractor may file insurance acceptable to the Agency covering more than one contract or permit. The coverage shall provide the following minimum limits:

Bodily Injury............................	$250,000 each person
	$500,000 each occurrence
	$500,000 aggregate products and completed operations.
Property Damage	$100,000 each occurrence
	$250,000 aggregate.

A combined single limit policy with aggregate limits in the amount of $1,000,000 will be considered equivalent to the required minimum limits.

Except as otherwise specified in 6-10, the Contractor shall save, keep, and hold harmless the Agency, its officers and agents from all damages, costs or expenses in law or equity that may at any time arise or be set up because of damages to property, or of personal injury received by reason of or in the course of performing work, which may be caused by any willful or negligent act or omission by the Contractor, any of the Contractor's employees, or any Subcontractor. The Agency will not be liable for any accident, loss or damage to the Work prior to its completion and acceptance, except as otherwise specified in 6-10.

All liability insurance policies shall bear an endorsement or shall have attached a rider whereby it is provided that, in the event of expiration or proposed cancellation of such policies for any reason whatsoever, the Agency shall be notified by registered mail, return receipt requested, giving a sufficient time before the date thereof to comply with any applicable law or statute, but in no event less than 30 days before expiration or cancellation is effective.

The cost of this insurance coverage shall be included in the Bid.

7-4 WORKERS' COMPENSATION INSURANCE.

Pursuant to Sections 1860 and 3700 of the Labor Code, the Contractor shall secure, pay for, and maintain in full force for the duration of the Contract, workers' compensation insurance. The Agency, its officers, employees and agents, will not be responsible for any claims in law or equity occasioned by failure of the Contractor to comply with this requirement.

Pursuant to Sections 1860 and 1861 of the Labor Code, the Contractor shall submit the following certification to the Engineer prior to execution of the Contract by the Board:

> "I am aware of the provisions of Section 3700 of the Labor Code which require every employer to be insured against liability for workers' compensation or to undertake self-insurance in accordance with the provisions of that code, and I will comply with such provisions before commencing the performance of the work of this contract."

7-5 PERMITS. The Agency will obtain, at no cost to the Contractor, all permits necessary to perform the Work in streets, highways, railways or other rights-of-way. The Contractor shall obtain and pay for all costs incurred for permits necessitated by its operations such as, but not limited to, those permits required for night work, overload, blasting, and demolition. For Private Contracts, the Contractor shall obtain all permits incidental to the Work or made necessary by its operations, and pay all costs incurred by the permit requirements.

The Contractor shall pay all business taxes or license fees that are required for the Work.

7-6 THE CONTRACTOR'S REPRESENTATIVE. Before starting the Work, the Contractor shall designate in writing a representative who shall have complete authority to act for it. An alternative representative may be designated as well. The representative or alternate shall be present at the Work site whenever work is in progress or whenever actions of the elements necessitate its presence to take measures necessary to protect the Work, persons, or property. Any order or communication given to this representative shall be deemed delivered to the Contractor. A joint venture shall designate only one representative and alternate. In the absence of the Contractor or its representative, instructions or directions may be given by the Engineer to the superintendent or person in charge of the specific work to which the order applies. Such order shall be complied with promptly and referred to the Contractor or its representative.

In order to communicate with the Agency, the Contractor's representative, superintendent, or person in charge of specific work shall be able to speak, read, and write the English language.

7-7 COOPERATION AND COLLATERAL WORK. The Contractor shall be responsible for ascertaining the nature and extent of any simultaneous, collateral, and essential work by others. The Agency, its workers and contractors and others, shall have the right to operate within or adjacent to the Work site during the performance of such work.

The Agency, the Contractor, and each of such workers, contractors and others, shall coordinate their operations and cooperate to minimize interference.

The Contractor shall include in its Bid all costs involved as a result of coordinating its work with others. The Contractor will not be entitled to additional compensation from the Agency for damages resulting from such simultaneous, collateral, and essential work. If necessary to avoid or minimize such damage or delay, the Contractor shall redeploy its work force to other parts of the Work.

Should the Contractor be delayed by the Agency, and such delay could not have been reasonably foreseen or prevented by the Contractor, the Engineer will determine the extent of the delay, the effect on the Work, and any extension of time.

7-8 WORK SITE MAINTENANCE.

7-8.1 General. Throughout all phases of construction, including suspension of the Work, and until acceptance per 6-8, the Contractor shall keep the Work site clean and free from rubbish and debris. Rubbish and debris collected on the Work site shall only be stored in roll-off, enclosed containers prior to disposal. Stockpiles of such will not be allowed.

When required by the Special Provisions, the Contractor shall provide a self-loading motorized street sweeper equipped with a functional water spray system. The sweeper shall clean all paved areas within the Work site and all paved haul routes at least once each working day

The Contractor shall ensure there is no spillage along haul routes. Any such spillage shall be removed immediately and the area cleaned.

Should the Contractor fail to keep the Work site free from rubbish and debris, the Engineer may suspend the Work per 6-3 until the condition is corrected.

7-8.2 Air Pollution Control. The Contractor shall not discharge smoke, dust, equipment exhaust, or any other air contaminants into the atmosphere in such quantity as will violate any Federal, State, or local regulations. The Contractor shall also abate dust nuisance by cleaning, sweeping and spraying with water, or other means as necessary. The use of water shall conform to 7-8.6.

7-8.3 Noise Control. Noise generated from the Contractor's operations shall be controlled as specified in the Special Provisions.

7-8.4 Storage of Equipment and Materials.

7-8.4.1 General. Materials and equipment shall be removed from the Work site as soon as they are no longer necessary. Before inspection by the Engineer for acceptance, the Work site shall be cleared of equipment, unused materials, and rubbish so as to present a satisfactory clean and neat appearance.

Excess excavated material shall be removed from the Work site immediately unless otherwise specified in the Special Provisions.

Forms and form lumber shall be removed from the Work site as soon as practicable after stripping.

7-8.4.2 Storage in Public Streets. Construction materials and equipment shall not be stored in streets, roads, or highways for more than 5 days after unloading unless otherwise specified in the Special Provisions or approved by the Engineer. All materials or equipment not installed or used in construction within 5 days after unloading shall be stored at a location approved by the Engineer.

Excavated material, except that which is to be used as backfill in the adjacent trench, shall not be stored in public streets unless otherwise specified in the Special Provisions or approved by the Engineer. Immediately after placing backfill, all excess excavated material shall be removed.

7-8.5 Sanitary Sewers.

7-8.5.1 General. The flow of sewage shall not be interrupted. Should the Contractor disrupt the operation of existing sanitary sewer facilities, or should disruption be necessary for performance of the Work, the Contractor shall bypass the sewage flow around the Work. Sewage shall be conveyed in closed conduits and disposed of in a sanitary sewer system. Sewage shall not be permitted to flow in trenches nor be covered by backfill.

Whenever sewage bypass and pumping is required by the Plans or Specifications, or the Contractor so elects to perform, the Contractor shall submit per 2-5.3 a working drawing conforming to 7-8.5.2 detailing its proposed plan of sewage bypass and pumping.

7-8.5.2 Sewage Bypass and Pumping Plan. The plan shall indicate the locations and capacities of all pumps, sumps, suction and discharge lines. Equipment and piping shall be sized to handle the peak flow of the section of sewer line to be bypassed and pumped. Equipment and piping shall conform to 7-10, the Plans, and the Special Provisions. Bypass piping, when crossing areas subject to traffic loads, shall be constructed in trenches with adequate cover and otherwise protected from damage due to traffic. Lay-flat hose or aluminum piping with an adequate casing and/or traffic plates may be allowed if so approved by the Engineer. Bypass pump suction and discharge lines that extend into manholes shall be rigid hose or hard pipe. Lay flat hose will not be allowed to extend into manholes. The Contractor shall provide a backup bypass pumping system in case of malfunction. The backup bypass system shall provide 100 percent standby capability, and be in place and ready for immediate use. Each standby pump shall be a complete unit with its own suction and discharge piping. In addition to the backup system, the Contractor shall furnish and operate vacuum trucks when required by the Plans or Special Provisions.

7-8.5.3 Spill Prevention and Emergency Response Plan. The Contractor shall prepare and submit per 2-5.3 a spill prevention and emergency response plan. The plan shall address implementation of measures to prevent sewage spills, procedures for spill control and containment, notifications, emergency response, cleanup, and spill and damage reporting.

The plan shall account for all storm drain systems and water courses within the vicinity of the Work which could be affected by a sewage spill. Catch basins that could receive spilled sewage shall be identified. Unless otherwise specified in the Special Provisions, these catch basins shall be sealed prior to operating the bypass and pumping system. The Contractor shall remove all material used to seal the catch basins when the bypass and pumping system operations are complete.

The Contractor shall be fully responsible for containing any sewage spillage, preventing any sewage from reaching a watercourse, recovery and legal disposal of any spilled sewage, any fines or penalties associated with the sewage spill imposed upon by the Agency and/or the Contractor by jurisdictional regulatory agencies, and any other expenses or liabilities related to the sewage spill.

7-8.6 Water Pollution Control.

7-8.6.1 General. The Contractor shall conform to all applicable local, state and Federal regulations and laws pertaining to water pollution control. The Contractor shall conduct and schedule its operations, and follow and implement best management practices in such a manner as to prevent water pollution. The Contractor shall also conform to the following requirements:

a) Sediments shall not be discharged to a storm drain system or receiving waters.

b) Sediments generated on the Work site shall be contained on the Work site using appropriate BMPs.

c) No construction-related materials, waste, spill, or residue shall be discharged from the Work site to streets, drainage facilities, receiving waters, or adjacent property by wind or runoff.

d) Non-storm water runoff from equipment, vehicle washing, or any other activity shall be contained within the Work site using appropriate BMPs.

e) Erosion shall be prevented. Erosion susceptible slopes shall be covered, planted or otherwise protected in a way that prevents discharge from the Work site.

7-8.6.2 Best Management Practices (BMPs). The Contractor shall implement and maintain such BMPs as are relevant to the Work, and as are specifically required by the Plans or Special Provisions.

The Contractor shall be responsible throughout the duration of the Contract for installing, constructing, inspecting, maintaining, removing and disposing of BMPs for wind erosion control, tracking control, erosion and sediment control, non-storm water control, and waste management and materials pollution control. Unless otherwise directed by the Engineer, the Contractor shall be responsible for BMP implementation and maintenance throughout any temporary suspension of the Work.

7-8.6.3 Storm Water Pollution Prevention Plan (SWPPP). When so specified in the Special Provisions, or if so required by a jurisdictional regulatory agency, the Contractor shall prepare and submit per 2-5.3 a storm water pollution prevention plan. The SWPPP shall conform to the requirements specified in the Special Provisions and those of the jurisdictional regulatory agency. The Notice of Intent will be filed by the Agency.

7-8.6.4 Dewatering. Dewatering shall be performed by the Contractor when specifically required by the Plans or Specifications, and as necessary for construction of the Work. Dewatering shall be performed in conformance with all applicable local, state and Federal laws and permits issued by jurisdictional regulatory agencies. Permits necessary for treatment and disposal of accumulated water shall be obtained by the Contractor or the Agency as specified in the Special Provisions. Accumulated water shall be treated prior to disposal if so specified in the Special Provisions or required by a permit. The Contractor shall submit a working drawing and related supporting information per 2-5.3 detailing its proposed plan and methodology of dewatering and treatment and disposal of accumulated water.

The plan shall identify the location, type and size of dewatering devices and related equipment, the size and type of materials composing the collection system, the size and type of equipment to be used to retain and, if required, treat accumulated water, and the proposed disposal locations. If the proposed disposal location is a sanitary sewer, the Contractor shall submit to the Engineer written evidence of permission from the owner. If the proposed disposal location is a storm drain system or receiving body of water, the Contractor shall submit written evidence of permission from the owner of the storm drain system and, if not obtained by the Agency, original signed permits from jurisdictional regulatory agencies or written evidence that such permits are not required.

7-8.6.5 Payment. Unless otherwise specified in the Special Provisions, payment for implementation and maintenance of BMPs shall be considered as included in the Contract Unit Price for each item in the Bid.

Payment for dewatering shall be as specified in the Special Provisions.

7-9 PROTECTION AND RESTORATION OF EXISTING IMPROVEMENTS. The Contractor shall be responsible for the protection of public and private property adjacent to the Work and shall exercise due caution to avoid damage to such property.

The Contractor shall repair or replace all existing improvements within the right-of-way which are not designated for removal (e.g., curbs, sidewalks, driveways, fences, walls, signs, utility installations, pavement, structures, etc.) which are damaged or removed as a result of its operations. When a portion of a sprinkler system within the right-of-way must be removed, the remaining lines shall be capped. Repairs and replacements shall be at least equal to existing improvements and shall match them in finish and dimension.

Maintenance of street and traffic signal systems that are damaged, temporarily removed or relocated shall be done in conformance with 307-2.

Trees, lawns, and shrubbery that are not to be removed shall be protected from damage or injury. If damaged or removed due to Contractor's operations, they shall be restored or replaced in as nearly the original condition and location as is reasonably possible. Lawns shall be reseeded and covered with suitable mulch.

The Contractor shall give reasonable notice to occupants or owners of adjacent property to permit them to salvage or relocate plants, trees, fences, sprinklers, and other improvements, within the right-of-way which are designated for removal and would be destroyed because of the Work.

All costs to the Contractor for protecting, removing, and restoring existing improvements shall be included in the Bid.

7-10 PUBLIC CONVENIENCE AND SAFETY.

7-10.1 Access.

7-10.1.1 General. The Contractor's operations shall cause no unnecessary inconvenience to the public or businesses in the vicinity of the Work. The Contractor shall have no greater length or quantity of Work under construction than can be properly prosecuted with a minimum of inconvenience to the public and other contractors engaged in adjacent or related work.

The Contractor shall provide continuous and unobstructed access to the adjacent properties unless otherwise specified in the Special Provisions or approved by Engineer. Work requiring traffic lane closures shall only be performed between the hours specified in the Special Provisions or shown on the traffic control plan. Traffic shall be permitted to pass through the Work site, unless otherwise specified in the Special Provisions or shown on the traffic control plan.

7-10.1.1.1 Vehicular Access. Vehicular access to residential driveways shall be maintained to the property line except when necessary construction precludes such access . If backfill has been completed to the extent that safe access may be provided and the street is opened to local traffic, the Contractor shall immediately clear the street and driveways and provide and maintain access.

7-10.1.1.2 Pedestrian Access. Pedestrian access shall be maintained unless otherwise approved by the Engineer.

7-10.2 Work Area Traffic Control.

7-10.2.1 General. Work area traffic control shall be as specified in the Special Provisions. The total length of the traffic control zone shall include buffer spaces, advance signing, striping transitions, and the Work site.

7-10.2.2 Traffic Control Plan (TCP).

7-10.2.2.1 General. If so specified in the Special Provisions, the Contractor shall submit a TCP in accordance with 2-5.3.

The sheets of the TCP shall display the title, phase identification, name of the firm preparing the TCP, name and stamp of the Registered Traffic or Civil Engineer, approval block for each jurisdictional agency, north arrow, sheet number, and number of sheets comprising the TCP. General notes and symbol definitions shall be included when required. Adequate dimensioning shall be provided to allow for proper field installation.

The TCP shall be drawn to a 1 inch = 40 feet scale on common size sheets, either 8-1/2 inches x 11 inches, 8-1/2 inches x 14 inches, 11 inches x 17 inches, or 2-foot x 3-foot plan sheets as dictated by the length of the Work.

The requirements in the Special Provisions shall govern the design of the proposed TCP.

7-10.2.2.2 Payment. Payment for work area control and preparation of the TCP, if so required, will be made as specified in the Special Provisions.

7-10.3 Haul Routes. Unless otherwise specified in the Special Provisions, haul routes shall be determined by the Contractor.

7-10.4 Safety.

7-10.4.1 Work Site Safety.

7-10.4.1.1 General. The Contractor shall provide safety measures as necessary to protect the public and workers within, or in the vicinity of, the Work site. The Contractor shall ensure that its operations will not create safety hazards.

The Contractor shall provide safety equipment, material, and assistance to Agency personnel so that they may properly inspect all phases of the Work.

When asbestos is being removed, the requirements of the CCR Title 8, Div. 1, Chapter 4, Subchapter 4 and Subchapter 7 shall be implemented.

7-10.4.1.2 Work Site Safety Official. The Contractor shall designate in writing a "Project Safety Official" who shall be at the Work site at all times, and who shall be thoroughly familiar with the Contractor's Injury and Illness Prevention Program (IIPP) and Code of Safe Practices (CSP). The Project Safety Official shall be available at all times to abate any potential safety hazards and shall have the authority and responsibility to shut down an unsafe operation, if necessary.

7-10.4.2 Safety Orders.

7-10.4.2.1 General. The Contractor shall have at the Work site, copies or suitable extracts of Construction Safety Orders, Tunnel Safety Orders, and General Industry Safety Orders issued by the State Division of Industrial Safety.

Prior to beginning any excavation 5 feet in depth or greater, the Contractor shall submit to the Engineer, the name of the "Competent Person" as defined in CCR, Title 8, Section 1504, in accordance with 2-5.3. The "Competent Person" shall be present at the Work site as required by Cal-OSHA.

7-10.4.2.2 Shoring Plan. Before excavating any trench 5 feet (1.5m) or more in depth, the Contractor shall submit in accordance with 2-5.3 a detailed working drawing (shoring plan) showing the design of the shoring, bracing, sloping, or other provisions used for the workers' protection . If the shoring plan varies from the shoring system standards, the shoring plan shall be prepared by a registered Structural or Civil Engineer. The shoring plan shall accommodate existing underground utilities. No excavation shall start until the Engineer has accepted the shoring plan and the Contractor has obtained a permit from the State Division of Industrial Safety. A copy of this permit shall be submitted to the Engineer in accordance with 2-5.3. If the Contractor fails to submit a shoring plan or fails to comply with an accepted shoring plan, the Contractor shall suspend work at the affected location(s). Such suspended work shall not be the basis of a claim for Extra Work and the Contractor shall not receive additional compensation or Contract time.

7-10.4.2.3 Payment. Payment for shoring shall be included in the Bid item provided therefor. Payment for compliance with the provisions of the safety orders and all other laws, ordinances, and regulations shall be included in the various Bid items.

7-10.4.3 Use of Explosives. Explosives may be used only when authorized in writing by the Engineer, or as otherwise specified in the Special Provisions.

Explosives shall be handled, used, and stored in accordance with all applicable regulations.

Prior to blasting, the Contactor shall comply with the following requirements:

a) The jurisdictional law enforcement agency shall be notified 24 hours in advance of blasting.

b) The jurisdictional fire department shall be notified 24 hours in advance of blasting.

c) Blasting activities and schedule milestones shall be included in the Contractor's construction schedule per 6-1.

For a Private Contract, specific permission shall be obtained from the Agency in writing, prior to any blasting operations in addition to the above requirements.

The Engineer's approval of the use of explosives shall not relieve the Contractor from liability for claims caused by blasting operations.

7-10.4.4 Hazardous Substances . An MSDS as described in CCR, Title 8, Section 5194, shall be maintained at the Work site for all hazardous material used by the Contractor.

Material usage shall be accomplished with strict adherence to California Division of Industrial Safety requirements and all manufacturer warnings and application instructions listed on the MSDS and on the product container label.

The Contractor shall notify the Engineer if a specified product cannot be used under safe conditions.

7-10.4.5 Confined Spaces.

7-10.4.5.1 Confined Space Entry Program (CSEP). The Contractor shall be responsible for implementing, administering and maintaining a CSEP in accordance with CCR, Title 8, Sections 5156, 5157 and 5158.

Prior to the start of the Work, the Contractor shall prepare and submit a CSEP in accordance with 2-5.3. The CSEP shall address all potential physical and environmental hazards and contain procedures for safe entry into confined spaces such as the following:

a) Training of personnel.

b) Purging and cleaning the space of materials and residue.

c) Potential isolation and control of energy and material inflow.

d) Controlled access to the space.

e) Atmospheric testing of the space.

f) Ventilation of the space.

g) Special hazards consideration.

h) Personal protective equipment.

i) Rescue plan provisions.

The submittal shall include the names of the Contractor's personnel, including each Subcontractor's personnel, assigned to the Work that will have CSEP responsibilities, their CSEP training, and their specific assignment and responsibility in carrying out the CSEP.

7-10.4.5.2 Permit-Required Confined Spaces. Entry into permit-required confined spaces as defined in CCR, Title 8, Section 5157 may be required as a part of the Work. Manholes, tanks, vaults, pipelines, excavations, or other enclosed or partially enclosed spaces shall be considered permit-required confined spaces until the pre-entry procedures demonstrate otherwise. The Contractor shall implement a permit-required CSEP prior to performing any work in a permit-required confined space. A copy of the permit shall be available at all times for review by the Contractor and the Engineer at the Work site.

7-10.4.5.3 Payment. Payment for the CSEP shall be included in the Bid items for which the CSEP is required.

7-10.5 Security and Protective Devices.

7-10.5.1 General. Security and protective devices shall consist of fencing, steel plates, or other devices as specified in the Special Provisions to protect open excavations.

7-10.5.2 Security Fencing. The Contractor shall completely fence open excavations. Security fencing shall conform to 304-3.5. Security fencing shall remain in place unless workers are present and construction operations are in progress during which time the Contractor shall provide equivalent security.

7-10.5.3 Steel Plate Covers. The Contractor shall provide steel plate covers as necessary to protect from accidental entry into openings, trenches, and excavations.

7-11 PATENT FEES OR ROYALTIES. The Contractor shall absorb in its Bid the patent fees or royalties on any patented article or process furnished or used in the Work. The Contractor shall indemnify and hold the Agency harmless from any legal action that may be brought for infringement of patents.

7-12 ADVERTISING. The names, addresses and specialties of Contractors, Subcontractors, architects, or engineers may be displayed on removable signs. The size and location shall be subject to the Engineer's approval.

Commercial advertising matter shall not be attached to or painted on the surfaces of buildings, fences, canopies, or barricades.

7-13 LAWS TO BE OBSERVED. The Contractor shall keep itself fully informed of State and national laws and County and municipal ordinances and regulations which in any manner affect those employed in the Work or the materials used in the Work or in any way affect the conduct of the Work. The Contractor shall at all times observe and comply with such laws, ordinances, and regulations.

7-14 ANTITRUST CLAIMS. Section 7103.5 of the Public Contract Code provides:

> "In entering into a public works contract or a subcontract to supply goods, services, or materials pursuant to a public works contract, the contractor or subcontractor offers and agrees to assign to the awarding body all rights, title, and interest in and to all causes of action it may have under Section 4 of the Clayton Act (15 U.S.C. Sec 15) or Cartwright Act (Chapter 2 [commencing with Section 16700] of Part 2 of Division 7 of the Business and Professions Code), arising from purchases of goods, services, or materials pursuant to the public works contract or subcontract. The assignment shall be made and become effective at the time the awarding body tenders final payment to the contractor, without further acknowledgment by the parties."

SECTION 8 - FACILITIES FOR AGENCY PERSONNEL

8-1 GENERAL. Facilities provided for Agency personnel shall be at suitable locations approved by the Engineer. Such facilities must be in a room, building, or trailer provided for this purpose with acceptable means for locking.

A Class "A" Field Office conforming to 8-2.1 shall be provided at any offsite plant facility furnishing pipe subject to Agency inspection during manufacture. A Field Laboratory conforming to 8-3.1 shall be provided at any offsite or Work site plant facility furnishing portland cement concrete or asphalt concrete pavement. Any other facilities for Agency personnel shall be provided only when required by the Specifications.

Offices and laboratories at plants may be used concurrently by inspection personnel of other agencies provided such use does not seriously conflict with Agency use. When facilities are shared in this manner, at least one locker provided with a hasp for a padlock must be available for the exclusive use of the Agency. Otherwise any facilities furnished are for the exclusive use of the Agency.

Facilities shall conform to the applicable codes, ordinances, and regulations of the local jurisdiction and of the State of California, and shall conform to current practice. The interior shall be paneled or suitably lined to provide a facility of good appearance.

The Contractor shall provide janitorial and other maintenance services in all types of facilities provided. Such services shall include the supply of the appropriate paper products and dispensers. Trash receptacles shall be provided and emptied by the Contractor at weekly intervals or sooner as required. The trash shall be removed from the Work site.

All costs to furnish, maintain, service, and remove the specified facilities at the Work site shall be included in the price in the Bid for such facilities. If no Bid item is provided in the Proposal, costs shall be included in other Bid items.

The first progress payment will not be approved until all facilities are in place and fully comply with the Specifications.

8-2 FIELD OFFICE FACILITIES.

8-2.1 Class "A" Field Office. The office shall have a minimum floor space of 175 ft^2 ($16m^2$), at least one door, and window area of not less than 22 ft^2 ($2m^2$). All doors and windows shall be provided with screens.

Furniture shall be provided as follows: one plan table, one standard 5-foot (1.5m) long double-pedestal desk with a drawer suitable for holding files, 2 chairs, one drafting stool, and one plan rack.

Electric power shall be provided to include a minimum of 4 duplex convenience outlets. The office shall be illuminated at the tables and desk. An outdoor lighting fixture with a 300-watt bulb shall be installed.

Heating and air conditioning of sufficient capacity shall be provided at no expense to the Agency. The Contractor shall provide drinking water within the office and integral sanitary facilities directly adjoining. Sanitary facilities shall include a toilet and wash basin with hot and cold running water.

Extended area, non-coin-operated telephone service shall be provided within the office area. The installation shall include sufficient extension cord to serve the plan table and desk.

8-2.2 Class "B" Field Office. The office shall be the same as class "A" except that integrated sanitary facilities and air conditioning are not required. A chemical toilet facility shall be provided adjacent to the office.

8-2.3 Class "C" Field Office. The office shall have a minimum floor space of 120 ft^2 ($11m^2$) of floor area. It shall be equipped with one 3-foot x 5-foot (0.9m x 1.5m) table, 4 chairs and one plan rack. It shall be adequately heated, ventilated, and lighted and 2 duplex convenience outlets shall be provided. Air conditioning, telephones, and sanitary facilities are not required.

8-3 FIELD LABORATORIES.

8-3.1 Offsite at Manufacturing Plant. Field laboratories shall conform to the requirements for a Class "C" Field Office specified in 8-2.3 except for the following:

a) Telephone service as specified in 8-2.1.

b) Chair.

c) Work table, 4 feet x 10 feet (1.2m x 3m), 3 feet (0.9m).

d) Sieves per 203-6.

e) Scales and weights.

f) Burner plate for heating samples.

g) Thermometer, with 200 to 400°F (90 to 260°C) degree range (asphalt concrete plants only).

h) Air meter for concrete in accordance with ASTM C231 of the type that indicates the percentage of air directly (precast concrete plants only).

Sampling and testing equipment shall be maintained in satisfactory operating condition by the Contractor or the plant owner. Laboratories shall be located immediately adjacent to and with full view of batching and loading operations.

8-3.2 At the Work Site. Field laboratories shall conform to 8-3.1, except that sieves, scales, weights, burner plates, sampling devices, pans, and thermometers will be furnished by the Agency at no expense to the Contractor. If air entraining admixtures are being used in the concrete on the Work, an air meter of the type specified in 8-3.1 shall be furnished by the Contractor.

8-4 BATHHOUSE FACILITIES. When the Plans or Special Provisions require bathhouse facilities, the following shall be provided:

a) One lavatory with hot and cold water.

b) One toilet in a stall.

c) One 3-foot (1m) trough-type urinal.

d) One enclosed shower at least 3 feet x 3 feet (1m x 1m) with hot and cold water.

e) One bench, 6 feet (2m) long.

f) Soap dispensers.

g) Toilet paper holders.

h) Paper towel cabinet.

i) Wastepaper receptacle.

These facilities shall be serviced and provided with necessary sanitary supplies.

These facilities shall be for the exclusive use of Agency personnel. However, a separate building need not be provided for this purpose if such facilities are located in a separate room in a building which includes other facilities.

8-5 REMOVAL OF FACILITIES. Field offices, laboratories, and bathhouse facilities at the Work site shall be removed upon completion of the Work. Buildings and equipment furnished by the Contractor at the Work site under the provisions of this subsection shall remain the property of the Contractor.

8-6 BASIS OF PAYMENT. All costs incurred in furnishing, maintaining, servicing, and removing field offices laboratories, or bathhouse facilities required at the Work site shall be included in the Bid item for furnishing such facilities. If such facilities are required by the Plans and no Bid item is provided in the Proposal, the costs shall be included in other Bid items. Such costs incurred in connection with offices and laboratories at plants shall be borne by the plant owners.

SECTION 9 - MEASUREMENT AND PAYMENT

9-1 MEASUREMENT OF QUANTITIES FOR UNIT PRICE WORK.

9-1.1 General. Unless otherwise specified, quantities of work shall be determined from measurements or dimensions in horizontal planes. However, linear quantities of pipe, piling, fencing and timber shall be considered as being the true length measured along longitudinal axis.

Unless otherwise provided in the Special Provisions, volumetric quantities shall be the product of the mean area of vertical or horizontal sections and the intervening horizontal or vertical dimension. The planimeter shall be considered an instrument of precision adapted to measurement of all areas.

9-1.2 Methods of Measurement. Materials and items of work which are to be paid for on basis of measurement shall be measured in accordance with methods stipulated in the particular sections involved.

9-1.3 Certified Weights. When payment is to be made on the basis of weight, the weighing shall be done on certified platform scales or, when approved by the Engineer, on a completely automated weighing and recording system. The Contractor shall furnish the Engineer with duplicate licensed weighmaster's certificates showing actual net weights. The Agency will accept the certificates as evidence of weights delivered.

9-2 LUMP SUM WORK. Items for which quantities are indicated "Lump Sum", "LS", or "Job", shall be paid for at the price indicated in the Bid. Such payment shall be full compensation for the items of work and all work appurtenant thereto.

When required by the Special Provisions or requested by the Engineer, the Contractor shall submit to the Engineer within 15 days after award of Contract, a detailed schedule in triplicate, to be used only as a basis for determining progress payments on a lump sum contract or designated lump sum Bid item. This schedule shall equal the lump sum Contract Price or Bid item price and shall be in such form and sufficiently detailed as to satisfy the Engineer that it correctly represents a reasonable apportionment of the lump sum.

9-3 PAYMENT.

9-3.1 General. The quantities listed in the Proposal will not govern final payment. Payment to the Contractor will be made only for actual quantities of Contract items constructed in accordance with the Contract Documents. Upon completion of the Work, if the actual quantities show either an increase or decrease from the quantities in the Proposal, the Contract Unit Prices will prevail except as otherwise specified in 3-2.2.

The unit and lump sum prices to be paid shall be full compensation for the items of work and all appurtenant work, including furnishing all materials, labor, equipment, tools, and incidentals.

Payment will not be made for materials wasted or disposed of in a manner not called for under the Contract. This includes rejected material not unloaded from vehicles, material rejected after it has been placed, and material placed outside of the Plan lines. No compensation will be allowed for disposing of rejected or excess material.

Payment for work performed or materials furnished under an Assessment Act Contract will be made as provided in particular proceedings or legislative act under which such contract was awarded.

Whenever any portion of the Work is performed by the Agency at the Contractor's request, the cost thereof shall be charged against the Contractor, and may be deducted from any amount due or becoming due from the Agency.

Whenever immediate action is required to prevent injury, death, or property damage, and precautions which are the Contractor's responsibility have not been taken and are not reasonably expected to be taken, the Agency may, after reasonable attempt to notify the Contractor, cause such precautions to be taken and shall charge the cost thereof against the Contractor, or may deduct such cost from any amount due or becoming due from the Agency. Agency action or inaction under such circumstances shall not be construed as relieving the Contractor or its Surety from liability.

Payment shall not relieve the Contractor from its obligations under the Contract; nor shall such payment be construed to be acceptance of any of the Work. Payment shall not be construed as the transfer of ownership of any equipment or materials to the Agency. Responsibility of ownership shall remain with the Contractor who shall be obligated to store any fully or partially completed work or structure for which payment has been made; or replace any materials or equipment required to be provided under the Contract which may be damaged, lost, stolen or otherwise degraded in any way prior to completion of the Work, except as otherwise specified in 6-10.

Warranty periods shall not be affected by any payment.

If, within the time fixed by law, a properly executed notice to stop payment is filed with the Agency, due to the Contractor's failure to pay for labor or materials used in the Work, all money due for such labor or materials will be withheld from payment to the Contractor in accordance with applicable laws.

At the expiration of 35 days from the date of acceptance of the performance of the Contract by the Board, or as prescribed by law, the amount deducted from the final estimate and retained by the Agency will be paid to the Contractor except such amounts as are required by law to be withheld by properly executed and filed notices to stop payment, or as may be authorized by the Contract to be further retained.

9-3.2 Partial and Final Payment. The Engineer will, after award of the Contract, establish a closure date for the purpose of making monthly progress payments. The Contractor may request in writing that such monthly closure date be changed. The Engineer may approve such request when it is compatible with the Agency's payment procedure.

Each month, the Engineer will make an approximate measurement of the work performed to the closure date and as a basis for making monthly progress payments, estimate its value based on Contract Unit Prices or in accordance with 9-2. When the Work has been satisfactorily completed, the Engineer will determine the quantity of work performed and prepare the final estimate.

From each progress payment, , not less than 5 percent will be deducted and retained by the Agency. The Agency will withhold not less than 5 percent of the total Contract amount until acceptance of the performance of the Contract by the Board.

No progress payment made to the Contractor or its Sureties will constitute a waiver of the liquidated damages specified in 6-9.

Pursuant to Section 22300 of the Public Contract Code, the Contractor may substitute securities for any monies withheld by the Agency to ensure performance under the Contract.

9-3.3 Delivered Materials. When provided for in the Special Provisions, and subject to the limitation and conditions therein, the cost of materials and equipment delivered but not incorporated into the Work will be included in the progress estimate.

9-3.4 Mobilization. When a Bid item is included in the Proposal for mobilization and subject to the conditions and limitations in the Special Provisions, the costs of work in advance of construction operations and not directly attributable to any specific Bid item will be included in the progress estimate. When no such Bid item is provided, payment for such costs will be considered to be included in the other items of work.

PART 2
CONSTRUCTION MATERIALS

SECTION 200- ROCK MATERIALS

200-1 ROCK PRODUCTS.

200-1.1 General. Rock products are defined as crushed rock, rock dust, gravel, sand, stone for riprap, or any combination thereof. All rock products shall be clean, hard, sound, durable, uniform in quality, and free of any detrimental quantity of soft, friable, thin, elongated or laminated pieces, disintegrated material, organic matter, oil, alkali, or other deleterious substance.

200-1.1.1 Testing. Sieve analysis shall be performed in accordance with California Test 202. Sand equivalent tests shall be performed in accordance with California Test 217. Unless otherwise specified, all percentages referred to in this Section 200 shall be determined by weight. All testing shall use the most current test methods.

200-1.1.2 Statistical Testing. Statistical testing shall conform to the following:

Whenever both individual test results and moving average requirements are specified, materials shall meet both requirements.

Individual samples tested prior to the first use of materials from each source, or prior to the first use of materials after any changes have been made in material processing procedures, shall conform to the limits specified for moving average.

Moving average shall be computed in accordance with 211-5.

200-1.2 Crushed Rock and Rock Dust. Crushed rock and rock dust shall be the product of crushing rock or gravel. The portion of the material that is retained on a 3/8 inch (9.5mm) sieve shall contain at least 50 percent of particles having 3 or more fractured faces. Not over 5 percent shall be pieces that show no such faces resulting from crushing. Of that portion which passes the 3/8 inch (9.5mm) sieve but is retained on the No. 4 (4.75mm) sieve, not more than 10 percent shall be gravel particles. Crushed rock will be designated by nominal size and shall conform to the following gradations:

TABLE 200-1.2 (A)

	Percentage Passing Sieves		
Sieve Size	**1" (25.0mm)**	**3/4" (19.0mm)**	**1/2" (12.5mm)**
1-1/2" (37.5mm)	100	-	-
1" (25.0mm)	90-100	100	-
3/4" (19.0mm)	30-60	90-100	100
1/2" (12.5mm)	0-20	30-60	90-100
3/8" (9.5mm)	-	0-20	20-60
No. 4 (4.75mm)	0-5	0-5	0-15
No. 8 (2.36mm)	-	-	0-5
ASTM C131 Test Grading	A	B	B

TABLE 200-1.2 (A) *(Continued)*

Sieve Size	3/8" (9.5mm)	1/4" (6.3mm)	3/16" (4.75mm)	Rock Dust
1/2" (12.5mm)	100	-	-	-
3/8" (9.5mm)	90-100	-	-	100
1/4" (6.3mm)	-	100	-	-
No. 4 (4.75mm)	30-60	75-100	100	90-100
No. 8 (2.36mm)	0-10	0-25	40-75	-
No. 16 (1.18mm)	-	0-5	0-10	-
No. 30 (600µm)	-	-	-	20-60
No. 200 (75µm)	-	0-2	0-2	5-20
ASTM C131 Test Grading	C	D	D	-

Crushed rock shall meet the following requirements:

TABLE 200-1.2 (B)

Test	Test Method No.	Requirements in percent
Percentage Wear	ASTM C131	
100 Revolutions		15 Maximum
500 Revolutions		52 Maximum

200-1.2.1 Screenings. Screenings when used as a cover aggregate for asphalt emulsion chip seals in accordance with 302-2.5, shall be composed of crushed rock and will be designated by the name of the size of screenings and shall conform to the following gradations in Table 200-1.2.1 (A):

All screenings in 200-1.2.1 and 600-2.6.1 shall meet the following requirements:

TABLE 200-1.2.1 (A)

Sieve Size	Percentage Passing Sieve			
	Coarse 1/2" x No. 4 (12.5 x 4.75mm)	Medium 3/8" x No. 6 (9.5 x 3.35mm)	Medium Fine 5/16" x No. 8 (8.0 x 2.36mm)	Fine 1/4" x No. 10 (6.3 x 2.00mm)
3/4" (19.0mm)	100			
1/2" (12.5mm)	90-100	100	-	-
3/8" (9.5mm)	50-80	90-100	100	100
No. 4 (4.75mm)	0-15	5-30	30-60	60-85
No. 8 (2.36mm)	0-5	0-10	0-15	0-25
No. 16 (1.18mm)	-	0-5	0-5	0-5
No. 30 (600µm)	-	-	0-3	0-3
No. 200 (75µm)	0-2	0-2	0-2	0-2

TABLE 200-1.2.1 (B)

Tests	Test Method No.	Requirements
Percentage Wear (100 revolutions)	ASTM C131	12 Maximum
Percentage Wear (500 revolutions)	ASTM C131	35 Maximum
Film Stripping	California 302	25 Maximum
Cleanness Value	California 227	80 Minimum
California Durability	California 229	52 Minimum

200-1.3 Gravel. Gravel shall be composed entirely of particles that have no more than one fractured face.

200-1.4 Coarse Aggregate for Portland Cement Concrete. Concrete aggregate shall be composed of gravel, crushed rock, or a blended mixture. All concrete aggregate shall be washed before delivery to the batching plant and shall conform to the following:

TABLE 200-1.4 (A)

Tests	Tests Method No.	Requirements
Cleanness Value	California 227	75 Minimum
Percentage Wear	ASTM C131	
100 revolutions		15 Maximum
500 revolutions		52 Maximum
Specific Gravity (Bulk saturated surface dry)	ASTM C127	2.58 Minimum[1]

1. Not more than 15 percent by weight shall be particles with a bulk specific gravity below 2.50.
2. Moving Average calculated in accordance with 211-5; no individual test result used shall be less than 71.

Concrete aggregate will be designated by number and shall conform to the following gradations:

TABLE 200-1.4 (B)

Sieve Size	Percentage Passing Sieve		
	No. 2	No. 3	No. 4
2" (50mm)	100	-	-
1-1/2" (37.5mm)	90-100	100	-
1" (25.0mm)	5-40	90-100	-
3/4" (19.0mm)	0-15	55-85	100
3/8" (9.5mm)	0-5	8-20	85-100
No. 4 (4.75mm)	-	0-5	0-30
No. 8 (2.36mm)	-	0-5	0-10
No. 200 (75μm)	0-2	0-2	0-2
ASTM C131 Test Grading	A	B	C

200-1.5 Sand.

200-1.5.1 General. Sand shall consist of natural or manufactured granular material, or a combination thereof, free of deleterious amounts of organic material, mica, loam, clay, and other substances not suitable for the purpose intended.

200-1.5.2 Sand for Asphalt Concrete. The sand shall conform to the gradation specified for asphalt concrete in 200-1.5.5.

200-1.5.3 Sand for Portland Cement Concrete. Sand for portland cement concrete shall be washed and shall conform to the gradation specified for portland cement concrete in 200-1.5.5 and the following quality requirements:

TABLE 200-1.5.3 (A)

Tests	Test Method No.	Requirements in percent
Organic Impurities	ASTM C40	Satisfactory[1]
Sand Equivalent	California 217	75 Minimum
Moving Average[3]		75 Minimum
Soundness[2]	California 214	10 Maximum

1. The resultant color of the testing solution shall not be darker than the ASTM C40 standard.
2. The soundness requirement will be waived, provided that the durability index, D_f, is 60 or greater, when determined by California Test 229.
3. Moving Average calculated in accordance with 211-5; no individual test result used shall be less than 70.

200-1.5.4 Sand For Air-Placed Concrete. Sand for air-placed concrete shall be washed and conform to the gradation for portland cement concrete in 200-1.5.5. The amount of deleterious substances shall not exceed the limits prescribed in ASTM C33.

200-1.5.5 Sand Gradations. The sand shall conform to the following gradations:

TABLE 200-1.5.5 (A)

Sieve Size	Percentage Passing Sieve		
	Asphalt Concrete	Portland Cement Concrete	Mortar
3/8" (9.5mm)	100	100	-
No. 4 (4.75mm)	-	95-100	100
No. 8 (2.36mm)	75-100	75-90	95-100
No. 16 (1.18mm)	-	55-75	70-95
No. 30 (600µm)	-	30-50	35-70
No.50 (300µm)	-	10-25	5-35
No. 100 (150µm)	-	2-10	0-10
No. 200 (75µm)	0-8[1]	0-5	0-5

1. May be exceeded to permit a maximum of 12 percent, provided the sand equivalent of the asphalt concrete sand is 35 or greater.

200-1.6 Stone for Riprap.

200-1.6.1 General. Stone for riprap shall be quarrystone or cobblestone. Quarrystone shall be angular and cobblestone shall be rounded. Stone shall be of such shape as to form a stable protection structure of the required section. Cobblestone shall not be used on slopes steeper than 1 vertical to 2 horizontal. Flat or elongated shapes will not be accepted unless the thickness of the individual pieces is at least 1/3 of the length.

Stone shall be sound, durable, hard, resistant to abrasion and free from laminations, weak cleavage planes, and the undesirable effects of weathering. It shall be of such character that it will not disintegrate from the action of air, water, or the conditions to be met in handling and placing. All material shall be clean and free from deleterious impurities, including alkali, earth, clay, refuse, and adherent coatings.

200-1.6.2 Grading Requirements. Stone for riprap shall be designated by class and conform to the following gradations:

TABLE 200-1.6 (A)

Rock Size	Percentage Larger Than			
	500 lb. (225 kg) Class	375 lb. (170 kg) Class	Light (90 kg) Class	Facing (35 kg) Class
1000 lbs (450 kg)	0-5	-	-	-
700 lbs (320 kg)	-	0-10	-	-
500 lbs (225 kg)	50-100	10-50	0-5	-
200 lbs (90 kg)	-	85-100	50-100	0-5
75 lbs (35 kg)	90-100	95-100	90-100	50-100
25 lbs (10 kg)	95-100	-	95-100	90-100
2.2 lbs (1 kg)	-	-	-	95-100

Note: The amount of material smaller than the smallest size shown in the table for any class shall not exceed the percentage limit as determined on a weight basis. Compliance with the percentage limits shown in the table for all other sizes of the individual pieces of any class of rock slope protection shall be determined by the ratio of the number of individual pieces larger than the specified size compared to the total number of individual pieces larger than the smallest size listed in the table for that class.

200-1.6.3 Quality Requirements. Visual evaluation of the quarry, including examination of blast samples and diamond drill core samples, suitable tests and service records may be used to determine the acceptability of the stone. The Contractor shall notify the Agency in writing of the intended source of stone at least 60 days prior to use. To ensure the required quality, stone may be subject to petrographic analysis or X-ray diffraction.

The material shall conform to the following requirements:

TABLE 200-1.6 (B)

Tests	Test Method No.	Requirements in percent
Apparent Specific Gravity	ASTM C127	2.50 Minimum
Absorption[1]	California 206	4.20 Maximum
Durability[1]	California 229	52 Minimum
Percentage Wear	ASTM C131	45 Maximum

1. Based on the formula below, absorption may exceed 4.2 percent if the Durability Absorption Ratio (DAR) is greater than 10. Durability may be less than 52 if DAR is greater than 24.

$$DAR = \frac{\text{Coarse Durability Index}}{\text{\% Absorption} + 1}$$

200-2 UNTREATED BASE MATERIALS.

200-2.1 General. Materials for use as untreated base or subbase shall be classified in the order of preference as follows:

Crushed Aggregate Base

Crushed Miscellaneous Base

Pulverized Miscellaneous Base

Processed Miscellaneous Base

Select Subbase

When base material without further qualification is specified, the Contractor shall supply crushed aggregate base. When a particular classification of base material is specified, the Contractor may substitute any higher classification, following the order of preference listed above, of base material for that specified. All processing or blending of materials to meet the grading requirement will be performed at the plant or source. The materials shall compact to a hard, firm, unyielding surface and shall remain stable when saturated with water.

200-2.2 Crushed Aggregate Base.

200-2.2.1 General. Crushed aggregate base shall consist entirely of crushed rock and rock dust conforming to the requirements of 200-1.1 and 200-1.2.

200-2.2.2 Grading. The aggregate shall be uniformly graded and shall conform to the following gradation:

TABLE 200-2.2.2

Sieve	Percentage Passing Sieve
1-1/2" (37.5mm)	100
3/4" (19.0mm)	90-100
3/8" (9.5mm)	50-80
No. 4 (4.75mm)	35-55
No. 30 (600μm)	10-30
No. 200 (75μm)	2-9
ASTM C131 Test Grading	B

200-2.2.3 Quality Requirements. The material shall conform to the following:

TABLE 200-2.2.3 (A)

Tests	Test Method No.	Requirements
R-value[1]	California 301	80 Minimum
Sand Equivalent	California 217	50 Minimum
Percentage Wear	ASTM C131	
100 revolutions		15 Maximum
500 revolutions		52 Maximum
Specific Gravity (Bulk saturated surface dry)	ASTM C127	2.58 Minimum[2]

1. The R-value requirement will be waived, provided the material has an SE of 55 or more.
2. Not more than 15 percent by weight shall be particles with a bulk specific gravity below 2.50.

The Engineer may waive percentage wear and specific gravity requirements, provided that the material has a minimum durability of 40 in accordance with California Test 229.

200-2.3 (Not Used).

200-2.4 Crushed Miscellaneous Base.

200-2.4.1 General. Crushed miscellaneous base shall consist of broken and crushed asphalt concrete or portland cement concrete and may contain crushed aggregate base or other rock materials. The material shall be free of any detrimental quantity of deleterious material as defined in 200-1.1. Material retained on the No. 4 (4.75mm) sieve shall contain no more than 15 percent gravel particles as defined in 200-1.3. The material may contain no more than 3 percent brick by weight as determined by California Test 202 as modified: Brick material retained on a No. 4 (4.75mm) sieve shall be identified visually and separated manually. Brick quantification shall be based on total weight of dry sample.

200-2.4.2 Grading. The material shall be uniformly graded and shall conform to one of the following gradations:

TABLE 200-2.4.2 (A)

Sieve Size	Percentage Passing Sieve	
	Coarse	Fine
2" (50.0mm)	100	
1-1/2" (37.5mm)	85-100	100
3/4" (19.0mm)	50-85	85-100
3/8" (9.5mm)		55-75
No. 4 (4.75mm)	25-45	35-60
No. 30 (600μm)	10-25	10-30
No. 200 (75μm)	2-9	2-9
ASTM C131 Test Grading	A	B

When there is a difference in specific gravity (bulk saturated surface dry per ASTM C127) of 0.2 or more between that portion retained and that portion passing a No. 4 (4.75mm) sieve, the grading will be modified by California Test 105.

200-2.4.3 Quality Requirements. This material shall conform to the following:

TABLE 200-2.4.3 (A)

Tests	Test Method No.	Requirements
R-value[1]	California 301	78 Minimum
Sand Equivalent	California 217	35 Minimum
Percentage Wear	ASTM C131	
100 revolutions		15 Maximum
500 revolutions		52 Maximum

1. The R-value requirement may be waived, provided the material has an SE of 40 or more.

The Engineer may waive the percentage wear requirements, provided the material has a minimum durability of 40 in accordance with California Test 229.

200-2.5 Processed Miscellaneous Base.

200-2.5.1 General. Processed miscellaneous base shall consist of broken or crushed asphalt concrete, portland cement concrete, railroad ballast, glass, crushed porcelain material, crushed rock, rock dust, or natural material. The material that is retained on a No. 4 (4.75mm) sieve shall contain no more than 75 percent gravel particles as defined in 200-1.3. The material shall be free of any detrimental quantity of deleterious material as defined in 200-1.1. The material may contain no more than 3 percent brick by weight as determined by California Test 202 modified as follows: Brick material retained on a No. 4 (4.75mm) sieve shall be identified visually and separated manually. Brick quantification shall be based on total weight of dry sample.

200-2.5.2 Grading. The material shall be uniformly graded and shall conform to 200-2.4.2.

200-2.5.3 Quality Requirements. This material shall conform to the following in Table 200-2.4.3 (A):

The Engineer may waive the percentage wear requirements, provided the material has a minimum durability of 35 in accordance with California Test 229.

200-2.6 Select Subbase

200-2.6.1 General. Select subbase shall consist of soil, mineral aggregates, asphalt concrete, portland cement concrete, or blends of these.

200-2.6.2 Grading. The material shall be uniformly graded and shall conform to one of the following gradations:

TABLE 200-2.6.2 (A)

Sieve Size	Percentage Passing Sieve	
	Coarse	Fine
1-1/2" (37.5mm)	100	-
1" (25.0mm)	-	100
No. 4 (4.75mm)	55-75	70-100
No. 16 (1.18mm)	30-75	40-90
No.50 (300μm)	15-40	20-60
No. 200 (75μm)	0-25	0-30

200-2.6.3 Quality Requirements. The material shall also conform to the following requirements:

TABLE 200-2.6.3 (A)

Tests	Test Method No.	Requirements
R-value[1]	California 301	60 Minimum
Sand Equivalent	California 217	20 Minimum

1. The R-value requirement may be waived, provided the material has an SE of 30 or more.

200-2.7 Disintegrated Granite.

200-2.7.1 General. Disintegrated granite shall be free from vegetable matter and other deleterious substances and shall be of such nature that it can be compacted readily under water and rolling to form a firm, stable base.

Disintegrated granite shall be any igneous rock, which has been weathered in place, or any sedimentary material principally derived from igneous rock.

200-2.7.2 Grading. The percentage composition by weight shall conform to Table 200-2.7.2(A).

TABLE 200-2.7.2 (A)

Sieve Size	Percent Passing
1-1/2 in (37.5mm)	100
1 in (25.0mm)	90-100
No. 4 (4.75mm)	50-100
No. 30 (600μm)	25-55
No. 200 (75μm)	5-18

200-2.7.3 Quality Requirements. Disintegrated granite shall also conform to Table 200-2.7.3(A).

TABLE 200-2.7.3 (A)

Test	Test Method No.	Requirements
R-value	California 301	73 Minimum
Sand Equivalent	California 217	30 Minimum

200-2.8 Pulverized Miscellaneous Base.

200-2.8.1 General. Pulverized miscellaneous base shall consist of asphalt concrete or Portland cement concrete pavement that has been pulverized in place to the dimensions shown on the Plans. Pulverized miscellaneous base may contain underlying base material or subgrade soil. The material shall be free of any detrimental quantity of deleterious material as specified in 200-1.1.

Pulverized miscellaneous base, if produced in place, shall only be produced through the operation of a machine specifically designed for pulverizing, stabilizing and blending of this material. Pulverized miscellaneous base shall not be produced by the operation of self-propelled drop hammers ("stompers"), breakers, steel padded rollers, or by ripping or excavation equipment.

200-2.8.2 Testing. The Agency will collect samples of the pulverized material in accordance with California Test 125. The Agency will perform a minimum of one gradation and one sand equivalent test per 500 cubic yards of material pulverized. The Contractor shall allow the Agency a minimum of 48 hours to complete each test and inform the Contractor of the results.

200-2.8.3 Grading. The material shall be uniformly graded and shall conform to Table 200-2.8.2 (A):

TABLE 200-2.8.2 (A)

Sieve Size	Percentage Passing Sieves
2" (50mm)	100
1" (25mm)	-
3/4" (19mm)	85-100
3/8" (9.5mm)	55-80
No. 4 (4.75mm)	35-60
No. 30 (600μm)	10-30
No. 200 (75μm)	2-9
ASTM C131 Grading	B

If, after testing by the Agency, the pulverized material does not conform to the gradation specified in Table 200-2.8.2 (A), a maximum of 35 percent rock products conforming to 200-1 may be blended into the pulverized material to correct the gradation. The Contractor shall determine the amount of rock products to be blended. The rock products shall be uniformly spread and blended over the area requiring correction of the gradation. The equipment used for blending shall conform to 200-2.8.1. The rock products shall be supplied, and blended at no additional cost to the Agency. The Agency will retest the corrected area in conformance with 200-2.8.2.

When there is a difference in specific gravity (bulk saturated surface dry conforming to ASTM C127) of 0.2 or more between that portion retained and that portion passing a No. 4 sieve, a modified grading will be required. The grading will be modified in accordance with California Test 105. The Agency will provide the Contractor with the modified grading required.

200-2.8.4 Quality Requirements. This material shall conform to the following:

TABLE 200-2.8.3 (A)

Tests	Test Method No.	Requirements
Sand Equivalent	California 217	40 Minimum
Percentage Wear	ASTM C131	
100 Rev		15 Maximum[1]
500 Rev		52 Maximum[1]
R-Value	California 301	78 Minimum[2]

1. The Engineer may waive the percentage wear requirements, provided the material has a minimum durability of 40 in accordance with California Test 229.
2. The R-Value may be waived if the SE is 45 or greater.

200-3 NOT USED.

200-4 LEAN CONCRETE BASE (LCB).

200-4.1 General. LCB shall consist of a mixture of:

a) portland cement,

b) aggregate,

c) water, and

d) at the option of the Contractor, water reducing and/or air-entraining admixtures.

The proportions of portland cement, water and other materials shall be determined by the Contractor.

The Contractor shall submit a mix design with laboratory results in accordance with 2-5.3.4. The mix design shall show the amount and type of portland cement, and the amount of water, the aggregate grading and source, and the type and manufacturer of any proposed admixtures.

The amount of portland cement shall not to exceed 300 pounds per cubic yard (178 kilograms per cubic meter). The compressive strength, when tested in accordance with California Test 548, shall be a minimum of 700 psi (4.8MPa) at 7 days.

200-4.2 Materials.

200-4.2.1 Portland Cement. Portland cement shall be Type II/V conforming to 201-1.2.1, added at a rate of 270 pounds per cubic yard (160 kilograms per cubic meter). SCMs shall not be substituted for portland cement.

200-4.2.2 Aggregate.

200-4.2.2.1 General. Aggregate shall conform to 200-2. Should the Contractor change the source or grading, the Contractor shall notify the Engineer, in writing, of the new source or grading. The Contractor shall make the aggregate available for sampling and testing at least 45 days prior to its intended use.

The 1-1/2 inch (38mm) maximum or the 1 inch (25mm) maximum may be used. Once a grading is selected, the grading shall not be changed without the Engineer's written approval.

200-4.2.2.2 Combined Aggregate Gradation. The combined aggregate gradation for the grading selected shall conform to Table 200-4.2.2.1 (A). The percentage retained shall be determined by California Test 202, modified by California Test 105 when there is a difference in specific gravity of 0.2 or more between the coarse and fine aggregate.

TABLE 200-4.2.2.1 (A)

Sieve Sizes	Percent Passing1			
	1-1/2" (37.5mm) Maximum		1" (25.0mm) Maximum	
	Moving Average[2]	Individual Test Results	Moving Average[2]	Individual Test Results
2" (50.0mm)	100	100	---	---
1 ½" (37.5mm)	90 -100	87-100	100	100
1" (25.0mm)	—	—	90 – 100	87 – 100
3/4" (19.0mm)	50 – 85	45 – 90	50 – 100	45 – 100
3/8" (9.5mm)	40 – 75	35 – 80	40 – 75	35 – 80
No. 4 (4.75mm)	25 – 60	20 – 65	35 – 60	30 – 65
No. 30	10 – 30	6 – 34	10 – 30	6 – 34
No. 200	0 – 12	0 – 15	0 – 12	0 – 15

1. Coarse aggregate is material retained on the No. 4 (4.75mm) sieve and fine aggregate is material passing the No. 4 (4.75mm) sieve.
2. Statistical testing shall conform to 400-1.1.3 as modified in 301-4.6.2.

200-4.2.2.3 Sand Equivalent. Aggregate shall not be treated with lime, portland cement or other chemicals before being tested. Aggregate shall have a sand equivalent value of not less than that shown in Table 200-4.2.2.2 (A).

TABLE 200-4.2.2.2 (A)

Moving Average[1]	30
Individual Test Results	27

1. Statistical testing shall conform to 400-1.1.3 as modified in 301-4.6.2.

If the test results of the tests for either or both aggregate grading and sand equivalent tests do not meet the requirements specified for the "moving average" but meet the "individual test" requirements, placement of the LCB may be continued for the remainder of the working day. Work shall not resume until tests indicate that the aggregate to be used complies with the requirements specified for the "moving average."

No single aggregate grading or sand equivalent test shall represent more than 500 cubic yards (400 cubic meters) or one day of production, whichever is smaller.

200-4.2.3 Water. Water shall conform to 201-1.2.3.

200-4.2.4 Chemical Admixtures. Chemical admixtures shall conform to 201-1.2.4. Water reducing admixtures shall be Type A or Type F.

The air content of LCB shall not exceed 4 percent. An air-entrainment reducing admixture shall be used when aggregate is produced from reclaimed material containing asphalt concrete or other material which would cause the air content to exceed 4 percent.

200-4.3 Proportioning, Mixing and Transporting. Proportioning shall conform to 201-1.3, except that the dividing of aggregate into sizes will not be required and the slump shall not exceed 3 inches (75mm).

Mixing and transporting shall conform to 201-1.4 and 201-1.5.

SECTION 201 - CONCRETE, MORTAR, AND RELATED MATERIALS

201-1 PORTLAND CEMENT CONCRETE.

201-1.1 Requirements.

201-1.1.1 General. Concrete shall consist of portland cement, concrete aggregates, water, and when specified or approved for use, chemical admixtures, and/or SCMs, fibers, color, and/or reclaimed concrete material. Concrete shall be specified by class, alternate class, special exposure, or compressive strength.

Concrete specified by compressive strength shall be designed by the Contractor in accordance with 201-1.1.4.

Chemical admixtures shall be used in accordance with 201-1.2.4 and at the manufacturer's specified dosage rate. Additional portland cement or SCMs may be used to obtain high early strength in concrete, except that the total amount of portland cement shall not exceed 700 lbs per yd^3 (415kg per m^3) unless otherwise approved by the Engineer.

Color concrete shall conform to 303-7.

Fiber reinforcement shall conform to 201-2.3 and be added at the rate specified in the Special Provisions.

Reclaimed concrete material conforming to 201-1.2.6 may be incorporated into concrete mixtures when so specified in the Special Provisions or approved by the Engineer.

When concrete is specified by class, alternate class, or special exposure, a mix design shall be submitted to the Engineer for approval in accordance with 2-5.3. The mix design shall specify the proportions of aggregate, water, and when applicable, SCMs, chemical admixtures, and reclaimed concrete material. The mix design shall also include the gradation and source of aggregate, the type and source of portland cement and SCMs, the brand and designation of chemical admixtures, and slump requirement. If so specified in the Special Provisions, the fiber type, fiber manufacturer and rate of addition, and the brand and amount of color, and the type of construction for which the concrete is to be used shall also be included in the mix design.

201-1.1.2 Concrete Specified by Class and Alternate Class. When the specified by class, concrete will be designated by a number, one or two letters, and a number. The first number is the weight of portland cement conforming to 201-1.2.1 in lbs per yd^3 (kgs per m^3), the first letter is the combined aggregate gradation conforming to 201-1.3.2 and the second letter (W) designates the required use of a water reducing admixture conforming to 201-1.2.4. The last number is the minimum compressive strength at 28 days in psi (MPa). A water reducing admixture conforming to 210-1.2.4 may be used in any concrete specified by class and is required in all 4000 psi (28MPa) compressive strength concrete specified by class.

When specified by alternate class, concrete will be designated by a number, three letters, and a number. The first number is the weight of cementitious material in lbs per yd^3 (kgs per m^3) which consists of 85 percent Portland cement conforming to 201-1.2.1 and 15 percent SCMs by weight. The first letter is the combined aggregate gradation, conforming to 201-1.3.2, the second letter (F) designates the required use of SCMs conforming to 201-1.2.5, the third letter (W) designates the required use of a water reducing admixture conforming to 201-1.2.4. The last number is the minimum compressive strength at 28 days in psi (MPa).

The concrete class, alternate class, and maximum slump for the various types of construction shall be as specified in Table 201-1.2.2(A).

TABLE 201-1.1.2 (A)

Type of Construction	Concrete Class[5, 6, 7] U.S. Standard Measures (Metric Units)	Alternate Class U.S. Standard Measures (Metric Units)	Maximum Slump Inches (Millimeters)
Street Surface Improvements			
Concrete Pavement (not integral with curb)	520-A-2500 (310-A-17)	494-AFW-2500 (295-AFW-17)	3 (75)
Curb, Integral Curb and Pavement, Gutter, Walk, Alley Aprons	520-C-2500 (310-C-17)	494-CFW-2500P[1] (295-CFW-17P[1]	4 (100)
Curb, Integral Curb and Pavement, Gutter, Walk, Alley Aprons	520-C-2500P[1] (310-C-17P[1])	494-CFW-2500 (295-CFW-17)	4 (100)
Extruded Curb, Curb and Gutter	520-C-2500 (310-C-17) 520-D-2500 (310-D-17)	494-CFW-2500 (295-CFW-17) 494-DFW-2500 (295-DFW-17)	2 (50) 2 (50)
Sewer & Storm Drainage Facilities			
Pipe Collars, Beam Support for Pipe, Pre-Cast Manhole Components, Catch Basins, Sidewalk Culverts	560-C-3250[2] (330-C-23[2]) 565-C-3250P[1] (335-C-23P[1])	532-CFW-3250[2] (315-CFW-23[2]) 537-CFW-3250P[1] (320-CFW-23P[1])	5 (125) 4 (100)
Sidehill Surface Drainage Facilities	500-C-2500 (295-C-17) 520-C-2500P[1] (310-C-17P[1]) 560-E-2500P[1] (330-E-17P[1])	475-CFW-2500 (280-CFW-17) 494-CFW-2500P[1] (295-CFW-17P[1]) 532-EFW-2500P[1] (315-EFW-17P[1])	3 (75) 4 (100) 4 (100)
Pipe Bedding and Encasement, Anchors and Thrust Blocks, Wall Support for Pipe[3]	450-C-2000[2] (265-C-14[2]) 565-E-2000P[1] (335-E-14P[1])	537-EFW-2000P[1] (320-EFW-14P[1])	4 (100) 6 (150)
Tunnel Backfill	480-C-2000 (285-C-14) 490-C-2000P[1] (290-C-14P[1])		5 (125) 5 (125)
Trench Backfill	100-E-100 (60-E-0.7)		5 (125)
Reinforced Structures			
Bridges, Buildings, Retaining Walls, and Tunnels	560-C-3250[2] (330-C-23[2]) 565-C-3250P[1] (335-C-23P[1]) 650-CW-4000[5] (385-C-28[5]) 660-CW-4000P[1, 5] (390-CW-28P[1, 5])	532-CFW-3250[2] (315-CFW-23[2]) 537-CFW-3250P[1] (320-CFW-23P[1]) 650-CFW-4000[5] (385-CFW-28[1]) 660-CFW-4000P[1,5] (390-CFW-28P[1,5])	4 (100) 4 (100) 4 (100) 4 (100)
Cast-In-Place piles	560-C-3250[2] (330-C-23[2]) 565-C-3250P[1] (335-C-23P[1])	532-CFW-3250[2] (315-CFW-23[2]) 537-CFW-3250P[1] (320-CFW-23P[1])	4 (100) 4 (100)

Table 201-1.1.2 (A) *Continued*

Type of Construction	Concrete Class[5,6,7] U.S. Standard Measures (Metric Units)	Alternate Class U.S. Standard Measures (Metric Units)	Maximum Slump Inches (Millimeters)
Channels and Boxes, Inverts	560-B-3250 (330-B-23) 565-B-3250P[1] (335-B-23P[1]) 650-BW-4000[5] (385-BW-28[5]) 660-CW-4000P[1,5] (390-CW-28P[1,5])	532-BFW-3250 (315-BFW-23) 537-BFW-3250P[1] (320-BFW-23P[1]) 650-BFW-4000[5] (385-BFW-28[5]) 660-CFW-4000P[1,5] (390-CFW-28P[1,5])	4 (100) 4 (100) 4 (100) 4 (100)
Walls and Deck	560-C-3250 (330-C-23) 565-C-3250P[1] (335-C-23P[1,5]) 650-CW-4000[5] (385-CW-28[5]) 660-CW-4000P[1,5] (390-CW-28P[1,5])	532-CFW-3250 (315-CFW-23) 537-CFW-3250P[1] (320-CFW-23P[1,5]) 650-CFW-4000[5] (390-CFW-28[5]) 660-CFW-4000P[1,5] (390-CFW-28P[1,5])	5 (125) 5 (125) 4 (100) 4 (100)
Miscellaneous			
Street Light and Traffic Signal Foundations, Survey Monuments	560-C-3250 (330-C-23)	532-CFW-3250 (315-CFW-23)	4 (100)
Fence and Guardrail Post Foundations	500-C-2500 (295-C-17)	475-CFW-2500 (280-CFW-17)	5 (125)
Concrete Not Otherwise Specified	560-C-3250 (330-C-23) 565-C-3250P[1] (335-C-23P[1])	532-CFW-3250 (315-CFW-23) 537-CFW-3250P[1] (320-CFW-23P[1])	5 (125) 4 (100)
Air Placed Concrete, Method B	650-D-3250P[1] (385-D-23P[1]) 650-E-3250P[1] (385-E-23P[1])	650-DFW-3250P[1] (385-DFW-23P[1]) 650-EFW-3250P[1] (385-EFW-23P[1])	4 (100) 4 (100)
Coarse Masonry Grout	610-E-2000G[4] (360-E-14G[4])	610-EF-2000G[4] (360-EF-14G[4]) 580-EFW-2000G[4] (345-EFW-14G[4])	10 (250) 10 (250)

1. Concrete mixes followed by a "P" have been designed to accommodate placement by a concrete pump using a 4-inch line. When placing conditions require use of a smaller line pump, a "CSP" gradation may be substituted for the "C" gradation with the approval of the Engineer. A pump mix may be substituted for a similar Class or Alternate Class mix and placed utilizing standard placement methods by the Contractor at its option. Said substitution, if made, shall be at the Contractor's expense.
2. Use "B" aggregate gradation when placing conditions permit.
3. Use limited to bedding concrete over which backfill will be placed not less than 40 hours after placement. For backfill after 24 hours, add 3 pints per 100 pounds of cement (31 milliliters per kilogram of cement) of calcium chloride. For backfill after 16 hours and removal of sheeting after 18 hours, use 390-C-26 (660-C-3750) with 3 pints per 100 pounds of cement (31 milliliters per kilogram of cement) calcium chloride solution.
4. Concrete mixes followed by a "G" have been designed to accommodate the grout requirements of Section 202.
5. A water reducing admixture conforming to 201-1.2.4 may be used in any concrete specified by class and is required in all 4000 psi (28MPa) compressive strength concrete specified by class.
6. Fibers conforming to 201-2.3 may be used in any concrete specified by class.
7. Color conforming to 303-7 may be added to any concrete specified by class.

201-1.1.3 Concrete Specified by Special Exposure. Concrete specified by special exposure, shall be designated by a number, followed by three letters and a number. The first number is the minimum weight of cementitious material in lbs per yd^3 (kg per m^3) as specified in Table 201-1.1.3 (A). The first letter is the combined aggregate gradation conforming to 201-1.3.2. The second and third letters (LE, ME, or SE) designate the level of exposure (Low Exposure, Moderate Exposure or Severe Exposure). The last number is the minimum compressive strength at 28 days in psi (MPa).

Concrete specified by special exposure shall contain Class F fly ash or other SCMs conforming to 201-1.2.5 and a water-reducing admixture conforming to 201-1.2.4, and shall conform to the water-cementitious material ratio specified in Table 201-1.1.3(A).

Concrete specified by special exposure shall be proportioned in accordance with ACI 318, Chapter 4, Durability Requirements. Air entraining admixtures may be incorporated at the manufacturer's recommended rates to produce 4 percent +/- 1 percent air content in the concrete mix. Admixtures containing calcium chloride are prohibited for use in concrete exposed to sulfates. The level of protection shall be as specified in the Special Provisions or shown on the Plans. A mix of more severe exposure may be substituted for a mix of less severe exposure.

TABLE 201-1.1.3 (A)

Special Exposure Mixes	Maximum Water – Cementitious Ratio[2]	Special Exposure U.S. Standard Measure (Metric Units)	Cementitious Material Requirement
LOW EXPOSURE – (% water soluble SO_3 in soil samples 0.10 to 0.20) (Sulfate SO_4 in water samples, PPM – 150 to 1,500)	0.50	650-BLE-4000P[1] (385-BLE-28P[1]) 650-CLE-4000P[1] (385-CLE-28P[1]) 740-DEL-4000P[1] (440-DLE-28P[1])	80% Type II or V portland cement with 20% Class F fly ash
MODERATE EXPOSURE – (% water soluble SO_3 in soil samples 0.20 to 2.00) (Sulfate SO_4 in water samples, PPM – 1500 to 10,000)	0.45	658-BME-4500P[1] (390-BME-31P[1]) 658-CME-4500P[1] (390-CME-31P[1]) 815-DME-4500P[1] (480-DME-35P[1])	80% Type II or V portland cement with 20% Class F fly ash
SEVERE EXPOSURE – (% water soluble SO_3 in soil samples 2.00 or more) (Sulfate SO_4 in water samples, PPM –10,000 or more)	0.40	750-BSE-5000P[1] (450-BSE-35P[1]) 750-CSE-5000P[1] (450-CSE-35P[1]) 875-DSE-5000P[1] (520-DSE-35P[1])	80% Type II or V portland cement with 20% Class F fly ash

1. Concrete mixes followed by a "P" have been designed to accommodate placement by a concrete pump using a 4-inch line.
2. Maximum slump is determined by placement conditions so long as water-cementitious ratio is not exceeded.

201-1.1.4 Concrete Specified by Compressive Strength. The Contractor shall determine the mix proportions of concrete specified in the Special Provision or shown on the Plans by its 28-day compressive strength within the minimum cement, maximum size coarse aggregate and admixture limitations specified herein or otherwise specified in the Special Provisions. Concrete specified by compressive strength shall contain not less than 560 pounds of cement, or cement and SCMs per cubic yard (330 kg of cement, or cement and SCMs per m^3), in accordance with 201-1.2.5, for concrete strengths of 3,250 psi (23MPa) or greater. The Contractor may submit mix designs specified by compressive strength in excess of 28 days with the approval of the Engineer. Such mix designs are not subject to the SCMs limitations of 201-1.2.5.

Calcium chloride may be used only with the approval of the Engineer. Admixtures proposed for use shall be evaluated in accordance with 201-1.2.4.

The proposed mix design shall be evaluated from tests of a field trial batch conforming to the size of load, materials, proportions, slump, mixing and placing equipment, and procedures to be used in the actual work or, with the approval of the Engineer, tests of a laboratory trial batch conforming to the materials, proportions, and slump to be used in the actual work. The trial batch procedure may be waived when test data of prior performance of the proposed mix design is presented by the Contractor and approved by the Engineer. The Contractor may utilize any strength data on file with the Agency for this purpose.

When approved by the Engineer, field trial batches may be placed in the Work at designated locations where concrete of a lower quality is specified. Concrete so placed will be considered for purpose of payment to be the type of concrete specified at that location.

Ten test cylinders shall be molded from the trial batch containing the maximum water content indicated by the mix design. Five of the cylinders shall be tested at 7 days in order to establish 7-day average compressive strength information, and the remaining 5 cylinders shall be tested at no more than 28 days after molding. The average compressive strength of the five 28 day cylinders for field trial tests shall be at least 600 psi (4MPa) greater than the specified strength. The average compressive strength of the five 28 day cylinders for laboratory trial batch tests shall be at least 1000 psi (7MPa) greater than the specified strength for specified strengths less than 3000 psi (21MPa), 1200 psi (8MPa) greater than the specified strength for specified strengths between 3000 psi and 5000 psi (21MPa and 36MPa), and 1400 psi (10MPa) greater than the specified strength for specified strengths greater than 5000 psi (36MPa). The minimum strength of any one cylinder shall not be less than the specified strength.

The placing of concrete specified by compressive strength shall not begin until the mix design has qualified in accordance with the aforesaid test criteria. Should the source of materials or established procedures change, new trial batches may be required.

201-1.1.5 Test for Portland Cement Concrete. Portland cement concrete shall be sampled and tested in accordance with the following ASTM and California Tests:

a) Sampling Fresh Concrete C172
b) Obtaining Drilled Cores C42
c) Molding and Curing Specimens C31
d) Compressive Strength C39
e) Flexural Strength C78
f) Slump C143
g) Air Content .... C173 or C231
h) Unit Weight Yield C138
i) Setting of Mortar .... C191 or C266
j) Mortar Cube Test .... California Test 515
k) Drying Shrinkage (with admixture) .... California Test 530

A compressive strength test shall consist of the average strength of 2 cylinders fabricated from a single load of concrete, except that if any cylinder shows evidence of improper handling, molding, or testing, it shall be discarded and the strength test shall consist of the strength of the remaining cylinder.

The Engineer will determine the frequency of sampling. The Contractor shall afford the Engineer all reasonable access, without charge, for the procurement of samples of fresh concrete at time of placement.

Concrete compressive strength tests representing concrete that has been placed shall attain the following 28-day strength: The average of any 3 consecutive strength tests shall be equal to or greater than the specified 28-day strength. Not more than 10 percent of the tests shall be less than the specified 28-day strength. No test shall be less than 85 percent of the specified 28-day strength.

Concrete represented by compressive strength tests that fail to meet the requirements of this subsection shall be removed from the Work. However, with the approval of the Engineer, the concrete represented by the failing compressive strength tests may be cored for strength testing. Coring shall commence within 5 days of notification by the Engineer. Drilled cores shall be obtained by the Contractor in the presence of the Engineer and tested at the Contractor's expense in accordance with ASTM C42 by a laboratory acceptable to the Engineer. A minimum of 3 cores shall be taken in each area represented by the failing 28-day compressive strength tests. If the average of 3 cores is at least 85 percent of the specified 28-day strength and no single core is less than 75 percent of the specified 28-day strength, the concrete represented may be accepted with no further action required. Additional testing of cores extracted from locations represented by erratic core strength results shall be permitted by the Engineer.

201-1.2 Materials.

201-1.2.1 Portland Cement. All cement to be used or furnished shall be:

a) Type II portland cement, or Type V portland cement conforming to ASTM C150. Additionally, Type II and Type V portland cements shall meet the optional requirements of ASTM C150, Table 2 for maximum equivalent alkalies ($Na_2O + 0.658K_2O$) of 0.60 percent, or Type IP (MS) portland-pozzolan cement conforming to ASTM C595, unless otherwise specified. Type IP (MS) cement shall meet the optional mortar expansion requirements of ASTM C595, Table 2 and contain no more than 20 percent pozzolan, which shall be inter-ground with clinker.

b) Type III low alkali portland cement conforming to the standard requirements of ASTM C150 with the additional optional requirements of ASTM C150, Table 2 for maximum equivalent alkalies ($Na_2O + 0.658K_2O$) of 0.60 percent may be used when approved by the Engineer.

c) Rapid-hardening hydraulic cement may be used when approved by the Engineer. When used, the concrete mix shall conform to the requirements of ASTM C928 and admixtures may be used when approved by the Engineer.

The Contractor shall furnish a Certificate of Compliance signed by the manufacturer identifying the cement and confirming compliance. Supporting test data will be furnished upon request by the Engineer.

Whenever suitable facilities are available for handling and weighing bulk cement, they shall be used. Otherwise, cement shall be delivered in original unopened sacks marked with brand, type, and weight.

Cement shall be stored in such a manner as to permit ready access for inspection and sampling and suitably protected against contamination and moisture. Should any cement show evidence of contamination or be otherwise unsuitable, the Engineer may reject it and require that it be removed from the site.

All portland cement used in concrete for any individual structure shall be of the same brand and type unless otherwise approved by the Engineer.

201-1.2.2 Aggregates. Aggregates shall conform to the requirements prescribed in 200-1 and shall be approved by the Engineer prior to use. Aggregate shall be of such character that it will be possible to produce workable concrete within the limits of slump and water content in 201-1.1.2 and 201-1.3.3.

Methods of handling materials resulting in segregation, degradation, or the combining of materials which results in failure to meet specifications shall not be permitted. The free moisture content of sand shall not exceed 8 percent at the time of batching.

Aggregates shall be nonreactive when tested in accordance with ASTM C289 and evaluated in accordance with Appendix X-1 of ASTM C33. Aggregates found to be potentially reactive may be used only upon written approval of the Engineer.

201-1.2.3 Water. Water used for concrete shall not contain deleterious substances. Water shall not contain an amount of impurities that will cause a change in the time of setting of portland cement of more than 25 percent nor a reduction in relative mortar strength at 7 and 28 days of more than 10 percent compared to results obtained with distilled water.

In conventionally reinforced concrete work, water shall not contain more than 1,000mg/L (ppm) of chlorides calculated as Cl, nor more than 1,000mg/L (ppm) of sulfates calculated as SO_4.

In prestressed concrete work, water shall not contain more than 650mg/L (ppm) of chlorides calculated as Cl, nor more than 800mg/L (ppm) of sulfates calculated as SO_4.

In nonreinforced concrete work, water shall not contain more than 2,000mg/L (ppm) of chlorides calculated as Cl, nor more than 1,500mg/L (ppm) of sulfates calculated as SO_4.

201-1.2.4 Chemical Admixtures. Chemical admixtures specified or approved by the Engineer shall conform to ASTM standards unless otherwise specified in the Special Provisions. Chemical admixtures shall be compatible with all other ingredients of the concrete mixture. Chemical admixtures shall be dispensed in liquid form and the quantity dispensed shall not vary more than 5 percent from the quantity specified. Approval to use an admixture shall not relieve the Contractor of the specified concrete requirements. Use of chemical admixtures shall conform to the trial batch requirements of 201-1.1.4, except for concrete specified by class, by alternate class, or by special exposure.

a) **Water-Reducing, Set-Retarding, and Hydration Stabilizing Admixtures.** Water-reducing (Type A), water-reducing set-retarding (Type D), and hydration stabilizing (Type B or D) admixtures shall conform to ASTM C494.

b) **Accelerating Admixtures.** When calcium chloride is approved for use, it shall conform to ASTM D98. Calcium chloride shall not be used in concrete containing prestressing steel. Calcium chloride shall not be used in concrete containing any reinforcing metal products, unless approved by the Engineer. Other accelerating admixtures shall conform to ASTM C494, accelerating (Type C), or water reducing and accelerating (Type E) admixtures.

c) **High Range, Water-Reducing Admixtures (HRWRA).** HRWRA, i.e., superplasticizers, shall conform to ASTM C494, water-reducing, high range admixtures (Type F), or water-reducing, high range, and retarding admixtures (Type G). The use of HRWRA shall conform to the manufacturer's recommendation for dosage, redose, and point of addition for each specific admixture. HRWRA may be added at the batch plant or at the Work site, as approved by the Engineer.

d) **Air-Entraining Admixtures.** Air-entraining admixtures shall conform to ASTM C260. Concrete containing air-entraining admixtures shall be sampled and tested in accordance with 201-1.1.5. The air content, at the point of discharge shall be within a tolerance of 1.5 percent of the value specified in ASTM C94.

e) **Specific Performance Admixtures (SPA).** SPA shall conform to ASTM C494, specific performance admixtures (Type S). The use of SPA shall conform to the manufacturer's recommendation for dosage, redose, and point of addition. SPA may be added at the batch plant or at the Work site, as approved by the Engineer.

201-1.2.5 Supplementary Cementitious Materials.

201-1.2.5.1 Definitions.

SCMs - Materials that are by products of other processes or natural materials. SCMs may or may not be furthered processed for use in concrete and distribute pozzolanic and/or cementitious reactions.

Pozzolans - Materials that possess little or no cementitious value, but are capable of reacting chemically with calcium hydroxide at ambient temperatures to form compounds with cementitious properties.

Cementitious Materials - Materials that react in the presence of water or calcium hydroxide to form complex silicates having the characteristics of cements.

201-1.2.5.2 General.

SCMs shall be cementitious and/or pozzolanic materials as specified in the Special Provisions.

Bulk SCMs shall be stored in silo storage facilities. SCMs shall be stored in such a manner as to permit ready access for the purposes of inspection, and sampling and provide suitable protection against contamination or moisture. The Engineer may reject SCMs that show evidence of contamination or are otherwise unsuitable, and require that they be removed from the storage site. Facilities shall be provided to discharge the SCMs into the cement hopper in accordance with 201-1.3.1.

SCMs used in concrete for any individual structure shall be from the same source in combination with the same source and type of portland cement, unless otherwise approved by the Engineer.

201-1.2.5.3 Fly Ash. Fly ash shall be incorporated into the alternate class mixes specified in 201-1.1.2, and special exposure mixes specified in 201-1.1.3. Fly ash may also be used in other mixes when approved by the Engineer. The amount of fly ash and portland cement used shall be based upon trial batches in accordance with 201-1.1.4.

The Contractor shall furnish a Certificate of Compliance conforming to 4-1.5 signed by the supplier. The certificate shall identify the type of fly ash, and state that the fly ash conforms to ASTM C618 and the Specifications. Supporting test data shall be furnished when requested by the Engineer. Testing and sampling procedures shall conform to ASTM C311.

a) **Class F Fly Ash.** Class F fly ash shall conform to ASTM C618 and the following:

Loss of ignition	4% maximum
SO_3 content	3% maximum
Moisture content	1% maximum

Test results for special exposure mixes shall be submitted to the Engineer in accordance with 2-5.3. The test results shall show the fly ash to be used is effective in contributing to sulfate resistance in conformance with ASTM C618, Table 3 (optional requirements) as tested in accordance with ASTM C1012. The data submitted shall be less than 6 months old.

Class F fly ash, as a percent by weight of total cementitious material, may exceed 20 percent, when approved by the Engineer.

b) **Class C Fly Ash**. Class C fly ash shall conform to ASTM C618 and the following:

Loss of ignition	2% maximum
SO_3 content	4% maximum
Moisture content	1% maximum

Class C fly ash, as a percent by weight of total cementitious material, shall not exceed 30 percent, unless otherwise approved by the Engineer.

Class C fly ash shall not be used where sulfate resistant concrete is required.

201-1.2.5.4 Class N Pozzolans. Class N pozzolans shall comply with the applicable requirements for the class specified herein and conform to ASTM C618.

Class N pozzolans are:

a) some diatomaceous earths;

b) opaline cherts and shales;

c) tuffs and volcanic ashes or pumicites, calcined or uncalcined; and

d) various materials requiring calcination to induce satisfactory properties, such as some clays and shales.

Class N pozzolans may be used in any mix when approved by the Engineer.

The Contractor shall submit a Certificate of Compliance conforming to 4-1.5, signed by the supplier. The certificate shall identify the type of pozzolan, and state that the pozzolan conforms to ASTM C618 and the Specifications. Supporting test data shall be furnished when requested by the Engineer. Testing and sampling procedures shall conform to ASTM C311.

Class N pozzolan, as a percent by weight of total cementitious material, may exceed 20 percent when approved by the Engineer.

Special exposure mixes specified in 201-1.1.3 shall be proportioned in accordance with ACI 318, Chapter 4. Class N pozzolan shall comply with ASTM C618 Table 3. All test data shall be made available upon the Engineer's request.

In any other mix, the amount of Class N pozzolan and portland cement used shall be determined by trial batches in accordance with 201-1.1.4.

201-1.2.6 Reclaimed Concrete Material. Reclaimed concrete material may be used in concrete mixtures in accordance with this section when approved by the Engineer. Reclaimed concrete material may be either:

a) Reclaimed plastic portland cement concrete (RPPCC)

or

b) Reclaimed non-plastic portland cement concrete materials

The Contractor is required to maintain suitable equipment to classify reclaimed concrete material and document its use in the proportioning of concrete mixtures. The addition and characteristics of reclaimed concrete material will be monitored to ensure the final portland cement concrete composite conforms to the specifications for its Class and use.

All mixtures incorporating reclaimed concrete material will be represented by mix designs in accordance with section 201-1.1.1. The Contractor shall evaluate all mix designs by laboratory or field trial batches. Each trial batch shall conform to the materials, proportions, and slump as proposed by the mix design. When approved by the Engineer, field trial batches may be placed in the Work at designated locations where concrete of lower quality is specified. Concrete so placed will be considered for the purpose of payment to be the type of concrete specified at that location. A minimum of 10 test cylinders shall be molded from the trial batch containing the maximum water content indicated by the mix design. Five of the cylinders shall be tested at 7 days to establish 7-day average compressive strength information. The remaining 5 cylinders shall be tested at no more than 28 days after molding. For field trial batches the average 28 day compressive strength shall be at least 600 psi (4MPa) greater than the specified strength. For laboratory trial batches the average 28-day compressive strength shall be at least 1000 psi (7MPa) greater for specified strengths less than 3000 psi (21MPa), 1200 psi (8MPa) greater for specified strengths between 3000 psi and 5000 psi (21MPa and 36MPa) and 1400 psi (10MPa) greater for specified strengths greater than 5000 psi (36MPa). The minimum strength of any one cylinder shall not be less than the specified strength. Changes in the source of materials or established procedures may require new trial batches. Changes in the quality of materials or failure to comply with the compressive strength requirements of 201-1.1.4 shall require new trial batches unless otherwise approved by the Engineer.

Reclaimed concrete material may not be used in special exposure mixtures and is not normally recommended for use in portland cement concrete where architectural aesthetics are a concern.

a) **Reclaimed Plastic Portland Cement Concrete (RPPCC).** A maximum of 15 percent by volume of reclaimed plastic portland cement concrete may be incorporated into fresh portland cement concrete. Each weighmaster certificate shall show the exact volume of RPPCC in addition to the weighmaster certificate requirements of section 201-1.4.3.

   RPPCC may be any un-hardened portland cement concrete provided its design strength is 2000 psi (14MPa) or greater, its constituent material conforms to section 201-1.2, and it has not attained or has been delayed from attaining initial set either by time or by the incorporation of set-delaying chemical admixtures. When set-delaying chemical admixtures are used, they shall be used at the manufacturer's recommended dosage rates and have a proven history of specifically maintaining and extending both plasticity and set. The Contractor shall maintain process documentation, mix designs, and supportive concrete test data and shall provide the information to the Engineer upon request.

   RPPCC shall be proportioned by volume in accordance with section 201-1.3. RPPCC may be added at any point during the proportioning process that results in a consistent, uniform, and homogeneous final product. For design and proportioning purposes, all RPPCC shall be considered as a 2000 psi (14MPa) mixture, consisting of 470 pounds (280kg) of cementitious material. Additional portland cement shall be added to achieve the minimum portland cement content and/or strength as required for a mixture's Class and use. The quantity and/or constituent materials of the RPPCC shall be monitored and proportioned such that the final portland cement concrete gradation conforms to the requirements of 201-1.3.2.

b) **Reclaimed Non-Plastic Portland Cement Concrete Materials.** Non-plastic portland cement concrete materials shall consist of an individual amount of or a combination of materials resulting from the reclaiming of portland cement concrete. Before reclamation, these materials shall conform to 201-1.2. The materials shall be designated as either reclaimed aggregates (RA) or reclaimed water (RW).

When crushed portland cement concrete is used as aggregate it shall, when combined with the non-reclaimed aggregate at the proposed percentage of use, conform to 201-1.2.2. A maximum of 30 percent RA by weight of total aggregate may be incorporated into fresh portland cement concrete. RA shall consist of crushed and graded concrete aggregates and/or a reclaimed naturally occurring aggregate. Reclaimed naturally occurring aggregates may contain minor residual amounts of portland cement concrete components as a result of reclamation. When 15 percent or less RA by weight of total aggregate is used, the requirements of 201-1.2.2 may be waived by the Engineer provided the final portland cement concrete gradation conforms to the requirements of 201-1.3.2.

RW may consist of non-deleterious amounts of hydrated and un-hydrated portland cement, admixtures, minor amounts of fly ash and fine aggregate. The reclamation process for RW shall include a mechanism to ensure uniformity and homogeneity of the RW. A maximum of 35 percent RW by weight of batch water may be incorporated into fresh portland cement concrete.

RA and RW will be proportioned by weight in accordance with 201-1.3. RA and RW may be added at any point during the proportioning process that results in a consistent, uniform, and homogeneous final product. The quantity and/or constituent materials of the RA shall be monitored and proportioned such that the final portland cement concrete gradation conforms to the requirements of 201-1.3.2.

201-1.3 Proportioning.

201-1.3.1 General. Aggregates and cement shall be proportioned by weight except that when the amount of concrete required for any one contract is 10 cubic yards (8 cubic meters) or less, the materials may be measured by volume. Materials that are proportioned by volume shall be measured in containers of known capacity.

Weigh hoppers shall be charged from bins located directly over them or from conveyor belts. When conveyor belts are used, there shall be a separate belt for each size aggregate. There shall be a moisture meter installed, accurate within 1 percent of actual moisture content, to indicate moisture in the sand.

Bulk cement shall be weighed in an individual hopper and kept separate from the aggregates until ingredients are released for discharge. The cement hopper shall be attached to a separate scale for individual weighing.

The amount of water added to the mixture shall be measured into the mixing drum through a valve with a positive cutoff. When water is measured by weight, it shall be weighed on a separate scale.

Whenever a portable batch plant is set up at a new location the scale assemblies shall be inspected and certified regardless of the date the scales were last tested.

Scales utilized in proportioning shall be either springless dial, multiple beam, or solid-state digital strain gage transducer. Scale graduation shall be no greater than the following:

Aggregate Scales ....25 lbs (10 kg)
Cement Scales ....5 lbs (2 kg)
Water Scales ....5 lbs (2 kg)

If a multiple beam type scale is used, the scale shall be provided with an indicator operated by the main beam which will give positive visible evidence of over or under weight. The indicator shall be so designed that it will operate during the addition of the last 400 pounds (200 kg) of any weighing. The over-travel of the indicator hand shall be at least 1/3 of the loading travel. Indicators shall be enclosed against moisture and dust.

Weighing equipment shall be insulated against vibration and movement of other operating equipment in the plant. When the entire plant is running, the scale reading at cutoff shall not vary from the weight designated by more than 1 percent for cement, 1 percent for water, 1-1/2 percent for any size of aggregate, nor 1 percent for the total aggregate in any batch.

201-1.3.2 Combined Aggregate Gradings. The aggregates, when proportioned and combined in accordance with an approved mix design, shall conform to the gradings shown in Table 201-1.3.2 (A). If the combined aggregate grading conforms to the grading of its specified class, the mix proportions shall be considered satisfied and the individual gradations required by 200-1.4 and 200-1.5 may be waived by the Engineer.

TABLE 201-1.3.2 (A)

Sieve Size	Percentage Passing Sieves					
	Grading A	Grading B	Grading C	Grading CSP	Grading D	Grading E
2" (50mm)	100	100	-	-	-	-
1-1/2" (37.5mm)	95-100	95-100	100	100	-	-
1" (25.0mm)	64-80	80-96	95-100	98-100	-	-
3/4" (19.0mm)	55-71	64-80	77-93	85-98	100	100
3/8" (9.5mm)	37-53	40-52	50-70	67-80	92-100	90-100
No. 4 (4.75mm)	32-42	35-46	39-51	46-65	42-60	60-80
No. 8 (2.36mm)	25-35	28-38	31-41	38-54	33-47	50-70
No. 16 (1.18mm)	18-28	21-31	22-32	27-40	22-38	33-53
No. 30 (600μm)	10-18	10-20	12-22	15-27	17-25	19-35
No.50 (300μm)	3-9	3-10	3-15	3-15	6-15	5-15
No. 100 (150μm)	0-4	0-4	0-5	0-7	1-6	2-7
No. 200 (75μm)	0-2	0-2	0-2	0-3	0-3	0-4

201-1.3.3 Concrete Consistency. The amount of water added at the mixer shall be regulated to take into account the free water in aggregates. Free water is defined as the total water minus the water absorbed by the aggregate in a saturated surface-dry condition.

The amount of water used in the mixture shall not exceed the amount necessary to permit practical placement and consolidation of the concrete. Total free water in the mixture shall not exceed an amount producing the maximum slump specified in 201-1.1.2, and shall not exceed amounts shown in the following:

TABLE 201-1.3.3 (A)

Aggregate Grading[1]	Pounds (Kilograms) Per Cubic Yard (Cubic Meter) of Concrete Slump Inches (millimeters)				
	1" (25mm)	2" (50mm)	3" (75mm)	4" (100mm)	5" (125mm)
A	270 (160)	280 (166)	290 (172)	300 (178)	310 (184)
B	275 (163)	285 (169)	295 (175)	305 (181)	315 (187)
C	290 (172)	300 (178)	310 (184)	320 (190)	330 (196)
D	320 (190)	335 (199)	350 (208)	365 (216)	375 (222)
E	335 (199)	350 (208)	365 (216)	380 (225)	395 (234)

1. When the coarse aggregate is composed solely of crushed rock, the above values may be increased up to 20 pounds (12 kilograms) of water per cubic yard (cubic meter).

When adverse or difficult conditions affect the placement of concrete, the Engineer may authorize a greater slump to be used, provided the cement is increased. Water shall be added at a ratio not to exceed 32 percent of added cement per cubic yard (cubic meter) of concrete, and such additional water and cement shall be provided at the Contractor's expense.

201-1.4 Mixing.

201-1.4.1 General. Machine mixing will be required in all cases other than those in which it would obviously prove to be impractical, in which case hand mixing is permitted. Mixing shall be commenced as soon as possible after the cement is placed in contact with the aggregates, but in no event shall the intervening period exceed 30 minutes.

All concrete mixers shall be of such design and construction and so operated to provide a properly mixed concrete with uniform distribution of ingredients. Mixers shall be maintained properly and be in working order and not have any aluminum parts which will have direct contact with concrete.

201-1.4.2 Paving and Stationary Mixers. Paving and stationary mixers shall be equipped with an accurate automatic timing device so designed and constructed as to lock the discharge lever. The regulation setting of this device shall be under the supervision of the Engineer. Water control equipment shall also be provided with each concrete mixer.

The proper proportions of aggregate, cement, and water for each batch of concrete shall be placed in the mixer and mixed for a period of not less than 1 minute (1-1/2 minutes for reinforced concrete).

The rotating speed at which the mixer shall be operated shall conform to that recommended by the manufacturer. The total volume of materials in any one batch shall not exceed the water level capacity of the mixer or the manufacturer's specified capacity of the mixer.

201-1.4.3 Transit Mixers. The type, capacity, and manner of operation of the mixing and transporting equipment for ready-mix concrete shall conform to the current "Standards for Operation of Truck Mixers and Agitators of the National Ready-Mixed Concrete Association" and the "Truck Mixer and Agitator Standards of the Truck Mixer Manufacturers Bureau." Transit mix concrete trucks shall be equipped with an automatic device for recording the number of revolutions of the drum during the mixing period. Each mixer and agitator shall have attached thereto in a prominent place, a metal plate or plates, installed by the manufacturer. The plate(s) shall be plainly marked with the manufacturer's designated capacity of the drum in terms of the volume of mixed concrete and the speed of rotation for the agitating and mixing speeds of the mixing drum or blades.

Each mixer shall have an identification number painted on it that can be easily read from the batching platform.

The total volume of materials introduced into the mixer shall not exceed the manufacturer's specified mixing capacity. If the concrete so mixed does not meet the uniformity requirements of 210-1.4.1, the amount of materials introduced into the mixer shall be reduced.

The drum of the mixer shall be completely emptied of any previously mixed load. The proper proportions of the required ingredients for each load of concrete shall be placed in the mixer and shall be mixed between 70 and 100 revolutions at the manufacturer's designated mixing speed unless otherwise approved by the Engineer. Additional revolutions of the drum shall be at the manufacturer's designated agitating speed.

When concrete is produced for pavement or concrete structures, all wash water shall be emptied from the mixer before any portion of the succeeding load is introduced, unless it has been measured. For all other work, the mixer shall be empty or may carry up to 10 gallons (40 liters) of water in the drum. This amount may be exceeded if the water is measured to an accuracy of 1 percent of the required total mixing water. Adequate control of ready-mixed concrete may require additional added water mixed into the batch at the discharge point. This water shall be mixed for a minimum of 30 revolutions at the manufacturer's designated mixing speed. Water shall not be added to the load during transit.

The total elapsed time between introduction of water at the batch plant and completely discharging the load shall not exceed 90 minutes. The Engineer may waive this limitation if the concrete is of such slump after the 90 minute time limit has been reached that it can be placed without the addition of water to the batch. Under conditions contributing to quick stiffening of the concrete a time limit less than 90 minutes may be specified by the Engineer.

The Engineer shall be provided with a legible certified weighmaster certificate which shall contain the following information:

a) Name of vendor.

b) Name of Contractor.

c) Work location.

d) Number of cubic yards in the load.

e) Concrete Class, Alternate Class, or Special Exposure mix designation.

f) Amount of water added at the plant (including water in fine aggregates).

g) Maximum allowable water.

h) Time and date of batching.

When concrete is specified by compressive strength, or when required by the Engineer, the certificate shall contain the following additional information:

a) Actual weights of cementitious materials and of each size of aggregate.

b) Brand and type of cement.

c) Brand, type and amount of admixture.

Space shall be provided on the certificate so that amount of water added at the Work site may be indicated.

201-1.4.4 Hand Mixing. Hand mixing is permitted when the amount of concrete required is less than 1 cubic yard (cubic meter). Hand-mixed concrete shall be mixed on a watertight platform or in a mortar box in batches not to exceed 1/3 cubic yard (cubic meter) each. The aggregates shall be spread in a uniform layer over which the required quantity of cement shall be evenly distributed. The entire batch shall be turned with shovels until the ingredients are thoroughly blended before adding water. After adding the proper amount of water, the batch shall again be turned with shovels until a uniform consistency is obtained. Methods of hand mixing which allow the loss of mixing water will not be permitted.

201-1.5 Transporting Batched Materials and Mixed Concrete. The compartments of trucks or other equipment used for the purpose of transporting proportioned dry aggregate and cement, or mixed concrete, shall be suitably constructed to protect and prevent loss or leakage of contents during charging, transit, or discharging.

201-1.6 Pervious Concrete.

201-1.6.1 General. Pervious concrete shall consist of portland cement, coarse concrete aggregate, water, and when approved by the Engineer or specified in the Special Provisions, chemical admixtures, fly ash, or reclaimed concrete material.

201-1.6.2 Materials. Materials shall consist of:

a) Coarse aggregate conforming to 200-1.4, No. 4 unless otherwise specified in the Special Provisions;

b) Portland cement conforming to 201-1.2.1, Type II, unless otherwise specified in the Special Provisions;

c) chemical admixtures conforming to 201-1.2.4;

d) fly ash conforming to 201-1.2.5, the content of which shall be less than 25 percent, and;

e) reclaimed concrete materials conforming to 201-1.2.6.

201-1.6.3 Mix Design. A mix design shall be submitted to the Engineer for approval in accordance with 2-5.3. The mix design shall show the mix identification number and the applicable proportions, weights, and quantities of portland cement, aggregate, water, and where specified fly ash, admixtures, and reclaimed concrete material. The mix design submittal shall also include the size and source of aggregate, the type and source of portland cement and fly ash, the brand and designation of admixtures, and the type of construction for which the concrete is to be used.

The minimum cementitious material shall be 630 lbs/yd^3 (10 Tonne/m^3). The maximum water-to-cementitious ratio shall be 0.30 lb/lb (136 g/g), unless otherwise specified in the Special Provisions. If so specified in the Special Provisions or approved by the Engineer, a hydration stabilizing admixture conforming to ASTM C494 Type B or a Type D admixture may be used. If so specified in the Special Provisions or approved by the Engineer, an air-entraining agent conforming to ASTM C260 may be used.

201-2 REINFORCEMENT FOR CONCRETE.

201-2.1 General. Reinforcement for concrete specified or approved by the Engineer shall conform to the appropriate ASTM standard. Bar, wire, and welded wire reinforcement shall conform to dimensions and details indicated on the Plans or otherwise prescribed. Before placed in any concrete work, it must be thoroughly cleaned. When steel, glass or synthetic fibers are used, the quantity and material type shall conform to details indicated on the Plans or in the Specifications, as approved by the Engineer. Approval to use any reinforcement shall not relieve the Contractor of the specified concrete requirements, including strength.

201-2.2 Steel Reinforcement.

201-2.2.1 Reinforcing Steel. Unless otherwise specified, reinforcing steel shall be either Grade 40 (300) or Grade 60 (400) billet steel conforming to ASTM A615/A615M. Steel bending processes shall conform to the requirements of Manual of Standard Practice of the Concrete Reinforcing Steel Institute. Bending or straightening shall be accomplished so that the steel will not be damaged. Kinked bars shall not be used.

201-2.2.2 Wire Reinforcement. Wire reinforcement shall in all respects fulfill requirements prescribed in ASTM A 82.

201-2.2.3 Welded Wire Reinforcement. Welded wire reinforcement shall conform to ASTM A185. The gage of the wire and the dimensions of the mesh shall be as shown on the Plans or in the Specifications. The welded wire reinforcements shall be so constructed as to retain its original shape and form during necessary handling. The effective cross-sectional area of the wire shall be equal to that specified or indicated on the Plans.

201-2.3 Fiber Reinforcement. Fiber-Reinforced Concrete or Air-Placed Concrete, Type B shall conform to ASTM C1116 material requirements and classifications. Concrete containing fibers shall conform to 201-1.1.4. Fiber Type (Type I, II, or III), fiber manufacturer and addition rate of fibers per cubic yard shall be included in the mix design approved by the Engineer. Type II and Type III fibers shall not be used to replace structural reinforcement, and shall be added at the batch plant.

201-2.3.1 Type I. Steel Fiber-Reinforced Concrete or Air-Placed Concrete, Type B. Type I shall contain stainless steel, alloy steel, or carbon steel fibers.

201-2.3.2 Type II. Glass Fiber-Reinforced Concrete or Air-Placed Concrete, Type B. Type II shall contain alkali-resistant glass fibers.

201-2.3.3 Type III. Synthetic Fiber-Reinforced Concrete or Air-Placed Concrete, Type B. Type III shall contain virgin homopolymer polypropylene fibers or other synthetic fibers.

201-2.4 Samples for Testing.

201-2.4.1 General. No reinforcing steel will be accepted until it has been approved by the Engineer. Samples shall be taken from bars selected by the Engineer and cut in the Engineer's presence. The Contractor shall furnish a certified mill test report for each heat or size of steel when required by the Engineer.

201-2.4.2 Reinforcing Steel Bars. Two sample bars, cut from different bars and 3 feet (1m) in length for sizes No. 3 (10M) through No.5 (16M) and 5 feet (1.5m) in length for size No. 6 (19M) and larger, shall be taken from each bar size and heat number delivered to the Work site on a cumulative tonnage basis in accordance with the following schedule:

TABLE 201-2.4.2 (A)

U.S. Standard Bar Sizes (Metric Sizes)	Bar Size Delivered To the Work Site Tons (Tonnes)
2 (6M)	0.5 (1/2)
3 (10M)	1 (1)
4 (13M)	2 (2)
5 (16M)	3 (3)
6 (19M)	4 (4)
7 (22M)	5 (5)
8 (25M)	7 (6)
9 (29M)	9 (8)
10 (32M)	11 (10)
11 (36M)	13 (12)
14 (43M)	20 (18)
18 (57M)	35 (32)

Note: At least two sample bars shall be taken from each bar size.

201-2.4.3 Wire Reinforcement. One sample consisting of two pieces, each 3 feet (1m) long, shall be taken from each lot of 2 tons (2 tonne) or less for each size of wire delivered to the Work site.

201-2.4.4 Wire Mesh Reinforcement. Two samples of a size suitable for testing shall be taken from each 3,000 square feet (280m^2) of fabric or fraction thereof.

201-3 EXPANSION JOINT FILLER AND JOINT SEALANTS.

201-3.1 General. This section specifies joint fillers and sealants to be used for treating joints in portland cement concrete.

All joints which are to be sealed shall be formed with filler. The filler shall be placed in correct position before concrete is placed against it. Holes or joints in the filler shall be filled with mastic to prevent the passage of mortar or concrete from one side of the joint to the other.

201-3.2 Premolded Joint Filler. Premolded joint filler material shall consist of premolded strips of a durable resilient material.

Unless otherwise specified, premolded joint filler shall be one of the following:

Preformed Expansion Joint Filler (Bituminous) ASTM D994

Nonextruding and Resilient Filler (Bituminous) ASTM D1751

Nonextruding and Resilient Filler (Non-bituminous) ASTM D1752

201-3.3 Polystyrene Joint Filler. Commercial quality expanded polystyrene foam blocks and planks shall be furnished by the Contractor and installed in-place as shown on the Plans. The foam shall be composed of non-interconnecting cells. Expanded polystyrene shall have a flexural strength of 35 psi (240 kPa) and a compressive yield strength of between 16 and 40 psi (110 kPa and 275 kPa) at 5 percent compression.

Surfaces of expanded polystyrene against which concrete is to be placed shall be faced with hardboard. Hardboard shall be 1/8 inch (3mm) minimum thickness conforming to Federal Specifications LLL-B-810, any type. Other facing materials may be used, provided they furnish equivalent protection. All boards shall be held in place by nails, waterproof adhesive, or other means approved by the Engineer.

201-3.4 Type "A" Sealant (Two-Part Polyurethane Sealant). The sealant shall be a polyurethane sealant furnished and placed in accordance with the Specifications for "Two-Component Machine-Mixed Polyurethane Sealant" (State Specification 8030-61J-01).

Polyethylene foam shall be commercial quality, with a continuous impervious, glazed top surface, suitable for retaining the liquid polyurethane sealant in the joint while hardening.

A primer, furnished by the manufacturer of the sealant, shall be applied to the sides of the groove and to all exposed vertical surfaces in the joint prior to placing the polyurethane sealant. The primer shall be dry prior to placing sealant. Contamination of the completed primer with foreign material will be cause for rejection of the primed surface.

201-3.5 Type "B" Sealant (Preformed Elastomeric Sealant). The sealant shall be a preformed elastomeric joint seal conforming to specifications of ASTM D268 and the following requirements:

The manufacturer shall designate the minimum uncompressed width of each size of seal to be furnished. Any seal which has a minimum uncompressed width, measured at any point in the height of the seal, less than that designated by the manufacturer shall not be used.

Seals delivered to the Work site that have dimensional or shape tolerances of such magnitude that the seal will not function as specified shall not be used.

The seal shall consist of a multi-channel, nonporous, homogeneous material furnished in a finished extruded form, and shall be furnished full length for each joint with no more than one shop splice in any 60 foot (18m) length of seal. Field splices shall not be used.

The adhesive used to splice the seal shall be an effective bonding agent and shall be resistant to water and ozone. The lubricant adhesive shall conform to State Specification 701-80-30. All abutting surfaces of shop splices shall be bonded together with adhesive. Shop splices shall have no visible evidence of bond failure.

After the seal has been installed in the joint, shop splices shall have no visible offset of the exterior surfaces and there shall be no evidence of bond failure at the abutting surfaces of the splices.

At all open ends of the seal, each cell of the elastomeric joint seal shall be filled to a depth of 4 inches (100mm) with commercial quality open cell polyurethane foam, or closed by other methods subject to Engineer approval.

Tests of elastomeric joint seals will be conducted in accordance with California Test 673.

201-3.6 Type "C" Sealant (Asphalt-Latex Emulsion Joint Sealant). Asphalt-latex emulsion joint sealant shall be designated for mixing and application by hand methods and shall be suitable for use at temperatures above 50°F (10°C). The sealing compound shall be an emulsion consisting of paving asphalt, emulsified with rubber latex in a suitable emulsifying agent. Rubber latex shall be natural rubber or synthetic latex containing approximately 40 percent solids. The resulting emulsion shall consist of a minimum of 55 percent paving asphalt and a minimum of 36 percent rubber latex.

A setting agent shall be provided in the form of a paste to be added at the rate of approximately 3 fluid ounces per gallon (25mL/L) of emulsion.

The joint sealant shall comply with the requirements of Table 201-3.6 (A).

TABLE 201-3.6 (A)

Determination	Test Method No.	Requirements	Remarks
Furol Viscosity at 77°F (25°C)	ASTM D88	80-250 Secs.	Before adding setting agent.
Sieve Test	ASTM D244	1% maximum	Before adding setting agent.
Penetration[1] at 77°F (25°C)	California 418	50-150	Immediately after mixing, the material is poured into a 6-ounce (175mL) deep ointment can and the specimen allowed to stand in air at a temperature of 77° ± 2°F (25° ± 1°C) for a period of 24 hours. It is then penetrated with a grease cone under a total load of 150 g in accordance with ASTM D217.
Elasticity[1]	California 418	70% minimum	Immediately after mixing, the material is placed in an aluminum pan and cured for 24 hours at 115° ± 2°F (46° ± 1°C). Specimen shall be tested in modified ductility mold per test method.
Total Solids[2]	California 418	70% minimum	3 to 5 g of freshly mixed material is placed in an aluminum pan and dehydrated in a suitable oven maintained at a temperature of 200° ± 3°F (93° ± 2°C) for a period of three hours.
Setting Time[2]	California 418	60 Min. Max.	Material is poured into a 6-ounce (175mL) deep ointment can and penetration checked each 15 minutes. Time is recorded when penetration is less than 200.

1. Mixing ratio of material: Add 10 ± 0.1 g of sodium fluosilicate powder to 200 g of emulsion component.
2. Mixing ratio of material: Add 12mL of setting agent component and 8 fluid ounces (24mL) of emulsion component.

201-3.7 Type "D" Joint Sealant (Hot-Poured Rubber-Asphalt Joint Sealant). Hot-poured rubber asphalt joint sealants shall conform to requirements of ASTM D1190. They shall be applied in accordance with the manufacturer's recommendation and 303-1.8.7.

201-3.8 Type "E" Joint Sealant (Polysulfide Polymer and Rubber Rod). Polysulfide polymer type joint sealant shall be a two-component, plastic, rubber-like, cold-applied joint sealant. This material shall resist the intrusion of foreign materials into the joint throughout repeated cycles of expansion and contraction.

The joint sealant shall be packaged in sealed containers identified by the name of the manufacturer, lot number, and date, and bear instructions for mixing and application. Accelerator containers shall be marked "A", and polysulfide containers shall be marked "B". The weight of component "A" in the mixture shall not be less than 10 percent of component "B". Minimum polysulfide content of the mixture shall be 20 percent by weight.

Upon opening the container, component "B" shall not exhibit more than a slight degree of "skinning" on the surface of the material.

The joint sealant shall comply with the following requirements when laboratory mixed materials are tested in accordance with California Test 413:

TABLE 201-3.8 (A)

Determination	Requirements
Viscosity, 5 minutes after mixing, poises (Pa·s)	200 to 350 (20 to 35)
Application time (pot life or time to reach 2000 poises [200 Pa·s] at 77°F [25°C], hours minimum)	1
Penetration at 77°F (25°C) after aging 24 hours at 77°F (25°C)	150 maximum
Nonvolatile content, percent	88 minimum

The following tests are to be performed on the mixed material after 96 hours of aging at 77°F (25°C) :

TABLE 201-3.8 (B)

Determination	Requirements
Penetration at 77°F (25°C)	50 to 120
Penetration at 158°F (70°C)	175 maximum
Resilience at 77°F (25°C), percent	70 min., 60 min.[1]
Resilience at 158°F (70°C), percent	60 min., 50 min.[1]
Bond to concrete, 100% extension, dry at -20°F (-29°C)	No failure
Bond to concrete, 50% extension, wet at -20°F (-29°C)	No failure

1. This requirement applied if the penetration at 77°F (25°C), after 96 hours aging at 77°F (25°C) is 100 to 129

The following test is to be performed on the sample from the resilience test above, oven-aged for 7 days at 158°F (70°C):

TABLE 201-3.8 (C)

Determination	Requirements
Resilience of oven-aged sample after 7 days at 77°F (25°C), percent	70 min., 60 min.[1]

1. This requirement applied if the penetration at 77°F (25°C), after 96 hours aging at 77°F (25°C) is 100 to 129

The rubber rod shall be a cellular synthetic rubber, either butyl or neoprene, of circular cross section with continuous skin impervious to water. It must comply with the following requirements:

TABLE 201-3.8 (D)

Determination	Requirements
Specific gravity (Bulk)	0.30 to 0.50
Surface hardness, Durometer (Fig. 1) ASTM D2240	60 to 85
Water absorption, by weight, one inch (25.4mm) head at 75°F (24°C), for 24 hours, percent	5 maximum
Resistance to ozone, ASTM D470, except 30 minutes exposure, at 70 ± 10mg/L ozone, at 85°F (29°C)	No cracks

201-3.9 Test Report and Certification. When requested by the Engineer, the Contractor shall provide at delivery time, certified copies of the vendor's test report. This shall indicate the vendor name, type of joint sealant, date and point of delivery, quantity, ticket number, lot number, and results of the required test. The test report shall be signed by an authorized representative of the vendor. The certified test reports and testing required in connection therewith shall be without cost to the Agency.

201-4 CONCRETE CURING MATERIALS.

201-4.1 Membrane Curing Compounds.

201-4.1.1 General. Curing compound shall consist of a liquid which, when applied to fresh concrete, will form a continuous membrane over the exposed surfaces of the concrete.

Curing compounds shall conform to the requirements of ASTM C309 Class B (Resin Type Only), except the loss of water shall not exceed 0.15 kilogram per square meter in 24 hours nor 0.45 kilogram per square meter in 72 hours when tested in accordance with ASTM C156.

Concrete curing compounds are designated by type as follows:

Type 1 - Clear or translucent without dye.

Type 1-D - Clear or translucent with fugitive dye.

Type 2 - White pigmented

Unless otherwise stated on the Plans or in the Specifications, Type 1-D shall be used, except that Type 2 shall be used for the top surface of bridge decks. Curing compounds which contain such coloration as to be easily visible following application shall be considered as meeting the fugitive dye requirement.

All compounds shall be furnished by the Contractor and shall be delivered ready-mixed in sealed original containers labeled in accordance with ASTM C309 and with the date of the manufacture. At the time of use, pigmented curing compounds shall be maintained in a thoroughly mixed condition. Containers of curing compound shall remain air-tight when not in use.

201-4.1.2 Application. Curing compound shall be applied in 2 coats according to the manufacturer's directions and 303-1.10. The direction of application of the second coat shall be perpendicular to the first. The second coat shall be applied when the first coat is dry to the touch, but not later than 4 hours after the first coat was applied.

The rate of application shall be such that the compound forms a continuous, unbroken film when applied to the work and shall not be less than the rate used when tested for water retention.

Application of the curing compound to the concrete shall commence as soon as the finished surface of the concrete reaches a uniformly damp appearance with no free water on the surface. At any point, the application rate shall be within 50 square feet per gallon ($1.2m^2/L$) of the nominal rate and the average application rate shall be within 25 square feet per gallon ($0.6m^2/L$) of the nominal rate specified when tested in accordance with California Test 535.

201-4.2 Sheet Curing Materials.

201-4.2.1 General. Sheet curing materials are designated by type as follows:

Type 1 - Waterproof Paper

Type 2 - Polyethylene Film

Type 3 - White-Burlap-Polyethylene Sheet

Type 4 - Burlap

Types 1, 2, and 3 sheet curing materials shall conform to the requirements of ASTM C171, Standard Specification for Sheet Materials for Curing Concrete.

201-4.2.2 Burlap. Burlap shall weigh not less than 10 ounces per linear yard, 40 inches wide ($305 g/m^2$).

201-4.3 Test Report and Certification. The Contractor shall provide, when requested by the Engineer, certified copies of the vendor's test report showing compliance with ASTM C309 and these specifications. The testing and the report required shall be supplied without cost to the Agency.

201-5 CEMENT MORTAR.

201-5.1 General. Cement mortar shall consist of portland cement, sand, and water. Cement and sand shall first be combined in the proper proportions, and then thoroughly mixed with the required amount of water.

Cement mortar shall be designated by class and proportioned by loose volume as shown in Table 201-5.1 (A).

TABLE 201-5.1 (A)

Mortar Designation Class	Proportions	
	Parts Cement	Parts Sand
A	1	1
B	1	1-½
C	1	2
D	1	2-½
E	1	3
F	1	3-½

The quantity of water to be used in the preparation of mortar shall be only that required to produce a mixture sufficiently workable for the purpose intended.

Mortar shall be used as soon as possible after mixing and shall show no visible signs of setting prior to use. Retempering of mortar will not be permitted.

201-5.2 Cement. Portland cement shall conform to 201-1.2.1.

201-5.3 Sand. Sand shall conform to requirements of 200-1.5.3. In proportioning the sand it shall be measured loose (without shaking or compacting) in measuring boxes or other suitable containers of known capacity.

201-5.4 Water. Water shall conform to 201-1.2.3.

201-5.5 Admixtures. No admixture shall be used in mortar unless approved by the Engineer.

201-6 CONTROLLED LOW STRENGTH MATERIAL (CLSM).

201-6.1 Requirements.

201-6.1.1 General. If approved by the Engineer, controlled low strength material (CLSM) may be used as a trench backfill, structural backfill, pipe bedding, or pipe backfill. CLSM shall consist of portland cement, aggregates, water, and fly ash. Chemical admixtures and other mineral admixtures may be used when approved by the Engineer.

The actual mix proportions and flow characteristics shall be determined by the producer of the CLSM to meet the Work site conditions and shall be approved by the Engineer. The mixture shall be workable and non-segregating. When air-entrainment is specified, it shall be evaluated with the actual materials to be used in the Work.

CLSM that is to be hand excavatable, shall contain aggregate no larger than 3/8 inch (9.5mm) and the 3/8 inch (9.5mm) aggregate shall comprise no more than 30 percent of the total aggregate content.

When minimum and/or maximum compressive strength is required, the strength shall be as shown on the Plans.

201-6.2 Materials

201-6.2.1 Portland Cement. Portland cement shall conform to the requirements of 201-1.2.1.

201-6.2.2 Aggregates. Aggregates shall conform to the requirements of 201-1.2.2, except as noted below. Aggregates shall be pretested in CLSM mixtures similar to those anticipated for the Work, confirming their ability to perform as required for the specific application. Aggregates not in conformance with 201-1.2.2 may be used when approved by the Engineer, providing the material has a minimum sand equivalent of 20, the percentage passing the No. 200 (75mm) sieve does not exceed 12 percent, and the fines are non-plastic.

201-6.2.3 Water. Water shall conform to the requirements of 201-1.2.3.

201-6.2.4 Admixtures. Chemical admixtures shall conform to the requirements of 201-1.2.4, and may consist of air-entraining agents, water-reducing admixtures, and other chemical additives. Chemical admixtures shall be approved by the Engineer.

Mineral admixtures other than fly ash may be used when approved by the Engineer.

201-6.2.5 Fly Ash. Fly ash shall conform to the requirements of 201-1.2.5, except it shall not be limited to 20 percent by weight of the total cementitious material. Class C fly ash shall not be used.

201-6.3 Proportioning.

201-6.3.1 General. Proportioning shall conform to requirements of 201-1.3.1. When CLSM is used underneath a paved public right of way, the mixture shall contain a minimum of 25 pounds per cubic yard (15 kilograms per cubic meter) of cement when using washed concrete sand.

201-6.4 Mixing.

201-6.4.1 General. Mixing shall conform to the requirements of 201-1.4, except the 90-minute time limit specified in 201-1.4 shall not apply. Unless otherwise specified, under conditions contributing to quick setting the Engineer may specify a time limit, not to exceed 2-1/2 hours.

201-6.5 Testing. CLSM shall be tested for plastic unit weight. Plastic unit weight shall not deviate more than ± 10 percent of theoretical unit weight shown on the approved mix design. Unit weight shall be determined in accordance with ASTM C138. When a compressive strength is specified on the Plans, a compressive strength test shall be performed as directed by the Engineer.

201-6.6 Placement.

201-6.6.1 General. CLSM may be placed by chutes, conveyors, buckets, pumps or tremies depending upon the application and accessibility.

For trench backfill, CLSM shall be placed continuously. To contain CLSM when filling long open trenches or open-ended structures in stages, the end points shall be adequately bulkheaded to prevent movement. Methods may include bulkheading with sandbags, earth dams, forms, or stiffer mixtures of CLSM.

For bedding, CLSM shall be placed in a manner to prevent flotation or displacement of the embedded item. Methods of preventing flotation or displacement may include placement of CLSM in lifts or, at strategic locations, placing sandbags, straps anchored into soil, faster setting CLSM, or lower slump CLSM over the embedded item.

201-6.7 Replacing Pavement.

201-6.7.1 General. Pavement may be placed directly upon the CLSM as soon as the surface will withstand the paving process without displacement or disruption. If the placement of the CLSM is not completed in time to allow permanent paving to be completed the same day, the Contractor shall prevent traffic contact with the CLSM until paving is completed.

201-7 NON-MASONRY GROUT.

201-7.1 General. Non-masonry grouts are composed of hydraulic cement, fine aggregate, and other ingredients. Quick-setting grouts may be used for setting posts and anchors in concrete or masonry. Non-shrink grout is typically used for setting bearing plates.

201-7.2 Quick-Setting Grout. Quick-setting grout shall be a high strength nonstaining grout approved by the Engineer. It shall reach an initial set within 90 minutes at 70°F (21°C) and reach minimum compressive strength of 2,500 psi (17.2MPa) within 24 hours. Shrinking shall be less than 0.1 percent when tested, using the test procedures of ASTM C596. The grout shall be mixed, handled and placed in accordance with manufacturer's instructions.

201-7.3. Non-Shrink Grout. Non-shrink grout shall be a high strength nonstaining grout approved by the Engineer and meeting the requirements of ASTM C1107. The grout shall be mixed, handled and placed in accordance with the manufacturer's instructions.

SECTION 202 - MASONRY MATERIALS

202-1 BRICK.

202-1.1 General. Brick shall be whole, sound, hard, burned, give a clear ringing sound when struck together, and be uniform in quality. Brick shall be culled or sorted before delivery to the Work site. Mortar used in brick construction shall be Class "D" conforming to 202-3.1. Grout used in brick construction shall conform to 202-3.2. Fine grout shall be used in spaces less than 2 inches (50mm) clear in any dimension. Coarse grout shall be used in spaces 2 inches (50mm) or larger in all horizontal directions.

202-1.2 Manhole Brick. Sewer manhole brick shall conform to ASTM C62, Grade MW, modified as follows:

a) The average compressive strength of 5 bricks shall not be less than 4,000 psi (28MPa) and the compressive strength of any individual brick shall not be less than 3,500 psi (24MPa).

b) The absorption of any individual brick shall not be more than 16 percent when submerged 24 hours in cold water.

c) Dimensions shall conform to the following:

TABLE 202-1.2 (A)

	Depth inches (mm)	Width inches (mm)	Length inches (mm)
Standard Size	2-1/2 (64)	3-7/8 (98)	8-1/4 (210)
Allowed Variation	± 1/4 (± 6)	± 3/8 (± 10)	± 1/2 (± 13)

d) Plaster for brick sewer structures shall be Class "D" mortar conforming to 201-5.1.

202-1.3 Building Brick. Building brick shall conform to ASTM C62, Grade MW. The size and texture shall be as shown on the Plans, as specified in the Special Provisions, or as approved by the Engineer.

202-1.4 Facing Brick. Facing brick shall conform to ASTM C216, Grade MW, Type FBS. The size, color, and texture shall be as shown on the Plans, as specified in the Special Provisions, or as approved by the Engineer.

202-2 CONCRETE BLOCK.

202-2.1 General. Mortar used in concrete block construction shall be Class "D" or "E" conforming to 202-3.1. Grout used in concrete block construction shall conform to 202-3.2. Fine grout shall be used in spaces less than 4 inches (100mm) clear in any dimension. Coarse grout shall be used in spaces 4 inches (100mm) or larger in all horizontal directions.

202-2.2 Masonry Units. Masonry units shall be made with sand-gravel aggregate and conform to ASTM C90. The net size of the units shall be as indicated on the Plans. Unless otherwise specified in the Special Provisions, all units shall be of the normal weight classification [oven-dry weight of concrete 125 pounds per cubic foot (2000 kilograms per cubic meter) or more].

202-3 Mortar, Grout, and Water.

202-3.1 Mortar. Mortar shall be as specified in 201-5.1 to which 1/4 to 1/2 part hydrated lime or lime putty has been added to portland cement mixtures. Mortar shall attain a minimum compressive strength of 1,800 psi (13MPa) in 28 days when tested in accordance with ASTM C109.

202-3.2 Grout.

202-3.2.1 General.

a) Grout shall attain a minimum compressive strength of 2,000 psi (14MPa) in 28 days.

b) Grout shall be tested in accordance with ASTM C1019.

202-3.2.2 Site-Mixed Grout.

a) Site-mixed grout shall be proportioned by volume.

b) Fine grout shall be 1 part portland cement and 2-1/4 to 3 parts sand to which 1/10 part hydrated lime may be added.

c) Coarse grout used in brick construction shall be 1 part portland cement, 2 to 3 parts sand, and not more than 2 parts No. 4 Concrete Aggregate.

d) Coarse grout used in concrete block construction shall be 1 part portland cement, 2 to 3 parts sand, and 1-3/4 to 2 parts No. 4 Concrete Aggregate.

202-3.2.3 Ready-Mixed Grout.

a) Ready mixed grout shall conform to 201-1.

b) Fine grout shall consist of a minimum of 750 pounds per cubic yard (445 kilograms per cubic meter) of portland cement, sand, and sufficient water to achieve a 10-inch (250mm) slump. Admixtures may be used as specified in the Special Provisions or as approved by the Engineer.

c) Coarse grout shall conform to Table 201-1.1.2 (A) for "Coarse Masonry Grout."

202.-3.3 Water. Water shall conform to 201-1.2.3. The quantity of water to be used in the preparation of the mortar or grout shall be the minimum required to produce a mixture sufficiently workable for the purpose intended.

SECTION 203 - BITUMINOUS MATERIALS

203-1 PAVING ASPHALT.

203-1.1 General. Paving asphalt shall be steam-refined asphalt produced from crude asphaltic petroleum or a mixture of refined liquid asphalt and refined solid asphalt. It shall be homogeneous and free from water and residues from distillation of coal, coal tar, or paraffin oil.

203-1.2 Testing Requirements. Paving asphalt shall be specified by performance grade and shall conform to the requirements in Table 203-1.2 (A).

TABLE 203-1.2 (A)

<table>
<tr><th>Grade</th><th>AASHTO Test</th><th>PG 64-10</th><th>PG 64-16</th><th colspan="2">PG 64-28</th><th>PG 70-10</th></tr>
<tr><td colspan="7">Original Binder</td></tr>
<tr><td>Flash Point, °C, minimum</td><td>T48</td><td>230</td><td>230</td><td colspan="2">230</td><td>230</td></tr>
<tr><td>Solubility, %, minimum</td><td>T44</td><td>99.0</td><td>99.0</td><td colspan="2">99.0</td><td>99.0</td></tr>
<tr><td>Viscosity, 135°C, Pa·s, maximum[a]</td><td>T316</td><td>3.0</td><td>3.0</td><td colspan="2">3.0</td><td>3.0</td></tr>
<tr><td>Dynamic Shear Test Temperature, °C</td><td>T315</td><td>64</td><td>64</td><td colspan="2">64</td><td>70</td></tr>
<tr><td>Dynamic Shear, 10 rad/s, G*/Sinδ, kPa, minimum</td><td>T315</td><td>1.00</td><td>1.00</td><td colspan="2">1.00</td><td>1.00</td></tr>
<tr><td colspan="7">RTFO Aged Binder (T240)</td></tr>
<tr><td>Mass Loss, %, maximum</td><td>T240</td><td colspan="2">1.00</td><td>1.00</td><td>1.00</td><td>1.00</td></tr>
<tr><td>Dynamic Shear Test Temperature, °C</td><td>T315</td><td colspan="2">64</td><td>64</td><td>64</td><td>70</td></tr>
<tr><td>Dynamic Shear, 10 rad/s, G*/Sinδ, kPa, minimum</td><td>T315</td><td colspan="2">2.20</td><td>2.20</td><td>2.20</td><td>2.20</td></tr>
<tr><td>Ductility, 25°C, 5 cm/min, cm, minimum</td><td>T51</td><td colspan="2">75</td><td>75</td><td>75</td><td>75</td></tr>
<tr><td colspan="7">RTFO and PAV Aged Binder (R28)</td></tr>
<tr><td>PAV Aging Temperature, °C</td><td>R28</td><td colspan="2">100</td><td>100</td><td>100</td><td>110</td></tr>
<tr><td>Dynamic Shear Test Temperature, °C</td><td>T315</td><td colspan="2">31[b]</td><td>28[b]</td><td>22[b]</td><td>34[b]</td></tr>
<tr><td>Dynamic Shear, 10 rad/s, G*•Sinδ, kPa, maximum</td><td>T315</td><td colspan="2">5000</td><td>5000</td><td>5000</td><td>5000</td></tr>
<tr><td>Bending Beam Test Temperature, °C</td><td>T313</td><td colspan="2">0</td><td>-6</td><td>-18</td><td>0</td></tr>
<tr><td>Creep Stiffness, MPa, maximum</td><td>T313</td><td colspan="2">300</td><td>300</td><td>300</td><td>300</td></tr>
<tr><td>m-value, minimum</td><td>T313</td><td colspan="2">0.300</td><td>0.300</td><td>0.300</td><td>0.300</td></tr>
</table>

a. This requirement may be waived by the Engineer if the supplier warrants that the paving asphalt can be adequately pumped and mixed at temperatures that meet all applicable safety standards.

b. If the PAV aged binder exceeds 5000 kPa at the designated test temperature, it will be deemed acceptable if it is less than 5000 kPa when tested at a 3°C higher test temperature.

203-1.3 Test Reports and Certification. Paving asphalt shall be supplied by Caltrans or other State Department of Transportation approved vendors or as specified in the Special Provisions. At delivery time, the supplying vendor shall deliver to the purchaser a certified copy of the test report. This report shall indicate the vendor's name, grade of paving asphalt delivered, date and point of delivery, quantity delivered, ticket number, purchase order number, and results of specified tests. The certified test report and the testing required in connection with the report shall be submitted to the Agency in accordance with 2-5.3.4.

Until the certified test report and samples of the material have been reviewed by the Engineer to determine their conformity with the specified requirements, the material to which such report relates and any work in which it may have been incorporated as an integral component, will be only tentatively accepted by the Agency. Final acceptance will be dependent upon the determination by the Engineer that the material involved conforms to the Specifications.

203-1.4 Temperatures. Paving asphalt shall not be heated during its manufacture, storage, or during construction so as to cause formation of carbonized particles. At no time shall the temperature be higher than 10°F (5°C) below the actual flash point of the paving asphalt, nor shall it be raised above 375°F (190°C) after loading into a tank car or truck for transport.

Unless otherwise specified in the Special Provisions, the various grades of paving asphalt shall be applied within the temperature range indicated in Table 203-1.4 (A).

TABLE 203-1.4 (A)

ASPHALT GRADE	PLANT MIXING TEMPERATURE °F (°C)		DISTRIBUTION APPLICATION TEMPERATURE °F (°C)	
	Minimum	Maximum	Minimum	Maximum
PG 70-10	300 (150)	350 (175)	285 (140)	350 (175)
PG 64-28	275 (135)	325 (160)	285 (140)	350 (175)
PG 64-16	275 (135)	325 (160)	285 (140)	350 (175)
PG 64-10	275 (135)	325 (160)	285 (140)	350 (175)

Paving asphalt shall be heated in such a manner that no steam or hot oils will be introduced into the paving asphalt during heating. The Contractor shall furnish and keep on-site an accurate thermometer suitable for determining the temperature of the paving asphalt.

203-1.5 Distributing Equipment. Distributing equipment shall conform to 203-2.5.

203-1.6 Measurement and Payment. For all volumetric quantities to be paid for at a Contract Unit Price, the unit of measurement shall be U.S. gallon (liter) at a temperature of 60°F (15°C).

In converting weight to volume, computations shall be based on Table 203-1.6 (A).

TABLE 203-1.6 (A)

Grade of Material	Gallons Per Ton At 60°F (Liters Per Tonne At 15°C)	Lbs. Per Gallon At 60°F (Grams Per Liter At 15°C)
PG 70-10	235 (981)	8.51 (1020)
PG 64-28	235 (981)	8.51 (1020)
PG 64-16	235 (981)	8.51 (1020)
PG 64-10	235 (981)	8.51 (1020)

203-2 LIQUID ASPHALT.

203-2.1 General. Liquid asphalt shall conform to the following classifications:

a) Slow curing products, designated by the letters SC, shall consist of natural crude oils or residual oils from refining of crude asphaltic petroleum, or products resulting from fluxing of a paving asphalt with a light oil.

b) Medium curing products, (MC), shall consist of a paving asphalt conforming to provisions in 203-1, fluxed or blended with a kerosene-type solvent.

c) Rapid curing products (RC), shall consist of a paving asphalt conforming to provisions in 203-1, fluxed or blended with a naphtha solvent.

203-2.2 Test Reports and Certifications. Test reports and certifications will be furnished in accordance with 203-1.3.

203-2.3 Temperatures. The asphalt shall not be heated during its manufacture, storage, or during construction so as to cause the formation of carbonized particles.

At no time after loading for transportation from the refinery to the purchaser, unless authorized by the Engineer, shall the temperature of the asphalt be raised above that given in the last column of the following table and in no case shall the temperature be higher than 10°F (5°C) below flash point. Unless authorized by the Engineer, no asphalt, except tack coats, shall be applied when the air temperature is lower than 40°F (5°C).

Unless otherwise specified, the various grades of asphalt shall be applied at a temperature range as indicated in table 203-2.3 (A):

TABLE 203-2.3 (A)

All Type of Liquid Asphalt	Pugmill Mixing Temp. for MC & SC Liquid Asphalts, °F (°C)		Distributor Application Temperature, °F (°C)	
	Minimum	Maximum	Minimum	Maximum
70	90 (30)	155 (70)	105 (40)	175 (80)
250	125 (50)	200 (90)	140 (60)	225 (110)
800	160 (70)	225 (105)	175 (80)	225 (125)
3000	200 (95)	260 (125)	215 (100)	290 (145)

The asphalt shall be heated in such a manner that no steam or hot oils will be introduced into the asphalt. The Contractor shall furnish and keep on the site an accurate thermometer suitable for determining the temperature of the asphalt.

203-2.4 Test Requirements. Liquid asphalt shall conform to the requirements in the following tables:

TABLE 203-2.4 (A): SLOW CURING PRODUCTS

Test Description	ASTM Test No.	GRADE			
		SC 70	SC 250	SC 800	SC 3000
Flash Point, C.O.C. °F (°C), Min.	D 92	150 (65)	175 (80)	200 (90)	225 (105)
Viscosity at 140°F (60°C) Kinematics, mm^2/s (cSt)	D 2170	70-140	250-500	800-1600	3000-6000
Distillation Total Dist. to 680°F (360°C) Vol. %	D 402	10-30	4-20	2-12	0-5
Kinematic Viscosity on Residue at 140°F (60°C) mm^2/s (cSt)	D 2170	400-7000	800-8500	2002-14000	4000-35000
Asphalt Residue of 100 Pen., % Min.	D 243	50	60	70	80
Ductility of Asphalt Residue at 77°F (25°C) mm, Min.	D 113	1000	1000	1000	1000
Solubility in trichloroethylene, % Min.	D 2042	99.5	99.5	99.5	99.5
Water, % Max.	D 95	0.5	0.5	0.5	0.5
Heptane Xylene Equivalent, % Max. (AASHTO T102)		35	35	35	35

TABLE 203-2.4 (B): MEDIUM CURING PRODUCTS

Test Description	ASTM Test No.	GRADE			
		MC 70	MC 250	MC 800	MC 3000
Flash Point (Tag Open Cup) °F (°C) Min.	D 3143	100 (40)	150 (65)	150 (65)	150 (65)
Viscosity at 140 °F (60 °C) Kinematic, mm^2/s (cSt)	D 2170	70-140	250-500	800-1600	3000-6000
Distillation Distillate [% of total distillate to 680°F (360°C)]	D 402				
To 437°F (225°C)		0-20	0-10	0-3	0-2
To 500°F (260°C)		20-60	15-55	0-35	0-75
To 600°F (315°C)		65-90	60-87	45-80	15-75
Residue from distillation to 680°F (360°C) Volume percent by difference, Min.		55	67	75	80
Tests on Residue from Distillation Penetration at 77°F (25°C), 0.1mm	D 5	120-250	120-250	120-250	120-250
Ductility, 77°F (25°C) mm, Min.[1]	D 113	1000	1000	1000	1000
Solubility in trichloroethylene, % Min.	D 2042	99.5	99.5	99.5	99.5
Water, % Max.	D 95	0.2	0.2	0.2	0.2
Heptane Xylene Equiv., % Max. (AASHTO T102)	—	35	35	35	35

1. If penetration of residue is more than 200 and its ductility at 77°F (25°C) is less than 1000, the material will be acceptable if the ductility at 60°F (15°C) is greater than 1000mm.

TABLE 203-2.4 (C): RAPID CURING PRODUCTS

Test Description	ASTM Test No.	GRADE			
		RC 70	RC 250	RC 800	RC 3000
Flash Point (Tag Open Cup) °F (°C), Min.	D 3143	—	80 (27)	80 (27)	80 (27)
Viscosity at 140°F (60°C) Kinematic mm^2/s (cSt)	D 2170	70-140	250-500	800-1600	3000-6000
Distillation Distillate [% of total distillate to 680°F (360°C), Min.]	D 402				
To 374°F (190°C)		10	—	—	—
To 437°F (225°C)		50	35	15	—
To 500°F (260°C)		70	60	45	25
To 600°F (315°C)		85	80	75	70
Residue from Distillation to 680°F (360°C) Volume percent by difference, Min.		55	65	75	80
Tests on Residue from Distillation Penetration at 77°F (25°C), 0.1mm	D 5	80-120	80-120	80-120	80-120
Ductility, 77°F (25°C) mm, Min.[1]	D 113	1000	1000	1000	1000
Solubility in trichloroethylene, % Min.	D 2042	99.5	99.5	99.5	99.5
Water, % Max.	D 95	0.2	0.2	0.2	0.2
Heptane Xylene Equiv., % Max. (AASHTO T102)	—	35	35	35	35

203-2.5 Distributing Equipment. Distributors shall be of the pressure type with insulated tanks and shall be equipped with the following:

a) A tachometer of the auxiliary wheel type which registers speed in feet (meters) per minute.

b) Charts and devices to provide for accurate and rapid determination and control of the amount of asphalt being applied.

c) A hose and nozzle attachment to be used for areas inaccessible to the spray bar.

d) A pressure gage for determining application pressure.

e) A thermometer for determining temperature of the asphalt.

Distributors and booster tanks shall be so maintained as to prevent dripping of asphalt from any part of the equipment.

Spray bars shall have a minimum length of 9 feet (2.7m). Spray bars and extensions shall be the full circulating type and be adjustable to permit varying height above the surface to be treated.

The nozzles attached to the bar shall be either conical or flat slotted. The distance center to center shall not exceed 6 inches (150mm). The valves which control the flow from nozzles shall be of a positive acting design so as to provide a uniform unbroken spread of asphalt. Valves shall be operated so that all may be simultaneously opened or closed. Each valve shall also be capable of similar independent control.

Spreading equipment shall be so designed and articulated that uniform application of the asphalt may be made ranging from 0.02 to 1.0 gallon per square yard (0.1 L/m^2 to 4.5 L/m^2) of surface and with a range of pressure from 25 to 75 psi (170kPa to 515kPa).

Equipment which fails to produce a satisfactory distribution of asphalt, as determined by the Engineer, shall not be used.

Application of asphalt will not be permitted when, as determined by the Engineer, the surface to be treated is damp or weather conditions are unsuitable.

203-2.6 Measurement and Payment. For all volumetric quantities to be paid for at Contract Unit Price, the unit measure shall be the U.S. gallon (liter) at a temperature of 60°F (15°C).

In converting weight to volume, computations are based on the following:

TABLE 203-2.6 (A)

Grade of Material	Gallons per Ton (Liters per Tonne) @ 60°F (15°C)	Pounds per Gallon (Grams per Liter) @ 60°F (15°C)
70	253 (1056)	7.90 (947)
250	249 (1039)	8.03 (962)
800	245 (1022)	8.16 (978)
3000	241 (1006)	8.30 (995)

All types SC, MC, and RC of the same grade shall be considered to have equal weights per volume.

203-3 EMULSIFIED ASPHALT.

203-3.1 General. Emulsified asphalt shall be composed of a paving asphalt base emulsified with water. Emulsified asphalt may also contain crumb rubber and/or latex. Emulsified asphalt shall remain homogeneous during storage.

203-3.2 Water. Water used in emulsifying paving asphalt shall be potable, free from harmful soluble salts and compatible with the other components.

203-3.3 Latex. Latex shall conform to 203-10.2.2.

203-3.4 Physical Properties.

203-3.4.1 General. Emulsified asphalt shall be anionic, cationic, or rubberized polymer modified emulsion (RPME).

203-3.4.2 Anionic Emulsified Asphalt. Anionic emulsified asphalt shall conform to the requirements shown in Table 203-3.4.2 (A).

TABLE 203-3.4.2 (A)

Test Description	ASTM Test Method No.	Rapid Setting				Slow Setting			
		RS-1		RS-2		SS-1		SS-1h	
		Min.	Max.	Min.	Max.	Min.	Max.	Min.	Max.
Furol Viscosity at 77°F (25°C), sec.	D7496	20	100			20	100	20	100
Furol Viscosity at 122°F (50°C), sec.	D7496			75	400				
Settlement[1] 5 days, %	D6930		5		5		5		5
Storage Stability[2] 1 day	D6930		1		1		1		1
Demulsibility[3] (5mL 0.02N $CaCl_2$), %	D6936	60		60					
Sieve Test (Retained on No. 20 (850μm),%	D6933		0.10		0.10		0.10		0.10
Cement Mixing Test, %	D6935						2.0		2.0
Residue from Distillation, %	D6997	55		63		57		57	
Penetration of residue at 77°F (25°C), 0.1mm[4]	D5	100[4]	200	100[4]	200	100	200	40	90
Solubility of residue,%	D2042	97.5		97.5		97.5		97.5	
Ductility of residue at 77°F (25°C), mm	D113	400		400		400		400	

1. The test requirements for settlement may be waived when the emulsified asphalt is used in less than 5 days time, or the purchaser may require that the settlement test be run from the time the sample is received until it is used, if the elapsed time is less than 5 days.
2. The 24-hour (1-day) storage stability test may be used instead of the 5-day settlement test.
3. The demulsibility test shall be made within 30 days from the date of shipment.
4. A harder base asphalt meeting current paving asphalt specifications may be specified with the provision that the test requirements on the Residue from Distillation test be waived.

203-3.4.3 Cationic Emulsified Asphalt. Cationic emulsified asphalt shall conform to the requirements shown in Table 203-3.4.3 (A).

TABLE 203-3.4.3 (A)

Test Description	ASTM Test MethodNo.	Rapid Setting		Medium Setting			Slow Setting		Quick Set
		CRS-1 Min. Max.	CRS-2 Min. Max.	CRS-2 Min. Max.	CMS-2 Min. Max.	CMS-2h Min. Max.	CSS-1 Min. Max.	CSS-1h Min. Max.	CQS-1h Min. Max.
Furol Viscosity at -77^{0}F (25^{0}C), sec.	D7496						20 100	20 100	15[6] 100[6]
Furol Viscosity at -122^{0}F (50^{0}C), sec.	D7496	20 100	100 400	50 450	50 450	50 450			
Storage Stability Test[1] 5 day %	D6930	5	5	5	5	5	5	5	5
Storage Stability Test[2] 1 day %	D6930	1	1	1	1	1	1	1	1
Demulsibility[3] 35mL 0.8% sodium dactyl sulfosuccinate, %	D6936	40	40						
Sieve Test (Retained on No. 20 [850 μm])	D6933	0.1	0.1	0.1	0.1	0.1	0.1	0.1	0.1
Coating ability and water resistance:	D244								
Coating, dry aggregate				Good	Good	Good			
Coating, after spraying				Fair	Fair	Fair			
Coating, wet aggregate				Fair	Fair	Fair			
Coating, after spraying				Fair	Fair	Fair			
Particle Charge Test	D7402	Positive	Positive	Positive	Positive	Positive	Positive	Positive[5]	Positive
Cement Mixing Test, %	D6935						2.0	2.0	
Oil distillate by vol. of emulsion, %	D6997	3	3	20	12	12			
Residue from Distillation, %	D6997	60	65	60	65	65	57	57	60
Penetration of residue, at 77^{0}F (25^{0}C), 0.1mm[4]	D5	100[4] 250	100[4] 250	100 250	100 250	40 90	100 250	40 90	45 80
Solubility of residue, %	D2042	97.5	97.5	97.5	97.5	97.5	97.5	97.5	97.5
Ductility of residue 77^{0}F (25^{0}C), mm	D113	400	400	400	400	400	400	400	400

1. The test requirements for settlement may be waived when the emulsified asphalt is used in less than 5 days time or the purchaser may require that the settlement test be run from the time the sample is received until it is used if the elapsed time is less than 5 days.
2. The 24-hour (1-day) storage stability test may be used instead of the 5-day settlement test.
3. The demulsibility test shall be made within 30 days from the date of shipment.
4. A harder base asphalt meeting current paving asphalt specifications may be specified with the provision that the requirements on the Residue from Distillation test be waived.
5. Must meet pH requirements of 6.7 maximum (ASTM E70) if the Particle Charge Test results are inconclusive.
6. Viscosity does not apply to latex or polymer modified emulsion.

203-3.4.4 Rubberized Polymer Modified Emulsion (RPME).

203-3.4.4.1 General. RPME shall consist of a mixture of emulsified asphalt, crumb rubber, and latex. RPME shall contain between 0.55 pounds per gallon (66 grams/liter) and 0.65 pounds per gallon (78 grams/liter) of crumb rubber. Latex shall constitute a minimum of 2 percent by weight of RPME.

Crumb rubber shall:

a) be granulated scrap tire rubber free from fabric, wires, and other contaminants

b) be dry and free flowing;

c) have a specific gravity between 1.15 and 1.20; and

d) have a gradation of 100 percent passing the No. 16 (1.18mm) sieve, 95 percent passing the No. 20 (900μm) sieve, and a maximum of 1 percent passing the No. 200 (75μm) sieve.

Calcium carbonate or talc may be added to a maximum of 4 percent by dry weight of crumb rubber to prevent the rubber particles from sticking together.

203-3.4.4.2 Composition. RPME shall conform to the requirements shown in Table 203-3.4.4.2 (A).

TABLE 203-3.4.4.2 (A)

Tests	ASTM Test Method	Requirement	
		Min.	Max.
Viscosity, 77°F (25°C), Brookfield, Model RVT #6 Spindle @ 10 RPM (Centipoise) @ 60 sec.	D 2196	4,000	12,000
Residue by Evaporation % (including fillers)	D6934	50	55
Sieve Test (% retained on No. 20 (850 μm) sieve)	D6933	---	2.0[1]
Penetration of Residue, 77°F (25°C), 0.1mm,	D 5	15	30
Solubility of Residue	D 2042	75	---
Weight lbs/gallon (g/L) 77°F± 1°F (25°C ± 5°C)	D 1475	8.33 lbs/gal (1000g/L)	8.75 lbs/gal (1050g/L)
Asphalt Content[2]	------	40	-----

1. Sieve test of original emulsion is 0.10 max.
2. Asphalt Content shall be determined by multiplying Residue by Evaporation by Solubility of Residue.

203-3.5 Certificate of Compliance. If so specified in the Special Provisions or requested by the Engineer, a Certificate of Compliance conforming to 4-1.5 shall be sent with each load of emulsified asphalt.

203-3.6 Temperature. Emulsified asphalt may be reheated, but at no time after loading shall the temperature be raised above 160°F (70°C). During reheating, emulsified asphalt shall be agitated to prevent localized overheating. Emulsified asphalt shall not be permitted to cool to a temperature less than 40°F (5°C).

Unless otherwise specified in the Special Provisions, emulsified asphalt shall be mixed and applied within the temperature range shown in Table 203-3.6 (A).

TABLE 203-3.6 (A)

Grade of Emulsified Asphalt	Pug Mill Mixing Temperature ^{0}F (^{0}C)		Application Temperature ^{0}F (^{0}C)	
	Min.	Max.	Min.	Max.
CQS-1h	50 (10)	130 (55)	77 (25)	130 (55)
RS-1, CRS-1			77 (25)	130 (55)
RS-2, CRS-2			110 (45)	160 (70)
SS-1, CSS-1	50 (10)	130 (55)	77 (25)	130 (55)
SS-1h, CSS-1h	50 (10)	130 (55)	77 (25)	130 (55)
CMS-2, CMS-2S, CMS-2h	50 (10)	140 (60)	100 (40)	160 (70)
RPME	60 (15)	140 (60)	---	---

The Contractor shall furnish and keep on the Work site a thermometer capable of accurately determining the temperature.

203-3.7 Distribution Equipment. Distribution equipment shall conform to 203-2.5 except that hand spraying by means of a hose or bar through a gear pump or air tank is acceptable for uniform application rates up to 0.10 gallon per square yard (0.45 liter per square meter) for work on flat surfaces or tacking of vertical edges.

203-3.8 Volumetric Measurement. For volumetric quantities, the unit of measurement shall be the U.S. gallon (liter) at a temperature of 60^{0}F (15^{0}C).

In converting weight to volume, computations shall be based on the following, for all grades of emulsified asphalt, except RPME:

Gallons per ton = 240 (Liters per Tonne = 1002)

Pounds per gallon = 8.33 (Liters per Tonne = 998)

For RPME:

Gallons per ton = 235 (Liters per Tonne = 962)

Pounds per gallon = 8.5 (Liters per Tonne = 1018)

203-4 NOT USED.

203-5 SLURRY SEAL.

203-5.1 General. Slurry seal shall be either emulsion-aggregate slurry (EAS) conforming to 203-5.4 or rubberized emulsion-aggregate slurry (REAS) conforming to 203-5.5.

203-5.2 Mix Design. The Contractor shall submit a mix design in accordance with 2-5.3.4 for each combination of emulsified asphalt grade and aggregate gradation to be used in the Work. Each mix design shall conform to ASTM D3910. Laboratory reports supporting each mix design shall be included with the submittals. Laboratory reports shall identify the aggregate source and supplier, emulsified asphalt supplier and all of the test results required in ASTM D3910 except for the cohesion test. When the use of slow-set emulsion is specified in the Special Provisions, the “set time” test is not required. In addition, each mix design shall include the following:

a) amount of emulsified asphalt or RPME in gallons (liters) per ton (Tonne) of aggregate,

b) amount of added water in gallons (liter) per ton (Tonne) of aggregate,

c) quantity and type of set control agents,

d) percent of latex, if so specified in the Special Provisions, and the

e) loose unit weight of aggregate (ASTM C29 with 0.1 ft^3 (3L) bucket).

When a mix design is more than 30 days old, it shall be supplemented with a Certificate of Compliance conforming to 4-1.5 that states the combined aggregate gradation is within ± 3 percent of the reference mix design based on a 30 day moving average or the average of a minimum of 10 of the most current laboratory results, whichever is greater. A mix design shall be reformulated if it is more than one year old or whenever the combined aggregate gradation changes from that in the previously submitted mix design by ± 3 percentage points on any sieve size shown in Table 203-5.3.2 (A). If the source of any aggregate or emulsified asphalt is changed, or the mix design or supporting laboratory reports are over one year old, a new mix design shall be submitted to the Engineer in accordance with 2-5.3.4.

Mix designs shall be based on the following:

a) for Type Fine, the content of RPME and water needed to produce a slurry seal with a maximum wet track abrasion test loss of 50 grams per square foot (540 grams per square meter) when tested in accordance with ASTM D3910;

b) for Type I, the content of emulsified asphalt and water needed to produce a Type I slurry seal with a maximum wet track abrasion test loss of 50 grams per square foot (540 grams per square meter) when tested in accordance with ASTM D3910; or,

c) for Type II or Type III, the content of emulsified asphalt and water needed to produce a slurry seal with a maximum wet track abrasion test loss of 60 grams per square foot (650 grams per square meter) when tested in accordance with ASTM D3910.

ASTM D3910 shall be modified to include the aggregate retained on the No. 4 (4.75mm) sieve for Type II and Type III slurry seals. Type III slurry seals shall use the 3/8 inch (9.5mm) template.

203-5.3 Aggregate.

203-5.3.1 General. Aggregate shall be 100 percent crushed rock conforming to 200-1.1, 200-1.2, and the requirements shown in Table 203-5.3.1 (A).

TABLE 203-5.3.1 (A)

Tests	ASTM Test Method	Requirements
Percentage Wear at 500 Revolutions[1]	C131	40% Maximum
Sand Equivalent	D2419	55 Minimum
Soundness(5 Cycles)[1]	C88	15% Maximum

1. To be run on the material retained on the No. 4 (4.75mm) sieve graded from the source.

203-5.3.2 Grading. The combined aggregate gradation shall be determined in accordance with ASTM C136 and conform to the requirements shown in Table 203-5.3.2 (A).

TABLE 203-5.3.2 (A)

	Percentage Passing Sieves							
	Type Fine		Type I		Type II		Type III	
Sieve Size	Min.	Max.	Min.	Max.	Min.	Max.	Min.	Max.
3/8 inch (9.5mm)	100	---	100	---	100	---	100	---
No. 4 (4.74mm)	100	---	100	---	90	100	70	90
No. 8 (2.36mm)	95	100	90	100	65	90	45	70
No. 16 (1.18mm)	75	92	65	90	45	70	28	50
No. 30 (600µm)	50	75	40	60	30	50	19	34
No. 50 (300µm)	35	50	25	42	18	36	12	25
No. 100 (150µm)	15	30	15	30	10	24	7	18
No. 200 (75µm)	10	20	10	20	5	15	5	15

203-5.4 Emulsion-Aggregate Slurry (EAS).

203-5.4.1 General. EAS shall be a stable mixture of cationic emulsified asphalt, aggregate, water, a set control agent, and when so required, latex. EAS shall be specified by combined aggregate gradation and emulsified asphalt grade, i.e. Type II-CQS-1h-EAS. The combined aggregate gradation and emulsified asphalt grade shall be as specified in the Special Provisions or shown on the Plans.

203-5.4.2 Materials.

203-5.4.2.1 Aggregate. Aggregate shall be Type I, Type II, or Type III conforming to 203-5.3.

203-5.4.2.2 Emulsified Asphalt. Emulsified asphalt shall be slow-set or quick-set, and, if so specified in the Special Provisions or shown on the Plans, contain latex. Slow-set emulsified asphalt shall be CSS-1h conforming to 203-3 unless otherwise specified in the Special Provisions or shown on the Plans. Quick-set emulsified asphalt shall be CQS-1h conforming to 203-3 unless otherwise specified in the Special Provisions or shown on the Plans.

The percentage of emulsified asphalt and residual asphalt content shall conform to the requirements shown in Table 203- 5.4.2.2 (A).

TABLE 203-5.4.2.2 (A)

	Test Method	Type I	Type II	Type III
Emulsified Asphalt %, by weight of dry aggregate.	—	17-20	14-18	11-15
Residual Asphalt Content, %by weight of dry aggregate[1].	ASTM D6307 or CT 382[1]	10 min.	7.5 min.	6.5 min.

1. Sample size shall be 500g minimum.

203-5.4.2.3 Water. Water shall conform to 203-3.2.

203-5.4.2.4 Latex. The addition and exact percentage of latex to be added shall be as specified in the Special Provisions or shown on the Plans. Latex for cationic emulsified asphalt shall conform to 203-3.3 except contain 65 ± 5 percent rubber solids.

203-5.4.2.5 Set Control Agents. Set control agents shall be either Type II or Type V portland cement conforming to 201-1.2.1, aluminum sulfate, or other material approved by the Engineer.

203-5.4.3 Mixing. Mixing shall conform to 302-4.

203-5.5 Rubberized Emulsion Aggregate Slurry (REAS).

203-5.5.1 General. REAS shall be a stable mixture of RPME, aggregate, water, and portland cement. REAS shall be specified by combined aggregate gradation, i.e. Type II-REAS. The combined aggregate gradation shall be as specified in the Special Provisions or shown on the Plans.

203-5.5.2 Materials.

203-5.5.2.1 Aggregate. Aggregate shall be Type Fine, Type I, Type II, or Type III conforming to 203-5.3.

203-5.5.2.2 Rubberized Polymer Modified Emulsion. RPME shall conform to 203-3.4.5. The percentage of RPME and residual RPME solids shall conform to the requirements shown in Table 203-5.5.2.2 (A).

TABLE 203-5.5.2.2 (A)

Tests	Test Method	Type Fine		Type I		Type II		Type III	
		Min.	Max.	Min.	Max.	Min.	Max.	Min.	Max.
RPME % by weight of dry aggregate[1]	ASTM D6307[2] or CT 382[2]	61	85	50	57	33	40	28	35
Residual RPME Solids[3]		31	47	26	31	17	22	15	19

1. Must meet Residual RPME Solids.
2. Sample size shall be 500g minimum.
3. Residual RPME Solids shall be determined by multiplying RPME % (ASTM D6307 or CT 382) by Residue by Evaporation of RPME % (ASTM D6934).

203-5.5.2.3 Water. Water shall conform to 203-3.2.

203-5.5.2.4 Portland Cement. Portland cement may be added to modify the viscosity and curing characteristics accordance with the approved mix design.

Portland cement shall be Type I/II or II/V conforming to 201-1.2.1 and shall not exceed 1.5 percent of the dry weight of the aggregate.

203-5.5.3 Central Plant Mixing.

203-5.5.3.1 General. Mixing at a central mixing plant shall conform to the following requirements:

a) Component materials conforming to 203-5.5.2 shall be stored separately at the plant.

b) Aggregate shall neither be stored nor transported in such a way that may cause segregation, degradation, or intermingling of different size aggregates.

c) Materials shall be proportioned by weight into the mixing tank. Volumetric proportioning will not be permitted.

d) The mixing tank shall be equipped with scales. The zero tolerance for the tank scales shall be 0.5 percent based on the total batch weight. The scales shall be calibrated and certified on a yearly basis in accordance with 4-1.7 or after every modification or repair.

e) The mixing tank shall be equipped with a full sweep mixer/agitator with blades or paddles of a sufficient size and number and operated at a speed sufficient to produce a homogeneous mix. Should the blades, paddles, or other parts of the mixer/agitator become worn to such an extent to adversely affect the quality of the mix they shall be promptly replaced. Insufficient mixing or agitation within the mixing tank shall be corrected by either a reduction in the volume of component materials or other adjustments.

f) Each batch shall be continuously mixed for 3 minutes or until all of the component materials are thoroughly blended, whichever is longer. The mixing time shall begin upon the introduction of the last component material. If the Engineer determines that the mixture is not thoroughly blended, the mixing time shall be increased.

g) Mixed REAS shall be stored at the central mixing plant site in storage tanks equipped with an agitator of a similar configuration to the agitator in the mixing tank. The agitator shall be capable of continuous operation.

203-5.5.3.2 Transporting. REAS shall be transported from the central mixing plant to the Work site in trucks specifically designed for this purpose equipped with an agitator. REAS shall be continuously agitated during transport.

203-5.5.4 Work Site Mixing. Mixing at the Work site in continuous-flow mixers shall conform to 302-4.

203-6 ASPHALT CONCRETE.

203-6.1 General. Asphalt concrete shall be the product of mixing mineral aggregate and a maximum of 15 percent reclaimed asphalt pavement (RAP) with paving asphalt conforming to 203-1 at a central mixing plant. Asphalt concrete containing greater than 15 percent RAP shall conform to 203-7. RAP shall conform to 203-7.2.2, except the viscosity of the paving asphalt recovered from the RAP in accordance with ASTM D1856 (Abson Recovery Method) will not be required. The content of the paving asphalt recovered from the RAP, if required in the Special Provisions, shall be determined in accordance with California Test 382.

Asphalt concrete shall be designated by class and grade (i.e., "C2-PG 64-10"). Asphalt concrete containing up to 15 percent RAP shall be identified by adding the suffix "RAP" to the class and grade (i.e., "C2-PG 64-10 RAP").

203-6.2 Job Mix Formula (JMF) and Mix Designs. The Contractor shall submit in accordance with 2-5.3 a JMF that summarizes each asphalt concrete mix design for each class and grade of asphalt concrete required to construct the Work. The JMF shall identify the source and individual grading of each material used to produce the mix design (including the percentage and individual gradation of any manufactured or natural sands), the combined aggregate gradation, the optimum binder content (OBC), void content, RAP percentage, stability value, plant identification, mix number, and the source and performance grade of the paving asphalt. Upon request, the asphalt concrete mix design test data represented by the JMF shall be immediately made available to the Engineer.

The asphalt concrete mix design method shall be Hveem with the OBC determined by California Test 367. The combined aggregate gradation shall be uniformly graded from coarse to fine when tested in accordance with ASTM C136 and C117 or California Test 202. The target S value, at the OBC, for the mix design shall be at least 2 points above the minimum specified in Table 203-6.4.3(A). The mix design shall conform to the requirements shown in Table 203-6.4.3(A) for the class shown on the Plans or specified in the Special Provisions. These values will be used for mix design only.

When the mix design is more than 30 days old, the JMF shall be supplemented with a mix certificate that shows that the combined aggregate gradation is +/- 3 percent from the referenced mix design based upon a 30 day moving average or a minimum of the 10 most current test results. If the combined aggregate gradation does not meet the previously submitted JMF by +/- 3 percentage points on any sieve even though it may be within the target values or if the source of any aggregate is changed, grade of binder is changed, or the mix design is over 2 years old, a new mix design shall be prepared and a new JMF shall be submitted to the Engineer for approval. Samples shall be obtained from the bins for batch-type plants. Individual samples shall be obtained from under the feeders for drum-type plants.

203-6.3 Materials.

203-6.3.1 Asphalt. The asphalt binder to be mixed with the aggregate shall be paving asphalt or liquid asphalt conforming to 203-1 or 203-2 and the grade shall be as shown on the Plans or specified in the Special Provisions.

203-6.3.2 Aggregate. Coarse aggregate shall be crushed rock conforming to 200-1.2.

Fine aggregate shall be sand, rock dust, mineral filler, or a blend of these materials. Sand shall conform to the requirements of 200-1.5.2. Rock dust shall conform to the requirements of 200-1.2.

If the fine aggregate is deficient in material passing the No. 200 (75μm) sieve, mineral filler conforming to requirements of 203-6.3.3 shall be added to meet the combined grading.

If the combined aggregate grading conforms to the grading of its designated class in 203-6.4.2, the individual grading required by 200-1.2 may be waived. Percentages for the combined gradings, within the specified limits, shall be of such uniformity that the material passing the indicated sieves during any day's run will not exceed the maximum variation as specified in 203-6.4.2.

203-6.3.3 Mineral Filler. Mineral filler shall consist of portland cement or mechanically reduced rock. Mechanically reduced rock shall conform to the grading in Table 203-6.3.3 (A) when tested in accordance with ASTM D422:

TABLE 203-6.3.3 (A)

Particle Size	Percentage
Passing 75μm Sieve	75 - 100
Finer than 50μm	65 - 100
Finer than 20μm	35 - 65
Finer than 10μm	26 - 35
Finer than 5μm	10 – 22

203-6.4 Asphalt Concrete Mixtures.

203-6.4.1 General. Acceptance of plant produced mixtures will be based upon gradation, binder content, and minimum stability values per Table 203-6.4.3 (A). Air void values are not to be used as acceptance criteria. The asphalt concrete shall be sampled in accordance with the requirements of California Test 125. The S-value test results shall conform to the moving average provisions of 400-1.1.3, except that no individual test result shall be found acceptable if it is more than 3 points below the minimum S-value listed in Table 203-6.4.3(A). The S-value result shall be the average of 3 tests from one sample representing 1000 tons or a single day's paving, whichever is less. If the range of stabilities for the 3 tests is more than 10, then the sample shall be discarded and a new sample shall be obtained and retested.

203-6.4.2 Combined Aggregates. Combined aggregates, after all processing except the adding of asphalt and mineral filler, shall have a minimum sand equivalent of 50 when tested by California Test 217.

When there is a difference in specific gravity (bulk saturated surface dry per ASTM C127 and C128) of 0.2 or more between that portion retained and that portion passing a No. 4 (4.75mm) sieve, the grading shall be modified as provided by California Test 105.

203-6.4.3 Composition and Grading. The grading of the combined aggregates and the percentage of asphalt binder shall conform to the requirements indicated in the following tabulations in which the percentages shown are based on the weight of dry aggregate only (outside):

TABLE 203-6.4.3 (A)

Sieve Size	Percentage Passing Sieves							
	A	B	C1	C2	D1	D2	E	F
	Dense Coarse	Dense Medium Coarse	Open Medium	Dense Medium	Open Fine	Dense Fine	Extra Fine	Channel Liner
1-1/2" (37.5mm)	100							
1" (25mm)	90-100	100						
3/4" (19.0mm)	78-90	87-100	100	100				
1/2" (12.5mm)	64-78	70-87	90-100	95-100	100	100		
3/8" (9.5mm)	54-68	55-76	72-88	72-88	90-100	95-100	100	100
No. 4 (4.75mm)	34-48	35-52	40-54	46-60	40-54	58-72	65-85	95-100
No. 8 (2.36mm)	25-35	22-40	18-34	28-42	20-32	34-48	45-65	70-84
No. 30 (600μm)	12-22	8-24	8-20	15-27	6-18	18-32	22-38	36-50
No.50 (300μm)	8-16	5-18	4-14	10-20	2-12	13-23	16-28	23-35
No. 200 (75μm)	3-6	0-7	1-6	2-7	0-5	2-9	6-12	6-12
Asphalt Binder %	4.5-5.5	4.5-5.8	4.6-6.0	4.6-6.0	4.8-6.5	4.8-6.5	5.8-7.8	8.0-10.0
Hveem Stability "S" Value (min.)	35	35	33	33	30	30		
Air Voids*	4%	4%	4%	4%	4%	4%		

* Per CTM 367

The Hveem stability ("S") value shall be per California Tests 304 and 366.

When highly absorptive aggregate with a K factor greater than 1.7 per California Test 303 is used, the maximum percentage for asphalt binder may be increased up to 2.0 percentage points over the values shown in the table. Percentages for the combined gradings, within the specified limits shall be of such uniformity that material passing the indicated sieves during any day's run will not exceed the following maximum variations:

No. 4 (4.75mm) sieve - 6 percentage points

No. 30 (600μm) sieve - 5 percentage points

No. 200 (75μm) sieve - 3 percentage points

The gradations in the above tabulations represent the limits which shall determine the suitability of aggregate for use from the sources of supply.

203-6.5 Aggregate Storing, Drying, and Screening.

203-6.5.1 General. The sand, rock dust, and various sizes of aggregates shall be stored separately at the plant. They shall be evenly fed to the dryer to ensure a uniform flow of properly combined aggregates. In placing materials in storage or in moving them from storage to feeder, no method shall be used which may cause segregation, degradation, or intermingling of different size aggregates. Materials not meeting the gradation requirements shall be discarded or reprocessed to comply with requirements of 203-6.3.2.

203-6.5.2 Batch Plant Method. The aggregates shall be dried and heated for a sufficient time in the dryer so that the moisture content will not be greater than 1 percent.

The dryer shall be provided with an approved heat indicating device in order that the temperature of the aggregate leaving the dryer may be determined. The heat indicating device shall be independently mounted from other plant components, accurate to the nearest 10°F (5°C), and shall be installed in such a manner that fluctuation of 10°F (5°C) in the aggregate temperature will be indicated within 1 minute.

After drying, the aggregates shall be evenly fed to the screens in such quantities as to maintain, in separate bins, a uniform grading of materials and proper balance in the amount of material. The operation of the screens shall be so controlled as to secure a thorough separation of aggregate sizes.

TABLE 203-6.5.2 (A)

Size of Openings	Bin No.
4-Bin Plants	
100% passing 1-1/2" (37.5mm) sieve	4
100% passing 5/8" (16.0mm) sieve	3
100% passing 3/8" (9.5mm) sieve	2
100% passing No. 4 (4.75mm) sieve	1
5-Bin Plants	
100% passing 1-1/2" (37.5mm) sieve	5
100% passing 7/8" (22.4mm) sieve	4
100% passing 5/8" (16.0mm) sieve	3
100% passing 3/8" (9.5mm) sieve	2
100% passing No. 4 (4.75mm) sieve	1

Screens having clear square openings shall be used to separate and classify materials for the hot storage bins. The aggregate passing these screens shall be separately stored in individual bins until proportioned into the mixer as described in 203-6.6.

The No.1 bin shall not contain more than 10 percent of material retained on the No. 8 (2.36mm) sieve. The remaining bins shall not contain more than 10 percent of material which will pass through the following sizes:

TABLE 203-6.5.2 (B)

Bin. No.	Sieve Size
2	No. 8 (2.36mm)
3	No. 4 (4.75mm)
4	3/8" (9.5mm)
5	3/8" (9.5mm)

Each bin shall be provided with an opening to prevent overflow into adjacent bins. All overflow material shall be discarded or returned to the appropriate storage area. All material fed to the No. 4 bin shall pass through the screen over that bin and the oversize rock shall be discarded or returned to the appropriate storage area.

If at any time there is a substantial change made in the cold feed to accommodate the demands of a different type of mixture, the hot storage bins shall be emptied and recharged with the correct materials. Discharged material may be returned to a storage area that contains aggregates of the approximate grading of the discharged material.

203-6.5.3 Dryer-Drum Method. When a dryer-drum is used, aggregate shall be fed directly to the mixer at a uniform rate.

At the time of spreading, mixtures produced in the dryer-drum shall not contain more than 1 percent moisture. Moisture content will be determined by California Test 310, 311, or 370.

A device shall be provided which indicates the temperature of the mixed material leaving the drum. The device shall be accurate to 10°F (5°C) and shall be installed in such a manner that a temperature change of 10°F (5°C) in the mixed material will be shown within 1 minute. The temperature indicator shall be located and maintained at the location where the proportioning operations are controlled.

203-6.6 Proportioning.

203-6.6.1 Batch Plant Method. All materials shall be proportioned by weight. Volumetric proportioning will not be permitted. The zero tolerance for aggregate scales, asphalt binder scales, and, when used, mineral filler scales shall be 0.5 percent based on the total batch weight of the aggregate.

The indicated weight of material drawn from storage for any material shall not vary from the preselected setting by more than the following percentages based on the total batch weight of the aggregate:

Aggregate	1.0 percent
Mineral Filler	0.5 percent
Asphalt Binder	0.1 percent

Automatic proportioning devices shall be operated so that all weight increments required for a batch are preset on the controls at the same time. The discharge from the several bins shall be interlocked so that only one bin can discharge onto a given scale at a time and that no new batch may be started until all weigh hoppers are empty, the scales are at zero and the discharge gate is closed. The interlock shall prevent the weigh box from discharging until the required quantity of each bin and the other weighing devices have been properly filled and weighed. The proportioning controls shall be equipped with means for inspection of the interlock tolerance settings. Instructions for determining these settings shall be posted at the control panel for the Engineer's use.

When bag house fines or mineral filler is used, it shall be proportioned by weight by a method that uniformly feeds the materials within 10 percent of the required amount.

203-6.6.2 Dryer-Drum Method. Asphalt binder shall be measured through a meter calibrated and certified for accuracy by one of the agencies designated in 4-1.7. The asphalt meter shall automatically compensate for changes in asphalt temperature. The system shall be capable of varying the rate of delivery of binder proportionate with the delivery of aggregate. During any day's run, the temperature of the binder shall not vary more than 50°F (30°C). The meter and lines shall be heated and insulated. The storage tank for the binder shall be equipped with a device for automatic plant cutoff when the level of binder in the tank is lowered sufficiently to expose the pump suction line

When bag house fines or mineral filler is used, it shall be proportioned by weight or volume by a method that uniformly feeds the material within 10 percent of the required amount and shall discharge the material directly into the mixer in proximity to where the binder is added.

The feeders for each material in the mixture and for combined aggregates shall be equipped with devices by which the rate of feed can be determined while the plant is in full operation.

The combined aggregate shall be weighed on a belt scale. The belt scale shall be of such accuracy that, when the plant is operating between 30 and 100 percent of belt capacity, the average difference between indicated weight of the material delivered and actual weight will not exceed 1 percent of the actual rate for three 3-minute runs. For any of the three individual 3-minute runs, the indicated weight of the material delivered shall not vary from the actual weight delivered by more than 2 percent of the actual weight. The actual weight of material delivered shall be determined by a vehicle platform scale.

The belt scale for the combined aggregate, other proportioning devices, and the binder proportioning meter shall be interlocked so that the rates of feed of the aggregates and binder will be adjusted automatically to maintain the binder ratio of dry aggregate designated by the Engineer. The plant shall not be operated unless this automatic system is operating and in good working condition.

Binder meters and aggregate belt scales used for metering the aggregates and binder into the mixer shall be equipped with rate of flow indicators to show the rates of delivery of asphalt and aggregates and resettable totalizers, so that the actual weight of binder and aggregate introduced can be determined. Rate of flow indicators and totalizer for like materials shall be accurate to within 1 percent when compared directly. The asphalt totalizer shall not register when the asphalt metering system is not delivering material to the mixer. The bin containing the fine aggregate, and bag house fines and mineral filler, if used, shall be equipped with a vibrating unit or other equipment which will prevent any hang-up of material while the plant is operating. Before the quantity of material in any one bin reaches strike-off capacity of the feed gate, a device shall automatically close down the plant.

The Contractor shall determine the moisture content of the aggregate at least once during each 2 hours of production and shall adjust the moistures control equipment accordingly.

An aggregate sampling device which will provide a 60 to 80 pound (25 to 40 kg) sample of the combined aggregate while the plant is in full operation shall be provided in advance of the point where the aggregate enters the dryer-drum mixer.

When bag house fines or mineral filler is used, a safe and suitable sampling device shall be installed in each feed line or surge tank preceding the proportioning device.

203-6.7 Mixing.

203-6.7.1 General. All aggregates, binder, and RAP, where applicable, shall be mixed in a batch or drier drum plant. The temperature of the completed mixtures, using paving asphalt, at the point of discharge from the plant shall not exceed 325°F (163°C) nor be less than 275°F (135°C). The temperature of the mixture at the Work site shall be as specified in 302-5.5.

Uniformity of distribution of binder shall be determined by an extraction test made in accordance with any of the following test methods or other published test methods approved by the Engineer:

ASTM D2172

California Test 382

ASTM D4125

The binder ratio, pounds of binder per 100 pounds of dry aggregate (kilograms of binder per 100 kilograms of dry aggregate), shall not vary by more than 5 percent above or 10 percent below the amount designated by the Engineer. This requirement shall apply to samples taken from a single batch at any location or operation designated by the Engineer.

203-6.7.2 Batch Plant Method. The mixer shall be of the twin-shaft pug mill type and operated at the speed recommended by the manufacturer. It shall be equipped with paddles of sufficient size and number to deliver a uniform mixture. Should the paddles or other parts of the pug mill become worn to such extent as to adversely affect quality of mixing or allow leakage from the discharge gate, they shall be promptly replaced.

The weight of material that may be mixed per batch shall not exceed the manufacturer's rated capacity of the mixer, nor exceed that which will permit complete mixing of all the materials. Dead areas in the mixer shall be corrected either by a reduction in the volume of materials or by other adjustments.

The entire batch shall be continuously mixed until all the materials are thoroughly blended. The batch mixing time shall begin on the charging stroke of the weigh hopper dump mechanism and shall end when discharge from the mixer has started. The mixer shall be equipped with a time lock mechanism which locks the mixer discharge gate for the mixing period and actuates an indicator light by the charging stroke of the weigh hopper charging mechanism. The indicator light shall be so located as to be visible from the operator's platform and from the Engineer's field laboratory.

The minimum mixing time shall be 35 seconds, of which 5 seconds shall be drop time for the aggregate and asphalt into the mixer. If drop time exceeds 5 seconds, the additional time will be added to the 35 seconds. If the Engineer determines that the mixture is not thoroughly blended, the mixing time shall be increased.

The mixer, weigh hopper, and sampling platforms shall be of ample size to provide safe and convenient access to the mixer and other equipment. Weighbox housings shall be provided with gates of ample size to permit ready sampling of the discharge of aggregates from each of the plant bins.

203-6.7.3 Dryer-Drum Method. Mixing shall continue for a sufficient time and at a sufficiently high temperature, that, at discharge from the mixer, the sizes of aggregates are uniformly distributed through the completed mixture and all particles are thoroughly and uniformly coated with binder.

The mixed material shall be discharged from the drum into the silo. The Contractor shall provide a means of diverting flow of material away from the silo, when starting and stopping the plant production, to prevent incomplete mixed portions of the mixture from entering the silo. The burner used for heating the aggregate shall achieve complete combustion of the fuel.

203-6.8 Asphalt Concrete Storage. Asphalt concrete shall be stored in insulated silos with heated discharge cones, unless the silo is being used for storage purposes. The storage silo shall be equipped to prevent segregation of the completed mixture as it is discharged into the silo. Asphalt concrete with hardened lumps in the mixture shall not be used.

203-6.9 Miscellaneous Requirements. Paving asphalt shall be added to the aggregate at a temperature conforming to the range of temperatures specified in 203-1.4.

Liquid asphalt shall be added to the aggregate at a temperature conforming to the range of temperatures specified in 203-2. The temperatures of the aggregates at the time of adding the asphalt binder shall not be in excess of maximum applicable pug mill mixing temperature specified in 203-2.3. The temperature of the aggregates may be designated by the Engineer.

A temperature-indicating device reading to 500°F (260°C) and accurate to 10°F (5°C) shall be fixed in the binder feed line or storage tank. An indicator shall be located and maintained where the proportioning operations are controlled. The discharge end of the binder circulating pipe shall extend to within 1 foot (0.3 m) of the bottom of the storage tank.

The Contractor shall provide a suitable sampling outlet in the binder feed lines connecting the plant storage tanks to the binder weighing system or spray bar. The sampling valve shall consist of a 1/2 inch or 3/4 inch (13mm or 19mm) valve constructed in such a manner that a 1 quart (1 liter) sample may be withdrawn slowly at any time during plant operations, and shall be placed in a nonhazardous location. A drainage receptacle shall be provided for flushing the valve prior to sampling. One gallon (4 liters) shall be drawn from the sampler prior to taking the sample.

The beds of trucks used to haul asphalt concrete may be coated with an EPA and Agency-approved bond breaker before loading. Amounts of bond breaker that form visible pools in the truck bed shall be removed prior to loading asphalt concrete. Beds of trucks hauling curb mixes may be sprinkled with mineral filler.

The temperature of asphalt concrete discharged into hauling vehicles shall not vary more than 20°F (10°C) for successive loads. When hauling time from mixing plant to the Work site exceeds 2 hours, or when atmospheric temperature is below 50°F (10°C), or when rain is falling along the haul route, asphalt concrete shall be completely covered with tarpaulins, which are fully secured, during transport. Asphalt concrete shall be delivered to the site of the Work, without segregation of the ingredients and within the temperature range specified in 302-5.5.

When dissimilar surface course mix characteristics are the result of multiple plant delivery, the Engineer may require delivery from only one plant for any one day's production.

203-7 RECYCLED ASPHALT CONCRETE-HOT MIXED.

203-7.1 General. Recycled Asphalt Concrete (RAC) shall be the product of mixing over 15 percent reclaimed asphalt pavement, virgin aggregates and paving asphalt and/or recycling agent.

RAC will be designated by class and grade, i.e., "C2-PG 64-10-RAC." The end product shall meet both the gradation and paving asphalt grade specified.

203-7.2 Materials.

203-7.2.1 Aggregate. New aggregate shall conform to 203-6.3.2.

203-7.2.2 Reclaimed Asphalt Pavement. Reclaimed asphalt pavement (RAP) is pavement containing asphalt and aggregates which has been processed to 1-1/2 inches (37.5mm) maximum size and is free of detrimental quantities of deleterious materials. The stored RAP shall be uniform in appearance and well graded from fine to coarse.

The RAC supplier shall perform sand equivalent tests on the unextracted RAP, and tests for RAP asphalt content, RAP asphalt viscosity or performance grade, and gradation of RAP aggregates on solvent extracted samples of RAP taken from stockpiles. Solvent extractions shall be performed in accordance with ASTM D2172 and the RAP asphalt shall be recovered in accordance with ASTM D1856 (Abson Recovery Method). The minimum sand-equivalent value of the unextracted RAP shall be 80, when determined in accordance with California Test 217.

The supplier shall maintain current records of the test results at the plant and make them available to the Engineer for its information and use in the approval of RAC mixes. The reports shall be made available prior to use of any RAC in the Work.

203-7.2.3 RAC Bituminous Materials. RAC bituminous materials shall conform to the following:

a) New asphalt shall be paving asphalt conforming to 203-1.

b) RAP asphalt shall be defined as the bituminous material present in RAP. The quantity of RAP asphalt shall be determined by solvent extraction of RAP. The quality of RAP asphalt shall be determined by tests performed on the asphaltic residues obtained from the Abson Recovery Method.

c) New binder shall be defined as the combined new asphalt and/or recycling agent added to RAP and new aggregates to create RAC.

d) RAC binder shall be defined as the total bituminous material present in RAC; consisting of the blend of RAP asphalt, new asphalt and/or recycling agent. The quality of the RAC binder shall be determined by tests performed on the asphaltic residues obtained from the Abson Recovery Method.

The test results of the RAC binder shall meet the RTFO and PAV test requirements in Table 203-1.2(A) for the PG grade specified.

203-7.2.4 Mineral Filler. Mineral filler shall conform to 203-6.3.3.

203-7.2.5 Recycling Agent. Recycling Agent (RA) shall comply with Table 203-7.2.5 (A):

TABLE 203-7.2.5 (A)[1]

TEST	ASTM Test Methods	RA 5 Min.	RA 5 Max.	RA 25 Min.	RA 25 Max.	RA 75 Min.	RA 75 Max.	RA 250 Min.	RA 250 Max.	RA 500 Min.	RA 500 Max.
Kinematic Viscosity at 140°F, cSt (60°C, mm²/s)	D2170 or D2171	200	800	1000	4000	5000	10000	15000	35000	40000	60000
Flash Point, COC, COC °F (°C), Min.	D92	400 (205)		425 (215)		450 (230)		450 (230)		450 (230)	
Saturated Wt. %, Max.	D2007	25		25		25		25		25	
Residue from RTFO Oven Test at 325°F (163°C)	D2872										
Viscosity Ratio[2] Max.	-	3		3		3		3		3	
RTFO Oven Weight Change, ±, %	D2872	4		3		2		2		2	
Specific Gravity	D70 or D1298	Report		Report		Report		Report		Report	

1. The acceptance of any recycling agent is subject to its ability to develop a RAC binder which will comply with the asphalt grade specified.
2. Viscosity Ratio = $\frac{\text{RTFO Viscosity at 140°F, cSt (60°C, mm}^2\text{/s)}}{\text{Original Viscosity at 140°F, cSt (60°C, mm}^2\text{/s)}}$

203-7.3 Recycled Asphalt Concrete Mixtures.

203-7.3.1 Combined RAC Aggregates. Combined aggregate and RAP, after all processing except the adding of new binder and mineral filler, shall have an unextracted minimum sand equivalent of 50 when tested by California Test 217.

When there is a difference in specific gravity (bulk saturated, surface dry per ASTM C127 and C128) of 0.2 or more between that portion retained and that portion passing a No. 4 (4.75mm) sieve, the grading shall be modified as provided by California Test 105.

203-7.3.2 Composition and Grading. The combined grading and RAC binder content shall conform to the following table. All percentages are based on weight of dry aggregate only (outside):

TABLE 203-7.3.2 (A)

	Percentage Passing Sieves									
CLASS	A-RAC		B-RAC		C1-RAC		C2-RAC		D2-RAC	
Sieve Size	Min.	Max.	Min.	Max.	Min.	Max.	Min.	Max.	Min.	Max.
1-1/2" (37.5mm)	100									
1" (25.0mm)	90	100	100	-	-	-	-	-	-	-
3/4" (19.0mm)	78	90	87	100	100	-	100	-	-	-
1/2" (12.5mm)	64	78	70	87	90	100	95	100	100	-
3/8" (9.5mm)	54	68	55	76	72	88	72	88	95	100
No. 4 (4.75mm)	34	48	35	52	40	54	46	60	58	72
No. 8 (2.36mm)	25	35	22	40	18	34	28	42	34	48
No. 30 (600μm)	12	25	8	24	8	20	15	27	18	32
No.50 (300μm)	8	16	5	18	4	14	10	20	13	23
No. 200 (75μm)	3	6	0	7	1	6	2	7	2	9
RAC binder %	4.5	5.7	4.5	6.5	4.6	6.6	4.6	6.6	4.8	7.2

When the amount of RAP is 15 percent or less of the total mix, the supplier shall maintain a job mix formula at the plant. The formula shall be based on current test data and approved by the Engineer.

When the amount of RAP to be added is over 15 percent of the total mix, a job mix formula, supporting test data, and blending charts shall be submitted to the Engineer for approval in accordance with 2-5.3. Blending charts shall be used to determine the grade of new paving asphalt. The supporting test data for RAC shall include the result for stability, swell and moisture vapor susceptibility. These tests are in addition to the tests for the RAP stockpile specified in 203-7.2.2.

TABLE 203-7.3.2 (B)

Test	California Test	Requirements
Stabilometer Value	366	35 Min.
Moisture Vapor Susceptibility (Stabilometer Value)	307	25 Min.
Swell, mils (μm)	305	30 (760) Max.

Percentages for the combined gradings, within the specific limits, shall be of such uniformity that the material passing the indicated sieves during any day's run will not exceed the following maximum variations:

No. 4 (4.75mm) sieve - 6 percentage points

No. 30 (600μm) sieve - 5 percentage points

No. 200 (75μm) sieve - 3 percentage points

203-7.4 Aggregate Storing, Drying and Screening.

203-7.4.1 General. New aggregate consisting of sand, rock dust, and various sizes of aggregates shall be stored separately at the plant and evenly fed to the dryer to insure a uniform flow of properly combined aggregates. In placing materials in storage or in moving them from storage to the feeder, no method shall be used which causes segregation, degradation, or the intermingling of different size aggregates. Materials not meeting the gradation requirements shall be discarded or reprocessed to comply with 203-6.3.2.

203-7.4.2 RAC Batch Plant Method. A conventional batch plant shall be modified to introduce the RAP at locations other then the dryer by:

a) providing a separate RAP storage facility, with direct access to the weigh hopper; or

b) providing for RAP introduction to the hot aggregate elevator; or

c) other method approved by the Engineer.

New aggregate shall be dried and heated for a sufficient time in the dryer so that the moisture content of the aggregate will not be greater than 1 percent.

The dryer shall be provided with an approved temperature indicating device to determine the temperature of the aggregate leaving the dryer. The device shall be mounted independent of the plant components, be accurate to the nearest 10°F (5°C), and shall be installed in such a manner that a temperature change of 10°F (5°C) in the aggregate will be indicated within 1 minute. The temperature indicator shall be located and maintained where the proportioning operations are controlled.

After drying, the aggregate shall be evenly fed to the screens in such quantities as to maintain, in separate bins, a uniform grading of the materials and a proper balance in the amount of material. The operation of the screens shall be controlled so as to secure a thorough separation of the aggregate sizes.

Screens having clear square openings shall be used to separate and classify materials for the hot storage bins. The aggregate passing these screens shall be separately stored in bins until proportioned into the mixer as specified in 203-7.5.

TABLE 203-7.4.2 (A)

Bin No.	Maximum Aggregate Size in Bin
	4-Bin Plants
4	100% passing 1-1/2" (37.5mm) sieve
3	100% passing 5/8" (16.0mm) sieve
2	100% passing 3/8" (9.5mm) sieve
1	100% passing No. 4 (4.75mm) sieve
	5-Bin Plants
5	100% passing 1-1/2" (37.5mm) sieve
4	100% passing 7/8" (22.4mm) sieve
3	100% passing 5/8" (16.0mm) sieve
2	100% passing 3/8" (9.5mm) sieve
1	100% passing No. 4 (4.75mm) sieve

The No. 1 bin shall contain not more than 10 percent of material retained on the No. 8 (2.36mm) sieve. The remaining bins shall contain not more than 10 percent of material which will pass through the following sieve sizes:

TABLE 203-7.4.2 (B)

Bin. No.	Sieve Size
2	No. 8 (2.36mm)
3	No. 4 (4.75mm)
4	3/8" (9.5mm)
5	3/8" (9.5mm)

Each bin shall be provided with an opening to prevent overflow into adjacent bins.

All overflow material shall be discarded or returned to the appropriate storage areas. All material fed to the No. 4 bin shall pass through the screen over that bin and the oversize rock shall be discarded or returned to appropriate storage area.

If at any time there is a substantial change made in the cold feed to accommodate the demands of a different type of mixture, the hot storage bins shall be emptied and recharged with the correct materials. Discharged material may be returned to a storage area that contains aggregates of the approximate grading of the discharged material, except when the hot storage bins contain RAP. Discharged material containing RAP shall be returned to a separate stockpile.

203-7.4.3 RAC Dryer-Drum Method. When producing RAC, new aggregate shall be fed directly to the mixer at a uniform rate. The RAP shall be introduced into the dryer-drum and combined with the hot, new aggregate in such a manner that the RAC is protected from direct contact with the burner flame by means approved by the Engineer. The new binder shall be introduced into the drum after the RAP and the new aggregate have been combined.

A device shall be provided which indicates the temperature of the mixed material leaving the drum. The device shall be accurate to 10°F (5°C) and shall be installed in such a manner that a temperature change of 10°F (5°C) in the mixed material will be shown within 1 minute. The temperature indicator shall be located and maintained where the proportioning operations are controlled.

The burner used for heating the aggregate shall achieve complete combustion of the fuel. The temperature indicator shall be located and maintained where the proportioning operations are controlled.

203-7.5 RAC Proportioning.

203-7.5.1 RAC Batch Plant Method. When introducing the RAP into the hot aggregate elevator the conveyors shall be equipped with belt scales with rate-of-flow indicators to show the rates of delivery of each of these ingredients. The belt scales shall be interlocked to maintain the proper proportion of RAP to new aggregate.

When introducing RAP from a separate storage facility, it shall be fed directly into the weigh hopper.

All materials shall be proportioned by weight. Volumetric proportioning will not be permitted. The zero tolerance for aggregate scales, asphalt binder scales, and, when used, mineral filler scales shall be 0.5 percent based on the total batch weight of the aggregate.

The indicated weight of material drawn from storage for any material shall not vary from the preselected setting by more than the following percentages based on the total batch weight of the aggregate:

Aggregate .... 1.0 percent
Mineral Filler.. 0.5 percent
Asphalt Binder.. 0.1 percent

Automatic proportioning devices shall be operated so that all weight increments required for a batch are preset on the controls at the same time. The discharge from the several bins shall be interlocked so that only one bin can discharge onto a given scale at a time and that no new batch may be started until all weigh hoppers are empty, the scales are at zero, and the discharge gate is closed. The interlock shall prevent the weigh box from discharging until the required quantity of each bin and the other weighing devices have been properly filled and weighed. The proportioning controls shall be equipped with means for inspection of the interlock tolerance settings. Instructions for determining these settings shall be posted at the control panel for the Engineer's use.

203-7.5.2 RAC Dryer-Drum Method. When producing RAC, the separate conveyor supplying the RAP to the dryer shall be equipped with a belt scale with rate of flow indicator. This belt scale shall be interlocked to maintain proper proportions of RAP to new aggregate.

New asphalt and RAP shall be measured through separate meters calibrated and certified for accuracy by one of the agencies designated in 4-1.7. The asphalt meter shall automatically compensate for changes in asphalt temperature. The system shall be capable of varying the rate of delivery of binder proportionate with the delivery of aggregate. The meter and lines shall be heated and insulated. The storage tanks for new asphalt and RAP shall be equipped with a device for automatic plant cutoff when the fluid level in the tank is lowered sufficiently to expose the pump suction line.

The system shall be capable of varying the rates of delivery of binder. During any day's run, the temperature of the binder shall not vary more than 50°F (30°C).

When bag house fines or mineral filler is used, it shall be proportioned by weight or volume by a method that uniformly feeds the material within 10 percent of the required amount. The material shall be discharged directly into the mixer where the binder is added.

The feeders for each material in the RAC shall be equipped with devices by which the rates of feed can be determined while the plant is in full operation.

The RAP and the combined new aggregate shall be weighed on separate belt scales. They shall be of such accuracy that, when the plant is operating between 30 and 100 percent of belt capacity, the average difference between the indicated weight of the material delivered and the actual weight delivered will not exceed 1 percent of the actual weight for three 3-minute runs. For any of the individual runs the indicated weight shall not vary from the actual weight by more than 2 percent. The actual weight of material delivered shall be determined by a vehicle platform scale.

The individual belt scales, for the RAP and the combined new aggregate, the proportioning meters for the new asphalt and RA, and the other proportioning devices, shall be interlocked so that the rates of feed of the RAP, new aggregate, new asphalt, and RA will be adjusted automatically to maintain the proper proportions. The plant shall not be operated unless this automatic system is operating and in good working condition.

Belt scales and proportioning meters shall be equipped with rate of flow indicators to show the rates of delivery of asphalt, RA, RAP, and aggregates and a resettable totalizer for determining the actual weight of asphalt, RA, RAP, and combined aggregates. Rate of flow indicators and totalizer for like materials shall be accurate within 1 percent when compared directly. The asphalt totalizer shall not register when the asphalt metering system is not delivering material to the mixer, and shall not be reset without approval of the Engineer. The bins containing the fine aggregate and mineral filler if used shall be equipped with a vibrating unit or other equipment which will prevent any hang-up of material while the plant is operating. Before the quantity of material in any one bin reaches the strike-off capacity of feed gate, a device shall automatically close down the plant.

The supplier shall determine the moisture contents of the RAP and aggregates at least once during each 2 hours of production and shall adjust the moisture control equipment accordingly.

A sampling device which will provide 60 to 80 pound (25 to 40 kg) samples of the RAP and the combined aggregate, while the plant is in full operation, shall be provided in advance of the point where these materials enter the dryer-drum mixer.

When bag house fines or mineral filler is used, a safe and suitable sampling device shall be installed in each feed line or surge tank preceding the proportioning device.

203-7.6 RAC Mixing.

203-7.6.1 General. Aggregates, RAP, recycling agent, and asphalt binder shall be mixed in a batch or dryer-drum plant. The temperature of the completed mixtures, using paving asphalt, at the point of discharge from the plant shall not exceed 325°F (163°C). The temperature of the mixture at the Work site shall be as specified in 302-5.5.

203-7.6.2 RAC Batch Plant Method. The mixer shall be of the twin-shaft pug mill type and shall be operated at the speed recommended by the manufacturer. It shall be equipped with sufficient paddles to deliver a uniform mixture. Should the paddles or other parts of the pug mill become worn to such an extent as to adversely affect the quality of mix or allow leakage from the discharge gate, they shall be promptly replaced.

The weight of material that may be mixed per batch shall not exceed the manufacturer's rated capacity of the mixer, nor exceed that which will permit complete mixing of all materials. Dead areas in the mixer, in which the material does not move or is not sufficiently agitated, shall be corrected either by a reduction in the volume of materials or by other adjustments.

The entire batch shall be continuously mixed until all the materials are thoroughly blended. The batch mixing time shall begin on the charging stroke of the weigh hopper dump mechanism and shall end when discharge from the mixer has started. The mixer shall be equipped with a time lock mechanism which locks the mixer discharge gate for the mixing period and actuates an indicator light by the charging stroke of the weigh hopper charging mechanism. The light shall be so located as to be visible from the operator's platform and from the Engineer's field laboratory.

If the plant has been modified to provide separate storage for the RAP, the RAP and the new aggregates shall be weighed together in the weigh hopper prior to introduction into the pug mill. These 2 materials shall be dry mixed for 10 seconds, or longer if necessary, to effect the heat transfer.

The minimum wet mixing time shall be 35 seconds, of which 5 seconds will be drop time for the RAP, new aggregate and new binder to enter the mixer. If the drop time exceeds 5 seconds, additional time shall be added to the 35 seconds. If the Engineer determines that the mixture is not thoroughly blended and the aggregate not fully coated, the mixing time shall be increased.

The mixer, weigh hopper and sampling platforms shall be of ample size to provide safe access to the mixer and other equipment. Weighbox housings shall be provided with gates of ample size to permit ready sampling of the discharge of aggregates from each of the plant bins.

203-7.6.3 RAC Dryer-Drum Method. RAC mixing shall continue for a sufficient time and at a sufficiently high temperature, that at discharge from the mixer, the sizes of aggregates are uniformly distributed through the completed mixture and all particles are thoroughly and uniformly coated.

The RAC shall be discharged from the drum into a storage silo. The supplier shall provide a means of diverting the flow of material away from the silo, when starting and stopping plant production, to prevent incompletely mixed materials from entering the storage silo.

203-7.7 RAC Storage. When RAC is stored, it shall be stored only in insulated silos with heated discharge cones, unless the silo is being used for surge purpose. The storage silo shall be equipped to prevent segregation of the completed mixture as it is discharged into the silo. RAC with hardened lumps in the mixture shall not be used.

203-7.8 RAC Miscellaneous Requirements. New binder shall be added to the aggregate and RAP at a temperature conforming to the range of temperatures prescribed in 203-1.4, with a minimum of 200°F (95°C) for RA.

A temperature indicating device reading to 500°F (260°C), accurate to 10°F (5°C), shall be fixed in the new asphalt and RA feed lines or storage tanks at suitable locations. The temperature indicator shall be located and maintained where the proportioning operations are controlled.

The discharge end of the new asphalt and RA circulating pipe shall extend to within 1 foot (0.3m) of the bottom of the storage tank.

The supplier shall provide sampling outlets in the new asphalt and RA feed lines connecting the plant storage tanks to the weighing system or spray bar. When new asphalt and RA are blended in a single line, a sampling outlet shall be provided. It shall be a 1/2 inch or 3/4 inch (13mm or 19mm) valve constructed so that a 1-quart (1 liter) sample may be withdrawn slowly at any time during plant operation. The sampling outlet shall be placed in an accessible, nonhazardous location. A container shall be provided for flushing the valve prior to sampling. One gallon (4 liters) shall be drawn from the valve prior to taking the sample.

The beds of trucks used to haul RAC may be coated with an EPA and Agency approved bond breaker before loading. Amounts of bond breaker that form visible pools in the truck beds shall be removed prior to loading RAC. Beds of trucks hauling curb mixes may be sprinkled with mineral filler.

The temperature of the RAC discharged into the hauling vehicles shall not vary more than 20°F (10°C) for successive loads. When the hauling time from the mixing plant to the Work site exceeds 2 hours, or when the atmospheric temperature is below 50°F (10°C), or when rain is falling along the haul route, RAC shall be covered with tarpaulins during transport. The tarpaulins shall completely cover the load and be firmly secured. RAC shall be delivered to the Work site without segregation of the ingredients and within the temperature range specified in 302-5.5.

At the time of spreading, RAC mixtures shall not contain more than 1 percent moisture. Moisture content will be determined by California Test 310, 311, or 370. RAC mixes shall be the product from one plant or source unless otherwise permitted by the Engineer.

203-8 ASPHALT PAINT. Shall conform to ASTM D41 or D43 and be furnished and applied to concrete surfaces as required and shown on the Plans.

203-9 SEALCOAT - ASPHALT BASED.

203-9.1 General. This specification applies to sealcoat intended to be used for sealing miscellaneous areas such as asphalt parking lots, playgrounds, and similar areas. Sealcoat material shall be a plant blended product composed of mineral aggregates uniformly distributed in a petroleum-based asphalt emulsion. The asphalt emulsion shall conform to 203-3. The sealcoat material shall contain non-asbestos fibers.

203-9.2 Testing Requirements. Sealcoat materials, undiluted except as noted, shall conform to the following requirements:

TABLE 203-9.2 (A)

Test	Specifications
Weight lbs. per gallon (grams per liter)	9.5 (1139) Min.
Nonvolatile component (%)[1]	60% Min. by weight
Mineral aggregate component	No. 20 (850μm) sieve – 100% passing
Working viscosity, diluted 4 parts product to 1 part water – ASTM D562	75 KREBS Min.
Dried film color	Black
Asphalt content	25% - 35% of nonvolatiles by weight

1. Weigh 10 grams of homogeneous product into a previously tarred, small ointment can lid. Place in a constant temperature oven at 325°F (163°C) for 1-1/2 hours. Cool, reweigh, and calculate nonvolatile components.

203-10 LATEX MODIFIED ASPHALT CONCRETE.

203-10.1 General. Latex modified asphalt concrete shall be the product of mixing latex, asphalt cement, and aggregate. Latex modified asphalt shall conform to 203-6 and the following requirements.

203-10.2 Materials.

203-10.2.1 Paving Asphalt. Paving asphalt shall be PG 64-10 conforming to 203-1, or as specified in the Special Provisions.

203-10.2.2 Latex. Latex shall be a water-based emulsified suspension of styrene butadiene rubber (SBR) in liquid form. The ratio of styrene butadiene shall be 70 to 30. The rubber solids shall be 70 ± 5 percent of the emulsified material unless otherwise directed by the Engineer.

At the time of delivery of each shipment of latex emulsion, the vendor supplying the material shall deliver to the purchaser a Certificate of Compliance conforming to 4-1.5 which includes, but is not limited to, the following information:

Actual percent of rubber solids

Ratio of styrene to butadiene

Temperature range or limits for product use

Unit weight of the emulsion

Recommended storage conditions

203-10.2.3 Aggregates. Aggregates shall conform to 203-6.3.2.

203-10.2.4 Composition and Grading. The class shall be C2 unless otherwise specified on the Plans or in the Special Provisions. Where Alternate Rock Products-Type S, Section 400, are specified, the class and grade shall be Type III-C3-PG 64-10 or as specified in the Special Provisions.

The amount of latex to be added shall provide 2 percent dry rubber solids, unless otherwise directed by the Engineer. The percentage of dry rubber solids shall be based on the weight of asphalt cement.

The exact proportions of aggregate and the amount asphalt binder for the mixture shall be determined by Contractor so as to incorporate the specified amount of latex.

If the mix design and/or its individual components have no previous record of use by the Agency, the Contractor shall submit a mix design with supporting test data at least 10 working days prior to initial use.

203-10.3 Proportioning. Latex may be added to the mixture by any method approved by the Engineer that will assure uniform distribution and accurate measurement of quantity of latex introduced. The latex shall be introduced to the mix at the same time as the introduction of asphalt.

The latex emulsion temperature at the time it is introduced to the mixture shall be as recommended by the manufacturer.

203-10.4 Mixing. The wet-mixing cycle for batch plant production shall be 50 seconds minimum.

Production rates of continuous mix plants shall be carefully regulated by the Contractor to ensure complete and uniform mixing.

203-10.5 Storage. Latex emulsion shall be stored in a closed bottom-draw vessel and agitated as recommended by the manufacturer's Certificate of Compliance.

A latex sampling outlet shall be provided in conformance to the requirements of 203-6.9.

203-10.6 Placement. Latex modified asphalt concrete placement shall conform to the requirements of 302-5 with the exception that pneumatic-tired rollers shall not be used.

203-11 ASPHALT RUBBER HOT MIX (ARHM)

203-11.1 General. Asphalt Rubber Hot Mix shall consist of a mixture of paving asphalt, asphalt modifier, crumb rubber modifier (CRM), and aggregate mixed in a central mixing plant.

203-11.2 Materials. The Contractor shall submit test reports and Certificates of Compliance, conforming to 4-1.5, for the paving asphalt, asphalt modifier, and CRM to be used. When requested by the Engineer, the Contractor shall also submit samples of the tested material.

203-11.2.1 Paving Asphalt. Paving asphalt used for asphalt-rubber shall be PG 64-16, conforming to 203-1. Performance graded paving asphalts other than PG 64-16 may be used if so specified in the Special Provisions.

203-11.2.2 Asphalt Modifier. The asphalt modifier shall be a resinous, high flash point, aromatic hydrocarbon compound and shall conform to the requirements shown in Table 203-11.2.2 (A). The asphalt modifier may be eliminated if approved by the Engineer.

TABLE 203-11.2.2 (A): REQUIREMENTS FOR ASPHALT MODIFIER

Property	ASTM Test Method	Value
Flash Point, C.L.O.C., °F (°C)	D 92	405 (207) min
Viscosity, cSt @ 212°F (100°C)	D 445	X ± 3*
Molecular Analysis		
Asphaltenes, percent by mass	D 2007	0.1 max
Aromatics, percent by mass	D 2007	55 min

* The symbol "X" is the viscosity of the asphalt modifier the Contractor proposes to furnish. The value "X" which the Contractor proposes shall be between the limits of 19 and 36 and shall be submitted in writing to the Engineer. Any proposed change requested by the Contractor in the value "X" shall require a new asphalt-rubber binder design.

203-11.2.3 Crumb Rubber Modifier (CRM). CRM shall consist of a combination of scrap tire CRM and high natural CRM. Scrap tire CRM shall consist of ground or granulated rubber derived from any combination of automobile tires, truck tires or tire buffings. The high natural CRM shall consist of ground or granulated rubber derived from materials that utilize high natural rubber sources. Whole scrap tire CRM shall be derived from whole scrap tires generated within the State boundaries of the user agencies. The Certificate of Compliance shall contain a statement confirming conformance with this requirement. The high natural CRM may consist of blended CRM.

CRM shall be ground or granulated at ambient temperature. Cryogenically produced CRM particles which can pass through the grinder or granulator without being ground or granulated shall not be used. Cryogenic separation, if utilized, shall be performed separately from and prior to grinding or granulating. Steel and fiber separation may employ any method.

CRM shall not contain more than 0.01 percent of wire by weight and shall be free of all other contaminants, except fabric. Fabric shall not exceed 0.05 percent by weight of CRM. A Certificate of Compliance certifying these percentages shall be furnished to the Engineer.

CRM shall be dry and free-flowing and not produce foaming when combined with the blended paving asphalt and asphalt modifier mixture. Calcium carbonate or talc may be added up to a maximum of 3 percent by weight of CRM to prevent CRM particles from sticking together. CRM shall have a specific gravity range from 1.1 minimum to 1.2 maximum as determined by ASTM D297. Scrap tire CRM and high natural CRM shall be delivered to the production site in separate bags and shall be sampled and tested separately. Scrap tire CRM material shall conform to the chemical analysis requirements shown in Table 203-11.2.3 (A).

TABLE 203-11.2.3 (A): CHEMICAL REQUIREMENTS FOR SCRAP TIRE CRM

Test	ASTM Test Method	Minimum	Maximum
Acetone Extract	D 297	6.0%	16.0%
Ash Content	D 297	—	8.0%
Carbon Black Content	D 297	28.0%	38.0%
Rubber Hydrocarbon	D 297	42.0%	65.0%
Natural Rubber Content	D 297	22.0%	39.0%

Scrap tire CRM shall be mixed at the production site with high natural CRM so that 75 percent ± 2 percent of the product used is derived from scrap tires and 25 percent ± 2 percent from materials that utilize high natural rubber sources. High natural rubber CRM may consist of blended CRM which, after blending, conforms to the chemical analysis requirements shown in Table 203-11.2.3(B).

TABLE 203-11.2.3 (B): CHEMICAL REQUIREMENTS FOR HIGH NATURAL CRM

Test	ASTM Test Method	Minimum	Maximum
Acetone Extract	D 297	4.0%	16.0%
Rubber Hydrocarbons	D 297	50.0%	–
Natural Rubber Content	D 297	40.0%	48.0%

With the approval of the Engineer, high natural CRM may be eliminated. If high natural rubber is not used, then the full amount of required CRM shall be scrap tire CRM. Asphalt rubber binder without high natural CRM shall be designated as Type I and ARHM manufactured with Type I binder shall be designated as ARHM-I.

203-11.2.3.1 CRM Gradations. CRM gradations shall conform to the requirements shown in Table 203-11.2.3.1 (A) when tested in accordance with ASTM C136, amended as follows:

Split or quarter 100 grams ± 5 grams from the representative CRM sample and dry to a constant weight at a temperature of not less than 135°F (57°C) nor more than 145°F (63°C) and record the dry sample weight. Place the CRM sample and 5.0 grams of talc (or calcium carbonate) in a 1 pint jar. Seal the jar and shake it by hand for a minimum of one minute to mix the CRM and the talc (or calcium carbonate). Continue shaking or open the jar and stir until particle agglomerates and clumps are broken and the talc (or calcium carbonate) is uniformly mixed.

A Rotap (or equivalent) test shaker shall be used for the sieve analysis. Place one rubber ball on each sieve. Each ball shall have a weight of 8.5 ± 0.5 grams, have a diameter of 24.5 ± 0.5mm, and shall have a Shore Durometer "A" hardness of 50 ± 5 in accordance with ASTM D2240. After shaking the combined material for 10 minutes ± 1 minute, disassemble the sieves. Any material adhering to the bottom of a sieve shall be brushed into the next finer sieve. Weigh and record the weight of the material retained on the No. 8 (2.36mm) sieve and leave this material on the scale or balance. Any observed fabric balls shall remain on the scale or balance and shall be placed together on the side of the scale or balance to prevent the fabric balls from being covered or disturbed when placing the material from finer sieves onto the scale or balance. The material retained on the next finer sieve shall be added to the scale or balance. Weigh and record that weight as the accumulative weight retained on that sieve. Continue weighing and recording the accumulated weight retained on the remaining sieves until the accumulated weight retained in the pan has been determined. Prior to discarding the CRM sample, separately weigh and record the total weight of fabric balls in the sample.

Determine the weight of material passing the No. 200 (75µm) sieve (or weight retained in the pan) by subtracting the accumulated weight retained on the No. 200 (75µm) sieve from the accumulated weight retained in the pan. If the material passing the No. 200 (75µm) sieve (or weight retained in the pan) has a weight of 5 grams or less, cross out the recorded number for the accumulated weight retained in the pan and copy the number recorded for the accumulated weight retained on the No. 200 (75µm) sieve and record that number (next to the crossed out number) as the accumulated weight retained in the pan. If the material passing the No. 200 (75µm) sieve (or weight retained in the pan) has a weight greater than 5 grams, cross out the recorded number for the accumulated weight retained in the pan, subtract 5 grams from that number and record the difference next to the crossed out number. The adjustment to the accumulated weight retrained in the pan is made to account for the 5 grams of talc (or calcium carbonate) added to the sample. For calculation purposes, the adjusted total sample weight is the same as the adjusted accumulated weight retained in the pan. Determine the percent passing based on the adjusted total sample weight and record to the nearest 0.1 percent.

TABLE 203-11.2.3.1 (A): GRADING REQUIREMENTS FOR CRM

Sieve Size	Scrap Tire CRM Percent Passing	High Natural CRM Percent Passing
No. 8 (2.36mm)	100	100
No. 10 (2.00mm)	98 – 100	100
No. 16 (1.18mm)	45 – 75	95 – 100
No. 30 (600µm)	2 – 20	35 – 85
No. 50 (300µm)	0 – 6	10 – 30
No. 100 (150µm)	0 – 2	0 – 4
No. 200 (75µm)	–	0 – 1

* CRM from more than one source may be used provided the combined CRM gradation meets the specified limits. No particles shall exceed a length of 5mm (3/16 inch) as measured on any axis.

203-11.2.4 Aggregate. The aggregate for ARHM shall meet the quality requirements specified in 200-1 for asphalt concrete.

203-11.3 Composition and Grading. Asphalt-rubber hot-mix gap-graded (ARHM-GG) shall be designated by type and class, i.e., ARHM-GG-C, and shall conform to the requirements shown in Table 203-11.3 (A).

TABLE 203-11.3 (A): REQUIREMENTS FOR ARHM-GG

SIEVE SIZE	CLASS		
	GG-B Min.- Max.	GG-C Min.- Max.	GG-D Min.- Max.
1" (25mm)	100	–	–
3/4" (19.0mm)	90-100	100	–
1/2" (12.5mm)	–	90-100	--
3/8" (9.5mm)	60-75	78-92	100
No. 4 (4.5mm)	28-42	28-42	28-42
No. 8 (2.36mm)	15-25	15-25	15-25
No. 30 (600μm)	5-15	5-15	5-15
No. 200 (75μm)	0-5	2-7	2-7
% Asphalt Rubber Binder by Weight of Dry Aggregate[1]	7.5-8.4	7.5-8.7	7.5-8.7
Air Voids % California Test 367	3-6	3-6	3-6
Stabilometer Value Min. California Test 304 and 306	25	23	23
Voids in Mineral Agg. Percent Min.[2]	18	18	18

1. Once the percent asphalt rubber binder is determined by the mix design, the production tolerance shall be ± 0.5% as determined by California Test Method 362, 379, or 382.
2. Percent voids in the mineral aggregate (VMA) is to be determined during the mix design process only and is to be calculated on the basis of ASTM bulk specific gravity as described in the Asphalt Institute MS-2 manual.

203-11.4 Mixing Asphalt and CRM. The paving asphalt and asphalt modifier shall be combined into a blended mixture that is chemically compatible with the crumb rubber modifier to be used. The blended mixture is considered to be chemically compatible when it conforms to the requirements for asphalt rubber binder (after reacting) shown in Table 203-11.4(A).

The asphalt modifier shall be proportionately added to the paving asphalt at the production site where the asphalt-rubber binder is blended and reacted. Asphalt modifier shall be added at an amount of 2.5% to 6.0% by weight of the paving asphalt based on the recommendation of the asphalt rubber binder supplier. The paving asphalt shall be at a temperature of not less than 375°F (190°C) nor more than 440°F (226°C) when the asphalt modifier is added. If the asphalt modifier is combined with the paving asphalt before being blended with the CRM, the combined paving asphalt and asphalt modifier shall be mixed by circulation for not less than 20 minutes. This premixing of asphalt modifier and the paving asphalt will not be required when all ingredients of the asphalt rubber binder are proportioned and mixed simultaneously. Asphalt modifier and paving asphalt shall be measured for proportioning with meters conforming to 203-6-6.

The proportions of the materials, by total weight of asphalt-rubber binder, shall be 80 percent ± 2 percent combined paving asphalt and asphalt modifier, and 20 percent ± 2 percent CRM. The temperature of the blended asphalt and modifier shall be between 375°F (190°C) and 440°F (226°C) when the CRM is added. The temperature shall not exceed 10°F (6°C) below the actual flash point of the mixture. The CRM shall be combined and mixed together in an asphalt-rubber mechanical blender conforming to 203-11.5. The combined asphalt and CRM shall be pumped into a storage/reaction tank or distributor truck conforming to meeting the requirements of 203-11.5. The required mixing/reaction time shall be 45 minutes minimum. The temperature of the asphalt-rubber mixture shall be between 375°F (190°C) to 425°F (218°C) during the reaction period. After reacting, the asphalt rubber binder shall conform to the requirements shown in Table 203-11.4 (A).

TABLE 203-11.4 (A): REQUIREMENTS FOR ASPHALT RUBBER BINDER

Test Parameter	Test Method	Specification Limit	
		Minimum	Maximum
Haake Field Viscosity @ 375°F (191°C), (Centipoise)	See 211-4	1500	4000
Cone Penetration @ 77°F (25°C),mm	ASTM D217	25	70
Resilient @ 77°F (25°C), % Rebound	ASTM D3407	18	—
Field Softening Point, °F (°C)	ASTM D36	52 (125)	74 (165)

The reacted asphalt rubber binder shall be maintained at a temperature of not less than 375°F (190°C) nor more than 425°F (218°C). If any of the material in a batch of asphalt rubber binder is not used within 4 hours after the 45-minute reaction period, heating of the material shall be discontinued. Any time the asphalt rubber binder cools below 375°F (190°C), and is then reheated, shall be considered a reheat cycle. The total number of reheat cycles shall not exceed 2. The material shall be uniformly reheated to a temperature of not less than 375°F (190°C)) nor more than 425°F (218°C) prior to use. Additional scrap tire CRM may be added to the reheated binder and reacted for a minimum of 45 minutes. The cumulative amount of additional scrap tire CRM shall not exceed 10 percent of the total binder weight. Reheated asphalt rubber binder shall conform to requirements shown in Table 203-11.4 (A).

When permitted by the Engineer, asphalt-rubber binder produced on another agency's project and defined here as "hold over material", may be used in the Work if the initial agency certifies the following:

a) The total weight and type of material being held over.

b) The amount of CRM contained within the holdover load on a percentage basis.

c) The grade of paving asphalt and asphalt modifier used and its source.

d) Date of original mixing.

e) Number of reheat cycles.

In no case, will more than 20 tons of holdover material be allowed to be transferred from one project to another. In all cases, the holdover material when blended with new asphalt-rubber binder, shall conform to the requirements shown in Table 203-11.4 (A).

203-11.5 Equipment for Production of Asphalt-Rubber. The Contractor shall utilize the following equipment for production of asphalt-rubber binder:

a) **Asphalt Heating Tank.** An asphalt heating tank equipped to heat and maintain the blended paving asphalt and asphalt modifier mixture at the necessary temperature before blending with the CRM. This unit shall be equipped with a thermostatic heat control device and a temperature reading device and shall be accurate to within ± 5°F (± 3°C) and shall be of the recording type.

b) **Blender Equipment.** A mechanical mixer for the complete, homogeneous blending of paving asphalt, asphalt modifier, and CRM. Paving asphalt and asphalt modifier shall be introduced into the mixer through meters. The blending system shall be capable of varying the rate of delivery of paving asphalt and asphalt modifier proportionate with the delivery of CRM. During the proportioning and blending of the liquid ingredients, the temperature of paving asphalt and the asphalt modifier shall not vary more than ± 25°F (± 14°C). The paving asphalt feed, the asphalt modifier feed and CRM feed, shall be equipped with devices by which the rate of feed can be determined during the proportioning operation. Meters used for proportioning individual ingredients shall be equipped with rate-of-flow indicators to show the rates of delivery and resettable totalizers so that the total amounts of liquid ingredients introduced into the mixture can be determined. The liquid and dry ingredients shall be fed directly into the mixer at a uniform and controlled rate. The rate of feed to the mixer shall not exceed that which will permit complete mixing of the materials. Dead areas in the mixer, in which the material does not move or is not sufficiently agitated, shall be corrected by a reduction in the volume of material or by other adjustments. Mixing shall continue until a homogeneous mixture of uniformly distributed and properly blended asphalt-rubber binder is produced. The Contractor shall provide a safe sampling device capable of delivering a representative sample of the completed asphalt-rubber binder of sufficient size to perform the required tests.

c) **Storage/Reaction Tank.** An asphalt-rubber binder storage/reaction tank equipped with a heating system that is equipped with a temperature reading device to maintain the proper temperature of the asphalt-rubber binder and an internal mixing unit capable of maintaining a homogeneous mixture of paving asphalt, asphalt modifier and CRM.

d) **Viscometers.** The Contractor shall supply a Haake Viscometer (or equivalent) conforming to 211-4 for use by the Engineer to verify the viscosity of the asphalt-rubber binder wherever a field laboratory is used. All asphalt concrete plants are required to have a field laboratory for use by the Engineer per 8-3.

The equipment shall be approved by the Engineer prior to use.

203-11.6 Mix Designs and Certifications. The optimum binder content for ARHM-GG mixes shall be determined by California Test Method 367 except that Step 2 regarding surface flushing shall not be used. Optimum binder content shall be determined by using a void content between 3 percent minimum to 6 percent maximum as approved by the Engineer. Compaction shall be in accordance with California Test Method 304 except for the following:

Mixing Temperatures:

Asphalt-rubber = 325°F to 360°F (163°C to 182°C)

Aggregate = 290°F to 325°F (143°C to 163°C)

Compaction Temperature = 290°F to 300°F (143°C to 149°C)

In addition to the formulations and certifications required in 203-11.2, 203-11.3 and 203-11.4 for asphalt-rubber, the Contractor shall furnish to the Engineer a mix design and samples of all materials to be used at least 10 working days before construction is scheduled to begin. The mix design and certifications shall include, but are not limited to, the following:

a) Combined aggregate gradation.

b) Individual bin gradations (hot for batch, cold for drum plant).

c) Percentage of each bin.

d) Asphalt rubber binder content.

e) Density.

f) Air Voids.

g) Voids in Mineral Aggregates (VMA).

h) Stability.

i) Aggregate source

j) Asphalt binder source

203-11.7 Mixing Binder With Aggregate. Mixing of the asphalt rubber binder with aggregate shall conform to 203-6.7 except that the temperature requirements of ARHM shall supercede the requirements in 203-6.7.

203-11.8 Storage. Storage of ARHM shall conform to 203-6.8.

203-11.9 Miscellaneous Requirements. Miscellaneous requirements shall conform to 203-6.9 except that the temperature of the asphalt rubber binder shall be 375°F (190°C) to 425°F (218°C) when added to the aggregate. The temperature of the aggregate at the time of adding the asphalt-rubber binder shall be 300°F (149°C) to 330°F (166°C).

203-12 ASPHALT RUBBER AND AGGREGATE MEMBRANE (ARAM).

203-12.1 General. ARAM surfacing or interlayer shall involve cleaning and preparing the existing surface, spreading asphalt rubber and cover aggregate, rolling, and sweeping.

The construction sequence of an ARAM shall be as follows:

a) The surface shall be cleaned.

b) Asphalt-rubber shall be applied.

c) Screenings shall be placed, rolled, and loose material removed. Only then will the ARAM surface be opened to traffic. A minimum of 48 hours shall elapse after placement of the screenings before a slurry seal conforming to 302-4 may be applied, unless otherwise directed by the Engineer.

Certified volume or weight slips shall be delivered to the Engineer for all materials supplied.

203-12.2 Materials. Materials used in the construction of ARAM shall be asphalt-rubber and screenings as specified below.

203-12.2.1 Asphalt Rubber. Asphalt-rubber shall conform to 203-11

203-12.2.2 Screenings. Screenings, when used as a cover aggregate for ARAM, shall be crushed rock conforming to the following gradations in Table 203-12.2.2 (A):

TABLE 203-12.2.2 (A): SCREENINGS

Sieve Size	Percentage Passing Sieve		
	Coarse 1/2" (12.5mm)	Medium 3/8" (9.5mm)	Fine 3/8" (9.5mm)
3/4" (19.0mm)	100	100	—
1/2" (12.5mm)	90 – 100	95 – 100	100
3/8" (9.5mm)	50 – 80	70 – 85	85 - 100
No. 4 (4.75mm)	0 - 15[1]	0 - 15[1]	5 - 20[1]
No. 8 (2.36mm)	0 – 5	0 – 5	
No. 16 (1.18mm)	—	—	—
No. 200 (75μm)	0 – 1	0 – 1	0 - 1

1. Lower percentages are desirable.

Screenings shall be medium 3/8 inch (9.5mm) unless otherwise specified. Screenings shall be preheated to between 260°F (127°C) to 325°F (163°C) and adequately coated with 0.70 percent to 1 percent PG 64-10 paving asphalt at a central mixing plant to prevent free dust. The exact amount of paving asphalt shall be recommended by the Contractor and approved by the Engineer. Screenings shall conform to the requirements of Table 200-1.2.1 (B).

203-13 CRUMB RUBBER MODIFIED ASPHALT CONCRETE GAP GRADED (CRUMAC-GG).

203-13.1 General. CRUMAC-GG shall be the product of mixing mineral aggregate, paving asphalt and crumb rubber modifier (CRM) at a central mixing plant. The Contractor shall submit test reports and certificates of compliance for the paving asphalt and the CRM at the time of delivery to the Work site.

203-13.2 Materials.

203-13.2.1 Paving Asphalt. Paving asphalt shall be PG 64-16 conforming to 203-1.

203-13.2.2 Crumb Rubber Modifier.

203-13.2.2.1 General. CRM shall be scrap tire CRM conforming to 203-11.2.3.

203-13.2.2.2 Gradation. The CRM gradation shall conform to 203-11.2.3.1 and Table 203-130.2.2.2 (A). The class of CRM gradation shall be the same as the class of aggregate gradation, i.e. CRM (CRUMAC-B) shall be specified with aggregate gradation CRUMAC-GG-B.

TABLE 203-13.2.2 (A)

GRADING REQUIREMENTS FOR CRM (CLASS B AND C)

SIEVE SIZE	CLASS	
	CRUMAC-B Min.-Max.	CRUMAC-C Min.-Max.
No. 10 (2.00mm)	98-100	100
No. 16 (1.18mm)	30-50	50-70
No. 30 (600μm)	3-13	10-20
No.50 (300μm)	0-6	3-8
No. 100 (150μm)	0-2	0-3

203-13.2.3 Aggregate. Aggregate for CRUMAC-GG shall conform to 203-6.3.2.

203-13.3 Composition and Grading. CRUMAC-GG will be designated by type and class, i.e., CRUMAC-GG-C. The grading of the combined aggregates shall conform to Table 203-11.3(A).

203-13.4 Mix Designs. Mix designs shall conform to 203-11.6. The proportions of paving asphalt and CRM, by weight, shall be 82 ± 2 percent paving asphalt and 18 ± 2 percent CRM. Once the total percent of paving asphalt and CRM is determined by the mix design, the production tolerance for the combined materials shall be ± 0.5 percent as determined by California Test 382.

203-13.5 Aggregate Storing, Drying, and Screening. Aggregate storing, drying, and screening shall conform to 203-6.5.

203-13.6 Proportioning. Proportioning shall conform to 203-6.6, except that proportioning of CRM shall be performed using an automatic batching system. The CRM feeder gate shall be operated by a switch or starter. The CRM feeder system shall be able to deliver CRM to the mixer at an accuracy of 0.1 percent of the total weight of mix. The Contractor shall submit to the Engineer, in writing, the method proposed to deliver the CRM to the mixture in accordance with 2-5.3. The method and equipment proposed for use shall be so designated and accessible that the Engineer can visually observe the materials being incorporated into the mixture. When batch-type plants are used, the CRM shall be proportioned by weight. When dryer-drum plants are used, the systems provided shall maintain positive interlock between the flow of CRM, paving asphalt and aggregate.

203-13.7 Mixing. Mixing shall conform to 203-6.7, except that the aggregate and the CRM shall be combined and mixed thoroughly for a minimum of 5 seconds prior to introducing the paving asphalt. The paving asphalt, aggregate, and CRM shall be mixed for a minimum of 35 seconds. If the Engineer determines that the mixture is not thoroughly mixed or that the aggregate and CRM are not fully coated with paving asphalt, the mixing time shall be increased. The temperature of the mix at discharge shall be a minimum of 320 °F (161 °C) and a maximum of 350 °F (177°C).

203-13.8 Storage. Storage of CRUMAC shall conform to 203-6.8.

203-13.9 Miscellaneous Requirements. Miscellaneous requirements shall conform to 203-6.9, except that CRUMAC shall be delivered to the Work site without segregation and within the temperature requirements specified in 302-11.2. No mix shall be delivered with segregation.

203-13.10 Acceptance. Acceptance of plant produced mixtures will be based upon gradation, combined paving asphalt and CRM content, and minimum stability values per Table 203-11.3 (A). Air voids will not to be used as an acceptance criteria. Sampling and S-value moving average requirements shall conform to 203-6.4.1.

203-14 TIRE RUBBER MODIFIED ASPHALT CONCRETE (TRMAC).

203-14.1 General. TRMAC shall be the product of mixing mineral aggregate and tire rubber modified paving asphalt at a central mixing plant.

203-14.2 Materials.

203-14.2.1 Tire Rubber Modified Paving Asphalt. Tire rubber modified paving asphalt shall consist of paving asphalt containing ground scrap tire rubber. The scrap tire rubber shall be incorporated into the paving asphalt such that a smooth and homogeneous composition results. Tire rubber modified paving asphalt shall be designated as MAC-10TR or MAC-15TR and shall conform to the requirements shown in Table 203-14.2.1 (A).

TABLE 203-14.2.1 (A)

Properties	Test Method No.	MAC-10TR	MAC-15TR
TESTS ON ORIGINAL ASPHALT:			
Minimum Recycled Whole Scrap Tire Rubber Content, %	CERTIFICATION	10.0	15.0
Penetration, at 25°C, 0.1mm, 100 grams, 5 seconds	ASTM D5	40-60	40-60
Penetration, at 4°C, 0.1mm, 200 grams, 60 seconds, Min.	ASTM D5	20	15
Dynamic Viscosity, at 140°F (60°C), Poise, Min.	ASTM D2171	5000	5000
Kinematic Viscosity, at 275°F (135°C), Centistokes, Max.	ASTM D2170	1000	1200
Flash Point, Cleveland Open Cup, °F (°C), Min.	ASTM D92	450 (232)	450 (232)
Softening Point, °F (°C), Min.	ASTM D36	127 (53)	127 (53)
Solubility in Trichloroethylene, %, Min.	ASTM D2042	97.5	97.5
TESTS ON RESIDUE FROM RTFO PROCEDURE:			
Penetration, at 25°C, 0.1mm, 100 grams, 5 seconds	ASTM D5	20-40	20-40
Penetration, at 4°C, 0.1mm, 200 grams, 60 seconds, Min.	ASTM D5	14	10
Dynamic Viscosity, at 140°F (60°C), Poise, Min.	ASTM D2171	20,000	20,000
Kinematic Viscosity, at 275°F (135°C), Centistokes, Max.	ASTM D2170	1500	2000
Percent of Original Penetration, at 25°C, 0.1mm, 100 grams, 5 seconds, Min.	ASTM D5	50	50

The Contractor shall submit in accordance with 2-5.3 a Certificate of Compliance conforming to 4-1.5.

203-14.2.2 Aggregate. Aggregate shall conform to 203-6.3.2.

203-14.2.3 Mineral Filler Mineral filler shall conform to 203-6.3.3.

203-14.3 Composition and Grading. TRMAC will be designated by class and tire rubber content for dense-graded mixes (i.e., "C2-TRMAC-10"); and by tire rubber content and type for gap-graded mixes (i.e., "TRMAC-15-GG-C").

The composition and grading of dense-graded mixes shall conform to 203-6.4.3.

Gap-graded mixes shall conform to the composition requirements shown in Table 203-14.3 (A) and the requirements for the grading of the combined aggregates shown in Table 203-11.3 (A).

TABLE 203-14.3 (A)

Properties	Requirements
% MAC-10TR or MAC-15TR by Weight of Dry Aggregate*	5.0 – 7.0
Air Voids % California Test 367	3 – 6
Stabilometer Value per California Test 304 and 366, Min.	23
Voids in Mineral Aggregate Percent, Min.	16

*Once the percent MAC-10TR or MAC-15TR is determined by the mix design, the production tolerance shall be ± 0.5% as determined by California Test method 362, 379, or 382.

203-14.4 Mix Designs. Mix designs for dense-graded mixes shall conform to 203-6.2. Mix designs for gap-graded mixes shall conform to 203-6.2 and the composition and grading requirements specified in 203-14.3.

203-14.5 Aggregate Storing, Drying, and Screening. Aggregate storing, drying, and screening shall conform to 203-6.5.

203-14.6 Proportioning. Proportioning shall conform to 203-6.6.

203-14.7 Mixing. Mixing shall conform to 203-6.7 except that the temperature of the aggregates at the time of adding tire rubber modified paving asphalt shall not exceed 350°F (176°C) nor be less than 300°F (149°C).

203-14.8 Storage. Storage shall conform to 203-6.8.

203-14.9 Miscellaneous Requirements. Miscellaneous requirements shall conform to 203-6.9.

203-14.10 Acceptance. Acceptance of plant-produced mixtures will be based upon gradation, tire rubber modified paving asphalt content, and the minimum stabilometer value shown in Table 203-14.3 (A). Air voids will not be used as an acceptance criteria. Sampling and S-value moving average requirements shall conform to 203-6.4.1.

203-15 POROUS ASPHALT CONCRETE.

203-15.1 General. Porous asphalt concrete shall be the product of mixing open-graded mineral aggregate with paving asphalt at a central mixing plant.

203-15.2 Materials.

203-15.2.1 Paving Asphalt. Paving asphalt shall:

a) conform to 203-1,

b) be one performance grade higher than that established by the State of California for the climate region of the Work site unless otherwise specified in the Special Provisions, and

c) contain 0.5 percent liquid antistrip by weight of paving asphalt.

203-15.2.2 Aggregate. Aggregate shall conform to 203-6.3.2.

203-15.2.3 Mineral Filler. Mineral filler shall conform to 203-6.3.3.

203-15.2.4 Liquid Antistrip. Liquid antistrip storage and method of introduction shall conform to the manufacturer's recommendations. Liquid antistrip shall be introduced into the paving asphalt by the paving asphalt producer. Liquid antistrip shall not be substituted for paving asphalt. Only one type or brand may be used at any one time during production. The total amine value shall be 325 minimum when tested in accordance with ASTM D2074.

203-15.3 Composition and Grading. The grading of the combined aggregates shall conform to the requirements shown in Table 203-15.3 (A) unless otherwise specified in the Special Provisions.

TABLE 203-15.3 (A)

Sieve Size	Percentage Passing 1/2"
3/4 in (19.0mm)	100
1/2 in (12.5mm)	95-100
3/8 in (9.5mm)	74-93
No. 4 (4.75mm)	24-41
No. 8 (2.36mm)	3-22
No. 16 (1.18mm)	0-10
No. 200 (75 μm)	0-3

Porous asphalt concrete shall conform to the requirements shown in Table 203-15.3 (B).

TABLE 203-15.3 (B)

Properties	Requirements
Paving Asphalt, %	6.0 % min.
Air Voids, %, California Test 309 *	18% min.
Draindown, ASTM D6390 or CT 368	0.30% max.

* Compact specimen and determine bulk specific gravity in accordance with CT 308 Method B.

203-15.4 Job Mix Formula (JMF) and Mix Designs. JMFs and mix designs shall conform to 203-6.2 and 203-15.3.

203-15.5 Aggregate Storing, Drying, and Screening. Aggregate storing, drying, and screening shall conform to 203-6.5.

203-15.6 Proportioning. Proportioning shall conform to 203-6.6.

203-15.7 Mixing. Mixing shall conform to 203-6.7.

203-15.8 Storage. Storage shall conform to 203-6.8. Porous asphalt concrete shall not be stored in a silo in excess of 2 hours.

203-15.9 Miscellaneous Requirements. Miscellaneous requirements shall conform to 203-6.9. Porous asphalt concrete shall be completely covered with tarpaulins during transport. The material temperature at the time of delivery to the Work site shall not exceed 275° F (135° C).

203-15.10 Acceptance. Acceptance will be based upon gradation and paving asphalt content.

SECTION 204 - LUMBER AND TREATMENT WITH PRESERVATIVES

204-1 LUMBER AND PLYWOOD.

204-1.1 Kinds.

204-1.1.1 General. Structural lumber shall be of the kinds and grades indicated on the Plans or in the Special Provisions. Proper allowance for shrinkage in the lumber shall be made by the Contractor where it is necessary to meet definite dimensions shown on the Plans. All sizes refer to nominal sizes. Rough and dressed sizes shall conform to the sizes set forth in the American Lumber Standards.

204-1.1.2 Douglas Fir. Unless otherwise specified, all lumber shall be Douglas fir and shall be selected as to grade and shall conform in all particulars to the Standard Grading Rules for Western Lumber, published by the Western Wood Products Association and approved by the American Lumber Standards Committee.

204-1.1.3 Redwood. Redwood lumber shall conform in all particulars to the Standard Specifications for Grades of California Redwood of the Redwood Inspection Service.

204-1.1.4 Plywood. Plywood shall be manufactured and graded in accordance with the rules of the American Plywood Association and the latest Product Standard for Softwood Plywood, Construction and Industrial, of the National Bureau of Standards.

204-1.2 Lumber Uses and Grades. Unless otherwise specified, lumber quality for the uses listed shall not be less than the grades in Table 204-1.2 (A).

TABLE 204-1.2 (A)

USES	GRADES
Major permanent construction, such as bridges.	"Select Structural" for main structural members. "No. 1" for beams, stringers, joists, and planks. "Select Structural" when redwood is specified.
Minor permanent construction, such as bulkheads, retaining structures, headers for bituminous pavement, wooden warning rails, posts for metal beam guard rails.	"No.1" for posts, joists and planks. "Construction" for boards.
Wooden warning rail posts, guide posts, sign posts.	"Select Structural" redwood. (No. 1 grade Douglas fir optional for wooden warning rail posts.)
Studs, headers, and wales for formwork. Form sheathing for non-showing surfaces of concrete.	"Standard", "No. 4 Common", or "No. 3"; any exterior-type grade of plywood optional for form sheathing.
All exposed surfaces of bridges, viaducts, over-crossing; soffits and sides of beams and girders; slabs between beams and girders; headwalls and endwalls of culverts or covering conduits; form sheathing for showing surfaces of retaining walls, channel walls, etc.	"Exterior B-B" (concrete form) grade of plywood.

204-1.3 Grade Marking.

204-1.3.1 Lumber. All lumber shall be grade marked by a lumber grading agency certified by the American Lumber Standards Committee.

204-1.3.2 Plywood. Each sheet of plywood shall bear the official stamp of a quality control agency stating the grade of the sheet.

204-2 TREATMENT WITH PRESERVATIVES.

204-2.1 General. All timber products to be treated shall conform to the requirements of the various sections of these specifications.

All preservatives shall comply with the applicable standards contained in the AWPA Book of Standards.

All cutting such as adzing, boring, chamfering, framing, graining, surfacing, trimming, etc., shall be done prior to treatment.

Wood shall be conditioned, seasoned, prepared, and treated by the pressure process in accordance with the applicable standards in the AWPA Book of Standards.

With the exception of southern yellow pine and ponderosa pine, all lumber and timber with a nominal thickness of 2 inches (50mm) and over shall be incised on all sides to ensure penetration in accordance with AWPA Standards. Lumber edge-worked to pattern (such as T&G) other than S4S need not be incised on the edges.

204-2.2 Wood Preservatives. The preservative used shall be one of the following:

a) Creosote or creosote solutions;

b) Oil-borne preservatives;

c) Water-borne preservatives;

d) Fire-retardant treatment (either exterior or interior type).

All preservatives and the selection of preservatives shall meet the requirements as listed in the AWPA Book of Standards. The retention and penetration for each preservative shall be equal to, or greater than, the minimums listed in the AWPA Standard.

Fire-retardant treated wood is lumber or plywood impregnated with chemicals. When tested in accordance with ASTM E84 for a period of 30 minutes, it shall have a flame spread of not over 25 and show no evidence of significant progressive combustion. Materials which may be exposed to weather shall maintain this fire-retardant classification when tested in accordance with the rain and weathering tests of ASTM E108. The materials shall bear identification showing the fire performance rating thereof and shall be further identified to indicate suitability for exposure to the weather. Such identification shall be issued by an approved agency having a series for inspection of materials at the factory.

Preservatives used for wood in contact with ground or fresh water shall be limited to:

a) Creosote or creosote solutions

b) Pentachlorophenol

c) Ammoniacal Copper Arsenate

d) Chromated Copper Arsenate

Preservatives used for wood subjected to marine (salt water) exposure shall be limited to:

a) Ammoniacal Copper Arsenate

b) Chromated Copper Arsenate

204-2.3 Field Treatment of Cut Surfaces. When cutting or drilling becomes necessary after plant treatment, the cut or drilled surfaces shall be given protection by field treatment in accordance with provisions of AWPA Standard M4.

204-2.4 Quality Control - Inspection. All materials treated shall be subject to inspection. The inspection and marking shall be in accordance with AWPA Standard M2.

Inspection shall be performed by any of the following:

a) The Engineer;

b) An American Wood Preservative License Bureau Inspection organization.

204-2.5 Handling and Protection of Treated Materials. Care shall be exercised to not damage the edges or abrade the surfaces to the extent of reducing the depth of treated wood or exposing any wood not penetrated. Cant hooks, peavies, sharp-pointed tools, and the use of metal slings without protective guards shall not be permitted for the handling of treated lumber. Treated piling may be handled with pointed tools, provided that side surfaces are not penetrated over 1/2 inch (13mm). The use of handling tools and loading devices will not be permitted in the groundline area of poles. Care shall be taken to assure that all damage such as abrasions, nail and spike holes, are thoroughly saturated with the field treating solution in accordance with AWPA Standard m 4.

Material that is stored at the Work site prior to its use shall be stacked neatly on skids to raise it from the ground, and shall be protected from the sun and weather.

SECTION 205 - PILES

205-1 TIMBER PILES.

205-1.1 General. Timber piles shall be Douglas fir, unless otherwise shown on the Plans or specified in the Special Provisions.

205-1.2 Quality. Piles shall conform to ASTM D25 for Class B piles and the requirements contained herein. Piles shall be cut from sound, live, close-grained trees and shall be free from large, loose, or unsound knots, scars, decay, holes, insect damage, barnacles, limnoria, or other forms of sea life, and other defects or imperfections that would materially impair their strength or durability.

All piles shall be machine-peeled and all inner skin shall be removed. All branch stubs and partially overgrown knots shall be neatly trimmed flush with the surface, and the butts and tips shall be sawed square with the longitudinal axis of the pile.

No cracks will be permitted in any pile. Splits, shakes, and checks will be permitted only to the extent provided herein. A crack is defined as a break across the grain of the wood extending from surface to surface through the pile; a shake is defined as a separation of the wood along the annual rings; and a check is defined as a lengthwise separation of the wood across the rings of annual growth. A through check or compound check is defined as a check extending from surface to surface, either through the pith center or shunted around by a shake.

Splits in piles shall not be longer than the butt diameter. The length of any shake or combination of shakes in the outer half of the radius of the butt of the pile, when measured along the curve of the annual ring, shall not exceed one-third of the circumference of the butt of the pile.

Any check that would impair the strength of the pile will be cause for rejection, but in no case will any check exceeding 3/4 inch (19mm) in width be allowed. Any compound or through check, regardless of length, will be cause for rejection.

Spiral grain shall not exceed one-half of a complete twist in any 20 feet (6m) of length.

Sound knots in piles 50 feet (15m) or less in length, and in three-quarters of the length from the butt of piles longer than 50 feet (15m), shall be no longer than 4 inches (100mm) or one-third of the diameter of the pile at the point where they occur, whichever is smaller. Sound knots in the remaining one-quarter of the length of piles longer than 50 feet (15m) shall be no larger than 5 inches (125mm) or one-half of the diameter of the pile at the point where they occur, whichever is the smaller. The size of a knot shall be its diameter measured at right angles to the length of the pile.

Unsound knots or knot clusters will not be permitted. A knot cluster is two or more knots grouped together, the fibers of the wood being deflected around the entire unit. A group of single knots, with fibers deflected around each knot separately, is not a cluster, even though the knots may be in close proximity. The sum of sizes of all knots in any foot of length of the pile shall not exceed twice the size of the largest single knot permitted.

Holes less than 1/2 inch (13mm) in average diameter will be permitted in piles, provided the sum of the average diameters of all holes in any 1 ft^2 ($0.1m^2$) of pile surface does not exceed 1-1/2 inches (38mm).

The presence of invisible internal checks, and shattering or water-burses, shall be determined by sounding with a suitable hammer, and a resultant hollow or shattered sound will be sufficient cause for rejecting the pile.

The requirements prescribed herein relative to cracks, splits, shakes, and checks shall apply equally to treated and untreated piles, and any such defects that develop or occur as a result of any treating operation may be cause for rejection.

No nails, spikes, or other metal shall be present in any timber pile unless specifically authorized by the Engineer.

205-1.3 Dimensional Requirements. The diameter of a pile at any section shall be considered as the average diameter at such section measured at right angles to the longitudinal axis of the pile; but in determining an average diameter, no single diameter that is more than 10 percent greater than the least diameter at the same section shall be used. The butt and tip diameters shall be such as to conform with the requirements indicated in the following tabulation and the diameter 3 feet (1m) from the butt shall not be smaller than 1 inch (25mm) less than that at the butt.

TABLE 205-1.3 (A)

Length Feet (Meters)	3 Ft (1m) From Butt				All Tip Minimum	
	Minimum		Maximum			
	Circumference inches (mm)	Approx. Dia. inches (mm)	Circumference inches (mm)	Approx. Dia. inches (mm)	Circumference inches (mm)	Approx. Dia. inches (mm)
Under 40 (12)	38 (965)	12 (305)	63 (1600)	20 (510)	25 (635)	8 (203)
40-54 (12-16)	38 (965)	12 (305)	63 (1600)	20 (510)	22 (559)	7 (178)
55-74 (17-22)	41 (1040)	13 (330)	63 (1600)	20 (510)	22 (559)	7 (178)
75-90 (23-27)	41 (1040)	13 (330)	63 (1600)	20 (510)	19 (483)	6 (152)
Over 90 (27)	41 (1040)	13 (330)	63 (1600)	20 (510)	16 (406)	5 (127)

The average diameter of the heartwood at the butt of Douglas fir piles that are to be treated shall be not less than 70 percent, and of those that are to be used untreated not less than 75 percent, of the average outside butt diameter of the pile.

A straight line from the center of the butt to the center of the tip pile shall lie entirely within the body of the pile.

Piles shall be free from short crooks in which the deviation from straightness in any 5 feet (1.5m) of length anywhere exceeds 2-1/2 inches (64mm). Short crooks shall also comply with the requirements for straightness.

The individual lengths in which timber piles are to be furnished shall be as stipulated by the Plans, Specifications, requisition, or purchase order issued in connection therewith. Unless otherwise prescribed, as many as 30 percent of the number furnished on any single order or in any single lot may run as much as 2 feet (0.6m) longer than the specified lengths.

205-2 STEEL PILES.

205-2.1 General. Steel piles furnished under this specification shall consist of structural steel shapes that fulfill the requirements prescribed for such material in ASTM A36 and shall conform to the details and dimensions shown on the Plans and Special Provisions relating directly thereto.

Splices in steel piles shall be made by a full penetration butt weld of the entire cross section. Care shall be taken to properly align adjacent sections so that the axis of the pile will be straight. The number of splices in the length of the pile shall be limited to 2. Splices in the top 10 feet (3m) of the piles will not be permitted. The locations of pile lugs, when used, shall be subject to the approval of the Engineer. All weldings shall be performed by qualified welding operators in accordance with 304-1.9.

205-2.2 Manufacturing Requirements. Material for steel piles shall not be made by the acid Bessemer process.

Steel sheet piling shall consist of standard interlocking sheet pile sections having positive interlocks in both longitudinal and transverse directions which are continuous throughout the entire length of the piece.

205-3 CONCRETE PILES.

205-3.1 General. The types of concrete piles covered by these specifications are precast, cast-in-place, and prestressed piles. The type to be used or furnished shall be as shown on the Plans or in the Special Provisions. Steel and concrete shall be placed in accordance with 303-1. Portland cement concrete and reinforcing steel shall conform to the provisions of 201-1 and 201-2.

Compressive strength tests for precast and prestressed piles shall be performed in accordance with ASTM C31.

205-3.2 Precast Piles. Precast concrete piles shall be of such quality that the finished piles can be handled and driven to required bearing without cracking or other damage which would impair their strength or durability. Concrete shall have a minimum strength of 4,000 psi (27.6MPa) at 28 days.

Concrete for precast concrete piles shall be cast in smooth, mortar-tight forms so supported as to prevent deformation or settlement during concrete placement or curing. The piles after being cast shall be cured by water, steam, curing compound, or other such methods of curing as may be approved by the Engineer. Curing shall be continued until specimens of the concrete from which the piles were cast attain a compressive strength of at least 4,000 psi (27.6MPa). Piles shall not be driven until completion of the specified curing.

The piles shall present true, smooth, even surfaces, free from honeycombs or voids and shall be sufficiently straight that a line stretched from butt tip along any face will not deviate nor be deflected for more than 1 inch (25mm) in 50 feet (15m) at any point. Defects in any pile may be accepted if repaired to the satisfaction of the Engineer.

Concrete piles may be cast the full length of the reinforcing bars provided that, after the piles have been driven, the concrete is removed to expose the steel as shown on the Plans.

205-3.3 Cast-in-Place Concrete Piles.

205-3.3.1 Metal-Cased Cast-in-Place Concrete Piles. Piles shall be cast in steel shells that have been previously driven to the penetration or bearing value required by the Engineer.

The shell shall be cylindrical and may be fluted, step-tapered, or uniformly tapered from butt to tip.

Shells that are driven without a mandrel shall be equipped with steel driving tips and shall be constructed of material conforming to the requirements prescribed in ASTM A252, Grade 2. They shall be of sufficient thickness (9 gage (3.8mm) minimum), strength, and rigidity to withstand distortion from driving, soil pressure, or the driving of adjacent piles. Continuous welds shall be used at all shell splices to develop the full strength of the section.

After being driven, but prior to placing of the reinforcing steel and concrete, the shells shall be examined for collapse or reduced diameter. The Contractor shall have available at all times a suitable light for the inspection of the shell throughout the entire length. Shells that are improperly driven, broken, or show partial collapse shall be replaced by and at the expense of the Contractor. Partial collapse of shells shall be interpreted to mean any collapse which reduces any diameter to less than 80 percent at any point. The replacement of the shell shall be made by withdrawing the entire shell and driving another in its place. Driving one shell within a shell is not permitted. If the withdrawal of the defective shell is impossible or impractical, as determined by the Engineer, the Contractor shall fill the defective shell with concrete and shall replace the defective pile with another pile driven alongside. Any enlargement of the footing required to accommodate such piling shall be at Contractor's expense.

Driven shells shall be clean and free of water before reinforcing steel and concrete are placed therein. Concrete shall be vibrated to within 5 feet (1.5m) of the tip of the shell.

205-3.3.2 Piles Cast in Drilled Holes. All holes for concrete piles cast in drilled holes shall be drilled dry to the tip elevations and diameter shown on the Plans or to the elevation determined by the Engineer. Any hole which shows less than one-half the diameter of the hole at the bottom shall be rejected. Suitable casings shall be furnished and placed when required to prevent caving of the hole before concrete is placed therein.

All loose material and water, existing at the bottom of the hole after drilling operations have been completed, shall be removed before placing concrete in the hole.

If the casing is to be removed, it shall be removed from the hole as concrete is placed therein. The bottom of the casing shall be maintained not more than 5 feet (1.5m) nor less than 1 foot (0.3m) below the top of the concrete during withdrawal and placing operations, unless otherwise permitted by the Engineer. Damage to the concrete during withdrawal and placing operations shall be avoided by hammering or vibrating the casing.

The Contractor shall ensure that the concrete in the hole is dense and homogeneous. Vibration of the concrete during placing will not be permitted. However, rodding may be required. After the hole has been filled with concrete, the concrete in the top 10 feet (3m) of the hole or for the length of the reinforcing, whichever is the greater, shall be vibrated.

The Contractor shall ensure that the concrete in the hole is dense and homogeneous. Vibration of the concrete during placing will not be permitted. However, rodding may be required. After the hole has been filled with concrete, the concrete in the top 10 feet (3m) of the hole or for the length of the reinforcing, whichever is the greater, shall be vibrated.

205-3.4 Prestresse Concrete Piles.

205-3.4.1 General. The manufacture of prestressed piles shall be performed in accordance with 303-3. Concrete shall have a minimum strength at 28 days of 5,000 psi (34.5MPa). Piles shall be of such quality that the finished piles can be handled and driven to required bearing without cracking or other damage which would impair their strength or durability. Piles shall present true, smooth, even surfaces free from honeycombs or voids and shall be sufficiently straight that a line stretched from butt to tip along any face will not deviate more than 1 inch (25mm) in 50 feet (15m) at any point. Piles showing defects in the upper 10 feet (3m) which reduce the cover over the steel to less than required, will be rejected. Defects in the remainder of the pile may be accepted if repaired in a manner satisfactory to the Engineer.

Prestressed forces shall not be transferred to the piles until the concrete has attained a strength of 3,500 psi (24.1MPa).

205-3.4.2 Handling and Driving. Prestressed concrete piles shall be lifted or supported only at the points shown on the approved shop drawings. Piles shall not be driven until they have attained a minimum compressive strength of 5,000 psi (34.5MPa) as determined by tests on concrete cylinders cast and cured under the same conditions as the piles.

SECTION 206 - MISCELLANEOUS METAL ITEMS

206-1 STRUCTURAL STEEL, RIVETS, BOLTS, PINS, AND ANCHOR BOLTS.

206-1.1 Requirements.

206-1.1.1 General. All steel, the class of which is not definitely designated herein, in the Special Provisions, or on the Plans, shall be structural steel and shall conform to the requirements of ASTM A 36. Steel manufactured by the acid Bessemer process shall not be used.

206-1.1.2 Certification. The Contractor shall furnish to the Engineer, before fabrication, a mill certified report (in duplicate) of the tests for each heat of steel or iron from which the material is to be fabricated. The certification shall contain the results of chemical and physical tests required by ASTM standards for materials.

206-1.1.3 Additional Tests. The Agency reserves the right to require or make additional mill and laboratory tests. The number of such additional tests will be limited as follows, except that in case of failure of the material to comply with the ASTM requirements, more tests may be made or the material rejected:

a) Structural steel - One complete test for each heat number or each 10 tons (9 tonnes) of identifiable stock

b) Rivets - One complete test for each size

c) Bolts - One complete test for each lot

"Identifiable stock" is material for which authentic records of the chemical and physical properties are available.

Test specimens shall be furnished cut and machined in accordance with the ASTM standards for the material to be tested, as referred to herein.

206-1.1.4 Stock Material. When the Contractor proposes to use material already in stock, it shall notify the Engineer of such intention at least 10 days in advance of beginning fabrication.

206-1.1.5 Mill Tolerance. Rolling and cutting tolerances, permissible variations in weight and dimensions, defects, and imperfections shall not exceed the limits contained in ASTM A6.

206-1.2 Structural Steel.

206-1.2.1 Stock Materials. The Contractor shall select the material intended for use from stock and place it in location apart from other stock material and accessible for inspection and sampling. It shall also select the material from as few heat numbers as possible and shall furnish certified mill test reports for each of the heat numbers. Two samples shall be taken by the Engineer from each heat number; one for the tension test and one for the cold-bend test. If the heat numbers cannot be identified, the Engineer may select random test specimens from the unidentifiable heats.

206-1.2.2 High-Strength Low-Alloy Structural Steel. The material shall conform to the requirements of ASTM A242, A440, A441, A606, A607, or A446 (Grades C, D, or E) as specified.

206-1.2.3 Copper Bearing Structural Steel. Copper bearing structural steel shall conform to requirements of ASTM A36, A440, A446, A570 or A611 as specified.

206-1.3 Rivets.

206-1.3.1 Stock Material. Rivets taken from identifiable stock may be accepted by the Engineer based on certified mill test reports.

Rivets from unidentifiable stock shall not be used except where shown on shop drawings.

206-1.3.2 High-Strength Structural Steel Rivets. The material shall conform to the requirements of ASTM A 502.

206-1.3.3 Structural Steel Rivets. The material shall conform to the requirements of ASTM A502, except that the test specimen shall be bent upon itself when performing the bend test.

206-1.4 Bolts.

206-1.4.1 Unfinished Bolts. The bolts shall have square heads and square nuts unless otherwise specified. The bolts shall be long enough to extend entirely through the nut but not more than 1/4 inch (6mm) beyond. Washers shall not be furnished unless specified. Bolts shall be of steel conforming to the requirements of ASTM A307.

206-1.4.2 High-Strength Bolts. Shall conform to ASTM A325.

206-1.4.3 Anchor Bolts. Shall be manufactured from steel conforming to ASTM A36 or A307.

206-1.5 Mild Steel Forgings for Structural Purposes.

Steel forgings shall conform to the requirements of ASTM A325. They shall be Class C forgings with a maximum carbon content of 0.35 percent and shall be given a thorough annealing. The metal shall have a minimum Brinell hardness of 130 and a maximum of 190, when tested in accordance with ASTM E10.

206-2 STEEL CASTINGS.

206-2.1 General. Steel castings shall be true to pattern in form and dimension and free from defects that would affect the service value of the casting. Minor defects which do not impair the strength of the casting may be repaired with the approval of the Engineer. Castings which have been repaired without the permission of the Engineer may be rejected. Chemical analysis shall be performed in accordance with ASTM E 30.

206-2.2 Test Specimens. When required by the Agency, test coupons shall be poured monolithically with the castings. If, in the opinion of the manufacturer, the design of the casting is such that test coupons should not be attached thereto, the test coupons shall be cast attached to separate blocks. Two coupons shall be cast to represent each lot. A lot shall be considered as all castings from a melt which constitutes a part of a charge. Coupons shall remain attached until properly identified by the Engineer. Where test coupons are cast separately from the castings, the Engineer shall be notified of the time of pouring so as to permit him to identify both coupons and castings. Coupons cast separately from the castings shall not be detached from the block to which they are fastened until identified by the Engineer.

Test specimens shall be furnished and machined in accordance with ASTM A 370 at the Contractor's expense.

If the results of the tests for any lot do not conform, the entire lot will be rejected.

206-2.3 High-Strength Steel Castings for Structural Purposes. Castings shall conform to ASTM A148, Grade 80-50, except that the steel shall contain not less than 0.60 percent of manganese and not less than 0.20 percent of silicon.

206-2.4 Mild-to-Medium Carbon-Steel Castings for General Application. Castings shall conform to ASTM A27, Grade 65-35. The metal shall have a minimum Brinell hardness number of 130 when tested in accordance with ASTM E 10.

206-3 GRAY IRON AND DUCTILE IRON CASTINGS.

206-3.1 General. This subsection provides material specifications for gray and ductile iron castings. Castings shall be coated or uncoated as specified in the Special Provisions or shown on the Plans.

206-3.2 Materials.

206-3.2.1 Castings for Vehicular Traffic Areas. Vehicular traffic areas shall be defined as streets, highways, alleys, parking lots, bike lanes, driveways and other areas intended for use by motor vehicles.

Gray iron castings shall conform to ASTM A48/A48M, Class 35B, or AASHTO M105, Class 35B. Ductile iron castings shall conform to ASTM A536/A536M, grade 80-55-06.

206-3.2.2 Castings for Non-Vehicular Traffic Areas. Non-vehicular traffic areas shall be defined as sidewalks, bike paths, and other areas intended for use by pedestrians and bicyclists.

Gray iron castings shall conform to ASTM A48/A48M, Class 30B, or AASHTO M105, Class 30B. Ductile iron castings shall conform to ASTM A536/A536M grade 65-45-12 or grade 80-55-06.

206-3.2.3 Railings, Railing Posts, and Wheel Guards. Gray iron castings shall conform to ASTM A48/A48M, Class 40B, or AASHTO M105, Class 40B. Ductile iron castings shall conform to the requirements specified in the Special Provisions or shown on the Plans.

206-3.2.4 Rockers, Rocker Plate Bearings, and Bearing Plates for Bridges. Gray iron castings shall conform to ASTM A48/A48M, Class 50B, or AASHTO M105, Class 50B.

206-3.3 Manufacturing and Finishing.

206-3.3.1 General. The dimensions of the finished casting shall be as shown on the Plans. The as-cast dimensions may vary by 1/2 the maximum shrinkage of the metal or +/- 1/16 inch per foot (+/- 5mm/m).

206-3.3.2 Manhole Frame and Cover Sets and Grates. Manhole covers and grates shall seat into the frame such that the top surface of the cover or grate is flush with the top surface of the frame. The cover or grate shall sit within 1/8 inch (3mm) of the frame at any point.

Circular manhole frames shall be fabricated with a machined horizontal bearing surface. Covers and grates shall seat firmly into the frame and not rock when rotated to any position in the frame. The cover or grate shall not sit higher than 1/16 inch (2mm) above the frame for a distance greater than 1/4 of the frame circumference.

Square and rectangular frames shall be fabricated with an as-cast bearing surface.

206-3.3.3 Railings, Railing Posts, and Wheel Guards. Castings shall be machined and finished as specified in the Special Provisions or shown on the Plans.

206-3.3.4 Rockers, Rocker Plate Bearings, and Bearing Plates for Bridges. Castings shall be machined and finished as specified in the Special Provisions or shown on the Plans.

206-3.4 Testing.

206-3.4.1 General. Tensile testing, proof-load testing, or both shall be performed as specified in the Special Provisions. No separate payment will be made for the material specimens, test bars, machining, testing, and inspection required in these specifications.

206-3.4.2 Tensile Testing.

206-3.4.2.1 General. Test bars shall be tested at a laboratory approved by the Engineer. The Engineer shall be notified of the time of the melt and the time of testing to permit identification of both the test bars or coupons and the castings. The Engineer may select the test bars to be tested and may be present during testing. The Contractor shall submit in accordance with 2-5.3 a Certificate of Compliance conforming to 4-1.5 with laboratory test results.

206-3.4.2.2 Gray Cast Iron Castings. Testing shall be performed in accordance with ASTM A48/A48M.

Test bars shall be provided and machined by the manufacturer to the dimensions specified for Type B test bars in AASHTO M105 Specimen B or ASTM A48/A48M.

Test bars cast on and with a casting shall be cast such that when removed, the breakage pattern remains on the member. Test bars shall be of sufficient size for the manufacturer to produce a machined test specimen with the dimensions specified for "Type B" test bars as shown in Table 3 of AASHTO M105 or ASTM A48/A48M. For lots of 30 or less, a minimum of 3 test bars shall be tested. For lots greater than 30, 10 percent of all the test bars shall be tested.

Test bars cast separately from the casting shall be machined and tested at the laboratory approved by the Engineer. The Certificate of Compliance shall state that the test bar was cast integral with the casting it represents and that the testing and results were obtained as specified.

206-3.4.2.3 Ductile Iron Castings. Testing shall be performed in accordance with ASTM A536/A536M as modified herein.

Test bars shall be machined by the manufacturer to the dimensions specified in ASTM A536/A536M. The size of the test bars shall be as specified in the Special Provisions. Test bars shall be marked for identification of the casting represented by the coupons and test bars.

At the time of manufacture, test bars shall be cast from the same melt. Test bars shall be machined and tested at the laboratory approved by the Engineer. The Contractor shall submit in accordance with 2-5.3 a Certificate of Compliance conforming to 4-1.5 with laboratory test results. The Certificate of Compliance shall state that the test bar was cast integral with the casting it represents and that the testing and results were obtained as specified.

206-3.4.2.4 Test Results. If any of the test bars fail, 3 additional test bars shall be tested. The Engineer may select the test bars. If any of the additional test bars fail, the lot shall be rejected.

206-3.4.3 Proof-Load Testing.

206-3.4.3.1 General. Proof-load testing shall be performed in accordance with AASHTO M306 as modified in these specifications. Testing shall be performed at a facility approved by the Engineer. The Engineer shall be notified a minimum of 2 weeks before the scheduled date of testing and may elect to be present.

Castings shall be loaded to the proof-load specified in the Special Provisions or shown on the Plans. The loads shall be concentrated on a 9-inch x 9-inch x 1 inch (230mm x 230mm x 25mm) minimum thickness steel plate placed on a 9-inch x 9-inch x 1/4 inch (230mm x 230mm x 6mm) thick rubber pad, with a shore D hardness of 30, centered on the assembled castings. Castings shall be supported in the same manner as they would be when placed in use. The specified load shall be applied by a calibrated testing machine and held for a period of 1 minute. Upon removal of the load, the castings shall be examined for cracks and permanent deformation. Castings subjected to proof-load testing shall be destroyed.

Additionally, for each lot of gray iron castings supplied with cast-on test bars, 3 assembled castings shall be proof-load tested. The castings may be selected by the Engineer. A lot shall be defined as consisting of no more than 100 complete units from the same foundry facility.

206-3.4.3.2 Test Results. The Contractor shall submit test results signed by a Registered Civil Engineer showing the ultimate strength of the castings conforms to the requirements of AASHTO M306.

In addition, each lot of gray iron castings with cast-on test bars shall conform to the following:

a) If each of the 3 assembled castings pass the test, the entire lot will be considered as complying with the load requirements.

b) If any of the 3 assembled castings fail to pass the test, then 5 additional assembled castings from the same lot shall be selected for testing.

c) If each of the 5 additional assembled castings pass the test, the remainder of the lot will be considered as complying with the Specifications, except for the previously failed casting(s).

d) If any of the 5 additional assembled castings fail, the entire lot will be rejected.

206-3.5 Workmanship. Before castings are removed from the foundry, they shall be smooth and cleaned by shot blasting and parting lines, gates, and risers ground flush. Castings that show injurious defects revealed by X-ray or machining operations shall be rejected.

Castings shall be rejected if an examination reveals cracks or permanent deformations exceeding 1/8 inch (3mm).

Castings shall be true to pattern in form and dimension, and properly fit together. Castings shall be filleted boldly at angles and the risers shall be sharp and true. Parting lines, gates, and risers shall be ground flush.

Castings with minor defects may be reworked and resubmitted for approval by the Engineer. Castings that have been repaired without the permission of the Engineer shall be rejected. Welding plugs shall not be used.

206-3.6 Weight and Dimensions. Castings shall be inspected for conformance to the required weight and dimensions. The weight of each component shall not vary by more than ± 5 percent from the weight specified in the Special Provisions or shown on the Plans.

Castings that fail to conform to the required weight or dimensions shall be rejected. All remaining castings in the lot shall be inspected and shall conform to the Specifications.

206-3.7 Acceptance. The basis of acceptance will be the test results from the tensile testing, proof-load testing or both, as specified in the Special Provisions, and conformance to the required workmanship and weights and dimensions.

If so specified in the Special Provisions, a Certificate of Compliance conforming to 4-1.5 shall be submitted to the Engineer in accordance with 2-5.3 in lieu of testing. The certificate shall state that the samples representing each lot have been tested, inspected, and conform to the Plans and Specifications. The certificate shall also state the country of origin of the castings. In addition, acceptance will be based on conformance to the required workmanship.

Castings that have been damaged prior to installation will be rejected.

206-4 BRONZE CASTINGS.

206-4.1 General. Bronze castings shall be true to pattern in form and dimension and free from defects that would affect the service value of the casting. Minor defects may be repaired with Engineer approval. Castings which have been repaired without Engineer approval may be rejected. Castings that show injurious defects revealed by machining operations or by X-ray will be rejected.

206-4.2 Testing Requirements. Chemical analysis shall be made in accordance with ASTM E54.

When required by the Agency, test coupons shall be provided as specified in 206-2.2 except specimens shall be prepared in accordance with ASTM B208.

206-4.3 Expansion and Bearing Plates. Expansion and bearing plates shall conform to the requirements of ASTM B22, Alloy C. The sliding contact faces shall be machined smooth to true planes. If practical, one plate shall be machined at right angles to other plates in the set.

206-4.4 Ornamental Tablets and Miscellaneous Castings.

206-4.4.1 General. Ornamental tablets, railings, miscellaneous ornaments, and fixtures shall conform to the chemical requirements of ASTM B584 Copper Alloy UNS No. C90300.

206-4.4.2 Bearing Plates. The letters shall be heavily raised and spaced carefully to secure a uniform and balanced effect over the entire area of the panel. The background of the letter panel shall have a finely pebbled surface. The model of the tablet shall be submitted to the Engineer for approval.

Castings shall be boldly filleted at angles, and the arrises shall be sharp and true.

The faces and edges of all lettering and ornaments shall be tooled sharp and clean. Beveled edges shall be tooled smooth and true. Outside borders shall be straight and true and thoroughly polished. Filling and other tool marks shall be removed.

Ornaments, lettering, and the beveled edges shall be given a fine satin, hand finish. Lettering, bevels, and rosettes shall be highlighted; leaves and scrolls slightly highlighted, but well-polished. The pebble background shall be finished in dark statuary bronze and polished.

206-5 METAL RAILINGS.

206-5.1 Metal Hand Railings Materials. The fabrication of metal hand railings shall be in accordance with 304-2.

Steel railing material shall be welded or seamless steel pipe conforming to ASTM A53, structural steel conforming to ASTM A36, or tubular sections of hot rolled-mild steel, conforming to ASTM A501.

The base metal for aluminum railing shall be ASA alloy designation 6063-T6. Pipe and tubing shall be extruded conforming to requirements of ASTM B429, plates and sheets shall be rolled conforming to ASTM B 209, and rods, bars, or shapes shall be extruded conforming to ASTM B221.

206-5.2 Flexible Metal Guardrail Materials. Unless otherwise provided on the Plans or in the Special Provisions, materials and construction for the railings shall conform to the following requirements:

The rail elements, terminal sections, bolts, nuts, and other fittings shall conform to the specifications of AASHTO M180, except as modified in this subsection. The edges and center of the rail element shall contact each post or block. Rail element joints shall be lapped not less than 12-1/2 inches (315mm) and bolted. The rail metal shall be open hearth, oxygen furnace, or electric furnace steel and, in addition to conforming to the requirements of AASHTO M180, shall withstand without cracking, a cold bend of 180 degrees around a mandrel of a diameter equal to two and one-half times the thickness of the plate.

Two certified copies of mill tests reports of each heat, from which the rail element is formed, shall be furnished to the Engineer.

The ends of each length of railing shall be fitted with terminal sections.

Workmanship shall be equivalent to commercial practice and all edges, bolt holes, and surfaces shall be free of torn metal, burrs, sharp edges, and protrusions.

Bolts shall have shoulders of such shape that will prevent them from turning.

Rail elements shall be fabricated for splicing at wood posts at intervals not to exceed 12.5 feet (4m).

The rail element shall have full bearing at joints. When the radius of curvature is 150 feet (50m) or less, the rail element shall be shaped in the shop.

Railing parts furnished under these specifications shall be interchangeable with similar parts regardless of source.

Unless otherwise specified, the rail elements, terminal sections, bolts, nuts, and other fittings shall be galvanized in accordance with 210-3.

Posts, including blocks, shall be "Construction" grade, Douglas fir, free of heart center.

The posts and blocks shall be pressure treated after fabrication with creosote, creosote-coal tar solution, creosote-petroleum solution (50-50), pentachlorophenol (oil borne), or copper-naphthenate (oil borne) as provided in 204-2.

206-6 CHAIN LINK FENCE.

206-6.1 General. All materials and fittings shall be new and all ferrous materials shall be coated in accordance with 210-3. Class 1A steel pipe shall additionally be coated in accordance with 210-4. When specified, Class 1 or Class 1A materials shall additionally be clad coated with PVC in accordance with 210-5. The base material for the manufacture of steel pipe used for posts, braces, top rails, and gate frames shall conform to the requirements of ASTM F1083, Schedule 40, for Class 1 or ASTM A569 for

Class 1A. Class 1A steel shall have a minimum yield strength of 50,000 psi (345MPa). All unit weights shall be subject to the standard mill tolerance of plus or minus 5 percent.

Posts shall be fitted with caps designed to fit securely over the posts and carry a top rail where specified. Posts shall have a total length of not less than the depth of the concrete footing, as specified herein, plus the length required above ground. Where no top rail is required, pipe posts shall be fitted with suitable caps. Caps will not be required for "C" or "H" section posts.

Top rails shall be furnished in random lengths of approximately 20 feet (6m) where required.

Barbed wire shall be installed on the fence only when specifically required by the Plans or Special Provisions. When required, it shall be installed on extension arms of a type specified under 206-6.6.

206-6.2 Materials for Posts, Rails and Braces. Materials for posts, rails, and braces shall conform to Table 206-6.2 (A).

TABLE 206-6.2 (A)

Use	Nominal[1] Type And Size	Actual O.D. inches (mm)	Weight lbs/ft (kg/m)	
			Class 1	Class 1A
End, corner, slope, and gate posts for single gates 6 feet (1.8m) or less in width and double gates 12 feet (3.6m) or less in width for fences less than 72 inches (1.8m) in height.	2 NPS	2.375 (60.3)	3.65 (5.43)	3.12 (4.64)
End, corner, slope, and gate posts for single gates 6 feet (1.8m) or less in width and double gates 12 feet (3.6m) or less in width for fences 72 inches (1.8m) or higher.	2-1/2 NPS	2.875 (73.0)	5.79 (8.62)	4.64 (6.91)
Gate posts for single swing gates over 13 ft (4m) but not over 18 ft (5.5m) in width and double swing gates over 26 ft (8m) but not over 36 ft (11m) in width.	3-1/2 NPS	4.0 (101.6)	9.11 (13.56)	— —
	3 NPS	3.5 (88.9)	— —	5.71 (8.50)
Gate posts for single swing gates over 13 feet (4m) but not over 18 feet (5.5m) in width and double swing gates over 26 feet (8m) but not over 36 feet (11m) in width.	6 NPS	6.625 (168.3)	18.97 (28.23)	—
Gate posts for single swing gates over 18 feet (5.5m) in width and double-swing gates over 36 feet (11m) in width.	8 NPS	8.625 (219.1)	24.70 (36.76)	—
Line posts for fences 72 inches (1.8m) or higher.	2 NPS	2.375 (60.3)	3.65 (5.43)	3.12 (4.64)
	1-7/8 x 1-5/8 in (48 x 41mm) C	—	2.15 (3.20)	— —
	2-1/4 x 1-7/8 in (57 x 48mm) H	—	4.10 (6.10)	
Line posts for fences less than 72 inches (1.8m) in height.	1-1/2 NPS	1.90 (48.3)	2.72 (4.05)	2.28 (3.39)
	17/8 x 1-5/8 in (48 x 41mm) H	— —	1.85 (2.75)	
	17/8 x 1-5/8 in (48 x 41mm) H	— —	2.80 (4.17)	—

TABLE 206-6.2 (A) *Continued*

Use	Nominal[1] Type And Size	Actual O.D. inches (mm)	Weight lbs/ft (kg/m)	
Top rails and braces.	1-1/4 NPS	1.660 (42.2)	2.27 (3.39)	1.83 (2.72)
	1-5/8 x 1-1/4 in (41 x 32mm) C	— —	1.35 (2.01)	— —
	1-1/2 x 1-1/4 in (38 x 32mm) H	— —	2.20 (3.27)	— —
Frames for gates.	1-1/2 NPS	1.900 (48.3)	2.72 (4.05)	2.28 (3.39)
Stiffeners for gates.	1-1/4 NPS	1.660 (42.2)	2.27 (3.39)	1.83 (2.72)

1. Nominal Pipe Size (NPS), a non-dimensional unit as defined in ASTM F1083.

206-6.3 Chain Link Fabric. Unless otherwise specified, shall conform to 206-3.1 or 206-6.3.2.

206-6.3.1 Galvanized Fabric. Chain link fabric shall conform to the requirements of ASTM A 392. The fabric shall be 11-gage (3.1mm) for all fence 60 inches (1500mm) or less in height and shall be 9-gage (3.8mm) for all fence over 60 inches (1500mm) in height, unless otherwise specified.

All chain link fabric shall be woven into approximately 2-inch (50mm) mesh and galvanized either prior to or after fabrication, unless otherwise specified by the Contract Documents. Fabric 60 inches (1500mm) or less in width shall have knuckled finish on the top and bottom edges. Fabric greater than 60 inches (1500mm) in width shall have knuckled finish on the top edges and twisted and barbed finish on the bottom edge. Barbing shall be done by cutting the wire on the bias.

206-6.3.2 Polyvinyl Chloride (PVC) Coated Fabric. This specification covers PVC coated chain link fabric coated before weaving. PVC coated fabric shall conform to ASTM F668. Fabric may be produced in two classes of wire defined as follows: Class 1 shall consist of PVC extruded or extruded and adhered to zinc-coated steel wire. Class 2 shall consist of PVC fusion-bonded to zinc-coated steel wire. PVC coating thickness shall be a minimum 15 mils (380μm) for Class 1 and 7 mils (180μm) for Class 2. The core wire for the fabric shall be 0.120 inches (3.0mm) for all fence 60 inches (1500mm) or less in height and shall be 0.148 inches (3.76mm) for all fence over 60 inches (1500mm) in height unless otherwise specified. The specified diameter is the metallic core wire diameter and the PVC coating shall not be considered when determining the diameter.

All chain link fabric shall be woven into approximately 2-inch (50mm) mesh. All fabric widths shall have knuckled finish on the top and bottom edges. At the time of fabrication, cut ends shall be covered with acrylic enamel. Acrylic enamel shall be a PVC resin in solution, consisting of high-level pigments, ultraviolet absorbers and solvent blends applied by brush or dabbing applicator.

206-6.4 Tension Wires and Fabric Ties. Tension wires shall be at least 7-gage (4.5mm) galvanized coil spring steel wire.

Ties used to fasten the fabric to posts, rails, and gate frames shall be not smaller than 11-gage (3.1mm) galvanized steel, 6 gage (4.9mm) aluminum wire, or approved noncorrosive metal bands.

Tension bars used in fastening fabric to end and corner posts and gate frames shall be galvanized high carbon steel bars not smaller than 3/16 inch (9.5mm) by 3/4 inch (19mm).

206-6.5 Truss or Tension Rods. Truss or tension rodsused in trussing gate frames and line posts adjacent to end, corner, slope or gate posts shall be adjustable 3/8 inch (9.5mm) diameter galvanized steel rod. When used in trussing line posts, adjustment shall be provided by means of galvanized turnbuckles or other suitable tightening devices.

206-6.6 Fittings. All required fittings and hardware shall be galvanized.

Couplings to connect the individual lengths of top rail shall be of the outside sleeve type and at least 7 inches (175mm) long. The bore of the sleeves shall be sufficiently true to maintain adjacent lengths of rail in alignment.

Extension arms for barbed wire shall be of a type that can be attached to the tops of the posts and carry three wires at approximately 5-1/2 inches (140mm) centers in a plane approximately 45 degrees from the vertical, inclined as shown on the Plans or as directed by the Engineer.

206-6.7 Barbed Wire. Barbed wire shall be 4-point pattern, composed of two strands of 12-1/2 gage (2.5mm) galvanized steel wire with barbs spaced 5 inches (125mm) apart and shall confirm to ASTM A121.

206-6.8 Repair of Damaged Coatings. All welds made after galvanizing shall be ground smooth and wire brushed to remove loose or burned zinc coating, after which the cleaned areas shall be prepared and neatly coated with 50-50 solder or as prescribed in 210-3.5. Repairs to abraded or otherwise damaged zinc coating shall be made in a similar manner.

206-6.9 Security Fencing. Security fencing shall be 5 feet (1.5m) high. Security fencing material shall be chain link fabric or welded wire fabric (6x6-W9xW9 minimum) and constructed using:

a) Tensioned fencing material with top and bottom tension wires securely fastened to driven steel posts or other equally rigid elements at a maximum spacing of 12 feet (3.7m); or

b) Untensioned fencing materials securely fastened to extended trench shoring elements at a maximum spacing of 8 feet (2.4m) and fastened to continuous top and bottom rails including toe plates constructed of nominal 2-inch x 4-inch (50mm x 100mm) lumber or equally rigid material.

Framed panels with suitable supporting elements fastened together to form a continuous fence may also be used.

SECTION 207 - PIPE

207-1 NONREINFORCED CONCRETE PIPE.

207-1.1 General. Nonreinforced concrete pipe to be furnished shall be as shown on the Plans, or as specified under the Bid item of work for the Contract.

Concrete pipe shall be extra-strength unless otherwise specified, shall be manufactured from portland cement concrete, and shall be so constructed that it will conform to the requirements described herein.

207-1.2 Materials. Materials used in manufacturing the pipe shall be as specified in ASTM C14/C14M with the following exceptions:

a) The portland cement shall be as specified in 201-1.2.1.

b) At least 28 percent of the aggregate by weight shall be larger than 9/10 inch (25mm) for pipes 12 inches (300mm) and larger in diameter.

c) All aggregates shall conform to 201-1.2.2.

207-1.3 General Requirements. The plane of the ends of the pipe, except for special shapes, shall be perpendicular to the longitudinal axis of the pipe. The interior surface shall be smooth and well-finished. Joints shall either be of the socket and spigot type or the tongue and groove type, as approved by the Engineer, and so constructed that, when laid, the pipe will form a continuous conduit with a smooth and uniform interior surface.

When shown on the Plans, the pipe shall have a gasketed joint. The gasket shall be seated in an accurately shaped groove on the spigot end of the pipe section and the gasket shall be of suitable cross section and size. Alternate joint details may be used with Engineer approval. The gasket shall be considered as the principal element in making the joint watertight. The gasket shall be manufactured from a synthetic rubber of neoprene base and shall conform to the requirements of 208-1.2 and 208-2.2.

The completed pipe shall be free from fractures, large or deep cracks, laminations, and surface roughness. Specimens which, when placed in a vertical position, do not give a metallic ring when struck with a hammer, or which exhibit any of the following defects, will be subject to rejection:

a) Indications of honeycomb or open texture or of imperfect mixing or molding.

b) Fractures or cracks passing through the wall or socket, except a single end crack less than 3 inches (75mm) measured transversely or 2 inches (50mm) measured longitudinally will not be deemed cause for rejection unless such defects appear in more than 5 percent of the number of sections inspected; in which event, the defective sections will be rejected.

c) Cracks sufficient to impair the strength, durability or serviceability of the pipe; a single crack in the body of the pipe, extending through one-half of the thickness of the wall and over 3 inches (75mm) in length; or two or more such cracks regardless of length.

d) Variations in dimensions in excess of those permissible, as set forth in the following tables.

e) Failure to meet the test requirements set forth herein.

f) In pipe designed to be straight, deviations greater than 3/8 inch per linear foot (30mm per linear meter).

The deviation shall be measured from a straight-edge on the concave side of the pipe.

207-1.4 Dimensions. Each straight pipe of all sizes and classes shall be not less than 3 feet (0.9m) in length unless otherwise specified for special purposes. The minimum length of 6-inch (150mm) Y's and T's shall be 18 inches (450mm) and 24 inches (600mm) for 8-inch (200mm) and larger Y's and T's. Other dimensions of pipe shall conform to the following tables:

TABLE 207-1.4 (A): STANDARD STRENGTH CONCRETE PIPE

1 Nominal Size inches (mm)	2 Thickness of Wall inches (mm)	3 Depth of Socket inches (mm)	4 Minimum Annular Space[1] inches (mm)
6 (150)	3/4 (19)	2-1/4 (57)	5/8 (16)
8 (200)	7/8 (22)	2-1/2 (64)	5/8 (16)
10 (250)	1 (25)	2-1/2 (64)	5/8 (16)
12 (300)	1-1/8 (29)	2-3/4 (70)	5/8 (16)
15 (375)	1-3/8 (35)	2-3/4 (70)	5/8 (16)
18 (450)	1-1/2 (38)	3 (76)	5/8 (16)
21 (525)	1-3/4 (44)	3-1/2 (89)	3/4 (19)

1. Applies to socket ends only.

TABLE 207-1.4 (B): EXTRA-STRENGTH CONCRETE PIPE

1 Nominal Size inches (mm)	2 Thickness of Wall inches (mm)	3 Depth of Socket inches (mm)	4 Minimum Annular Space[1] inches (mm)
6 (150)	3/4 (19)	2-1/4 (57)	5/8 (16)
8 (200)	7/8 (22)	2-1/2 (64)	5/8 (16)
10 (250)	1 (25)	2-1/2 (64)	5/8 (16)
12 (300)	1-1/2 (38)	2-3/4 (70)	5/8 (16)
15 (375)	1-7/8 (48)	2-3/4 (70)	5/8 (16)
18 (450)	2-3/8 (60)	3 (76)	5/8 (16)
21 (525)	2-3/4 (70)	3-1/2 (89)	3/4 (19)

Note: The annular space is the space between the inside of the socket and the outside of the spigot of the pipe as placed in the socket. The minimum annular space, as shown in column 4, is measured at the outer end of the socket. The space at the bottom of the socket, known as the caulking space, shall be not less than 1/4 inch (6mm).

1. Applies to socket ends only.

TABLE 207-1.4 (C): LIMITS OF PERMISSIBLE VARIATION IN STANDARD STRENGTH AND EXTRA-STRENGTH CONCRETE PIPE

Nominal Size inches (mm)	Variation from Indicated Length (±) inches (mm)	Depth of Socket (–) inches (mm)	Thickness of Wall (–) inches (mm)	Inside Diameter of Bore inches (mm)		Length of Two Opposite Sides inches (mm)
				Minimum	Maximum	
6 (150)	1/2 (13)	1/4 (6)	1/16 (2)	5-13/16 (150)	6-3/16 (160)	3/8 (10)
8 (200)	1/2 (13)	1/4 (6)	1/16 (2)	7-13/16 (200)	8-1/4 (210)	3/8 (10)
10 (250)	1/2 (13)	1/4 (6)	1/16 (2)	9-13/16 (250)	10-1/4 (260)	3/8 (10)
12 (300)	1/2 (13)	1/4 (6)	1/16 (2)	11-13/16 (300)	12-5/16 (312)	3/8 (10)
15 (375)	1/2 (13)	1/4 (6)	1/8 (3)	14-3/4 (375)	15-5/16 (390)	3/8 (10)
18 (450)	1/2 (13)	1/4 (6)	1/8 (3)	17-3/4 (450)	18-3/8 (465)	3/8 (10)
21 (525)	1/2 (13)	1/4 (6)	1/8 (3)	20-11/16 (525)	21-7/16 (545)	3/8 (10)

Note: The minus sign (–) alone indicates the plus variation is not limited; the plus and minus sign (±) indicates variations in both excess and deficiency.

207-1.5 Marking. Each pipe shall be marked clearly and legibly to show the class of pipe, date of manufacture, and the name or trademark of the manufacturer.

207-1.6 Test Requirements.

207-1.6.1 General. Before pipe is delivered to the Work site for use in any work, test pipe shall meet the requirements of the hydrostatic pressure test and the loading test described herein. Tests shall be made at the point of manufacture and shall be made under the supervision of the Engineer.

All testing equipment shall be calibrated by an agency approved by the Engineer at intervals not to exceed 6 months.

207-1.6.2 Selection of Test Pipes. Each lot of non-reinforced concrete pipe is defined as not more than 450 sections of pipe of one size and class. All the pipe shall be manufactured from the same mix of concrete, have the same cure, and be from 1 day's manufacture.

The Engineer will select at random and have tested one pipe for each 50 pipes or fraction thereof in each lot. The minimum number of pipes tested for any lot shall be 5. The pipes selected for test shall be sound and have dimensions consistent with these specifications.

The Contractor shall furnish the test pipes without charge and shall provide adequate equipment and facilities for conducting tests. Unless otherwise specified, the Contractor shall bear all costs.

207-1.6.3 Hydrostatic Pressure Test. The hydrostatic pressure test shall precede the loading test by not more than 3 hours.

When the pipe is subjected to an internal hydrostatic pressure of 10 psi (70kPa) for the time shown in the following table, the accumulated moisture on the exterior surface shall not run down the sides in such quantity that will exceed 0.34 fluid ounce (10mL). Each 10 square inches (6500mm^2) of moisture appearing on the exterior surface shall be considered to be 0.03 fluid ounce (1mL) of runoff.

TABLE 207-1.6.3 (A)

Thickness of Wall	Testing Time (Minutes)
Up to and including 1 inch (25mm)	7
Over 1 inch (25mm) and including 1-1/2 in (38mm)	9
Over 1-1/2 inches (38mm) and including 2 in (50mm)	12
Over 2 inches (50mm) and including 2-1/2 in (60mm)	15
Over 2-1/2 inches (63mm) and including 3 in (75mm)	18
Over 3 inches (75mm)	21

207-1.6.4 Loading Test. The loading test shall be the three-edge bearing conforming to ASTM C14.

TABLE 207-1.6.4 (A)

pounds per linear foot (kN per linear meter)

Size Inches (mm)	Standard Strength	Extra Strength
6 (150)	1100 (16.1)	2000 (29.2)
8 (200)	1300 (19.0)	2000 (29.2)
10 (250)	1400 (20.4)	2000 (29.2)
12 (300)	1500 (21.9)	2250 (32.8)
15 (375)	1750 (25.5)	2750 (40.1)
18 (450)	2000 (29.2)	3300 (48.2)
21 (525)	2200 (32.1)	3850 (56.2)

The net inside length of pipe from the bottom of the socket to the spigot end of the pipe shall be used as the divisor to calculate the load per unit length.

207-1.6.5 Acceptance or Rejection of Pipe. When all the pipe tested passes the required tests, the entire lot of pipe will be acceptable. When 2 test pipes fail, the entire lot will be rejected.

When only one test pipe of the first group selected fails, 5 additional pipes shall be tested. If none of the additional pipes fail, the entire lot will be acceptable. If one fails, the entire lot will be rejected.

207-1.7 Perforated Pipe. Perforated pipe shall conform to the requirements of this section except the hydrostatic pressure test will not be required, and the strength requirements shall be reduced 10 percent from the values given in 207-1.6.4.

The pipe shall be shop perforated with perforations symmetrically located within a maximum arc of 160 degrees. Perforations shall have a total open area of at least 0.3 square inches per linear foot (630mm^2 per linear meter) of pipe, with a minimum of one perforation per linear foot (3 perforations per linear meter) ; except for joint areas. Perforations shall be either holes or slots. Diameter of the holes may vary from 1/4 inch (6mm) minimum to 3/8 inch (10mm) maximum; the width of the slots may vary from 3/16 inch (5mm) minimum to 5/16 inch (8mm) maximum; the length of the slot shall not exceed 4 inches (100mm).

207-1.8 Curing.

207-1.8.1 General. The manufacturer shall provide adequate steam plant, enclosures, piping, and other facilities for curing the pipe. The enclosures shall be such that the humidity shall be maintained so as to keep the pipe surfaces moist at all times. The temperature shall be maintained continuously between 120°F and 150°F (50°C and 65°C).

207-1.8.2 Curing Requirements. The pipe shall be cured under either of the following methods:

a) After the pipe has been manufactured, it shall be placed in enclosures and saturated steam applied for at least 6 hours, starting not sooner than 1 hour nor more than 10 hours, after completion of manufacture. Starting not later than 1 working day after completion of the steam curing, the pipe shall be kept constantly and thoroughly wet for at least 7 days. When an approved temperature-time recorder is used, 1 hour of steam may be substituted for each 4 hours of the required water cure.

b) Starting not later than 24 hours after the pipe has been manufactured, it shall be kept constantly and thoroughly wet for at least 14 days.

207-1.9 Inspection. Sections of pipe to be inspected shall be so situated at the manufacturer's plant as to provide the Engineer with free accessibility for inspection and marking. In no case shall the pipe be stacked for inspection to a height that would require the Engineer to climb or use a ladder.

At the place of manufacture, the Engineer will indicate acceptance of the pipe for delivery to the Work site by marking the pipe with the Agency's stamp. Such acceptance, however, shall be considered a tentative acceptance. Final acceptance will be made only when the project has been completed.

If pipe is rejected subsequent to its manufacture, the mark placed thereon by the Engineer shall be defaced.

207-2 REINFORCED CONCRETE PIPE (RCP).

207-2.1 General. These specifications apply to reinforced concrete pipe intended to be used for the construction of storm drains, sewers, and related structures.

The size, type, and D-load of the concrete pipe to be furnished shall be as shown on the Plans or in the Special Provisions.

Prior to the manufacture of the RCP, three sets of prints of the pipeline layout diagrams shall be furnished to the Engineer per 2-5.3, except transparencies will not be required. Catch basin connector pipes need not be included in the layout; but, in lieu thereof, a list of catch basin connector pipes shall accompany the layout. The connector pipe list shall include size and D-load of pipe, station at which pipe joins mainline, number of sections of pipe, length of sections, and type of sections (straight, horizontal bevel, vertical bevel, etc.). The diagrams and lists submitted will be used by the Agency for reference only, and their use shall in no way relieve the Contractor of its responsibility for correctness. The Engineer may waive the pipeline layout and connector pipe list requirement.

Unless otherwise specified, RCP shall be either wet-cast, spun, or machine-made.

Plants and processes not previously qualified by the Agency may require initial qualification by the Agency using any or all of the following tests described in ASTM C497/C497M and as listed below. Upon qualification, no additional tests will be required unless changes have been made in the equipment and procedures, or an increase in the largest aggregate size is made from the previous approved mix designs. The manufacturer shall provide qualifying data upon request by the Engineer:

a) **Three-Edge Bearing Test.** Two tests minimum on each of three different pipe sizes per 207-2.9.2.

b) **Core Strength Test.**

1) Minimum allowable adjusted compressive strength: 4,000 psi (27.6MPa).

2) Number of tests: 2 each from 3 different pipes.

3) Unit weight to be used as a baseline for future reference.

c) **Absorption Test Per ASTM C497/ASTM, C497M, Section 7, Method A.**

1) Obtain 2 core samples (one each from the middle area of two pipes for each mix design).

2) Maximum allowable absorption value: 9 percent.

d) **Hydrostatic Test.** One test sample selected by the Engineer shall meet the following criteria:

1) 24-hour presoak (manufacturer's option).

2) Hydrostatic pressure of 13 psi (90 kPa) minimum shall be maintained for 20 minutes.

3) End bulkheads or internal plugs may be used at the manufacturer's option. Leakage at the bulkhead of plugs is allowed if leakage does not interfere with the test.

4) Allowable leakage: none. Moisture appearing on the exterior surface of the pipe in the form of beads adhering to the surface will not be considered leakage. The tests may be repeated after the 24-hour presoak at the manufacturer's option to determine if test pipe stops leaking. Pipe which stops leakage after the presoak will be considered to have passed.

e) **Visual Inspection.** Pipe inspection shall include the following:

1) 207-2.5 (Joints).

2) Interior surface finish and textures.

3) Reinforcing steel placement and twist, per 207-2.4.2 and 207-2.4.3. Concrete from a portion of one piece of green pipe shall be stripped or raked to determine cage twist, location, clearance, and voids. Voids around reinforcing steel cage caused by cage twist will not be acceptable. Cage twist resulting from pipe production in excess of 1/4 inch (6mm) in 8 feet (2.4m) will not be acceptable.

4) Longitudinally cut pipe. One pipe shall be longitudinally sawcut in half using equipment which will not damage the concrete or reinforcing steel. The exposed surface shall be inspected for voids adjacent to the circumferential steel. Voids shall be considered continuous if a 1/16 inch (1.5mm) diameter pin can be inserted 1/4 inch (6mm) deep.

This test will be acceptable if no more than 10 percent of the circumferential bars exposed have continuous voids.

Plant inspection shall include cage manufacturing, curing process, batching equipment and process, aggregate and cement storage, concrete mix designs, and product handling.

The interior surface of the pipe shall be smooth and well-finished. Joints shall be of such type and design and so constructed as to be adequate for the purpose intended so that, when laid, the pipe will form a continuous conduit with a smooth and uniform interior surface.

Sockets and spigots shall be free from any deleterious substance or condition which might prevent a satisfactory mortar bond at the joints.

If the Engineer determines that the forms, end rings, or form gaskets used in the manufacture of the pipe are inadequate, the Contractor shall replace or repair said equipment to the satisfaction of the Engineer.

Pipe of greater strength than that specified or cover greater than required in 207-2.4.2 may be furnished at the Contractor's option, and its expense, provided such pipe conforms in all other aspects to the applicable provisions of these specifications.

The Contractor shall furnish, install, and maintain stulls or other devices in the pipe as may be necessary to meet the limitations on cracks specified herein.

207-2.2 Materials. Unless otherwise approved by the Engineer, no materials shall be used in manufacturing the pipe other than water, water reducing admixture, portland cement, mineral aggregates, and steel conforming to ASTM C76/ASTM, C76M with the following requirements:

a) The Portland cement shall conform to 201-1.2.1.

b) Aggregate shall conform to the reactivity requirements in 201-1.2.2.

c) Fly ash shall conform to the quality requirements in 201-1.2.5.

Aggregate shall be so graded, proportioned, and thoroughly mixed in a batch mixer so as to produce a homogeneous concrete mixture of such quality that pipe will conform to the test and design requirements of these Specifications. The proportion of cementitious material (portland cement and fly ash) shall not be less than 560 pounds per cubic yard (330 kg/m^3) of concrete.

207-2.3 Dimensions.

207-2.3.1 Length. The nominal length shall be not less than 8 feet (2.4m) except as otherwise specified or required for bends or special joints.

Variations in laying lengths of 2 opposite sides of pipe shall not be more than 1/8 inch per foot (10mm per m) of diameter with a maximum of 5/8 inch (16mm) in any length of pipe.

207-2.3.2 Cast Pipe Wall Thickness. Cast pipe wall thickness shall conform to the following:

TABLE 207-2.3.2 (A)

Nominal Internal Diameter inches (millimeters)	Minimum Barrel Thickness inches (millimeters)
12 (300)	2 (50)
15 (375)	2-1/4 (57)
18 (450)	2-1/4 (57)
21 (525)	2-1/4 (57)
24 (600)	3 (76)
27 (675)	3 (76)
30 (750)	3 (76)
33 (825)	3 (76)

For pipe larger than 33 inches (825mm) in diameter, the minimum wall thickness permitted will depend upon the strength of the concrete used. If the concrete has a strength of 4,500 psi (31.0MPa) in 28 days, any wall thickness not less than 1/12 the internal diameter is permitted. If the concrete has a strength less than 4,500 psi (31.0MPa) but more than 4,000 psi (27.6MPa) at 28 days, a wall thickness 1 inch (25mm) greater than 1/12 the internal diameter will be required. The concrete strength shall be determined in accordance with ASTM C31 and C39.

207-2.3.3 Spun Pipe Wall Thickness. Spun pipe wall thickness shall conform to the following:

TABLE 207-2.3.3 (A)

Nominal Diameter inches (millimeters)	Minimum Wall Thickness inches (millimeters)
12 (300)	2 (50)
15 (375)	2 (50)
18 (450)	2 (50)
21 (525)	2 (50)
24 (600)	2-1/2 (63)
27 (675)	2-5/8 (66)
30 (750)	2-3/4 (69)
33 (825)	2-7/8 (73)
36 (900)	3 (76)
39 (975)	3-1/8 (85)
42 (1050)	3-3/4 (95)
45 (1125)	3-7/8 (98)
48 (1200)	4 (101)

The thickness of the barrel of pipe larger than 48 inches (1200mm) in diameter shall not be less than 1/12 the internal diameter of the pipe.

207-2.3.4 Machine-Made Pipe Wall Thickness. The minimum wall thickness for all pipe sizes shall be as specified in ASTM C76/C76M Wall B (1/12 the internal diameter plus 1 inch (25mm)) using 4,000 psi (27.6MPa) minimum strength concrete.

207-2.4 Reinforcement.

207-2.4.1 General. The reinforcement shall be a cage fabricated of bars or wire. Circumferential reinforcement shall be in the amount and type shown on Plans, or that required to sustain the specified test loads. Longitu dinal reinforcement shall be sufficient to make the cage rigid and to support the circumferential reinforcement firmly in place in the forms during placing and consolidation of the concrete.

Pipe which is to be jacked shall have a circular cage reinforcement.

In pipe larger than 96 inches (2400mm) diameter, the minimum size of circumferential reinforcement for the inside circular cage shall be 7/16 inch (11mm) diameter.

Fastenings (supports) and/or retractable mechanical devices, approved by the Engineer, shall be used for holding the cage rigidly in place in the form in its elliptical or circular shape. In wet-cast and machine-made pipe these fastenings shall be spaced not closer than 2 feet (0.6m) center-to-center along the longitudinal reinforcement except for pipe having a nominal length of 4 feet (1.2m) or less.

All reinforcing steel shall be clean and free from loose rust, scale, paint, grease, form oil, or other foreign matter.

Splices shall be butt-welded or lap-welded a minimum of 6 diameters. Nonwelded splices shall be lapped a minimum of 20 diameters for deformed bars and 40 diameters for plain bars or cold-drawn wire. Nonwelded lapped splices shall be wired tightly. Welds shall develop not less than 75 percent of the minimum specified ultimate strength of the bars or wires being welded.

Elliptical cages may be specifically fabricated or deformed from a circular cage to the required elliptical dimensions.

Cages for machine-made pipe shall be circular.

Upon request, the Contractor shall furnish data to the Engineer indicating lot number, wall thickness, and the size, spacing, and positioning of reinforcement for any pipe manufactured.

207-2.4.2 Location of Reinforcement. Measurements of position, except for concrete cover, shall apply to the center of the bar or wire.

For pipe with wall thickness greater than 2-1/4 inches (57mm), the required minimum cover measured between reinforcement surface and pipe surface, and the permitted cover tolerances are given in the following table:

TABLE 207-2.4.2 (A)

Reinforcement	Storm Drain Pipe Nominal Diameter			Sewer Pipe Nominal Diameter	
	24 to 36 inches (600 to 900mm)	39 to 96 inches (975 to 2400mm)	Greater than 6 inches (2400mm)	96 inches (2400mm) or less	Greater than 96 inches (2400mm)
Circumferential inches (mm)	3/4 ± 1/4 (19 ± 6)	3/4 ± 1/4 (19 ± 6)	1-1/4 ± 1/4 (32 ± 6)	1 ± 1/4 (25 ± 6)	1-1/4 ± 1/4 (32 ± 6)
Longitudinal inches (mm)	3/8 (10) minimum	1/2 (13) minimum	1/2 (13) minimum	1 (25) minimum	1 (25) minimum

Note: In lined sewer pipe, the tabulated cover shall be increased as necessary to ensure that the distance from any reinforcement to embedded lugs or locking extensions of the liner is at least 1/4 inch (6mm).

The required covers shown in the table are minimums. Certain conditions such as high velocities or corrosive environments, may require greater covers as indicated on the Plans or in the Special Provisions.

The tabulated tolerances shall not apply where the Plans show cover over circumferential steel which differs from the values shown above.

For pipe with a wall thickness of 2-1/4 inches (57mm) or less, the circumferential reinforcement shall be located at the center of the wall. The cover over longitudinal reinforcement shall be at least 3/8 inch (10mm).

207-2.4.3 Placement of Reinforcement. The location of the minor axis of elliptically reinforced pipe shall be marked so that an imprint will be left in the interior of the concrete.

The circumferential reinforcement shall be placed in the body of pipe in such a manner that the longitudinal reinforcement of the cage will hold the last hoop or coil 3/4 inch (19mm) minimum from the end plates or rings. Both ends of the circumferential reinforcement shall be finished off with a complete hoop of reinforcement.

For storm drain pipe, the longitudinal reinforcement may extend to the base plate or ring to act as a cage support. For sanitary sewer pipe, the longitudinal reinforcement shall have a minimum cover of 3/4 inch (19mm) from the end faces of the pipe.

Circumferential reinforcement may be included in the joint projections at the option of manufacturer, and shall be included when required by Plans or Special Provisions.

Where the wall reinforcement does not extend into the joint projection, the longitudinal distance to the last circumferential reinforcement from the inside shoulder of the bell or shoulder of the spigot shall be a maximum of 1-3/4 inches (44mm), but not less than 3/4 inch (19mm) from any face.

207-2.5 Joints. Joints shall be designed to be self-centering. Unless otherwise specified, joints in concrete pipe shall be of the tongue-and-groove mortar type of joint.

When pipe joints of the reinforced concrete collar type or of rubber-gasketed type are specified or indicated on the Plans, joint details shall be submitted to the Engineer for approval before commencing pipe manufacture.

Pipe with beveled ends for use around curves, the radii of which are shown on the Plans, shall be provided where necessary. Either one or both ends shall be beveled a maximum of 5 degrees as may be required to provide well-fitting joints.

If required by the Engineer, the pipe shall be "match-marked" to meet specified laying tolerances at the place of manufacture and laying diagrams shall be furnished to the Contractor.

207-2.6 General Manufacturing Requirements.

207-2.6.1 Wet-Cast Pipe. Wet-cast RCP shall be manufactured by placing the concrete into stationary, vertical, cylindrical metal forms. During placing of each batch the concrete shall be vibrated continuously with internal or external mechanical vibrators operating at a minimum rate of 6,500 vibrations per minute.

207-2.6.2 Spun Pipe. Spun RCP shall be manufactured by introducing the concrete into a rotating horizontal cylindrical metal form. After the concrete materials have been mixed, they shall be promptly placed in the forms and spun on a horizontal axis. If, for any reason, the work of filling the forms is interrupted long enough for the concrete to take its initial set, any partly filled form shall be emptied and the concrete rejected. While the concrete is being placed in the forms, they shall be revolved on a horizontal axis at a speed that will ensure uniformity of aggregates. After all the concrete has been placed in the forms, they shall be revolved at proper speed and duration to secure a dense concrete with smooth interior surface. Water and laitance collecting on the surface of the concrete shall be removed, and the interior surface of the pipe shall be troweled and finished to the form of a true cylinder.

207-2.6.3 Machine-Made Pipe. Machine-made RCP shall be manufactured by placing the concrete between vertical cylindrical forms or by placing the concrete into a vertical form and the interior surface formed with one or more roller packing heads and a long bottom-trowel rotating in opposite directions. The concrete shall be mixed to a uniform consistency and the reinforcing cage shall be held by fastening (supports) and/or retractable mechanical devices to ensure correct cage position.

Concrete placed between inner and outer forms shall be continuously vibrated throughout the manufacturing cycle and after all concrete has been placed in the forms, axial pressure shall be applied simultaneously with vibration to further densify the concrete, or the concrete shall be placed against the outer jacket and densified by continuous vibration during the manufacturing cycle. On completion of densification, the pipe may be removed immediately from the forms.

207-2.7 Curing.

207-2.7.1 Steam Curing Facilities. The manufacturer shall provide adequate enclosures, steam plant, piping, and other facilities for curing pipe. The enclosures shall be such that the temperature and humidity can be controlled to keep the pipe surfaces moist at all times and temperature maintained continuously between 80°F and 170°F (27°C and 77°C). Bulkhead curing is permissible, but will only be given credit for a maximum of 12 hours of the steam cycle.

207-2.7.2 Curing Procedures.

General. Wet-cast, machine-made, and spun pipe shall be cured by steam or water, or a combination of both, as described in the following paragraphs:

a) **Steam Curing.** Steam may be applied as soon as the pipe is enclosed, but not later than 10 hours after completion of concrete placement. The temperature within the enclosure shall not be raised above 100°F (38°C) by the use of steam for the first hour; thereafter the temperature shall not be increased at a rate greater than 40°F (22°C) per hour. The temperature shall be maintained continuously between 80°F and 170°F (27°C and 77°C) for 28 hours.

For wet-cast and spun pipe, 6 hours of steam (at least 5 hours at 80°F (27°C) or higher) is required before forms may be removed. The period necessary to remove the forms may be included in the 28 hours, as long as it does not exceed 4 hours. Any time more than 4 hours shall be added to the total steam cycle. The time lapse during form removal shall not exceed 1 working day.

b) **Combination Curing.** At any time after 6 hours of the steam cycle, the steam may be stopped and the cure continued using water applied in such a manner that the outside surfaces of the pipe is kept moist. Water may be substituted for steam on a basis of 4 hours of water being equal to 1 hour of steam. The manufacturer shall notify the Engineer prior to using water in lieu of steam.

c) **Water Curing.** Pipe may be water cured by any method that will keep the outside surface moist for 4 consecutive days. Pipe to be given a total water cure may not be stripped from the forms until 20 hours after concrete placement or until the concrete has reached a compressive strength of 1,500 psi (10.3MPa), whichever occurs first.

d) **Alternate Curing Based On Compressive Strength.** Pipe cured by any of the above methods may have the water or steam stopped when the concrete strength reaches 3,000 psi (20.7MPa). The strength shall be determined in accordance with ASTM C31 and C39.

e) **Alternate Curing Procedures.** The manufacturer may request approval of alternate methods of curing that differ from the procedures specified above. The alternative method shall be specified in writing to the Engineer and fully substantiated by test data.

207-2.8 Causes for Rejection. Inspection of pipe as may be deemed necessary by the Agency will be made at the place of manufacture, and unless it can be repaired as provided in 207-3.3, or as approved by the Engineer, pipe may be rejected for any of the following reasons:

a) A piece of any size broken out of the pipe.

b) Defects that indicate imperfect mixing or molding.

c) Any crack extending entirely through the wall of the pipe and having a longitudinal or transverse length greater than the wall thickness of the pipe.

d) Any shattering or flaking of concrete at a crack.

e) A deficiency greater than 1/4 inch (6mm) from the specified wall thickness of pipe 30 inches (750mm) or smaller in diameter, or a deficiency greater than 5 percent (6 percent) from the specified wall thickness of pipe larger than 30 inches (750mm) in internal diameter, except that the deficiency may be 8 percent (7 percent) adjacent to the longitudinal joint, provided that additional deficiency does not lie closer than 20 percent of the internal diameter to the vertical axis of the pipe and does not extend along the circumference for a distance greater than 20 percent of the internal diameter.

The deficiencies in wall thickness permitted herein do not apply to gasket contact surfaces in gasketed joint pipe. Dimensions and tolerances of such contact surfaces shall be submitted for approval.

f) Internal diameter of the pipe varying from the true circle of the specified diameter by more than 1 percent (for projects in metric units, pipe having less than the specified diameter, or exceeding it by more than 2.6 percent).

g) The roundness of the pipe varies from a true circle of the actual internal diameter by more than one percent at any location along the barrel.

h) Rock pockets and water pockets in any pipe.

i) Exposure of any reinforcement arising from misplacement thereof.

j) Evidence of cage twist or misplacement of reinforcement.

k) Delamination of the concrete.

l) Surface defects indicating honeycomb or open-texture.

m) Separations or "blisters".

n) Slumped or sagged concrete.

o) For sewer pipe, any crack showing two visible lines or separation for a continuous length of 2 feet (0.6m) or more, or an interrupted length of 3 feet (0.9m) or more anywhere in evidence, both inside and outside, except where such cracks occur during the external loading test specified herein.

p) Any continuous crack or concrete separation having a surface width of 0.010 inch (0.255mm) or more and extending for a length of 12 inches (300mm) or more, regardless of depth or position in the wall of the pipe.

q) The pipe fails the D-load bearing strength test.

The imperfections and variations as causes for rejection in sewer and storm drain pipe, as specified herein, shall apply to pipe for which design details are indicated on the Plans as well as for pipe which is specified by D-load. The procedure of the Engineer for marking the pipe with the Agency's stamp at the place of manufacture shall not be considered a final acceptance of the pipe.

Pipe shall be considered ready for transporting to the Work site when it conforms to the specified requirements for curing, testing, and inspection.

207-2.9 Basis for Acceptance.

207-2.9.1 General. The basis for acceptance shall be by one of the following, as designated in the Special Provisions:

a) The D-load bearing strength test, compliance with these specifications, inspection of the pipe manufacture and inspection of the completed pipe.

b) The structural design details, materials, tests, inspection of the pipe manufacture and inspection of the completed pipe.

c) Acceptance of certification of compliance with these specifications.

207-2.9.2 D-load Bearing Strength Test. Pipe to be D-load tested shall be selected at random by the Engineer at the point of manufacture. One pipe will be selected for each lot, or fraction thereof, of the pipe to be furnished for the Work.

For the purpose of these specifications, a lot is defined as 400 feet (122m) but no more than 50 sections of pipe, or fraction thereof, of one size and class manufactured on consecutive working days. If the 400 feet (122m), but no more than 50 sections, of pipe are not made on consecutive working days, then only those made on consecutive working days shall be considered a lot. If an interruption in manufacturing occurs, the Engineer may permit the pipe made after the interruption to be included in the lot, provided that the interruption lasts less than 7 calendar days. A new lot number will be assigned if any change occurs in size or spacing of reinforcing steel, in the concrete mix, or in the curing method.

The Contractor shall furnish the test pipe without charge and shall provide adequate equipment and facilities for conducting tests and shall bear all expense in connection therewith, all tests being under the supervision of the Engineer. All testing equipment shall be calibrated at intervals not to exceed 6 months, by an agency approved by the Engineer.

Test pipe shall conform in all other respects to the applicable requirements specified herein. Pipe shall be tested by the 3-edge bearing test as prescribed in ASTM C497/C497M.

TABLE 207-2.9.2 (A): ALLOWABLE CRACK WIDTH

Pipe Wall Thickness inches (mm)	Allowable Crack Width in inches (mm) Pipe Diameters up to 96 inches (2400mm)					
	Concrete Cover on Transverse Reinforcement in inches (millimeters)					
	0.75" (19mm)	1.25" (32mm)	1.75" (44mm)	2.25" (57mm)	2.75" (70mm)	
2.5 (63)	0.010 (0.255)	0.015 (0.380)				
3.0 (76)	0.010 (0.255)	0.014 (0.355)				
3.5 (89)	0.010 (0.255)	0.013 (0.330)	0.018 (0.455)			
4.0 (101)	0.010 (0.255)	0.013 (0.330)	0.017 (0.430)			
4.5 (114)	0.010 (0.255)	0.012 (0.305)	0.015 (0.380)	0.020 (0.510)		
5.0 (127)	0.010 (0.255)	0.012 (0.305)	0.015 (0.380)	0.018 (0.455)		
5.5 (140)	0.010 (0.255)	0.012 (0.305)	0.014 (0.355)	0.017 (0.430)	0.021 (0.535)	
6.0 (152)	0.010 (0.255)	0.012 (0.305)	0.013 (0.330)	0.016 (0.405)	0.019 (0.485)	
6.5 (165)	0.010 (0.255)	0.012 (0.305)	0.013 (0.330)	0.015 (0.380)	0.018 (0.455)	
7.0 (178)	0.010 (0.255)	0.011 (0.280)	0.013 (0.330)	0.015 (0.380)	0.017 (0.430)	
7.5 (191)	0.010 (0.255)	0.011 (0.280)	0.013 (0.330)	0.014 (0.355)	0.017 (0.430)	
8.0 (203)	0.010 (0.255)	0.011 (0.280)	0.013 (0.330)	0.014 (0.355)	0.016 (0.405)	
8.5 (216)	0.010 (0.255)	0.011 (0.280)	0.013 (0.330)	0.014 (0.355)	0.015 (0.380)	
9.0 (229)	0.010 (0.255)	0.011 (0.280)	0.013 (0.330)	0.013 (0.330)	0.015 (0.380)	
9.5 to 10.5 (241 to 267)	0.010 (0.255)	0.011 (0.280)	0.013 (0.330)	0.013 (0.330)	0.014 (0.355)	
Pipe Wall Thickness inches (mm)	Allowable Crack Width in inches (mm) Pipe Diameters Greater than 96 inches (2400mm)					
	Concrete Cover on Transverse Reinforcement in inches (millimeters)					
		1.25" (32mm)	1.75" (44mm)	2.25" (57mm)	2.75" (70mm)	3.25" (83mm)
8.0 (203)		0.010 (0.255)	0.011 (0.280)	0.012 (0.305)	0.014 (0.355)	0.016 (0.405)
8.5 (216)		0.010 (0.255)	0.011 (0.280)	0.012 (0.305)	0.014 (0.355)	0.015 (0.380)
9.0 (229)		0.010 (0.255)	0.011 (0.280)	0.012 (0.305)	0.013 (0.330)	0.015 (0.380)
9.5 (241)		0.010 (0.255)	0.011 (0.280)	0.012 (0.305)	0.013 (0.330)	0.015 (0.380)
10.0 (254)		0.010 (0.255)	0.011 (0.280)	0.012 (0.305)	0.013 (0.330)	0.014 (0.355)
10.5 (267)		0.010 (0.255)	0.011 (0.280)	0.012 (0.305)	0.013 (0.330)	0.014 (0.355)
11.0 (279)		0.010 (0.255)	0.011 (0.280)	0.012 (0.305)	0.013 (0.330)	0.014 (0.355)
11.5 (292)		0.010 (0.255)	0.011 (0.280)	0.012 (0.305)	0.012 (0.305)	0.013 (0.330)
12.0 (305)		0.010 (0.255)	0.011 (0.280)	0.012 (0.305)	0.012 (0.305)	0.013 (0.330)
12.5 (318)		0.010 (0.255)	0.011 (0.280)	0.011 (0.280)	0.012 (0.305)	0.013 (0.330)
13.0 (330)		0.010 (0.255)	0.011 (0.280)	0.011 (0.280)	0.012 (0.305)	0.013 (0.330)
13.5 (343)		0.010 (0.255)	0.011 (0.280)	0.011 (0.280)	0.012 (0.305)	0.013 (0.330)

1. Concrete cover is measured between reinforcement surface and concrete surface.
2. The tabulated crack width dimensions are measured in inches (millimeters).
3. Sewer pipe shall have an allowable crack width of 0.01 inch (0.25mm) regardless of concrete cover.

The required strength of the pipe specimens undergoing the bearing test will be designated in terms of D-load. Such designation indicates the actual load in pounds per linear foot (newtons per meter of length) of pipe, divided by the inside diameter of the pipe in feet (millimeters). The pipe shall withstand the required test load before a crack having a width as indicated in the following table, measured at close intervals, occurs throughout a length of 1 foot (300mm) or more. The crack shall be considered to be at the indicated width when the point of the measuring gauge will, without forcing, penetrate it 1/16 inch (1.5mm) at close intervals throughout the specified distance of 1 foot (300mm).

The load shall be applied at a uniform rate not to exceed 2,000 pounds per minute per foot-length (29.2 kN per minute per meter of length) of pipe for the first 80 percent of the required load and then at a uniform rate not to exceed 500 pounds per minute per foot-length (7.3 kN per minute per meter of length) of pipe for the remainder of the test.

The test specimens shall be surface dry when tested.

The length on which the test load is computed shall be determined by measuring the inside length of the barrel of the pipe from the bottom of the socket to the end of the spigot. The length of a beveled pipe shall be the average length of the inside of the barrel of the pipe, measured from the bottom of the socket to the end of the spigot.

If the tested specimen of a designated lot passes the test, all of the pipe of that lot shall be considered as complying with the requirements.

If the tested specimen of a designated lot fails to pass the test, then 5 additional specimens from that same lot shall be selected for testing.

If the 5 additional specimens pass, the total number of that lot to be furnished shall be considered as complying, except the one previous failing specimen.

If any of the five additional specimens fail, the entire lot shall be rejected; or may be downgraded except those specimens which passed the first time.

The Contractor may test specimens of a rejected lot individually to determine whether they may comply with the requirements for acceptance.

207-2.9.3 Structural Design Basis. Where structural details of the pipe are shown on the Plans, the manufacture of pipe shall be checked by making the appropriate tests on the concrete placed in the pipe forms, by inspection of the steel reinforcing cages that are to be used in the pipe, and by inspection of the fabrication of the pipe.

Concrete in pipe shall attain a minimum compressive strength of 4,500 psi (31.0MPa) at 28 days.

207-2.9.4 Downgrading of Pipe. For the purpose of these specifications, "downgraded" pipe shall be defined as pipes which are to be used under loads less than those for which they were designed.

Pipes manufactured in accordance with these specifications which have not met their designed test loads may be downgraded by the Engineer and used provided that:

a) Enough load tests are made to establish the load under which they may be used. The number of tests to be made shall be as determined by the Engineer and may require the testing of each section.

b) They comply with the testing and inspection requirements of these specifications.

Individual specimens of pipe which will require major repair, or which have numerous hairline cracks extending the full length of the section on the inside of the pipe at the minor axis or on the outside of the pipe at the major axis, may be tested for downgrading purposes only at the discretion on the Engineer.

207-2.9.5 Acceptance of Stockpiled Pipe. Pipe may be used from stockpiles only when approved in advance by the Engineer, and provided the pipe meets all other requirements for pipe with the exception of inspection of manufacture.

For the purpose of these specifications, "stockpiled" pipe shall be defined as pipe manufactured in quantity which will meet requirements of this section, but was not manufactured for use in specific projects. However, pipe which has been rejected by another agency will not be considered as "stockpiled" pipe, nor will such pipe be accepted.

207-2.10 Marking. The date of manufacture, size and D-load, lot number, manufacturer's identification mark, and where elliptical reinforcement is used, a 4-inch (100mm) high "T" marking the location of the minor axis of the reinforcement shall be legibly painted or stamped on the inside of each pipe. If the Plans or Specifications require a minimum interior clearance for the reinforcing steel, the minimum clearance shall also be painted or stamped on the inside of the pipe.

Sections of pipe to be inspected shall be so situated at the manufacturer's plant as to provide the Engineer with free accessibility for inspection and marking. In no case shall the pipe be stacked to a height that would require the Engineer to climb or use or a ladder to properly inspect the pipe.

At the place of manufacture, the Engineer will indicate acceptance of the pipe for delivery to the Work by marking the pipe with the Agency's stamp. Such acceptance, however, shall be considered a tentative acceptance. Final acceptance will be made only when the Work has been completed.

If pipe is rejected subsequent to its manufacture, the mark placed thereon by the Engineer shall be defaced.

207-2.11 Inspection Platform. For machine-made pipe, a permanently installed inspection platform shall be provided. The platform shall be located such that the inspector can observe the cage and concrete during placement and manufacturing.

207-3 LINED REINFORCED CONCRETE PIPE.

207-3.1 General. These specifications apply to pipe manufactured with a plastic lining. The plastic lining material shall be tested in accordance with 211-2 and conform to 210-2. Pipe not installed within 180 days from the date of pipe manufacture shall have both ends of the pipe covered with an opaque material to prevent ultra-violet degradation of the plastic liner. The Contractor shall not install any lined RCP that is more than 2 years old from the date of manufacture.

Such pipe shall conform to the applicable provisions in 207-2, except that the causes for rejection shall be as listed herein, and repair of pipe will be permitted only within limits set forth in this subsection.

All such pipe shall be manufactured with Type II cement unless otherwise specified in the Special Provisions. All cement shall contain not more than 0.6 percent by weight of alkalis calculated as Na_2O plus 0.658 K_2O.

Chairs or spacers between the reinforcement and forms or base rings shall be stainless steel or a nonferrous material approved by the Engineer. Spacers may only come in contact with the liner plate if they are made of plastic and have a flat base plate of sufficient size to prevent puncture of the liner plate.

207-3.2 Causes for Rejection. Lined pipe may be rejected for any of the following reasons:

a) Exposure of any wires, positioning spacers, or chairs used to hold the reinforcement cage in position, or steel reinforcement in any surface of the pipe, except for holding rods in end projections.

b) Transverse reinforcing steel found to be in excess of 1/4 inch (6mm) out of specified position after the pipe is molded.

c) Any shattering or flaking of concrete at a crack.

d) Air bubble voids (bugholes) on the interior and exterior surfaces of the pipe exceeding 1/4 inch (6mm) in depth unless pointed with mortar or other approved material.

e) Unauthorized application of any wash coat of cement or grout.

f) A deficiency greater than 1/4 inch (6mm) from the specified wall thickness of pipe 30 inches (750mm) or smaller in internal diameter.

g) A deficiency greater than 6 percent (5 percent) from the specified wall thickness of pipe larger than 30 inches (750mm) in internal diameter, except that the deficiency may be 8 percent (7 percent) adjacent to the longitudinal form joint, provided that the additional deficiency does not lie closer than 20 percent of the internal diameter to the vertical axis of the pipe and does not extend along the circumference for a distance greater than 20 percent of the internal diameter of the pipe.

The deficiencies in wall thickness permitted herein do not apply to gasket contact surfaces in gasketed joint pipe. Dimensions and tolerances of such contact surfaces shall be submitted for approval.

h) Internal diameter of the pipe varying from the true circle of the specified diameter by more than 1 percent (for projects in metric units, pipe having less than the specified diameter, or exceeding it by more than 2.6 percent) or interior surfaces reworked after placing of the concrete. The variation in internal diameter permitted herein does not apply to gasket contact surface in gasketed joint pipe. Tolerances at such contact surfaces shall be submitted for approval.

i) A water pocket (identified by tapping the internal surface of the pipe) which is greater than 30 inches (750mm) in length or wider than three times the specified wall thickness. Repair of such defective areas not exceeding these limits shall be made as described in 207-3.3.

j) A piece broken from the end projections of the pipe which has circumferential length exceeding 60 degrees of the circle, or extends into the body of the pipe, or extends into the gasket contact surfaces of gasketed joint pipe for a circumferential length in excess of 6 inches (150mm) measured at the midpoint of the gasket contact surface on the bell end, and at the inner shoulder of the gasket groove at the spigot end. If 2 or more pieces are broken from an end projection, the total length of such broken pieces on any end shall not exceed 90 degrees of the circle; and there shall be a distance of at least 9 inches (230mm) of sound concrete between breaks. The total length of broken pieces that extends into the gasket contact surfaces of gasketed joint pipe shall not exceed a circumferential length of 6 inches (150mm).

If less than 9 inches (230mm) of sound concrete exists between two individual breaks, the 2 breaks shall be considered as one continuous break. Repair of such defects not exceeding above limits shall be made by Method III as described in 207-3.3.3. Unsound portions of end projections shall be removed, and if pieces removed do not exceed the above limits, the pipe may be similarly repaired.

k) Defects that indicate imperfect molding of concrete; or any surface defect indicating honeycomb or open-texture (rock pockets) greater in size than an area equal to a square with a side dimension of 2-1/2 times the wall thickness or deeper than 2 times the maximum graded aggregate size; or a local deficiency of cement resulting in loosely bonded concrete, the area of which exceeds in size the limits of area described in 9) and 10) above when the defective concrete is removed. Repair of such defects not exceeding these limits shall be made as provided in 9) and 10) above. Sand rings occurring at the ends of the pipe shall be repaired for the full circumference.

l) Any of the following cracks:

1) A crack having a width of 0.01 inch (0.255mm) or more throughout a continuous length of 12 inches (300mm) or more.

2) Any crack extending through the wall of the pipe and having a length in excess of the wall thickness.

3) Any crack showing two visible lines of separation for a continuous length of 2 feet (0.6m) or more, or an interrupted length of 3 feet (0.9m) or more anywhere in evidence, both inside and outside, except where such cracks occur during the external loading test specified in 207-2.9.

When required by the Engineer, any crack which is 0.01 inch (0.255mm) wide or wider and is not a cause for rejection, shall be filled with neat cement grout composed of cement mixed with water to a fluid consistency.

207-3.3 Repair of Imperfections.

207-3.3.1 Method I - Repair by Hand-Placed Mortar.

a) **Preparation of Surfaces to be Repaired.** Unsound or imperfect concrete shall be removed by chipping. Edges where concrete has been chipped out shall be sharp and squared with the surface, leaving no feathered edges. The chipped area shall be washed with water to remove all loose material and concrete dust.

Surfaces within the trimmed areas shall be kept wet for several hours, preferably overnight, before the repair replacement is made. All surfaces in areas to be repaired shall be damp, but not wet, when the material is applied.

b) **Placement of the Mortar.** The mortar used for the repair shall contain the same proportions of cement and sand as the mix from which the pipe was made.

This mortar shall be pre-shrunk by mixing it to a plastic consistency as far in advance of its use as possible. Trial mixes shall be made and aged to determine the longest period the mortar's use can be delayed while retaining sufficient plasticity to permit good workmanship.

Immediately prior to the application of the mortar, the damp surface of the area to be repaired shall be scrubbed thoroughly with a small quantity of neat cement grout, using a wire brush. Remaining loose sand particles shall be swept away before application of the mortar.

In applying the mortar, it shall be compacted into the space to be filled, care being taken to eliminate air pockets and to secure bond at the edges. The surfaces shall be shaped and finished to correspond with the adjacent surface of the pipe.

c) **Curing.** The newly repaired surfaces shall be kept damp for 24 hours after the repair is completed. A membrane coating of an approved white-pigmented sealing compound shall then be applied.

207-3.3.2 Method II - Repair by Pneumatically Applied Mortar (PAM).

a) **General.** PAM shall not be used when the repair extends to a depth greater than the embedment of the reinforcing steel. Such repairs shall be made with pre-shrunk mortar.

b) **Preparation of Surface to be Repaired.** Surfaces to which PAM is to be applied shall be prepared in the same manner as described in 207-3.3.1 a) except that the edges of the area from which unsound or imperfect concrete is removed shall be beveled so as not to entrap rebound.

c) **Placement of Mortar.** No rebound shall be included in the repair. The pipe shall be turned so that the area being repaired is at the side of the pipe in a near vertical position to permit rebound to fall clear.

The mortar used for the repair shall contain the same properties of cement and sand as the mix from which the pipe was made.

Before repairing grooved concrete spigots, the snap ring shall be replaced and retained in position until the repair has attained sufficient strength to assure no damage to the gasket groove by its removal.

Areas repaired with PAM shall be filled in excess of the dimension required and then carefully trimmed to correspond with adjacent surfaces.

d) **Curing.** Surfaces to which PAM has been applied shall be cured in the same manner as described in 207-3.3.1 c).

207-3.3.3 Method III - Bonding Mortar Repairs With Epoxy Resin Adhesives.

Unsound or imperfect concrete shall be removed by chipping. If hand placed mortar is to be used, the edges shall be left sharp and square with the surface. If PAM is to be used, the edges shall be beveled.

The area to be repaired shall be kept dry. Loose material and concrete dust remaining after the chipping operation shall be removed by means of an air jet.

Epoxy resins previously approved for such use by the Engineer shall be used in the manner prescribed by the Engineer. The prepared area shall be primed with the epoxy resin compound, care being taken to ensure intimate contact with the base material. Mortar shall be applied before the epoxy resin compound sets. Mortar shall be applied by either Method I or II as described in 207-3.3.1 and 207-3.3.2.

207-4 CONCRETE CYLINDER PIPE.

207-4.1 General. These specifications apply to concrete cylinder pipe intended for use in water supply pipelines and distribution systems that carry water under pressure. Concrete cylinder pipe may be furnished in pipe diameters of 10 inches (250mm) or larger for operating pressures from 100 psi (690 kPa) to 400 psi (2,760 kPa).

207-4.2 Design, Manufacture, and Tests. Concrete cylinder pipe shall be designed, manufactured and tested in accordance with AWWA C303. The average circumferential stress in the steel cylinder and bar or wire reinforcement of the pipe at design pressure shall not exceed 16,500 psi (114MPa) nor 50 percent of the minimum yield strength of the steel used in the cylinder.

207-4.3 Fittings. Fittings for concrete cylinder pipe shall conform to requirements of Federal Specification SS-P-381.

207-5 REINFORCED CONCRETE PRESSURE PIPE.

207-5.1 General. These specifications apply to three types of reinforced concrete pressure pipe, 2 of which are not prestressed (and one prestressed), with internal diameters of 12 inches (300mm) and larger, to be used in the transmission and distribution systems that carry water under pressure as shown on the Plans.

207-5.2 Manufacture and Tests. Reinforced concrete pressure pipe and fittings shall be manufactured and tested to conform to one of the following specifications:

a) AWWA C300 for the steel bar reinforcement and cylinder type in pipe diameters 24 inches (600mm) and larger, for design pressures of 40 to 260 psi (275 kPa to 1,800 kPa), and for external loading conditions as may be designated on the Plans or in the Special Provisions.

b) AWWA C301 for the prestressed steel wire reinforcement and cylinder type, in pipe diameters of 30 inches (750mm) and larger, for design pressure up to 350 psi (2,410 kPa), and for external loading conditions as may be designated on the Plans or in the Special Provisions.

c) AWWA C302 for the steel reinforcement (without cylinder) type, in pipe diameters 12 inches (300mm) and larger, for design pressures of not more than 45 psi (310 kPa), and for external loading conditions as may be designated on the Plans or in the Special Provisions.

207-6 (Not Used).

207-7 (Not Used).

207-8 VITRIFIED CLAY PIPE (VCP).

207-8.1 General. Except as modified in this subsection, vitrified clay pipe and fittings including perforated pipe shall be extra strength or high strength manufactured in accordance with ASTM C700.

207-8.2 Manufacturing Requirements.

207-8.2.1 General. All pipe and fittings shall be clearly marked with the name or trademark of the manufacturer, the location of the plant, and the strength designation. All standard length straight pipe as defined in 207-8.2.2 shall, in addition to the above, be marked with a manufacturer's date code. All fabricated bends and/or bevels shall be manufactured from pipe meeting all requirements of the pipe specifications for the Work.

207-8.2.2 Dimensions and Tolerances. The pipe diameter shall not vary from a true circle by more than 3 percent of nominal diameter. Except for special purposes, the minimum standard length of straight pipe, exclusive of socket depth, shall be 40 inches (1000mm). Pipe shall not deviate from straight by more than 1/16 inch per foot (5mm per 1m of length).

207-8.2.3 Imperfections. Imperfections in pipe and fittings containing blisters, cracks, and chips in excess of the limits herein will be rejected; however, certain cracks and chips meeting the requirements of 207-8.2.4 may be repaired in accordance with 207-8.6.

a) **Blisters.** For pipe of nominal sizes 3 to 18 inches (75 to 450mm), blisters shall not exceed 3 inches (75mm) in any direction, and no blister or pimple shall project more than 1/8 inch (3mm) above the surface of the pipe.

   For pipe of nominal sizes over 18 inches (450mm), no blister shall exceed in any direction, 2 inches per foot (170mm per meter) of internal diameter, and no blister or pimple shall project above the surface of the pipe more than 1/8 inch per foot (10mm per meter) of internal diameter.

   Pipe shall have no broken blisters.

b) **Cracks.** There shall be no cracks passing through the barrel or socket except that a single crack at the spigot end of the pipe not exceeding 75 percent of the depth of the socket, or a single circumferential crack in the socket not exceeding 3 inches (75mm) in length or a single crack not exceeding 2 inches (50mm) in the axial direction is permitted.

c) **Chips.** Chips on the interior surface shall not exceed 2 inches (50mm) in length, 1 inch (25mm) in width, and a depth of 1/4 of the barrel thickness, but not to exceed 1/4 inch (6mm). A single pipe shall contain no more than 2 such defects.

207-8.2.4 Repairable Imperfections.

a) **General.** Structurally sound clay pipe larger than 15 inch (375mm) size, may be repaired as provided in 207-8.6.

   Repairs of any type at the spigot or socket, shall be limited to one for each 60 degrees of circumference, and a maximum of four at either end. Repaired pipe shall not be used for fabricated fittings unless the repaired pipe is tested. Molded fittings may be repaired within the scope of the Specifications.

b) **Cracks.** The following longitudinal cracks parallel to the axis caused by shrinkage or drying and not more than 1/32 inch (0.8mm) wide may be repaired:

   1) A crack on the exterior of the spigot that does not penetrate the entire barrel thickness and does not exceed 50 percent of the depth of the socket in length.

2) A crack in the socket of the pipe that does not penetrate the entire thickness and does not exceed 75 percent of the depth of the socket in length.
3) A crack that penetrates the entire thickness of the socket and does not exceed 50 percent of the depth of the socket in length.
4) A crack on the interior of the socket and in the shoulder on the exterior of the socket which does not exceed 3 inches (75mm) in length and does not penetrate more than 20 percent of the wall thickness.

c) **Surface Chips.** Surface chips located on the exterior of the spigot, the interior or exterior of the socket, or on the shoulder of the socket may be repaired, provided:

1) The length of the circumference of the chip does not exceed twice the barrel thickness.
2) The width is not greater than 50 percent of the socket depth measured parallel to the axis.
3) The depth is not greater than 25 percent of the wall thickness measured perpendicularly to the axis.

d) **Full Depth Chips.** Full depth chips located on the socket may be repaired provided the length of the chip does not exceed twice the barrel thickness or the width does not exceed 25 percent of the socket depth.

207-8.3 Fittings and Stoppers. Fittings shall be made to such dimensions as will accommodate the joint system specified. Y- and T-branch fittings shall be furnished with spurs securely fastened by the manufacturer to the barrel of the pipe. There shall be no projection on the inner surface of the barrel.

T-branch fittings shall have their axis perpendicular to the longitudinal axis of the pipe. The axis of the spur on Y-branch fittings shall be 45-degrees from the longitudinal axis of the pipe. The barrel of each spur shall be of sufficient length to permit the proper jointing of the connecting pipe.

Stoppers furnished for installation in branch fittings and ends of pipe left unconnected shall be strong enough to sustain all applied construction and in-place loads, including field pressure tests. Stoppers for pipe shall be one of the following: polyethylene (PE), polyurethane, polypropylene, acrylonitrilebutadiene-styrene (ABS), polyvinyl chloride (PVC), ozone-resistant synthetic rubber, clay discs, or other material approved by the Engineer. The Contractor shall retest within 60 days prior to installation any stopper that is more than 180 days old from the date of manufacture to ensure compliance with the requirements of the Specifications. The Contractor shall not install any stopper that is more than 2 years old from the date of manufacture.

207-8.4 Joints. Joints for vitrified clay pipe are specified in 208-2. The maximum bevel of the ends of pipe to be laid on a curve is 4 degrees. Plain-end pipe shall not be beveled.

Each joint within vertical and horizontal curves shall be constructed using factory fabricated mitered or beveled pipe or by deflecting joints. In no case shall joints be deflected more than allowed in 306-1.2.3. Shop and layout drawings for mitered or beveled pipe shall be submitted to the Engineer for review and approval.

207-8.5 Testing Requirements.

207-8.5.1 General. Before a lot of pipe is acceptable for use, test pipes selected from the lot shall meet the requirements of the hydrostatic pressure and bearing tests described herein. The tests shall be made at the point of manufacture, under the supervision of the Engineer, and at no cost to the Agency. Fittings will be subject to the requirements of the hydrostatic pressure test only.

When less than 50 pipes are to be furnished to the Work, the Engineer may waive the sampling, testing and acceptance requirements of this subsection and accept the manufacturer's written certification that the pipe meets the quality and strength specified. All pipe will be subject to visual inspection and acceptance by the Engineer after delivery to the Work site.

All testing equipment shall be calibrated by a testing agency acceptable to the Engineer at intervals not to exceed 12 months or following repair of the hydraulic system or modification or relocation of the equipment.

Each lot of vitrified clay pipe is defined as not more than 500 sections of pipe for one size and class. The Engineer shall select for testing the following number of samples:

TABLE 207-8.5.1 (A)

No. of Sections in Lot	No. of Sections for Testing
Less than 41	2
41 - 60	3
61 - 80	4
81 - 250	5
251 - 300	6
301 - 350	7
351 - 400	8
401 - 450	9
451 - 500	10

The pipe selected for testing shall be sound and shall meet the dimensions and tolerances of the specifications.

207-8.5.2 Hydrostatic Pressure Test. Shall be carried out in accordance with the applicable requirements of ASTM C301 and shall precede the bearing test by not more than 3 hours.

207-8.5.3 Bearing Tests. Shall be carried out in accordance with the 3-edge bearing method requirements of ASTM C301. The pipe shall withstand the minimum 3-edge bearing test loads indicated in the following table:

TABLE 207-8.5.3

Nominal Diameter inches (mm)	Extra Strength Pipe lbs/ft (kN/m)	High Strength Pipe lbs/ft (kN/m)
3 (75)	2000 (29.2)	2200 (32.1)
4 (100)	2000 (29.2)	2200 (32.1)
6 (150)	2000 (29.2)	2200 (32.1)
8 (200)	2200 (32.1)	2400 (35.0)
10 (250)	2400 (35.0)	2600 (37.9)
12 (300)	2600 (37.9)	2900 (42.3)
15 (375)	3100 (45.2)	3400 (49.6)
18 (450)	3600 (52.5)	4000 (58.4)
21 (525)	4200 (61.3)	4600 (67.1)
24 (600)	4800 (70.1)	5300 (77.3)
27 (675)	5200 (75.9)	5700 (83.2)
30 (750)	5500 (80.3)	6100 (89.0)
33 (825)	5800 (84.6)	6400 (93.4)
36 (900)	6300 (91.9)	6900 (100.7)
39 (975)	6600 (96.3)	7300 (106.5)
42 (1050)	7000 (102.2)	7700 (112.4)

207-8.5.4 Acceptance. When all test pipes meet the hydrostatic pressure and bearing tests, the entire lot of pipe is acceptable. When 2 test pipes fail, the entire lot will be rejected. When one pipe fails, a second group of 2 pipes must pass the tests, otherwise the entire lot will be rejected.

207-8.6 Clay Pipe Repair.

207-8.6.1 Repair Methods and Materials.

a) **General.** All surfaces to be repaired shall be clean and dry. All unsound material at lumps or blisters shall be ground smooth and flush with adjacent surfaces. Cracks shall be grooved 1/8 to 1/4 inch (3mm to 6mm) wide and 1/8 to 1/4 inch (3mm to 6mm) deep for the full length of the crack. All unsound material at chips, flakes, pits, and spalls shall be removed and edges shall be 1/16 inch (1.5mm) minimum below adjacent surfaces. There shall be no feather edges.

Prepared areas shall be cleaned of dust and other loose particles and then filled with repair material compounded to provide properties most desirable for sewerage service. Repair material shall resist bacterial attack and attack by chemicals or combinations of chemicals normally present in domestic and industrial sewage.

Repair material shall be mixed, applied, and cured as recommended by the manufacturer and approved by the Engineer, and shall have a color contrasting with the color of pipe to be repaired. If necessary to produce a contrast in color, carbon black in a small quantity may be added to the repair material. The repair material shall be subject to adhesion and chemical testing as required by the Engineer to determine its suitability for use.

b) **Adhesion Test.** Vitrified clay bars 1 inch (25mm) square in cross section and approximately 8 inches (200mm) in length, compounded of the same materials as the vitrified clay pipe and fired to clay pipe manufacturing temperature shall be used in preparing test specimens. The bars shall have a modulus of rupture of not less than 1,600 psi (11.0MPa) when tested in flexure with third-point loading.

The bars shall be cut through at the midpoint and then bonded with the repair material. Following a 7-day maximum cure period at ambient room temperature, the bonded bars shall be tested in flexure with third-point loading.

The average modulus of rupture of 5 test bars bonded with the repair material shall not be less than 1,600 psi (11.0MPa).

Five additional test bars bonded with the repair material and immersed for 60 days in water at ambient room temperature shall have an average modulus of rupture not less than 1,500 psi (10.3MPa).

c) **Chemical Tests.** Each specimen of repair material shall lose not more than 2 percent of its weight after being immersed in the solutions listed in 208-2.3.2 (Note 3 in table) for a period of 30 days and being reconditioned as indicated.

207-8.6.2 Inspection of Repairs. All pipe to be repaired shall be inspected by the Engineer after preparation for repair, again after repair has been made. Repairs made without prior inspection will be rejected. The Engineer may require retesting of any repaired pipe to demonstrate its soundness. The Agency shall be reimbursed for all costs incurred for inspection and testing of repaired pipe.

207-9 IRON PIPE AND FITTINGS.

207-9.1 General. This subsection specifies ductile iron pipe for water and other liquids, and cast iron soil pipe. Cast iron soil pipe shall not be used for mainline gravity sewers or for any pressure piping used to convey sewage.

207-9.2 Ductile Iron Pipe for Water and Other Liquids.

207-9.2.1 General. Ductile iron pipe for water and other liquids shall be furnished in the sizes, classes, grades, or nominal thicknesses and joint types shown on the Plans or specified in the Special Provisions.

Ductile iron pipe shall comply with AWWA C 151.

207-9.2.2 Pipe Joints. Iron pipe joints shall comply with the following requirements for the types shown on the Plans or specified in the Special Provisions:

TABLE 207-9.2.2 (A)

Type of Joint	Specifications
Rubber Gasket Push-on Joint	AWWA C 111
Mechanical Joint	AWWA C 111
Flanged Joint	AWWA C 110
Flanged Joint (Threaded Flanges)	AWWA C 115

Rubber gasket material shall conform to 208-1.2.

207-9.2.3 Fittings. All ductile iron and cast iron rubber gasket, push-on, mechanical, and flanged joint fittings for iron pipe shall be manufactured in accordance with AWWA C 110.

207-9.2.4 Lining and Coating. Unless otherwise specified, the internal surfaces of iron pipe and fittings shall be lined with a uniform thickness of cement mortar then sealed with a bituminous coating in accordance with AWWA C 104. The outside surfaces of ductile iron pipe and fittings for general use shall be coated with a bituminous coating 1 mil (25μm) thick in accordance with AWWA C 151 or AWWA C 110.

207-9.2.5 Inspection and Certification. The manufacturer shall furnish a certified statement that the pipe and fittings have been manufactured and tested in accordance with these specifications.

207-9.2.6 Polyethylene Encasement for External Corrosion Protection. When loose polyethylene encasement for the protection of iron pipes, cast or ductile iron fittings, valves, and appurtenances is specified in the Plans or in the Special Provisions, it shall be furnished and installed in accordance with the requirements of AWWA C 105.

207-9.3 Cast Iron Soil Pipe and Fittings. Cast iron soil pipe and fittings shall comply with ASTM A74.

207-10 STEEL PIPE.

207-10.1 General. These specifications apply to steel pipe for the transmission and distribution of water under pressure. Steel pipe shall be of the size, type, and cylinder wall thickness or pressure class shown on the Plans or specified in the Special Provisions.

207-10.2 Fabricated Steel Pipe.

207-10.2.1 General. Fabricated steel pipe shall consist of butt or offset butt electrically welded straight or spiral-seam steel cylinders, shop fabricated from plates or sheets in accordance with AWWA C 200.

Steel pipe to be cement-mortar lined may also be fabricated to the requirements of Federal Specification SS-P-385 a.

Pipe with 4-inch (100mm) nominal diameter, when required, shall conform to the requirements of AWWA C 200.

Four-inch (100mm) pipe, when required, shall conform to the requirements of AWWA C 200.

Prior to fabrication of pipe, the Contractor shall in accordance with 2-5.3, submit complete shop fabrication drawings of all pipe, pipe specials and joint details. The Contractor shall also submit a schedule of pipe marks accompanied by a plan showing the field location of each mark.

Steel plates or sheets used in the manufacture of fabricated steel pipe shall comply with the physical and chemical requirements of ASTM A570, Grades 30 or 33, for steel sheets or strip, or to ASTM A283 Grades C or D, or A36 for steel plates, or be of other steel of equal quality if approved by the Engineer.

The Contractor shall submit to the Engineer a certified laboratory report stating the type of steel, and the physical and chemical properties for each heat number of the steel used in the fabrication of the pipe.

After fabrication, but prior to the application of the specified lining and coating, straight pipe cylinders of all sizes shall be successfully hydrostatically tested to 75 percent of the specified minimum yield point of the steel sheets or plates used in the manufacture of the pipe in accordance with Section 3.5 of AWWA C 200.

207-10.2.2 Design Criteria. The cross-sectional area of steel in the wall of the pipe cylinder shall be computed on the basis of a minimum design safety factor of 2 of the ratio of yield point stress for that particular type steel used in the fabrication of the pipe, to the maximum design working stress in the pipe cylinder wall.

Steel cylinders shall have a wall thickness of not less than 14 gage (1.9mm).

207-10.2.3 Diameter. Fabricated steel pipe shall have a minimum net inside diameter, after application of the interior protective lining, equal to the nominal diameter of the pipe shown on the Plans or in the Special Provisions, with a permissible tolerance of minus 1/4 inch (6mm).

207-10.2.4 Length. Unless otherwise specified, fabricated steel pipe shall be manufactured in lengths to fit the pipeline alignment shown on the Plans, subject to a maximum pipe length of 42 feet (13m). Pipe may be furnished in lengths less than the manufacturer's standard length to facilitate the laying of the pipe on curved alignments.

The Contractor shall fabricate and furnish all necessary short pipe sections required for pipeline closures and the proper location of any special section, fitting or pipeline appurtenance shown on the Plans.

207-10.2.5 Joints. Joints for fabricated steel pipe shall be one of the following types as specified on the Plans or in the Special Provisions and shall comply with the requirements of Section 3.7 of AWWA C 200.

a) Bell-and-spigot ends with rubber gaskets.

b) Lap joints for field welding.

c) Plain ends fitted with butt straps for field welding.

d) Ends prepared for mechanical coupled field joints.

e) Plain ends fitted with flanges.

Joint tolerances shall be as specified in Section 3.7.9 of AWWA C200.

For pipe less than 27 inches (675mm) in nominal diameter, butt straps for butt strap joints shall be furnished with a 4 inch (100mm) diameter hand hole, complete with screwed cap or plug, suitable for use in "pointing" the interior joint lining after field installation of the joint.

All joints, except flanged and mechanical coupling joints, shall be designed for a maximum interior pipe lining gap joint of 1/2 inch (13mm) after joint assembly, measured from the ends of the lining of the two pipe sections being joined.

The laying of bell-and-spigot, rubber-gasket-joint pipe on curved alignment by means of unsymmetrical closure of spigot into bell will be permitted. The amount of "pull" and the method of achieving the "pull" shall be in accordance with the manufacturer's recommendation, but the deviation from the normal lining gap at the joint shall not exceed a 3/4 inch (19mm) "pull" on the outside of the curve, and a 1/4 inch (6mm) "push" on the inside of the curve, with no "push" to be permitted until the full 3/4 inch (19mm) "pull" has been used.

Where the allowable deflection of the pipe by means of unsymmetrical closure of joints is not sufficient to fit the curved alignment of the pipe, as shown on the Plans, special sections consisting of beveled ends or multiple-piece bends shall be fabricated. In no case shall the pipe bell or spigot composing the joint between sections of pipe be cut or altered in any manner.

207-10.2.6 Rubber Gaskets. Rubber gaskets and gasket lubricant shall be furnished with rubber gasket joint pipe in sufficient quantity for the amount of pipe ordered. Rubber gaskets shall be new and made in conformance with section 3.4 of AWWA C 300. Gasket material shall conform to 208-1.2.

207-10.2.7 Special Sections. Special sections, including curves, tees, branches, manifolds, outlets, reducers, and enlargers shall be fabricated as shown on the Plans or in the Special Provisions. Special sections shall be fabricated from sections of previously lined and coated straight pipe, of the kind being furnished under the Contract or of specially rolled, lined, and coated pipe for fittings to meet job conditions shown in the Plans or in the Special Provisions.

Bends shall be required on curves which have a shorter radius than can be accommodated by beveled pipe ends or by pushing and pulling the joint. Bends, tees, crosses, and wyes shall be fabricated and dimensioned in accordance with AWWA C 208.

The courses of special sections shall be joined by butt welding after which the pipe protective lining and coating shall be repaired to a like condition as existed on the previously lined and coated straight pipe section from which the special was fabricated.

Bends, angles, reducers, enlargers, and other special pipe sections made by certified welding operations from previously tested pipe sections need not be subjected to a hydrostatic test. All special pipe sections not fabricated from tested pipe shall be hydrostatically tested after fabrication as specified in 207-10.3.2 for straight pipe. Sections that cannot be tested in a testing machine may be prepared for testing by welding a head on the open ends, or as may otherwise be approved by the Engineer. After testing, the ends shall be reconditioned.

Any welds which have not been subject to a hydrostatic test shall be tested by the air-soap or dye-check method.

207-10.2.8 Welding. Welding of fabricated steel pipe shall conform to the requirements of AWWA C 200.

Patching of pipe where test specimens have been taken will not be permitted.

If provided for in the Special Provisions, the Contractor shall furnish the Engineer with a certified laboratory report stating the results of the welding tests required in this subsection.

207-10.3 Mill-type Steel Water Pipe.

207-10.3.1 General. Mill-type steel pipe shall consist of furnace-welded, electrically welded, or seamless pipe produced to meet finished pipe specifications in accordance with AWWA C 200.

Acceptable grades of pipe under this standard shall be:

ASTM A53 Grades A & B

ASTM A135 All grades

ASTM A139 All grades

ASTM A134 Using only grades of steel shown in 10.2.1

If required in the Special Provisions, the Contractor shall submit to the Engineer a certified laboratory report stating the type of pipe and the physical and chemical properties for each heat number of the steel used in the manufacture of the pipe.

207-10.3.2 Dimensions, Weights, and Test Pressure. The inside and outside diameters, wall thickness, weight, test pressures, and weight and dimension tolerances for mill-type steel pipe shall conform to ANSI B36.10 or AWWA C 200.

207-10.3.3 Length. Mill-type steel pipe shall be furnished in single random lengths, double random lengths, or in specified cut lengths.

For specified cut lengths, the actual length of the pipe shall not vary from the specified length by more than 1/8 inch (3mm).

For special random lengths, the average length shall not be less than 17-1/2 feet (5.3m) and no piece shall be shorter than 9 feet (2.7m). Double random lengths shall have an average length of not less than 35 feet (10.7m) with not more than 10 percent of the lengths shorter than 26-1/4 feet (8.0m) and no piece shorter than 14 feet (4.3m).

For specified average lengths in excess of 20 feet (6.1m), not more than 10 percent of the lengths shall be shorter than 75 percent of the average length specified and no length shall be shorter than 40 percent of the specific average length.

207-10.3.4 Joints.

a) **General.** Joints for mill-type steel pipe shall be one of the following types as shown on the Plans or in the Special Provisions.

b) **Mechanical Couplings.** Pipe ends for mechanical couplings shall be square cut or beveled with all burrs removed. All outside surfaces of the pipe ends, where the mechanical coupling will be seated, shall be free of all indentations, projections, or roll marks to ensure a watertight joint. Pipe ends for mechanical couplings shall have tolerances within the limits required by the manufacturer of the mechanical coupling specified.

c) **Field Butt-Welded Joints.** Pipe with a wall thickness 15/64 inch (6.0mm) or greater, intended for field butt welding, shall have the pipe ends beveled on the outside or inside or both sides, as shown on the Plans or in the Special Provisions. Bevels shall have an angle of 30 degrees, with a maximum plus tolerance of 5 degrees and no minus tolerance, measured from a line drawn perpendicularly to the axis of the pipe. The width of the root face at the end of the pipe shall be 1/16 inch ± 1/32 inch (1.5mm ± 0.8mm).

d) **Flanged Joints.** Welding flanges shall be forged steel conforming to ASTM A 181 and faced and dimensioned in accordance with ANSI B16.5 for the pressure class shown on the Plans or in the Special Provisions. Class ANSI B12.5 shall be the minimum class flange permitted for water distribution purposes. Threads for screwed flanges and companion pipe ends shall be dimensioned in accordance with ANSI B2.1.

Pipe ends for welding neck flanges shall be beveled.

207-10.3.5 Special Sections. These include elbows, returns, tees, reducers, and crosses which shall be manufactured from mill-type steel pipe in accordance with ANSI B16.9.

Special sections manufactured from previously tested pipe need not be tested hydrostatically but any welds not previously hydrostatically tested shall be tested by the air-soap or dye-check method.

207-10.3.6 Marking of Pipe and Special Sections. Each length of pipe and each special section shall be legibly marked by paint stenciling, die stamping, or hot-roll marking to show the manufacturer's name or mark, size and weight of the pipe or special section, and the type of steel from which the pipe or special section was made.

207-10.4 Protective Lining and Coating for Steel Pipe.

207-10.4.1 General. The interior and exterior surfaces of all steel water pipe shall be protected from corrosion by one of the following methods, as shown on the Plans or in the Special Provisions.

207-10.4.2 Cement-Mortar Lining and Coating. Cement mortar lining and coating shall comply with the requirements of AWWA C 205, or shall be fabricated, lined, and coated to the requirements of Federal Specification SS-P-385a.

The interior and exterior surfaces of all pipe to which cement mortar will be applied shall be thoroughly cleaned prior to the application of the cement mortar. Loose scale, rust, and all accumulations of dirt or debris shall be removed. Oil and grease shall be removed by a volatile solvent.

When a curing compound is used to facilitate curing of the cement mortar lining, it shall be applied to the cement mortar within 1 hour after placing the mortar. The curing time of cement-mortar so cured shall be 7 days. Sealing compound used shall be in conformance with Section 2.5 of AWWA C 205.

The results of cleaning, lining, and coating of special sections shall be equivalent to the results of similar work on straight pipe sections and shall conform to the requirements of Sections 4.5 and 5.7 of AWWA C 205.

207-10.4.3 Cement Mortar Lining and Asphalt Mastic Coatings. Cement mortar lining shall comply with the requirements of 207-10.4.2.

Materials, manufacture, and application of asphalt protective coatings for pipelines (mastic systems) shall comply with "Asphalt Protective Coating for Pipelines" of the Asphalt Institute, Construction Series No. 96, Specifications M-2.

Mastic systems for pipelines shall consist of an asphalt prime coat followed by a coating of a dense, impervious, essentially voidless mixture of asphalt, mineral aggregate, and mineral filler which may include asbestos fiber. The finished mastic coating shall be painted with a water-resistant whitewash.

A Grade 3 mastic system shall be furnished under these specifications.

Unless otherwise specified, the minimum thickness of the asphalt mastic coating shall be as shown in Table 207-10.4.3 (A).

TABLE 207-10.4.3 (A)

Diameter inches (Millimeters)	Coating	
	Thickness[1] in (mm)	Tolerance in (mm)
4 thru 8 (100 thru 200)	3/8 (10)	+1/8 (+3.2)
10 thru 12 (250 thru 300)	7/16 (11)	+1/8 (+3.2)
Over 12 (300)	1/2 (13)	+1/8 (+3.2)

1. Subject to the provisions of Section 2.8 of the Asphalt Institute "Specifications M-2".

207-10.4.4 Coal Tar Enamel Lining and Coating. Coal-tar enamel protective lining and coating shall comply with the requirements of AWWA C 203.

The interior and exterior surfaces of the pipe shall be thoroughly cleaned by grit blasting immediately prior to the application of the coal-tar protective system.

The interior surfaces of the pipe shall first be coated with a coal-tar primer followed by a uniform lining of hot coal-tar enamel. Thickness of the lining shall be 3/32 inch (2.4mm) with an allowable variation of ± 1/32 inch (0.8mm).

The exterior surfaces of the pipe shall be protected with one coat of coal-tar primer followed by one or more coats of hot coal-tar enamel. A single layer of 15 pounds per 100 ft^2 (73 kg per 100m^2) asbestos coal-tar saturated felt or fibrous glass mat shall be bonded into each coat of hot coal-tar enamel. The coating system shall be finished with a single wrap of heavy paper followed by a coat of water-resistant whitewash.

The thickness of each exterior coal-tar enamel coating shall be 3/32 inch (2.4mm) with an allowable variation in thickness not to exceed ± 1/32 inch (0.8mm) (excluding the wrapping).

The number of exterior hot coal-tar enamel coats and the type of exterior wrapping shall be as shown on the Plans or in the Special Provisions.

The results of cleaning, lining and coating of special sections shall be equivalent to the results of similar work on straight pipe sections. Methods deviating from the prescribed procedures of AWWA C 203 shall require the approval of the Engineer. If the shape precludes spinning, the lining and exterior coating shall be applied by hand daubers and hand wrapping as specified in Section 3.11 of AWWA C 203.

Whenever possible, all welding operations for straight pipe sections, including the welding of flanges, shall be completed prior to the application of the coal-tar enamel lining and coating.

Coal-tar enamel lining and coating shall be held back from the ends of all joints to be field welded. Field-welded joints on coal-tar enamel lined and coated pipe less than 27 inches (675mm) in diameter shall have a 4-inch (100mm) diameter hand hole complete with a screwed cap or plug to permit priming and daubing of the interior pipe joint upon completion of the field welding.

207-10.4.5 Cement Mortar Lining and Asphalt Coating and Wrapping. Cement mortar lining shall comply with 207-10.4.2.

The exterior surfaces of the pipe shall be thoroughly cleaned by grit blasting followed immediately by a coat of hot asphalt primer. The pipe shall then be coated with one or more coats of hot asphalt. A single layer of 23-1/2-pounds per 100 ft^2 (115 kg per 100m^2) asbestos felt or 40-pounds per 100 ft^2 (195 kg per 100m^2) rag felt or fiberglass wrap shall be machine wrapped into each coat of hot asphalt. The coating system shall be finished with a single wrap of heavy paper followed by a coat of water resistant whitewash.

The thickness of each exterior asphalt coating shall be 3/32 inch (2.4mm) with an allowable variation in thickness not to exceed ± 1/32 inch (0.8mm) excluding the wrapping.

The number of exterior hot asphalt coats and the type of exterior wrap shall be as shown on the Plans or in the Special Provisions.

The results of coating special sections shall be equivalent to the results of similar work on straight pipe sections. If the shape precludes machine application, the coating of specials shall be applied by hand daubing and wrapping.

207-10.4.6 Preparation of Pipe Ends (Lined and Coated Pipe). The pipe protective lining and coating shall be held back from the socket and spigot ends respectively, for all socket-and-spigot rubber gasket, welded lap, and driven field joint pipe in accordance with the manufacturers' standard practices. The pipe protective coating shall be held back from the ends of all butt strap, mechanical coupling, and flanged joint pipe for a sufficient distance to permit field assembly of the particular joint.

The protective lining shall terminate at the pipe ends of all pipe except where otherwise specified herein.

All noncoated surfaces shall be cleaned and given a brush coat of a suitable rust preventative material to protect said surfaces from pitting or rusting. The rust preventative material used shall be nontoxic and of such properties that the quality of the field joint or the lining and coating to be applied to the joint will not be adversely affected.

207-11 CORRUGATED STEEL PIPE AND PIPE ARCHES.

207-11.1 General. Corrugated steel pipe, pipe arches, nestable pipe, slotted pipe, spiral rib pipe, spiral rib pipe arches, and coupling bands shall be manufactured and inspected in conformance with AASHTO M36/ASTM A 760/A760M or AASHTO M245/ASTM A 762/A762M and as specified herein. The size, type, and metal thickness of the pipe to be furnished shall be shown on the Plans or in the Special Provisions.

Corrugated steel pipe arches and spiral rib pipe arches shall consist of round pipe which has been re-formed to multi-centered pipe having an arch shaped top with a slightly curved integral bottom. Nominal diameter shall be the minimum inside dimensions of the round pipe. Pipe arch dimensions shall be in accordance with AASHTO M36/ASTM A760/A760M or AASHTO M245/ASTM A762/A762M.

207-11.2 Materials.

207-11.2.1 General. The material for corrugated steel pipe, pipe arches, spiral rib pipe, and spiral rib pipe arches shall be zinc coated (galvanized) or aluminum coated (AL-T-2) or polymer precoated steel sheet conforming to AASHTO M218/ASTM A444/A444M or AASHTO M274/ASTM A819 or AASHTO M246/ASTM A742/A742M. Material Safety Data Sheets (MSDS) are available from the pipe manufacturers.

The manufacturer of corrugated steel pipe, pipe arches, spiral rib pipe, and spiral rib pipe arches shall furnish to the Engineer a Certificate of Compliance stating that the materials furnished comply in all respects with these specifications. Additional testing, as required by the Engineer, shall be done at the Contractor's expense.

207-11.2.2 Coupling Bands. The coupling bands shall be of either one-, two-, or three-piece construction and shall be of the same material as the pipe. The coupling bands for the pipe larger than 42 inches (1050mm) shall be of two- or three-piece construction. Coupling bands shall be formed metal not more than three standard culvert sheet thicknesses lighter than that of the pipe to be connected, but shall be not less than 0.052 inch (1.32mm) or more than 0.109 inch (2.77mm) thick. The minimum width of the coupling bands for pipe ends other than flanged shall be 10-1/2 inches (265mm) for annular corrugations and rerolled ends and 12 inches (300mm) for helical corrugations except for Hugger bands for pipes 42 inches (1050mm) or less in diameter installed in a buried condition without watertight joints, which may be 4 inches (100mm). Dimpled coupling bands shall not be used without prior written approval of the Engineer. All coupling band connection hardware shall be galvanized in accordance with 210-3 or electroplated in accordance with ASTM A164, Type RS, or ASTM B633, Class Fe/Zn 5. Bolts and nuts for all types of coupling bands shall conform to the requirements of ASTM A 307.

The installed distance between corrugated pipe ends shall be no greater than 1-1/2 inches (38mm) and the maximum difference in diameter between pipe ends to be joined shall be 1/2 inch (13mm) for pipe sizes 48 inches (1200mm) and under and 1 percent of the diameter for larger pipe sizes.

When required by the Plans or the Special Provisions, watertight joints shall be provided by the use of approved sealant or gasket materials. Gasket material shall conform to 208-1.2. These materials shall be neoprene expanded rubber or sheet rubber gaskets, "O" ring rubber gaskets, butyl rubber base joint sealant, or other approved materials. Sheet rubber gaskets shall be at least 7 inches (175mm) wide and 3/8 inch (9.5mm) thick and shall conform to the requirements of ASTM D1056 with Grades 41 to 43 inclusive, unless otherwise specified. "O" rings shall conform to ASTM C443 and shall not be used on helical corrugations. Test for watertight joints shall conform to 306-1.4.6.

207-11.3 Fabrication.

207-11.3.1 General. Unless otherwise specified, corrugated steel pipe, pipe arches, spiral rib pipe and spiral rib pipe arches, shall be furnished with pipe ends cut perpendicularly to the longitudinal axis of the pipe. Pipe ends shall be cut uniformly and shall be fabricated so that the pipe can be effectively joined with coupling bands conforming to 207-11.2.2.

Corrugated steel pipe may be fabricated either by riveting, resistance spot welding or using a helically corrugated steel pipe with a continuous helical lock seam or a continuous helical welded seam paralleling the corrugation.

Spiral rib pipe shall be fabricated with helical ribs projecting outwardly with a continuous lock seam or welded seam paralleling the rib.

207-11.3.2 Fabrication by Spot Welding. Pipe fabrication by spot welding shall conform to the requirements of AASHTO M36/ASTM A760/A760M. Pipes fabricated by resistance spot welding shall be the full circle type with lap joint construction. The center of each spot weld shall be at least its radius plus 1/4 inch (6.5mm) from the edge of the steel sheets.

Spot welding shall be performed in such a manner that the exterior surfaces of 90 percent or more of the spot welds on a length of pipe show no evidence of melting or burning of the base metal; and the base metal is not exposed when the area adjacent to the electrode contact surface area is wire brushed. Discoloration of the spot weld surfaces will not be cause for rejection.

207-11.3.3 Fabrication by Continuous Helical Seam.

a) **Helical Corrugations and Pitches.** Helically corrugated steel pipe and spiral rib pipe shall be fabricated using corrugated profiles and continuous helical pitches as shown in Table 207-11.3.3 (A).

TABLE 207-11.3.3 (A)

Diameters inches (Millimeters)	Nominal Pitch[1] inches (Millimeters)	Max. Pitch[1] inches (Millimeters)	Rib Width inches (Millimeters)	Min. Depth inches (Millimeters)	Seam Pitch[1] inches (Millimeters)
6 thru 12 (150 thru 300)	1-1/2 (38)	1-7/8 (48)	—	1/4 (6.5)	12 (305)
12 thru 96 (300 thru 2400)	2-2/3 (68)	2-7/8 (73)	—	1/2 (13)	24 (610)
42 thru 120 (1050 thru 3000)	3 (75)	3-1/4 (83)	—	1 (25)	21 (533)
42 thru 120 (1050 thru 3000)	5 (125)	5-1/4 (133)	—	1 (25)	29-1/2 (749)
18 thru 72 (450 thru 1800)	11-1/2 (292)	11-3/4 (298)	3/4 (19)	1 (25)	—

1. Pitch shall be measured at right angles to the direction of the corrugation.

For spiral rib pipe and spiral rib pipe arches, the corrugations shall be essentially rectangular ribs projecting outwardly from the pipe wall. The width of the ribs shall be 3/4 inch (19mm) plus 2 times the wall thickness ± 1/8 inch (3mm). The height of the ribs shall be a minimum of 0.90 inch (23mm). The radius bend of the metal at the corners of the ribs shall be not less than 0.10 inch (2.55mm) or greater than 0.17 inch (4.30mm). All rib measurements will be made outside to outside except for the inside radius bend. The rib spacing shall be 11-1/2 inches (292mm) ± 1/2 inch (13mm) center to center.

b) **Continuous Lock Seam Pipe.** Pipe shall be fabricated with the lock seam parallel to the corrugation, and may be used for full circle and equivalent pipe-arch sizes. The lock seam shall be formed in the tangent element of the corrugation profile with its center near the neutral axis, and shall meet the following requirements:

1) The edges of the sheets within the cross-section of the lock seam shall lap at least 5/32 inch (4mm) for the pipe 10 inches (250mm) or less in diameter and at least 5/16 inch (8mm) for pipe greater than 10 inches (250mm) in diameter, with an occasional tolerance of minus 10 percent of lap width allowable.
2) The lapped surface shall be in tight contact.
3) There shall be no excessive angularity on the interior of the 180-degree fold of metal at the lock seam which will cause any visual cracks in the sheet. Roller indentations shall not cause any cracks in the sheet or a loss of metal-to-metal contact within the seam.
4) Tensile specimens cut from production pipe normal to and across the lock seam shall develop the strength as tabulated in Table 207-11.3.3 (B).
5) Continuous lock seam pipe shall be sampled and tested in accordance with AASHTO T 249.

TABLE 207-11.3.3 (B)

Pipe Sheet Thickness inches (mm)	Gage No. (U.S. Std.)	Minimum Lock Strength lb/in (kN/m) of width
0.064 (1.63)	16	425 (74)
0.079 (2.01)	14	650 (114)
0.109 (2.77)	12	875 (153)
0.138 (3.51)	10	1,100 (193)
0.168 (4.21)	8	1,500 (263)

c) **Continuous Welded Seam Pipe.** Pipes fabricated with a continuous helical weld seam parallel to the corrugations may be used for full circle and equivalent pipe arch sizes. The welding process shall be so controlled that the combined width of the weld and adjacent spelter coating burned by the welding does not exceed three times the thickness of the metal. If the spelter is burned outside of these limits, the weld and burned spelter shall be treated as specified in 207-11.4. When pipe is fabricated with aluminum coated sheets, the weld seam shall be metalized during the fabrication of the pipe with an aluminum alloy wire to completely cover and protect the damaged coated area. Testing for welded seam quality control shall conform to AASHTO T241. This test may be waived for pipe manufactured with factory reformed ends.

207-11.4 Repair of Damaged Galvanizing or Aluminizing. When the metallic coated surface has been burned by gas or arc welding, all surfaces of the welded connections shall be thoroughly cleaned by wire brushing and all traces of the welding flux and loose or cracked zinc or aluminum removed, after which the areas shall be repaired as specified in 210-3.5. In addition, damaged aluminized surfaces shall be given an asphalt mastic coating per 207-11.5.2.

207-11.5 Coatings, Linings, and Pavings.

207-11.5.1 General. When required by the Special Provisions, corrugated steel pipe, coupling bands, and fittings shall be coated, lined, or paved. Each section of pipe and fittings shall have the nominal metal thickness painted on the inner surface so that the metal thickness can readily be identified. Coupling bands need not be coated on the interior surface.

207-11.5.2 Coatings.

a) **General.** Coatings shall be applied to one or both pipe surfaces to a minimum thickness as specified herein. Any appearance of pinholes, blisters, cracks, excessive whiskering (stalactites), or lack of bond shall be cause for rejection.

b) **Bituminous Coatings.** Hot-applied bituminous coating shall be applied to both surfaces and shall conform to AASHTO M190, Type A/ASTM A849, Class B. Minimum coating thickness shall be 0.050 inch (1.3mm). Where concrete lining or paving is to be placed on aluminum coated steel, that portion of the conduit in contact with the concrete shall be coated with asphalt mastic coating per 207-11.5.2 prior to placement of the concrete. With the approval of the Engineer, the Contractor or manufacturer may add (anti-corrosion) additives to the concrete mix in lieu of asphalt coating.

c) **Asphalt Mastic Coating.** Cold-applied asphalt mastic coating shall conform to the requirements of AASHTO M243, except that asbestos fiber shall not be used. The asphalt mastic material shall be applied to the surface specified to a minimum thickness of 0.050 inch (1.3mm).

d) **Polymeric Coating.** The polymeric coating shall comply with the requirements of ASTM A849 Class P, and may be plant applied after pipe fabrication. Polymeric coatings shall be applied to the surfaces specified to a minimum thickness of 0.010 inch (0.25mm).

207-11.5.3 Linings.

a) **General.** Linings shall be applied to the interior of the pipe to produce a smooth surface.

b) **Bituminous Linings.** Hot-applied bituminous linings shall conform to the requirements of AASHTO M190, Type D/ASTM A849, Class M. Minimum lining thickness shall be 1/8 inch (3.2mm) over the tops of the crests of the inside corrugations.

c) **Concrete Lining.** The concrete lining shall be plant applied to produce a homogeneous, non-segregated, smooth lining with a minimum thickness of 1/8 inch (3mm) over the crest of the inside corrugations, in accordance with ASTM A849, Class C.

The concrete shall comply with 201-5, except that sand gradation requirements will be waived. Fly ash in accordance with 201-1.2.5 is permitted. The 28-day strength of the concrete shall be a minimum of 5,000 psi (34.5MPa). Cracks in the lining greater than 1/16 inch (1.5mm) shall be repaired in accordance with 207-3.3. Interior joints shall be mortared in accordance with 306-1.2.4.

207-11.5.4 Pavings.

a) **General.** Paving material shall cover at least 25 percent of the circumference of round pipe and 40 percent of the circumference of pipe arches and shall be placed symmetrically, left and right of the bottom centerline of the pipe.

b) **Bituminous Paving.** Hot-applied bituminous paving shall conform to the requirements of AASHTO M190, Type C/ASTM A849, Class M.

c) **Concrete Paving.** The concrete paving shall be Class 600-E-3250 (360-E-23). Steel reinforcing, when specified, shall conform to 201-2. The concrete shall be placed to the thickness specified, after the pipe has been backfilled.

207-11.6 Nestable Pipe. Nestable pipe shall be fabricated in two separate half-circle sections and the sections shall be firmly joined together, all in accordance with Federal Specification MIL-P-236E. The longitudinal joint of the nestable pipe sections may be either Type I, flanged, or Type II, notched, as specified in MIL-P-236E.

207-11.7 Slotted Pipe. The corrugated steel pipe used in the slotted drain shall meet the requirements of AASHTO M36/ASTM A760/A760M and shall be galvanized or aluminized coated (AL-T-2) steel. The diameter and gage shall be as shown on the Plans.

The grates shall be manufactured from ASTM A570/A570M, Grade 36 steel. The spacers and bearing bars (sides) shall be 3/16 inch (4.763mm) material ± 0.0075 inch (0.190mm).

The spacers shall be on 6-inch (152mm) centers and welded on both sides to each bearing bar (sides) with four 1-1/4 inch (32mm) long, 3/16 inch (5mm) fillet welds on each side of the bearing bars.

The Engineer may require tensile strength tests on the grate if the grate is not in compliance with the above. If tensile strength tests are required, minimum results for an in-place spacer pulled perpendicular to the bearing bar shall be:

T = 12,000 pounds (53.4 kN) for 2-1/2 inch (64mm) grate

15,000 pounds (T = 66.7 kN) for 6-inch (152mm) grate

The grates shall be trapezoidal with a 1-3/4 inch (44mm) opening in the top and 30-degree slanted spacers unless shown otherwise on the Plans. If vertical grates are shown on the Plans, the grates shall be vertical (straight sides) with a 1-3/4-inch (44mm) opening in the top and 30-degree slanted spacers. The grate shall be 2-1/2 inches (64mm) or 6 inches (152mm) high as shown on the Plans.

The grate shall be galvanized in accordance with ASTM A123 except with a 2 oz/ft^2 (605 g/m^2) galvanized coating.

The grate shall be fillet welded with a minimum 1 inch (25mm) long weld to the corrugated steel pipe on each side of the grate at every other corrugation.

The corrugated steel pipe shall have a minimum of two re-rolled annular ends. The slotted drain bands shall be a modified Hugger band to secure the pipe and prevent infiltration of the backfill. When the slotted drain is banded together, the adjacent grates shall have a maximum 3-inch (76mm) gap.

207-11.8 Underdrains. The pipe for underdrains shall be of the full-circle type and perforations in the pipe may be drilled or punched. The perforations may be located either in the inside crests or in the flat tangent portion of the corrugations, but not in both locations in the same length of pipe.

Steel band couplers shall be as specified in 207-11.2.2. Sleeve type couplings may be substituted for the band couplers. The couplings may be either plastic or galvanized steel, suitable for holding the pipe firmly in alignment without the use of sealing compound or gaskets. The couplings shall not distort under normal conditions of use.

207-11.9 Siphons. Pipe and couplings shall be watertight and shall conform to 207-11.2.

The pipe for siphons shall be of such lengths that the number of field connections will be held to a minimum. The outside seams of riveted, spot welded or continuous helical lock seam pipe shall be continuously soldered. Rivets and spot welds on circumferential seams shall be spaced at approximately 2-1/2 inches (65mm) on center with maximum of 3 inches (76mm) on center.

Prior to backfilling, the pipe for siphons shall be subject to the hydrostatic test as specified in 306-1.4.6.

207-11.10 Pipe Downdrains. Pipe and couplings for downdrains shall be watertight in accordance with 207-11.2. When required by the Plans or the Special Provisions, the Engineer may require testing in accordance with 306-1.4.6.

Joints for downdrains shall comply with the following joint property requirements:

TABLE 207-11.10 (A): TENSILE STRENGTH (PULL APART STRENGTH)

0 thru 42 inches (0 thru 1050mm) pipe diameter	5,000 lbs (22.3 kN)
42 thru 84 inches (1050 thru 2200mm) pipe diameter	10,000 lbs (44.5 kN)

207-12 STRUCTURAL STEEL PLATE PIPE AND ARCHES.

207-12.1 General. The sizes, thicknesses, and dimensions of structural steel plate pipe, arches, and pipe arches shall be as designated on the Plans or in the Special Provisions.

207-12.2 Materials.

207-12.2.1 General. Steel sheets, plates, bolts, and nuts shall conform to AASHTO M 167. In accordance with the requirements of Section 4, the fabricator of the structural steel products shall furnish to the Engineer a Certificate of Compliance stating that the materials furnished comply with these requirements. Galvanized surfaces which are damaged shall be repaired in accordance with 210-3.5.

207-12.2.2 Identification. When plates of 2 dissimilar thicknesses are involved in one cross-section of an installation, the thickness of structural steel plates will be identified on the Plans, or in the proposal in accordance with the following:

Each installation will be designated by size and symbol indicating the number and thickness of plates required. Thus, "(4x0.109-1x0.168)" ["(4x2.77-1x4.27)"]) will be used to designate an installation for one plate length composed of four 0.109 inch (2.77mm) thick plates and one 0.168 inch (4.27mm) thick plate, with the heaviest thickness to be placed in the invert. This designation does not prevent the Contractor from using fewer or more plates, provided the minimum thickness requirements are met.

207-12.3 Distorting. When required by the Plans or the Special Provisions, circular structural steel plate pipe shall be distorted, either at the fabricating shop or in the field.

Plates distorted in the fabricating shop shall provide an increase in the vertical diameter of the pipe, after assembly, of approximately 5 percent for the full length. Plates shall be marked to assure that they will be placed in proper position.

The method for distorting plates in the field shall conform to details shown on the Plans and to 304-4.

207-12.4 Coatings. When required by the Plans or the Special Provisions, a protective coating shall be applied to structural steel plate pipe and arches, including all appurtenant fasteners and fittings. Bituminous coating shall conform to 207-11.5.2. Asphalt mastic coating shall conform to 207-11.5.4. Asphalt mastic coating may be field applied after assembly of the structure.

When bituminous coatings are shop applied, each plate shall have the plate thickness on the inner surface so that the plate thickness can be readily identified.

207-13 CORRUGATED ALUMINUM PIPE AND PIPE ARCHES.

207-13.1 General. Corrugated aluminum pipe, pipe arches, and connectors shall be manufactured and inspected in conformance with the requirements of AASHTO M 196, M 197, and as specified herein. The size, type, and gage of the pipe to be furnished shall be as shown on the Plans or in the Special Provisions.

Corrugated aluminum pipe arches shall consist of corrugated aluminum pipe which has been re-formed to multicentered pipe having an arch-shaped top with a slightly curved integral bottom. The minimum radius of any part of the pipe section shall be not less than 3 inches (76mm).

The specifications contained herein for pipe shall also apply to pipe arches.

Nominal diameter, as referred to in AASHTO M 196 and M 197, shall be defined as meaning the minimum inside dimension of the pipe.

207-13.2 Materials.

207-13.2.1 General. Corrugated aluminum sheets covered by this section shall be fabricated from Alloy Alclad 3004 with Temper H-34 and shall conform to ASTM B209/B209M and the following mechanical properties:

TABLE 207-13.2.1 (A)

Thickness inches (millimeters) Property	0.051 to 0.113 (1.30 to 2.87)	0.114 to 0.249 (2.88 to 6.32)
Tensile Strength (minimum)	31,000 psi (215MPa)	31,000 psi (215MPa)
Yield Strength (2% offset) (minimum)	24,000 psi (165MPa)	24,000 psi (165MPa)
Elongation in 2 inches (50mm)	4% min.	5% min.

The gages and thickness referred to in this specification are as follows:

TABLE 207-13.2.1 (B)

Thickness inches (mm)	Gage Number (U.S. Standard)
0.060 (1.52)	16
0.075 (1.91)	14
0.105 (2.67)	12
0.135 (3.43)	10
0.165 (4.19)	8

The cladding thickness shall be 5 percent of the total composite thickness.

Rivets shall conform to ASTM B316M/B316 for Alloy 6053 with Temper T-4 and the following physical properties:

TABLE 207-13.2.1 (C)

Tensile Strength (minimum)	25,000 psi (170MPa)
Yield Strength (2% Offset) (minimum)	14,000 psi (95MPa)
Shear Strength (minimum)	15,000 psi (105MPa)
Elongation in 2 inches (50mm) (minimum)	16%

The corrugations for all pipe, measured at right angles to the direction of the corrugation, shall conform to the dimensions in the following table:

TABLE 207-13.2.1 (D)

Diameter[1] inches (mm)	Width[2] inches (mm)	Depth (Min.)[2] inches (mm)	Type of Corrugation
8 and larger (200 and larger)	2-2/3 (68)	1/2 (13)	Annular
36 and larger (900 and larger)	3 (76)	1 (25)	Annular
6 through 10 (150 through 250)	1-1/2 (38)	1/4 (6)	Helical
12 and larger (300 and larger)	2-2/3 (68)	1/2 (13)	Helical
36 and larger (900 and larger)	3 (76)	1 (25)	Helical

1. Inside diameter of pipe shall not vary more than the following:
   6 inch through 18 inch (150mm through 450mm) in diameter, 1/4 inch (6mm) maximum
   21 inch through 24 inch (525mm through 600mm) in diameter, 3/8 inch (9.5mm) maximum
   24 inch (600mm) and over in diameter, 1/2 inch (13mm) maximum.
2. Minimum and maximum width and depth of corrugation shall conform to AASHTO M196 and M197.

The lengths and thickness of sheet, width of laps, and computed weight per lineal foot of finished corrugated aluminum pipe and arches shall be as specified in Table 4 of AASHTO M196. The dimensions of the corrugated aluminum pipe arch shall be as shown in Table 7 of AASHTO M196.

207-13.2.2 Connecting Bands. The connecting bands shall conform to the requirements of AASHTO M196, except the minimum width of band for 12-inch (300mm) and larger pipe shall be 12 inches (300mm). Minimum width of band for pipe less than 12 inches (300mm) shall be 7 inches (175mm) and minimum width of band for 1 inch x 3-inch (25mm x 75mm) corrugations shall be 14 inches (355mm). The base metal of connecting bands shall be the same base metal as that of the pipe. The gage of the connecting bands for corrugated pipe and pipe arches may be 2 standard-use thicknesses lighter than that used for the pipe, but not less than 0.060 inch (1.52mm) thick.

The band couplers shall be connected with galvanized steel bolts of not less than 1/2 inch (13mm) diameter.

207-13.2.3 End Finish. The ends of 0.060 inch (1.52mm) and 0.075 inch (1.91mm) thickness installations shall be reinforced where shown on the Plans. The reinforcement shall consist of an aluminum band of at least 0.135 inch (3.43mm) material at least 6 inches (150mm) wide, on at least the outer 12 inches (300mm) of 0.060 inch (1.52mm) and 0.075 inch (1.91mm) pipe shall be at least 0.075 inch (3.43mm) material.

207-13.3 Fabrication.

207-13.3.1 General. At the option of the Contractor, corrugated aluminum pipe may be fabricated by riveting or by using a helically corrugated metal pipe with a continuous helical lock seam paralleling the corrugation.

207-13.3.2 By Riveting. Pipe fabricated by riveting shall conform to AASHTO M 196.

207-13.3.3 By Continuous Lock Seam. Pipe fabricated with a continuous helical lock seam parallel to the corrugations shall conform to the requirements of AASHTO M 196 and as specified herein.

207-13.4 Underdrains. Pipe for underdrains shall be of the full circle type and perforations in the pipe may be drilled or punched in accordance with AASHTO M 196. The perforations may be either in the inside crests or in the flat tangent portion of the corrugations but not in both locations in the same length of pipe.

Sleeve-type couplings may be substituted for the band couplers. The couplings may be either plastic or aluminum, suitable for holding the pipe firmly in alignment without the use of sealing compound or gaskets. The couplings shall not distort under normal conditions of use. Minimum thickness for underdrains shall be 0.060 inch (1.52mm).

207-13.5 Pipe Downdrains. Joints for downdrains shall be watertight in accordance with 207-11.2.2. When required by the Special Provisions, the Engineer may require testing in accordance with 306-1.4.6.

Joints for downdrains shall comply with the following joint property requirements:

TABLE 207-13.5 (A): TENSILE STRENGTH (PULL APART STRENGTH)

0 to 42 inches (0 to 1050mm) pipe diameter	5,000 pounds (22.3 kN)
45 to 84 inches (1125 to 2100mm) pipe diameter	10,000 pounds (44.5 kN)

207-13.6 Bituminous Coating. When required by the Plans or Special Provisions, pipe and connecting bands shall be protected, both inside and outside, with a bituminous coating. The bituminous coating shall conform to the requirements of AASHTO M190 and as hereinafter specified.

The minimum thickness of bituminous material for all coated pipe measured on the crest of the corrugations shall be 0.05 inch (1.3mm).

In paving the invert of aluminum pipes, the bituminous material shall cover the crests to a minimum depth of 1/8 inch (3mm) and the width of paving shall at least cover one third of the periphery of pipe arches and one quarter of the periphery of circular pipes.

The bituminous material shall conform to Paragraph 4 of AASHTO M190, except that it shall be at least 90 percent soluble in cold carbon disulfide.

When corrugated aluminum pipes are to have a bituminous coating, the fabrication requirements specified in AASHTO M196 shall be altered so that the rivet heads inside the pipe will be in the valley of the corrugations.

Damaged bituminous coatings shall be repaired by the Contractor at its expense by applying bituminous material conforming to the provisions of this subsection.

207-13.7 Repair of Damaged Alclad Coating. Alclad coatings which have been damaged shall be repaired. The damaged area shall be thoroughly cleaned by wire brushing. The cleaned area shall be painted with 2 coats of zinc oxide-zinc dust paint conforming to the requirements of Military Specification MIL-P-15145. The paint shall be properly mixed with a suitable vehicle in the ratio of one part zinc oxide to four parts zinc dust by weight.

207-14 STRUCTURAL ALUMINUM PLATE PIPE AND ARCHES.

207-14.1 General. The sizes, thicknesses, and dimensions of structural aluminum plate pipe, arches, and pipe arches shall be as designated on the Plans or in the Special Provisions.

207-14.2 Materials.

207-14.2.1 General. Aluminum sheets, plates, bolts, and nuts shall conform to AASHTO M 219. The fabricator of the structural aluminum products shall furnish to the Engineer a Certificate of Compliance stating that the materials furnished comply with these requirements. Each plate shall be marked on the inner surface so that the plate thickness can be readily identified after installation.

207-14.2.2 Identification. When plates of 2 dissimilar thicknesses are involved in one cross-section of an installation, the thickness of structural aluminum plates will be identified in accordance with 207-12.2.2.

207-14.2.3 Distorting. When required by the Plans or the Special Provisions, circular structural aluminum plate pipe shall be distorted, either at the fabricating shop or in the field.

Plates distorted in the fabricating shop shall provide an increase in the vertical diameter of the pipe, after assembly, of approximately 5 percent for the full length. Plates shall be marked to ensure that they will be placed in proper position.

Distorting of plates in the field shall conform to 304-4.3.

207-14.2.4 Coatings. When required by the Plans or the Special Provisions, a protective coating shall be applied to structural aluminum plate pipe and arches, including all appurtenant fasteners and fittings. Bituminous coating shall conform to 207-11.5.2 b). Asphalt mastic coating shall conform to 207-11.5.2 c). Asphalt mastic coating may be field applied after assembly of the structure.

207-15 ABS SOLID WALL PIPE.

207-15.1 General. This subsection applies to ABS plastic solid wall pipe for use as sanitary sewers, storm drains, and house connection sewers. Pipe, fittings, and joints shall comply with ASTM D2751 except as modified herein. Minimum wall thickness shall correspond with SDR 35. Joints for sanitary sewers, except for house connection sewers, shall be gasketed joints.

Joint solvent cement shall be an ABS cement conforming to ASTM D2235. Gaskets shall conform to the requirements of 208-4.

207-15.2 Material Composition and Testing. The pipe and fittings shall be made of ABS plastic which shall meet the minimum cell classification of 1-3-3, 3-2-2, or 2-2-3 as defined in ASTM D1788 and having the chemical composition as follows:

Acrylonitrile-butadiene-styrene (ABS) pipe - plastics containing polymers or blends of polymers, or both, in which the minimum butadiene content is 6 percent, the minimum acrylonitrile content is 15 percent, the minimum styrelene or substituted styrene content or both, is 15 percent and the maximum content of all other monomers is not more than five parts by weight per 100 parts of ABS resin. Additives and fibers, including but not limited to stabilizers, antioxidants, colorants, etc., shall not exceed 10 parts by weight per 100 parts of ABS resin.

Material shall meet or exceed the values and properties below:

TABLE 207-15.2 (A)

Property	ASTM Test Method	Minimum Value Based on Cell Classification		
		1-3-3	3-3-2	2-2-3
Izod impact @ 23° ± 2°C foot-lb/inch (J/m) of notch	D256	1 (53)	3 (160)	2 (107)
Deflection Temperature under load 264 psi (1820 kPa)	D648			
°F		190	180	190
°C		87	82	87
Tensile stress at yield , psi (MPa)	D638	5000 (34.5)	4000 (27.6)	5000 (34.5)
Specific Gravity:	D792			
Minimum		1.0	1.0	1.0
Maximum		1.2	1.2	1.2

207-15.3 Chemical Resistance (Pickle Jar Test). The Engineer may at any time direct the manufacturer to obtain samples of the compound and to prepare test specimens in accordance with ASTM D1987. These specimens shall comply with the minimum property values shown below and also with the applicable ASTM requirements.

Tensile, impact and weight change exposure specimens shall be immersed in the solutions specified in Table 211-2 (A) for a period of 112 days. At 28-day intervals, selected specimens shall be removed, washed, surface dried, and tested.

The weight change specimens shall be 2 inches (50mm) in diameter and may be molded discs or discs cut from the pipe wall. They shall be conditioned in a mechanical convection oven for 7 days at 110° ± 4°F (43° ± 2°C), then cooled in a desiccator for 3 hours at 73°± 4°F (23° ± 2°C), weighed, and then immersed in the solutions in Table 211-2 (A). At 28-day intervals, selected specimens shall be removed, washed, surface dried, and weighed. These same specimens shall be reconditioned in a mechanical convection oven for 7 days at 110° ± 4°F (43° ± 2°C), then cooled in a desiccator for 3 hours at 73°± 4°F (23° ± 2°C) and weighed again. If any specimen fails to meet these requirements at any time, the material will be rejected.

TABLE 207-15.3 (A)

Property	ASTM Test Method	Initial Values	Values After 112-Day Exposure
Tensile stress at yield, psi (MPa), minimum	D 638	5,000 (34.5)	5,000 (34.5)
Impact Strength foot-lbs/inch (J/m) of notch, minimum	D 256 Method A 1/2 x 1/8 x 2-1/2" (Size 12.7 x 3.17 x 63.5mm)	2 (107)	2 (107)
Weight Change	D 543		
Unconditioned			± 1.5% maximum
Conditioned			± 1.0% maximum

207-15.4 Pipe Acceptance. At the time of manufacture each lot of pipe and fittings shall be inspected for defects, and tested for impact, stiffness, and flattening in accordance with ASTM D2751.

When testing subsequent to manufacture, the impact requirement shall be excluded. For the flattening requirement, the percentage reduction in pipe diameter shall be not less than 15 percent for pipe marked SDR 23.5 or lower, and not less than 25 percent for pipe marked with higher SDR numbers. The stiffness requirement is unchanged.

A pipe lot shall consist of all pipe having the same marking number. The lot test specimen shall have a minimum length of 4 feet (1.2m).

207-15.5 Marking. Pipe shall have a home mark to indicate full penetration of the spigot when the joint is made. Pipe shall be marked at 5-foot (1.5m) intervals or less with a marking number which identifies the manufacturer, SDR, size, machine, date, and shift on which the pipe was produced.

207-15.6 Installation. Bedding shall conform to 306-1.2.1. Backfill shall conform to 306-1.3. Field inspection shall conform to 306-1.2.12.

207-15.7 Installation Time Limit. The Contractor shall retest within 60 days prior to the installation of all pipe and fittings that are more than 180 days old from the date of manufacture to ensure compliance with the requirements of the Specifications. The Contractor shall not install any pipe that is more than 2 years old from the date of manufacture.

207-16 ABS OR PVC COMPOSITE PIPE.

207-16.1 General. This subsection applies to ABS or PVC composite pipe for use as sanitary sewers, storm drains, and house connection sewers. Pipe, fittings, and joints shall comply with ASTM D2680, except as modified herein.

The pipe shall consist of two concentric extruded thermoplastic tubes integrally connected by webs to form a circular truss. The longitudinal void spaces shall be filled with inert material. The maximum average ID of the pipe, as determined by ASTM D2122, shall be:

TABLE 207-16.1 (A)

Nominal Size inches (millimeters)	Max. Average ID inches (millimeters)
6 (150)	5.81 (148)
8 (200)	7.90 (201)
10 (250)	9.88 (251)
12 (300)	11.83 (301)
15 (375)	14.80 (376)

Joint solvent cement shall be in accordance with 207-15.1 for ABS and 207-17.3.3 for PVC.

207-16.2 Material Composition and Testing. The ABS resin compound used in the manufacture of ABS composite pipe shall conform to the requirements of 207-15.2. The PVC resin shall conform to the requirements of 207-17.2.2.

207-16.3 Chemical and Physical Testing. ABS resins shall conform to 207-15.3. PVC resins shall conform to 207-17.5.

207-16.4 Pipe Acceptance. Each lot of pipe and fittings shall be inspected for defects and tested for stiffness and deflection in accordance with ASTM D2680. Installation time shall conform to 207-15.7.

A pipe lot shall consist of all pipe having the same marking number. The lot test specimen shall be a minimum length of 4 feet (1.2m).

207-16.5 Marking. Pipe shall have a home mark to indicate full penetration of the spigot when a joint is made. Pipe shall be marked at 5-foot (1.5m) intervals or less with a marking number which identifies the manufacturer, size and machine, date, and shift on which the pipe was made.

207-16.6 Repair. There shall be no discontinuity of the pipe inner wall. Ruptures in the pipe outer wall may be repaired if the damage is limited to an area that can be encompassed by a 3-inch (75mm) diameter circle superimposed over the damage. Cell filler repair is unnecessary. A solvent welded repair patch of the same material as the pipe, at least equal to the thickness of the pipe outer wall, shall extend at least 1 inch (25mm) beyond the damage. When damage exceeds these limits, the damaged section shall be cleanly removed.

207-16.7 Installation. Bedding shall conform to 306-1.2.1. Backfill shall conform to 306-1.3. Field inspection shall conform to 306-1.2.12.

207-17 PVC PLASTIC PIPE.

207-17.1 General. This subsection applies to the requirements for unplasticized PVC plastic pipe for sanitary sewers, storm drains, and house connection sewers. Pipe, fittings, couplings, and joints shall conform to the requirements listed in Table 207-17.1 (A), except as otherwise modified by the Plans or the Special Provisions.

TABLE 207-17.1 (A)

Nominal Size inches (mm)	ASTM	Wall Thickness Min.
4 - 15 (100 - 375)	D 3034	SDR 35
18 - 30 (450 - 750)	F 679	"T-1" only

Joints for sanitary sewers, except house connection sewers, shall be gasketed joints.

207-17.2 Manufacturing Requirements.

207-17.2.1 Identification Marks. All pipe, fittings, and couplings shall be clearly marked at intervals not to exceed 5 feet as follows:

a) Nominal pipe diameter.

b) PVC cell classification.

c) Company, plant, shift, ASTM, SDR, and date designation.

d) Service designation or legend.

For fittings and couplings, the SDR designation is not required.

207-17.2.2 Cell Classification. Pipe shall be made of PVC plastic having a cell classification of 12454 or 13364, as defined in ASTM D1784. The fittings shall be made of PVC plastic having a cell classification of 12454, or 13343. PVC compounds of other cell classifications shall be in conformance with 207-17.5. Additives and fillers, including but not limited to stabilizers, antioxidants, lubricants, colorants, etc., shall not exceed 10 parts by weight per 100 of PVC resin in the compound.

207-17.3 Joining Systems.

207-17.3.1 General. All pipe shall have a home mark on the spigot end to indicate proper penetration when the joint is made.

The socket and spigot configurations for the fittings and couplings shall be compatible to those used for the pipe.

207-17.3.2 Elastomeric Gasket Joints. Pipe with gasketed joints shall be manufactured with a socket configuration which will prevent improper installation of the gasket and will ensure that the gasket remains in place during the joint operation. The gasket shall be manufactured from a synthetic elastomer and shall conform to the requirements of 208-1.2 and 208-4.

207-17.3.3 Solvent Cement Joints. Pipe with solvent cement joints shall be joined with PVC cement conforming to ASTM D2564.

207-17.3.4 Injection Sealed Joints. Pipe with injection sealed joints shall be sealed with a PVC adhesive compound. The compound shall conform to the requirements of ASTM D2564 and shall have a minimum viscosity of 50,000 centipoise (50 pascal-second). The internal diameter of the socket shall be uniform with a locking taper at the base and an outer seal ring attached to the end. The socket shall have an injection port to inject the adhesive and an exhaust port on the opposite side to allow air to escape from the annular space.

207-17.4 Test Requirements.

207-17.4.1 General. Pipe, fittings, and couplings shall meet the requirements of the section titled "Requirements" of ASTM D3033, D3034, or F679 ("T-1" wall only). During production of the pipe, the manufacturer shall perform the specified tests for each pipe marking. A Certificate of Compliance by the manufacturer indicating compliance with specification requirements shall be delivered with the pipe. The Certificate of Compliance shall include the test result data. The PVC compound shall also meet the chemical resistance requirements of 207-17.5.

207-17.4.2 Acceptance. The basis for acceptance will be the inspection of pipe, fittings, and couplings; the tests specified in 207-17.4.1; and compliance with the Specifications. When the pipe is delivered to the Work site, the Engineer may require additional testing to determine conformance with the requirements of pipe flattening, impact resistance, pipe stiffness, and extrusion quality. Installation time shall conform to 207-15.7.

207-17.4.3 Selection of Test Pipe. When testing is required by the Engineer, one test pipe shall be selected at random by the Engineer from each 1,200 feet (370m) or fraction thereof of one test pipe per lot. A lot shall be defined as pipe having the same identification marking. The length of specimen for each selected pipe shall be a minimum of 8 feet (2.4m).

207-17.5 Chemical Resistance and Physical Testing. PVC shall be tested in accordance with 211-2 and conform to Table 207-17.5(A).

TABLE 207-17.5 (A)

Property	ASTM Test Method	Values (Initial and After 112-Day Exposure)		
		Cell Classification		
		12454	13343	13364
Tensile stress at yield, psi (MPa), min.	D 638	7000 (48.3)	6000 (41.4)	6000 (41.4)
Impact Strength ft-lbs/inch (J/m) of notch, min.	D 256 Method A Size 1/2 x 1/8 x 2-1/2" (12.7 x 3.17 x 63.5mm)	0.65 (35)	1.5 (80)	1.5 (80)
Weight Change	D 543			
Unconditioned		± 1.5% max	± 1.5% max	± 1.5% max
Conditioned		± 1.0% max	± 1.0% max	± 1.0% max

207-17.6 Installation. Bedding shall conform to 306-1.2.1. Backfill shall conform to 306-1.3. Field inspection shall conform to 306-1.2.12.

207-18 ANNULAR HIGH-DENSITY POLYETHYLENE PIPE WITH SMOOTH INTERIOR, CORRUGATED EXTERIOR, WITH BELL AND SPIGOT JOINTS (TYPE S).

207-18.1 General. These specifications apply to high density polyethylene, annular corrugated pipe and applicable fittings and coupling systems intended for use in the construction of gravity flow storm drains, culverts, and subsurface drains. The size, type, and material properties of the pipe to be furnished shall be as shown on the Plans or as specified in the Special Provisions.

207-18.2 Materials.

207-18.2.1 Extruded Pipe and Fittings and Blow Molded Fittings. Pipe and fittings shall be made of virgin HDPE that conforms to the requirements of AASHTO M294. The residue from ignition shall not exceed 30 percent as determined by ASTM D2584 except that the furnace temperature shall be 840° +/- 45° F (450° +/- 25° C).

207-18.2.2 Reworked Plastics. In lieu of virgin HDPE, the manufacturer may use clean reworked plastics, provided that it meets the cell class requirements specified in AASHTO M294. Reworked plastics shall be from a processor's own production that has been reground, palletized, or solvated after having been previously processed by molding, extrusion, etc.

207-18.3 Nominal Diameter, Wall Thickness, and Pipe Stiffness. The inside diameter, minimum wall thickness and minimum pipe stiffness shall be as specified in Table 207-18.3(A). The nominal diameter for each pipe shall be determined in accordance with ASTM D2122. The diameter shall not vary from a true circle by more than 2 percent of the nominal diameter. The wall thickness and inside diameter for each production pipe shall be determined in accordance with AASHTO M294. Wall thickness of Type S corrugated polyethylene pipe shall be the thickness of the inner liner measured between corrugated valleys. Pipe stiffness values shall be determined in accordance with 207-18.5.2.

TABLE 207-18.3(A)

Diameter in (mm)	Minimum Wall Thickness in (mm)	Minimum Pipe Stiffness psi (kPa)
12 (300)	0.035 (0.89)	50 (345)
15 (375)	0.039 (1.00)	42 (290)
18 (450)	0.050 (1.27)	40 (275)
24 (600)	0.059 (1.50)	34 (235)
30 (750)	0.059 (1.50)	28 (195)
36 (900)	0.069 (1.70)	22 (150)
42 (1050)	0.070 (180)	20 (140)
48 (1200)	0.070 (180)	18 (125)

Tolerance on the above specified inside diameter shall be 4.5 percent oversize and 1.5 percent undersize, but not more than 1-1/2 inch (37mm) oversize.

207-18.4 Joints, Fittings, and Connections.

207-18.4.1 Water-Tight Joints. Pipe shall be joined with bell and spigot joints and unless otherwise shown on the Plans or specified in the Special Provisions, be water-tight. Pipe shall incorporate a gasket conforming to 208-4. Water-tight joints shall be tested at a pressure of at least 11 psi (74kPa) in conformance with ASTM D3212. The manufacturer shall clearly identify watertight joints on the pipe delivered to the Work site.

207-18.4.2 Silt-Tight Joints. Pipe shall be joined with bell and spigot joints and shall at a minimum be silt-tight and leak resistant. The pipe shall incorporate a gasket conforming to 208-4. Silt-tight joints shall conform to ASTM D3212, modified to a test pressure of at least 2 psi (14kPa).

207-18.4.3 Fittings. Corrugated fittings include in-line joint fittings, such as couplings and reducers, and branch or complimentary assembly fittings such as tees, wyes, and end caps. Fittings shall be installed in accordance with the Plans and Special Provisions.

a) Fittings shall not reduce or impair the overall integrity or function of the pipe.

b) Fittings shall not reduce the inside diameter of the pipe being joined by more than 1/2 inch (12mm). Reducer fittings shall not reduce the cross-sectional area of the smaller size.

c) Fabricated fittings shall be welded on the interior and exterior of all junctions.

d) Fittings shall conform to 207-18.4.2 unless otherwise specified in the Special Provisions.

207-18.5 Testing.

207-18.5.1 General. For testing purposes a production lot size shall consist of all pipe having the same lot marking number, but shall not exceed a total of 50 lengths per day.

207-18.5.2 Pipe Stiffness. A minimum of 3 pipe specimens per production run shall be tested for pipe stiffness (PS) in accordance with ASTM D2412 modified as follows:

a) The test specimens shall be a minimum of one diameter length.

b) Position the first specimen in the loading machine with an imaginary line connecting the two seams formed by the corrugation mold (end view) parallel to the loading planes, when applicable. The specimen shall lie flat on the plate within 1/8 inch (3mm). The specimen may be straightened by hand bending at room temperature to accomplish this. The first location shall be used as a reference point for rotation and testing of the two other specimens. Rotate the other specimens 45 and 90 degrees, respectively, from the reference point. Each specimen shall be tested in one position only.

c) The deflection indicator shall be readable and accurate to +/- 7.9 x 10^{-4} inches (+/- 0.2mm).

d) The beginning point for deflection measurement shall be at a load of 4.5 +/- 1.1 lbs. (20 +/- 5 Newtons). This point shall be considered as the origin of the load deflection curve.

207-18.5.3 Pipe Flattening. The specimens specified in 207-18.5.2 shall be flattened until the vertical inside diameter is reduced by 20 percent. The rate of loading shall conform to 207-18.5.2. The specimens shall not exhibit any cracking, splitting, or delamination. Wall buckling is indicated by reverse curvature in the pipe wall accompanied by a decrease in load carrying ability of the pipe. Any decrease or downward deviation in the pipe stiffness test curve shall be considered a wall buckling point.

207-18.5.4 Brittleness. Test pipe specimens shall be tested in accordance with ASTM D2444, except 6 specimens shall be tested, or 6 impacts shall be made on one specimen. In the latter case, successive impacts shall be separated by 120 +/- 10 degrees for impacts made on one circle, or at least 12 inches (300mm) longitudinally for impacts made on one element. Impact points shall be at least 6 inches (150mm) from the end of the specimen. A tup (type B) shall be used, with a mass of 10 pounds (4.5 kg). The height of drop shall be 10 feet (3.0m). A flat plate specimen holder shall be used. Specimens shall be conditioned for 24 hours at a temperature of 24.8 +/- 1.8° F (–4 +/- 2° C), and all tests shall be conducted within 60 seconds of removal from this atmosphere. The center of the falling tup shall strike on a corrugation crown for all impacts.

207-18.5.5. Environmental Stress Cracking of Finished Pipe. Pipe shall be tested for environmental stress cracking in accordance with ASTM D1693 modified as follows:

a) Three specimens shall be tested.

b) Each specimen shall consist of a 90-degree arc length of pipe.

c) Bend the specimens to shorten the inside chord length 20 +/- 1 percent and retain in this position using a suitable holding device. Determine the arc chord measurement (B) of the specimen under test as follows:

Place the bent specimen in a container of suitable size and cover completely with a preheated wetting agent at 122 +/- 1.8° F (50 +/- 2° C). Maintain this temperature for 24 hours, and then remove the sample and inspect immediately. The wetting agent used in this test shall be 100 percent nonylphenoxy poly (ethyleneoxy) ethanol.

207-18.5.6 Slow Crack Growth Resistance. Basic resin compounds shall be tested for stress crack resistance in accordance with ASTM F2136 modified as follows:

a) The applied stress for the NCLS test shall be 600 psi (4100kPa).

b) The notched depth shall be 20 percent of the nominal thickness of the specimen.

c) The average failure time of the five test specimens shall not exceed 24 hours with no single test specimen's failure time less than 17 hours.

207-18.5.7 Joint Integrity. Joint integrity shall be tested in accordance with the following:

a) Each fitting or coupling shall be assembled in accordance with the manufacturer's recommendations. Use pipe samples at least 12 inches (300mm) in length. Assemble a specimen at least 21 inches (600mm) in length with the connection at the center.

b) Load the connected pipe and fitting between parallel plates at the rate of 1/2 inch (12mm) per minute until the vertical inside diameter is reduced by at least 20 percent of the nominal diameter of the pipe.

c) Inspect for damage while at the specified deflection and after removal of the load.

d) Measure the maximum radial distance between pipe and fittings or between bell and spigot, during test and after the load is removed.

Pipe connections shall not separate to create a gap exceeding 3/16 inch (5mm) when measured in the radial direction between pipe and coupling or between bell and spigot portions of the pipe. The gap measurement shall be taken at the gasket or hinge point. The test shall be conducted on a fully assembled joint, including the gasket. Fittings shall not crack or delaminate.

207-18.5.8 Cell Classification. Cell classification of pipe and fittings shall conform to 207-18.2.1.

207-18.6 Marking. Pipe and fittings shall be clearly marked at intervals of 11.5 feet (3.5m) or less. Markings shall display the manufacturer's name and/or trademark, nominal size, plant designation code, and date of manufacture.

207-18.7 Pipe Acceptance or Rejection. Pipe shall be free of cracks, holes, delaminations, foreign inclusions, blisters, or other defects that would, due to their nature, degree or extent, have a deleterious effect on pipe performance as determined by the Engineer. Prior to installation, damaged pipe shall either be repaired or field cut to remove the damaged portion as approved by the Engineer and in accordance with the manufacturer's recommendations.

207-19 POLYETHYLENE (PE) SOLID WALL PIPE.

207-19.1 General. Polyethylene (PE) plastic solid wall pipe and liner for use in gravity flow sanitary sewers, storm drains, and house connection sewers shall comply with ASTM D3350 or ASTM F714. Unless otherwise specified, pipe shall conform to SDR 21. Fittings shall comply with ASTM D2683 or D3261.

207-19.2 Material Composition. Pipe and fittings shall be made from PE resins complying with ASTM D1248, Type III, Class C, Category 5, Grade P34, and ASTM D3350, and which shall further meet the requirements as listed in the following table:

TABLE 207-19.2 (A)

Property	Value	ASTM Test
Density (g/cm^3)	0.941-0.959	D 1505
Melt Index (g/10 minutes)	0.15 max.	D 1238 cond. E.
Flexural Modulus psi (MPa)	120,000 (827) minimum	D 790
Tensile strength at yield psi (MPa)	3,200 (22.1) minimum	D 638
Elongation at break	800%	D 638
Brittleness temperature °F (°C)	-180° (-118°) max.	D 746
Environmental Stress Crack Resistance F_o (hrs.)[1] Test Condition "C"	192 min.	D 1693
Hydrostatic Design Basic psi (MPa) @ 73°F (23°C)	1600 (11.0)	D 2837
Color	2% Carbon Black min.	

1. F_o indicates no failures.

Additives and fillers including, but not limited to, stabilizers, antioxidants, lubricants, colorants, etc., shall not exceed 5 parts by weight per 100 of PE resin in the compound. The Engineer may require a Certificate of Compliance by the manufacturer that the test results conform to the requirements of the Specifications.

207-19.3 Pipe Acceptance. At the time of manufacture, each lot of pipe, liner, and fittings shall be inspected for defects and tested for Elevated Temperature Sustain Pressure in accordance with ASTM F714. Installation time limit shall conform to 207-15.7.

At the time of delivery, the pipe shall be homogeneous throughout, uniform in color, free of cracks, holes, foreign materials, blisters, or deleterious faults.

For testing purposes, a production lot shall consist of all pipe having the same marking number. It shall include any and all items produced during any given work shift and must be so identified as opposed to previous or ensuing production.

207-19.4 Marking. Pipe shall be marked at 5-foot (1.5m) intervals or less with a coded number which identifies the manufacturer, SDR, size, material, machine, date, and shift on which the pipe was extruded.

At the end of the production shift, during which a production lot has been extruded, the marking code on the pipe shall be changed to indicate that said time intervals have elapsed and that a new production shift has begun.

Fittings shall be marked with the name of the manufacturer or its logo, the size, and the material from which they were molded or fabricated.

207-19.5 Chemical Resistance and Physical Testing. PE resins shall be tested in accordance with 211-2 and conform to Table 207-19.5 (A).

TABLE 207-19.5 (A)

Property	ASTM Test Method	Initial Value	Value After 112 Days Exposure
Tensile stress at yield, psi (MPa) min.	D 638	3,200 (22.1)	3,200 (22.1)
Impact Strength, ft-lb/inch (J/m) min.	D 256 Method A Size 1/2 x 1/8 x 2-1/2" (12.7 x 3.17 x 63.5mm)	3.5 (187)	3.5 (187)
Weight Change	D 543		
Unconditioned			± 1.5% max.
Conditioned			± 1.0% max.

The Engineer may, at any time, direct the manufacturer and/or Contractor to obtain compound samples and to obtain compression molded test specimens in accordance with ASTM D1928. These specimens shall comply with the minimum property values in Table 207-19.5 (A).

207-19.6 Installation. Bedding shall conform to 306-1.2.1. Backfill shall conform to 306-1.3.

207-20 CENTRIFUGALLY CAST FIBERGLASS REINFORCED PLASTIC MORTAR (CCFRPM) PIPE.

207-20.1 General. These specifications apply to CCFRPM pipe to be used for the construction of direct bury gravity sanitary sewers. storm drains, and related structures. All pipes. joints, and fittings shall conform to ASTM D3262. Unless otherwise specified, the minimum pipe stiffness shall be 46 psi (318 kPa) when tested in accordance with ASTM D2412. The size, type, and stiffness of the pipe to be furnished shall be as shown on the Plans or in the Special Provisions.

207-20.2 Materials. The amount, location, and orientation of the chopped glass-fiber reinforcement shall be specifically designed for each application. The glass shall be a commercial grade of E-type glass fibers with a finish compatible with the resin used. The sand shall be a minimum 98 percent silica kiln-dried and graded. The polyester wall resin shall be an isophthalic, orthophthalic, or other approved resin with a minimum tensile elongation of 2 percent. A vinyl ester liner resin shall be used to meet the chemical requirements of 207-20.5. Designation per ASTM D3262 shall be Type 1, Liner 2, Grade 3, and a minimum pipe stiffness of 46 psi (318 kPa) or greater, unless a higher value is indicated in the Plans or in the Special Provisions. Elastomeric sealing gaskets shall conform to the requirements of ASTM F477.

207-20.3 Dimensions.

207-20.3.1 Length. The nominal length shall be 20 feet (6m) except as otherwise specified or required for bends or special joints, with a minimum length of 5 feet (1.5m). At least 90 percent of the total footage of each size and class of pipe, excluding special order lengths, shall be furnished in nominal length sections.

207-20.3.2 Minimum Wall Thickness and Weight for Class SN 46 (Minimum Pipe Stiffness of 46 psi (318 kPa)). The outside diameter, minimum wall thickness, and weight are shown in Table 207-20.3.2 (A).

TABLE 207-20.3.2 (A)

Nominal Inside Diameter inches (mm)	Maximum Outside Diameter inches (mm)	Minimum Wall Thickness inches (mm)	Average Weight lbs/ft (kg/m)
12 (300)	13.4 (340)	0.30 (7.7)	11 (16)
18 (450)	19.5 (495)	0.42 (10.7)	21 (31)
20 (500)	21.6 (549)	0.46 (11.7)	25 (37)
24 (600)	25.8 (655)	0.54 (13.8)	35 (52)
30 (750)	32.0 (813)	0.66 (16.8)	54 (80)
36 (900)	38.3 (973)	0.78 (19.9)	77 (115)
42 (1050)	44.5 (1130)	0.90 (22.9)	104 (155)
48 (1200)	50.8 (1290)	1.02 (26.0)	134 (199)
54 (1350)	57.1 (1450)	1.14 (29.0)	169 (251)
60 (1500)	62.9 (1598)	1.26 (32.0)	209 (311)
66 (1650)	69.2 (1758)	1.38 (35.1)	249 (371)
72 (1800)	75.4 (1915)	1.50 (38.1)	298 (443)
78 (1950)	81.6 (2073)	1.62 (41.2)	352 (524)
84 (2100)	88.7 (2253)	1.76 (44.7)	414 (616)
90 (2250)	94.3 (2395)	1.86 (47.3)	468 (696)
96 (2400)	100.6 (2555)	1.98 (50.3)	532 (792)
102 (2550)	108.0 (2743)	2.13 (54.1)	611 (909)

207-20.4 Joints. Pipes and fittings shall be field connected with a coupling or bell-and-spigot joint containing a compressed elastomeric gasket. Elastomeric gasket material shall conform to 208-1.2. Joints shall meet the performance requirements of ASTM D4161. Maximum allowable joint angular deflections for curves are shown in Table 207-20.4 (A).

TABLE 207-20.4 (A)

Nominal Diameter inches (millimeters)	Maximum Joint Deflection Angle (Degrees)
12 to 14 (300 to 350)	4
16 to 20 (400 to 500)	3
24 to 36 (600 to 900)	2
42 to 48 (1050 to 1200)	1.5
54 to 60 (1350 to 1500)	1
66 to 78 (1650 to 1950)	0.75
84 to 102 (2100 to 2550)	0.50

207-20.5 Chemical Resistance and Physical Testing. All pipe furnished under this subsection shall be tested in accordance with 211-2 and conform to Table 207-20.5(A). In addition, the pipe shall be tested as follows:

a) **Properties.** In accordance with ASTM D3262.

b) **Accelerated Aging of Pipe Using A Deflected Condition.** Pipe rings shall be tested for retention of properties after exposure for 112 days in a strained deflected condition to solutions specified in Table 207-20.5 (B).

Tests for initial properties to comply with Table 207-20.5 (A) shall be conducted on pipe rings and specimens cut from pipe rings all prior to their exposure.

Initial properties to be used in evaluation of physical property retention for each exposed pipe ring shall be determined from specimens cut from each ring after exposure. Initial property test specimens shall be taken from ring locations that were 45 degrees either side of the top position in the exposure test.

The deflection of pipe rings during the exposure shall be based on achieving a specific strain as shown in Table 207-20.5 (C).

c) **Test Procedures.** The following procedures shall be used for performing the accelerated aging test on pipe specimens:

1) Select 2 pipe diameters for testing.
2) Obtain 16 pipe rings each 12 inches (300mm) long from both diameters.
3) One ring of each diameter shall be tested for initial properties in accordance with 211-2 and meet or exceed the values in Table 207-20.5 (A).
4) Weigh the remaining pipe rings and test for pipe stiffness per ASTM D2412.
5) Determine the appropriate test deflection of each pipe ring in accordance with Table 207-20.5 (C).
6) Deflect each pipe ring in accordance with ASTM D3681 procedures to the determined test deflection, dam the ends, and immediately begin the test exposure. Depth of test solution shall be 1/3 of pipe diameter.
7) Three pipe rings of each diameter and pipe joint section shall be exposed for 112 days to each test solutions specified in Table 207-20.5 (B).
8) After exposure, immediately remove rings from deflections restraints, clean, and visually inspect (note any changes).
9) Weigh each ring and test for pipe stiffness.
10) Determine initial strength and elongation properties for each pipe ring in accordance with Table 207-20.5 (A).
11) Determine exposed properties of each pipe ring in accordance with Table 207-20.5 (A).
12) Compute changes in pipe stiffness, weight, and properties for each ring.
13) Compute average retentions for each test solution and evaluate for conformance to the requirements of Table 207-20.5 (A).

TABLE 207-20.5 (A)

Property	Initial and After Exposure[1]	After Exposure in strained deflected condition
Minimum Axial Tensile Strength	100% of initial	90% of initial
Minimum Axial Tensile Modulus	350,000 psi (2413MPa) (secant tangent method)	90% of initial
Minimum Axial Elongation at break	0.5%	90% of initial
Weight change	+ 1.5%	+ 1.5%
Minimum Pipe Stiffness	46 psi (0.32MPa)	90% of initial

1. After 112 days in chemical solutions listed in 211-2.

TABLE 207-20.5 (B)

Chemical Solutions	Concentration
Sulfuric Acid	20%
Sodium Hydroxide	5%
Nitric Acid	1%
Sodium Hypochlorite	1%
Bacteriological	BOD not less than 700 PPM

1. Volumetric percentages: Actual concentration of reagent must be corrected to 100%.
2. After 112 days in chemical solutions listed on Table 207-20.5 (B) in the stressed condition given in Table 207-20.5 (C).
3. May be 85% of initial if after exposure value exceeds 46 psi (0.32MPa).

TABLE 207-20.5 (C)

Pipe Stiffness psi (MPa)	Minimum Strain (%)	Test Deflection
46 (0.32)	0.39t/D[1]	9.1%
72 (0.50)	0.34t/D[2]	7.9%

1. Linearly interpolate minimum strain values and test deflections for other values of pipe surface.
2. t and D are the nominal local wall thickness and the inside diameter.

207-20.6 Marking. Each pipe section shall be marked at both ends inside and every 5 feet (1.5m) on the outside to identify the manufacturer, manufacturer number (identifies factory location, date of manufacture, shift, and sequence), nominal diameter, pipe stiffness, ASTM D3262 and designation, and lot number.

207-20.7 Pipe Acceptance or Rejection. The pipe shall be free of cracks, holes, delaminations, foreign inclusions, blisters, or other defects that would, due to their nature, degree, or extent, have a deleterious effect on the pipe performance as determined by the Engineer. Prior to installation, damaged pipe shall be either repaired or field cut to remove the damaged portion as approved by the Engineer. For testing purposes, a production lot shall consist of all pipe having the same lot marking number, but shall not exceed a total of 50 pipes. Pipe length, wall thickness, joint dimensions, pipe stiffness, and deflection characteristics shall be verified by inspection and testing in accordance with 4-1.3, 4-1.4, and ASTM D3262 requirements for each lot. Installation time limit shall conform to 207-15.7.

207-20.8 Installation. Bedding shall conform to ASTM D3839 and 306-1.2.1. Backfill shall conform to 306-1.3. Field inspection shall conform to 306-1.2.12.

207-21 VITRIFIED CLAY MICROTUNNELING PIPE.

207-21.1 General. This specification is for vitrified clay microtunneling pipe to be used for sanitary sewers, storm drains, and related structures. Vitrified clay microtunneling pipe shall be manufactured in accordance with 207-8 and ASTM C1208, except as modified herein.

207-21.2 Dimensions and Tolerances. For typical tolerances, see Table 207-21.2 (A). Since the physical properties of clay pipe vary at different manufacturing locations, the minimum wall thickness of the pipe shall be included with the submittals.

a) **Length.** Typical lengths shall be 4, 6, or 8 feet (1.5, 2, or 2.5m).

b) **Straightness.** Pipe shall not deviate from straight by more than 0.05 inch per linear foot (4mm per linear meter) when the maximum offset is measured from the concave side of the pipe. Measurements shall be taken by placing a straightedge along the concave side of the full length of the pipe barrel, excluding the joint, and measuring the maximum distance between the straightedge and concave side of the pipe.

c) **Roundness.** The diameter shall not vary from a true circle by more than 2 percent of the nominal diameter. The out-of-round dimension is the difference between the maximum and minimum diameters measured at any one location along the barrel.

d) **End Squareness.** The ends of the pipe shall be perpendicular to the theoretical longitudinal axis within 0.004 inch per inch (4mm per meter) of outside diameter.

TABLE 207-21.2 (A): DIMENSIONS AND TOLERANCES

Nominal Diameter inches (mm)[1]	Maximum Out-of-Round inches (mm)	Maximum Out-of-Square inches (mm)[2]	Minus Tolerance from Lying Length, inch/foot (mm per m)	Minimum Allowable Diameter inches (mm)[3]
8 (200)	0.16 (4)	0.04 (1.2)	0.25 (21)	7.6 (192)
10 (250)	0.20 (5)	0.06 (1.4)	0.25 (21)	9.5 (240)
12 (300)	0.24 (6)	0.06 (1.6)	0.25 (21)	11.4 (289)
16 (400)	0.30 (8)	0.09 (2.2)	0.25 (21)	15.2 (385)
18 (450)	0.36 (9)	0.09 (2.3)	0.25 (21)	17.2 (437)
20 (500)	0.38 (10)	0.10 (2.5)	0.25 (21)	18.9 (480)
21 (525)	0.42 (11)	0.10 (2.7)	0.25 (21)	20.0 (509)
24 (600)	0.43 (12)	0.12 (3.0)	0.38 (32)	23.1 (586)
27 (675)	0.54 (14)	0.13 (3.4)	0.38 (32)	25.9 (658)
30 (750)	0.60 (15)	0.14 (3.6)	0.38 (32)	28.7 (730)
33 (825)	0.70 (16)	0.16 (4.2)	0.38 (32)	31.5 (801)
36 (900)	0.72 (18)	0.18 (4.5)	0.38 (32)	34.6 (878)
39 (975)	0.80 (20)	0.19 (4.9)	0.38 (32)	37.5 (953)
42 (1050)	0.84 (21)	0.21 (5.2)	0.38 (32)	40.5 (1028)

1. The outside diameter of the delivered pipe shall not vary by more than 2.5% for outside diameters up to and including 24 inches (600mm), and 2.0% for outside diameters greater than 24 inches (600mm).
2. These are based on the outside diameter and are approximate. Consult with the pipe manufacturer for the specific dimensions.
3. No fixed "+" tolerance limit.

207-21.3 Imperfections.

a) **Fractures or Cracks.** There shall be no fractures or cracks passing through the barrel which are visible to the unaided eye.

b) **Chips, Fractures, or Blisters.** Chips, fractures, or blisters on the pipe shall not exceed 2 inches (50mm) laterally in any surface dimension and shall not exceed a depth of 1/8 the minimum thickness of the barrel.

207-21.4 Repair Methods and Materials. Repairs, if any, shall be made in accordance with 207-8.6.

207-21.5 Fittings and Stoppers. 207-8.3 is not applicable to microtunnel-installed vitrified clay pipe.

207-21.6 Joints. Joints for vitrified clay pipe shall consist of a seat, an elastomeric sealing element, a sleeve, and a compression disc.

a) **Seat.** The seat is either formed or recess-ground on the pipe ends.

b) **Elastomeric Sealing Element(s).** The elastomeric sealing element is an elastomeric gasket configuration and location may vary by pipe manufacturer. The sealing element(s) shall conform to 208-2.

c) **Sleeve.** The sleeve is an element which bridges between the pipe sections and shall be made of non-corrosive materials which, in conjunction with the sealing element(s), forms a joint which meets the test requirements of 207-21.7. Stainless steel sleeves shall be AISI Type 316.

d) **Compression Disc.** The compression disc is a flat disk that forms a continuous ring of contact with the ends of the pipe and which functions to distribute the jacking forces which develop during pipeline installation. The width of the compression disc shall not exceed the maximum wall thickness of the pipe, not extend into the flow, nor inhibit the installation of the sleeve onto the joint of the pipe.

207-21.7 Testing Requirements.

207-21.7.1 General. Testing shall be in accordance with 207-8.5 and ASTM C1208.

207-21.7.2 Joint Test. Testing shall be in accordance with ASTM C1208.

207-21.7.3 Hydrostatic Pressure Test. Testing shall be in accordance with ASTM C1208.

207-21.7.4 Bearing Strength Test. Testing shall be in accordance with ASTM C1208. Pipe shall meet Table 207-8.5.3 (A) for Extra Strength pipe.

207-21.7.5 Compressive Strength Test. Testing shall be in accordance with ASTM C1208. Pipe material shall have a minimum compressive strength of 7,000 psi (48MPa).

207-21.7.6 Acid Resistance Test. Testing shall be in accordance with ASTM C301.

207-21.7.7 Acceptance. Pipe acceptance shall be in accordance with 207-8.5.4.

207-21.8 Inspection. Pipe shall be inspected in accordance with 4-1.3 and ASTM C1208.

207-21.9 Installation and Field Acceptance Testing. The installation of pipe shall be in accordance with ASTM C1208. Exercise care to maintain proper alignment during installation.

207-22 CENTRIFUGALLY CAST FIBERGLASS REINFORCED PLASTIC MORTAR (CCFRPM) MICROTUNNELING PIPE.

207-22.1 General. This specification applies to CCFRPM microtunneling pipe to be used for sanitary sewers, storm drains, and related structures. Such pipe shall conform to 207-20 except as modified herein. The minimum pipe stiffness shall be 120 psi (830kPa) when tested in accordance with ASTM D2412.

207-22.2 Materials. Materials shall conform to 207-20.2, except that the minimum pipe stiffness shall be 120 psi (830kPa). Pipe material shall have a minimum compressive strength of 10,000 psi (69MPa).

207-22.3 Dimensions. Dimensions and sizes of pipe shall conform to the requirements of Table 207-22.3.

207-22.3.1 Length. The nominal lengths shall be 6, 10, or 20 feet (2, 3, or 6m). At least 90 percent of the total footage of each size and class of pipe, excluding special order lengths and pipes sampled for quality control testing, shall be furnished in nominal length sections.

207-22.3.2 Minimum Wall Thickness. The outside diameter and minimum wall thickness are shown in Table 207-22.3. The allowable jacking load for microtunnel pipe shall be determined from the pipe wall thickness where the cross-section is reduced (the bottom of the spigot gasket groove).

TABLE 207-22.3

Nominal Diameter inches (mm)	Outside Diameter inches (mm)[1]	Nominal Inside inches (mm)[2]	Minimum Pipe Wall inches (mm)[2]
18 (450)	19.5 (495)	17.0 (432)	1.12 (28.4)
20 (500)	21.6 (549)	19.0 (483)	1.15 (29.2)
24 (600)	25.8 (655)	23.1 (587)	1.22 (31.0)
28 (700)	30.0 (762)	27.1 (688)	1.30 (33.0)
30 (750)	32.0 (813)	28.7 (729)	1.52 (38.6)
33 (825)	34.0 (864)	30.6 (777)	1.54 (39.1)
36 (900)	38.3 (973)	34.8 (884)	1.59 (40.4)
42 (1050)	44.5 (1130)	40.8 (1036)	1.67 (42.4)
48 (1200)	50.8 (1290)	46.7 (1186)	1.88 (47.8)
51 (1275)	53.9 (1369)	49.7 (1262)	1.94 (49.3)
54 (1350)	57.1 (1450)	52.8 (1341)	2.00 (50.8)
57 (1425)	60.0 (1524)	55.5 (1410)	2.06 (52.3)
60 (1500)	62.9 (1598)	58.3 (1481)	2.11 (53.6)
63 (1575)	66.0 (1676)	61.3 (1557)	2.15 (54.6)
66 (1650)	69.2 (1758)	63.9 (1623)	2.44 (62.0)
69 (1725)	72.5 (1842)	67.1 (1704)	2.50 (63.5)
72 (1800)	75.4 (1915)	69.9 (1775)	2.56 (65.0)
78 (1950)	81.6 (2073)	75.8 (1925)	2.94 (68.8)
82 (2050)	87.0 (2210)	80.7 (2050)	2.94 (74.7)
84 (2100)	88.7 (2253)	82.3 (2090)	3.00 (76.2)
90 (2250)	94.3 (2395)	87.6 (2225)	3.13 (79.5)
96 (2400)	99.5 (2527)	92.6 (2352)	3.23 (82.0)
102 (2550)	108.0 (2743)	100.7 (2558)	3.42 (86.9)

1. The outside diameter of the pipe delivered to the Work site shall not vary by more than ± 0.1% for each pipe size.
2. For standard tonnage ratings.

207-22.3.3 Tolerances.

a) **Length.** Laying length tolerance shall be ± 1/2 inch per foot (± 42mm per m).

b) **Straightness.** Pipes shall not deviate from straight by more than 0.006 inch per linear foot (0.5mm per linear meter). Measurements shall be taken by measuring gaps between the pipe wall and a straightedge placed along any longitudinal line on the exterior surface of the pipe.

c) **Roundness.** The diameter shall not very from a true circle by more than ± 0.5 percent of its nominal diameter. The out-of-round dimension is the difference between the maximum and minimum diameters measured at any one location along the barrel.

d) **End Squareness.** All points around each end of the pipe shall fall within 1/16 inch (1.5mm) to a plane perpendicular to the longitudinal axis of the pipe.

207-22.4 Joints. Joints shall conform to 207-20.4, except that bell and spigot joints shall have a flush exterior. Joints at tie-ins, when needed, may utilize fiberglass gasket-sealed closure couplings which have an outside diameter larger than the pipe. The maximum allowable joint angular deflections for curves shown in Table 207-20.4 (A) shall not apply.

207-22.5 Accelerated Aging and Physical Testing. Testing shall conform to 207-20.5, and values used shall reflect actual pipe stiffness.

207-22.6 Marking. Marking shall conform to 207-20.6. The maximum allowable jacking load shall also be indicated.

207-22.7 Pipe Acceptance or Rejection. Pipe shall conform to 207-20.7.

207-22.8 Field Inspection. Field Inspection per 306-1.2.12 is required when: 1) pressure grouting is specified or otherwise implemented; or 2) the allowable rate of return is exceeded; 3) a void, loss of face control, or surface disruption has occurred; or 4) the actual jacking force exceeds the maximum allowable jacking force. The deflection test or equivalent means shall be performed from access shaft for entire length of affected section.

207-23 REINFORCED CONCRETE MICROTUNNELING PIPE.

207-23.1 General. This specification applies to reinforced concrete microtunneling pipe manufactured for the construction of storm drains, sewers, water pipelines, and related structures. Such pipe shall comply with 207-2 for reinforced concrete pipe; 207-3 for lined reinforced concrete pipe; or 207-5 for AWWA C300 Reinforced Pressure Pipe, Steel Cylinder Type, or AWWA C302 Reinforced Pressure Pipe, Noncylinder Type, except as modified herein. A minimum concrete compressive strength of 5,000 psi (34.5MPa) is required.

207-23.2 Materials. All materials used shall conform to the requirements of 201 and 207-2.2, except that the Contractor shall submit written certification from the pipe manufacturer that no calcium chloride has been used in the manufacturer of the pipe.

207-23.3 Dimensions and Tolerances.

a) **Length.** Pipe shall have a nominal laying length of 8 to 24 feet (2.4 to 7.3m) unless otherwise specified. Variations in laying lengths of 2 opposite sides of pipe shall not be more than 1/16 inch per foot (5mm per m) with a maximum of 3/8 inch (10mm) in any length of pipe for diameters up to and including 72-inch (1800mm), or 1/2 inch (13mm) for larger diameters. The underrun in length of a section of pipe shall not be more than 1/8 inch per foot (10mm per m) with a maximum of 1/2 inch (13mm) in any length of pipe.

b) **Wall Thickness.** The minimum wall thickness for all pipe sizes shall be as specified in ASTM C76, Wall B.

c) **Inside Diameter.** The inside diameter of 12-inch to 24-inch (300mm to 600mm) pipe shall not vary more than 1/4 inch (6mm). The internal diameter of 27-inch (675mm) and larger pipe shall not vary more than 1 percent or 5/8 inch (16mm), whichever is less.

d) **Outside Diameter.** The outside diameter of 12-inch to 24-inch (300mm to 600mm) pipe shall not vary more than 1/4 inch (6mm). The outside diameter of 27-inch (675mm) and larger pipe shall not vary more than 1 percent or 5/8 inch (16mm), whichever is less.

e) **Straightness.** Pipes shall not deviate from straight by more than 1/16 inch per linear foot (5mm per linear meter). Measurements shall be taken by measuring gaps between the pipe wall ends and a straightedge placed along any longitudinal line on the exterior surface of the pipe.

f) **Roundness.** The outside diameter of the pipe shall not vary from a true circle by more than 1 percent.

g) **End Squareness.** The ends of the pipe in contact with the jacking pads shall be perpendicular to the longitudinal axis of the pipe with a maximum of 1/2 inch (13mm), measured with a square and a straight edge across the end of the pipe. The bearing surfaces shall be smooth and free of projections.

207-23.4 Reinforcement. Pipe shall have concentric circular cage reinforcement conforming to the requirements of 207-2.4 for reinforced concrete pipe or 207-5 for reinforced concrete pressure pipe. The total area of reinforcement provided shall be at least 1/8 in^2 per ft (265mm^2 per m) in each direction. When 2 reinforcement cages are used, the longitudinal reinforcement shall be divided approximately equal between the two cages, and only the longitudinal reinforcement on the outer cage need extend into the bell. Supports between the reinforcement and the forms that are to be exposed in the finished pipe shall be made of stainless steel, plastic or plastic coated steel.

207-23.5 Joints. Pipe shall have rubber-gasket type, concrete or steel joints. Tongue and groove joints are not acceptable. Joint details shall be submitted to the Engineer for approval before commencing pipe manufacturing. For flush bell and spigot joints, the slope of the longitudinal gasket contact surfaces of the joint with respect to the longitudinal axis of the pipe shall not exceed 2 degrees. The ends of the pipe shall be so formed that, when the pipes are joined, they shall make a continuous and uniform line of pipe with a smooth and regular surface. The rubber gasket shall conform to the requirements of 208-3 and 306-1.2.4 d).

207-23.6 Repair. Unsound or imperfect concrete shall be repaired per the applicable sections of 207-2, 207-3, 207-4 or other methods approved by the Engineer.

207-23.7 Pipe Acceptance or Rejection. Pipe shall conform to 207-2.9 and 207-3.2.

207-24 STEEL MICROTUNNELING PIPE.

207-24.1 General. This specification applies to steel microtunneling pipe manufactured for the construction of storm drains, sewers, water pipelines, and related structures. Such pipe shall comply with 207-10 except as modified herein.

207-24.2 Materials. All materials used shall conform to the requirements of 207-10.2.

207-24.3 Dimension and Tolerances.

a) **Length.** The nominal length shall conform to the requirements of 207-10.2.4. Pipe shall have a nominal laying length of 8 to 40 feet (2.4m to 12m) unless otherwise specified. Variations in laying lengths of 2 opposite sides of pipe shall not be more than with a maximum of 3/8 inch (10mm) in any length of pipe for diameters up to and including 72 inches (1800mm), or 1/2 inch (13mm) for larger diameters. The underrun in length of a section of pipe shall not be more than 3/8 inch per foot (10mm per m) with a maximum of 1/2 inch (13mm) in any length of pipe.

b) **Wall thickness.** Plate and sheet thickness shall conform to 306-2.3. For plate, the maximum allowable thickness variation shall be 0.010 inch (0.254mm) under the ordered thickness. For sheet, the maximum allowable thickness variation shall be as tabulated in ASTM A568, ASTM A635, or 0.010 inch (0.254mm) under the ordered thickness,

c) **Diameter.** Internal diameter shall conform to 207-10.2.3. The outside circumference of the pipe shall not vary by more than 1 percent but not to exceed 3/4 inch (19mm) from the nominal outside circumference. The circumference of the pipe ends shall not vary by more than 0.196 inch (5mm) under, or 0.393 inch (10mm) over the required outside circumference.

d) **Straightness.** Finished cylinder shall not deviate by more than 1/8 inch (3mm) from a 10-foot (3m) long straight edge held against the cylinder.

e) **Roundness.** The outside diameter of the pipe shall not vary from a true circle by more than 1 percent. The out-of-round of the pipe ends shall be determined by measuring the ring diameter at a minimum three locations equally spaced on the contact surface. The out-of-round shall be the difference between the high and low readings. The maximum out-of-round shall not exceed 0.5 percent of the average of the low and high readings or 1/4 inch (6mm), whichever is less.

f) **End Squareness.** For pipe that is to be butt welded in the field, the end of the pipe shall be perpendicular to the longitudinal axis of the pipe with a maximum deviation of no more than 1/8 inch (3mm) measured with a square and a straight edge across the end of the pipe.

207-24.4 Joints. Joints shall conform to the requirements of 306-2.3. Connecting adjacent pieces of steel pipe shall be achieved by field butt welding. For casing application, integral press fit connectors may be used as approved by the Engineer.

207-24.5 Protective Lining and Coating. Exterior coatings, if required, shall be the type that minimize skin friction between the exterior of the pipe and the soil. Such an exterior coating shall be epoxy-based polymer concrete, fusion-bond epoxy, or other product that provides a hard, smooth surface. Strict adherence to the coating manufacturer's recommendations for surface preparation and application procedure shall be required with holiday detection and repair. If the steel pipe is field welded, the Contractor shall submit a procedure for approval to the Engineer to field repair the coating which maintains the coating integrity and minimizes the repair and curing time.

Exterior coating of steel pipe, used as encasement, is not required if a wall thickness increase of at least 0.063 inch (1.5mm) is specified.

Interior lining of steel pipe should be shop-applied coating such as liquid epoxy, polyurethane, cement mortar, or appropriate material for the desired service. Consideration should also be given to line-in-place procedures.

207-24.6 Inspection and Acceptance. Pipe shall be inspected and accepted in accordance with AWWA C 200.

SECTION 208 - PIPE JOINT TYPES AND MATERIALS

208-1 GENERAL.

208-1.1 Material Designation. The type of pipe joint shall be designated on the Plans. The Agency may require testing of the joint materials for compliance with these requirements prior to delivery to the jobsite.

208-1.2 Installation Time Limit. The Contractor shall retest within 60 days prior to installation any pipe gasket that is more than 180 days old from the date of manufacture to ensure compliance with the requirements of the Specifications. The Contractor shall not install any pipe gasket that is more than 2 years old from the date of manufacture.

208-2 JOINTS FOR CLAY PIPE.

208-2.1 General. Joints other than those specified below, made of approved materials and meeting the chemical and bacteriological resistance requirements contained herein, may be submitted to the Engineer for approval. The installation of joint material shall conform to 208-1.2.

208-2.2 Type "D" Joints (Synthetic Rubber Coupling with Corrosion-Resistant Shear Ring for Plain-End Clay Pipe, 12 inch (300mm) Maximum). The joint shall consist of 3 parts: a circular synthetic rubber sleeve of 2 stainless steel compression bands with stainless steel nut and bolt type tightening devices and an injection molded Acrylonitrile-Butadiene-Styrene (ABS), Polyethylene (PE), or Polyvinylchloride (PVC) shear ring.

208-2.2.1 Sealing Components. The compression bands shall be fabricated from AISI Type 316 stainless steel with stainless steel nut and bolt (any AISI series 300) type tightening devices and meeting the requirements of ASTM A240. The sleeve shall be fabricated in a configuration approved by the Engineer. It shall be made of a synthetic rubber molded to form a smooth surface, free of pits, cracks, air marks, porosity, and air pockets.

208-2.2.2 Testing of the Synthetic Rubber Sealing Component. Test specimens that are exposed to various chemical and bacteriological environments shall be conditioned in the same manner, both before and after exposure, prior to testing.

a) **Initial Physical Requirements.** The initial physical requirements shall conform to Table 208-2.2.2 (A):

TABLE 208-2.2.2 (A)

Property	Value	ASTM Test Method No.
Tensile strength, at 73.4° ± 3.6°F (23° ± 2°C), psi (kPa), min.	1000 (6900)	D 412 (Die C)
Elongation at break, 73.4° ± 3.6°F (23° ± 2°C), %, min.	200	D 412 (Die C)
Shore Durometer, Type A (1 sec. reading, min.)	60	D 2240
Compression set (after exposure to 158° ± 3.6°F [70° ± 2°C], for 22 hours), % max.	20	D 395 (Method B)
Water absorption (after immersion at 73.4° ± 1.8°F [23° ± 1°C] for 28 days), %, max.	4	D 570

b) **Physical Requirements after Accelerated Aging.** The physical requirements after accelerated aging shall conform to Table 208-2.2.2 (B).

TABLE 208-2.2.2 (B)

Property	Value	ASTM Test Method No.
Tensile strength (after exposure to 300 psi (2100 kPa) oxygen at 158° ± 1.8°F [70° ± 1°C] for 96 hours), % of initial, min.	70	D 572
Elongation at break (after exposure to 300 psi (2100 kPa) oxygen at 158° ± 1.8°F [70° ± 1° C] for 96 hours), % of initial, min.	70	D 572
Ozone resistance (after exposure to 100 pphm ozone for 50 hours at 104° ± 2°F [40° ± 1°C])	No cracks or crazing	D 518 (Procedure C) and D 1149

c) **Chemical Resistance and Physical Testing.** After exposure to the chemical solutions specified in Table 211-2 (A), test specimens shall be tested in accordance with 211-2 and meet the physical requirements specified in Table 208-2.2.2 (C).

TABLE 208-2.2.2 (C)

Property	Value	ASTM Test Method No.
Tensile strength at 73.4° ± 3.6°F (23° ± 2°C), psi (kPa), min.	800 (5500)	D 412 (Die C)
Elongation at break at 73.4° ± 3.6°F (23° ± 2°C), % min.	150	
Shore Durometer, Type A at 75° ± 5°F (23° ± 2°C) (1 second reading) point change, max.	15	D 2240
Compression set (after exposure to 158° ± 1.8°F [70° ± 1°C] for 22 hours), % max.		D 395 (Method B)
a. Chemical exposures	25	
b. Bacteriological exposure (unconditioned surface dry)	30	
Weight change, % max. (approx. specimen size 1.0 x 3.0 x 0.1 inch) (25 x 75 x 2.5mm)	–1.0, +5.0	

208-2.2.3 Circular Plastic Shear Ring Component and Testing Requirements. The plastic shear ring shall be injection molded from an ABS, PE, or PVC resin meeting the requirements specified below. The ABS shear ring material shall conform to 207-15.2. The PE shear ring shall conform to 207-19.2 except that the hydrostatic design and color requirements need not be satisfied and crack resistance failure maximum percentage be zero.

The PVC shear ring shall conform to 207-17.2.2 and 207-17.4.1.

The shear rings shall be homogeneous throughout, uniform in color and free of cracks, holes, foreign materials, blisters, or deleterious faults.

Markings on all shear rings shall indicate the name of the manufacturer and Type "D" joint.

208-2.2.4 Laboratory Test of Joint. An assembled joint shall provide sufficient resistance to shear loading to allow a load of 150 lbf per inch (26.3 N per mm) of nominal diameter to be uniformly applied over an arc of not less than 120 degrees and a longitudinal distance of 12 inches (305mm) immediately adjacent to one edge of the sleeve coupling. The assembled pipe shall rest on three supports. A support shall be located at each extreme end of the assembly. The third support shall be placed immediately adjacent to the coupling. The shear load shall be placed on the unsupported end of the pipe, immediately adjacent to the coupling. There shall be no visible leakage when tested with an internal hydrostatic pressure of 10 psi (70 kPa) for 10 minutes. The joint including the plastic shear ring for the 6-inch to 12-inch (150mm to 300mm) diameter pipe, inclusive, shall exhibit sufficient flexibility when joined to allow a maximum deflection of 3 degrees in any direction. The deflected joint shall show no visible leakage when subjected to the same shear load as indicated in the previous paragraph and when tested under an internal hydrostatic pressure of 10 psi (70 kPa) for 10 minutes.

During these tests, the ends of the tested pipe shall be restrained only in the amount necessary to prevent longitudinal movement. Upon completion of the above tests, the joint shall be disassembled and if any component has failed, that joint will be rejected. Failures shall include, but not be limited to the following: breaks, flange cracks, gouges, tears, and deformation beyond reuse.

208-2.2.5 Joint Acceptance. When all components of this joint meet the requirements as set forth in these specifications and when one percent of the joints are successfully laboratory tested, then the joints will be accepted.

208-2.3 Type "G" Joints (Polyurethane).

208-2.3.1 General. The Type "G" joint shall consist of polyurethane elastomer sealing components, one bonded to the outside of the spigot and the other bonded to the inside of the socket. The sealing components shall be shaped, sized, bonded, and cured to uniform hardness to form a tight seal of the joint when it is assembled. The sealing components shall resist attack by bacteria and chemicals or combinations of chemicals normally present in domestic or industrial waste sewage.

208-2.3.2 Polyurethane Sealing Components. The polyurethane sealing component material shall be tested in accordance with 211-2. The number of samples to be tested shall be designated in the Special Provisions.

TABLE 208-2.3.2 (A)

Property	Values[1]		Test Method and Conditions
Tensile strength psi (kPa) min.	4125 (600)	2425 (350)	ASTM D412 (Die C) 73.4 ± 3.6°F (23° ± 2°C)
Elongation, % min.	70	70	ASTM D412 (Die C) 73.4° ± 3.6°F (23° ± 2°C)
Compression set, % max.	3	3	ASTM D395, Method B, 24 hours 73.4° ± 3.6°F (23° ± 2°C)
Shore durometer	Not less than value designated for the joint design by the manufacturer		ASTM D2240, Type A, 5-second reading, 32° to 80°F (0° to 27°C)
Water absorption, % max. (Weight gain)	3.5	3.5	ASTM D570, after immersion for 28 days at 73.4° ± 1.8°F (23° ± 1°C)
Volatile loss, % max. (Weight loss)	1	1	After 28 days in mechanical convection oven at 150° ± 3°F (66° ± 2°C)
Adhesive strength, psi (kPa) min			Before immersion in accordance with Note 2. After immersion in water at 73.4° ± 3.6°F (23° ± 2°C) in accordance with Note 2
Original	2425 (350)	2425 (350)	
Final	1725 (250)	1725 (250)	
Chemical resistance (See Note 3)			After exposure to each of the chemical environments for 112 days as described in Note 3
Weight change, % max.	1.5	1.5	
Tensile strength, psi (kPa)	2975 (430)	1800 (260)	
Shore durometer, change max.	± 15	± 15	
Compression set, % max.	5	5	
Bacteriological resistance (See Note 3)			After exposure to bacteriological environment for 112 days as described in Note 3.
Weight change, % max.	2	2	
Tensile strength, psi (kPa) min.	3450 (500)	2000 (290)	
Shore Durometer			
Before reconditioning, loss max.	15	15	
After reconditioning, change max.	5	5	
Compression set			
Before reconditioning, % max.	3	3	
After reconditioning, % max.	3	3	

1. The configuration of the jointing system determines the necessary physical properties of the polyurethane joint material. The columns of values in this table represent properties of polyurethane that, in conjunction with specific joint configurations, will provide functionally equal acceptable jointing systems. The values for a single product shall be all from one of the columns.
2. Adhesion test specimens shall be clay blocks 1/2 inch (13mm) thick and 1 inch square (25mm x 25mm), of the same composition of materials and fired at the same vitrifying temperatures as sewer pipe.
   The clay block shall be placed flat in the center of a mold 7 inches (175mm) long, 1 inch (25mm) wide, and 1/2 inch (13mm) deep. The edges of the clay block at right angles to the longitudinal axis of the form shall be coated with the adhesive and the form on each side of the block shall be filled to a depth of approximately 1/4 inch (6mm) with the sealing component compound. Curing of this last specimen shall simulate the curing process at the pipe manufacturing plant.
   At the end of the immersion period, samples shall be removed, surface dried, and immediately pulled in tension at the rate of 20 inches (500mm) per minute to determine the final tensile strength of the bonded interfaces. The specimens retained for controls shall be pulled at the same time to determine the original strength of the bonded interfaces.
3. Exposure environments for bacteriological and chemical resistance tests per Table 211-2 (A). At the end of the exposure period, specimens shall be washed and reconditioned before testing.

208-2.3.3 Type "G" Joint Acceptance Tests. Joint tests shall be performed in accordance with ASTM C425. The Engineer may perform additional acceptance tests on one joint for each 100 pipes in a lot (or fraction of a lot) for each size. For this additional testing the pipe joints shall not leak when subjected to the shear load test described in ASTM C425, except that the internal hydrostatic pressure shall be 10 psi (70 kPa) for pipe sizes up to 18 inches (450mm) and 5 psi (35 kPa) for pipe sizes 21 inches (525mm) and larger before and after being deflected the following distances at the sealing components of the joint for a period of 10 minutes.

TABLE 208-2.3.3 (A)

Nominal Pipe Size inches (mm)	Deflection inches (mm)
4 - 10 (100 - 250)	5/8 (16)
12 - 21 (300 - 525)	3/4 (19)
24 - 30 (600 - 750)	7/8 (22)
33 - 42 (825 - 1050)	1 (25)

If one of the selected joints fails the test, the lot shall be rejected unless 2 additional joints selected from the lot pass the test.

208-3 GASKETS FOR CONCRETE PIPE. Unless otherwise specified, gaskets shall be manufactured from a synthetic elastomer. The compound shall contain not less than 50 percent by volume of first-grade synthetic rubber. The remainder of the compound shall consist of pulverized fillers free of rubber substitutes, reclaimed rubber, and deleterious substances. The installation of gaskets shall conform to 208-1.2.

Gaskets shall be extruded or molded and cured in such a manner as to be dense, homogeneous and of smooth surface, free of pitting, blisters, porosity and other imperfections. The tolerance for any diameter or profile dimension measured at any cross section shall be ± 1/32 inch (0.8mm).

When required by the Engineer, the Contractor shall furnish test samples of gaskets from each batch used in the work. Gasket material shall meet the following:

TABLE 208-3 (A)

Property	Value	ASTM Test Method
Tensile strength, psi (MPa) min.	1500 (10.3)	D 412
Elongation at break (% min.)	350	D 412
Shore durometer, Type A (Pipe manufacturer shall select value suitable for type of joint).	40 to 65[1]	D 2240
Compression set (constant deflection) max. % of original deflection.	16	D 395 Method B
Tensile strength after oven aging (96 hours, 158°F [70°C]) % of tensile strength before aging.	80	D 573
Increase in Shore durometer hardness after oven aging. Maximum increase over original Shore durometer.	10	D 2240
Physical Requirements after exposure to ozone concentration (50 pphm. 72 hours, 104°F [40°C], 20% extension).	No Cracks	D 1149

1. This applies only to the sealing component of the gasket.

No more than one splice will be permitted in a gasket. A splice shall be made by applying a suitable cement to the ends and vulcanizing the splice in a full mold. The splice shall show no separation when subjected to the following tests:

a) Elongation Test. The part of the gasket which includes the splice shall withstand 100 percent elongation with no visible separation of the splice. While in the stretched position, the gasket shall be rotated in the spliced area a minimum of 180 degrees in each direction in order to inspect for separation.

b) Bend Test. The portion of the unstretched gasket containing the splice shall be wrapped a minimum of 180 degrees and a maximum of 270 degrees around a rod of a diameter equal to the cross section diameter of the gasket.

208-4 GASKETS FOR THERMOPLASTIC PIPE. Gaskets shall be manufactured from a synthetic elastomer conforming to the requirements of 208-3. The installation of gaskets shall conform to 208-1.2.

208-5 TYPE "Z" JOINT.

208-5.1 General. Except as modified herein, Type "Z" Joints shall conform to Type "D" Joints per 208-2.2. This joint shall be used as a field closure coupling, repair coupling, or OD transition coupling for identical or dissimilar pipe materials for gravity sewers and drains. The joint shall be applied to 4-inch through 12 inch (100mm through 300mm) diameter pipe sizes only.

208-5.2 Components. The joint shall consist of three components as follows:

a) A circular synthetic rubber sleeve meeting the physical, chemical and bacteriological requirements of the Type "D" Joint per 208-2.2.2. The sleeve shall conform to 208-1.2.

b) Two stainless steel compression bands (AISI 316) with stainless steel nut and bolt (any AISI Series 300) type tightening devices and meeting the requirements of ASTM A 240.

c) A stainless steel shear band shall wrap around the joint a minimum of 380 degrees. Welded to the shear band shall be two stainless steel nut and bolt tightening devices or worm drive tightening devices fabricated from any 300 Series stainless steel and meeting the requirements of the compression bands above. The minimum shear band thickness shall be 0.012 inches (0.30mm). Minimum shear band width shall be 2.120 inches (54.0mm) for 4-inch and 6-inch (100mm and 150mm) diameter pipe sizes and 2.437 inches (62.0mm) for 8-inch through 12-inch (200mm through 300mm) diameter pipe sizes.

208-5.3 OD Transitions. Pipe joints where the pipe OD's have a differential of 3/8 inch (9mm) or greater shall require an OD transition coupling. OD transition coupling shall be comprised of the components per 208-5.2 plus a synthetic rubber bushing meeting the requirements of 208-2.2.2. The bushing shall compensate for the differential in pipe OD's.

208-5.4 Testing Requirements. The Type "Z" Joint, with or without bushing, shall meet the test requirements for Type "D" Joints per 208-2.2.4.

208-6 PIPE TO MANHOLE FLEXIBLE COUPLINGS.

208-6.1 General. The joint shall consist of a flexible connector designed to produce a positive watertight connection for pipes entering precast manholes and other concrete structures.

208-6.1.1 Seal. The connector shall be in accordance with ASTM C923/C923M so that a positive seal is made between the connector and the manhole wall and between the connector and the pipe. The seal between the connector and the manhole wall may be made by either mechanical means or by casting the connector integrally with the manhole wall. The seal between the connector and the pipe may be made by mechanical means or by compression of the resilient material against the outside of the pipe.

208-6.1.2 Parameters. The connector shall withstand 23 feet (10 psi) (7.1m (70 kPa)) of hydrostatic pressure and be capable of sustaining an axial deflection of at least 7 degrees in any direction. The test methods and requirements shall be in accordance with ASTM C923 (C923M), Section 7.

208-6.2 Materials. The gaskets shall be manufactured from a synthetic elastomer and shall contain not less than 50 percent by volume of first-grade synthetic rubber. All rubber gaskets shall be either molded or extruded and cured in such a manner that any cross-section shall be dense, homogeneous, and free of porosity blisters, pitting, and other imperfections. The gaskets shall comply with the physical requirements prescribed by ASTM C923/C923M Table 1 when tested in accordance with the referenced ASTM. However, the chemical resistance shall be tested in accordance with 211-2 and meet the weight change per Table 210-2.4.1 (A) of these specifications. Metal components shall be fabricated from AISI

Type 316 stainless steel for all bands, and the nut and bolt shall be AISI Type 305 stainless steel. Gaskets shall conform with the installation time requirements in 208-1.2.

208-6.3 Performance Test Requirements. A performance test of the connector to be used shall be made at least once when the manhole producer begins using a pipe to manhole connector system. The test methods and requirements shall be made in accordance with ASTM C923/C923M, Section 7. The connector shall be marked clearly by the manufacturer with his trade name and size designation or part number. This shall be visible on the gasket when installed in the manhole.

208-6.4 Plant Fabrication Method. The flexible connector shall be installed in accordance with the specific instructions of the manufacturer.

208-6.5 Installation Instructions. The manhole manufacturer shall provide installation instructions to the Contractor. The connector shall have all foreign matter removed and shall be inspected to ensure that there are no defects in the rubber or splice.

SECTION 209 - STREET LIGHTING AND TRAFFIC SIGNAL MATERIALS

209-1 GENERAL. The following material specifications set forth the requirements for components installed in conjunction with the construction of street lighting and traffic signal systems.

Manufacturer's warranties, guaranties, instruction sheets and parts lists supplied for materials used in the Work shall be submitted to the Engineer in accordance with 2-5.3.

209-2 REFERENCE SPECIFICATIONS. In addition to the requirements of the Plans, Specifications, and Special Provisions, components shall conform, where applicable, to the following:

a) NEMA
b) UL
c) EIA
d) RUS
e) Aluminum Association Publication 30
f) ANSI
g) AASHTO "Standard Specification for Structural Support for Highway Signs, Luminaires and Traffic Signals"
h) ETL
i) IEEE
j) IMSA
k) MIL SPEC
l) AWS D1.1 "Structural Welding Code"
m) FCC
n) California Administrative Code, Title 8, Chapter 4, Subchapter 5
o) NEC

209-3 COMPONENTS FOR STREET LIGHTING AND TRAFFIC SIGNAL SYSTEMS.

209-3.1 General. The following material Specifications set forth the requirements for components that are common to street lighting and traffic signal systems. All materials furnished and installed shall be new, except materials specified on the Plans or in the Special Provisions to be reused.

209-3.2 Anchor Bolts, Nuts, and Washers. Anchor bolts shall conform to ASTM A307 and shall be tested in accordance with 206-1.1.3. Each anchor bolt shall be round, have a minimum of 8 inches (200mm) of thread, and be provided with 2 nuts and 2 flat 1/4 inch (6mm) thick washers.

Nuts shall be symmetrically formed with the hole centered and at right angles to the face. Nuts shall be tapped to fit a corresponding thread on the anchor bolt such that the nut can be run the entire length of the thread by the fingers without undue forcing, and without play or rocking.

Anchor bolts, nuts, and washers shall be galvanized by the hot-dip process conforming to 210-3, or cadmium plated with Type NS coating conforming to ASTM A165.

209-3.3 Standards.

209-3.3.1 General. Standards shall have an aluminum identification plate attached with stainless steel rivets or screws, as specified on the Plans. Each Standard shall have a handhole in the base and a handhole cover. The handholes shall conform to the details shown on the Plans or Standard Plans. The handhole cover shall be securely attached to the Standard with tamper-resistant hardware or as detailed on the Plans or specified in the Special Provisions.

209-3.3.2 Straightness. For street lighting and pedestrian Standards the maximum deviation shall not exceed the tolerance specified in Table 209-3.3.2 (A) when measured with the Standard in the vertical position. For traffic signal standards the maximum deviation shall not exceed the tolerance specified in Table 209-3.3.2 (A) when measured with the Standard in the horizontal position.

TABLE 209-3.3.2 (A)

Length of Standard – Excluding Base		Maximum Allowable Deviation From String Line[1]
Over	Equal to or Less	
---	21 feet (6.4m)	1/2 inch (12mm)
21 feet (6.4m)	25 feet (7.9m)	3/4 inch (19mm)
25 feet (7.9m)	35 feet (10.7m)	1 inch (25mm)
35 feet (10.7m)	40 feet (12.2m)	1-1/4 inches (32mm)
40 feet (12.2m)	---	As specified in the Special Provisions.

1. The maximum deviation shall be measured from a string line on the face of the Standard, in a plane passing through the longitudinal axis.

Short crooks in the Standard shall not exceed 1/4 inch (6mm) deviation from the centerline of the Standard for each 5 feet (1.5m) of length.

Offsets or jogs due to mold extensions or joints shall not exceed 1/16 inch (1.6mm) in thickness along the surface of the Standard.

209-3.3.3 Metal Standards.

209-3.3.3.1 General. The top of each Standard shall be equipped with an ornamental cap which shall be securely held in place by a 3/4 inch (19mm) diameter hex head machine bolt. Provisions shall be made for substituting 3/4 inch (19mm) diameter steel insulator pin. The cap and cap support surface shall have sufficient strength to transfer to the Standard, from a point 5 inches (125mm) above the top of the cap, a 950 lbs (4,250 N) horizontal loading.

Metal standards shall withstand, without permanent deformation, a maximum horizontal load of 950 lbs (4,250 N) applied to the center of the cap that is attached to the Standard. In addition, metal standards shall withstand, without exceeding a deflection of 2-5/16 inches (60mm) a normal horizontal load of 370 lbs (1,650 N) applied to the cap that is attached to the top of the Standard.

Exposed edges of plates that comprise the base assembly shall be finished smooth. Exposed corners shall be rounded unless otherwise shown on the Plans. Slots or drilled holes shall have a tolerance of 0 to + 1/8 inch (3mm).

Metal standards shall be equipped with anchor bolt covers made of metal of the same type as that used for the shafts or as specified on the Plans or Standard Plans. Anchor bolt covers shall be equipped with all necessary fittings and hardware for securing the bolt covers to the Standard.

209-3.3.3.2 Steel Standards. Unless otherwise specified in the Special Provisions, steel standards shall be fabricated from sheet steel of weldable grade having minimum yield strength, after fabrication, of 40,200 psi (276MPa). When a single-ply 5/16 inch (8mm) thick steel standard is specified, a 2-ply steel standard with an equivalent section modulus may be substituted unless otherwise specified in the Special Provisions. Certified test reports that verify conformance to this minimum yield strength shall be submitted to the Engineer in accordance with 2-5.3.4.

Steel standards may be fabricated from full length sheets or shorter sections. Each section shall be fabricated from not more than 2 pieces of sheet steel. Where 2 pieces are used, the longitudinal welded seams shall be directly opposite each other. When the sections are butt welded together, the longitudinal welded seams on adjacent sections shall be placed to form continuous straight seams from the base to the top. In addition, butt welded sections shall be strengthened by inserting a welded sleeve at each joint. The sleeve shall be fabricated from steel 1/8 inch (3mm) nominal thickness or thicker of the same composition as the steel used in the Standard. The sleeve shall have a minimum length of 1 inch (25mm). The sleeve shall be centered at the joint and have the same taper as the Standard such that the outside of the sleeve is in full contact throughout its length and circumference. The weld metal at the transverse joint shall extend to the sleeve. No transverse joint shall occur within 3 inches (75mm) of mast arm fittings.

All welds shall be continuous and conform to the requirements of 209-2. Welds joining the shafts of Standards to their base plates shall be as shown on the Plans or Standard Plans. However, alternative weld joint details may be approved by the Engineer. Approval of alternative weld joint details will be contingent upon the proposed weld joint passing both weld procedure and nondestructive testing as deemed necessary by the Engineer. All costs of the weld procedure and nondestructive testing shall be borne by the Contractor.

Longitudinal welds in steel tubular sections will be tested by the Agency in conformance with the requirements in California Test 664. The sampling frequency shall be as directed by the Engineer. Welds may be made by the electric resistance welding process. Exposed welds shall be ground flush with the base metal.

Steel standards shall be galvanized by the hot-dip process conforming to 210-3, or cadmium plated with Type NS coating conforming to ASTM A165. If specified in the Special Provisions or shown on the Plans, steel standards shall be painted in accordance with 210-1.

209-3.3.3.3 Aluminum Standards. Aluminum standards shall be fabricated from seamless tubing conforming to "6063-T6 wrought aluminum alloy of the Specifications of the Aluminum Association" or the specifications in Table 209-3.3.3.3 (A).

TABLE 209-3.3.3.3 (A)

ITEM	ASTM
Castings	B26/B26M
Luminaire Arm	B490-90a
Spun Shaft	B241/B241M
Square Extruded Shaft	B429

The wall thickness of the shaft shall be a minimum of 1/4 inch (6mm). Aluminum standards shall be supplied with a mill finish, and be uniform and commercially sound in conformance with ASTM B209M.

209-3.3.4 Fiberglass Standards. Fiberglass standards shall consist of fiberglass-reinforced thermosetting plastic poles. Fiberglass standards shall be hollow, tapered or with tapered sections, be non-conductive and chemically inert. Fiberglass standards shall be in conformance with AASHTO and ANSI specifications in 209-2. The Contractor shall submit to the Engineer a Certificate of Compliance from the manufacturer in accordance with 2-5.3. The certificate shall include a copy of all applicable test reports. The test reports shall be signed and stamped by the licensed engineer who supervised the tests. The certificate shall also comply with 4-1.5 and certify that the Standards were manufactured in accordance with an Agency-approved testing and quality control program.

Fiberglass standards shall be constructed of continuous fiberglass filaments combined with thermosetting polyester and ultraviolet-resistant resin. The fiberglass and resin ratio shall contain at least 65 percent glass and 35 percent resin by weight. The resin shall be pigmented as required by the Standard Plans or Special Provisions and be of uniform color throughout the entire body of the Standard. The glass filament shall be helically wound under tension at angles to provide axial strength. The finish shall be smooth.

Fiberglass standards shall be flame resistant in accordance with ASTM D635. In addition fiberglass standards shall be reinforced in areas of handholes, conduit entrance openings, and mast arm connections.

Fiberglass standards shall be direct burial or have an anchor base as shown on the Plans or the Standard Plans. The base shall be bonded to the pole with an adhesive recommended by the manufacturer, and coated with an aliphatic-type acrylic-modified polyurethane finish. Each Fiberglass standard shall be equipped with a removable aluminum or galvanized steel pole top cap.

An aliphatic-type acrylic-modified polyurethane coating shall be applied to the exterior of each fiberglass standard. The coating shall be semi-gloss, weather resistant and match the color of the resin specified in the Special Provisions. The coating shall have a minimum of 3 mils (75μm) dry film thickness. A 1 quart (1 liter) can of the coating that matches the specified color shall be furnished as specified in the Special Provisions. The polyurethane coating shall be tested by for adhesion in accordance with the requirements of ASTM D3359, Method A and shall have a scale rating of 5A. The adhesion testing shall be conducted before and after the accelerated weathering evaluation by the manufacturer. The Contractor shall provide the Engineer with a copy of the test results from the manufacturer in accordance with 2-5.3. The test reports shall be signed and stamped by the licensed engineer who supervised the tests.

Finished surfaces shall be capable of withstanding a minimum of 2,500 hours of accelerated weathering when tested by the Agency, in accordance with the requirements of ASTM G 53. Testing lamps shall be UV-B (313nm wavelength). The testing cycle shall consist of 4 hours of ultraviolet (UV) exposure at 140°F (60°C), followed by 4 hours of condensate exposure at 100°F (40°C). After testing, the finished surface of fiberglass standards shall exhibit the following:

a) Fiber Exposure – None

b) Crazing – None

c) Checking – None

d) Chalking – Very slight

e) Change in color - May dull slightly.

Each fiberglass standard shall be spiral-wrapped in its entirety with a weatherproof wrap for protection during shipping and storage.

209-3.4 Mast Arms. Mast arms shall be fabricated from the same material specified for the adjoining Standards, except for fiberglass standards. Mast arms for fiberglass standards shall be fabricated from aluminum. Mast arms shall be fabricated in conformance with 209-2, 209-3.3.3.2 or 209-3.3.3.3 depending on the material the mast arm is fabricated from. Mast arms shall be smoothly curved to the approximate configuration shown on the Plans or Standard Plans. A smooth curving arm is required. Mast arms shall have an aluminum identification plate attached with stainless steel rivets or screws, as specified on the Plans.

Exposed welds, except fillet and fatigue resistant welds on top of the mast arms shall be ground flush with the base metal.

Mast arms shall be furnished complete with all necessary fittings and hardware for attachment of the arm to the adjoining Standard.

209-3.5 Conduit.

209-3.5.1 General. Conduit shall be as shown on the Plans or specified in the Special Provisions.

209-3.5.2 Rigid Metallic Conduit. Conduit and conduit fittings shall be galvanized by the hot-dip, electrodepositing, or metallizing process in accordance with 210-3. Conduit shall conform to Underwriters' Laboratories, Inc., publication UL 6 for Rigid Metallic Conduit, and shall bear the UL label on each length. The zinc coating shall be tested in conformance with ASTM A 239.

209-3.5.3 Galvanized Pipe. Galvanized pipe used as conduit shall conform to the following:

a) ASTM A53 Grade A.

b) Only standard lengths shall be supplied to the Work site.

c) Need not be UL listed.

209-3.5.4 Rigid Non-Metallic Conduit. Rigid non-metallic conduit shall conform to the requirements of the UL Standard for Rigid Non-Metallic Conduit, Publication UL 651 (PVC Schedule 80) and UL 651B (HDPE). Rigid non-metallic conduit connections shall be of the solvent weld type.

209-3.6 Pull Tape or Rope. Pull tape or rope shall be as specified in the Special Provisions and used to install wires or cables. Pull tape shall consist of a flat, woven, lubricated, soft-fiber polyester tape with a minimum tensile strength of 1,800 lbs. (8,000 N) and shall have a painted sequential measurement marking at least every 3 feet (1m).

Pull rope shall be a 1/4 inch (6mm) diameter polyethylene rope with a minimum tensile strength of 1,800 lbs (8,000 N).

209-3.7 Pull Boxes. Pull boxes, covers, and extensions shall be approved by the Agency or as shown on the Plans or Standard Plans and/or specified in the Special Provisions.

When a ballast, transformer or other device is to be placed in a non-metallic pull box, the pull box shall be provided with recesses for a hanger.

209-3.8 Splice Insulation.

209-3.8.1 General. Splice insulation shall be as specified herein or as specified in the Special Provisions.

209-3.8.2 Splice Insulation Tape and Pads. Splice insulation shall consist of layers of vinyl chloride electrical insulating tape, conforming to ASTM D2301, Type I.

Low-voltage tape shall be UL or ETL listed and shall be the following types:

a) Self-fusing, oil and flame-resistant, synthetic rubber.

b) Pressure-sensitive, adhesive, polyvinyl chloride, 150µm minimum thickness.

Tape for insulating splices in high-voltage (over 600 V) circuits shall be designed for use on 5kV circuits and shall be resistant to ozone, corrosion and water. Insulating pads shall be composed of a laminated 1/16 inch (2mm) thickness of electrical grade polyvinyl chloride and a 1/8 inch (3mm) thickness of butyl splicing compound with removable liner.

209-3.8.3 Heat-Shrink Tubing. Heat-shrink tubing shall be medium or heavy wall thickness, irradiated polyolefin tubing containing an adhesive mastic inner wall. The minimum wall thickness prior to contraction shall be 1/32 inch (800µm). When heated, the inner wall shall melt and fill all crevices and interstices of the object being covered while the outer wall shrinks to form a waterproof insulation. Each end of the tube or open end of the end cap of heat-shrink tubing shall, after contraction, overlap the conductor insulation at least 1-1/2 inches (38mm). Heat-shrink tubing shall conform to the requirements of UL 468D and ANSI C119.1 for extruded insulated tubing at 600 V. Heat-shrink tubing shall also meet the following requirements:

a) **Shrinkage Ratio:** 33 percent, maximum, of supplied diameter when heated to 257°F (125°C) and allowed to cool to 77°F (25°C).

b) **Dielectric Strength:** 140kV per 3/8 inch (10mm), minimum.

c) **Resistivity:** 10^{13} Ω-mm, minimum.

d) **Tensile Strength:** 2000 psi (14MPa), minimum.

e) **Operating Temperature:** -40°F to 194°F (-40°C to 90°C) [275°F (135°C) Emergency].

f) **Water Absorption:** 0.5 percent, maximum.

209-3.9 Bonding and Grounding. Bonding jumpers shall be copper wire or copper braid of the same cross sectional area as No. 6 AWG conductor for multiple circuit lighting systems and No. 8 AWG conductor or larger for series and all other street lighting and traffic signal systems. The cross sectional area of the jumper shall match the cross sectional area of the conductors in the circuit, or shall be as shown on the Plans or specified in the Special Provisions. Equipment grounding conductors shall be color coded to code requirements or shall be bare.

Bonding bushings to be installed on metal conduit shall be of the galvanized or zinc alloy type.

Ground electrodes shall be one-piece, 100 inches (2.5m) long galvanized steel rod or pipe no less than 1 inch (25mm) in diameter, or of copper-coated (minimum thickness of rod coating 10 mils (300µm)) steel rod not less than 1/2 inch in diameter by 8 feet long (12mm in diameter by 2.4m long).

209-3.10 Service.

209-3.10.1 General. Service conduit shall conform to the requirements of the serving utility and shall not be less than 2 inches (50mm) in diameter.

Where a kilowatt-hour meter is required, a meter socket with sealing ring, as approved by the serving utility, shall be furnished and installed. Where a meter socket is installed, the meter enclosure shall be provided with factory installed test bypass facilities as required by the serving utility.

Service equipment enclosures shall be galvanized or if so specified on the Plans, Standard Plans or in the Special Provisions, the enclosures shall be provided with a factory applied rust resistant prime coat and finish coat instead of galvanizing.

209-3.11 Circuit Breakers. Circuit breakers shall be of the quick-break type and also be capable of either automatic or manual operation. The operating mechanism shall be enclosed and shall be trip-free from the operating handle on overload. Circuit breakers shall be trip-indicating, have the frame size plainly marked, and have the trip rating clearly indicated on the operating handle. Overload tripping breakers shall not be influenced by an ambient temperature range from 0°F (–18°C) to 122°F (50°C). Multiple-pole breakers shall be the internal trip type. All circuit breakers shall be listed by UL or ETL.

209-4 STREET LIGHTING SYSTEM MATERIALS.

209-4.1 Reinforced Concrete Standards.

209-4.1.1 General. The manufacture of, and all material used in the construction of reinforced concrete standards shall conform to 201-1 and 201-2. Portland cement concrete shall be as specified in the Special Provisions. Shop drawings and a concrete mix design shall be submitted to the Engineer for approval in accordance with 2-5.3.

Reinforced concrete standards shall have a uniform color and surface appearance.

Reinforced concrete standards shall be constructed of pre-stressed, centrifugally spun, exposed aggregate, portland cement concrete. They may have black and white marble aggregate or natural exposed aggregate, and be direct burial or anchor base type. The cross section shall not deviate more than 3/16 inch (5mm) from the shape and dimensions shown on the Plans, Standard Plans or approved shop drawings. The length shall not deviate more than 1 percent of its total length or a maximum of 3 inches (75mm) from the dimensions shown on the Plans, Standard Plans or approved shop drawings. Maximum deviation from a stringline at any point shall not exceed 1/32 inch per foot (2.6mm per meter) of length. Reinforced concrete standards shall be sufficiently straight when in a vertical position such that the maximum deviation from a string line on the face of the Standard in a plane passing through the longitudinal axis shall not exceed the tolerance listed in Table 209-3.3.2 (A).

A central opening or duct shall be formed throughout the entire length. The opening or duct shall be 1-1/4 inches (32mm) in diameter or as shown on the Standard Plans or approved shop drawings. The opening or duct shall be free from sharp projections or edges which might damage wire or cable. The base shall be cored to the dimensions shown on the Standard Plans or approved shop drawings.

Reinforced concrete standards used with post-top luminaires shall be terminated with a post-top adapter with a tenon as shown on the Plans, Standard Plans or approved shop drawings.

209-4.1.2 Reinforcement. Reinforced concrete standards shall be reinforced with at least 4 prestressed strands. The prestressed steel strand shall consist of an uncoated 7-wire stress relieved strand conforming to 201-2. Each strand shall be mechanically anchored to the base plate or base-reinforcing cage depending on type of base specified. Strands shall be pretensioned to develop the design strength specified in the Special Provisions or on the Standard Plans. However, in no case shall the stress in the strands immediately after transfer exceed 70 percent of the minimum specified ultimate strength of the strands. Deformed bars may be used in addition to the prestressed strands depending on the size and loading.

Longitudinal reinforcing shall be spirally wrapped with No. 11 AWG (3mm) or larger steel wire on a variable pitch, dependent upon the loading. The maximum pitch shall not exceed 6 inches (150mm). Transformer bases cast integral with the shaft shall be reinforced with deformed steel bars securely attached to the base anchor lugs or plate and prestressing cables.

Four galvanized deformed bar studs at least 1 foot (300mm) long shall be extended 1 inch (25mm) above the top of the reinforced concrete standard for attachment of a bracket arm or tenon when required. The metal components of the reinforced concrete standard shall conform to the requirements in Table 209-4.1.2 (A).

TABLE 209-4.1.2 (A)

MATERIAL	ASTM
Prestressed Steel	A 416
Deformed Bar	A615 Grade 40 or Grade 60/A706
Base Plate	A 36
Base Anchor Lugs	A 36

209-4.1.3 Casting of Concrete Standards.

Reinforced concrete standards shall be cast in rigid steel molds such that the relief in the design of the concrete standard is sharp and distinct. Longitudinal steel reinforcement shall be securely anchored to the top and bottom of the mold plates and shall be placed to provide 3/4 inch (19mm) minimum concrete cover except within 2 feet (600mm) of the top where the minimum cover shall be 5/8 inch (16mm).

Reinforced concrete standards shall be cured in accordance with the following phases:

a) Mold Cure:

Reinforced concrete standards shall be cured in their molds until the concrete has attained a set sufficiently hard to prevent deformation or slipping of cable strands.

b) After the mold cure is complete, the remaining cure shall conform to 207-2.7.

After sufficient curing, the entire outside surface shall be blasted to remove cement laitance and develop the surface texture.

209-4.1.4 Finish of Concrete Standards. The finish shall be without cracks or crazing, rock pockets or any discoloration or ridges and shall have a uniform surface (without objectionable mold marks) and texture throughout the entire length.

Surface defects extending less than 5/8 inch (16mm) in the shell may be repaired without the prior approval of the Engineer. The repair shall be made by a method that will produce a permanent bond and a surface that blends into the color and surface appearance of the remainder. Surface defects extending greater than 5/8 inch (16mm) into the shell shall not be repaired prior to inspection by the Engineer. Reinforced concrete standards containing voids or insufficient cover over prestressing cables, reinforcing steel or a blocked central opening or duct will not be accepted.

209-4.1.5 Acceptance and Rejection of Concrete Standards. Reinforced concrete standards shall be labeled or marked in a permanent fashion to indicate the manufacturer, and month and year of fabrication. Markings shall be located within 2 feet (600mm) of the bottom but not on the hand-hole cover. Markings shall be visible through the handhole opening.

A lot shall be defined as 100 reinforced concrete standards or less of the same design, produced in sequence on consecutive working days, and bearing the specified markings. The standards composing each lot shall be uniform in color and surface appearance. A new lot number shall be assigned if there is any change in the size, type, or spacing of reinforcing or prestressing steel, concrete mix, or curing method. The frequency of sampling and testing shall be as specified in the Special Provisions. The Contractor shall notify the Engineer, in writing, at least 5 days prior to any contemplated change in supplier or source of raw materials, and shall obtain approval of any changes before the standards are delivered to the Agency.

209-4.2 Wire/Conductors.

209-4.2.1 General. Wire shall be a single strand of drawn metal that may or may not be insulated. Conductor shall be a single (solid) or multiple (stranded) strands metal that are insulated. Cable shall be multiple wires or conductors that are insulated and enclosed in an insulating cable jacket. Copper wire shall conform to ASTM B3 and B8. Wire sizes shall be based on the American Wire Gage (AWG). The wire diameter shall not be less than 98 percent of the specified AWG diameter.

All conductors shall have clear, distinctive and permanent markings on the outer surface throughout the entire length showing the manufacturer's name or trademark, insulation type letter designation, wire size, and voltage rating.

Conductor insulation shall be a solid color or basic colors with a permanent colored stripe unless otherwise specified in the Special Provisions. Solid or basic colors shall be homogenous throughout the full depth of insulation except for the clear nylon layer, if present.

209-4.2.2 Series Circuits Conductors. Wire for series street lighting systems shall be No. 8 AWG solid copper wire. The wire shall be insulated with a110 mils (2800μm) thickness of black polyethylene insulation per Standard S-61-402 of the Insulated Power Cable Engineers Association, and designated for operation at 5,000V. The Contractor shall submit a Certificate of Compliance in accordance with 2-5.3.3 and 4-1.5, for all 5,000V (6.6A) series lighting conductors.

Where isolating transformers or external (remote) ballasts are used, the secondary wires, in the Standard, from the transformer or external (remote) ballast to the luminaire shall be insulated No. 10 AWG solid copper wire. The insulation for these wires shall be black, rated, and UL approved for 600V operation. The insulation shall be standard THW or THWN/THHN grade polyvinyl chloride, conforming to the applicable provisions of ASTM D2219 and D2220.

209-4.2.3 Multiple Circuits Conductors. Conductors for multiple circuits shall be of the type, size and insulation colors specified in Table 209-4.2.3(A) or on the Plans. The insulation shall be THW or THHN/THWN, or as specified on the Plans, grade polyvinyl chloride, conforming to the applicable provisions of ASTM D2219 and D2220. At any point the minimum thickness of any THW or THHN/THWN insulation shall be 55 mils (1400μm) for conductor sizes No. 14 AWG to No. 10 AWG, inclusive and 59 mils (1500μm) for No. 8 AWG to No. 2 AWG inclusive.

TABLE 209-4.2.3 (A)

Multiple Circuits	Function	Identification				Size
		Insulation Colors		Band Symbols[c]		
		Base	Stripe[a]			
Multiple Service (Stranded)	Ungrounded Line 1	Black	None	NBR		8
	Ungrounded Line 2	Red	None	NBR		8
	Grounded/Neutral	White	None	NBR		8
Street Lighting Pull Box to Pull Box[b] (Stranded)	Ungrounded Line 1 (odd)	Black	None	NBR		8
	Ungrounded Line 2 (even)	Red	None	NBR		8
	Grounded/Neutral	White	None	NBR		8
Street Lighting Pull Box to Luminaire[b] (Solid)	Ungrounded Line 1 (odd)	Black	None	NBR		10
	Ungrounded Line 2 (even)	Red	None	NBR		10
	Grounded/Neutral	White	None	NBR		10
Lighting Control From Luminaire's with PEU's to Luminaire's without PEU's (Solid)	Ungrounded to PEU	Black	None	C1		10
	Switching leg of PEU to terminal strip.	Red	None	C2		10
	Switching leg of PEU to unswitched Luminaire[d].	Purple	None	NBR		8/10[d]

NBR = No Band Required PEU = Photoelectric Unit

a. Band conductors in each pull box and near ends of termination point.

b. Both wires between external high intensity discharge (HD) ballast lamp shall be black No. 10 AWG Solid wire.

c. "S" if circuit is switched on line side of service equipment by utility.

d. No. 8 AWG Stranded conductor from pull box to pull box and No. 8 AWG solid wire in the Standards.

209-4.3 Splicing Units and Terminal Blocks.

209-4.3.1 Disconnecting Splice Connector (High Voltage).

209-4.3.1.1 Connector Case. The case or shell shall be formed of a neoprene or plastic composition having a volume resistivity not less the 3.94 x 10^{12} ohms per inch (10^{13} ohms per cm). The diameter of the case shall not exceed 2-1/2 inches (63mm) and the length shall not exceed 10 inches (250mm). The case shall withstand a load of 200 lbs (900 N) applied normal to the axis of the case without rupture, cracking, or permanent deformation.

The two halves of the case shall lock together with screw threads forming a watertight joint and be separated by a neoprene "O" ring under compression when the connector is assembled. The wires that are to be spliced shall attach to the connector by a self threading lug made from a No. 8 AWG solid copper wire. The connecting lug shall be slotted at 90 degrees to provide a pressure fit between the male and female parts. All metal parts shall be cadmium plated.

Each half of the case shall be completely filled with a sealing compound complying with 209-4.3.1.2 and held in place with a wire/cable entrance plug designed to accommodate any cable having an overall diameter of between 11/32 inch (9mm) and 5/8 inch (15mm).

209-4.3.1.2 Sealing Compound. The sealing compound for the cable connector case shall be a viscous paste composed of oil and an inert mineral filler. The compound shall be waterproof and resistant to oxidation, weathering and biological attack. The volume resistivity shall not be less than 3.94 x 10^{11} ohms per inch (10^{12} ohms per cm).

When the sealing compound is exposed to air for 18 hours in a film 1/8 inch (3mm) thick at a temperature of 70°F ± 2°F (21°C ± 1.1°C) absorption of water shall not exceed 0.02 fl. oz. per inch^2 (0.6mL per cm^2) of surface area in contact with the water.

When allowed to dry in contact with air, the lineal shrinkage of the exposed surface of the compound shall not exceed 0.2 percent.

209-4.3.1.3 Electrical Characteristics. When two wire/cable ends are connected and sealed in the connector case with sealing compound, the unit shall sustain an electrical stress of 11,000 V at 70°F± 2° F (21°C± 1.1°C), without breakdown, after immersion in water for 18 hours.

209-4.3.2 Fused Splice Connectors. Fused disconnect connectors shall have no exposed metal parts except for the head of a stainless steel assembly screw. The head of the stainless steel assembly screw shall be recessed a minimum of 1/32 inch (800μm) below the top of a plastic boss which surrounds the head.

Fused splice connectors shall completely enclose the fuse and shall protect the fuse against damage from water and weather. The contact between the fuse and the fuseholder shall be spring pressure.

Fuses shall be standard midget, ferrule type, with "Slow Blow" feature and shall be 3/8 inch x 1-1/2 inches (10mm x 38mm).

209-4.3.3 Terminal Blocks. Terminal blocks shall have a minimum of 12 positions with terminals rated at size No. 8 AWG or larger, to accept the field wires indicated on the Plans.

209-4.4 High Pressure Sodium Luminaires.

209-4.4.1 General. High-pressure sodium luminaires shall be the enclosed cutoff type with a horizontal burning lamp unless otherwise specified in the Special Provisions or on the Plans. Each luminaire shall consist of a housing, a reflector, a refractor or lens, a lamp socket, an internal (integral) ballast, starter, a terminal strip and a lamp.

Each mast arm mounted luminaire shall be furnished without a photoelectric unit receptacle unless otherwise specified in the Special Provisions or on the Plans. If a photoelectric unit receptacle is included, a rain tight shorting cap shall be installed. If a luminaire housing is provided with a hole for the receptacle, the hole shall be closed, covered and sealed with weatherproof material.

The housing shall be fabricated from aluminum. Housings that are painted shall withstand a 1000-hour salt spray test as specified in ASTM B117. Other metal parts of the housing shall be fabricated from metal that equals or exceeds the corrosion resistance and finish of the metal used for the housing. The housing shall be equipped with a slip-fitter capable of mounting on a 2 inches (50mm) pipe tenon and of being adjusted 5 degrees from the axis of the tenon. The clamping brackets of the slip-fitter shall not bottom out on the housing bosses when adjusted within the ± 5 degree range.

No part of the slip-fitter mounting brackets shall develop a permanent set in excess of 1/64 inch (400μm) when the four 3/8 inch (10mm) diameter cap screws used for mounting are tightened to 9.62 ft·lbs (13 N·m).

Luminaires to be mounted on horizontal mast arms, shall be tested in conformance with California Test 611, and be capable of withstanding cyclic loading in (G = Acceleration of Gravity):

a) A vertical plane at a minimum peak acceleration level of 3.0 G peak-to-peak sinusoidal loading (same as 1.5 G peak) with the internal ballast removed, for a minimum of 2 million cycles without failure of any luminaire parts, and

b) A horizontal plane perpendicular to the direction of the mast arm at a minimum peak acceleration of 1.5 G peak-to-peak sinusoidal loading (same as 0.75 G peak) with the internal ballast installed, for a minimum of 2 million cycles without failure of any luminaire parts, and

c) A vertical plane at a minimum peak acceleration level of 1.0 G peak-to-peak sinusoidal loading (same as 0.5 G peak) with the internal ballast installed, for a minimum of 2 million cycles without failure of any luminaire parts.

The surface of each reflector shall be specular and shall be protected by either an anodized finish or a silicate film.

Each refractor or lens shall be mounted in a frame that is hinged to the housing and secured with a spring-loaded latch. Each refractor shall be made of glass or polycarbonate plastic as specified in the Special Provisions. Each lens shall be made of heat and impact resistant glass.

The optical system, consisting of the reflector, refractor or lens, lamp socket and lamp, shall be contained in a sealed chamber. The seal shall consist of a gasket between the reflector and refractor or lens and a gasket between the reflector and lamp socket. The chamber shall have provision for a filtered flow of air in and out due to lamp heat. Filtering shall be accomplished by either a separate filter or a filtering gasket.

Each lamp socket shall be a porcelain enclosed mogul-multiple type. The shell shall contain integral lamp grips to assure electrical contact under conditions of normal vibration. The socket shall be mounted in the luminaire in a manner to permit pre-setting a variety of specified light distribution patterns. The socket shall be rated for 1500W and 600V, and shall be rated for a 4KV pulse.

When the ballast, starter and terminal strip are mounted on a down-opening door, the door shall be hinged and secured to the luminaire separately from the refractor or flat lens frame. The door shall be removable and replaceable. The door shall be secured to the housing in a manner to prevent its accidental opening when the refractor or flat lens frame is opened.

Field conductors connected to the luminaire shall terminate on a barrier type terminal block secured to the housing. Terminal screws shall be captive and equipped with wire grips for conductors up to No. 6 AWG. Each terminal position shall be clearly identified.

The minimum light distribution for each luminaire shall be as shown on the isolux diagrams on the Plans. The maximum brightness of each cutoff luminaire, with the lamp indicated shall be as specified in Table 209-4.4.1 (A).

TABLE 209-4.4.1 (A)

Lamp ANSI Code No.	Lamp Wattage	Maximum Brightness (cd/m^2)
S55	150	140
S66	200	140
S50	250	175
S67	310	210
S51	400	260

Brightness readings specified are based on the use of a brightness meter with an acceptance angle of 1.5 degrees. When measured on the 90-degree and 270-degree lateral angle line, the maximum brightness shall not exceed the above specified brightness when the meter is located at a horizontal distance of 120 feet (37m) and a vertical distance of 7 feet 6 inches (2.3m) between the luminaire and the meter, or at an angle of 3 degrees 35 minutes from the horizontal to the line between the luminaire and the meter. Measurements shall be made from both the 90 degree line and the 270 degree line and averaged. The lamp for each test shall be operated at the wattage necessary to produce a light output as specified in Table 209-4.4.1 (B).

TABLE 209-4.4.1 (B)

Lamp Wattage	Lumens
150	16,000
200	22,000
250	27,000
310	37,000
400	50,000

209-4.5 Lamp Ballasts.

209-4.5.1 General. The input voltage for ballasts and ballast types shall be as shown on the Plans or as specified in the Special Provisions. The ballast shall be plainly marked as to its operating voltage and wattage characteristics. Noise generated by the ballast or in combination with the starting aid shall be minimized. The ballast, including starting aid, shall incorporate protection against normal lamp failure modes. Ballasts for luminaires to be mounted on mast arms, brackets or lowering assemblies shall be located within the luminaire housing.

Each ballast shall be designed for the type, characteristics and wattage of the lamp it is to operate and it shall provide lamp starting and waveforms, voltage and current specified in the Special Provisions. Ballasts shall also be designed to operate under outdoor conditions, either mounted internally (integral) within the luminaire or lighting fixture or mounted externally (remote) to the luminaire. Ballasts shall provide reliable starting and operation at ambient temperatures down to -22°F (–30°C) for the rated life of the lamp.

The ballast shall be designed for continuous operation at ambient air temperatures from -4°F (-20°C) to 149°F (65°C) without reduction in ballast life. Ballasts shall have a design life of not less than 100,000 hours. Ballast shall be designed to operate for a minimum of 180 cycles of 12 hours on and 12 hours off, with the lamp circuit in an open or short-circuited condition and without measurable reduction in the operating requirements in 209-4.4.1.

Heat-generating components shall be mounted so as to use the portion of the luminaire they are mounted on as a heat sink. Capacitors shall be located as far as practicable from heat-generating components or shall be thermally shielded to limit the case temperature to 167°F (75°C).

Transformers and inductors shall be resin-impregnated for protection against moisture. Capacitors, except those in starting aids, shall be encased in metal and hermetically sealed.

Ballasts shall be tested in conformance with ANSI C82.6-1980.

209-4.5.2 Internal Ballasts. Each internal ballast shall consist of separate components, each of which shall be capable of being easily replaced. A starting aid which is encapsulated will be considered as a single component. Each component shall be provided with screw terminals, NEMA tab connectors or a single multi-circuit connector. All conductor terminals shall be identified as to the component terminal to which they are connected.

Internal ballasts shall be tested within the luminaire and shall be rated at a minimum temperature of 221°F (105°C) or as specified in the Special Provisions.

209-4.5.3 Remote Ballasts. Ballasts to be located remote (external) from the luminaire shall be the submersible type and shall conform to 209-4.8.3. All components, including starting aids, shall be enclosed in a single housing. Ballast leads shall extend a minimum of 12 inches (300mm) from the housing. Steel housings shall be galvanized or painted. Ballast housings shall be clearly labeled to indicate lamp type, lamp wattage and input voltage.

209-4.5.4 Starting Aids. Starting aids, if required, shall be interchangeable between ballasts of the same wattage and manufacturer without adjustment.

209-4.5.5 High Pressure Sodium Lamp - Constant Wattage (CW) or Constant Wattage Auto-Regulator (CWA) – Ballasts. CW and CWA ballasts shall have a ballast characteristic curve which will intersect both of the lamp-voltage limit lines between the wattage limit lines and remain between the wattage limit lines throughout the full range of lamp voltage. This requirement shall be met not only at the rated input voltage of the ballast, but also at the lowest and highest input voltage that the ballast is rated. Throughout the rated lifetime of the lamp, the ballast curve shall fall within the limits of lamp voltage and wattage specified in the Special Provisions.

209-4.5.6 Regulator Type Ballasts.

209-4.5.6.1 General. Regulator type ballasts shall be lag-type or lead-type conforming to the following:

a) For nominal input voltage and lamp voltage, the ballast design center shall not vary more than 7.5 percent from rated lamp wattage.

b) The ballast shall be designed such that a capacitance variance of ± 6 percent will not cause more than a ± 8 percent variation in lamp wattage regulation throughout the rated lamp life for nominal input voltage.

c) The lamp current crest factor shall not exceed 1.8 for input voltage variation of ± 10 percent at any lamp voltage from initial through life.

209-4.5.6.2 Lag-Type. Lag-type regulator ballasts shall have the primary and secondary windings electrically isolated and, when operated with the appropriate lamp, shall have the following characteristics and maintain the following lamp operation:

a) The power factor shall be not less than 90 percent throughout the life of the lamp at nominal line voltage with a nominally rated reference lamp.

b) Lamp wattage regulation spread at any lamp voltage from nominal through life shall not vary by more than 18 percent for a ± 10 percent input voltage variation.

209-4.5.6.3 Lead-Type. Lead-type regulator ballast (CWA) shall when operated with the appropriate lamp, have the following characteristics and shall maintain the following lamp operation:

a) The power factor shall be not less than 90 percent when the ballast is operated at nominal line voltage with a nominally rated reference lamp.

b) Lamp wattage regulation spread at any lamp voltage from nominal through life shall not vary by more than 30 percent for ± 10 percent input voltage variation.

209-4.5.7 Autotransformer or Reactor Type Ballast. Each nonregulating reactor, autotransformer, or high reactance ballast shall, when operated with the appropriate lamp, have the following characteristics and maintain the following lamp operations:

a) The power factor shall be not less than 90 percent when the ballast is operated at nominal line voltage with a nominally rated reference lamp.

b) Lamp wattage regulation spread at any lamp voltage from nominal through life shall not vary by more than 25 percent for ± 5 percent input voltage variation.

c) For nominal input voltage and lamp voltage, the ballast design center shall not vary more than 7.5 percent from rated lamp watts.

d) The lamp current crest factor shall not exceed 1.8 for an input voltage variation of ± 5 percent at any lamp voltage from initial through life.

209-4.5.8 Physical Requirements. External leads for multiple to multiple and series to multiple secondary connections shall be Type USE No. 10, rated 600VAC. Primary conductors for series to multiple ballasts shall be rated for use on 5000V AC circuits.

Ballast leads shall extend a minimum of 12 inches (300mm) from the case. The ballast insulation shall be NEMA 365°F (185°C) or as specified in the Special Provisions.

Series to multiple ballasts shall withstand the application of 12,000V AC from core to primary coil and from coil to coil for a one minute period. Series to multiple ballasts secondaries and multiple-to-multiple ballast shall withstand the application of 2,200V AC from core to coils and, for multiple units only, from coil to coil for a one minute period. These tests shall be made immediately after operation of the ballast at full load for 24 hours.

Non-submersible ballast shall be provided with metal half-shell coil protection, moisture resistant, synthetic varnish impregnated windings and shall be suitable for outdoor operation in a rain tight enclosure.

Each ballast to be installed in a pull box shall be the submersible type and be provided with handle and a hanger.

209-4.6 High Pressure Sodium Lamps. High pressure sodium lamps shall conform to ANSI Standard C 78 "Lamp Specifications, Physical and Electrical Characteristics of High-Intensity Discharge Lamps," when tested in conformance with the requirements in ANSI Standard C78.388, "Methods of Measurement of High Pressure Sodium Lamp Characteristics." High-pressure sodium lamps shall have a minimum average rated life of 24,000 hours.

209-4.6.1 Lamp Size Identification Targets. Targets shall be:

a) Constructed of noncorrosive material that will provide a durable and legible surface that is designed to endure for the life of the luminaire,

b) have a stable color background and black block numerals,

c) be 3 inches (75mm) square, and

d) have the identification numbers and background color shown in Table 209-4.6.1(A) and Table 209-4.6.1(B) respectively. The identifying numerals shall be 2 inches (50mm) high, minimum, and have a stroke width of 1/4 inch (6mm).

TABLE NO. 209-.4.6.1(A)

Identifying Numeral	Lamp Wattage
1	18
3	35
5	50 or 55
7	70
9	90
10	100
13	135
15	150 – 55V
15H	150 – 100V
17	175
17B	175 Position Oriented
18	180
20	200
21	215

TABLE NO. 209-.4.6.1(A) *Continued*

25	250
25B	250 Position Oriented
30	300
31	310
36	360
40	400
40B	400 Position Oriented
70	700
XI	1000

TABLE NO. 209-4.6.1(B)

Background Color	Lamp Type
GOLD	High Pressure Sodium
GREEN	High Pressure Sodium/Mercury Ballast
LIGHT BLUE	Mercury
RED	Metal Halide
PURPLE	Incandescent
AQUA	Fluorescent

209-4.7 Photoelectric Controls.

209-4.7.1 General. Photoelectric controls shall consist of a photoelectric unit which plugs into an EEI-NEMA twist lock receptacle integral with the luminaire. Photoelectric controls, if required in the Special Provisions or on the Plans, shall be capable of switching multiple lighting systems directly.

209-4.7.2 Photoelectric Unit. Photoelectric units shall provide an output in response to changing light levels. The response level shall remain stable throughout the life of the control unit. Components shall not require periodic replacement.

Photoelectric units shall have a "turn-on" between 10 and 50 lux and a "turn off" at between 1.5 and 5 times "turn-on." Measurements shall be by the procedures specified in "EEI-NEMA Standards for Physical and Electrical Interchangeability of Light–Sensitive Control Devices Used in the Control of Roadway Lighting".

Photoelectric units shall also conform to the following:

a) The supply voltage rating shall be 60 Hz, 105-130 V, 210-240 V, or 1-5-240 V, as specified.

b) The load rating shall be 1000 W minimum, incandescent, mercury or fluorescent.

c) The operating temperature range shall be from -20°F (-29°C) to 149°F (65°C).

d) The power consumption shall be less than 10 W.

e) The unit shall be housed in a weatherproof enclosure.

f) The base of the unit shall be provided with a three prong, EEI-NEMA standard, twist-lock plug mounting.

g) Be equipped with a "fail-on" feature.

209-4.7.3 Contactor. The contactor shall have contacts rated to switch the specified lighting load and shall be normally open, unless otherwise specified in the Special Provisions. The contactor shall be either the mechanical armature type or the mercury displacement type. The contacts of the mechanical armature type contactor shall be either fine silver, silver alloy, or superior alternative material. The contactor shall have a minimum rating of 30A, per contact, inductive load.

209-4.7.4 Terminal Blocks. Terminal blocks shall be rated at a minimum of 30A, 600V or as specified in the Special Provisions. Terminal blocks shall be molded from phenolic or nylon material and shall be the barrier type with plated brass screw terminals and integral type marking strips.

209-4.8 Transformers.

209-4.8.1 General. Multiple to multiple and series to multiple transformers shall be of the single-phase, dry type designed for operation on a 60 Hz supply.

209-4.8.2 Electrical Requirements. Transformer ratings shall be 120/480V, 240/480V, 480/120V or 480/240V for multiple to multiple units or as shown on the Plans. Transformer ratings for series to multiple units shall be 6.0A/120V or 6.6A/4.8kV or as shown on the Plans. Secondary 480V windings shall be center tapped. Volt-ampere ratings shall be as shown on the Plans. Transformer efficiency shall exceed 95 percent for multiple to multiple units and 80 percent for series to multiple units.

Secondary voltage regulation and tolerance shall be ± 3 percent from half load to full load for multiple to multiple units and +10 percent (maximum) at no load to ± 3 percent at full load for series to multiple units.

Transformers shall have a decal showing a connection diagram. The diagram shall show either color coding or tagging of wires with primary (H1, H2) or secondary (X1, X2) markers, and shall also show the primary and secondary voltage and volt-ampere rating.

209-4.8.3 Submersible Type Transformers. Submersible type transformers shall be securely encased in a rugged corrosion resistant, watertight case and shall withstand a 5-day test submerged in 24 inches (600mm) of salt water (2 percent salt by mass) with 12-hour on and off periods. The operating periods shall be at full load. Leads of submersible transformers shall be brought out through one or more sealed hubs and shall be secured in a manner that will withstand a 101 lbs force (450 N) static pull without loosening or leaking.

209-5 TRAFFIC SIGNAL MATERIALS.

209-5.1 Steel Pedestrian Standards and Pedestals for Controller Cabinets. All steel pedestrian standards and pedestals for controller cabinets shall be constructed of 1/8 inch (3mm) or thicker galvanized steel, 4 inches (100mm) schedule 40 galvanized steel pipe, or size 103 conduit. Steel pedestrian standards and pedestals shall be constructed in conformance with 209-3.3, with the top designed for post top slip-fitting. Galvanized steel pipe shall conform to ASTM A53. Push button posts shall also be manufactured of 2-1/2 inches (64mm) schedule 40 galvanized steel pipe conforming to ASTM A53.

209-5.2 Pedestals. Pedestals shall have an aluminum identification plate attached with stainless steel rivets or screws, as specified on the Plans.

All welds shall be continuous and conform to the requirements of 209-2, 209-3.3.3.2 or 209-3.3.3.3 depending on the material the pedestal is fabricated from. Exposed welds shall be ground flush with the base metal.

Pedestals shall be galvanized by the hot-dip process conforming to 210-3, or cadmium plated with Type NS coating conforming to ASTM A165. If specified in the Special Provisions or shown on the Plans, pedestals shall also be painted in accordance with 210-1.

Pedestals shall be furnished complete with all necessary fittings and hardware for attachment of the controller cabinet to the adjoining pedestal.

209-5.3 Conductors and Cable.

209-5.3.1 General. Copper wire shall conform to ASTM B3 (solid) and 209-4.2. Wire sizes, other than wires used in loop detector lead-in cable, shall be based on the American Wire Gage (AWG). The wire diameter shall not be less than 98 percent of the specified AWG diameter. Conductors used in loop detector lead-in cable shall conform to the requirements in ASTM B286. A Certificate of Compliance shall be submitted by the Contractor in conformance with 2-5.3 and 4-1.5 for each type of cable used in the Work.

Unless otherwise specified on the Plans or in the Special Provisions, Table 209-5.3.1 (A) shall be used to identify the conductors to be used in the Work.

TABLE 209-5.3.1 (A)

Circuit	Signal Phase Or Function	Identification			Size [f]
		Insulation Colors		Band Symbols[d]	
		Base	Stripe [a]		
Vehicle Signals [c]	2, 6	Red, Yellow, Brown	Black	2.6	14
	4, 8	Red, Yellow, Brown	Orange	4.8	14
	1, 5	Red, Yellow, Brown	None	1.5	14
	3, 7	Red, Yellow, Brown	Purple	3.7	14
Pedestrian Signals [c]	2p, 6p	Red, Brown	Black	2p, 6p	14
	4p, 8p	Red, Brown	Orange	4p, 8p	14
	1p, 5p	Red, Brown	None	1p, 5p	14
	3p, 7p	Red, Brown	Purple	3p, 7p	14
Pedestrian Push Buttons[c]	2p, 6p	Blue	Black	P-2, P-6	14
	4p, 8p	Blue	Orange	P-4, P-8	14
	1p, 5p	Blue	None	P-1, P-5	14
	3p.7p	Blue	Purple	P-3, P-7	14
Traffic Signal Controller Cabinet [b]	Underground between Service Switch & Cabinet	Black	None	CON-1	6
		Red	None	CON-2	6
Sign Lighting [e]	Ungrounded Line 1	Black	None	SL-1	10
	Ungrounded Line 2	Red	None	SL-2	10
Grounded and Common	Pedestrian Push Buttons	White	Black	NBR	14
	Signals	White	None	NBR	10
	Sign Lighting	White	None	NBR	12
Railroad Pre-Emption		Black	Black	R	14
Spares		Black	None	NBR	14

NBR = No Band Required

a. On overlaps, insulation is striped for first phase in designation, e.g. Phase (2 & 3) insulation is striped as for phase 2.

b. Band for overlap and special phases as required.

c. These requirements do not apply to signal cable.

d. Band conductors in each pullbox and near each termination point. On signal light circuits, a single band may be placed around 2 or 3 ungrounded conductors comprising a phase.

e. Conductors between ballasts and sign lighting lamps shall be No. 16 AWG and color shall correspond to the ballast leads.

f. The No. 14 AWG conductor shall be solid copper in a 16 mils (410μm) minimum polyvinyl chloride insulation encased in a 5 mils (130μm) minimum clear nylon jacket. The No. 10 conductor shall be solid copper in a 22 mils (560μm) minimum polyvinyl chloride insulation encased in a 8 mils (200μm) minimum clear nylon jacket.

All single conductors, multi-conductor cables, and lead-in cables, shall have clear, distinctive and permanent markings on the outer surface throughout the entire length showing the manufacturer's name or trademark, insulation type letter designation, conductor size, voltage rating and the number of conductors. The multi conductor cable jacket shall be black polyethylene binder sheath, rated for 600V and 167°F (75°C), or black polyvinyl chloride conforming to IMSA 9-1. Filler material, if used to make the multi-conductor round, shall be polyethylene.

Conductor insulation shall be of a solid color or of basic colors with a permanent colored spiral stripe under the nylon jacket, as detailed in Table 209-4.2.3 (A) unless otherwise specified in the Special Provisions. Solid or basic colors shall be homogeneous through the full depth of insulation. Identification stripes shall be a continuous spiral over the entire length of the conductor. For conductor size No. 2 and larger, the insulation may be black and the ends of the conductors shall be taped with electrical insulating tape of the required color for a minimum of 20 inches (508mm).

209-5.3.2 Materials.

209-5.3.2.1 General. Individual wires in the cable shall be solid copper with Type THWN or polyvinyl chloride insulation as specified in the Special Provisions, and shall conform to the requirements in 209-4.2.1 and ASTM B286. The minimum thickness of Type THWN insulation, at any point, shall be 12 mils (300μm) for No. 14 conductor and 16 mils (400μm) for No. 10 conductor. The minimum thickness of the polyvinyl chloride insulation, at any point shall be 16 mils (400μm) for No. 14 AWG conductor and 22 mils (560μm) for No. 10 AWG conductor. No. 14 AWG conductors shall have a 5 mils (127μm) minimum thickness clear nylon jacket and the No. 10 AWG conductors shall have a 8 mils (203μm) clear nylon jacket.

209-5.3.2.2 Three-Conductor Cable. Three-conductor cable shall be used for pedestrian push buttons and a spare.

Three conductor cable shall consist of 3 No. 14 AWG wires. The cable jacket shall have a minimum average thickness of 43 mils (1100μm) and a minimum thickness of 35 mils (900μm). The nominal outside diameter of the cable shall not exceed 3/8 inch (10mm). The color code of the wires shall be blue/black stripe, blue/orange stripe and white/black stripe.

209-5.3.2.3 Five-Conductor Cable. Five-conductor cable shall consist of 5 No. 14 AWG wires. The cable jacket shall have a minimum average thickness of 43 mils (1100μm) and a minimum thickness of 35 mils (900μm). The nominal outside diameter of the cable shall not exceed 1/2 inch (12mm). The color code of the wires shall be red, yellow, brown, black, and white.

209-5.3.2.4 Nine-Conductor Cable. The nine-conductor signal cable shall consist of 8 No. 14 AWG wires and 1 No. 12 AWG wire. The cable jacket shall have a minimum average thickness of 59 mils (1500μm) and shall have a minimum thickness of 47 mils (1200μm). The nominal outside diameter of the cable shall not exceed 11/16 inch (17mm). The color for the No. 12 AWG wire shall be white. The color code for the No. 14 AWG wires shall be as specified in Table 209-5.3.2.4 (A).

TABLE 209-5.3.2.4 (A)

Primary Color	Color Code
Red	Yellow/Black strip
Yellow	Brown/Black strip
Brown	Black
Red/Black strip	White/Black strip

209-5.3.2.5 Twelve-Conductor Cable. Twelve-conductor cable shall be used for vehicle signals, pedestrian signals, spares, and the signal common or as specified in the Special Provisions. Twelve conductor cable shall consist of 11 No. 14 AWG wires and 1 No. 12 AWG wire. The cable jacket shall have a minimum average thickness of 59 mils (1500μm) and shall have a minimum thickness of 47 mils (1200μm). The nominal outside diameter of the cable shall not exceed 3/4 inch (19mm). The color code of the No. 12 AWG wire shall be white. The color code and functional connections for the No. 14 AWG wires shall be as specified in Table 209-5.3.2.5 (A), unless otherwise specified in the Special Provisions or approved by the Engineer.

TABLE 209-5.3.2.5 (A)

Color Code	Termination	Phase
Red	Vehicle Signal Red	2, 4, 6, or 8
Yellow	Vehicle Signal Yellow	2, 4, 6, or 8
Brown	Vehicle Signal Green	2, 4, 6, or 8
Red/Black stripe	Vehicle Signal Red	1, 3, 5, or 7
Yellow/Black stripe	Vehicle Signal Yellow	1, 3, 5, or 7
Brown/Black stripe	Vehicle Signal Green	1, 3, 5, or 7
Black/Red stripe	Spare, or Ped. Signal use as required for Vehicle Signal red or “Don’t Walk”	----
Black/White stripe	Spare, or use as required for Yellow	----
Black	Spare, or Ped. Signal use as required for Vehicle Green or “Walk”	----
Red/White stripe	Ped. Signal “Don’t Walk”	----
Brown/White stripe	Ped. Signal “Walk”	----

209-5.3.2.6 Twenty Eight -Conductor Cable. Twenty eight-conductor signal cable shall consist of 27 No. 14 AWG wires and 1 No. 10 AWG wire cabled with polyethylene fillers, if necessary to make the cable round. Clear binder tape shall be used to wrap the cable core. The black polyvinyl chloride cable jacket shall have a minimum average thickness of 80 mils (2000μm) and shall have a minimum thickness at any point of 60 mils (1600μm). The cable jacket shall have printed on the outside the manufacturers name, the type of insulation for each wire, the number of wires of each size, the size of the wires in the cable and the voltage rating. Sequential distance markings shall also be shown on the surface of the jacket, printed at 2-foot (600mm) intervals. The nominal outside diameter of the cable shall not exceed 7/8 inch (23mm). The voltage rating shall be 600V and the temperature rating shall be 167°F (75°C). Signal commons in each cable shall be kept separate except at the signal controller. Each cable shall be labeled in each pull box, “C1” or “C2”. The cable identified as “C1” shall be used for signal Phases 1, 2, 3 and 4. The cable identified “C2” shall be used for signal Phases 5, 6, 7 and 8. The “C1” and “C2” cable may also be used as shown on the Plans or as specified in the Special Provisions. The reel size shall allow for up to 2000 ft (610m) of cable maximum with a 2-3/4 inches (70mm) arbor hole.

The color code for the No. 10 AWG wire shall be white. The color code and functional connections for the No. 14 AWG wires shall be as shown in Table 209-5.3.2.6 (A), unless otherwise specified in the Special Provisions.

TABLE 209-5.3.2.6(A)

Color Code	Termination	Phase
Red/Black stripe	Vehicle Signal Red	2 or 6
Yellow/Black stripe	Vehicle Signal Yellow	2 or 6
Brown/Black stripe	Vehicle Signal Green	2 or 6
Red/Orange stripe	Vehicle Signal Red	4 or 8
Yellow/Orange stripe	Vehicle Signal Yellow	4 or 8
Brown/Orange stripe	Vehicle Signal Green	4 or 8
Red/Silver stripe	Vehicle Signal Red	1 or 5
Yellow/Silver stripe	Vehicle Signal Yellow	1 or 5
Brown/Silver stripe	Vehicle Signal Green	1 or 5
Red/Purple stripe	Vehicle Signal Red	3 or 7
Yellow/Purple stripe	Vehicle Signal Yellow	3 or 7
Brown/Purple stripe	Vehicle Signal Green	3 or 7
Red/2 Black stripes	Ped. Signal DON'T WALK	2 or 6
Brown/2 Black stripes	Ped. Signal WALK	2 or 6
Red/2 Orange stripes	Ped. Signal DON'T WALK	4 or 8
Brown/ 2 Orange stripes	Ped. Signal WALK	4 or 8
Red/2 Silver stripes	Overlap A, C, Red	OLA, OLC
Brown/2 Silver stripes	Overlap A, C Green	OLA, OLC
Red/2 Purple stripes	Overlap B, D Red	OLB, OLD
Brown/2 Purple stripes	Overlap B, C Green	OLB, OLD
Blue/Black stripe	Ped. Push Button	2 or 6
Blue/Orange stripe	Ped. Push Button	4 or 8
Blue/Silver stripe	Overlap A, C Yellow	OLA(y), OLC(y)
Blue/Purple stripe	Overlap B, D Yellow	OLB(y), OLD(y)
White/Black stripe	Ped. Push Button Common	---
Black/Red stripe	Railroad Preemption	---
Black	Spare	---

209-5.3.3 Interconnect Cable.

209-5.3.3.1 Direct Wire. Interconnect cable shall consist of 7 No.14 AWG tinned copper wires as shown on the Plans or specified in the Special Provisions. Each wire shall be insulated with 13 mils (330μm) minimum nominal thickness color coded, polypropylene material.

The interconnect cable jacket shall be black, high density polyethylene, rated for a minimum of 300V at 140°F (60°C), and shall have a nominal wall thickness of 40 mils (1000μm) minimum. The cable jacket or the moisture-resistant tape directly under the outer jacket shall be marked with the manufacturer's name, insulation type designation, number of wires and wire size, and voltage and temperature ratings.

209-5.3.3.2 Filled Telephone Cable. Filled telephone cable shall consist of paired No. 22 AWG solid annealed copper wires. The cable shall be polyethylene insulated and aluminum shielded, conforming to the REA Specification PE-39 for filled telephone cable. The cable sizes shall be 12, 25, 50, and 75 pair.

209-5.3.3.3 Telephone. Telephone lines shall be No. 14 AWG stranded, twisted, pair copper wire having 600V insulation and overall shield and jacket.

209-5.3.4 Fiber Optic Cable.

209-5.3.4.1 General. Fiber optic cable shall be a 6 optical glass fiber cable or as specified in the Special Provisions. The optical fibers shall be multi-mode graded index optical glass or as specified in the Special Provisions, with a nominal fiber core diameter of 50 microns and a nominal fiber cladding diameter of 125 microns. The fiber shall have a numerical aperture of 0.2.

209-5.3.4.2 Performance. The minimum optical performance of each fiber measured at wave lengths of both 850 nanometers and 1300 nanometers shall have a maximum attenuation of 3.5 decibels per kilometer at 858 nanometers and 1.0 decibel per kilometer at 1300 nanometers, and shall have a minimum bandwidth of 800 megahertz-kilometers.

209-5.3.4.3 Construction. The cable shall be constructed using 6 filled, loose-tube, color-coded dual-layer buffer tubes stranded (reverse oscillation) around a steel central member. The 6 color coded fibers shall be contained in a buffer tube and the remaining buffer tubes shall be natural or white. A layer of aramid yarn (e.g. Kevlar) shall hold the tubes in position around the strength member and provide additional tensile strength. The buffer tube diameter shall be a nominal 250 microns. The color code for the fibers and the buffer tubes shall be as shown on the Plans.

209-5.3.4.4 Jacket. The filled cable core shall be covered with a medium density polyethylene jacket. The color of the jacket shall be as shown on the Plans. This outer jacket shall be abrasion and crack resistant, non-nutrient to fungus, electronically non-conductive and compatible with all cable components that it may come in contact with. The jacket shall be free from holes, splits, blisters or other imperfections.

209-5.3.4.5 Identification. Each length of cable shall be permanently identified by specifying the manufacturer and type of cable at intervals not greater than 6 feet (2m) along the outside of the outer jacket. Each length of cable shall be permanently marked with length markings at intervals not greater than 3 feet (1m).

209-5.3.4.6 Reels. The cable shall be wound on standard reels in a manner which provides access to both ends of the cable for testing while the cable is still on the reel.

209-5.4 Controllers.

209-5.4.1 General. Controllers, controller cabinets and auxiliary equipment shall be as specified in the Special Provisions, Plans, or Standard Plans.

209-5.4.2 Wiring. Conductors in controller cabinet wiring shall be No. 22 AWG or larger, with a minimum of 19 strands. Conductors shall conform to Military Specification MIL-W-16878D, Type B or better. The insulation shall have a minimum thickness of 10 mils (250μm) and shall be nylon jacketed polyvinyl chloride, irradiated cross-linked polyvinyl chloride, polyhalocarbon or polychloro-alkene, except for conductors No. 14 AWG and larger which may be UL Type THHN.

Flat cable may be used in lieu of individual conductors. Cable shall be constructed of No. 28 AWG or larger wires. Conductor insulation shall be rated at 300V and shall rated be for use at 221°F (105°C). Cables shall be provided with strain relief.

Conductors used in controller cabinet wiring shall conform to the following color-code requirements:

a) The grounded conductor of a circuit shall be identified by a continuous white or natural gray color.

b) The equipment grounding conductor shall be identified by a continuous green color or by a continuous green color with one or more yellow stripes.

c) Ungrounded conductors shall be identified by any color not specified in 1 or 2 above.

Conductors used in cabinet wiring shall terminate with properly sized captive or spring spade type terminals, or shall be soldered to a through-panel solder lug on the rear side of the terminal block. Crimp-style connectors shall be applied with a tool that is designed to prevent opening of the handles until the crimp is complete.

An equipment grounding conductor bus shall be provided in each controller cabinet. The bus shall be grounded to the cabinet and shall be connected to the metal conduit system or other approved ground with a No. 8 or larger, grounding conductor.

With all of the cabinet equipment in place and connected, the resistance between the grounded conductor terminal bus and the equipment grounding conductor bus shall be 50 MΩ minimum, when measured with an applied voltage of 150V DC.

If a direct current (DC) circuit is to be grounded, the circuit shall be connected to the equipment ground only.

Two or more terminal blocks shall be provided for field connections. Terminals blocks for field connections shall be installed within 22 inches (550mm) of the face on the cabinet and shall be oriented for screwdriver operation from the door opening. All terminals shall be a minimum of 5 inches (125mm) above the foundation.

No more than 3 conductors shall be brought to any one terminal. Two flat metal jumpers, straight or U shaped, may also be placed under a terminal screw. At least 2 full threads of all terminal screws shall be fully engaged when the screw is tightened. No live parts shall extend beyond the barrier.

Wiring diagrams shall conform to 307-4.3.

209-5.4.3 Cabinets.

209-5.4.3.1 General. Controller cabinets shall be rainproof and conform to the dimensions shown on the Plans or the Standard Plans. The cabinet top shall be crowned 1/2 inch (12mm.) or slanted to the rear to prevent standing water.

Cabinets and doors shall be fabricated of either 1/16 inch (1.8mm) minimum thickness cold rolled steel, 1/16 inch (1.8mm) minimum thickness stainless steel or 1/8 inch (3mm) minimum thickness aluminum. All exterior seams shall be continuously welded.

Exterior welds shall be ground smooth. Edges shall be filed to a radius of 1/32 inch (800µm) minimum.

Metal shelves or brackets shall be provided to support the controller unit and auxiliary equipment.

Machine screws and bolts shall not protrude beyond the outside wall of the cabinet.

Conduit shall enter the bottom and towards the front of the cabinet unless otherwise shown on the Plans.

A pliable seal, composed of caulking compound or mastic, shall be placed between each controller cabinet and the concrete foundation to prevent water, dust and dirt from entering the cabinet.

209-5.4.3.2 Steel Cabinets. Cabinets fabricated from steel shall be finished with a polymer or enamel coating system conforming to Color No. 14672 of Federal Standard 595B. Coatings shall be commercially smooth, substantially free of flow lines, paint washout, streaks, blisters and other defects that would impair serviceability or distract from the general appearance. Coatings shall conform to the following requirements:

a) Coating Hardness: The finish shall have a pencil lead hardness of HB minimum using an Eagle Turquoise pencil.

b) Salt Spray Resistance: Undercutting of the film of the coating system shall not exceed 1/8 inch (3mm) average, from lines scored diagonally and deep enough to expose the base metal, after 336 hours exposure in a salt spray cabinet in conformance with ASTM B117.

c) The cabinet shall suffer no coating loss when subjected to the following:

1. Two test specimens 4 inches x 8 inches (100mm x 200mm), of the same material and coating as the cabinet supplied, shall be furnished by the Contractor for the tests. The samples shall be furnished by the Contractor at no additional cost to the Agency.

2. Two 9-inch (225mm) diagonal scratches exposing bare metal will be made on one specimen. The specimen will be soaked in de-mineralized water for 192 hours. A 1 inch (25mm) wide strip of masking tape will be tightly affixed to the scratched surface and removed with one quick motion; evidence of blistering, softening, or peeling of the paint or coating from the base metal shall be cause for rejection. Testing will be in conformance with California Test 645, except that conformance to the 180 degree bend test will not be required.

d) Metal preparation shall conform to the 3-step iron phosphate conversion coating bonderizing technique.

e) The inside walls, doors, and ceiling of the housing shall be finished the same as the out side.

209-5.4.3.3 Stainless Steel Cabinets. Cabinets fabricated from stainless steel shall conform to the following:

a) Annealed or quarter hard stainless steel sheet shall be used and shall conform to ASTM A666 for Type 304 Grades A or B.

b) Welding shall be by the gas tungsten arc welding (GTAW) process using bare stainless steel welding electrodes. Electrodes shall conform to American Welding Society (AWS) A5.9 for ER308 chromium-nickel bare arc welding electrodes.

c) Procedures, welders and welding operators for welding on stainless steel shall conform to AWS C5.5.

d) Exposed, exterior surfaces shall be ground or brushed to a 0.6 μm to 1.3 μm finish using iron-free abrasives or stainless steel brushes.

e) All overlapping exterior seams shall meet the requirements of NEMA Type 4 enclosures.

f) After grinding or brushing, no rust discoloration shall show when subjected to the following:

1. Forty-eight hours of exposure in a salt spray cabinet in conformance with ASTM B117; and

2. Twenty-four hours of exposure in a tap water spray cabinet with the water temperature between 100°F (38°C) and 113°F (45°C). Any cabinet which shows any rust discoloration anywhere on its surface after the test will be rejected.

g) Cabinets that have been rejected because of surface discoloration may be cleaned, passivated and resubmitted for testing.

209-5.4.3.4 Aluminum Cabinets. Cabinets fabricated from aluminum sheet shall conform to the requirements in ASTM B209/B209M for 5052-H32 aluminum sheet and conform to the following:

a) Welding shall be done by the gas metal arc welding (GMAW) process using bare aluminum welding electrodes. Electrodes shall conform to AWS A5.10 for ER5356 aluminum alloy bare welding electrodes.

b) Procedures, welders and welding operators for welding on aluminum shall be qualified in conformance with AWS B3.0 and C5.6.

c) Surfaces shall be finished in conformance with Military Specification MIL-A-8625C "Anodic Coatings for Aluminum and Aluminum Alloys" for a Type II, Class 1 coating, except that anodic coating shall have a minimum thickness of 20μm and a minimum coating weight of 40μg/mm^2. Anodic coating shall be sealed in a 5 percent aqueous solution of nickel acetate (pH 5.0 to 6.5) for 15 minutes at 207°F (97°C). Prior to applying anodic coating the cabinets shall be cleaned and etched as follows:

1. Immerse in inhibited alkaline cleaner such as Oaklite 61 A or Diversey 909, or equivalent, 6 to 8 oz./gal (45 to 60 g/L), 160°F (71°C) for 5 minutes.

2. Rinse in cold water.

3. Etch in a solution of 0.39 oz (11 g) of sodium fluoride, plus 4 to 6 oz (30 to 45 g) of sodium hydroxide per gal (liter) of distilled water at 140°F to 149°F (60°C to 65°C) for 5 minutes.
4. Rinse in cold water.
5. Desmut in a 50 percent, by volume, nitric acid solution at room temperature for 2 minutes.
6. Rinse in cold water.

209-5.4.3.5 Doors. The cabinet shall have a single front door equipped with a lock. The door width shall be as shown on the Standard Plans. When the door is closed and latched, the door shall be locked. The handle shall have provision for locking in the closed position. The handle shall have a minimum length and steel shank as shown on the Standard Plans. The handle shall be fabricated of cast aluminum or of zinc-plated or cadmium-plated steel. The cabinet frame shall be designed so that the latching mechanism will hold tension on and form a firm seal between the door gasketing and frame. Locks shall be the solid brass, 6-pin tumbler rim type, with rectangular, spring-loaded bolts. Locks shall be left hand, and rigidly mounted with stainless steel machine screws approximately 2 inches (50mm) apart. Keys shall be removable in the locked and unlocked positions, and 2 keys shall be furnished with each cabinet. The front position of the lock shall extend 1/8 inch to 1/4 inch (3mm to 6mm) beyond the outside surface of the door.

The latching mechanism shall be a 3-point cabinet latch with nylon rollers. The center catch and pushrods shall be zinc-plated or cadmium-plated steel. Pushrods shall be turned edgewise at the outer supports and shall be 1/4 inch (6mm) x 3/4 inch (19mm) minimum. Nylon rollers shall have a minimum diameter of 3/4 inch (19mm) and shall be equipped with ball bearings.

Cadmium plating shall conform to the requirements in Military Specification MIL-QQ-416b. Zinc plating shall conform to 210-3.

Doors shall be equipped with 3 bolt butt hinges. Each hinge shall have a fixed pin. Doors larger than 22 inches (550mm) or 6 feet^2 (0.56m^2) in area shall be equipped with catches to hold the door open at both 90-degrees and 180-degrees, plus or minus 10 degrees. Catches shall be 3/8 inch (9mm) diameter, minimum plated steel rods. Catches shall be capable of holding the door open at 90-degrees in a 56 mph (90km/h) wind at an angle perpendicular to the plane of the door.

A police panel shall be mounted on the door as shown on the Standard Plans and shall be equipped with a lock keyed to a master police key. Two keys shall be furnished with each cabinet for the police lock. Each police key shall have a shaft at least 1-3/4 inches (45mm) in length. Police panels shall not be furnished for controller assemblies that do not control traffic signals.

Door hinges, pins and bolts shall be made of stainless steel. Hinges on aluminum cabinets may be aluminum with a stainless steel hinge pin. Hinges shall be bolted to the cabinet. Hinge pins and bolts shall not be accessible when the door is closed.

Gasketing shall be provided on all door openings and shall be dust-tight. Gaskets shall be permanently bonded to the metal. The mating surface of the gasketing shall be covered with a silicone lubricant to prevent sticking of the mating surfaces.

Details of alternative designs shall be submitted in conformance with 2-5.3.3.

Metal shelves or brackets shall be provided to support the controller unit and auxiliary equipment.

Machine screws and bolts shall not protrude beyond the outside wall of the cabinet.

Conduit shall enter the bottom and toward the front of the cabinet unless shown otherwise on the Plans.

A pliable seal, composed of caulking compound or mastic, shall be placed between each controller cabinet and the concrete foundation to prevent water, dust and dirt from entering the cabinet.

209-5.4.3.6 Cabinet Ventilation. Each controller cabinet shall be provided with one of the following as specified in the Special Provisions:

a) Eight screened, raintight vent holes 1/2 inch (12mm) in diameter or larger, in the lower side or bottom of the cabinet.

b) Louvered vents with a permanent metal mesh.

c) Four-ply woven polypropylene air filter held firmly in place that will pass 100 cfm ($2.83m^3$/min.) of air.

Each controller cabinet shall be equipped with an electric fan with ball or roller bearings and a capacity of at least 100 cfm ($2.83m^3$/min.). The vents shall be able to pass this volume of air.

The fan shall be thermostatically controlled and shall be manually adjusted to turn on between 90°F (32°C) and 150°F (65°C) with a differential of not more than 10°F (6°C) between automatic turn on and turn off. The cabinet fan circuit shall be fused at 125 percent of the ampacity of the fan motor installed.

The fan and cabinet vent holes shall be located with respect to each other so as to direct the bulk of the air flow over the controller or through the ventilating holes of the controller where those holes exist.

209-5.4.3.7 Cabinet Accessories. The following accessories shall be furnished with each cabinet assembly as required by the Plans, Standard Plans, or Special Provisions.

a) **Labels.** A permanent printed, engraved or silk screened label shall be provided for the following equipment and for all other removable items of equipment:

1. Receptacles for relays and switching devices.
2. Switches, fuses and circuit breakers.

Labels shall conform to the designations on the cabinet wiring diagram. Labels for all shelf-mounted equipment shall be on the face of the shelf below the item. Labels for wall-mounted equipment shall be below the item.

b) **Convenience Receptacle.** A convenience receptacle shall be mounted in a readily accessible location inside the cabinet. The convenience receptacle shall be a duplex, 3-prong, NEMA, Type 5-15R grounding outlet and shall conform to UL Standard 943.

c) **Lighting Fixture.** The lighting fixture shall be a florescent lighting fixture, mounted on the inside top of the cabinet near the front edge. The fixture shall be provided with an F15T8, cool white lamp operated from a normal power factor UL or ETL listed ballast. The "On-Off" switch for the lighting fixture shall be either of the following:

1. A toggle switch mounted on the inside door panel.
2. A door-actuated switch that turns the light on when the door is open, and off when the door is closed.

d) **Surge Arrestor.** The surge arrestor shall reduce the effects of power line voltage transients and shall have ratings as follows:

Recurrent peak voltage	184V
Energy rating maximum	20 J
Power dissipation average	0.85 W
Peak current for pulses less than 7 ps	1250 A

Standby current shall be one milliampere or less for 60 Hz sinusoidal input.

e) **Terminal Blocks.** Terminal blocks shall be rated 600VAC, minimum, and shall be provided with nickel, silver or cadmium plated brass binder head screw terminals.

Heavy duty terminal blocks shall be rated at 20A and shall be provided with 12 poles with No. 10 x 5/16 inch (8mm) nickel plated brass binder head screws and nickel plated brass inserts. Each pole position shall be provided with 2 terminal positions. The terminal blocks shall be the barrier type, with shorting bars in each of the 12 positions, and shall be provided with integral type marking strips.

Light duty terminal blocks shall be rated at 5A and shall be provided with 12 poles with No. 6 x 1/8 inch (3mm) binder head screws. Each pole position shall be provided with one terminal position.

209-5.4.4 Components.

209-5.4.4.1 Toggle Switches. Toggle switches shall have poles as required and shall be rated at 200 percent of circuit current for circuits of 10A or less and 125 percent of circuit current for circuits over 10A.

209-5.4.4.2 Cartridge Fuses. Cartridge fuses shall be installed in panel mounted fuseholders. Fuse type and rating shall be as recommended by the fuse manufacturer for the type of load being protected.

209-5.4.4.3 Circuit Breakers. Circuit breakers shall conform to 209-3.11 except that the circuit breakers shall have a minimum interrupting capacity of 5000A, RMS.

209-5.4.4.4 Connectors. Connectors used for interconnecting various portions of circuits together shall be designed and constructed for the application involved. Connectors shall be designed to provide positive connection of all circuits, and easy insertion and removal of mating contacts. Connectors shall be permanently keyed to prevent improper connection of circuits.

Connectors, or devices plugging into connectors, shall be provided with positive means to prevent any individual circuit from being broken due to vibration, pull on connecting cable or similar disruptive force.

209-5.4.5 Dual 5-Way Active Data Bridge.

209-5.4.5.1 General. Dual 5-way active data bridges shall be provided in the controller assemblies when shown on the Plans or specified in the Special Provisions. The bridge shall provide a 4-wire transmission interconnection between a dedicated common port and four multiple ports. The bridge shall incorporate a splitter channel and a combiner channel. Multiple inputs shall be connected to a common output in the combiner channel and in the splitter channel a common input shall be connected to multiple outputs. Splitter and combiner shall be separate and independent allowing operation in full-duplex data transmission applications.

The bridge shall be equipped with a front-panel-accessible potentiometer with continuous adjustment of cross-bridge gain or loss within two switch-selectable loss/gain ranges suitable for most data applications; -30dB to –10dB and –10dB to +10dB. Either range shall be independently selectable for each channel. The level to which each channel is adjusted shall be common to all cross-bridge port combinations in that channel (that is, if the splitter-channel potentiometer is set for +3dB gain, all common-to-multiple-port paths in the splitter shall receive a +3dB level increase).

All ports of the bridge, both common and multiple, shall be characterized by a balanced 600Ω terminating impedance. The module's active circuit shall allow unused multiple ports to be left unterminated without affecting the transmission response of the ports in use.

Input, output and line monitor jacks on the front panel of the bridge shall allow isolated-module isolated-facility, and cross-bridge measurement of transmission parameters in both the splitter and combiner channels.

The bridge shall be powered from input voltage of –22 to –56VDC at a maximum current of 60mA.

Each dual 5-way active bridge unit shall be individually packaged in a housing. All circuitry shall be solid state, constructed on removable industry standard circuit boards with plug-in edge connectors.

All test jacks, edge connectors and external plug connectors shall be of a type and material suitable for use in the above stated environment without deterioration of electrical connection for the useful life of the equipment. The physical size of the case shall not exceed 105 ft^3 ($3700cm^3$) (approximately 7 inches x 8 inches x 4 inches (175mm x 200mm x 100mm) and shall be suitable for mounting to the frame of a field cabinet.

Mounting holes shall be provided for attaching the unit to one side of the frame of a relay rack. Input and output ports shall be provided for five full duplex telephone circuits with input and output ports labeled as to function. Level control or switches and level adjustment potentiometers along with the input and output level test jacks shall provide external adjustments without removal of the housing.

209-5.4.5.2 Application. The bridge shall be used to interconnect several 4-wire data modems to a common data channel or data link. The common data link shall be terminated at the distant end into a central processor unit (CPU) or computer. A bridge shall be used to provide the central transmission arrangement or "hubbing" network, that extends data transmission to outlying terminals.

Tandem bridge arrangements to expand the number of multiple ports of a 4-wire data hubbing network shall be effected by direct connection of one multiple port in each channel of the first bridge to the common port of the same channel of a second bridge to provide a 4-wire data bridge with one common and 7 multiple ports.

209-5.4.5.3 Circuit Description. The combiner channel of the bridge shall consist of a variable-gain integrated-circuit operational amplifier with an input summing circuit. The summing circuit shall add the transmission energy from all multiple input ports at a low-impedance summing point to provide input port isolation. The output of the amplifier shall be connected to a transformer for balanced connection to the facility.

The splitter channel shall consist of a transformer-input-variable-gain, integrated-circuit operational amplifier and a power amplifier. The output of the power transformer shall be a low impedance and shall drive eight 300Ω resistors to drive impedance at the multiple output circuits and provide isolation between output circuits.

The multiple ports in the splitter and combiner channels of the bridge shall be isolated and balanced via an output transformer and matched precision resistors in each channel.

209-5.4.5.4 Electrical Requirements. Electrical requirements shall be as specified in Table 209-5.4.5.4(A).

TABLE 209-5.4.5.4(A)

Splitter channel loss/gain	-30 to +10dB usable range
Combiner channel loss/gain	-30 to +10dB usable range
Maximum output level (overload point)	Splitter: +5Dbm Combiner: +12dBm
Level change with loading	1dB maximum, one port to all ports loaded.
Noise	20dBm maximum
Input port impedance	Splitter (multiple ports): 600Ω balanced. Combiner (common port): 600Ω balanced.
Output port impedance	Splitter (multiple ports): 600Ω balanced. Combiner (common port): 600Ω balanced.
Harmonic distortion	Splitter: less than one percent at + 3dBm Combiner: less than one percent at + 8 dBm
Frequency response	± 1 dBm re 1000Hz level, 300 to 5000Hz
Delay distortion	Less than 75μs, 400 to 3000Hz
Cross-port coupling loss (crosstalk)	Greater than 55dB
Mounting	One position
Operating environment	-19°F to 129°F (-7°C to 54°C), humidity to 95% (no condensation)

209-5.4.5.5 Testing and Troubleshooting. Complete testing and troubleshooting instructions, circuit diagrams and pictorial component location and identification guides shall be provided with each bridge.

209-5.5 Traffic Signal Faces and Fittings.

209-5.5.1 Signal Faces. Each vehicle signal face shall be of the adjustable type conforming to ITE Publication ST-008B.

Metal signal sections shall conform to the requirements of California Test 666. Any fracture within the housing assembly or a deflection of more than one-half the lens diameter of the signal section during the wind load test will be considered a structural failure.

Plastic signal sections shall conform to the requirements of California Test 605. Any fracture within the housing assembly, or deflection of more than 10 degrees in either the vertical or horizontal plane after the wind load has been removed from the front of the signal face, or a deflection of more than 6 degrees in either the vertical or horizontal plane after the wind load has been removed from the back of the signal face will be considered structural failure.

Vehicle signal faces, except arrow and "X" faces, shall conform to the requirements of California Test 604.

Adjustment shall permit rotation of 360 degrees about a vertical axis. The number and type of sections shall be as specified in the Special Provisions or shown on the Plans. Each vehicle signal face shall be installed at the location and mounted in the manner shown on the Plans.

Unless otherwise shown on the Plans, all vehicle signal faces shall contain 3 sections arranged vertically; red-top, yellow-center, and green-bottom.

All new vehicle signal faces, except programmed visibility type, installed at any one intersection shall be of the same manufacturer and of the same material.

Top openings of signal faces shall be sealed with neoprene gaskets.

209-5.5.2 Optical Units. Each optical unit shall consist of a lens, a reflector or reflector assembly, a lamp receptacle and a clear traffic signal lamp conforming to the following:

a) Lenses, reflectors, reflector assemblies, lamp receptacles, lamps, wiring and light distribution shall conform to ITE Publication ST-008B.

b) Each lens shall be of the best quality glass, true to color and free of imperfections.

c) Reflectors shall conform to ITE Publication ST-008B except that the reflectors shall be made of silvered glass or of specular aluminum with an anodic coating. Reflector ring holders shall be made of cast aluminum.

A single piece formed metal reflector ring holder may be used.

209-5.5.3 Signal Sections. Each signal section housing shall be either die-cast or permanent mold-cast aluminum conforming to ITE Publication ST-088B or, when so specified in the Special Provisions shall be structural plastic.

Signal sections shall conform to the following:

a) Maximum height shall be 10 inches (250mm) for 8-inch (200mm) sections and 14 inches (375mm) for 12-inch (300mm) sections.

b) Housings shall be provided with a one-piece, hinged, square-shaped door designed to permit access to the section for relamping without the use of tools. The door shall be secured by a method that will hold the door closed during the loading tests specified in these Specifications. The lens shall be mounted in the door in a watertight manner.

c) Exposed hardware, such as hinge pins and door latching devices, shall be manufactured of Type 304 or 305 stainless steel. Interior screws and fittings shall be manufactured of stainless steel or steel with a corrosion resistant plating or coating.

d) An opening shall be provided in the top and bottom of each section to receive a 1-9/16 inches (40mm) pipe. Sections sized 8 inches (200mm) and 12 inches (300mm) produced by an individual manufacturer shall be capable of joining to form a signal face in any combination. This interchangeability is not required between metal and plastic signal sections.

e) Gaskets, including those for the door, lens, reflector and lampholder, shall be made of a material that is not affected when installed in a signal section with a metal or plastic housing that is operated continuously for 336 hours.

209-5.5.4 Metal Signal Sections. Each metal signal section shall be provided with a metal visor and/or backplates as specified in the Special Provisions. Visors shall conform to 209-5.5.8. Backplates shall conform to 209-5.5.10.

209-5.5.5 Plastic Signal Sections. Housing shall be either molded in one piece or shall be fabricated from two or more pieces joined into a single piece. The plastic shall have ultraviolet stability, be unaffected by the heat of the lamp used and be self-extinguishing. Housing and doors shall be colored throughout and shall be black matching Color No. 17038, 27038 or 37038 of Federal Standard 595B.

Each plastic signal face shall be provided with plastic or metal visor as specified in the Special Provisions. Visors shall conform to 209-5.5.8. Plastic signal sections shall also be provided with plastic backplates as specified in the Special Provisions. Plastic backplates shall conform to 209-5.5.10.2.

A serrated nylon washer shall be inserted between each plastic signal section and the metal mounting assembly. Each washer shall be not less than 5/32 inch (4mm) nor more than 1/4 inch (6mm) thick. Serrations shall match those on the signal section and the mounting assembly.

209-5.5.6 Fittings. Each section in a face shall be joined to the adjacent section by one of the following methods:

a) A minimum of 3 machine screws for 8-inch (200mm) sections and 4 machine screws for 12 inch (300mm) sections installed through holes near the front and rear of the housings. Machine screws shall be No. 10 and each shall be provided with a nut, flat washer and lock washer.

b) Two machine screws (each with a nut, flat washer and lock washer) through the 1- 9/16 inches (40mm) pipe opening. The fastening system shall consist of 2 large flat washers to distribute the load around the pipe opening and 3 carriage bolts, each with a nut and lock washer. Minimum size of the machine screws shall be No. 10. Minimum size carriage bolts shall be 1/4 inch (6mm).

The supporting section of each signal face supported solely at the top or bottom shall be provided with reinforcement.

Reinforcing plates shall be either sheet aluminum, galvanized steel or cast aluminum. Each plate shall be not less than 7/64 inch (2700μm) thick and shall have a hole concentric with the 1-9/16 inches (40mm) pipe mounting hole in the housing. Sheet aluminum reinforcing plates shall be placed both inside and outside the housing. Galvanized steel reinforcing plates shall only be placed inside, and aluminum reinforcing plates shall be placed only on the outside of the housing. Reinforcing plates placed on the outside of the housing shall be finished to match the color of the signal face and mounting hardware. A minimum of 3 No. 10 machine screws shall be installed through holes in the plates and matching holes in the housing. Each screw shall have a round or binder head and shall be provided with a nut and lock washer.

Where a signal face is to be supported by a Type MAS side attachment slip-fitter inserted between 2 sections, a spacer or spacers shall be placed between the 2 sections. The vertical dimensions of the spacers shall permit proper seating of the serrations between the slip-fitter and the 2 sections. Holes in spacers shall align with the front holes in the section housings. In addition to the fastening through the large opening in the housings, the 2 sections shall be joined with at least 2 machine screws through holes near the front of the housings and the spacers, and through matching holes in a reinforcing plate installed in each housing. Machine screws shall be No. 10 size. Spacers shall be made of the same material as the signal housings. Reinforcing plates and machine screws shall be as specified herein.

Reinforcing plates will not be required where the housing is provided with reinforcing webs connecting the rear of the housing with the top, bottom and sides.

Holes for machine screws shall be either cast or drilled during fabrication of the signal section. Each hole shall be surrounded by a 1/8 inch (3mm) minimum width boss to permit contact between signal sections about the axis of the hole.

209-5.5.7 Electrical Components. Lamp receptacles and wiring shall conform to ITE Publication ST – 008B. The metal portion of the medium base lamp socket shall be brass, copper or phosphor bronze.

Each lamp receptacle shall be wired with a conductor, connected to the shell of the receptacle, with white insulation, and a conductor, to the bottom or end terminal of the receptacle, with a black insulation or insulation color-coded.

Conductors shall, in turn, be connected to a terminal block mounted inside at the back of the housing. The terminal block shall have sufficient screw type terminals or NEMA type tab connectors to terminate all field conductors and lamp conductors independently. The terminals to which field conductors are attached shall be permanently identified or conductors shall be color-coded to facilitate field wiring.

Lamp receptacle conductors shall be No. 18, or larger, 600V, appliance wiring material (AWM), with 1/32 inch (800μm) thickness insulation rated for 221°F (105°C) or with insulation conforming to Military Specification MIL-W 16878 D, Type B, with vinyl nylon jacket rated for 239°F (115°C). The manufacturer's name or trademark, conductor size, insulation type letter designation and temperature rating shall be marked on the insulation or a Certificate of Compliance in conformance with 4-1.5 shall be submitted by the Contractor with each shipment of traffic signal faces in accordance with 2-5.3.

209-5.5.8 Visors. Each signal section shall be provided with a removable visor conforming to ITE Publication ST-008B. Visors shall be classified, on the basis of lens enclosure, as full circle, tunnel (bottom open), or cap (bottom and lower sides open). Unless otherwise specified in the Special Provisions, visors shall be the tunnel type.

Visors shall be a minimum of 9-7/16 inches (240mm) in length for nominal 12-inch (300mm) round lenses and 7-1/16 inches (180mm) in length for nominal 8-inch (200mm) round lenses, with a downward tilt between 3 degrees and 7 degrees.

Metal visors shall be made from 3/64 inch (1.2mm) minimum thickness, aluminum alloy sheets.

Plastic visors shall conform to the following:

a) Visors shall be either formed from sheet plastic or assembled from one or more injection, rotational or blow molded plastic sections.

b) Sections shall be joined using thermal, chemical or ultrasonic bonding or with aluminum rivets and washers permanently colored to match the visor.

c) Visors shall be of black homogeneous colored material with lusterless finish.

Each visor shall be secured to its door in a manner that will prevent its removal or permanent deformation when the wind load specified in California Test 605 for plastic visors or California Test 666 for metal visors is applied to the side of the visor for 24 hours.

209-5.5.9 Directional Louvers. Where shown on the Plans, directional louvers shall be furnished and installed in the signal visors. Directional louvers shall be so constructed as to have a tight fit in the signal visors. Dimensions of louvers and arrangements of vanes shall also be as shown on the Plans.

The outside cylinder may be constructed of 1/32 inch (800μm) nominal thickness, or thicker, sheet steel. The vanes may be constructed of 1/64 inch (400μm) nominal thickness, or thicker, sheet steel. Alternatively, the cylinder and vanes may be constructed of 5052-H32 aluminum alloy of equivalent thickness.

209-5.5.10 Backplates.

209-5.5.10.1 General. Where shown on the Plans, backplates shall be furnished and installed on signal faces. Dimensions, materials and installation details shall be as shown on the Plans. No background light shall show between the backplate and the signal face or between sections.

209-5.5.10.2 Plastic Backplates. Plastic backplates shall be either formed from sheet plastic or assembled from extruded molded or cast sections. Backplates shall be of black homogeneous colored material with a lusterless finish. Each plastic backplate shall be secured to the plastic signal face in a manner that will prevent its removal or permanent deformation when the wind-load test specified in 209-5.5.8 is applied to either the front or the rear of the signal face. The permanent deformation of any portion of the backplate shall not exceed 5 degrees forward or backward after the wind loading has been applied for 24 hours.

a) Sections shall be factory joined using:
   1) Solvent cement recommended by the manufacturer,
   2) Aluminum rivets and washers painted or permanently colored to match the backplate, or
   3) No. 10 machine screws with washer, lock washer and nut, painted to match the backplate.

b) When a plastic backplate requires field assembly, it shall be joined with a minimum of 4 No. 10 machine screws at each field assembled joint. Each machine screw shall be provided with an integral or captive flat washer, a hexagonal head slotted for a standard screwdriver, and either:
   1) A locking nut with an integral or captive flat washer, or
   2) A nut, flat washer, and lock washer.

Machine screws, nuts, and washers shall be stainless steel or steel with a zinc or black oxide finish.

In lieu of the screws shown on the Plans, plastic backplates may be fastened to the plastic signal face using self-threading No. 10 steel screws. The screws shall have an integral or captive flat washer and a hexagon head slotted for a standard screwdriver, and shall be stainless steel or steel with a zinc or black oxide finish.

209-5.5.10.3 Metal Backplates. Where a metal backplate consists of 2 or more sections, the sections shall be fastened with rivets or with aluminum bolts peened after assembly to prevent loosening.

209-5.5.11 Circular Light Emitting Diode (LED) Signal Modules.

209-5.5.11.1 General. LED signal modules shall consist of an assembly that utilizes light emitting diodes as the light source. Modules shall be from the same manufacturer and each size shall be the same model. Modules shall be single, self-contained devices, not requiring on-site assembly for installation into the existing traffic signal housings. Modules shall fit into traffic signal section housings built in conformance with the ITE publication ST-008B without modification to the housing. Special tools shall not be required for their installation. Installation of LED signal modules shall only require removal of the optical unit components (lens, gaskets, lamp, lamp socket and reflector) from the traffic signal section.

Each LED signal module shall be designed to be installed in the door frame of the signal housings specified in the Contract Documents. The module shall be sealed securely in the door frame with a one-piece ethylene propylene rubber (EPDM) gasket and shall be weather tight.

LED signal modules shall be sealed units with 2 conductors for connecting to power, a printed circuit board, a power supply, a lens and gasket and shall be weatherproof after installation and connection. The circuit board and power supply shall be contained inside the module. Circuit boards shall conform to Chapter 1, Section 6, of the "Transportation Electrical Equipment Specifications" published by the State of California, Department of Transportation or as specified in the Special Provisions.

Conductors for LED signal modules shall be 3 feet (1m) in length, with terminals attached, and shall conform to 209-5.3.

Connections shall be to the terminal block in the signal face or shall utilize an adapter that screws into the medium base lamp socket. Contacts shall be brass. Splices will not be allowed.

The lens shall be integral to the module, shall be convex with a smooth outer surface and shall be made of ultraviolet stabilized plastic or glass. The lens shall be capable of withstanding exposure to ultraviolet light for a minimum period of 48 months without exhibiting evidence of deterioration.

LED's shall utilize aluminum indium gallium phosphate (AllnGaap) technology and shall be the ultra bright type or equivalent rated for 100,000 hours of continuous operation from -42°F (–40°C) to 165°F (+74°C).

Each individual LED shall be wired such that physical damage or the failure of 1 LED will result in the loss of not more than 5 percent of the LED signal module light output.

Maximum power consumption requirements for LED signal modules shall be specified in Table 209-5.5.11.1(A).

TABLE 209-5.5.11.1(A)

LED Signal Module	Power Consumption at	
	77°F (25°C)	165°F (74°C)
12 inch (300mm) Circular	25.0 Watts	30.0 Watts
8 inch (200mm) Circular	15.0 Watts	18.0 Watts
12 inch (300mm) Arrow	15.0 Watts	18.0 Watts

LED signal modules shall be rated for a minimum useful life of 48 months.

209-5.5.11.2 Physical and Mechanical Requirements. LED signal modules shall:

a) Have a maximum mass of 4.4 pound avoirdupois (2.0 kg).

b) Be rated for use in the operating temperature range of -42°F (-40°C) to +165°F (+74°C).

c) Be protected against dust and moisture intrusion in accordance with NEMA Standard 250-1991 for Type 4 enclosures.

d) Be manufactured to withstand mechanical shock and vibration from winds and other sources as specified in the Special Provisions.

e) Have the manufacturer's name, trademark, model number, serial number, lot number and the month and year of manufacture permanently marked on the back of each module.

f) Have prominent and permanent vertical markings for correct indexing and orientation within a signal housing. The markings shall consist of an "UP" arrow, or the word "UP" or "TOP".

209-5.5.11.3 Operating Characteristics. The following operating characteristics shall be identified and submitted to the Engineer in accordance with 2-5.3:

a) Rated voltage,

b) Power consumption and

c) Volt-ampere (VA).

Enclosures, excluding the lens, containing either the power supply or electronic components of LED signal modules shall be made of UL94VO flame retardant materials.

Lens may be tinted or use transparent film or material with similar characteristics to enhance "ON/OFF" contrasts. The use of tinting or other materials to enhance "ON/OFF" contrasts shall not affect chromaticity and shall be uniform across the face of the lens. If a polymeric lens is used, a surface coating or chemical surface treatment shall be used to provide front surface abrasion resistance.

209-5.5.11.4 Photometric Requirements. The minimum initial luminous intensity values for LED signal modules shall conform to Section 11.94 of ITE publication ST-008B at 77°F (25°C). Luminous intensity values shall meet or exceed 85 percent of the standard light output values specified in the ITE VTCSH, after 48 months of continuous use over the temperature range of -42°F (-40°C) to 165°F (+74°C) in a traffic signal operation.

Measured chromaticity coordinates shall conform to Section 8.04 and Figure 1 of the ITE VTCSH over the temperature range of -42°F (-40°C) to 165°F (+74°C).

209-5.5.11.5 Electrical Requirements.

a) LED signal modules shall operate over a voltage range from 95VAC to 135VAC at a frequency of 60 Hz ± 3 Hz.

b) Fluctuations of line voltage shall have no visible effect on the luminous intensity of the indications. Rated voltage for the measurements shall be 120V.

c) LED signal modules and associated on-board circuitry shall conform to FCC Title 47, SubPart B, Section 15 regulations concerning the emission of electronic noise.

d) Circuitry shall include voltage surge protection to withstand high-repetition noise transients as specified in Section 2.1.6 of NEMA Standard TS2-1992.

e) Circuitry shall prevent perceptible flicker over the specified voltage range.

f) LED signal modules shall provide a power factor of 0.90 or greater while operating throughout the temperature range of -42°F (-40°C) to 165°F (+74°C).

g) Total harmonic distortion (current and voltage) induced into an AC power line shall not exceed 20 percent while operating throughout this temperature range.

h) Wiring and terminal blocks shall conform to Section 13.02 of the ITE VTCSH. Two secured color coded 600V, 20 AWG minimum, jacketed wires shall be provided for electrical connection for each LED signal module. The wires shall conform to the NEC, rated for service at or greater than 221°F (105°C).

LED signal modules shall be operationally compatible with the controller assemblies (solid state load switches, flashers and conflict monitors) shown on the Plans or specified in the Special Provisions.

209-5.5.11.6 Arrow LED Signal Modules. Arrow LED signal modules shall conform to Section 9.01 of the ITE VTCSH for arrow lenses. In addition LED's shall be spread evenly across the illuminated portion of the arrow area. Each arrow LED signal section indication shall provide an average luminous intensity of 511 cd/ft^2 (5500cd/m^2). Measurements shall be performed at the rated operating voltage of 120VAC.

209-5.5.11.7 Testing Requirements. LED signal modules tested or submitted for testing shall be representative of typical average production units. Circular LED signal modules shall be tested in conformance with the requirements in California Test 604. Optical testing shall be performed with the LED signal module mounted in a standard traffic signal section but without a visor or hood attached to the signal section. The number of units tested (sample size) shall be determined by the quantity of each models in the shipment. The sample size shall conform to the American National Standards Institute/Acceptance Sampling in Quality Control, ANSI/ASQC Z1.4.

Sampling parameters to be used for the random sample testing, if any, shall be as specified in the Special Provisions. The specified parameters may be tested on the sample.

Acceptance or rejection of the shipment shall conform to ANSI/ASQC Z1.4 for shipments which are sampled randomly. Upon rejection of a shipment the Contractor shall arrange for pick-up of the rejected shipment at no cost to the Agency.

209-5.5.11.8 Certificate of Compliance. The Contractor shall submit a Certificate of Compliance from the manufacturer conforming to 4-1.5 in accordance with 2-5.3. The certificate shall certify that the LED signal modules comply with the requirements in these Specifications. The certificate shall also include a copy of the applicable test reports on the LED signal modules.

209-5.5.12 Programmed Visibility Vehicle Signal Sections.

209-5.5.12.1 General. Programmed visibility vehicle signal sections shall provide a nominal 11-3/4 inch (300mm) diameter circular or arrow indication. Color and arrow configuration shall conform to ITE Publication ST-008B. Each signal section shall be provided with:

a) A cap visor.

b) An adjustable connection that permits incremental tilting from 0 to 10 degrees above or below the horizontal while maintaining a common vertical axis through couplers and mounting. Terminal connections shall permit external adjustments about the mounting axis in 5 degree increments.

Signal sections shall be mountable with ordinary tools and capable of being serviced without tools.

Adjustments shall be preset at 4 degrees below the horizontal, unless otherwise specified in the Special Provisions.

209-5.5.12.2 Programmed Visibility. The Contractor shall program the signal section as recommended by the manufacturer, shown on the Plans or specified in the Special Provisions. The visibility of each indication shall be capable of adjustment or programming within the section. When programmed, the indication shall be visible only in those areas or lanes to be controlled, except that during dusk and darkness a faint glow to each side will be permissible.

Prior to programming, each signal section with a yellow indication shall provide:

a) A minimum luminous intensity of 2500 cd on the optical axis.

b) A maximum intensity of 100 cd at 15 degrees at from 0.5 to 2 degrees horizontal from the axis.

c) A maximum of 10 cd at from 2 to 15 degrees horizontal from the axis.

Under the same conditions, the intensity of the red indication and the green indication shall be a minimum of 19 to 38 percent, respectively, of the yellow indication.

Each signal face or each signal section shall include integral means for regulating its luminous intensity between limits in proportion to the individual background luminance. Lamp intensity shall not be less than 97 percent of uncontrolled intensity at 10,000 lux and shall reduce to 15 ± 2 percent of maximum intensity at less than 10 lux. The dimming device shall operate over an applied voltage range of 95 to 135 V, 60 Hz and a temperature range of –42°F (–40°C) to 165°F (74°C).

209-5.6 Pedestrian Signal Sections.

209-5.6.1 General. Pedestrian signal sections shall be as shown on the Plans or as specified in the Special Provisions. Message symbols shall be white "WALKING PERSON" and Portland orange "UPRAISED HAND" conforming to the requirements in the ITE "Pedestrian Traffic Control Signal Indications" and the "Manual of Uniform Traffic Control Devices. The height of each symbol shall be not less than 9-3/4 inches (250mm) and the width of each symbol shall not be less than 6-1/2 inches (163mm). Each housing including the front screen, shall have maximum overall dimensions of 18-1/2 inches (463mm) wide, 19 inches (475mm) high and 11-1/2 inches (288mm) deep. All new pedestrian signal sections installed at any one intersection shall be the same manufacturer and type.

The luminance of the "UPRAISED HAND" symbol shall be 349 cd/sf (3750 cd/m^2) minimum and the luminance of the "WALKING PERSON" symbol shall be 493 cd/sf (5300 cd/m^2), minimum, when tested in conformance with California Test 606. The uniformity ratio of an illuminated symbol shall not exceed 4 to 1 between the highest and the lowest luminance area. The luminance difference between a nonilluminated symbol and the background around the symbol shall be less than 30 percent when viewed with the visor and front screen in place and at a low sun angle.

Brightness measurements for pedestrian signal sections designed for an incandescent lamp shall be made when the signal is equipped with an A-21 traffic signal lamp operated at a voltage to produce 665 lm.

209-5.6.2 Components.

209-5.6.2.1 General. Each pedestrian signal section shall consist of a housing, 2-color message plate, a reflector assembly, one incandescent lamp or one LED module conforming to 209-5.5.11, with sockets and a front screen.

209-5.6.2.2 Message Plate. Each message plate shall be one piece and shall be made of 1/8 inch (3mm) minimum thickness, ultraviolet stabilized polycarbonate plastic or 3/16 inch (5mm) tempered glass. The symbols shall be applied to the inside smooth surface of the message plate.

209-5.6.2.3 Reflector Assembly. Each reflector assembly shall consist of a double reflector or 2 single reflectors. Reflectors shall conform to ITE Publication ST-008B. Reflectors shall be made of either aluminum or plastic. Each completed reflector shall, when operated with the appropriate lamp and lens, provide the message brightness specified.

Plastic reflectors shall consist of molded or vacuum-formed plastic, with a vacuum-deposited aluminum reflecting surface. The plastic used shall not distort when the reflector is used with the lamp of the wattage recommended by the manufacturer of the section. In addition, the UL nonmechanical loading temperature of the material shall not be exceeded by more than 50°F (10°C) in conformance with UL 746B.

209-5.6.2.4 Front Screens. Front screens shall be provided on each signal face as shown on the Plans or specified in the Special Provisions. The front screen shall not fracture, separate at the welds or compress more than 1/8 inch (3mm) when a 3-inch (75mm) diameter, 4-pound (1.8kg) ball is dropped onto the screen from a height of 4 feet (1.2m). The screen will be laying in a horizontal position and supported on its edges for this test. The message plate will be removed from the pedestrian signal housing, when the pedestrian housing is used to support the front screen during the test, so there will be no back support for the screen.

The screen and frame shall be fabricated from aluminum anodized flat black, or finished with lusterless black exterior grade latex paint formulated for application to properly prepared metal surfaces, or shall be fabricated from flat black plastic.

One of the following types of screens shall be provided as specified in the Special Provisions:

a) Honeycomb screens. Honeycomb screens shall consist of:

1. Aluminum 3/16 inch (5mm) hexagon cells, 3/8 inch (10mm) thick or,
2. Plastic 3/8 inch (10mm) squares, 1/2 inch (12mm) thick with a wall thickness of 1/16 inch (1500μm).
3. A clear front cover of 1/8 inch (3mm) minimum thickness acrylic plastic sheet or 1/16 inch (1.5mm) minimum thickness polycarbonate plastic.

Honeycomb screens shall be installed tilting downward, at an angle of 15 ± 2 degrees, out from the top. Honeycomb screens shall completely cover the message plate. The honeycomb screen and cover shall be held firmly in place by stainless steel or aluminum clips or stainless steel metal screws.

b) Eggcrate or Z-crate screens. Eggcrate or Z-crate screens shall consist of one 1-1/2 inches (38mm) deep eggcrate or Z-crate screen of 1/32 inch (800μm) nominal thickness polycarbonate. The eggcrate or Z-crate screen shall be mounted in a frame constructed of 1/32 inch (800μm) minimum thickness aluminum alloy or polycarbonate.

Eggcrate or Z-crate screens shall be installed parallel to the face of the message plate and shall be held in place by stainless steel screws. The visor described in 209-5.6.2.6 is not required with this type screen.

209-5.6.2.5 Housing. Pedestrian signal housings shall conform to 209-5.5.3.

209-5.6.2.6 Visors. Visors shall be constructed of a material similar to the housing. The top of the visor shall extend a minimum length of 6 inches (150mm) at the top and 5 inches (125mm) at the bottom when measured from the front surface of the housing. The front shall be normal to the top.

209-5.6.3 Finish. The exterior of each housing and visor and the interior of the visors shall be painted in conformance with 210 or as specified in the Special Provisions.

209-5.6.4 Controls. Pedestrian signals sections shall be capable of being controlled by solid-state switching devices specified for traffic signal controller assemblies.

209-5.6.5 Terminal Blocks. Each pedestrian signal section shall be provided with a terminal block conforming to 209-4.3.3 or as specified in the Special Provisions. All field wiring shall connect to this terminal block.

209-5.6.6 LED Pedestrian Signal Section "Upraised Hand" Modules.

209-5.6.6.1 General. In addition to the requirements in 209-5.5.11 these LED modules shall conform to the following specifications. The installation of the LED modules shall not require modification of a standard lamp socket or reflector. LED modules shall be designed as retrofit replacements for existing optical units and shall not require special tools for installation.

Maximum power consumption requirements for these LED modules shall be 15W at 77°F (25°C) and 18W at 165°F (74°C).

The luminance of the "Upraised Hand" symbol shall be a minimum of 348 cd/ft^2 (3,750 cd/m^2). The color of the "Upraised Hand" shall be Portland orange conforming to the requirements of ITE "Pedestrian Traffic Control Signal Indications" and the MUTCD. The height of each symbol shall be not less than 10 inches (250mm) and the width of each symbol shall not be less than 6-1/2 inches (163mm).

The uniformity ratio of an illuminated symbol shall not exceed 4 to 1 between the highest luminance area and the lowest luminance area.

LED modules shall be rated for a minimum useful life of 48 months.

209-5.6.6.2 Physical and Mechanical Requirements. In addition to the requirements in 209-5.5.11 the physical and mechanical requirements for these LED modules shall conform to the following specifications.

LED modules shall not require a specific mounting orientation or have a variance in light output, pattern or visibility for any mounting orientation. Installation of LED pedestrian signal modules shall only require removal of the lamp.

LED modules shall be rated for use in the operating temperature range of –42°F to 165°F (–40°C to +74°C).

209-5.6.6.3 Photometric Requirements. The minimum initial luminous intensity values shall be 347 cd/ft^2 (3750 cd/m^2). Luminous intensity values shall meet or exceed 85 percent of 347 cd/ft^2 (3750 cd/m^2) after 48 months of continuous use over the temperature range of –42°F to 165°F (–40°C to +74°C) in a traffic signal operation.

The measured chromaticity coordinates shall conform to 5.3.2.1 and Figure C of the ITE VTCSH while operating throughout the temperature range of –42°F to 165°F (–40°C to +74°C).

209-5.6.6.4 Electrical Requirements. Electrical requirements for LED modules shall conform to 209-5.5.11.5 or as specified in the Special Provisions.

209-5.6.6.5 Testing. The LED pedestrian signal modules tested or submitted for testing shall be representative of the typical average production units. Modules shall be tested in accordance with California Test 606. Optical testing shall be performed with the module mounted in a housing as specified in the Special Provisions or on the Plans but without a visor or hood attached to the housing.

209-5.6.6.6 Quality Assurance Testing. Quality assurance testing may be performed by the Agency on each new light emitting diode pedestrian signal module. The number of units tested (sample size) shall be determined by the quantity of each model in the shipment. The sample size shall conform to the requirements in the American National Standards Institute/Acceptance Sampling in Quality Control. ANSI/ASQC Z1.4.

The Engineer will determine the sampling parameters to be used for the random sample testing. The parameters specified in the Special Provisions may be tested on each sample.

The basis of acceptance or rejection of the shipment shall be the criteria specified in ANSI/ASQC Z1.4 for shipments that are sampled randomly. Upon rejection of the shipment the Contractor shall arrange for pick-up of the shipment at no cost to the Agency.

209-5.6.6.7 Certificate of Compliance. A Certificate of Compliance shall be submitted in accordance with 209-5.5.11.8.

209-5.7 Signal Mounting Assemblies. Signal mounting assemblies shall consist of Size 41 standard steel pipe or galvanized conduit, necessary fittings, slip-fitters and terminal compartments. Pipe fittings shall be ductile iron, galvanized steel, aluminum alloy Type AC-84B No. 380, or bronze. Mast arm slip-fitters, post top slip-fitters and terminal compartments shall be cast bronze or hot-dip galvanized ductile iron. After installation, any exposed threads of galvanized conduit brackets and areas of the brackets damaged shall be cleaned with a wire brush and painted with 2 applications of an approved unthinned zinc-rich primer (organic vehicle type) conforming to 210. Aerosol cans shall not be used.

Each terminal compartment shall be fitted with a terminal block containing a minimum of 12 poles, each with 2 screw type terminals. Each screw type terminal shall be designed to accommodate at least five No. 14 conductors. A cover shall be provided on the compartment to give ready access to the terminal block. Where used to bracket mount signals, the terminal compartment shall be designed to bolt securely to a pole or standard.

The horizontal dimension of the mounting assembly members between the vertical centerline of the terminal compartment or slip-fitter and the vertical centerline of each signal face shall not exceed 11 inches (275mm), except where required to provide proper signal face alignment or permit programming of programmed visibility signal faces unless otherwise directed by the Engineer. Mounting assemblies shall be watertight and free of sharp edges or protrusions which might damage conductor insulation. Mounting assembly members shall be either plumb or level, symmetrically arranged and securely assembled. Each mounting assembly shall be provided with positive locking, serrated fittings that, when mated with similar fittings on the signal faces shall prevent the faces from rotating. Fitting shall permit fastening at increments of not more than 7 degrees. Construction shall be such as to permit all conductors to be concealed.

For post-top mounting of signals, a slip-fitter shall be used. The slip-fitter shall fit over a 4-1/2 inches (113mm) outside diameter pipe or tapered standard end. Each slip-fitter shall be provided with cadmium-plated steel set screws, arranged as shown on the Plans. Each slip-fitter used to post-top mount signals with brackets shall be provided with an integral terminal compartment.

Each mounting assembly shall be oriented to provide maximum horizontal clearance to the adjacent roadway.

209-5.8 Detectors.

209-5.8.1 General. Vehicle detectors shall be the type or types shown on the Plans. All sensor units, control units and amplifiers shall conform to the specifications in the Special Provisions. The units shall not be affected by transient voltages when tested in conformance with the requirements in California Test 667. After a power interruption the units shall return to normal operation within one minute.

Each unit shall be provided with a light or meter, for each output circuit, to indicate when the detector is detecting a vehicle. Each detector shall operate over the range of voltages from 100V to 135V at 60 Hz.

Circuitry shall be solid-state, except relays with normally closed contacts may be used for the output circuit. Units shall use printed circuit boards designed to facilitate identification of components. This shall be done by either part identification markings or by providing a pictorial diagram showing the physical location and identification of each component. Each printed circuit board shall conform to the following minimum quality requirements:

a) NEMA FR-4 glass cloth base epoxy resin board, 1/4 inch (6mm) minimum thickness,

b) Organic masking,

c) Gold plated contacts.

Units shall be designed to provide vehicle detection without readjustment from 0°F to 160°F (–18°C to 71°C).

209-5.8.2 Inductive Loop Detectors.

209-5.8.2.1 General. The term “inductive loop detector” shall be defined as a complete installation consisting of a loop or group of loops installed in the roadway, as shown on the Plans, lead-in cable and sensor unit with a power supply installed in a controller cabinet.

209-5.8.2.2 Sensor Units. “Card” type sensor units shall conform to the requirements in “Traffic Signal Control Equipment Specifications” issued by the State of California Department of Transportation, and all addenda thereto current at the time of advertising the Contract. Shelf-mounted sensor units shall conform to the requirements in Section 11 of the NEMA Standards Publication No. TS.1.

Capacitors or inductors necessary for loop tuning shall not be mounted external to the sensor unit.

209-5.8.2.3 Conductors. Conductors for inductive loop detectors shall be continuous and unspliced and shall conform to the Special Provisions or the following:

a) Type 1 loop wire shall be Type RHW-USE neoprene-jacketed or Type USE cross-linked polyethylene insulated No. 12 stranded copper wire. The minimum insulation thickness at any point shall be 1/32 inch (800μm) or

b) Type 2 loop wire shall consist of a conductor inside of plastic tubing. The conductor shall be Type THWN or Type XHHW, No. 14 stranded copper wire. The tubing shall be polyethylene or vinyl, rated for use at 221°F (105°C) and shall be resistant to oil and gasoline. The tubing shall have a maximum outside diameter of 1/4 inch (6mm) and a minimum wall thickness of 1/32 inch (800μm). The conductors shall not be spliced inside the tubing.

Conductor for loop detector lead-in cable shall be 2 No. 16 AWG (19 x 29) stranded tinned copper. Loop detector lead-in cable shall conform to the calculated cross sectional area of ASTM Designation B 286, Table 1. The lead-in cable shall conform to the Special Provisions or the following:

a) Type B lead-in cable shall be insulated with 1/64 inch (400μm) of high-density polyethylene. The conductors shall be twisted together with at least 6 turns per 3 feet (meter) and the twisted pair shall be protected with a copper or aluminum polyester shield. A No. 20 AWG, minimum, copper drain ground wire shall be provided and connected to the equipment ground within the cabinet. The cable shall be provided with a high-density polyethylene or high-density polypropylene outer jacket with a nominal thickness of 1/32 inch (800μm). An amorphous interior moisture penetration barrier of nonhydroscopic polyethylene or polypropylene fillers shall be provided, or

b) Type C lead-in cable shall conform to the International Municipal Signal Association's (IMSA) Specifications No.50-2. A No. 20 AWG minimum copper drain ground wire shall be provided and connected to the equipment ground within the cabinet.

All detector lead-in cable connections and/or terminations shall be soldered.

209-5.8.2.4 Sealants. Sealants shall be packaged in containers clearly labeled "detector loop sealant." The label shall include the batch and lot number of the manufacturer. Sealants for filling slots shall conform to one of the following:

a) Elastomeric Sealant.

Elastomeric sealant shall be a polyurethane material of a composition that will within its stated shelf life, cure only in the presence of moisture. Sealant shall be suitable for use in both asphalt concrete and portland cement concrete. The cured sealant shall have the performance characteristics shown in Table 209-5.8.2.4 (A).

TABLE 209-5.8.2.4 (A)

Specifications	ASTM Designation	Requirement
Hardness (indentation) at 77°F (25°C) and 50% relative humidity (Type A Model 1700 only)	D2240 Rex.	65-85
Tensile Strength, pulled at 20 inches (508mm) per minute	D412 Die C	500 psi (3.45MPa) min.
Elongation: pulled at 508mm per minute	D412 Die C	400%, min.
Flex at -40°F (–40°C): 0.6mm Free Bend (180°) over a 1/2 inch (12mm) Mandrel.	---	No cracks
Weather Resistance: Weatherometer 350 h cured 7 days at 77°F (25°C) @ 50% relative humidity.	D822	Slight chalking
Salt Spray Resistance: 28 days at 100°F (38°C) with 5% NaCl, Die and pulled at 20 inch (508mm) per minute.	B117	500 psi (3.45MPa) min. tensile, 400% min. elongation
Dielectric Constant over a temperature range of -22°F (–30°C)	D 150	Less than 25% change

b) Asphaltic Emulsion Sealant.

Asphaltic emulsion sealant shall conform to the State of California's Specification 8040-41A-15 and shall only be used for filling slots in asphalt concrete pavement. This material shall not be used in slots which exceed 5/8 inch (16mm) in width or where the slope causes the material to run from the slot. The material shall not be thinned in excess of the manufacturer's recommendations and shall not be placed when the air temperature is less than 45°F (7°C).

c) Hot-Melt Rubberized Asphalt Sealant.

Hot-melt rubberized asphalt sealant shall be in solid form at room temperature and fluid at an application temperature of 375°F (190°C) to 400°F (205°C). Fumes from the material shall be non-toxic. Sealant shall be suitable for use in both asphalt concrete and portland cement concrete.

Performance characteristics of the cured hot-melt rubberized asphalt sealant shall be as shown in Table 209-5.8.2.4 (B).

TABLE 209-5.8.2.4(B)

Specification	ASTM Designation	Requirement
Cone Penetration, 77°F (25°C) 5.3 oz (150g), for 5 sec.	D3407, Sec. 5	1/8 inch (3mm) max.
Flow, 40°F (60°C 1)	D3407, Sec. 6	3/16 inch (5mm) max.
Resilience, 77°F (25°C)	D 3407, Sec. 8	25% min.
Softening Point	D 36	180°F (82°C) min.
Ductility, 77°F (25°C) 2 inch (50mm/min.) min.	D113	11 13/16 inches (300mm) min.
Flash Point, Cleveland Open Cup, °F (°C)	D92	550°F (288°C)
Viscosity, Brookfield Thermosel, No. 27 Spindle, 20 rpm, 374°F (190°C)	D4402	2.5-3.5 Pa·s

209-5.8.3 Magnetic Detectors.

209-5.8.3.1 General. Each magnetic detector amplifier shall be a low noise, high gain amplifier designed for use with all types of magnetic detector sensing units as specified by the Agency in the Special Provisions. Sensitivity of units shall be readily adjustable by a knob on the front panel. An indicator light shall be provided to indicate detector operation.

The sensing elements shall be no larger than 2-1/4 inches (56mm) in diameter and shall contain no moving parts or transistors. The element casing shall be constructed of non-ferrous materials suitable for use in the environment that it will operate in and shall be sealed to prevent the entrance of moisture. The element shall be shaped to facilitate pushing into conduit and shall not have any sharp edges along its length. The overall length of the sensing elements shall not exceed 21 inches (525mm).

209-5.8.3.2 Performance Characteristics. Each magnetic detector shall be capable of being activated by a voltage induced in the coil of the sensing element by the passage of a vehicle at any speed from 3 to 81 mph (5km/hr to 130km/hr). Any vehicle passing within 18 inches (450mm) of either end of the sensing elements shall provide an output closure.

209-5.8.3.3 Amplifier Unit Construction. Each amplifier shall be provided with an integral power supply. Each amplifier shall be designed to provide ease of maintenance with all electronic components readily accessible. All input and output circuits for each amplifier shall enter via a single MS connector. Connector circuitry for the amplifiers shall be as shown in Table 209-5.8.3.3 (A).

TABLE 209-5.8.3.3 (A)

MS Connector Circuit	18 – 8 Pin
Magnetic Detector (-)	A
Magnetic Detector (+)	B
120 V, (AC+)	C
Chassis Ground	D
Detector Common	E
Output N.O.	F
Output N.C.	G
Grounded Conductor, (AC-)	H

All controls, indicator lights, fuse holders and connectors shall be mounted on the front of the amplifier.

209-5.8.3.4 Construction Materials. The cable from the pull box adjacent to the magnetic detector sensing element to the field terminals in the controller cabinet shall be the type specified for inductive loop detectors.

209-5.9 Pedestrian Push Button Assemblies. Pedestrian push button assembly housings shall be either die-cast or permanent mold cast aluminum or, when specified in the Special Provisions, shall be ultraviolet stabilized, self-extinguishing, structural plastic. Plastic housings shall be black matching Color No. 17038, 27038 or 37038 of Federal Standard No.595B or as specified in the Special Provisions and shall be colored throughout. Assemblies shall be rainproof and shockproof in any weather conditions.

Switches shall be the phenolic enclosed precision snap-acting type, single-pole, double-throw, switching unit, with screw type terminals, rated 15A at 125VAC, and shall conform to the following:

a) Have a stainless steel plunger actuator and be provided with a U-frame to permit recessed mounting in the push button housing.
b) Have an operating force of 9oz (2.5N) to 13oz (3.6N) and a minimum release force of 4oz (1N).
c) Pre-travel shall be 1/64 inch (400μm) maximum.
d) Overtravel shall be 1/32 inch (800μm) minimum.
e) Differential travel shall be 0.01mm to 0.05mm.
f) The actuator shall have a minimum diameter of 2 inches (50mm).

Pedestrian push button signs shall be porcelain enameled metal or structural plastic.

SECTION 210 - PAINT AND PROTECTIVE COATINGS

210-1 PAINT.

210-1.1 General Requirements. Paint shall be homogeneous, free of contaminants, and of a consistency suitable for the use for which it is specified. The pigment shall be finely ground and properly dispersed in the vehicle according to the requirements of the paint; and this dispersion shall be of such nature that the pigment does not settle appreciably, does not cake or thicken in the container, or become granular or curdled. Paint and paint materials shall be delivered to the Work site in new, unopened, airtight containers, appropriately identified with the manufacturer's name, date of manufacture, type of paint or paint material, State Specification number, and lot or batch number.

No paint shall be used until at least 7 days have elapsed from the date of manufacture.

210-1.2 Testing. All paint and paint materials shall be sampled and tested prior to use. All tests will be conducted in accordance with the methods specified in ASTM or methods set forth in Federal Standard 141. In the absence of any such methods, other suitable methods may be designed and utilized by the Engineer.

Lots or batches of paint of proprietary brand, which have been previously sampled and tested by the Agency and approved as conforming with these specifications, may be used without further testing, if permitted by the Engineer. For the purpose of these specifications, proprietary brands of paint and paint materials are construed to mean those conforming to the requirements of these specifications which are produced for distribution through regular wholesale and retail outlets.

210-1.3 Paint Coats. Paint coats shall consist of pre-treatment when specified, primer, and finish coats in that order, in accordance with 210-1.5.

210-1.4 Paint Materials. Paint materials shall conform in all respects to the requirements of the reference specifications indicated for such material in the table of paint systems shown below.

210-1.5 Paint Systems. Unless otherwise specified, the paint systems to be used shall correspond with the following table and the California Department of Transportation specifications:

TABLE 210-1.5 (A)

Surface to be Painted	Pre-Treatment/Surface Preparation[1]	Primer[1]	Finish Coats[1]
Galvanized Metal	Vinyl Wash	Zinc Dust – Zinc Oxide Fed. Spec. TT-P- 641 Type II	Aluminum White Enamel White Tint Base[2] Aluminum Green Tan Burnt Sienna
Structural Steel – General Use	Commercial Blast Per SSPWC 310-2.5	Per Manufacturer's Recommendations	
Wood Structures, Wood Primer and Surfaces		Wood Primer	Exterior White[2]
Signal Standards	Vinyl Wash	Yellow	Yellow
Signal Heads and Mounting	Vinyl Wash		Olive Green[2]
Backplates, Visors, Louvers	Vinyl Wash		Black[2]

1. Refer to Section 91, State of California Department of Transportation (CalTrans) Standard Specifications unless otherwise specified.
2. Finish color shall be in accordance with Specifications or as determined by the Engineer.

210-2 PLASTIC LINER.

210-2.1 General. Plastic liner sheet, joint, corner, and weld strips shall be manufactured from a polyvinyl chloride compound that meets the properties specified herein and is approved by the Engineer. The material shall be suitable for use as protective liner and joint systems in pipe or other structures. Copolymer resin will not be permitted.

At any time during the manufacture or prior to the final acceptance of the Work, utilizing material qualified under 210-2.3, the Engineer may sample specimens taken from any part of the assembled or unassembled product for testing of physical properties per 211-2.

Changes in the compound formulation may be permitted only after 12 months if prior notice is given to the Engineer, and the Contractor proves that the new formulation meets or exceeds all requirements specified herein and is approved by the Engineer. The Engineer shall be notified 7 days prior to any approval testing as to the type of material to be tested.

All plastic liner sheets including locking extensions, joint, corner, connecting, and welding strips shall be free of cracks, cleavages, or other defects adversely affecting the protective characteristics of the material. The Engineer may authorize the repair of such defects by approved methods.

Plastic liner older than 180 days from the date of manufacture shall not be used in lined reinforced concrete pipe as specified in 207-3.

210-2.1.1 Adhesive Products. The Engineer shall approve adhesive products and application procedures to be used in the installation of the plastic prior to their use. Adhesive products intended for use inside joined plastic-lined pipe or cast-in-place structures shall be nonflammable.

210-2.1.2 Cleaners. The Engineer shall approve cleaners used in the installation of the liner prior to their use. Cleaners shall be nonflammable, shall be water soluble or water dispersible, and shall not be detrimental to the plastic liner.

210-2.2 Details and Dimensions.

210-2.2.1 Approval of Details. Liner sheet, strip, and other accessory pieces shall conform to the requirements shown on the Plans and specified in the Specifications.

210-2.2.2 Thickness of Material. The minimum thickness of sheet and strip shall be as noted in Table 210-2.2.2 (A):

TABLE 210-2.2.2 (A)

Material	Thickness mils (mm)
Sheet, integral locking extensions	65 (1.65)
Sheet, plain	94 (2.39)
Joint strip	75 (1.91)
Weld strip	94 (2.39)

210-2.2.3 Material Sizes. Sheets of liner used for pipe shall be sized to provide the coverage required by the Plans. Joint strips shall be 4 ± 0.25 inches (100 ± 6mm) in width and shall have each edge beveled prior to application. Welding strips shall be 1 ± 0.125 inch (25 ± 3mm) in width. All welding and outside corner strips shall have edges beveled at time of manufacture.

210-2.2.4 Locking Extensions. All liner to be embedded in concrete shall have integral locking extensions.

Locking extensions shall be of the same material as the liner; shall be integrally bonded, molded, or extruded with the sheets; and shall have an approved cross section with a minimum height of 0.375 inch (9.5mm) and a minimum web thickness of 0.085 inch (2.2mm). Locking extensions shall be 2-1/2 ± 0.250 inches (65 ± 6mm) apart for pipe and structure applications. The locking extensions shall be such that when they are embedded in concrete, the liner will be held permanently in place and meet the pull-out requirements of 210-2.3.4. There shall be a minimum of one locking extension embedded in each continuous surface, except any continuous surface of 4 inches (100mm) or more shall have a minimum of two locking extensions, or the material shall be bonded in place. Locking extensions shall be parallel and continuous except where interrupted for joint flaps, weep channels, strap channels, and for other purposes shown on the Plans or where permitted by the Engineer.

A flexible liner sheet edge which will be the lower terminal edge in the structure or pipe shall not extend beyond the base of the final locking extension more than 3/8 inch (10mm).

Liner shall be bonded to concrete surfaces with an adhesive if shown on the Plans or specified in the Specifications.

210-2.2.5 Provisions for Strap Channels. If required, strap channels shall be) wide maximum and formed by removing the locking extensions so that a maximum of 3/16 inch (4.7mm) remains.

210-2.2.6 Flaps. When transverse flaps are specified or required, they shall be fabricated by removing locking extensions so that a maximum of 0.032 inch (0.80mm) of the base of the locking extensions remains on the sheet.

210-2.3 Tests.

210-2.3.1 General. Liner materials shall be tested for physical and chemical properties prior to approval by the Engineer. Samples taken from sheets, joints, or weld strips of material representative of those to be furnished shall be tested to determine material properties. Determination of physical properties shall be in accordance with the tests specified in 210-2.4 or 210-2.5. Properties specified in 210-2.4 or 210-2.5 shall be submitted to the Engineer per 2-5.3.4.

210-2.3.2 (Not Used).

210-2.3.3 Chemical Resistance Test (Pickle Jar Test). Test specimens shall be tested in accordance with 211-2.

210-2.3.4 Pull Test for Locking Extensions. Liner locking extensions embedded in concrete shall withstand a test pull of at least 100 pound per linear inch (18N per linearmm), applied perpendicularly to the concrete surface for a period of 1 minute, without rupture of the locking extensions or withdrawal from embedment. This test shall be made at a temperature between 70°F to 80°F (21°C to 27°C) inclusive.

210-2.3.5 Shop Welded Joints. Shop-welded joints, used to fuse individual sections of liner together, shall be at least equal to the minimum requirements of the liner for thickness, corrosion resistance, and impermeability. Welds shall show no cracks or separations and shall be tested for tensile strength. Tensile strength measured across the welded joint in accordance with ASTM D412 using Die B shall be at least 2,000 psi (13.8MPa). Test temperatures shall be 77°F ± 5°F (25°C ± 3°C) and the measured minimum width and thickness of the reduced section shall be used.

210-2.3.6 Spark Test. All liner shall be shop tested for holes with a spark tester set to provide from 15,000 to 20,000 volts. Sheets having holes shall be satisfactorily repaired in the shop prior to shipment from the manufacturer's plant. Welders qualified in accordance with 311-1.2.2 shall make all repairs.

210-2.4 Flexible PVC Liner.

210-2.4.1 Flexible PVC Liners for Structures, Manholes, and Pipes. Flexible PVC plastic liner sheets, joint assembly components, corner and weld strips shall be tested in accordance with 211-2 and shall conform to the requirements in Table 210-2.4.1(A).

TABLE 210-2.4.1 (A)

Property	Initial[1]	Exposure[1]
Tensile Strength ASTM D412	2,200 psi (15MPa)	2,100 psi (14.4MPa)
Elongation at break ASTM D412	200%	200%
Hardness, Shore durometer, Type D ASTM D2240	Within 1 sec. 50-60	± 5[3]
	Within 10 sec. 35-50	± 5[3]
Weight change	—	± 1.5%[3]

1. For 112 days in chemical solutions listed in 211-2. All above values are minimum required except for hardness and weight, which is the maximum permissible gain or loss in weight.
2. Except that a single thickness of material shall be used.
3. With respect to initial test result.

210-2.4.2 Shop-Welded Joints. Tensile strength measured across the welded joint in accordance with ASTM D412 using Die B shall be at least 2,200 psi (15MPa) minimum.

210-2.5 Rigid PVC Liners.

210-2.5.1 Rigid PVC Liners for Structures, Manholes, and Pipes. Rigid PVC plastic liner sheets, joint assembly components corner and connecting strips for pipe, manhole, and linings for structures shall be tested in accordance with 211-2 and shall conform to the requirements in Table 210-2.5.1 (A).

TABLE 210-2.5.1 (A)

Property	Initial[1]	After Exposure[1]
Tensile Strength ASTM D638	6,500 psi (44MPa)	80% of initial
Tensile Modulus ASTM D638	408,000 psi (2810MPa)	N/A
Elongation at break ASTM D638	25%	N/A
Hardness, Shore Durometer, Type D ASTM D2240[2]	Instantaneous 70	N/A
Flexural Strength ASTM D790	13,000 psi (90MPa)	N/A
Flexural Modulus ASTM D790	455,000 psi (3140MPa)	N/A
Heat Deflection ASTM D648	160°F at 264 psi (71°C at 1.8MPa)	N/A
Izod Impact ASTM D256	1.2 ft-lbs/inch (1.63 Joules/25mm) of notch	N/A
Variable Height Impact Test ASTM D4226		
W/.25 Hemispherical Dart		
Procedure A[2]	1.6 in-lbs/mil (7.2 Joules/mm) thickness	N/A
Procedure B[3]	2.8 in-lbs/mil (12.8 Joules/mm) thickness	N/A
W/.125 Conical Dart		
Procedure A[2]	1.7 in-lbs/mil (7.7 Joules/mm) thickness	80% of initial
Procedure B[3]	2.9 in-lbs/mil (13.1 Joules/mm) thickness	80% of initial
Weight Change		± 1.5%

1. All above values are minimum required except for hardness and weight, which is the maximum permissible gain or loss in weight.
2. Minimum Failure Energy.
3. Ductile/Brittle Transition.
4. With respect to initial test result.

210-3 GALVANIZING.

210-3.1 General. Zinc used for galvanizing shall be grade Prime Western conforming to ASTM B6. Except as otherwise specified, materials shall be galvanized by the hot-dip, mechanical, or electrode positing process.

210-3.2 Requirements of Coating. The minimum weight of coating and other requirements shall be as shown in the following table. If there is a conflict between the ASTM and minimum weight columns, the minimum weight column shall apply. The weight shown is ounces per square foot (grams per square meter) of surface area. The weight of coating shall be determined in accordance with ASTM A90, modified to determine the coating of each surface separately. All surfaces, when tested separately, shall meet the minimum requirements.

TABLE 210-3.2 (A)

Material	ASTM	Minimum Weight of Coating oz/ ft^2 (g/m^2)
Steel products including structural shapes, tie rods, handrails, manhole steps, and miscellaneous items.	A 123 A 153 B 633 B 695	2.00 (610) 2.00 (610) 2.00 (610) 2.00 (610)
Hardware including cast, rolled, pressed and forged articles.	A 153	2.00 (610)
	B 633	2.00 (610)
	B 695	2.00 (610)
Bolts, screws, nuts and washers	A 153	1.25 (381)
	B 633	1.25 (381)
	B 695	1.25 (381)
CSP culverts and underdrains	A 444	1.00 (305)
Chain link fence fabric, tie wire only.	A 392	1.20 (366)
Steel pipe (includes fence posts, braces, and rails)		
Class 1 (see Note 1)	F1083	1.80 (550)
Class 1A (pipe only, see Notes 2 and 3).		1.00 (305)
All other chain link fence articles	A 123	1.80 (550)
Iron or steel wire fencing	A 116	0.80 (244)
Steel or iron sheets	A 525	1.20 (366)
Barbed wire	A 121	0.80 (244)
Electrolier standards, 7-gage steel and over	A 386	2.00 (610)
Electrolier standards, under 7-gage steel	A 386	1.50 (458)

1. "C" and "H" section fence posts shall be hot-dipped galvanized after forming.
2. Class 1A pipe shall, in addition to the galvanized exterior coating indicated, have coatings of chromate conversion and acrylic urethane in accordance with Subsection 210-4.
3. The interior surface coating for Class 1A pipe shall have 1 ounce per square foot (305 g/m^2) of hot-dipped galvanized coating or a zinc-rich organic a minimum of 0.3 mils (8μm) thick. The coating shall have a minimum zinc powder loading of 87 percent by weight. The interior coated surface shall have a demonstrated ability to resist 650 hours of exposure to salt fog with a maximum of 5 percent red rust when tested in accordance with ASTM B 117.

210-3.3 Workmanship. The zinc coating shall adhere tenaciously to the surface of the base material. The finished product shall be free from blisters and excess zinc, and the coating shall be even, smooth, and uniform throughout. Machine work, die work, cutting, punching, bending, welding, drilling, thread cutting, straightening, and other fabricating shall be done as far as is practicable before the galvanizing. All members, nuts, bolts, washers, etc., shall be galvanized before a structural unit is assembled. All uncoated spots or damaged coatings shall be cause for rejection.

Products that are warped or distorted to the extent of impairment for the use intended shall be rejected.

210-3.4 Test Coupons. Test coupons for determining the quantity and quality of the galvanizing shall be of such size and shall be wired to the materials to be galvanized before immersion so as to represent the amount of coating deposited on the finished product.

Nondestructive tests for uniformity of coating may be made by the Engineer with a magnetic instrument in accordance with ASTM E 376.

210-3.5 Repair of Damaged Zinc Coatings. Zinc coating which has been field or shop cut, burned by welding, abraded, or otherwise damaged to such extent as to expose the base metal, shall be repaired and recoated by one of the following methods approved by the Engineer.

210-3.5.1 Hot-Dip Process. The damaged areas shall be thoroughly stripped and cleaned and a coating of zinc shall be applied by the hot-dip process per 210-3.1 and 210-3.2.

210-3.5.2 Metalizing Process. The damaged area shall be repaired per ASTM A780, Annex A3, and the following requirements:

a) The damaged areas shall be thoroughly cleaned by blasting with sharp sand or steel grit per 310-2.5.1b).

b) The blasted area shall lap the undamaged zinc coating at least 1/2 inch (15mm).

c) Zinc wire containing not less than 99.98 percent zinc shall be used in the metalizing operation.

d) A zinc coating shall be applied to the damaged area with a metalizing gun to a thickness of not less than 5 mils (130μm) on the damaged area, and shall taper to zero thickness at the edge of the blasted undamaged section.

210-3.5.3 Zinc Dust Paint. When zinc surfaces have small areas of abrasion which occur after shop application of zinc coating, zinc dust paint may be used to repair these areas when approved by the Engineer. The damaged area shall be thoroughly cleaned by wire brushing and traces of welding flux and loose or cracked zinc coating removed prior to painting. The cleaned area shall be painted with a minimum of two coats of an unthinned zinc paint to provide a total minimum film thickness of 8 mils (205μm). The zinc dust paint shall conform to the requirements of ASTM A 780, Annex A3, except that it shall have a 90 percent minimum dry film content of zinc dust by weight. The method of application shall be approved by the Engineer.

210-3.5.4 Zinc Based Solders. The damaged areas shall be repaired using zinc alloy solders per ASTM A 780, Annex A3. Zinc solder shall be deposited until a minimum thickness of 5 mils (130μm) is applied to the damaged area.

210-4 CHROMATE CONVERSION AND CLEAR ACRYLIC URETHANE COATINGS.

210-4.1 General. Class 1A Steel pipe for chain link fencing used for posts, braces, top rail, and gate frames shall have an exterior chromate conversion coating placed on the galvanized coating followed by a clear acrylic urethane coating.

210-4.2 Chromate Conversion Coating. The chromate conversion coating weight shall be a minimum of 23 g/m^2. The weight of chromate coating shall be determined by stripping the acrylic urethane coating with a neutral epoxy polyurethane stripper, determine the total weight of chromate and zinc coating in accordance with ASTM A90; then determine the percent of chromate and percent of zinc in the total metallic coating by atomic absorption spectrophotometer.

210-4.3 Clear Acrylic Urethane Coating. The coating shall be manufactured from high grade raw materials which produce a crosslinked acrylic polyurethane coating.

The acrylic urethane coating shall be a minimum of 0.3 mil (8μm) thick. The thickness shall be determined in accordance with ASTM G12. The gage shall be calibrated on the bare metal substrate after chemically removing the urethane and underlying zinc coatings. The total thickness of the urethane and zinc coatings shall then be determined. Chemically strip the urethane and determine the thickness of the zinc coating only. The difference between the thickness of the zinc and the total thickness of the urethane coating and zinc is the thickness of the urethane coating.

Alternate test methods may be used when approved by the Engineer.

210-4.4 Test of Coating System. These tests shall be performed with the entire coating system in place.

The exterior clear-coated surface of the pipe shall have a demonstrated ability to resist 1,000 hours of exposure to salt fog with a maximum of 5 percent red rust when tested in accordance with ASTM B117.

There shall be no film cracking of the clear finish coat after 500 hours exposure in a weatherometer in accordance with ASTM G23, Type E, or EH carbon arc weatherometer, or G 26, Type B, or BH xenon arc weatherometer. There shall be no blistering or cracking of the clear finish coat after 500 hours of exposure to 100 percent relative humidity in accordance with ASTM D2247.

210-4.5 Repair of Coatings. Repairs of damaged areas for Class 1A steel pipe shall be done by first repairing the galvanized coating in accordance with 210-3.5. In addition, a clear acrylic lacquer such as acrylon, shall be used as a final coating on the repaired area.

210-5 POLYVINYL CHLORIDE (PVC) COATINGS.

210-5.1 General. This specification covers PVC coatings for posts, frames, gates, and fittings. PVC coating thickness shall be a minimum of 15 mils (380μm).

210-5.2 Properties. The requirements for PVC coating shall be as shown in the following table:

TABLE 210-5.2 (A)

Property	ASTM Test No.	Requirements
Specific Gravity of Fused Resin	D 792	1.30 ± 0.030
Hardness,		
Shore A	D 2240	88 ± 2
Shore D	D 2240	36 ± 2
Tensile Strength, psi (MPa)	D 651	2500 ± 100 (17.2 ± 0.2)
Elongation, %	D 638	250 ± 10
Tear Strength, lbs/in (N/mm)	D 1004	440 ± 20 (77 ± 3)
Brittleness Temperature	D 746	0 ± 4°F (-18 ± 2°C)
Weathering Test Method, Langley Melting Range	E 838	395°F to 415°F (202°C to 213°C)
Dielectric Strength Range V/mil (V/μm)	D 419	800 to 1000 (32 to 40)

210-5.3 Workmanship. All surfaces to be PVC coated shall be thoroughly cleaned, removing all foreign matter to ensure noncontaminated surfaces. The substrate shall be primed, preheated to a temperature of 450°F (232°C) and with 0.5 mil (13μm) thickness primer. Primer shall be composed of a mixture of acrylic, phenolic, and epoxy. All pipe and fittings shall be immersed in a fluidized-bed system for PVC coating. Gate frames shall be fabricated and welded prior to the application of zinc substrate and PVC coatings.

210-5.4 Repair of Damaged PVC Coatings. Damaged PVC coatings, which have been field or shop cut, burned, abraded, or otherwise damaged, shall be repaired and recoated by the following method: the damaged area shall be thoroughly cleaned by wire brushing, then wiped clean with a non-oil-base solvent such as methyl ethyl ketone or acetone solvent. The cleaned area shall be painted with 5-mil (130μm) thickness of acrylic enamel. Acrylic enamel shall be a PVC resin in solution, consisting of high level pigments, ultraviolet absorbers, and solvent blends applied by brush in a minimum of three coats.

SECTION 211- MATERIAL TESTS

211-1 COMPACTION TESTS.

211-1.1 Laboratory Maximum Density. The following method shall be used for compaction tests unless otherwise specified:

Laboratory maximum densities will be performed in accordance with ASTM D1557.

The Engineer may modify ASTM D1557 at his option to calculate relative compaction based on adjusted laboratory maximum wet density calculated as follows:

Da = (100 Dm) / (100 ± Wa)

Da = Adjusted laboratory maximum wet density.

Dm = Maximum wet density per ASTM D1557.

± Wa = Percent change in moisture content from field moisture to laboratory optimum moisture. Use minus when field moisture content is higher than laboratory optimum moisture content. Use plus when field moisture content is lower than laboratory optimum moisture content.

211-1.2 Field Density. Field density of soil shall be determined by ASTM D2922 or ASTM D1556.

211-1.3 Relative Compaction. The words Relative Compaction shall mean the ratio of the field dry or wet density to the laboratory maximum dry or adjusted wet density, respectively, expressed as a percentage.

211-2 CHEMICAL RESISTANCE (PICKLE JAR) TEST. This test is used to determine the physical properties and weight change of specimens of materials used in sewers after exposure to chemical solutions. Specimens of composite materials shall be seal coated on 2 adjacent edges of their 4 edges and not seal coated on the inner or outer surface. Specimens of non-composite materials shall not be seal coated. Test specimens shall be conditioned in a mechanical convection oven for 7 days at a constant weight and at a temperature of 110° F ± 3° F (43° C ± 3° C) and subsequently cooled for 3 hours in a desiccator. This conditioning shall be performed before and after submersion of the test specimens in the solutions shown in Table 211-2 (A) for a period of 28, 56, 84 and 112 days at 77° F ± 5° F (25° C ± 3° C). Test specimens shall be provided as follows, unless otherwise approved by the Engineer.

a) Specimens For Non-Composite Materials - Provide 55 each ASTM D638 specimens and 164 each ASTM D543 specimens measuring 1 inch x 3 inches x 0.125 inch thick (25mm x 76mm x 3mm).

b) Specimens For Composite Materials - Provide 55 each ASTM D638 specimens and 164 each ASTM D543 specimens measuring 1 inch x 3 inches x 0.125 inch to 0.25 inch thick (25mm x 76mm x 3mm to 6mm).

The allowable percent weight gain or loss for these specimens shall be as specified for the respective type of material.

TABLE 211-2 (A)

Chemical Solution	Concentration[1]	Tolerance	Check Concentration	Replace Chemical Solution
Sulfuric Acid (H_2SO_4)	20%	± 0.1%	NA[2]	NA
Sodium Hydroxide (NaOH)	5%	± 0.2%	56 days	When < 4.8%
Ammonium Hydroxide (NH_4OH)	5%	± 1.0%	28 days	When < 4.0%
Nitric acid (HNO_3)	1%	± 0.1%	28 days	When Turbid
Ferric Chloride ($FeCl_3$)	1%	± 0.2%	28 days	When Turbid
Sodium Hypochlorite (NaOCl)	1%	± 0.1%	NA	28 days
Soap	0.10%	NA	NA	28 days
Detergent (Linear alkyl benzyl sulfonate or LAS)	0.10%	NA	NA	28 days
Bacteriological (BOD)	≥ 700 ppm	NA	NA	7 days

1. Volume percentages – Actual concentration of reagent must be corrected to 100%.
2. NA – Not Applicable.

If required by the type of material and the manufacturer, specimens thicker than 0.25 inch (6mm) will be accepted and evaluated. The same number of ASTM D638 AND D543 specimens, as specified in 211-2a or 211-2b above, will be required. The allowable weight change value for the 112-day test period for these thicker specimens shall conform to the values shown in Table 211-2 (B).

At 28-day intervals, specimens shall be removed from each chemical solution and tested. If any specimen fails to meet the 112-day requirement specified for the material being tested before completion of the 112-day exposure, the material will be rejected.

TABLE 211-2 (B)

Thickness of Sample Inches (mm)	Weight Change 1,2
0.375 (9.5)	± 0.75 %
0.50 (12.7)	± 0.63 %
0.75 (19.1)	± 0.50 %
1.0 (25.4)	± 0.44 %
1.5 (38.1)	± 0.38 %

1. With respect to initial test results.
2. Allowable percent weight gain or loss for the thickness between the designated values can be a straight line interpretation.

If required by the type of material being evaluated, additional size and quantity specimens may be required in order to meet the testing and ASTM requirements in the section for the material.

Five of the ASTM D638 specimens and 16 of the ASTM D543 specimens for the type of material being tested shall be submersed in each of the 9 solutions listed in Table 211-2 (A). At 28-day intervals, 4 specimens shall be removed from each chemical solution, tested for weight change and the percent weight gain or loss shall be recorded. Required physical property testing shall be obtained at 0 and 112 days. The chemical strength of each solution shall be tested and replaced if not in compliance with the required values shown in Table 211-2 (A). If any specimen fails to meet the 112-day requirement specified for the material being tested before completion of the 112-day exposure, the material will be rejected.

The Contractor shall furnish a Certificate of Compliance conforming to 4-1.5, signed by an authorized representative of the testing laboratory, identifying the product, the test results for each of the 9 solutions and confirming conformance with the Specifications. Supporting test data shall be furnished upon request by the Engineer.

The chemical resistance test is a qualification test only. Requalification is required only when the compound formulation changes.

211-3 INFILTRATION CAPACITY TEST.

211-3.1 Test Apparatus. Test apparatus shall consist of a:

a) One gallon (4L), minimum size, water container with a spout. The spout shall be able to produce a stream with a circular cross section, the diameter of that which is large enough to discharge the entire contents of the container in 20 seconds or less;

b) Stopwatch capable of indicating elapsed time to the nearest second;

c) Tape measure of at least 36 inches (1000mm) that is graduated in 1/4 inch (6mm) increments or smaller.

211-3.2 Water. Water shall be free of suspended solids. The volume of water shall be determined to 2 significant figures.

211-3.3 Test Procedure. The testing procedure shall be as follows:

a) Place a pre-measured amount of water into the container.

b) Pour the water onto the surface in one spot. Control the discharge rate by manually adjusting the angle of the spout so that the diameter of the pool of water is between 10 to 30 inches (250mm to 760mm). Empty the container holding the spout over the spot until the pool of water vanishes.

c) Start the stopwatch when the water initially touches the concrete surface and stop it when the pool disappears from the surface.

d) Measure the longest dimension (d1) of the dampened area. Measure the width (d_2) of the pool perpendicular to d_1.

e) Repeat this procedure at a minimum of 4 separate locations.

211-3.4 Infiltration Capacity. Infiltration Capacity shall be calculated as follows:

a) The formula for U.S. Standard Measures shall be: $IC=(V)(3{,}326{,}400)/(\pi)(d_1)(d_2)(t)$ inches per hour.

b) The formula for SI Units shall be: $IC=(V)(14.4X10^6)/(\pi)(d_1)(d_2)(t)$ cm per hour.

Where:

IC is Infiltration Capacity

V is the volume of water in gallons or liters

d_1 and d_2 are the dimensions that were determined in part 211-3.3

π is approximately 3.14159

211-3.5 Test Report. The test report shall include:

a) The time and date of testing.

b) The name and affiliation of the person performing the test.

c) The location of the Work site.

d) The location of each test site tested within the Work site.

e) The volume of water used at each test site.

f) The length of the two measurements taken at each test site.

g) The discharge time for the water at each test site.

h) The Infiltration Capacity at each test site.

i) The average Infiltration Capacity for the Work site.

211-4 HAND HELD VISCOMETER TEST.

211-4.1 Test Apparatus. Test apparatus shall consist of the following:

a) Viscometer: Haake Model VT – 02 rotational viscometer or equivalent.

b) Spindle: Rotor No. 1, diameter 24mm ± 0.1mm, height 53mm ± 0.1mm or equivalent.

c) Thermometer: Digital with metal-jacketed probe, 1°F (1°C) precision.

d) Sample Container: 1 gallon metal can with wire bale.

e) Standard Fluids: per fluid manufacturer's recommendations, calibrate viscometer in absolute viscosity (centipoises).

211-4.2 Calibration. The viscometer shall be calibrated in accordance with the manufacturer's recommendations. The accuracy of the viscometer shall be verified by comparing the viscosity results obtained with the hand held viscometer to 3 separate calibration fluids of known viscosity ranging from 1000 to 5000 centipoises. The viscometer will be considered accurate if the values obtained are within 300 centipoises of the known viscosity. The known viscosity value shall be based on the fluid manufacturer's standard test temperature or the test temperature versus viscosity correlation table provided by the fluid manufacturer. The viscometer calibration may be performed at an offsite laboratory and a Certificate of Compliance conforming to 4-1.5 verifying the calibration shall be submitted to the Engineer.

211-4.3 Test Procedure. The test procedure shall be in accordance with the following:

a) Obtain a representative sample of asphalt rubber binder from an appropriate sample valve during production. Prior to sampling, run a one-gallon (4 liter) sample of binder through the sampling valve and discard the sample. Place a clean one-gallon (4 liter) sample can under the sampling valve and fill the container to approximately 85 percent full.

b) Transport the sample immediately to the testing location. The testing location shall be close to the sampling location to avoid undue temperature loss. The temperature of the binder prior to sampling shall be a minimum of 375°F (190°C). The sample container shall then be placed on a smooth, level support for testing.

c) The binder shall be continuously stirred using an appropriate metal stirring rod. Use of the temperature probe is permitted. Insert the temperature probe into the binder and check the temperature. Stirring shall continue until the binder reaches a temperature of 375°F ± 5°F (190°C ± 3°C).

d) The viscometer spindle shall be cleaned in a suitable solvent and free of any binder material prior to the test. While holding the viscometer over the sample container insert the spindle of the viscometer into the sample container near the side of the container and turn the viscometer on. While the spindle is rotating, move the spindle in and out of the binder 3 times to bring the spindle up to the temperature of the binder. Continue stirring the binder.

e) Determine the viscosity of the binder at 375°F ± 5°F (190°C ± 3°C) as follows:

   i. In one continuous operation, remove the spindle vertically from the binder, turn off the spindle rotation after the binder stops dripping from the spindle (after heating), and immediately insert the spindle back into the center of the binder sample until fully immersed.

ii. Wait 5 seconds to fill the spindle. While holding the viscometer level, t[illegible]ndle on, watch the needle on the viscometer dial and record the maximum value [illegible] on the dial.

iii. Record the test temperature and maximum viscosity. (The viscometer s[illegible]aintained and operated in accordance with the manufacturer's instructions. However, this test method shall apply if there are any differences in the instructions for the determination of the viscosity.)

f) Report the following information for the viscosity test:

i. Name of the technician performing the viscometer test,

ii. date and location of the plant,

iii. asphalt rubber binder supplier,

iv. binder test temperature and viscosity,

v. spindle size and rotational speed in revolutions per minute, and viscometer model and serial number.

211-5 MOVING AVERAGE COMPUTATIONS.

The moving average shall be computed as follows:

The moving average shall be rounded to the same number of significant figures as are reported for individual test results. When the figure to be dropped is less than 5, round down, if greater than 5, round up, and if it is 5, round up or down to the even number.

The moving average shall be continuous for the individual material processing or batch plant. In determining a moving average for a material property, all of the individual test results that represent material actually used in the Work shall be used in the calculation. The test results shall be entered into the calculation sequence in the chronological order that the Work is performed. The first individual test result shall start the moving average and shall meet the moving average requirements. Until more than 4 test results are available, the moving average shall be the numerical average of the individual test results. When more than 4 test results are available the moving average shall be determined by multiplying the last moving average by 4, adding the new result to this product and then dividing this sum by 5.

In computing the moving average, whenever an upper calculation limit value for an individual test is stated in the Specifications, the upper calculation limit value shall be used in the calculation in lieu of any actual individual test results which exceed said upper calculation limit value.

SECTION 212 - LANDSCAPE AND IRRIGATION MATERIALS

212-1 LANDSCAPE MATERIALS.

212-1.1 Topsoil.

212-1.1.1 General. Topsoil shall be designated as Class A (imported), Class B (selected) or Class C (unclassified), as specified herein. The Engineer shall determine the suitability of topsoil prior to use. Topsoil shall be transported from the source to its final position unless stockpiling is specified.

212-1.1.2 Class "A" Topsoil. Class "A" topsoil shall be from a source outside the limits of the project selected by the Contractor and in compliance with the requirements specified herein. The Engineer may make such inspections and perform such tests as deemed necessary to determine that the material meets the requirements.

At least 15 days before scheduled use, the proposed source of topsoil must be submitted to the Engineer for approval. The Contractor shall submit a written request for approval which shall be accompanied by a written report of a testing agency registered by the State for agricultural soil evaluation which states that the proposed source complies with these specifications. Class "A" topsoil shall have the same relative composition and structure, a friable sandy loam character, and be free of roots, clods and stones larger than 1 inch in greatest dimension, pockets of coarse sand, noxious weeds, sticks, brush, and other litter. It shall not be infested with nematodes or other undesirable insects and plant disease organisms.

Class "A" topsoil shall meet the following additional requirements:

1) **Gradation Limits.** Sand, 50 to 80 percent; clay 20 percent maximum; and silt, 30 percent maximum. The sand, clay, and silt gradation limits shall be as defined in ASTM D422.
2) **Permeability Rate.** Not less than 1/2 inch (13mm) per hour nor more than 2 inches (50mm) per hour when tested in accordance with ASTM D2434, California Test 220, or other approved methods.
3) **Agricultural Suitability.** The topsoil shall be suitable to sustain the growth of the plants specified.

212-1.1.3 Class "B" Topsoil. Class "B" is defined as material which is obtained from sources and in the quantities designated on the Plans or in the Specifications and which requires transport to the designated landscape areas. Such designated sources of the Class "B" topsoil may be within or outside the Work limits.

The cost of stripping the surface of vegetation and debris at the designated locations and processing of the material to a finely divided state, before it is spread, shall be included in the price in the Bid for hauling and placing.

Except as provided above, Class "B" topsoil shall be considered selected material in accordance with 300-2.7, 300-2.8, and 300-2.9.

212-1.1.4 Class "C" Topsoil. Class "C" topsoil is defined as soil found in place in the designated landscape area, including soil compacted in place as part of the earthwork specified for the Work.

212-1.2 Soil Fertilizing and Conditioning Materials.

212-1.2.1 General. Fertilizing materials shall comply with the applicable requirements of the State Food and Agricultural Code. All fertilizing materials shall be packaged first grade, commercial quality products identified as to source, type of material, weight and manufacturer's "guaranteed analysis". Fertilizing material shall not contain toxic ingredients or fillers in quantities harmful to human life, animals or plants.

When required by the Engineer, the Contractor shall furnish a Certificate of Compliance stating that the material substantially conforms to the Specifications.

212-1.2.2 Manure. Manure shall be the product of yard fed cattle; free of weed seed, straw, or other inert material; and aged at least 3 months. The manure shall have been processed by grinding and screening and shall be of a consistency that will readily spread with a mechanical spreader.

Manure may be supplied in bulk if the source is approved in advance by the Engineer.

212-1.2.3 Commercial Fertilizer. Commercial fertilizer shall be a pelletized or granular product having a chemical analysis as specified on the Plans or in the Special Provisions. Commercial fertilizer shall be free-flowing material delivered in unopened sacks. Material which becomes caked or otherwise damaged shall not be used.

212-1.2.4 Organic Soil Amendment. Organic soil amendment shall be selected from Type 1, 2, or 3 products as described herein.

Type 1 organic soil amendment shall be a ground or processed wood product derived from redwood, fir, or cedar sawdust, or from the bark of fir or pine, treated with a non-toxic agent to absorb water quickly, and shall comply with the following requirements:

TABLE 212-1.2.4 (A)

Gradation: Sieve Size	Percent Passing (minimum)
1/4 inch (6.3mm)	95
No. 8 (2.36mm)	80
No. 35 (500μm)	30

Nitrogen Content (%, dry weight)

Redwood	0.4 – 0.6%
Fir	0.56 – 0.84%
Cedar	0.56 – 0.84%
Fir bark	0.8– 1.2%
Pine bark	0.8 – 1.2%

Salinity Maximum saturation extract conductivity: 2.50 millisiemens per centimeter (6.35 millimhos/inch) at 77°F (25°C).

Wettability When one teaspoon of tap water is applied to 4 cubic inches (1 cm^3 of top water is applied to 15 cm^3) (volumetric ratio of 1:15) of the air-dry product, the material shall become completely damp in a period not exceeding 2 minutes. Any wetting agent added shall be guaranteed non-phyto-toxic at the rate used.

Type 2 organic soil amendment shall be a relatively dry friable organic composite derived from sewage sludge processed for agricultural use. It shall contain at least 1 percent nitrogen by dry weight and comply substantially with the gradation for Type 1 soil amendment.

Type 3 soil amendment shall be hay and stable bedding which has been processed and used as the growing medium for the commercial production of mushrooms. It shall contain at least 1 percent nitrogen by dry weight and comply substantially with the gradation for Type 1 soil amendment.

212-1.2.5 Mulch. Mulch shall be designated by Type in accordance with the requirements herein. Mulch shall be packaged in bales or bags unless the Engineer approves a bulk source in advance of delivery to the Work site.

a) **Type 1 mulch** (ground wood product), shall comply with the requirements for Type 1 organic soil amendment.

b) **Type 2 mulch** (sewage sludge product), shall comply with the requirements for Type 2 organic soil amendment.

c) **Type 3 mulch** (mushroom compost) shall comply with the requirements for Type 3 organic soil amendment.

d) **Type 4 mulch** (peat), shall be brown compressed sphagnum or hypnum.

e) **Type 5 mulch** (fir bark chips), shall be fir bark chips in the gradation specified.

f) **Type 6 mulch** (straw), shall be either threshed new straw or stable bedding material derived from rice, oats or barley. Straw in an advanced state of decomposition will not be acceptable.

212-1.3 Seed. Seed shall be fresh, clean, new crop seed, mechanically premixed to specified proportions.

Seed shall be delivered to the site in original unopened containers bearing the dealer's "guaranteed analysis" and germination percentage, and a certificate or stamp or release by a County agriculture commissioner. Any seed tagged "warning, hold for inspection" shall be inspected and released by the agriculture commissioner of the County within which the seeds are to be planted.

212-1.4 Plants.

212-1.4.1 General. Plants shall be inspected and approved at the nursery by the Engineer prior to shipment to the planting site.

All plants shall have a growth habit normal to the species and shall be sound, healthy, vigorous and free from insect pests, plant diseases, sun scalds, fresh bark abrasions, excessive abrasions, or other objectionable disfigurements. Tree trunks shall be sturdy and well "hardened off." All plants shall have normal well-developed branch systems, and vigorous and fibrous roots systems which are neither root- nor pot-bound and are free of kinked or girdling roots.

Other than the normal side pruning during the growth period, no pruning shall be done prior to inspection at the nursery.

212-1.4.2 Trees. All trees shall be of the specified height and crown to the last division of the terminal leader and diameter. The height shall be measured from the root crown. The diameter shall be measured 6 inches (150mm) above the root crown. The height of palm trees shall be measured from the groundline to the base of the growing bud. The tree shall stand reasonably erect without support.

212-1.4.3 Shrubs. Shrubs shall be of the specified type and size, selected from high quality, well-shaped nursery stock.

212-1.4.4 Flatted Plants. Groundcover plants and other flatted plants shall be grown and remain in the flats until transplanted at the site. The soil and spacing of the plants in the flat shall ensure the minimum disturbance of the root system at time of transplanting.

212-1.4.5 Sod and Stolons (turf grass). Turf grass stolons shall be fresh, clean, living sections of runners of hybrid Bermuda grass or hybrid bent grass as designated in the Contract Documents. They shall be free of turf disease, insects, or weeds, and capable of healthy vigorous growth.

For mechanical or hand spreading, Bermuda grass stolons shall be 1 to 4 inches (25mm to 100mm) long and bent grass 4 to 8 inches (100mm to 200mm) long. Stolons to be planted in a slurry mixture as described in 308-4.8 shall be supplied in shorter sections as required.

212-1.4.6 Cuttings. Cuttings shall be fresh stock cut with a sharp hand tool from the stems of healthy vigorous plants of the species specified. If not otherwise specified, the length of cuttings shall be in accordance with the best horticultural practice.

212-1.5 Headers, Stakes, and Ties.

212-1.5.1 General. Lumber for landscape work shall be construction heart rough redwood in the sizes specified. Galvanized steel pipe shall be as specified in 212-2.1. Nails, lag screws, and miscellaneous hardware shall be galvanized commercial quality material. Miscellaneous fabricated metal items shall be made from steel conforming to ASTM A36.

212-1.5.2 Headers and Stakes. Headers shall be 2 inches x 4 inches (50mm x 100mm) except that two 1 inch x 4-inch (25mm x 100mm) boards shall be supplied for laminations on turns and curves. Header stock shall be supplied in lengths of at least 10 feet (3m). Stakes for headers shall be pointed 2 inches x 4 inches (50mm x 100mm), at least 18 inches (450mm) long. Joint splicing lumber shall be 1 inch x 4 inches (25mm x 100mm), 2 feet (0.6m) long.

212-1.5.3 Tree Stakes. The type of tree stake shall be as designated in the Special Provisions. The tree support stakes shall be 10 feet (3.1m) long.

Guy wire shall be No. 12 BWG (2.64mm) zinc-coated iron. Plastic ribbon tie material shall be 1 inch (25mm) wide with a minimum tensile strength of 500 pounds (2 kN).

Deadman stakes shall be either 2 inches x 4 inches (50mm x 100mm) redwood or 3/4 inch (19mm) diameter steel pipe 3 feet (1m) long. Covers for wire shall be garden hose, 1/2 inch (13mm) minimum diameter.

212-2 IRRIGATION SYSTEM MATERIALS.

212-2.1 Pipe and Fittings.

212-2.1.1 General. The type of pipe materials and fittings shall be as designated on the Plans or in the Special Provisions and shall comply with the following:

212-2.1.2 Steel Pipe. Steel pipe shall be galvanized standard weight (Schedule 40) complying with the requirements of ASTM A120. Steel pipe shall be jointed with galvanized, threaded, standard weight malleable iron fittings and couplings.

212-2.1.3 Plastic Pipe for Use with Solvent Weld Socket or Threaded Fittings. Plastic pipe shall be rigid unplasticized polyvinyl chloride PVC 1220 (Type 1, Grade 2), conforming to ASTM D1785. Plastic pipe marked with product standard PS-21-70 conforms to the ASTM requirements. The minimum gaskets shall be rigid unplasticized polyvinyl chloride pressure rating shall not be less than the working pressures indicated therein for the schedule and sizes listed.

Schedule 40 pipe shall be used for installation on the discharge side of control valves and Schedule 80 pipe shall be used for continuously pressurized pipe on the supply side of control valves. Schedule 80 only, shall be supplied when threaded joints are specified or otherwise permitted by the Engineer.

Fittings and couplings for plastic pipe shall be threaded or slip-fitted tapered socket solvent weld type. Threaded adapters shall be provided with socket pipe for connections to threaded pipe. Plastic pipe fittings and couplings shall be PVC I or PVC I/II material supplied in the same schedule size specified for the pipe. The type of plastic material and schedule size shall be indicated on each fitting or coupling. Fittings and couplings shall comply with the following specifications:

TABLE 212-2.1.3 (A)

Socket Fittings	
Schedule 40	ASTM D2466
Schedule 80	ASTM D2467
Threaded Fittings	
Schedule 80	ASTM D2464

212-2.1.4 Plastic Pipe for Use with Rubber Ring Gaskets. Plastic pipe for use with rubber ring PVC 1120 (Type 1, Grade 1), manufactured in accordance with ASTM D2241. Plastic pipe marked with product standard PS 22-70 conforms to ASTM requirements. Pipe shall be supplied with plain ends or with an integral thickened expanded bell with rubber ring groove. Couplings for plain end pipe shall be of the single rubber ring type with solvent weld socket on one end or shall be of the double ring type.

Rubber ring gaskets shall be of a synthetic rubber supplied in accordance with the requirements of ASTM D1869.

Pipe shall be furnished in the following Standard Dimension Ratios (SDR) and Pressure Ratings:

160 psi (1100 kPa) SDR 26

200 psi (1380 kPa) SDR 21

212-2.1.5 Copper Pipe. Copper pipe shall be Type K in accordance with ASTM B88M/ASTM B88. Copper pipe shall be jointed with the appropriate solder type wrought copper fittings for 2-1/2 inch (67mm) and smaller sizes. Cast brass fittings shall be used for sizes over 2-1/2-inch (67mm).

212-2.2 Valves and Valve Boxes.

212-2.2.1 General. Valves shall be of the size, type, and capacity designated on the Plans or in the Special Provisions and shall comply with the requirements specified herein.

All valves except garden valves shall be capable of satisfactory performance at a working pressure of 200 psi (1380 kPa). Valves shall be designed to permit disassembly to replace sealing components without removal of the valve body from the pipeline.

212-2.2.2 Gate Valves. Gate valves in sizes 2 inches (50mm) and smaller shall be all bronze double disc wedge type with integral taper seats and nonrising stem.

Sizes 2-1/2 inches (60mm) and larger shall be iron body, brass trimmed with other features the same as for 2 inch (50mm).

212-2.2.3 Manual Control Valves. Manual control valves z shall be brass or bronze, and shall be straight or angle pattern globe valves, full opening, key operated with replaceable compression disc and ground joint union on the discharge end.

212-2.2.4 Remote Control Valves. Remote control valves shall be electrically or hydraulically operated. They shall be brass or bronze with accurately machined valve seat surfaces, equipped with flow control adjustment and capability for manual operation. They shall be made so that they may be readily disassembled for servicing.

212-2.2.5 Garden Valves. Garden valves shall be brass or bronze except for the handle. They shall have a replaceable compression disc, and shall be 3/4 inch (19mm) straight-nosed, key operated and pressure rated for operation at 150 psi (1035 kPa).

212-2.2.6 Quick-coupling Valves and Assemblies. Quick-coupling valves shall be brass or bronze with built-in flow control and self-closing valve and supplied in 3/4 inch (19mm) size unless otherwise required. When a quick-coupler assembly is specified, it shall consist of the valve, quick coupler connection and hose swivel.

212-2.2.7 Valve Boxes. Valve boxes and covers shall be precast portland cement concrete.

212-2.3 Backflow Preventer Assembly. The backflow preventer assembly shall consist of a backflow preventer unit and related components conforming to the governing code requirements.

212-2.4 Sprinkler Equipment. Sprinkler heads, bubbler heads and spray nozzles shall be of the types and sizes shown on the Plans. Such equipment shall be brass, bronze and stainless steel except for minor components. Equipment of one type and flow characteristic shall be from the same manufacturer and all equipment shall bear the manufacturer's name and identification code in a position where they can be identified in the installed position.

Fixed head sprinklers shall have a one-piece housing with provision for interior parts replacement. Pop-up sprinklers shall rise at least 1 inch (25mm) during operation.

Shrubbery and bubbler heads shall be adjustable from full flow to shutoff.

212-3 ELECTRICAL MATERIALS.

212-3.1 General. The Contractor shall furnish and install all electrical equipment and materials required for a complete electrical system.

All equipment and materials shall comply with the requirements of the governing code and the serving utility and shall be approved and identified by Underwriters Laboratories, Inc. (UL).

212-3.2 Conduit and Conductors.

212-3.2.1 Conduit. Conduit shall be galvanized steel conforming to the applicable provisions of 209-3.5.

212-3.2.2 Conductors. Line voltage conductors shall be supplied in the sizes and types shown on the Plans and shall be THW or THWN, 600-volt insulation rating, conforming to the applicable provisions of ASTM D2219 and D2220.

Low voltage control conductors shall be Type UF and supplied in the sizes shown on the Plans or in accordance with the control equipment manufacturer's recommendation, and shall be UL approved for direct burial installation.

212-3.3 Controller Unit. The type of control unit shall be as called for on the Plans. It shall be fully automatic, with provisions for manual operation, sized to accommodate the number of stations or control valves included in the system. Outdoor models shall be housed in vandal-proof and weatherproof enclosure with locking cover.

SECTION 213 - ENGINEERING FABRICS

213-1 PAVEMENT FABRIC MATERIAL.

213-1.1 General. Pavement fabric material shall be nonwoven, needlepunched polyester or polypropylene materials conforming to Table 213-1.1 (A).

TABLE 213-1.1 (A)

Property	ASTM Test No.	Requirements
Weight, oz./yd^2 (g/m^2)	D 5261	4.1 to 5.5 (139 to 187)
Grab Tensile Strength 1 inch (25mm) grip, lbf (N)	D 4632	101 (449) min.
Elongation at Break, %	D 4632	40 min., 100 max.
Fabric Thickness, Mils (μm)	D 5199	30 to 50 (760 to 1270)
Asphalt Retention, oz./ft^2 (g/ m^2)	D 6140	2.90 (886)min.
Grab Tensile Strength After Asphalt Saturation 1 inch (25mm Grip, lbf (N)	D 4632	200 (890) min.
Elongation at Break, % After Asphalt Saturation	D 4632	40 min., 70 max

Pavement fabric material shall be treated by heat or other processes approved by the Engineer causing the fibers on one side only to become bonded together, forming a glazed, delamination-free surface.

213-1.2 Identification. Pavement fabric material shall be shipped in rolls wrapped with a covering that protects against ultra violet radiation, abrasion, dust, dirt, mud, debris, and other deleterious forces and substances. Each roll in the shipment shall be marked or tagged to identify the manufacturer, type, length, width, date and place of manufacture, and production identification number. Pavement fabric material shall be free from defects or flaws.

213-1.3 Sampling and Test Compliance.

A Certificate of Compliance conforming to 4-1.5 shall accompany each shipment and be submitted to the Engineer prior to installation. The certificate shall include name of the manufacturer, chemical composition, product description, lot number, test results, and signature of an authorized official.

A unit shall be 600 yd^2 (500m^2) or one roll, whichever is less. A lot shall be the number of units produced by a single machine on a single shift without interruption but shall not exceed 1,000 units. The number of units tested within a lot shall be equal to, but not less than, the cube root of the units in that lot (fractions of a number to be rounded to the next higher whole number). A minimum of 8 tests shall be performed in each of the principal directions of each unit tested. In the event of any test failure, the entire lot will be considered as rejected.

213-1.4 Storage and Handling. Pavement fabric material shall be stored on clean, dry surfaces, free of foreign substances such as grease, oil, paint, epoxy, cement, or any other substance that could have a deleterious effect on the fabric. When stored in outside areas, pavement fabric material shall be kept 1 foot (300mm) minimum above ground level. Pavement fabric material shall be protected from exposure to ultraviolet rays. The Contractor shall not remove the protective covering until just prior to installation. Open rolls shall be covered by a waterproof covering. No hooks, tongs or other sharp tools or instruments shall be used in handling. Unloading of fabric material shall be performed:

a) by placing slings under the rolls; or

b) by using a pole inserted through a hollow core, provided the pole extends 1 foot (300mm) minimum beyond each end of the core, and lifting and handling devices are attached to only that portion of the pole located outside the ends of the core; or

c) by hand.

213-2 GEOSYNTHETICS.

213-2.1 General. This section applies to nonwoven and woven geosynthetics. Products include nonwoven and woven geotextiles, biaxial, and uniaxial geogrids.

Nonwoven geotextiles shall be needle punched or needle entangled and shall consist of long chain polymeric filaments of polypropylene, polyester, or nylon. The fabric shall be a stable network of fibers which retain their positions relative to each other. Heat bonding on one side is acceptable.

Woven geotextiles shall consist of long chain polymeric monofilaments, slit film tapes, or multifilaments of tape and nonwoven yarn of polypropylene, polyester or nylon. The fabric shall be woven into a stable network and the edges of the fabric shall be selvedged or surged in such a way that fabric will not unravel or fray during installation or usage.

Geogrids, uniaxial and biaxial, shall be manufactured using polyolefins (high density polyethylene [HDPE] and/or polypropylene [PP] or polyester [PET]). The geogrid reinforcement elements shall consist of a regular network of integrally connected longitudinal and transverse polymer tensile elements with aperture geometry sufficient to permit significant mechanical interlock with the surrounding soil, aggregate or other material.

213-2.2 Physical Properties. Nonwoven and woven geotextiles shall meet the requirements for the type indicated in table 213-2.2 (A). Geogrids shall meet the requirements for the type indicated in Table 213-2.2 (B).

213-2.3 Identification. Geotextiles shall be furnished in rolls wrapped with protective covering to protect them against ultraviolet radiation and abrasion. Torn wrappers shall be repaired within 48 hours, using an approved protective covering. Each roll of fabric shall be marked or tagged to identify the manufacturer, type length, width, and production identification number. Geogrid identification shall follow ASTM D4873.

213-2.4 Sampling and Test Compliance. Sampling and testing of geosynthetics shall conform to the requirements of ASTM D4354. Specification conformance for geosynthetics shall conform to the requirements of ASTM D4759.

213-2.5 Storage and Handling. Storage and handling of geosynthetics shall conform to the requirements of ASTM D4873.

TABLE 213-2.2 (A): NON WOVEN

Property	Test Reference	Type[1]		
		90N	180N	250N
Grab Strength[2] lbs (N), Min.	ASTM D4632	90 (400)	180 (800)	250 (1110)
Elongation, Minimum (at peak load) %, Max.	ASTM D4632	50	50	50
Puncture Strength, lbs (N), Min.	ASTM D3787	45 (200)	80 (355)	115 (510)
Permitivity, Sec.[1], Min.	ASTM D4491	0.7	0.7	0.7
Burst Strength psi (kPa), Min.	ASTM D3786	180 (1240)	320 (2205)	360 (2485)
Toughness, lbs (N), Min.	% Elongation x Grab Strength	5,500 (24,500)	10,000 (44,500)	14,000 (62,500)
Ultraviolet Resistance % Strength Retained @ 500 Weatherometer Hours	ASTM D4355	70	70	70

1. N – Nonwoven.
2. Minimum roll average in the weakest principal direction.

TABLE 213-2.2 (A): WOVEN

Property	Test Reference	Type[1]			
		90WS	200WS	270WS	200WM
Grab Strength[2] lbs (N), Min.	ASTM D4632	90 (400)	200 (890)	270 (1200)	200 (890)
Elongation, Maximum, at peak load %, Max.	ASTM D4632	25	25	25	25
Puncture Strength, lbs (N), Min.	ASTM D3787	30 (135)	70 (310)	110 (490)	100 (445)
Permitivity, Sec.[1], Min.	ASTM D4491	0.02	0.02	0.02	0.04
Burst Strength, psi (kPa), Min.	ASTM D3786	200 (1380)	400 (2760)	500 (3450)	400 (2760)
Ultraviolet Resistance % Strength Retained @ 500 Weatherometer Hours	ASTM D4355	70	70	70	70

1. WS – Slit Film; WM – Monofiliment.
2. Minimum roll average in each direction.

TABLE 213-2.2 (B): UNIAXIAL

Property	Test Reference	Type[1]				
		P1	P2	P3	P4	P5
Tensile Strength lbs/ft (kN/m), Min.	GRI GG-1[2]	2700 (39.4)	4400 (64.2)	6900 (100.7)	9000 (131.3)	10800 (157.6)
Tensile Strength @ 2% strain lbs/ft (kN/m), Min.	GRI GG-1[2]	550 (8)	1000 (14.6)	1800 (26.3)	2330 (34)	2740 (40)
Tensile Strength @ 5% strain lbs/ft (kN/m), Min.	GRI GG-1[2]	1165 (17)	2000 (29.2)	3700 (54)	4450 (65)	5400 (78.8)
Initial Modulus lbs/ft (kN/m)	GRI GG-1[2]	27500 (400)	50000 (730)	90000 (1310)	116500 (1700)	137000 (2000)
Junction Strength, lbs/ft (kN/m), Min.	GRI GG-2[2]	2430 (35)	3960 (58)	6210 (91)	8100 (118)	9720 (142)
Long-Term Allowable Strength, lbs/ft (kN/m)	GRI GG-4[3]	809 (11.8)	1602 (23.4)	2424 (35.4)	3203 (46.7)	4026 (58.7)

1. Primary reinforcement geogrid. All property values are listed for the machine direction.
2. GRI GG-1 conducted at 10%/minute with 2 aperture or 8 inches (200mm) minimum gage length.
3. Minimum reduction factors are as follows: installation damage reduction factor for HDPE of 1.05 and 1.1 for PET geogrids, durability reduction factor of 1.1 for HDPE and 1.3 for PET geogrids

TABLE 213-2.2 (B): BIAXIAL

Property	Test Reference	Type[1]	
		S1	S2
Ultimate Tensile Strength lbs/ft (kN/m), Min.	GRI GG-1[1]	1300 (19)	1975 (28.8)
Tensile Strength @ 2% strain lbs/ft (kN/m), Min.	GRI GG-1[1]	450 (6.57)	600 (8.75)
Tensile Strength @ 5% strain lbs/ft (kN/m), Min.	GRI GG-1[1]	920 (13.4)	1340 (19.5)
Flexural Stiffness mg-cm	ASTM D1388	250000	750000
Junction Strength, lbs/ft (kN/m), Min.	GRI GG-2[1]	1170 (17.1)	1778 (26)
Torsional Stiffness[3], cm-kg/degree	U.S. Army COE[3]	3.2	6.5

1. Secondary reinforcement geogrid. All property values are listed for the cross machine direction with the exception of flexural stiffness, measured in the machine direction, and torsional stiffness which is measured at the junction.
2. GRI GG-1 conducted at 10%/minute with 2 aperture or 8 inches (200mm) minimum gage length.
3. Resistance to in-plane rotation movement measured by applying a 20 cm-kg moment to the central junction of a 225mm by 225mm specimen restrained at its perimeter and measured in units of cm-kg/deg.

SECTION 214 - TRAFFIC STRIPING, CURB AND PAVEMENT MARKINGS, AND PAVEMENT MARKERS

214-1 GENERAL. The following material specifications set forth the requirements for glass beads, paint, thermoplastic material, and reflective and non-reflective pavement markers for use in traffic striping (striping) and curb and pavement markings (markings).

Manufacturer's warranties, guaranties, and instruction sheets for materials used in the Work shall be submitted to the Engineer in accordance with 2-5.3.4.

State Specifications referred to in these specifications are available from the Caltrans Division of Engineering Services, Materials Engineering and Testing Services, Office of Roadway Materials Testing, Chemical Testing Branch (Caltrans transportation laboratory).

214-2 TEST REPORTS AND CERTIFICATES OF COMPLIANCE. The Contractor shall submit to the Engineer certified copies of the manufacturer's test reports and MSDS in accordance with 2-5.3 and 7-10.4.3. Test reports shall indicate the name of the manufacturer, type of material, date of manufacture, quantity, State Specification number, manufacturer's lot or batch number, and results of the required tests. Test reports shall be signed by an authorized representative of the manufacturer.

In lieu of test reports the Contractor may provide the Engineer with a Certificate of Compliance in accordance with 2-5.3.4 and 4-1.5. The certificate shall certify that the paint or thermoplastic material conforms to the Specifications and that the paint was manufactured to the same formulation and process that has passed Caltrans testing.

214-3 GLASS BEADS. Glass beads shall conform to State Specification 8010-004 (Type II).

214-4 PAINT FOR STRIPING AND MARKINGS.

214-4.1 General. Paint for striping and markings shall conform to the requirements shown in Table 214-4.1 (A).

TABLE 214-4.1 (A)

Paint Type	Color	State Specification No.[1]
Waterborne Traffic Line, Type I (Rapid Dry)	White, Yellow and Black	PTWB-01
Acetone-Based	White, Yellow and Black	PT-150VOC(A)
Waterborne Traffic Line	Blue, Red and Green	Federal Specification No. TT-P-1952D

1. Copies of the State Specifications may be obtained from the Caltrans transportation laboratory.

The Contractor may select waterborne or acetone-based paint, depending on the weather conditions and local air pollution control regulations, subject to approval by the Engineer. Thinning of paint will not be allowed.

Paint shall contain at least 12 percent titanium dioxide in white paint.

Glass beads conforming to 214-3 shall be mechanically applied at a rate of 5 pounds per gallon (2 kilograms of beads per liter) paint to the surface of each coat of prior to setting. Application shall conform to 314-4.3.5.

214-4.2 White, Yellow, and Black Traffic Line Paint. White, yellow, and black traffic line paint shall dry to a "no traffic pickup" condition in 30 seconds. The "no traffic pickup" time shall be determined in accordance with ASTM D711. The paint shall be completely dry in not more than 3 minutes when preheated to the temperature recommended by the manufacturer before application.

214-5 THERMOPLASTIC MATERIAL FOR TRAFFIC STRIPING AND MARKINGS.

214-5.1 General. Thermoplastic material shall conform to State Specification PTH-025SPRAY, PTH-02HYDRO, or PTH-02ALKYD as specified in the Special Provisions.

Glass beads conforming to 214-3 shall be added directly to the combined pigment, filler, and resin. Thermoplastic material shall contain at least 20 percent by weight of glass beads in the white and yellow paints and at least 12 percent titanium dioxide in the white.

Thermoplastic material shall be applied in a molten state by mechanical means. Prior to setting, an additional application of glass shall be mechanically applied to the surface of the molten thermoplastic material in conformance with 314-4.4.4. Thermoplastic material, when applied at a temperature range of 400^0F to 425^0F (200^0C to 220^0C) and a thickness of 90 to 125 mils (2300 to 3200μm) for traffic markings and 60 to 70 mils (1500 to 1800μm) for traffic striping, shall set to bear traffic in 2 minutes or less when the atmospheric temperature is 50^0F (10^0C) or greater and not more than 10 minutes when the air atmospheric temperature is 90^0F (32^0C) or greater. Thermoplastic material shall dry to "no traffic pick-up" in accordance with ASTM D711.

214-6 PAVEMENT MARKERS.

214-6.1 Types of Markers. Pavement markers shall conform to one or more of the following types:

a) Type A - Non-Reflective White.
b) Type AY - Non-Reflective Yellow.
c) Type C - Red-Clear Retroreflective.
d) Type D - Two-Way Yellow Retroreflective.
e) Type G - One-Way Clear Retroreflective.
f) Type H - One-Way Yellow Retroreflective.
g) Type I - Two-Way Blue Retroreflective.

214-6.2 Sampling and Acceptance.

214-6.2.1 Sampling. A representative sample size for each lot shall be selected in accordance with the requirements shown in Table 214-6.2.1 (A). For sampling purposes the lot size shall not exceed 25,000 markers.

Table 214-6.2.1 (A)

Lot Size	Sample Size
10000 or less	20
10,001-25,000	40

214-6.2.2 Test Specimens. Three test specimens shall be randomly selected from the sample for each test except as otherwise specified in 214-6.2.3, and tested for conformance with these Specifications. If any one of the 3 specimens fails to comply with the requirements of these Specifications, 6 additional specimens will be tested. The failure of any one of these 6 specimens shall be a basis for rejection of the entire lot or shipment represented by the sample.

214-6.2.3 Reflectance. The entire sample of retroreflective pavement markers shall be tested for reflectance. The failure of 10 percent or more of the original sampling shall be a basis for rejection.

214-6.2.4 Re-samples. At the discretion of the Engineer, a re-sample may be taken consisting of twice the number of samples originally taken. Tolerances for re-samples shall be in the same ratio as specified in 214-6.2.2.

214-6.3 Non-Reflective Pavement Markers.

214-6.3.1 General. Non-reflective pavement markers shall be either ceramic or plastic as specified in the Special Provisions or shown on the Plans.

The bottom of each marker shall be free from gloss or glaze and shall have between 50 and 200 integrally formed protrusions or indentations approximately 50 mils (1.25mm) high projecting from the surface in a uniform pattern. The tips of the protrusions shall not deviate more than 50 mils (1.25mm) from a flat surface. Each protrusion shall have a face parallel to the bottom of the marker and shall project approximately 40 mils (1mm). The area of each parallel face shall be between 0.01 and 0.065 square inches (6 and 42 square millimeters) and the combined area of these faces shall be between 2.2 and 4 square inches (1420 and 2580 square millimeters). The sides of each protrusion may be tapered. This taper shall not exceed 15 degrees from perpendicular to the marker bottom. Markers manufactured with protrusions whose diameter is less than 1/8 inch (3mm) may have an additional taper not exceeding 30 degrees from perpendicular to the marker bottom and exceeding no more than 1/2 the total height of the protrusion.

The top surface of each marker shall be convex and the radius of curvature shall be between 3-1/2 inches (90mm) and 6 inches (150mm), except that the radius within 1/2 inch (15mm) of the nearest the edge may be less. Any change in curvature shall be gradual. The top and sides shall be smooth and free of mold marks, pits, indentations, air bubbles, or other objectionable marks or discolorations that will affect the adhesion or appearance.

214-6.3.2 Ceramic Non-Reflective Pavement Markers. Ceramic non-reflective pavement markers shall consist of a heat-fired, vitreous, ceramic base and a heat-fired opaque glazed surface to produce the properties required in these Specifications. Markers shall be produced from any suitable combination of intimately mixed clays, shales, talcs, flints, feldspars, or other inorganic materials which conform to the material requirements. The markers shall be thoroughly and evenly matured and free from defects which affect appearance or serviceability.

214-6.3.3 Plastic Non-Reflective Pavement Markers. Plastic non-reflective pavement markers shall be either polypropylene or ABS plastic as specified in the Special Provisions.

214-6.3.4 Requirements. Non-reflective pavement markers shall conform to the requirements shown in Table 214-6.3.4 (A). Testing shall be performed in accordance with California Test 669.

TABLE 214-6.3.4 (A)

Test[1]	Test Description	Requirement
A	Bond Strength	700 psi (4826 kPa), min.
B	Glaze thickness	7 mils (178μm), min.
C	Hardness	6 Mob, min.
D	Luminance factor, Type A white markers only, glazed surface.	75, min.
E	Yellowness index, Type A white markers only, glazed surface.	7, max.
F	Color-yellow, Type AY, yellow markers only, The chromaticity coordinates shall be within a color box defined in CTM 669.	Pass
G	Compressive strength	1500 lbs (680 kg) min.
H	Water absorption	2.0% max.
I	Artificial weathering, 500 hours exposure yellowness index	20, max.

1. Tests A, B, C, and H do not apply to plastic markers.

214-6.4 Retroreflective Pavement Markers.

214-6.4.1 General. Retroreflective pavement markers shall be of the prismatic, reflector-type consisting of a methyl methacrylate or suitably compounded shell with a mixture of an inert thermosetting compound and filler material. The exterior surface of the shell shall be smooth and contain 1 or 2 methyl methacrylate prismatic reflector faces of the number and color specified in the Special Provisions or shown on the Plans.

The infrared curves of the compounded methyl methacrylate or ABS shells shall match approved curves on file in the Caltrans transportation laboratory.

The retroreflective lens shall not contain any voids or air space and the back of the lens shall be metalized.

The shell shall be fabricated in a manner that will provide a mechanical interlock between the thermosetting compound and the shell. The thermosetting compound shall bond directly to the back side of the metalized lens surface.

The base shall be flat (the deviation from a flat surface shall not exceed 50 mils (1.25mm), rough textured, and free from gloss or substances which may reduce its bond to the adhesive.

The color of the reflector faces, when illuminated by the white light from a sealed-beam automobile headlight as defined in the Society of Automotive Engineers (SAE) Standard J 578, shall be clear, yellow, blue or red color as specified in the Special Provisions or shown on the Plans. Off-color reflection shall be a basis for rejection. The daylight color of the marker body shall be compatible with the color of the primary lens, and shall be subject to approval by the Engineer.

214-6.4.2 Requirements. Retroreflective pavement markers shall conform to the requirements shown in Table 214-6.4.2 (A). Testing shall be performed in accordance with California Test 669.

TABLE 214-6.4.2 (A)

Test Description	Requirement		
Bond strength[2]	500 psi (3400kPa), min.		
Compressive strength[3]	2000 lbs. (900kg), min.		
Abrasion resistance, marker shall meet the respective specific intensity minimum requirements after abrasion.	Pass		
Water Soak Resistance	No delamination of the body or lens system of the marker nor loss of reflectance.		
	Specific Intensity		
Reflectance	Clear	Yellow	Red
0^0 Incidence Angle, min.	3.0	1.5	0.75
20^0 Incidence Angle, min.	1.2	0.60	0.30
After one year field evaluation	0.30	0.15	0.08

1. Failure of the marker body or filler material prior to reaching 500 psi (3400kPa) shall constitute a failing bond strength.
2. Deformation of more than 125 mils (3200μm) at a load of less than 2000 lbs. (3400 kPa) or delimitation of the shell and the filler material of more than 125 mils (3200μm) regardless of the load required to break the marker shall be a basis for rejection of the markers.

Pavement markers to be placed in pavement recesses shall conform to the above requirements for retroreflective pavement markers except that the minimum compressive strength requirement shall be 1200 pounds (540 kg).

214-6.6 Storage. Pavement markers shall be stored indoors and shall be protected from any source of moisture both during shipment to, and at, the Work site. Pavement markers shall be stored at a temperature high enough to preclude moisture condensation. At the time of placement, both the pavement markers and their packaging shall be dry.

214-7 ADHESIVES FOR PAVEMENT MARKERS.

214-7.1 General. This specification covers high viscosity paste rapid-set and standard-set epoxy adhesives formulated primarily for use in bonding pavement markers to Portland cement concrete and asphalt concrete pavement, and hot-melt bituminous adhesives. The type of adhesive to be applied shall be as specified in the Special Provisions or shown on the Plans.

214-7.2 Epoxy Adhesives.

214-7.2.1 General. Epoxy adhesives shall be rapid-set or standard-set.

214-7.2.2 Rapid-Set Epoxy Adhesive. This specification covers a high viscosity paste, rapid-set epoxy formulated primarily for use in bonding pavement markers to Portland cement concrete and asphalt concrete pavement. Rapid-set epoxy adhesive shall conform to ASTM C881, Type IV, Viscosity Grade 3, Classes B and C, except that the gel time may be shorter than 30 minutes. Components shall conform to the requirements shown in Table 214-7.2.2 (A).

TABLE 214-7.2.2 (A)
COMPONENT A

Ingredients	Parts by Weight
Epoxy Resin[1]	90.00
Orthocresol Glycidyl Ether[2]	10.00
Titanium Dioxide ASTM Designation D 476	3.00
Talc[3]	50.00
Oleeophilic Fumed Silica[4]	4.50[9]

COMPONENT B

Ingredients	Parts by Weight
High Functionality Polymercaptan Harder[5]	60.00
2,4,6-Tri (Dimethylaminomethyl) Phenol[6]	7.00
Polysulfide Polymer[7]	35.00
Furnace Black[8]	0.10
Talc[3]	52.00
Oleophilic Fumed Silica[4]	3.50[9]
Silicone Anti-Foam, Type DB 100, 100% Solids	0.005

1. Di Glycidyl ether of bisphenol A, viscosity 100-160 poise (10-16 Pa·s) at 25^{0}C; weight per epoxide equivalent, 180-200. Gardner 19 maximum.
2. Viscosity at 25^{0}C, 5-10 Centipoises (0.005 to 0.01 Pa·s). Weight per gallon (liter), 9.00-9.10 (1.08-1.09 Kg). Weight per epoxy equivalent, 180-200.
3. Specify Gravity........ 2.68 to 2.96
Oil Absorption, ASTM D281........ 26 to 33
pH .... 8.9 to 9.6
Hegman Rating........ 3 to 5
Particle Shape .... Platey
Maximum Particle Size, micrometers........ 55
Percentage passing U.S. 45μm (No. 325 Screen), Min........ 99
Dry Brightness, Min........ 93
4. High purity fumed silica treated with a silicone oil, with the following properties: Appearance, fluffy while powder; surface area, N_2 B.E.T. method 70 ± 15M^2/gram; pH 4 grams dispersed in 100 milliliters of 20/80 volume mixture of ethyl alcohol and distilled water, 4.7; weight %, carbon, 5.0 minimum; ignition loss (dry basis) 2 hours at 1000^{0}C, 6 to 7; specify gravity, 1.8.
5. Liquid polymercaptan resin, viscosity 100-130 poise at 25^{0}C, 0.973; specific gravity 1.14-1.16: mercaptan value, 3.6 meg/gram. Color, Gardner 1933, 1 maximum. Infrared curve shall match the curve on file in the Caltrans transportation laboratory.
6. Formula weight, 265; specific gravity at 250/25^{0}C, 0.973; refractive index 1,514 at 25^{0}C; distillation range, 96% at 130^0 to 160^{0}C (0.5-1.5mm); flash point. Tag Open Cup, 300^{0}F minimum; water content, 0.08% maximum.
7. Specific gravity, 1.24-`1.30 at 20^0/20^{0}C, viscosity, 0.7-1.2 Pa·s (700-1200 centipoises). Brookfield at 25^{0}C, pH water extract 6.0-8.0; moisture content, 0.1% maximum; pour point , -10^{0}C (15^{0}F); average molecular weight, 1000; flash point ^{0}C (^{0}F). Cleveland Open Cup, 199^{0}C (390^{0}F) minimum; sulfur content, percent, 36-40; color, Hellige, 9-12. The product shall be a difunctional mercaptan made from 98 mole percent of bis (2-chloroetlyl) formal and 2 mole percent of trichloropropane.
8. Surface area, square meters/gram, 115/130; particle diameter nanometers, 18-30; pH, 7.0-8.5, fixed carbon (moisture free), percent, 96-98; volatile matter, percent, 1-4; oil absorption, stiff paste end-point; CCSi/gram, 0.80-0.90.
9. A range of 4.0 to 5.0 parts is permitted in the A Component and 3.0 to 4.0 parts in the B Component, to achieve the required viscosity and thixotropy. Small preproduction batches should be made to determine the oleophilic silica level best suited for manufacturing equipment used.

214-7.2.3 Standard-Set Epoxy Adhesive. This specification covers a high viscosity paste, standard-set epoxy formulated primarily for use in bonding pavement markers to Portland cement concrete and asphalt concrete pavement. Standard-set epoxy adhesive shall conform to ASTM C881, Type IV, Viscosity Grade 3, Classes B or C, except that the gel time may be shorter than 30 minutes. Components shall conform to the requirements shown in Table 214-7.2.3 (A).

TABLE 214-7.2.3 (A)
COMPONENT A

Ingredients	Parts by Weight
Epoxy Resin[1]	87.00
Orthocresol Glycidyl Ether[2]	13.00
Titanium Dioxide ASTM Designation D 476	3.00
Oleeophilic Fumed Silica[3]	6.50[8]
Talc[4]	34.00

COMPONENT B

Ingredients	Parts by Weight
N-Ameinoethyl Piperazine[5]	23.20
Nonylphenol[6]	52.00
Furnace Black[7]	0.10
Oleophilic Fumed Silica[3]	6.50[8]
Silicone Anti-Foam, Type DB 100, 100% Solids	0.005

1. Di Glycidyl ether of bisphenol A, viscosity 10-16 Pa·s (100-160 poise) at 25^{0}C; weight per epoxide equivalent 180-200. Gardner 1933.3 maximum.
2. Aliphatic mono functional reactive ether, derived from an aliphatic alcohol. Viscosity at 1-15 Centipoises. Weight per epoxide equivalent 220-250. Specific gravity 0.88-0.95.
3. High purity fumed silica treated with a silicone oil, with the following properties: Appearance, fluffy while powder; surface area, N_2 B.E.T. method 70 ± 15M^2/gram; pH 4 grams dispersed in 100 milliliters of 20/80 volume mixture of ethyl alcohol and distilled water, 4.7; weight % carbon 5.0 minimum; ignition loss (dry basis) 2 hours at 1000^{0}C, 6 to 7; specify gravity, 1.8.
4. Specify Gravity .... 2.68 to 2.96
Oil Absorption, ASTM D281 .... 26 to 33
pH .... 8.9 to 9.6
Hegman Rating .... 3 to 5
Particle Shape .... Platey
Maximum Particle Size, micrometers 55
Percentage passing U. S. 45μm (No. 325 Screen), Min 99
Dry Brightness, Min 93
5. Color (APHA) 50 maximum; amine value 1250-1350 based on titration which reacts with the 3 nitrogens in the molecule; appearance clear and substantially free of suspended matter.
6. Color (APHA) 50 maximum; hydroxyl number 245-255; distillation range, 25^{0}C at 760 millimeters first drop 295 minimum, 5% 298 minimum, 95% 325 maximum; water, % (K.F.) 0.05 maximum.
7. Surface area, square meters/gram, 115/130; particle diameter nanometers, 18-30; pH, 7.0-8.5, fixed carbon (moisture free), percent, 96-98; volatile matter, percent, 1-4; oil absorption, stiff paste end-point; CCSi/gram, 0.80-0.90.
8. A range of 6.0 to 7.0 parts is permitted in the A Component and B Component to achieve the required viscosity and shear ratio.

All tests shall be performed in accordance with California Test 425.

214-7.2.4 Mixing. Any material which shows evidence of crystallization or a permanent increase in viscosity or settling of pigments which cannot be readily re-dispersed with a paddle shall not be used. Prior to beginning mixing, Components A and B shall be at a temperature between 60^{0}F (15^{0}C) and 85^{0}F (30^{0}C), unless otherwise specified in the Special Provisions. Any heating of the adhesive components shall be done by application of indirect heat. Immediately prior to mixing, each component shall be thoroughly mixed with a paddle. Separate paddles shall be used to stir each component. Immediately prior to use, Components A and B shall be mixed together in the ratio specified by the manufacturer. Mixed adhesives shall have a uniform gray color without black or white streaks. No solvent shall be added. Mixed adhesives shall be used before thickening has begun.

214-7.2.5 Requirements of Combined Components. Components A and B, when combined, shall conform to the requirements shown in Table 214-7.2.6 (A). Testing shall be performed in accordance with California Test 434.

TABLE 214-7.2.6 (A)

Property	Requirement
Gel time, minutes, maximum, at 77^0 F (25^0 C)	30
Bond Strength to Concrete, Time, minutes (maximum) to reach not less than 200 psi.	
at 77^0 F ± 2^0 F (25^0 C ± 1^0 C)	35
at 50^0 F ± 2^0 F (10^0 C ± 1^0 C)	45
Slant Shear Strength	
2 days at 77^0 F ± 2^0 F (25^0 C ± 1^0 C), psi	1,000
14 days at77^0 F ± 2^0 F (25^0 C ± 1^0 C), plus water soak, psi	1,500
Tensile Adhesion and Cohesion	
Ceramic marker bottom, psi	700 min
Ceramic marker bottom, including post cure, psi	700 min
Retroreflective pavement marker bottom, psi	500 min
Color of mixed epoxy	Gray
Glass transition temperature, Tg, samples conditioned at 77^0 F (25^0 C) for 24 hours, ASTM D4065.	86^0 F (30^0 C)

214-7.3 Hot-Melt Bituminous Adhesives.

214-7.3.1 General. This specification covers a bituminous-type, hot-melt adhesive used for the placement of pavement markers. The adhesive shall be capable of bonding ceramic and plastic pavement markers to PCC and asphalt concrete pavement when the roadway surface and pavement marker temperatures are in the range of 40^{0}F to 160^{0}F (5^{0}C to 70^{0}C). The composition must be such that its properties will not deteriorate when heated to and applied at temperatures up to 425^{0}F (220^{0}C) using either air or oil-jacketed melters.

214-7.3.2 Material Requirements. Filler material used in hot-melt bituminous adhesives shall be Type PC, Grade III, calcium carbonate conforming to ASTM D1199 and the requirements shown in Table 214-7.3.2 (A).

TABLE 214-7.3.2 (A)

Sieve Size	Percentage Passing
No. 100	100
No. 200	95
No. 325	75

Hot-melt bituminous adhesive material, when homogeneously mixed with filler material, shall conform to the requirements shown in Table 214-7.3.2 (B).

TABLE 214-7.3.2 (B)

Property	Min.	Max.	Test Method No.
Softening Point, ^{0}F (C^{0})	200 (93)	230 (110)	ASTM D36
Penetration, 100 g, 5 seconds, 77 ^{0}F (25^{0} C)	10	20	ASTM D5
Flow, inch (mm)	---	0.2 (5)	ASTM D3407[1]
Heat Stability Flow, inch (mm)	---	0.2 (5)	See Note 2
Brookfield Thermosel Viscosity, Centipoise, No. 27 Spindle, 20 RPM, 400 ^{0}F (200^{0}C)	3,000	6,000	ASTM D4402
Flash Point, C.O.C, ^{0}F (^{0}C)	550 (290)		ASTM D92
Property	**Min.**	**Max.**	**Test Method No.**
Recommended Pouring Temp., ^{0}F (^{0}C)	400 (200)	425 (220)	---
Shelf Life, Years	---	2	----
Filler Content, percent by weight (Insoluble in 1,1,1 Trichlorethane	65	75	ASTM D2371

1. Flow shall be determined according to Section 6, Flow, of ASTM D3407 with the exception that the oven temperature shall be 158° ± 2°F (70°C ± 1°C) and sample preparation shall be according to Section 7.1 of ASTM D5.
2. Heat Stability Flow shall be determined according to Flow with the exception that 1000 grams of adhesive shall be placed in a loosely-covered quart can, heated to 220°C (425°F) and maintained at this temperature for 4 hours to preparing the sample panel.

SECTION 215 NOT USED

SECTION 216 – PRECAST REINFORCED CONCRETE BOX

216-1 GENERAL. These specifications cover materials for single-cell precast reinforced concrete box (PRCB) sections intended for the conveyance of storm water.

The span, rise, and design earth cover shall be as shown on the Plans or specified in the Special Provisions.

Three sets of prints of the PRCB layout diagrams and 2 sets of shop drawings shall be submitted to the Engineer in accordance with 2-5.3, except one reproducible print of the layout diagrams will not be required. The layout diagrams will be used by the Engineer for reference only, and their use shall in no way relieve the Contractor for its responsibility for accuracy. The Engineer may waive the PRCB layout diagrams requirement.

216-2 MATERIALS.

216-2.1 Portland Cement Concrete (PCC). PCC for PRCB sections shall have a compressive strength of 5000 psi minimum and conform to 201-1.1.4 unless otherwise specified in the Special Provisions.

PCC shall conform to the following:

a) Portland cement shall conform to 201-1.2.1.

b) Aggregate shall conform to the reactivity requirements specified in 201-1.2.2.

c) The combined aggregate gradation shall include aggregate of 3/4 inch (18mm) maximum diameter unless otherwise approved by the Engineer.

d) Water shall conform to 201-6.2.3.

e) Chemical admixtures shall conform to 201-1.2.4.

f) Cementitious material(s) shall be Portland cement or Portland cement and fly ash.

g) The water-cementitious material(s) ratio shall not exceed 0.53 by weight.

h) Proportioning shall conform to 201-1.3.

i) Mixing shall conform to 201-1.4.

216-2.2 Fly Ash. Fly ash shall be Class F conforming to 201-1.2.5 unless otherwise specified in the Special Provisions.

216-2.3 Reinforcement.

216-2.3.1 Deformed Welded Wire Reinforcement. The diameter of any deformed wire in finished deformed welded wire reinforcement shall conform to ASTM A496 or AASHTO M32 or M221. Deformed welded wire reinforcement shall conform to ASTM A497 or AASHTO M55 or M221.

216-2.3.2 Deformed Bars. Deformed bars shall be Grade 60, billet-steel bars conforming to 201-2.2.1.

216-2.4 Leveling Bed Material. Leveling bed material shall be sand, crushed aggregate or crushed miscellaneous base, native free-draining granular material having a sand equivalent of not less than 30, or the material specified in the Special Provisions.

216-3 FABRICATION.

216-3.1 General. PRCB sections shall be fabricated as shown on the Plans.

PRCB sections of greater strength than that specified may be furnished at the Contractor's option, and at its own expense. The interior surfaces of PRCB sections shall be smooth.

216-3.2 Joints. Joints for PRCB sections shall be fabricated with tongue and groove ends.

Outer cage transverse reinforcement shall be placed in the top and bottom slabs at the groove portion of the joint.

216-3.3 Reinforcement.

216-3.3.1 General. Reinforcement shall be deformed welded wire reinforcement conforming to 216-2.3.2 or deformed bars conforming to 216-2.3.3 unless otherwise specified in the Special Provisions. Other details shall be as shown on the Plans.

Before reinforcement is placed it shall be free of mortar, oil, dirt, excessive mill scale and scabby rust, and any other coating of any nature that would destroy or reduce its ability to bond.

216-3.3.2 Area of Steel Reinforcement. The area of steel reinforcement shall be as shown on the Plans or Standard Plans. Steel areas greater than those shown on the Plans shall not exceed 25 percent, unless the Engineer is provided with calculations verifying that the required ductile response is maintained. The calculations shall be prepared by a Registered Civil or Structural Engineer and submitted to the Engineer in accordance with 2-5.3.

If deformed steel bars are used, the steel area shall be increased to account for the difference in steel yield strength, steel spacing, concrete cover, and crack control between the welded wire reinforcement and steel bars.

216-3.3.3 Placement. Reinforcement placement shall conform to the details shown on the Plans. Reinforcement shall be firmly and securely held in position by wiring at intersections and splices and by using plastic or ferrous metal chairs, spacers, metal hangers supporting wires, or other devices of sufficient strength to resist crushing under applied loads. Wooden or aluminum supports shall not be used. Placement on layers of fresh concrete as the work progresses will not be permitted. Tack welding of reinforcing bars will not be permitted.

Welded wire fabric shall be rolled flat before placing concrete, unless otherwise shown on the Plans or Standard Plans.

216-3.3.4 Splicing. Splices in reinforcing bars shall be constructed using lap splices.

Splicing of reinforcing bars will not be permitted at locations shown on the Plans or Standard Plans as a "No-Splice Zone." At the option of the Contractor, reinforcing bars may be continuous at locations where splices are shown on the Plans. The locations of splices, except where shown on the Plans, shall be determined by the Contractor.

Unless otherwise shown on the Plans, splices in adjacent reinforcing bars at any particular section shall be staggered. The minimum distance between staggered lap splices or mechanical lap splices shall be the same as the length required for a lap splice in the largest bar.

216-3.4 Curing.

216-3.4.1 General. Curing shall be for a length of time sufficient for the PCC to develop the specified compressive strength in 28 days or less. Any one of the following methods of curing or combinations thereof may be used:

a) Steam Curing.

   Steam curing shall conform to ASTM C1433/C1433M. PRCB sections shall be low pressure, steam-cured by a system capable of maintaining a moist atmosphere.

b) Water Curing.

   Water curing shall conform to ASTM C1433/C1433M. PRCB sections shall be kept moist continuously.

c) Membrane Curing.

   A sealing membrane conforming to ASTM C309 shall be applied and left intact until the specified PCC compressive strength is attained. The temperature of the PCC at the time of application of the membrane shall be within ± 100F (± 60C) of the atmospheric temperature. All surfaces shall be kept moist prior to the application of the membranes and shall be damp when the membrane is applied.

216-3.5 Forms. Forms shall have sufficient rigidity to be capable of maintaining section dimensions within the permissible tolerances. Form surfaces which will come into contact with PCC shall be constructed of smooth non-porous material.

216-3.6 Lifting Holes or Devices. Lifting holes or devices may be cast into, or attached to, PRCB sections. Shop drawings shall be submitted in accordance with 2-5.3.3 if so specified in the Special Provisions.

216-4 TESTING REQUIREMENTS.

216-4.1 Test Specimens. Test specimens shall conform to 201-1.1.5 as modified to allow the use of Section 11 of ASTM C497 or AASHTO T280. The Engineer shall be notified before testing is started.

216-4.2 Compression Testing of Cylinders.

216-4.2.1 General. Compression testing of cylinders shall conform to 201-1.1.5. A minimum of 3 test cylinders shall be prepared from each PCC mix used for each lot of PRCB sections. A production lot for PRCB shall be the lesser of one day's production, 400 feet (120m) or 50 units.

216-4.2.2 Acceptance. When the average compressive strength of all cylinders tested is equal to or greater than the specified compressive strength of the PCC, and not more than 10 percent of the cylinders tested have a compressive strength less than the specified compressive strength, and no cylinder tested has a compressive strength less than 80 percent of the specified compressive strength, the lot will be accepted.

When the compressive strength of the cylinders tested does not conform to the aforementioned acceptance criteria, acceptance of the lot shall be determined in accordance with 216-4.3.

216-4.3 Compression Testing of Cores.

216-4.3.1 General. Cores shall be obtained and tested for compressive strength in accordance with 201-1.1.5.

216-4.3.2 Core Holes. Core holes shall be plugged and sealed using high-strength, non-shrink grout. The compressive strength of the repaired core holes shall meet or exceed the compressive strength requirements of the PRCB section.

216-5 PERMISSIBLE VARIATIONS.

216-5.1 Internal Dimensions. Internal dimensions shall not vary more than 1 percent from the dimensions shown on the Plans or Standard Plans. The haunch dimensions shall not vary more than 1/4 inch (6mm) from the dimensions shown on the Plans.

216-5.2 Slab and Wall Thickness. Slab and wall thicknesses shall not be less than that shown on the Plans or Standard Plans by more than 5 percent or 3/16 inch (5mm), whichever is greater. A thickness greater than that shown on the Plans will not be a cause for rejection.

216-5.3 Length of Opposite Surfaces. The length of opposite surfaces shall not vary more than 1/8 inch/foot of internal span. The maximum variation shall not exceed 5/8 inch (15mm) for all sizes through an internal span of 7 feet (2100mm), and 3/4 inch (18mm) for an internal span greater than 7 feet (2100mm), except where beveled ends for laying of curves are specified.

216-5.4 Length of Section. The under-run in length shall not be more than 1/8 inch/foot of length with a maximum of 1/2 inch (12mm) in any PRCB section.

216-5.5 Position of Reinforcement. The maximum variation in the position of the reinforcement for slab and wall thicknesses of 5 inches (125mm) or less shall be ± 3/8 inch (9mm), and for thicknesses greater than 5 inches (125mm) shall be ± 1/2 inch (12mm).

The cover over the reinforcement shall not be less than 5/8 inch (15mm), as measured to the internal surface or the external surface of the slab except as follows. The cover over the reinforcement for the external surface of the top slab for PRCB sections with less than 2 feet (600mm) of cover shall not be less than 1-5/8 inches (40mm). The aforementioned minimum cover limitations do not apply at the mating surfaces of a joint.

Hooks and bends shall conform to ACI 318.

216-6 MARKINGS. The following information shall be legibly marked on each PRCB section by indentation or waterproof paint:

a) PRCB section span,

b) rise,

c) table number,

d) maximum and minimum design earth cover,

e) specification designation,

f) interior invert cover,

g) date of manufacture, and

h) the word "top" lettered on the inside top surface.

216-7 CAUSES FOR REJECTION. Inspection of PRCB as may be deemed necessary by the Engineer will be made at the place of manufacture. Individual PRCB sections may be rejected due to any of the following unless repairs are made and approved by the Engineer:

a) Fractures or cracks with widths exceeding 0.10 inch (3mm).

b) Mixing and molding defects, honeycombed or open texture that would adversely affect the function of the PRCB section,

c) Failure to meet the permissible variations specified in 216-5.

d) Exposure of any reinforcement arising from misplacement thereof.

216-8 BASIS OF ACCEPTANCE. The basis of acceptance shall be by one of the following as specified in the Special Provisions:

a) Compliance with these Specifications, inspection of the manufacturing, and inspection of the completed PRCB sections.

b) Acceptance of a Certificate of Compliance conforming to 4-1.5.

Such acceptance, however, shall be considered a tentative acceptance. Final acceptance will only be made when the Work is completed.

PART 3
CONSTRUCTION METHODS

SECTION 300 - EARTHWORK

300-1 CLEARING AND GRUBBING.

300-1.1 General. Clearing and grubbing operations preceding construction of debris dams and basins shall comply with 300-6.2.

Clearing and grubbing shall consist of removing all natural and artificial objectionable materials from the right-of-way in construction areas, road approaches, material sites within the right-of-way, areas through which ditches and channels are to be excavated, and such other areas as may be shown on the Plans. This work shall be performed in advance of grading operations and in accordance with the requirements herein specified, subject to erosion control requirements. Demolition of buildings and structures, other than foundations or slabs, shall be as shown in the Plans or in the Specifications.

The natural ground surface shall be cleared of all vegetable growth, such as trees, logs, upturned stumps, roots of downed trees, brush, grass, weeds, and all other objectionable materials, within the limits of construction.

Grubbing shall extend to the outside excavation and fill slope lines, except that where slopes are to be rounded, the areas shall extend to the outside limits of slope rounding. Within the limits of clearing, all stumps, roots 1-1/2 inches (38mm) in diameter or larger, buried logs, and all other objectionable material shall be removed 3 feet (1m) below the existing ground surface or subgrade, whichever is deeper.

No payment will be made to the Contractor for clearing and grubbing outside the stated limits, unless such work is authorized by the Engineer.

Trees and plants that are not be removed shall be fully protected from injury by the Contractor at its expense. Trees shall be removed in such a manner as not to injure standing trees, plants, and improvements which are to be preserved.

Tree branches which hang within 13.5 feet (4.1m) above finished roadway grade or within 9 feet (2.7m) above finished sidewalk or parkway grade shall be cut off to the boles in a workmanlike manner. The Contractor shall remove additional tree branches under the direction of the Engineer, in such a manner that the tree will present a balanced appearance. Scars resulting from the removal of branches shall be treated with a heavy coat of an approved tree sealant.

300-1.2 Preservation of Property. Existing improvements, adjacent property, utility and other facilities, and trees and plants that are not to be removed shall be protected from injury or damage as provided for in 7-9.

300-1.3 Removal and Disposal of Materials.

300-1.3.1 General. All materials removed shall be disposed of outside of the right-of-way, unless burning is permitted. Burning shall be done only if permitted by local regulations and at such times and in such manner as to prevent the fire from spreading to areas adjoining the right-of-way. In case burning precedes construction operations, the piles may be placed in the most convenient location at the side of the right-of-way and beyond slope lines where they may be burned without damage to the surrounding area. No accumulation of flammable material shall remain on or adjacent to the right-of-way. The roadway and adjacent areas shall be left with a neat and finished appearance.

300-1.3.2 Requirements.

a) **Bituminous Pavement.** Bituminous pavement shall be removed to clean, straight lines. Saw cutting of edges to be joined is optional. Where only the surface of existing bituminous pavement is to be removed, the method of removal shall be approved by the Engineer, and a minimum laying depth of 1 inch (25mm) of new pavement material shall be provided at the join line. Where bituminous pavement adjoins a trench, the edges adjacent to the trench shall be trimmed to neat straight lines before resurfacing to ensure that all areas to be resurfaced are accessible to the rollers used to compact the subgrade or paving materials.

b) **Concrete Pavement.** Concrete pavement shall be removed to neatly sawed edges. Saw cuts shall be made to a minimum depth of 1-1/2 inches (38mm). If a saw cut in concrete pavement falls within 3 feet (1m) of a construction joint, cold joint, expansion joint, or edge, the concrete shall be removed to the joint or edge. The edges of existing concrete pavement adjacent to trenches, where damaged subsequent to saw cutting of the pavement, shall again be saw cut to neat, straight lines for the purpose of removing the damaged pavement areas. Such saw cuts shall be either parallel to the original saw cuts or shall be cut on an angle which departs from the original saw cut not more than 1 inch (25mm) in each 6 inches (150mm).

c) **Concrete Curb, Walk, Gutters, Cross Gutters, Driveway, and Alley Intersections.** Concrete shall be removed to neatly sawed edges with saw cuts made to a minimum depth of 1-1/2 inches (38mm). Concrete sidewalk or driveway to be removed shall be neatly sawed in straight lines either parallel to the curb or at right angles to the alignment of the sidewalk. No section to be replaced shall be smaller than 30 inches (750mm) in either length or width. If the saw cut in sidewalk or driveway would fall within 30 inches (750mm) of a construction joint, expansion joint, or edge, the concrete shall be removed to the joint or edge, except that where the saw cut would fall within 12 inches (300mm) of a score mark, the saw cut shall be made in and along the score mark. Curb and gutter shall be sawed to a depth of 1-1/2 inches (40mm) on a neat line at right angles to the curb face.

300-1.4 Payment. The lump sum Bid price, or the Contract Unit Price per acre, bid for clearing and grubbing shall include full compensation for removal and disposal of all the resulting materials.

When the Contract does not include a pay item for clearing and grubbing as specified above, and unless otherwise provided in the Special Provisions, full compensation for any necessary clearing and grubbing required to perform the construction operations specified shall be included in the Bid price for other items of work and no additional compensation will be allowed therefore.

300-2 UNCLASSIFIED EXCAVATION.

300-2.1 General. Unclassified excavation shall consist of all excavation, including roadways, unless separately designated.

300-2.2 Unsuitable Material.

300-2.2.1 General. Material that is unsuitable for its planned use shall be excavated and disposed of as directed by the Engineer.

The removal and disposal of unsuitable material will be paid for at the Contract Unit Price for unclassified excavation if the removal of such material is shown on the Plans or specified in the Special Provisions.

If the removal of unsuitable material is not shown on the Plans or specified in the Special Provisions, the removal and disposal of such material will be paid for at the Contract Unit Price for unclassified excavation. However, if the removal and disposal of such material involves a substantial change in the character of the work, an adjustment in payment will be made per 3-2.4.

300-2.2.2 Wet Material. If required excavated material is unsatisfactory for the specified use on the project solely because of high moisture content, the Contractor may be directed by the Engineer to either process the material to reduce the moisture content to an optimum condition, or to remove the material and replace it with suitable material. If such high moisture content is not the result of any action on the part of the Contractor, or inaction in protecting the work during the course of the Contract, the work involved will be paid for in accordance with 3-2 or 3-3. Otherwise, the Contractor shall submit to the Engineer for approval, a plan for drying or removing and replacing the wet material, and such work and material shall be at the expense of the Contractor.

300-2.3 Overshooting. Excessive blasting will not be permitted. Material outside the authorized cross section which may be shattered or loosened because of blasting shall be removed at the Contractor's expense. The Contractor shall discontinue any method of blasting which leads to overshooting, is hazardous to the public, or is destructive to property or natural features.

300-2.4 Slides and Slipouts. Material outside the planned excavation limits which is unstable and constitutes a potential slide as determined by the Engineer, material which has come into the planned excavation limits, and material which has slipped out of new or old fills shall be excavated to designated lines or slopes either by benching or in such manner as directed by the Engineer. Such material may be used in the construction of an unclassified fill or disposed of as directed by the Engineer.

The removal and disposal of slide and slipout material as above specified, not resulting from overshooting as specified in 300-2.3 and not resulting from any act or failure to act on the part of the Contractor, will be paid for at the Contract Unit Price for unclassified excavation for the quantities involved.

However, if due to the character of the work, the removal and disposal of such material is not properly compensable at the Contract Unit Prices for unclassified excavation, the work may be paid for as provided in 3-2 or 3-3.

Payment will be made only for those quantities of slide or slipout material which are actually removed as ordered by the Engineer.

300-2.5 Slopes. Excavation slopes shall be finished in conformance with the lines and grades shown on the Plans. All debris and loose material shall be removed. When completed, the average plane of the slopes shall conform to the slopes indicated on the Plans and no point on the completed slopes shall vary from the designated plane by more than 6 inches (150mm) measured at right angles to the slope. Where excavation is in rock, no point shall vary more than 2 feet (600mm) from the designated plane of the slope. In no case shall any portion of the slope encroach so as to interfere with the planned use of the facility.

The tops of excavation slopes and the ends of excavations shall be rounded where shown on the Plans, and these quantities will not be included in the quantities to be paid for as excavation. This work will be considered as a part of finishing slopes and no additional compensation will be allowed therefore.

300-2.6 Surplus Material. Unless otherwise shown on the Plans or in the Special Provisions, no surplus excavated material may be disposed of within the right-of-way. The Contractor shall make all arrangements for disposal of the material at offsite locations and shall, upon request, file with the Engineer the written consent of the owner of the property upon which it intends to dispose of such material.

Quantities of surplus material, if shown on the Plans or in the Special Provisions, are approximate only. The Contractor shall satisfy itself that there is sufficient material available for the completion of the fills before disposing of any indicated surplus material inside or outside the right-of-way. Any shortage of material, caused by premature disposal of the indicated surplus material by the Contractor, shall be replaced by it and no compensation will be allowed for such replacement.

300-2.7 Selected Material. Selected material encountered in excavation within the right-of-way shall be used as shown on the Plans, in the Special Provisions, or as directed by the Engineer. Topsoil excavated within the limits of the Work may be considered as a selected material only for the purpose of backfilling areas to be planted.

300-2.8 Measurement. The following earthwork operations will be measured as unclassified excavation for the quantities of material involved:

a) Excavating the roadway prism including public and private road approaches;

b) Connections and driveways;

c) Excavating unsuitable material when shown on the Plans or in the Special Provisions;

d) Excavating slides and slipouts not resulting from overshooting;

e) Excavating surplus material;

f) Excavating selected material and topsoil from within the limits of the Work and removing such materials from stockpiles when stockpiling is ordered;

g) Excavating local borrow.

Measurement of material removed from required stockpiles will be based on the volume it occupies in its final position after compaction.

Excavation in excess of the planned or authorized cross section will not be paid for, except as provided in 300-2.2. and 300-2.4. The Contractor shall backfill and compact unauthorized excavated areas to the original ground elevation or authorized section at its expense.

Care shall be exercised to prevent excavating below grade. Areas excavated below grade shall be filled with suitable material and compacted by the Contractor at its expense.

300-2.9 Payment. Payment for all unclassified excavation will be made at the Contract Unit Price per cubic yard (cubic meter). Payment for unclassified excavation shall include compensation for excavating, sloping, rounding tops and ends of excavations, loading, disposing of surplus material, stockpiling, and hauling it to its final location.

Where required by the Plans or the Special Provisions or where directed by the Engineer, the excavating and stockpiling of selected material will be paid for at the Contract Unit Price for unclassified excavation. Removing such selected material from the stockpile and placing it in its final position will also be paid for at the Contract Unit Price for unclassified excavation. The Contractor may stockpile material; however, no separate payment will be made for excavating material from an optional stockpile and placing it in its final position.

No separate payment will be made for excavating topsoil temporarily stockpiled along the top of slopes and placing it in its final position on the slope for erosion control planting work, whether or not required by the Contract Documents or by the Engineer.

300-3 STRUCTURE EXCAVATION AND BACKFILL.

300-3.1 General. Structure excavation shall consist of the removal of material for the construction of foundations for bridges, retaining walls, headwalls, culverts, or other structures, and other excavation designated on the Plans or in the Specifications as structure excavation. Excavation and backfill for underground conduit construction, including box conduit, shall be in accordance with 306.

Structure backfill shall consist of furnishing material, if necessary, and placing and compacting backfill material around structures to the lines designated on the Plans.

Structure excavation and structure backfill shall include the furnishing of all materials and equipment; the construction or installation of all cofferdams and other facilities which may be necessary to perform the excavations and to place and compact the backfill; and the subsequent removal of such facilities, except where they are required or permitted by the Plans or the Special Provisions to remain in place.

300-3.2 Cofferdams. Cofferdams for foundation construction shall be carried well below the bottom of the footings and shall be well braced and reasonably watertight. The interior dimensions of cofferdams shall provide sufficient clearance inside the wales for constructing forms and driving piles and to permit pumping outside the forms.

If in the judgment of the Contractor, the clearance provided on the Plans between the outside line of the footing and any pile or interior wall or surface is not sufficient to permit the driving of piles or building of forms, it may provide such necessary clearance by constructing the cofferdam sufficiently large to provide such clearances as may be deemed necessary. Any such enlargement in excess of 1 foot (0.3m) outside the dimensions of the footing as shown on the Plans shall be considered as being for the sole purpose of expediting the work of the Contractor and such excavation and backfill shall be at the Contractor's expense.

Cofferdams which are tilted or moved out of position by any cause during the process of sinking shall be plumbed or enlarged to provide the necessary clearance and proper pier location and such work shall be at the Contractor's expense.

In tidal waters or in streams at a time of probable flood, cofferdam walls shall be vented at low water elevation to ensure equal hydrostatic head both inside and outside of the cofferdam during the period of pouring and setting of seals.

No shoring will be permitted in cofferdams which will induce stress, shock, or vibration in the permanent structure.

When permitted by the Engineer, cross struts or bracing may extend through foundation concrete. Such struts or bracing below low water will be permitted to remain in place, except in navigable streams or, when shown on the Plans or in the Special Provisions to be removed. Struts or bracing above low water shall be removed and the resulting space filled with concrete of the same mix as that specified for the surrounding concrete.

For substructure work, the Contractor shall, in accordance with 2-5.3, submit Working Drawings showing the proposed method of cofferdam construction and other details left open to choice or not fully shown on the Plans. The type and clearance of cofferdams, insofar as such details affect the character of the finished work, will be subject to the approval of the Engineer, but other details of design will be left to the Contractor, who shall be responsible for the successful construction of the work.

After completion of the substructure, the cofferdams with all sheeting and bracing shall be removed to at least 2 feet (0.6m) below the level of the streambed by the Contractor at its expense and such removal shall be performed in a manner that will not disturb or mar the finished concrete or masonry.

300-3.3 Foundation Material Treatment. When footing concrete or masonry is to rest upon rock, the rock shall be fully uncovered and the surface thereof shall be removed to a depth sufficient to expose sound rock. The rock shall be roughly leveled off or cut to approximate horizontal and vertical steps, and shall be roughened. Seams in the rock shall be grouted under pressure or treated as the Engineer may direct and the cost thereof will be paid for as Extra Work per 3-3.

When no piles are used and footing concrete or masonry is to rest on an excavated surface other than rock, care shall be taken not to disturb the bottom of the excavation. Final removal of the foundation material to grade shall not be made until just before the concrete or masonry is placed. Except when overexcavation is directed by the Engineer, excavation below grade shall be replaced at the Contractor's expense with the same class of concrete specified for the structure and at the time the concrete for the structure is being placed. Where it is determined by the Engineer that it will not be detrimental to the structure, the Contractor may backfill above grade with not less than 90 percent relative compaction and then trim to the specified grade.

Where the original ground is below the specified elevation for footings, the Contractor shall backfill to 6 inches (150mm) above grade with not less than 90 percent relative compaction and then excavate to the prescribed grade prior to placing concrete.

The excavation for piers and abutments shall be completed to the bottom of the footings before any piles are driven therein, and excess material remaining in the excavation after pile driving shall be removed to the elevation of the bottom of the footings.

When piles are used and ground displacement results from pile driving operations, the Contractor shall at its expense excavate or backfill the footing area to the grade of the bottom of the footing as shown on the Plans with structure backfill material.

300-3.4 Inspection. Whenever any structure excavation is completed, the Contractor shall notify the Engineer who will make an inspection of the foundation. No concrete or masonry shall be placed until the foundation has been approved by the Engineer.

300-3.5 Structure Backfill.

300-3.5.1 Requirements. Structure backfill shall not be placed until the structure has been inspected by the Engineer and approved for backfilling. No backfill material shall be deposited against the back of concrete abutments or concrete retaining walls, until the concrete has developed not less than the specified 28-day compressive strength. Backfill at the inside of bridge wingwalls shall be placed before railing bases on the wingwalls are constructed.

Material used for structure backfill shall have a sand equivalent of not less than 20 and shall have the following grading:

TABLE 300-3.5.1 (A)

Sieve Size	Percent Passing
4" (100mm)	100
No. 4 (4.75mm)	35 - 100
No. 30 (600μm)	20 - 100

Structure backfill shall be placed in accordance with 300-4.5 and shall be mechanically compacted to a minimum relative compaction of 90 percent.

Where consolidation of structure backfill by jetting is permitted by the Plans or the Special Provisions, the following conditions shall apply:

a) Backfill material shall be placed and consolidated in layers not exceeding 4 feet (1.2m) in thickness.

b) The jetting shall be performed without softening the embankment and in a manner that excess water will not be impounded.

c) Jetting methods shall be supplemented by the use of vibratory or other consolidation equipment when necessary to obtain 90 percent relative compaction.

d) The upper 3 feet (0.9m) below finished subgrade shall be mechanically compacted in roadway areas.

300-3.5.2 Pervious Backfill. Pervious backfill material shall be placed behind bridge abutments, wingwalls, and retaining walls as shown on the Plans and in accordance with the following requirements.

Pervious backfill material shall consist of gravel, crushed gravel, crushed rock, natural sands, manufactured sand, or combinations thereof and shall conform to the following grading requirements:

TABLE 300-3.5.2 (A)

Sieve Size	Percent Passing
3/4" (19mm)	100
3/8" (9.5mm)	80 - 100
No. 100 (150µm)	0 - 8
No. 200 (75µm)	0 - 3

That portion of filter material passing a No. 4 (4.75mm) sieve shall have a sand equivalent of not less than 60.

All weep holes shall be backed with 2 cubic feet (0.06m^3) of material conforming to the requirements for No. 3 Concrete Aggregate as specified in 200-1.4; securely tied in a burlap sack; and placed in such a manner that the backing covers the weep holes and extends at least 12 inches (300mm) above the bottom of the opening. An 8 inch (200mm) square section of 1/4 inch (6mm) galvanized or aluminum screen having a minimum wire diameter of 0.03 inch (0.8mm) shall be firmly attached at the back of each weep hole before the material is placed.

Pervious backfill material shall be placed in layers along with and by the same methods specified for structure backfill. Pervious backfill material at any one location shall be approximately the same grading, and, at locations where the material would otherwise be exposed to erosion, shall be covered with at least a 1 foot (0.3m) layer of earthy material approved by the Engineer.

300-3.6 Payment. Unless otherwise provided in the Special Provisions or the Bid, no payment will be made for structure excavation or backfill. The cost thereof shall be considered as included in the price bid for the construction or installation of the item to which such excavation or backfill is incidental or Bid appurtenant. Unless otherwise shown on the Plans, the quantity of the structure excavation, where paid for as a separate item or not, shall be that volume in place included within the vertical plane 1 foot (0.3m) outside of and parallel with the outermost horizontal dimensions of the structure and the surface of the existing ground, final ground surface, or proposed street subgrade, whichever is lower, and the footing subgrade. Structure backfill will be measured to the finished ground surface, or to the proposed street subgrade, whichever is lower, within the limits specified above.

300-4 UNCLASSIFIED FILL.

300-4.1 General. Unclassified fill shall consist of all fill unless separately designated. Construction of unclassified fill includes preparing the area on which fill is to be placed, and the depositing, conditioning, and compacting of fill material.

Rocks, broken concrete, or other solid materials which are larger than 4 inches (100mm) in greatest dimension shall not be placed in fill areas where piles are to be placed or driven.

Clods or hard lumps of earth of 6 inches (150mm) in greatest dimension shall be broken up before compacting the material in fill, except when the fill material originating from the Work contains large rocks, boulders, or hard lumps (such as hardpan or cemented gravel which cannot be broken readily) over 12 inches (300mm) in greatest dimension, such materials may be incorporated in the fill. The location and depth of its placement in the fill and the method to be used shall be subject to the approval of the Engineer.

Fills shall not be constructed when material is frozen or a blanket of snow prevents proper compaction.

300-4.2 Preparation of Fill Areas. Areas over which fills are to be placed shall first be cleared and grubbed in accordance with the provisions of 300-1. The areas shall then be scarified to provide a bond between the existing ground and the material to be deposited thereon.

When fills are designated to be placed over existing surface improvements which are to remain in place, 4-inch (100mm) drainage holes shall be made through the structure on 5 foot (1.5m) centers each way or the pavement shall be broken in a grid pattern of 5 feet (1.5m) each way.

300-4.3 Other Fill Materials. Brick rubble, broken asphalt pavement, and broken concrete originating from the Work may be disposed of at the Contractor's expense off the Work site, or incorporated in the fill in conformance with 300-4.1. Unless otherwise provided, no such materials may be imported from outside the limits of the Work.

300-4.4 Benching. When fill is to be made and compacted against hillsides or existing fill, the slopes of original hillsides and old or new fills shall be horizontally benched to key the fill material to the underlying ground. A minimum of 2 feet (0.6m) normal to the slope of the original hillside and old or new fill shall be removed and recompacted, as the fill is brought up in layers, to ensure that the new work is constructed on a firm foundation free of loose or disturbed material. For the purpose of this subsection, fill slopes or hillsides shall be defined as steeper than 1 vertical in 4 horizontal. Slopes 1 vertical in 4 horizontal or flatter shall be prepared in conformance with 300-4.2.

Material thus cut shall be recompacted along with new fill material at the Contractor's expense, unless additional excavation is ordered by the Engineer. The additional excavation will be measured and paid for as unclassified excavation.

300-4.5 Placing Materials for Fills. Fill material shall be placed in horizontal layers of depths compatible to the material being placed and the type of equipment being used. Each layer shall be evenly spread and moistened or aerated, as necessary. Unless otherwise approved by the Engineer, each layer spread for compaction shall not exceed 8 inches (200mm) of compacted thickness.

Unless otherwise permitted by the Engineer, each layer of fill material shall cover the length and width of the area to be filled before the next higher layer of material is placed. The top surface of each layer shall be approximately level, but with a crown or crossfall of at least 1 vertical in 50 horizontal, but no more than 1 vertical in 20 horizontal, to provide adequate drainage at all times during the construction period.

When fill material contains by volume over 25 percent of rock larger than 6 inches (150mm) in greatest dimension, the fill below a plane 3 feet (1m) from finished grade may be constructed in layers of a loose thickness before compaction up to the maximum size of rock in the material, but not exceeding 3 feet (1m) in thickness. The interstices around the rock in each layer shall be filled with earth or other fine material and compacted.

Broken portland cement concrete and bituminous type pavement obtained from the Work excavation will be permitted in the fill with the following limitations:

a) The maximum dimension of any piece used shall be 6 inches (150mm).

b) Pieces larger than 4 inches (100mm) shall not be placed within 12 inches (300mm) of any structure.

c) Pieces larger than 3 inches (75mm) shall not be placed within 12 inches (300mm) of the subgrade for paving.

d) "Nesting" of pieces will not be permitted.

300-4.6 Application of Water. At the time of compaction, the moisture content of fill material shall be such that the specified relative compaction will be obtained and the fill will be firm, hard, and unyielding. Fill material which contains excessive moisture shall not be compacted until the material is dry enough to obtain the required relative compaction.

Subject to the approval of the Engineer, the Contractor may use chemical additives in water to be applied to fill material. Where reasonable grounds exist for concern that the use of a chemical additive will be in any way detrimental, the Engineer may prohibit its use.

The furnishing and application of such additives shall be at the Contractor's expense.

300-4.7 Compaction. Each layer of earth fill shall be compacted by approved tamping or sheepsfoot rollers, pneumatic-tired rollers, or other mechanical means acceptable to the Engineer, to such extent as will produce the specified relative compaction. At locations where it would be impractical because of inaccessibility to use such compacting equipment, fill layers shall be compacted to the specified requirements by hand-directed compaction equipment.

Unless otherwise specified, each layer of earth fill shall be compacted to a relative compaction of at least 90 percent.

When soil types, or a combination of soil types, are encountered which tend to develop densely packed surfaces as a result of spreading or compacting operations, the surface of each layer of fill shall be sufficiently roughened after compaction to ensure bond to the succeeding layer.

300-4.8 Slopes. Fill slopes shall be finished in conformance with the lines and grades shown on the Plans. When completed, the average plane of the slopes shall conform to the slopes indicated on the Plans and no point on the completed slopes shall vary from the designated plane by more than 6 inches (150mm) measured at right angles to the slope.

300-4.9 Measurement and Payment. The Contract Unit Price for unclassified fill shall include full compensation for the all grading, shaping, compacting or consolidating, or other work that is required under this subsection. The quantities used in determining payment for unclassified fill shall be those of the completed fills within the limits shown on the Plans or as directed by the Engineer.

300-5 BORROW EXCAVATION.

300-5.1 Local Borrow. Local borrow shall consist of material excavated and used in the construction of fills, for use as selected material or for other construction purposes. Local borrow shall be obtained by widening cuts or by excavating from other sources outside the planned or authorized cross section within the right-of-way and within the limits of the Work. Local borrow shall be material which is excavated from sources specified or designated by the Engineer. The sources of material to be excavated shall be approved in advance by the Engineer. Local borrow shall be excavated to the lines and grades established by the Engineer.

300-5.2 Imported Borrow. Imported borrow shall consist of material required for construction of fills, and unless otherwise specified in the Special Provisions, the Contractor shall make arrangements for obtaining imported borrow and shall pay all costs involved.

The Contractor shall notify the Engineer sufficiently in advance of opening any borrow site so that adequate time will be allowed for testing the material and establishing cross section elevations and measurements of the ground surface.

300-5.3 Placing and Compacting. Borrow shall be placed and compacted as specified in 300-4.

The Contractor shall satisfy itself that there is sufficient space available in fill locations for placing any excavated material before placing imported borrow. Any excess excavation which develops as a result of placing imported borrow in advance of completing excavations shall be disposed of at the Contractor's expense in accordance with the provisions of 300-2.6. A corresponding reduction in the quantity of imported borrow to be paid for will be made, for which the Contractor will have no claim for compensation.

300-5.4 Measurement and Payment. Quantities of local borrow will be measured and paid as specified for unclassified excavation in 300-2.8 and 300-2.9.

Quantities of imported borrow will be measured and paid in accordance with the Special Provisions.

300-6 EARTHWORK FOR DEBRIS DAMS AND BASINS.

300-6.1 General. Earthwork for debris dams and basins shall include clearing, stripping, excavation, fill, backfill, grading, and disposal of excess excavated material.

300-6.2 Clearing and Grubbing. Unless otherwise specified, the entire area of all rights-of-way shall be cleared of all trees and brush by cutting at the ground surface. All main root clusters shall be entirely removed from areas upon which improvements are to be constructed, and in addition, those areas upon which compacted fills are to be constructed shall be grubbed as follows:

Individual roots of diameters between 1/2 inch (13mm) and 1-1/2 inches (38mm) shall be removed to a depth of not less than 1 foot (0.3m) below natural ground. Larger roots shall be removed down to the level at which the root diameter is 1-1/2 inches (38mm) or less. Regardless of diameter, no roots need to be removed to a depth greater than 3 feet (1m).

All bulking of soil resulting from the grubbing operations shall be compacted and holes filled and compacted to subgrade for subsequent compacted fills.

300-6.3 Stripping. When stripping is indicated on the Plans or in the Special Provisions, the Contractor shall strip the soil from the designated areas to the depths specified.

The material obtained from stripping operations shall be disposed of away from the Work site unless otherwise specified.

Soil loosened below the stripping depth specified shall be compacted. Soil removed below stripping depth shall be replaced and compacted to subgrade. All such filling and compacting shall be at the Contractor's expense.

300-6.4 Basin Excavation. Unless otherwise specified, material obtained from the basin excavation shall be used for compacted fills. The Engineer will designate the exact limits of basin excavation and the depths thereof in order to obtain a material suitable for use in the compacted fills. Rocks over 6 inches (150mm) in greatest dimension will not be permitted in compacted fills and shall be disposed of away from the Work site unless otherwise specified.

300-6.5 Compacted Fills. Unless otherwise specified, the relative compaction of all fills shall be at least 90 percent.

Compacted fills shall be constructed of materials obtained from the onsite excavation unless otherwise specified.

The work of preparing subgrades, placing fill materials, watering, and compacting shall be performed only in the presence of the Engineer.

On hillsides and at abutments where the existing natural slope is steeper than 1 vertical in 4 horizontal, the existing ground shall be benched as the fill is brought up in layers. The benches shall be approximately horizontal and shall extend below the surface of the cleared and stripped ground a minimum depth of 2 feet (0.6m) normal to the slope unless otherwise specified. The material cut from the slope shall be incorporated in the fill.

Before placing the materials for the compacted fills, the subgrade therefore shall be moistened, compacted, and scarified in accordance with the requirements hereinafter set forth for subsequent layers of fill. The Engineer may determine the locations at which each load of fill shall be placed in order to obtain the best possible blending of materials. The fill material shall be placed in approximately horizontal, evenly distributed layers not exceeding 8 inches (200mm) in depth, except that where the Contractor clearly demonstrates that it can attain the required relative density with the type of equipment being used, a greater lift may be authorized. Unless otherwise permitted by the Engineer, each layer of fill material shall cover the full length and width of the entire area to be filled before the next higher layer of material is placed, and each layer shall be sufficiently scarified, after compaction, to provide bond with the succeeding layer. The top surface of each layer shall have sufficient crown to provide adequate drainage for water at all times during the construction period.

Before rolling or tamping, sufficient water shall be evenly applied to each layer of loose material so as to provide proper moisture content for satisfactory compaction to the specified relative compaction. The material shall be disc harrowed, or otherwise similarly worked, as the water is applied. The moisture content at the time of compaction shall be subject to the approval of the Engineer. In case any layer of fill shall prove to be too wet to permit the attainment of the specified relative compaction, the compacting work shall be delayed until the material has dried sufficiently to permit the attainment of said relative compaction.

After each layer has been spread, worked, and properly moistened, it shall be compacted by approved tamping or sheepsfoot rollers, pneumatic-tired rollers, mechanically operated hand tampers, or other mechanical means acceptable to the Engineer, to such extent as will produce the specified relative compaction.

Compacted fill which is to become subgrade for concrete channel or basin facing slabs, spillways, or other hydraulic structures, shall be overfilled sufficiently to permit the trimming thereof to an even and firm subgrade for the concrete to be placed thereon. No direct payment will be made for such overfill. Any costs involved therefore shall be included in the price bid for the compacted fill.

300-6.6 Payment. Clearing and grubbing shall be paid for on a lump sum or per acre (hectare) basis as provided in the Special Provisions.

Stripping, excavation, and compacted fill shall be measured and paid for on a cubic yard (cubic meter) basis as provided for in the Special Provisions.

300-7 EARTHWORK FOR CHANNELS.

300-7.1 General. Earthwork for channels shall consist of clearing, stripping, excavation, fill, backfill, grading, and disposal of excess excavated and removed material.

Channels for the purposes of this section shall mean open rectangular concrete channels and lined or unlined trapezoidal channels.

300-7.2 Stripping. When stripping is indicated on the Plans or otherwise specified, the Contractor shall strip the soil from the designated areas to the depths specified.

The material obtained from stripping operations shall be disposed of away from the site unless otherwise specified.

Soil loosened or removed below the stripping depth specified shall be replaced and compacted to not less than 90 percent relative compaction. All such filling and compacting shall be at the Contractor's expense.

300-7.3 Excavation. Excavation in an open cut for lined channels may be made so as to place concrete directly against excavated surfaces, provided the faces of the excavation are firm, hard, and unyielding; are such as will stand or can be made to stand without sloughing; and are at all points outside the concrete lines shown on the Plans.

Excavated surfaces becoming subgrade for lined channels, or subdrainage material, shall be excavated to the lines indicated on the Plans. Excavation made below subgrade shall be backfilled and compacted to a relative compaction of not less than 90 percent or, with the approval of the Engineer, with concrete or other materials being placed. However, no payment will be made for such overexcavation or material used for the backfill.

Where it becomes necessary to excavate beyond the normal lines of excavation in order to remove boulders or other interfering objects, the voids remaining after the removal of such boulders or interfering objects shall be backfilled as hereinafter specified, or as otherwise approved by the Engineer.

When the void is below the subgrade for reinforced concrete channel, suitable material approved by the Engineer shall be compacted to a relative compaction of not less than 90 percent or as specified. With the approval of the Engineer, concrete of the same mix as that used in the concrete channel may be used.

The removal of all boulders or other interfering objects and the backfilled of voids caused by such removals shall be at the expense of the Contractor and no direct payment for the cost of such work will be made by the Agency. The cost of such work shall be included in the prices bid for the various items of work.

If, during the progress of excavation, material is encountered which, in the opinion of the Engineer, is unsuitable for subgrade for the structure to be constructed thereon, the Engineer will direct the Contractor to excavate beyond the pay lines shown on the Plans. However, the suitability of subgrade will be determined by the Engineer on the basis of its ability to withstand the load of the proposed channel and not upon the capacity to withstand the loads which may be placed thereon by the Contractor's equipment. Should the Contractor be directed to excavate beyond the pay lines shown on the Plans, said pay lines will be extended to include such excavation. The pay lines for subdrainage material, if used, will be adjusted accordingly.

Materials used or work performed by the Contractor beyond the Agency's requirements for stabilization of the subgrade, so that it will withstand the loads which may be placed upon it by the Contractor's equipment, shall be at the Contractor's expense.

300-7.4 Fill and Backfill. Unless otherwise specified, the material obtained from the Work excavations will be suitable for use as fill or backfill, provided that all organic material, rubbish, debris, and other objectionable material contained therein is first removed. However, rocks, boulders, broken portland cement concrete, and bituminous type pavement obtained from the Work excavations will be permitted in the backfill or fill with the following limitations:

a) The maximum dimension of any piece used shall be 6 inches (150mm).
b) Pieces larger than 4 inches (100mm) shall not be placed within 12 inches (300mm) of any structure.
c) Pieces larger than 3 inches (75mm) shall not be placed within 12 inches (300mm) of the subgrade for paving.
d) "Nesting" of pieces will not be permitted.

Unless otherwise specified, the relative compaction of all fill and backfill shall be 90 percent.

300-7.5 Grading. Grading of unlined channels, levees, and access roads shall conform to the following tolerances:

A vertical tolerance of zero above and 3 inches (75mm) below the specified grade will be allowed for:

a) The channel bottom.
b) The channel sideslopes in both cut and fill.
c) The levee and access road sideslopes in cut.

A vertical tolerance of zero below and 3 inches (75mm) above the specified grade will be allowed on:

1) The top surface of levees and access roads in both cut and fill.
2) The levee and access road sideslopes in fill.

Regardless of the construction tolerances specified, the excavation and grading shall be performed so that the finished surfaces are in uniform planes with no abrupt breaks in the surface.

The construction tolerances specified herein for grading are solely for purposes of field control.

300-7.6 Measurement and Payment. Clearing, stripping, and excavation shall be measured and paid for on a cubic yard (cubic meter) basis or as specified in the Special Provisions.

Compacted fill shall be measured and paid for as provided in 300-4.9.

All costs involved in backfilling and grading shall be included in the prices in the Bid for the applicable items of earthwork.

300-8 GEOTEXTILES FOR DRAINAGE.

300-8.1 Trench Drains. Geotextiles for trench drains shall be placed in accordance with the following provisions:

300-8.1.1 Placement. Fabric shall be placed in the trench in accordance with the Plans. The fabric shall be placed loosely and seated firmly into the corners. If heat-bonded fabric is used, the bonded side shall be placed to the inside of the trench and the fuzzy side shall face the outside of the trench, against the native soil. Overlapping, if necessary, shall be a minimum of 12 inches (300mm). Damaged fabric shall be repaired at Contractor's expense by placing new fabric that meets overlap requirements over the damaged area. Fabric shall be covered as soon as possible after being placed, but not later than 7 calendar days after placement. Fabrics left uncovered for more than 7 calendar days shall be removed and rejected.

Trench sides and bottom shall be excavated to provide a smooth surface, free of obstructions and debris. After placement of granular fill, the 2 edges of the geotextile protruding at the top of the trench shall be overlapped 12 inches (300mm) on top of the granular fill, and then soil or other materials required by the Plans shall be compacted in the trench to the required grade.

300-8.1.2 Measurement and Payment. Geotextiles shall be measured for payment by the square yard (square meter) of fabric placed, not including any fabric for overlaps or splices.

300-9 GEOTEXTILES FOR EROSION CONTROL.

300-9.1 Bank and Shore Protection. Geotextiles for bank and shore protection shall be placed in accordance with the following provisions:

300-9.1.1 Placement. Prior to placement of fabric, the Contractor shall construct a subgrade in accordance with the Plans and Specifications. The fabric shall be placed loosely (not in a stretched condition), aligned, and placed in a manner to minimize wrinkling. Adjacent borders and ends of fabric shall be overlapped a minimum of 18 inches (460mm) or stitched when using nonwoven fabrics with a grab tensile strength of 90 pounds (400 N) or woven fabrics with a grab tensile strength of 200 pounds (890 N) or less. Borders and ends shall be overlapped 36 inches (900mm) or stitched for nonwoven fabrics with a grab tensile strength greater than 90 pounds (400 N) or woven fabrics with a grab tensile strength greater than 200 pounds (890 N).

If the fabric is overlapped, the upstream or higher panel shall overlap the downstream or lower panel. When stitched, the seam shall have seam breaking strength of not less than 80 percent of the minimum required fabric strength. The size and composition of the stitching material and stitching pattern shall be approved by the Engineer. The stitching yarn shall be of a contrasting color.

Anchoring of the fabric at terminal ends and top and bottom of the slope shall be accomplished through the use of key trenches or aprons as shown on the Plans. If the geotextile is placed in a vertical direction on the slope, there shall be no end joints between rolls. If heat bonded fabric is used, the fuzzy side shall be placed against the soil with the heat bonded side up.

The fabric shall be secured with pins placed on 6 foot (1.8m) centers at the midpoint of the overlaps unless a key trench is used. Pins at 6-foot (1.8m) centers shall be placed along the top edge of the slope. Spacing of pins shall be reduced to eliminate tearing of the fabric. Pins shall be 12 inches (300mm) long or of sufficient length to prevent pin movement.

The outer stone cover shall be thick enough to completely prevent penetration of sunlight, unless a bedding layer of aggregate particles greater in size than the openings in the outer stone cover is installed first.

Fabric shall be covered as soon as possible after being placed, but not later than 7 calendar days after placement. Fabric left uncovered for more than 7 calendar days shall be removed and rejected.

Equipment or vehicles shall not be operated on the fabric. Damaged fabric shall be repaired, at the Contractor's expense, by placing new fabric over the damaged area in a manner that meets the overlap requirements for horizontal placement. Vertically placed fabric shall be replaced in its entirety.

300-9.1.2 Measurement and Payment. Measurement and payment shall conform to 300-8.1.2.

300-10 GEOTEXTILES FOR SEPARATION.

300-10.1 Subgrade Enhancement. Geotextiles for subgrade enhancement shall be placed in accordance with the following provisions:

300-10.1.1 Placement. During grading operations, care shall be taken not to disturb or scarify the subgrade. This may require use of lightweight dozers, etc., for low-strength soils such as saturated, cohesionless, or low cohesion soils.

Once the subgrade along a particular segment of road alignment has been prepared, the geotextile shall be unrolled in line with the placement of the new aggregate. The fabric shall not be dragged across the subgrade, and the entire fabric shall be placed and rolled out as smoothly as possible. If heat bonded fabric is used, the fuzzy side shall be placed against the soil, with the bonded side up.

Parallel rolls or ends of fabric shall be overlapped 24 inches (600mm) or sewn if required by the Plans or Special Provisions. When indicated on the Plans or Special Provisions that soils have a CBR less than 5 or an R-value less than 20, a 36 inch (900mm) overlap shall be used.

The fabric shall be secured with pins placed on 6 foot (1.8m) centers at the midpoint of all overlaps and at the edges to maintain them during construction activities. Spacing of pins shall be reduced if necessary to eliminate tearing and movement of fabric.

For roadways, fabric widths shall be selected such that overlaps of parallel rolls occur at lane lines. Overlaps shall not be placed along anticipated main wheel path locations.

Overlaps at the end of rolls shall be in the direction of the aggregate placement with the previous roll on top.

When fabric intersects an existing pavement area, the fabric shall extend to the edge of the old system, and the fabric shall be anchored.

The aggregate base material shall be placed on the fabric in such a manner and thickness that wheel rutting of aggregate over the fabric is limited to 1/2 inch (13mm). Unless otherwise specified, the minimum thickness shall be 12 inches (300mm). Lightweight dozers shall be used if necessary. Equipment shall not be allowed directly on the fabric.

Before covering, the conditions of the fabric will be observed by the Engineer to determine that no holes or rips exist in the fabric. All such occurrences shall be repaired by placing a new layer of fabric extending beyond the defect in all directions a distance equal to the minimum overlap required for adjacent rolls.

300-10.1.2 Measurement and Payment. Measurement and payment shall conform to the requirements of 300-8.1.2.

300-11 STONEWORK FOR EROSION CONTROL.

300-11.1 General. Stone for erosion control shall be in conformance with 200-1.6 and the following provisions. When shown on the Plans, stonework shall be shall be concreted as specified herein.

300-11.2 Placing Stone. Stone for erosion control shall be placed in accordance with the following method:

A footing trench shall be excavated along the toe of the slope, as shown on the Plans. The larger stones shall be placed in the footing trench. Stones shall be placed with their longitudinal axis normal to the embankment face, and arranged so that each stone above the foundation course has a 3-point bearing on the underlying stones. The foundation course is the course placed on the slope in contact with the ground surface.

Bearing on smaller stones which may be used for chinking voids will not be acceptable. Nesting of the smaller stones used for chinking voids will not be permitted. Placing of stones by dumping will not be permitted.

Local surface irregularities of the slope protection shall not vary from the planned slope by more than 1 foot (0.3m) measured at right angles to the slope.

300-11.3 Concrete Stone Slope Protection. Stone for concreted stone slope protection shall be placed in accordance with 300-11.2.

300-11.3.1 Concrete. Concrete for concreted stone slope protection shall be 520-C-2500P (310-C-17P) and conform to the requirements of 201-1, except that the slump of the concrete shall be adjusted to provide the penetration shown in Table 300-11.3.1(A).

TABLE 300-11.3.1(A)

Class	500 lb (225 kg)	375 lb (170 kg)	Light (90 kg)	Facing (35 kg)
Minimum Penetration in inches (mm)	14 (350)	12 (300)	10 (250)	8 (200)

300-11.3.2 Placing Concrete. The surfaces of the stone to be concreted shall be cleaned of adhering dirt and clay and then moistened. The concrete shall be placed in a continuous operation for any day's run at one location. Concrete shall be brought to the place of final deposit by buckets or pump. Other placement methods shall be approved by the Engineer prior to use.

In no case shall concrete be permitted to flow on the slope protection a distance greater than 10 feet (3m).

Immediately after depositing, the concrete shall be spaded and rodded into place with suitable tools until the minimum penetration is that shown in Table 300-11.3.1.

After the concrete has been placed, the stones shall be thoroughly brushed so that their top surfaces are exposed. The outer stones shall project 1/3 to 1/4 their diameter above the concrete surface. After completion of any 10 foot (3m) strip, no person or load shall be permitted on the surface for a period of at least 24 hours, or longer if so ordered by the Engineer.

Concreted stone slope protection shall be cured as provided in 303-1.10.

300-11.4 Measurement and Payment. Stone and stonework for riprap and erosion control will be measured and paid for per ton (tonne) of stone in place.

Concrete and concrete placement will be measured and paid for per cubic yard (cubic meter) in place.

SECTION 301 - TREATED SOIL, SUBGRADE PREPARATION, AND PLACEMENT OF BASE MATERIALS

301-1 SUBGRADE PREPARATION.

301-1.1 General. This section shall govern the preparation of natural, filled, or excavated roadbed material prior to the placement of subbase or base material, pavement, curbs and gutters, driveways, sidewalks, or other roadway structures.

301-1.2 Preparation of Subgrade. Scarifying and cultivating will be required for dry soils which are impervious to the penetration of water, for soils which contain excessive amounts of moisture which may result in unstable foundations, for soils which are nonuniform in character which may result in nonuniform relative compactions and subsequent differential settlements of finished surfaces, or when pavement is to be placed directly on the roadbed material. Unsuitable material found below the processing depth for subgrade specified herein shall be treated in accordance with 300-2.2, Unsuitable Material.

After rough grading has been completed, when scarifying and cultivating are required, the roadbed shall be loosened to a depth of at least 6 inches (150mm). The loosened material shall then be worked to a finely divided condition and all rocks larger than 3 inches (75mm) in diameter shall be removed. The moisture content shall be brought to minus 1 percent of optimum or wetter by the addition of water, by the addition and blending of dry suitable material, or by the drying of existing material. The material shall then be compacted by approved equipment to the specified relative compaction.

Uniform pervious soils, that allow the immediate penetration of water or uniform impervious soils which will allow the penetration of water to a depth of at least 6 inches (150mm) after the addition of a suitable wetting agent, will not require scarifying and cultivating unless a condition previously set forth in this subsection requires such processing. When scarifying and cultivating are not required, the moisture content of the top 6 inches (150mm) of the subgrade material shall be brought to optimum by the addition of water at the surface, and the material shall be compacted by approved equipment to the specified relative compaction.

301-1.3 Relative Compaction. Except in alleys, when pavement is to be placed directly on subgrade material, the top 6 inches (150mm) of subgrade material shall be compacted to a relative compaction of 95 percent. When base or subbase material, curb, gutter, alley pavement, driveways, or sidewalks are to be placed on the subgrade material, the top 6 inches (150mm) of such subgrade material shall be compacted to a relative compaction of 90 percent.

After compaction and trimming, the subgrade shall be firm, hard, and unyielding.

301-1.4 Subgrade Tolerances. Subgrade for pavement, sidewalk, curb and gutter, driveways, or other roadway structures shall not vary more than 0.02 foot (6mm) from the specified grade and cross section. Subgrade for subbase or base material shall not vary more than 0.04 foot (12mm) from the specified grade and cross section. Variations within the above specified tolerances shall be compensating so that the average grade and cross section specified are met.

301-1.5 Grading of Areas Not To Be Paved. Roadway areas where "grade only" is called for on the Plans shall be graded to meet tolerances for base subgrade. The surface shall be constructed to a straight grade from the finish pavement or curb elevations shown on the Plans to the elevation of the existing ground at the extremities of the area to be graded.

301-1.6 Adjustment of Manhole Frame and Cover Sets to Grade. Utility manhole and vault frames and covers within an area to be paved or graded will be set by the owners thereof to finished grade. Sewer and storm drain manhole frames within the area to be paved or graded shall be set to finish grade by the Contractor. Manholes in asphalt concrete pavement shall be set to finish grade in accordance with provisions of 302-5.8. In the case of portland cement concrete pavement, manhole frames shall be set to finish grade before paving. Repaving required as a result of reconstructing or adjusting all manhole and vault frames and covers to grade shall be the responsibility of the Contractor and the cost thereof shall be included in the bid item for pavement.

The Contractor shall remove all debris from the interior of manholes and shall clean all foreign material from the top of the frames and covers.

301-1.7 Payment. Payment for preparing a subgrade will be considered as included in the item of work for which the subgrade is prepared.

Payment for grading operations in areas designated as "grade only" will be considered as included in the Bid price for excavation or fill.

Payment for adjusting manhole frames and covers to grade, where the difference between the lowest point of manhole removal and final elevation of the top of the frame is less than 15 inches (375mm) or where the adjustment is accomplished by adjustment rings only, will be made at the Contract Unit Price for adjusting each manhole frame.

Payment for adjusting manhole frames and covers to grade, where the difference between the lowest point of manhole removal and final elevation of the top of the frame is 15 inches (375mm) or more, will be made at the Contract Unit Price per linear foot (meter) for reconstructing each manhole.

If no payment provision for manhole adjustment or reconstruction is made, payment for such work will be as provided in 3-3.

301-2 UNTREATED BASE.

301-2.1 General. Untreated base shall be constructed of material conforming to 200-2.

301-2.2 Spreading. Imported aggregate bases shall be delivered to the roadbed as uniform mixtures and each layer shall be spread in one operation. Segregation shall be avoided and the base shall be free from pockets of coarse or fine material.

Aggregate bases shall be deposited on the roadbed at a uniform quantity per linear foot (meter), which quantity will provide the required compacted thickness within the tolerances specified herein without resorting to spotting, picking up, or otherwise shifting the aggregate base material. At the time aggregate base is spread, it shall have a moisture content sufficient to obtain the required compaction. Such moisture shall be uniformly distributed throughout the material.

Where the required thickness is 6 inches (150mm) or less, the base material, except "Pulverized Miscellaneous Base", may be spread and compacted in one layer. Where the required thickness is more than 6 inches (150mm) the base material, except "Pulverized Miscellaneous Base", shall be spread and compacted in 2 or more layers of approximately equal thickness, and the maximum compacted thickness of any one layer shall not exceed 6 inches (150mm). Each layer shall be spread and compacted in a similar manner.

The use of motor graders will be permitted during depositing, spreading and compacting operations, except when self-propelled spreaders are specified. Pulverized Miscellaneous Base shall be spread and compacted in one or more layers. The maximum compacted layer shall not exceed 10 inches (250mm).

When the subgrade for aggregate base consists of cohesionless sand and written permission is granted by the Engineer, a portion of the aggregate base may be dumped in piles upon the subgrade and spread from the dumped material in sufficient quantity to stabilize the subgrade. Segregation of aggregates shall be avoided and the material as spread shall be free from pockets of coarse or fine material.

301-2.3 Compacting. Rolling shall always be commenced along the edge of the area to be compacted and the roller shall gradually advance toward the center of the area to be compacted.

Rollers shall be operated along lines parallel or concentric with the centerline of the road being constructed, and no material variation therefrom will be permitted. All rollers must be maintained in good mechanical condition.

The relative compaction of each layer of compacted base material shall not be less than 95 percent, except in the areas back of curb (under sidewalks and driveways). Compaction in the excepted areas shall be as specified in 211-1 with each layer of compacted base material having a minimum relative compaction of 90 percent.

The surface of the finished aggregate base at any point shall not vary more than 0.02 foot (6mm) above or below the grade established by the Engineer.

Base which does not conform to the above requirements shall be reshaped or reworked, watered, and thoroughly recompacted to conform to the specified requirements.

301-2.4 Measurement and Payment. Quantities of untreated base will be measured as shown in the Bid. The volumetric quantities of base material shall be those compacted in place within the limits of the dimensions shown on the Plans.

The weight of material to be paid for will be determined by deducting (from the weight of material delivered to the Work) the weight of water in the material (at the time of weighing) in excess of 1 percent more than the optimum moisture content. No payment will be made for the weight of water deducted as provided in this subsection.

301-3 PORTLAND CEMENT TREATED MIXTURES.

301-3.1 Soil-Cement.

301-3.1.1 General. Soil-cement shall consist of soil, portland cement, and water which are uniformly mixed, compacted, finished, and cured in such a manner that the in-place soil-cement mixture forms a dense, uniform mass conforming to the lines, grades, and cross sections shown on the Plans. Soil-cement mixtures shall also include cement treated base and cement modified soil.

301-3.1.2 Materials. Portland cement and water shall conform to the provisions of 201-1.2.1 and 201-1.2.3, respectively, or as otherwise specified. Cement content and moisture content shall be specified.

Soil to be treated may consist of in-place material, select borrow material, a combination of these materials, or select aggregate material proportioned as directed by the Engineer or as specified. The soil shall consist of material not larger than 2 inches (50mm) in diameter and with not less than 55 percent passing the No. 4 (4.75mm) sieve. The material shall not be deleterious in its reaction with cement.

301-3.1.3 Equipment. Soil-cement shall be constructed utilizing a machine or a combination of machines that will produce results meeting all the requirements herein. Such machines shall be approved by the Engineer prior to use.

301-3.1.4 Preparation. Prior to importing materials to be treated, the soil subgrade shall be prepared as provided in 301-1.

Subgrade material that is to be treated in place shall be shaped and rolled to specified cross section prior to the scarification and treatment. No further subgrade preparation will be necessary.

The soil subgrade shall be firm enough to support construction equipment without displacement. Unsuitable soils shall be removed in accordance with 300-2.2.

301-3.1.5 Cement Application, Mixing, and Spreading. Mixing of the soil, cement, and water shall be accomplished either by the mixed-in-place or the central-plant-mixed method. No cement or soil-cement mixture shall be spread when the aggregate or subgrade is frozen or when the air temperature is less than 40°F (5°C) in the shade. The finished soil-cement shall be protected against freezing.

After mixing is complete, the soil shall be so pulverized that 100 percent by dry weight passes a 1 inch (25mm) sieve, and a minimum of 80 percent passes a No. 4 (4.75mm) sieve, exclusive of gravel or stone retained on these sieves.

301-3.1.6 Mixing In Place. The specified quantity of cement shall be spread uniformly on the soil to be treated. Cement that has been spread and then tracked or displaced by traffic or other operations shall be replaced before mixing is started to provide a uniform coverage in all areas.

For the windrow method of mixing, the material shall be shaped and sized uniformly prior to the addition of cement. The tops of soil windrows shall be flattened or slightly trenched to receive the cement. Windrows shall be limited to such size that all material can be passed through the mixing machine in one operation.

Cement shall be spread by an approved cement spreader in such a manner that cement content shall not vary from the specified rate of application by more than 10 percent. Cement may be supplied in sacks and spread by hand when specified or when approved by the Engineer. When hand spreading is used, the sacks shall be spaced at uniform intervals and then emptied, following which the cement shall be spread in a layer of uniform thickness. After the cement has been spread, it shall be uniformly mixed with the soil.

Water shall be applied under pressure by means of controlled distributing equipment which will produce a completed mixture with a uniform moisture content. Leakage of water from equipment will not be permitted. Care shall be exercised to avoid the addition of any excess water.

Mixing shall continue until a uniform mixture of soil, cement, and water has been obtained. If equipment is used that requires more than one pass of the mixer, at least one pass shall be made before any water is added to the material.

301-3.1.7 Central-Plant Mixing. Soil-cement mixed in central mixing plants shall use an approved pugmill or continuous flow mixer. Soil, cement, and water shall be stored separately.

The rate of feed of soil, cement, and water shall be within 3 percent of the amount of each material designated by the Engineer or as specified.

Water shall be so proportioned that the Engineer may readily verify the amount of water in a batch or the rate of flow for continuous mixing.

The charge in a batch mixer or the rate of feed to a continuous mixer shall not exceed that which will permit complete mixing of all of the material. Dead areas in a mixer, in which the material does not move or is not sufficiently mixed, will not be permitted.

Mixing shall continue until a homogeneous mixture of uniformly distributed and properly coated aggregates of unchanging appearance is produced. Cement content shall not vary more than 5 percent from that specified.

The soil-cement mixture shall be transported from the mixing plant to the site in clean equipment provided with suitable protective devices to prevent material loss and significant moisture change. The total elapsed time between the addition of water to the mixture and the start of compaction shall not exceed 45 minutes.

301-3.1.8 Placing, Compacting, and Finishing. Soil-cement shall be uniformly compacted to at least 95 percent of relative compaction.

The mixture shall be placed on the moistened subgrade, or previously completed soil-cement, using mechanical spreading equipment that will produce layers of such width and thickness that it will compact to the required dimensions of the completed soil-cement layers.

The mixture may be spread and compacted in one layer where the required thickness is 8 inches (200mm) or less. Where the required thickness is more than 8 inches (200mm), the mixture shall be spread and compacted in 2 or more layers of approximately equal thickness, provided that the maximum compacted thickness of any one layer does not exceed 8 inches (200mm).

Compaction shall commence within 30 minutes after the mixture is placed on the grade and shall proceed continuously until completed. Final compaction of the mixture to the specified density shall be completed within 2-1/2 hours after the application of water during the mixing operation.

When two or more layers of soil-cement are to be placed, the surface which will be in contact with succeeding layers shall be kept continuously moist for 7 days or until the placement of the subsequent layer. Any loose material on the surface of the completed layer shall be removed and the surface moistened immediately before placement of the next layer. No standing water will be permitted.

At the start of compaction, the mixture shall be in a uniform, loose condition throughout its full depth.

During finishing operations, the surface of the soil-cement shall be shaped to the required lines, grades, and cross sections and shall be kept moist. The finished surface of the soil-cement shall conform to the requirements of 301-1.4.

301-3.1.9 Curing. After placement and compaction of the soil-cement is completed, it shall be protected against drying and from traffic for 7 days.

Curing shall be moist (water fogging), bituminous seal, or other method approved by the Engineer. If moist curing is used, exposed surfaces of the soil-cement shall be kept continuously moist with a fog spray for 7 days. If a bituminous curing is used, it shall consist of liquid asphalt or emulsified asphalt meeting the requirements of 203.

The bituminous curing seal shall be applied in sufficient quantity to provide a continuous membrane over the soil at a rate of between 0.10 and 0.20 gallon per square yard (0.45 L/m^2 and 0.90 L/m^2) of surface with the exact rate determined by the Engineer. It shall be applied as soon as possible after the completion of final rolling. The surface shall be kept moist until the seal is applied. At the time the bituminous material is applied, the soil surface shall be dense, shall be free of all loose and extraneous material, and shall contain sufficient moisture to prevent excessive penetration of the bituminous material.

301-3.1.10 Construction Joints. Construction joints shall have vertical faces and shall be made in thoroughly compacted material. Additional mixture shall not be placed against the construction joint until the joint has been approved by the Engineer. The face of the cut joint shall be clean and free of deleterious material and shall be kept moist until the placing of the adjacent soil-cement.

301-3.1.11 Repair. If the soil-cement is damaged, it shall be repaired by removing and replacing the entire depth of affected layers in the damaged area. Feathering will not be permitted for repair of low areas.

301-3.1.12 Measurement and Payment. Cement treated soils, base, and subbase will be paid for by the square yard (square meter) or cubic yard (cubic meter) in place as shown on the Plans or as directed by the Engineer. The Contract Unit Price shall include full compensation for mixing, spreading, shaping, compacting, trimming, and curing.

301-3.2 Plastic Soil-Cement.

301-3.2.1 General. Plastic soil-cement shall consist of soil, portland cement, and water. It shall be proportioned, uniformly mixed to the consistency of plastering mortar, finished and cured in such a manner that the completed plastic soil-cement in place forms a dense uniform mass conforming to the lines and grades shown on the Plans.

301-3.2.2 Mixing and Finishing. Mixing of the soil, cement, and water shall be accomplished by a pugmill or rotary drum mixer.

The soil, cement, and water shall be proportioned as specified and uniformly mixed to a consistency similar to that of plastering mortar.

The plastic soil-cement shall be placed on the prepared grade, consolidated, and finished to the lines and grades shown on the Plans.

Finished, plastic soil-cement shall be kept moist and cured as specified in 301-3.1.9.

301-3.2.3 Measurement and Payment. Measurement and payment of plastic soil-cement will be in accordance with 301-3.1.12.

301-3.3 Cement-Treated Base.

301-3.3.1 General. The cement-treated base shall conform to the requirements of 301-3.1 and 301-3.3.

Cement-treated base shall consist of a combination of specified base material and portland cement, uniformly mixed, moistened, and compacted as described herein, and shaped to conform to the lines, grades, and cross sections shown on the Plans.

301-3.3.2 Mixed In Place. Water distributing equipment shall be provided with a metering device for introducing water at the time of mixing.

Should the Contractor elect to perform road-mixing operations off the roadbed at a designated location, the preparation of the material for mixing and the mixing of base material, cement, and water shall conform to the applicable provisions specified herein for preparing and mixing the materials on the roadbed. When the materials are road-mixed off the roadbed, the device for loading the mixed material into the transporting vehicle shall be so constructed and so operated that no untreated material will be picked up. The time required for loading and hauling the material shall be taken into account when determining the amount of material to be mixed at any one time for placement and compaction within the time specified in 301-3.1.8.

301-3.3.3 Central-Plant Mixing. Base material, cement, and water shall be mixed in a rotary drum or pugmill mixer with the materials proportioned by batch weights. If a continuous type of mixer is used, the materials shall be proportioned by volume.

If the Contractor elects to use a continuous type mixer, the correct proportions of each aggregate size introduced into the mixer shall be drawn from the storage bins by means of a continuous feeder through adjustable calibrated gates, which gates shall supply the correct amount of aggregate in proportion to the cement and water, and the gates shall be so arranged that the proportion of each aggregate size can be separately adjusted. The mixer shall be equipped with metering devices which will introduce the cement and water into the mixer in the specified proportions. The metering devices and feeder shall be interlocked and so synchronized as to maintain a constant ratio between cement, water, and aggregate. Storage bins shall be equipped with overflow chutes for each compartment. A positive signal system shall be provided to indicate when the level of material in each bin approaches the strikeoff capacity of the feed gate. The plant shall not be permitted to operate unless this signal system is in good working condition. The plant shall be equipped with facilities to weigh samples for calibrating gate openings.

Water may be proportioned by weight or volume. The quantity of water added to the mixture shall be adjusted to produce optimum moisture content. The addition of water shall be made under conditions which will permit an accurate determination of the quantity of water utilized.

301-3.3.4 Placing, Compacting, and Finishing. Materials mixed at a location off the roadbed shall be deposited by means of approved spreading equipment. Dumping in piles upon the subgrade will not be permitted.

The surface of the finished cement-treated base at any point shall vary not more than 0.02 foot (6mm) above or below the grade established by the Engineer.

The mixed materials shall be spread for the full width of the subgrade or base under construction, either by one spreader or by several spreaders operating in echelon across the subgrade, unless traffic conditions require part-width construction. Should only one spreader be used, not more than 45 minutes shall elapse between the time of placing material in adjacent lanes at any location without trimming the longitudinal joint.

Where traffic or other conditions make part-width construction of a base necessary, a windrow of shoulder material or soil shall be placed and compacted to form a choker to restrain the inner edge of the base during compacting operations. The choker shall be constructed to the same elevation as that of the compacted base, and shall be completed in advance of the spreading of the treated material. The toe of the choker shall be not less than 3 inches (75mm) outside the finished trimming line of the compacted section of base material. The use of side forms, or other methods which will satisfactorily retain the base material during compacting operations, will be permitted in lieu of a choker.

After a part-width section has been completed, the longitudinal joint against which additional material is to be placed shall be trimmed to a vertical edge at the neat line of the section. Choker material and material cut away in trimming shall be used only in the construction of adjacent shoulders or otherwise disposed of along the roadway, unless suitable for incorporation in the Work.

The use of self-propelled graders will be permitted for spreading the mixed material on the roadbed and for trimming.

301-3.3.5 Curing. No equipment or traffic shall be permitted on the finished surface during the first 3 days after the application of the curing seal.

In cases where the Contractor has permitted equipment or traffic to operate on a cement-treated base, or subbase, the Contractor shall, before the paving material is placed on the surface, completely remove any other material which has been spread to protect the treated surface and shall apply additional curing seal to any areas where the curing seal has been destroyed. Full compensation for furnishing, spreading, and removing such material and for furnishing and applying the additional curing seal will be considered as included in the price bid for cement-treated base and no additional compensation will be made therefore.

301-3.3.6 Measurement and Payment of Cement-treated Base. Measurement and payment of cement-treated base will be in accordance with 301-3.1.12.

301-4 LEAN CONCRETE BASE (LCB).

301-4.1 General. This subsection specifies the methods to be used to place LCB conforming to 200-4.

301-4.2 Subgrade. Immediately prior to placement, the subgrade shall conform to 301-1. The subgrade shall be free of loose or extraneous material and shall be sufficiently dampened to ensure that no water is absorbed from the LCB.

301-4.3 Placement. Placement shall conform to 302-6.3.

Unless otherwise shown on the Plans or specified in the Special Provisions, LCB shall be placed in widths of not less than 12 feet (4m) separated by construction joints. LCB placed monolithically in widths greater than 26 feet (8m) shall be placed with a longitudinal weakened plane joint offset not more than 3 feet (1m) from the centerline of the width being constructed. Longitudinal weakened plane joints shall conform to 302-6.5.4.

When PCC pavement is to be placed over LCB, longitudinal construction joints and longitudinal weakened plane joints in the LCB shall not be within 1 foot (300mm) of longitudinal weakened plane joints nor longitudinal construction joints in the PCC pavement.

Areas of the subgrade that are lower than the grade established on the Plans shall be filled with LCB.

LCB shall not be mixed nor placed while the atmospheric temperature is below 35° F (2° C), and shall not be placed on frozen ground.

301-4.4 Spreading, Compacting and Shaping. Spreading, compacting, and shaping shall conform to 301-3.3.

In advance of curing operations, LCB to be surfaced with asphalt concrete pavement shall be textured with a drag strip of burlap, a broom, or a spring steel tine device capable of producing scoring in the finished surface. The scoring shall be parallel with or transverse to the centerline.

LCB to be surfaced with PCC pavement shall not be textured and shall be finished to a smooth surface, free of mortar ridges and other projections, before curing compound is applied.

The finished surface of LCB shall be free from porous areas and not vary at any point more than 5/8 inch (16mm) above or below the grade shown on the Plans.

301-4.5 Curing. LCB shall be cured with a pigmented curing compound conforming to 201-4. Curing compound shall be applied to areas to be surfaced with PCC pavement in 2 separate applications. Each application shall be applied at a rate of approximately 1 gallon per 150 square feet (1 liter per 4 square meters).

Curing compound shall be applied before the atmospheric temperature falls below 40° F (40 C).

Areas to be covered with PCC pavement that are not covered by the fourth day after the curing compound is applied shall be given a subsequent application (or applications, if necessary) so that each application occurs no more than 4 days in advance of placing the overlying PCC pavement. LCB surfaces shall be cleaned of all foreign material prior to the application of curing compound.

Curing compound shall be applied at a rate of approximately 1 gallon per 200 square feet (1 liter per 5 square meters) for each subsequent application.

Damage to the curing compound or the LCB shall be promptly repaired by the Contractor at the Contractor's expense, as directed by the Engineer.

301-4.6 Acceptance.

301-4.6.1 Material. LCB will be tested for acceptance in accordance with 201-1.1.5.

301-4.6.2 Placement. Hardened LCB with a surface lower than 5/8 inch (16mm) below the grade shown on the Plans shall be removed and replaced with LCB, or if permitted by the Engineer, low areas shall be filled with paving material as follows:

a) When the surface pavement is PCC, low areas shall be filled with PCC at the time and in the same operation the PCC pavement is placed.

b) When the surface pavement is asphalt concrete, the low areas shall be filled with asphalt concrete at the time and in the same operation that asphalt concrete pavement is placed.

Hardened LCB with a surface higher than 5/8 inch (16mm) above the grade shown on the Plans shall be removed and replaced with LCB, or if permitted by the Engineer, high areas may be ground until the surface conforms to the tolerances specified in 301-4.4. Grinding shall be performed with diamond blade or carborundum blade grinding equipment. Ground areas to be surfaced with PCC pavement shall be cleaned of all foreign material and grinding residue as soon as all free water has left the surface. Curing compound conforming to 301-4.5 shall be applied at a rate of approximately 1 gallon per 150 square feet (1 liter per 4 square meters).

301-4.7 Measurement. LCB will be measured by the cubic yard (cubic meter). The volume will be calculated on the basis of the dimensions shown on the Plans adjusted by the amount of any change approved by the Engineer. LCB used to fill depressed areas of subgrade will not be measured for payment.

301-4.8 Payment. Payment for LCB will be made at the Contract Unit Price per cubic yard (cubic meter).

No separate payment will be made for bringing the surface of the LCB within tolerance. Payment will be considered as included in the Contract Unit Price per cubic yard (cubic meter).

301-5 LIME-TREATED SOIL.

301-5.1 General. Lime treatment shall consist of mixing a soil with a specified percentage of hydrated lime or quicklime and water; spreading, compacting, and curing the mixture to the lines, grades, and dimensions shown on the Plans or as specified herein.

301-5.2 Lime. Lime shall be either a commercial dry hydrated lime or quicklime, conforming to ASTM C51.

When sampled by the Engineer at the point of delivery, the lime shall meet the following quality requirements:

TABLE 301-5.2 (A)

Sieve Sizes	Percent Passing	
	Hydrated Lime	Quicklime
No. 4 (4.75mm)	100	100
No. 30 (600μm)	95 - 100	
No. 100 (150μm)		0 - 20
No. 200 (75μm)	75 - 100	
Test Method	ASTM C110[1]	ASTM C136[2]
Chemical	**Percent of Chemical (Min.)**	
$Ca(OH)_2$	85	119
CaO		80
Test Method	California Test 414 or ASTM C25	

1. The sample shall be washed for 15 ± 1 mins.
2. Dry sieving only.

The lime shall be protected from moisture exposure until used and shall be sufficiently dry to flow freely. Prior to spreading, a certificate of compliance and a certified copy of the shipping weight shall be submitted to the Engineer with each delivery.

Lime may be mixed with water to form a slurry prior to being added to the soil. Mixing of the lime and water may be accomplished by any method which will produce a uniform consistency. When once established, the proportioning method may not be changed without the approval of the Engineer.

301-5.3 Water. All water used must be fresh and potable.

301-5.4 Preparation. When material to be treated is to be imported to the Work site, the subgrade shall be prepared as provided in 301-1 prior to placing such imported material.

If in-place material is to be treated, it shall be shaped and rolled to the section specified prior to scarifying and treating. No subgrade preparation shall be necessary provided that it is firm and able to support the construction equipment to the satisfaction of the Engineer. Where needed, stabilization of the earth subgrade shall be as directed by the Engineer.

The material to be treated shall contain no rocks larger than 2 inches (50mm) in diameter. Removing and disposing of rocks larger than 2 inches (50mm), when in excess of 5 percent of the total material to be treated, will be paid as Extra Work as provided in 3-3.

301-5.5 Spreading Lime. Lime shall be spread by a system, approved by the Engineer, which will uniformly distribute the required amount of lime for the full width of the pass or on the windrow. When lime slurry is used, the distributor truck shall be equipped with a recirculating pump or other device to prevent settling of the lime prior to depositing on grade. The lime spread rate shall not vary more than 10 percent from the designated spread rate. All lime shall be mixed the same day as spread.

If the material to be treated is windrowed, the top of the windrow shall be flattened or trenched to receive the lime. No traffic other than water trucks and the mixing equipment will be allowed to pass over the spread lime until after completion of mixing.

Lime or lime-treated material shall not be spread or mixed while the atmospheric temperature is below 35°F (2°C) or when wind conditions are such that blowing lime can have adverse effect on traffic or adjacent property.

301-5.6 Mixing. The mixing shall be accomplished with a traveling mixer, approved by the Engineer, which shall be equipped with a system capable of introducing water at a controlled rate during mixing in order to produce a completed mixture with a uniform moisture content. Mixing operations shall be performed in such a manner as to produce a uniform mixture of the lime, water, and the materials being treated, free of streaks and pockets of lime. Nonuniformity of color reaction, when the treated material is tested with the standard phenolphthalien-alcohol indicator, shall be considered as evidence of inadequate mixing. Water shall be added during the mixing process until the water content of the mixture is approximately 2 percent above the optimum moisture content of the lime-treated soil. Sufficient water shall be added during initial mixing to slake all the quicklime.

The final mixture shall not contain more than 5 percent untreated dirt clods larger than 1 inch (25mm) in diameter.

At the conclusion of mixing, suitable protective measures shall be taken to minimize evaporation loss and to prevent excessive wetting.

301-5.7 Spreading and Compacting. After the mixing operation is complete, a curing period of not less than 16 nor more than 72 hours is required before the lime-treated material is spread and compacted. The completed mixture shall be spread to the plan, grade, and cross section.

The thickness of a compacted layer shall not exceed 8 inches (200mm). Where the thickness is more than 8 inches (200mm), the mixture shall be spread and compacted in two or more approximately equal layers. The moisture content shall be maintained to achieve compaction.

Where used directly under pavement, lime-treated soil shall be compacted to not less than 95 percent relative density. This finished surface shall not vary more than 0.02 foot (6mm) from the specified grade and cross section.

In other cases, the relative density shall be 90 percent minimum. The finished surface shall not vary more than 0.04 foot (12mm) from the specified grade and cross section.

Initial compaction shall be by means of sheepsfoot or segmented-wheel rollers. Final rolling shall be by means of steel-tired or pneumatic-tired rollers. Areas, inaccessible to rollers, shall be compacted to the required compaction by other means satisfactory to the Engineer.

At the end of each day's work, a vertical-faced construction joint shall be made in thoroughly compacted material normal to the centerline of the roadbed. Additional mixture shall not be placed until the construction joint has been approved by the Engineer.

301-5.8 Curing. The surface of the compacted layer shall be kept moist until covered by other base material or application of a curing seal of emulsified asphalt. If a curing seal is used, it should be applied as soon as possible after completion of final rolling, at a rate of between 0.10 and 0.20 gallon per square yard (0.45 L/m^2 and 0.90 L/m^2), the exact rate to be determined by the Engineer.

No equipment or traffic shall be permitted on lime-treated material for 72 hours after curing seal is applied, unless otherwise permitted by the Engineer.

301-5.9 Safety Program. The Contractor shall provide to the Engineer for review a detailed safety program for the protection of the workers and public, covering precautions to be exercised and emergency treatment to be available on the Work site. The program shall include protective equipment for eye, mouth, nose, and skin protection; and a first-aid kit with an eyeball wash. Said protective equipment shall be available on the Work site during spreading and mixing operations. This program shall be provided and agreed upon before the lime spreading begins. The Contractor shall actively enforce said program for the protection of its work force and others in the construction area. Adequate care must be taken to avoid quicklime contact during spreading and slaking operations.

301-5.10 Measurement and Payment. Lime-treated soil will be paid for by the square yard or cubic yard (square meter or cubic meter) in place as shown on the Plans or as directed by the Engineer. The Contract Unit Price per square yard or cubic yard (square meter or cubic meter) shall include payment for breaking-up, mixing, spreading, compacting, trimming, and curing treated soil. Lime will be paid for by the ton (tonne) when spread at the rate prescribed on the Plans or directed by the Engineer. Payment for lime will include furnishing and spreading lime.

SECTION 302 - ROADWAY SURFACING

302-1 COLD MILLING OF EXISTING PAVEMENT.

302-1.1 General. The Contractor shall cold mill existing pavement as shown on the Plans and specified in the Special Provisions. The type of pavement and depth to be cold milled shall be as shown on the Plans or specified in the Special Provisions. The presence of pavement fabric, rubberized material (i.e. ARHM, REAS, ARAM's, etc.), or steel reinforcement within the depth to be cold milled shall be noted on the Plans or in the Special Provisions. The surface after cold milling will be uniformly grooved or ridged unless otherwise specified in the Special Provisions. The outside lines of the milled pavement shall be neat and uniform.

The milled pavement shall be true to grade and cross section. When the straightedge specified in 302-5.6.2 is laid on the finished surface parallel to the centerline of the roadway, the surface shall not vary from the edge of the straightedge more than 3/8 inch (10mm) at any point, except at intersections or at changes of grade. Any areas that are not within tolerance shall be brought to grade within 1 working day following initial cold milling.

Cold milling operations shall be performed without damage to the remaining pavement. Whenever cold milling is adjacent to existing portland cement concrete curbs, gutters or pavement, the Contractor shall protect these improvements from damage. Any portland cement concrete curbs, gutters or pavement damaged during cold milling operations shall be repaired as directed by the Engineer at the Contractor's expense. Any portland cement concrete curbs, gutters or pavement that is cracked or displaced shall be removed and replaced at the Contractor's expense. Replaced sections of portland cement concrete curb, gutter or pavement shall be a minimum of 5 feet (1.52m) in length or to the next joint.

302-1.2 Milling Machine. Milling machines shall be specially designed for cold milling of asphalt concrete, portland cement concrete, or a combination of asphalt and portland cement concrete pavement. Milling machines shall conform to the following:

a) The cutting drum shall be a minimum of 60 inches (1500mm) wide and shall be equipped with carbide-tipped cutting teeth placed in a variable pattern to produce the desired finish.

b) Be self-propelled and capable of removing the pavement to the depth shown on the Plans.

c) Be equipped with a conveyor system that will immediately convey the milled material into a transport vehicle for disposal as specified in the Special Provisions.

d) Have the capability of spraying water at the cutting drum to minimize dust.

e) Be designed so that the operator can observe the milling operation, at all times, without leaving the controls.

f) Be adjustable for slope and depth.

g) Be able to deep cut, in one pass, to the maximum depth recommended by the manufacturer without producing fumes or smoke.

The Contractor shall provide smaller machines if required to cold mill areas that are inaccessible to the larger machine and to provide the surface specified in the Special Provisions.

302-1.3 Cold Milling to Specified Elevations.

302-1.3.1 General. Milling to specified elevations is the controlled removal of a portion of the existing pavement and underlying base or substrate. The finished elevations and depth of cut shall be as shown on the Plans or specified in the Special Provisions.

302-1.3.2 Milling Machine. Milling machines used for milling to specified elevations shall conform to 302-1.2. The machines shall also be equipped with automatic grade controls that reference the existing pavement elevations or independent grade references. The references shall be those required to achieve the specified elevations shown on the Plans.

302-1.4 Profile Milling.

302-1.4.1 General. Profile milling is the controlled removal of a portion of the existing pavement to a nominal depth using longitudinal grade controls to remove surface irregularities in the pavement and improve ride ability. The grade shall be as shown on the Plans.

302-1.4.2 Equipment. Milling machines used for profile milling operations shall conform to 302-1.2. The Contractor shall also use a minimum 20-foot (6m) paving ski with spring loaded feet attached to the bottom at not more than 18-inch (0.5m) increments. The upper portion of the ski shall be one-piece and manufactured such that the ski does not flex or bend by more than 3/16 inch (5mm) when supported off of the surface of the pavement by an attachment located at the ski's longitudinal center of gravity. The grade control system of the milling machine shall be referenced to the center of the ski. During profile milling operations, the center of the ski shall be on a line coincident with the transverse centerline of the milling machine's cutting drum. A ski shall be attached to each side of the milling machine cutting drum during the first pass, and on one side of the milling machine on subsequent, adjacent passes with a joint matching grade control on the other side.

The resultant milled surface shall not deviate from the grade shown on the Plans, using the straightedge specified in 302-5.6.2, by more than 1/4 inch (6mm) at any point.

302-1.5 Full-Depth Milling.

302-1.5.1 General. Full depth milling is the removal of the full depth of the existing pavement as shown on the Plans or specified in the Special Provisions. When full depth milling is specified, the Contractor shall continuously control the depth of milling to stay no more than 1/2 inch (12mm) below the full depth of the existing pavement. In areas of resurfaced trenches, individual excavations or bore holes, the required depth of milling shall be the same as that of the adjacent pavement. The Contractor shall remove existing asphalt concrete overlay from gutters adjacent to any area specified to be cold milled, as directed by the Engineer.

302-1.5.2 Equipment. Milling machines used for full depth milling of pavement shall conform to 302-1.2 and be capable of milling a minimum depth of 10 inches (250mm) in a single pass.

302-1.6 Cold Milling of Composite Pavements. Composite pavements are pavements that consist of more than one material overlaid with another (e.g. portland cement concrete overlaid with asphalt concrete). The thickness of each existing pavement material shall be shown on the Plans. The area and depth of a composite pavement to be cold milled shall be as shown of the Plans.

302-1.7 Work Site Maintenance. Work site maintenance shall conform to 7-8. A motorized street sweeper shall follow within 50 feet (15m) of the cold milling machine unless otherwise approved by the Engineer.

302-1.8 Disposal of Millings. Unless otherwise specified in the Special Provisions all material removed shall be considered the property of the Contractor and shall be disposed of by the Contractor.

302-1.9 Traffic Signal Loop Detectors. The Contractor shall not mill within 12 inches (300mm) of any existing loop detectors that are shown to be protected in place on the Plans or in the Special Provisions. Traffic signal loop detectors that were shown to be protected in place but are damaged or removed shall be replaced in conformance with 7-9.

302-1.10 Pavement Transitions. Structures and vertical joints within the cold-milled areas that are transverse to through traffic shall be ramped in conformance with 7-10 and 306-1.5.1. Ramps shall be constructed the same day as the existing pavement is cold milled and removed prior to placement of the permanent paving.

302-1.11 Measurement. Cold milling will be measured by the linear foot (meter) along the edge of the transverse joint lines, adjacent curb, gutter, or cross gutter, or by the square foot (square meter) as shown in the Bid. The amount of each type of pavement cold milled (asphalt, concrete, or composite pavement) shall be measured separately.

302-1.12 Payment. The Contract Unit Price for cold milling each type of pavement shall include cold milling, construction and removal of pavement transitions, disposal of millings, and all other necessary work.

302-2 CHIP SEAL.

302-2.1 General. A chip seal shall consist of an application of emulsified asphalt and screenings to an existing roadway surface. Streets shall be chip sealed to the edge of the existing pavement, excluding inverted shoulders except when shown on the Plans. Manholes, valves, survey monuments, or miscellaneous frames and covers shall not receive a chip seal. If any frame and cover is sealed, it shall be cleaned to the satisfaction of the Engineer.

Chip seals are defined by type, which identifies the size of aggregate and the rate of application of emulsified asphalt and screenings. Unless otherwise specified, a medium type chip seal shall be applied.

TABLE 302-2.1 (A)

Chip Seal Types	Size of Screenings
Fine	1/4" x No. 10 (6.3 x 2.00mm)
Medium Fine	5/16" x No. 8 (8.0 x 2.36mm)
Medium	3/8" x No. 6 (9.5 x 3.35mm)
Coarse	1/2" x No. 4 (12.5 x 4.75mm)
Double	
1st Application	1/2" x No. 4 (12.5 x 4.75mm)
2nd Application	1/4" x No. 10 (6.3 x 2.00mm)

302-2.2 Surface Preparation. The Contractor shall remove any raised pavement markers within the area to be sealed, and the pavement surface shall be prepared as required by the Plans or Specifications. Immediately before applying the emulsified asphalt, the surface shall be clean and dry. Cleaning shall be performed by sweeping, flushing, or other means necessary to remove all loose particles of paving, all dirt, and all other extraneous material.

302-2.3 Application and Spreading. Chip seals shall only be applied when:

a) Atmospheric temperature is above 60°F (16°C) and pavement temperature is above 55°F (13°C).

b) The pavement is clean and dry.

c) Wind conditions are such that a uniform emulsion coverage can be achieved.

d) Rain is not imminent.

e) Sufficient screenings shall be on hand before the emulsified asphalt is applied.

Application of emulsified asphalt shall be stopped before the distributor tank is empty to assure designated application rate.

Stockpiling of screenings prior to placing will be permitted; however, any contamination resulting during storage or from reloading operations will be cause for rejection. Stockpiling of screenings within the right-of-way will require the approval of the Engineer.

The emulsified asphalt shall be covered with screenings before setting or "breaking" of the emulsion occurs. Operating the chip spreader at speeds which cause the chips to roll over after striking the bituminous covered surface will not be permitted. The speed of the distributor shall be governed by the speed of the chip spreader.

Screenings shall be spread by means of a self-propelled chip spreader, equipped with a mechanical device which will spread the screenings at a uniform rate over the full width of a traffic lane in one application.

The spreader shall be capable of reducing the aggregate spreading width to accommodate variable widths.

Trucks for hauling screenings shall be tailgate discharge and shall be equipped with a device to lock onto the hitch at the rear of the aggregate spreader. Haul trucks shall also be compatible with the aggregate spreader so that the dump bed will not push down on the spreader when fully raised or have too short an apron resulting in aggregate spillage while dumping into the receiving hopper. Screenings shall be surface damp at the time of application, but excess water on the aggregate surface will not be permitted. Screenings shall be redampened in the vehicles prior to delivery to the spreader when directed by the Engineer.

All join edges shall be sweep clean of excess screenings prior to the adjacent application of emulsified asphalt material. Precautions shall be taken to avoid "skips" and "overlaps" at joins and to protect the surfaces of adjacent structures from being spattered or marred. Correction of any such defects shall be performed at the Contractor's expense. All transverse joins shall be made by placing building paper over the ends of the previous applications, and the joining application shall start on the building paper. The paper shall be removed and disposed of as required by 7-8.1. The longitudinal join between adjacent applications of screenings shall coincide with the line between designated traffic lanes.

Application of emulsified asphalt shall be discontinued sufficiently early in the day to permit the termination of traffic control prior to darkness.

302-2.4 Emulsified Asphalt. Emulsified asphalt shall be Type CRS-2 per 203-3 and shall be applied by distributing equipment per 203-2.5 and 203-3.

The application rate of emulsified asphalt shall be within the following ranges. The exact rates will be determined by the Engineer.

TABLE 302-2.4 (A)

Chip Seal Types	Application Rate gal/yd^2 (L/m^2)
Fine	0.15 to 0.30 (0.68 to 1.36)
Medium Fine	0.20 to 0.35 (0.91 to 1.58)
Medium	0.30 to 0.40 (1.36 to 1.81)
Coarse	0.30 to 0.40 (1.36 to 1.81)
Double	
1st Application	0.20 to 0.35 (0.91 to 1.58)
2nd Application	0.20 to 0.30 (0.91 to 1.36)

The distribution of emulsified asphalt shall not vary more than 15 percent transversely, nor more than 10 percent longitudinally from the specified rate of application as determined by California Test 339.

302-2.5 Screenings. Screenings shall conform to 200-1.2.1. The spread rate of screenings for the various types of chip seals shall be within the following ranges. The exact rate will be determined by the Engineer.

TABLE 302-2.5 (A)

Chip Seal Types	Spread Rate lbs/yd^2 (kg/m^2)
Fine	12 to 20 (6.5 to 10.9)
Medium Fine	16 to 25 (8.7 to 13.6)
Medium	20 to 30 (10.9 to 16.3)
Coarse	23 to 35 (12.5 to 19.0)
Double	
1st Application	23 to 30 (12.5 to 16.3)
2nd Application	12 to 20 (6.5 to 10.9)

302-2.6 Finishing. After the screenings have been spread, any piles, ridges, or uneven distribution shall be carefully removed to ensure against permanent ridges, bumps or depressions in the completed surface. Additional screenings shall be spread in whatever quantities may be required to prevent picking up by the rollers or traffic.

Initial rollings shall commence immediately following spread of screenings. The compaction of screenings shall be accomplished by a minimum of three self-propelled, pneumatic-tired rollers meeting the requirements of 302-5.6.1, except that tires shall be inflated to 100 psi (690 kPa) and the operating weight shall be 5,000 pounds (2300 kg) per tire. The rolling equipment shall maintain a distance of not more than 200 feet (60m) behind the chip spreader on the first pass. The rollers shall operate at a maximum speed of 5 miles per hour (8km/hr). There shall be at least three complete coverages by the pneumatic-tired rollers (one initial and two secondary) to embed particles firmly into the emulsified asphalt.

After completion of pneumatic-tired rolling and the emulsified asphalt has broken, traffic will be permitted to travel over the chip seal if necessary. Guide vehicles shall be used to limit traffic speeds on the chip seal to 15 mph (25 km/hr) for a period of 2 to 4 hours as determined by the Engineer.

Sweeping shall be a multi-step operation following placement of the screenings. A power broom shall be used to remove loose material without dislodging aggregate set in the asphalt. The initial sweeping shall be a light brooming performed before the roadway may be opened to unguided traffic but no sooner than 3 hours following placement, with a second sweeping completed the day following placement. Additional sweeping shall be performed as necessary during the 5-day period following placement of the chip seal. If because of high temperatures or other causes there is dislodgement of cover aggregate, sweeping shall be discontinued until such time as there will be satisfactory retention of cover aggregate. Final sweeping shall be done and all loose aggregate shall be removed prior to acceptance.

Within 24 hours after placement, the Contractor shall also start removing loose aggregate from parkways, sidewalks, and intersecting streets. Both operations shall continue until all excess or loose aggregate is removed from the roadway surface and abutting right-of-way. The aggregate shall be disposed of by the Contractor at no cost to the Agency.

The sweeping operations shall be accomplished without the use of gutter brooms or steel-tined brooms unless approved by the Engineer.

Blotter material, either fine screenings or concrete sand, may be required immediately after the initial pass of the rollers or after sweeping and opening to traffic so as to prevent bleeding and pickup of the aggregate.

The completed surface shall present a uniform appearance and shall be thoroughly compacted, and free from ruts, humps, depressions, or irregularities due to an uneven distribution of emulsified asphalt or screenings.

302-2.7 Weighmaster Certificates. The Contractor shall supply the Engineer with licensed weighmaster's certificates of weight for all materials delivered to the Work during the course of each day. Materials so certified as being delivered for use in the Contract shall be used only in the chip seal. The Contractor shall also present weighmaster's certificates for the amount of such materials remaining unused at the completion of the work at no cost to the Agency. Payment will be determined by deducting the amount of unused materials from the total amount of materials delivered, all as shown on the licensed weighmaster's certificates. The certificates shall be presented to the Engineer at the time of delivery.

302-2.8 Measurement and Payment. Emulsified asphalt and screenings will be paid for at the Contract Unit Price per square yard (square meter). Such price shall include full compensation for specified surface preparation, removals, sweeping, and sanding if necessary, and for doing all the work involved in constructing the chip seal complete in place.

302-3 (Not Used).

302-4 SLURRY SEAL SURFACING.

302-4.1 General. Slurry seal surfacing shall consist of mixing, spreading, and application of slurry seal conforming to 203-5 at the Work site. The combined aggregate gradation and emulsified asphalt grade shall be as shown on the Plans or specified in the Special Provisions.

302-4.2 Aggregate Stockpile. The Contractor shall make arrangements for and provide an aggregate stockpile site a minimum of 3 working days prior to starting the Work. The stockpile site shall be clean and free from objectionable material.

302-4.3 Continuous-Flow Mixers.

302-4.3.1 General. Slurry seal mixed at the Work site shall be mixed in hydraulic or mechanical continuous flow mixers (mixers). The Contractor shall have a minimum of 2 fully-operational mixers available for use at the Work site at all times.

Mixers shall be multi-blade or spiral continuous flow units capable of accurately proportioning and mixing aggregate, water, emulsion, portland cement and set control agents, and discharging the mixture to a spreader box on a continuous basis. Mixers shall be equipped with a counter to count the revolutions of the aggregate belt head pulley, a metering device capable of measuring the quantity of water in gallons (liters), and a separate metering device capable of measuring the quantity of emulsified asphalt used to produce each load of slurry seal.

Each emulsion pump shall have its own identification number as determined by the Contractor. Variable rate emulsion pumps shall not be used. The identification number shall be affixed in a location visible to the Engineer.

302-4.3.2 Hydraulic Continuous-Flow Mixers. The drive shaft of the hydraulic motor shall be connected directly to the emulsion pump and have the same or a parallel shaft connected to the hydraulic motor that drives the aggregate belt head pulley.

302-4.3.3 Mechanical Continuous-Flow Mixers. The emulsion pump and aggregate belt head pulley shall be driven off the same drive shaft.

302-4.3.4 Calibration.

302-4.3.4.1 General. Each continuous-flow mixer to be used in the Work shall be calibrated annually by the Contractor. The calibration procedure shall conform to the ISSA "Inspector's Manual MA-1" except as modified in 302-4.3.4.2. A Registered Civil Engineer in the State of California shall sign and seal the calibration report.

The calibration shall be valid for one year from the date the procedure was performed.

Re-calibration will be required if:

a) the emulsion pump or the sprocket drive ratio is changed, and/or

b) the aggregate source is changed and the loose unit weight, as determined by 302-4.3.4.2, varies by more than 5 percent from the loose unit weight determined at the time of calibration of the mixer.

Slurry seal surfacing work shall not begin until the calibration report has been reviewed and accepted by the Engineer.

302-4.3.4.2 Procedure. The ISSA procedure shall be modified to require a minimum sample size of 6,000 pounds (2,700 kilograms) of aggregate for each gate opening and each test run. The calibration of the emulsion pump shall require a minimum of 25 revolutions of the aggregate belt. The difference in the weight of aggregate delivered shall not vary by more than 2 percent from the average of 3 test runs. The number of revolutions of the head pulley shall be the same for each test run for the gate openings being tested. The emulsion calibration need only be verified for one gate setting. The emulsion delivery rate shall not vary by more than 2 percent from the average of 3 test runs. The Contractor may elect to use larger samples of aggregate or emulsion.

The Contractor shall reproduce and complete the "Emulsion Calibration Worksheet" in the Appendix of the ISSA "Inspector's Manual MA-1." In addition to the information required on the worksheet, the Contractor shall also provide the following:

a) the type and grade of emulsified asphalt,

b) the aggregate source and type,

c) the proposed gate setting based upon the approved mix design,

d) the date of calibration, and

e) the loose unit weight of aggregate determined in accordance with ASTM C29 with a minimum 0.1 ft^3 (3 L) bucket.

The Contractor shall also include the following information on all worksheets and graphs:

f) the identification number for the sprocket and emulsion pump, and

g) the model of the vehicle and type of mixer used for the calibration.

The graph showing pounds (kilograms) of aggregate vs. gate openings inches (millimeters) shall also be signed and dated by the Registered Civil Engineer that signed and sealed the calibration report. A copy of the gate opening curve and calibration data shall be kept with each mixer for review by the Engineer.

302-4.4 Verification Testing. The Contractor shall adjust each continuous-flow mixer to be used in the Work to produce slurry seal conforming to the approved mix design(s) required in 203-5.2. The Contractor shall allow 2 working days prior to the start of the Work for verification testing of each continuous flow and plant mixer. The Contractor shall provide field samples at the time of verification testing for extraction tests (ASTM D6307), consistency tests, and modified wet track abrasion tests (ASTM D3910). When the field samples conform to the Specifications, the Engineer will notify the Contractor it may start the Work. If not, additional tests shall be performed at the Contractor's expense until an acceptable mix is produced.

302-4.5 Scheduling, Public Convenience and Traffic Control. In addition to the requirements of 7-10, the Contractor shall comply with the following:

At least 5 working days prior to commencing the Work, the Contractor shall submit its proposed schedule to the Engineer for approval. Based upon the approved schedule, the Contractor shall, at least 48 hours in advance, notify residents and businesses of the Work and post temporary "No Parking" signs. Requests for changes in the approved schedule shall be submitted to the Engineer for approval at least 3 working days before the street is scheduled to be sealed.

Streets where slurry seal is scheduled to be applied shall be closed from the time the Contractor begins to clean the street surface until the Engineer determines the slurry seal has achieved sufficient set to be opened to traffic.

Rock dust or sand shall be spread, as directed by the Engineer, to eliminate tracking or damage and to provide vehicular or pedestrian crossings. Rock dust or sand used for this purpose shall conform to 200-1.2 and 200-1.5, respectively.

302-4.6 Emulsion-Aggregate Slurry (EAS).

302-4.6.1 General. EAS shall conform to 203-5.4. Field adjustments to the set control agents may be made in accordance with the approved mix design. The amount of set control agents to be included shall be that amount necessary to ensure that quick-set EAS can support vehicular traffic within 60 minutes after the completion of application.

302-4.6.2 Mixing. Mixing shall be performed on the Work site by the use of continuous flow mixers conforming to 302-4.3.

302-4.6.3 Application Temperature. EAS shall not be applied if either the pavement or the ambient temperature is less than 50^0F (10^0C) and falling, but may be applied when the pavement and ambient temperatures are both above 45^0F (7^0C) and rising.

302-4.6.4 Aggregate Application Rate.

302-4.6.4.1 General. The aggregate application rate shall conform to the requirements shown in Table 302-4.6.4.1 (A) unless otherwise specified in the Special Provisions or shown on the Plans.

TABLE 302-4.6.4.1 (A)

Aggregate Type	Aggregate Application Rate	
	MINIMUM	MAXIMUM
Type I	8 lbs/yd^2 (4.3 kg/m^2)	10 lbs/yd^2 (5.4kg/m^2)
Type II	12 lbs/yd^2 (6.5 kg/m^2)	15 lbs/yd^2 (8.1kg/m^2)
Type III	20 lbs/yd^2 (10.8 kg/m^2)	25 lbs/yd^2 (13.5 kg/m^2)

302-4.6.4.2 Corrective Action. When the aggregate application rate is less than the minimum shown in Table 302-4.6.4.1 (A), the Contractor shall apply additional EAS to the nonconforming areas as necessary to conform to the Specifications.

When the aggregate application rate exceeds the maximum shown in Table 302-4.6.4.1(A), the nonconforming material shall be removed and replaced, or be left in place at no additional cost to the Agency, as determined by the Engineer.

302-4.7 Rubberized Emulsion Aggregate Slurry (REAS).

302-4.7.1 General. REAS shall conform to 203-5.5.

302-4.7.2 Mixing. REAS shall be mixed by one of the following methods as specified in the Special Provisions:

a) in a continuous flow mixer conforming to 302-4.3 at the Work site, or

b) at a central mixing plant conforming to 203-5.5.3.

302-4.7.3 Transporting. Transporting of REAS mixed at a central mixing plant shall conform to 203-5.5.3.2.

302-4.7.4 Work Site Storage. REAS mixed at a central mixing plant may be stored at the Work site in tanks specifically designed for this purpose and which are equipped with an agitator similar to that in a central mixing plant. The agitator shall be capable of continuous operation.

302-4.7.5 Application Temperature. REAS shall not be applied if either the pavement or the ambient temperature is less than 55^{0}F (13^{0}C) and falling, but may be applied when the pavement and ambient temperatures are both above 50^{0}F (10^{0}C) and rising.

302-4.7.6 Application Rate.

302-4.7.6.1 Continuous-Flow Mixer. The aggregate application rate of REAS mixed on the Work site in a continuous-flow mixer shall conform to the requirements shown in Table 302-4.7.6.1 (A) unless otherwise specified in the Special Provisions or shown on the Plans.

TABLE 302-4.7.6.1 (A)

Type	Aggregate Application Rate	
	Minimum	Maximum
Type Fine REAS	2.8 lbs/yd^2 (1.5 kg/m^2)	3.4 lbs/yd^2 (1.8 kg/m^2)
Type I REAS	4.4 lbs/yd^2 (2.4 kg/m^2)	5.2 lbs/yd^2 (2.8 kg/m^2)
Type II REAS	7.5 lbs/yd^2 (4.1 kg/m^2)	10.0 lbs/yd^2 (5.4 kg/m^2)
Type III REAS	15.7 lbs/yd^2 (8.5 kg/m^2)	21.8 lbs/yd^2 (14.7 kg/m^2)

302-4.7.6.2 Central Mixing Plant. The application rate of REAS mixed at a central mixing plant shall conform to the requirements shown in Table 302-4.7.6.2 (A) unless otherwise specified in the Special Provisions or shown on the Plans.

TABLE 302-4.7.6.2 (A)

Type	Application Rate (REAS)	
	Minimum	Maximum
Type Fine REAS	5.0 lbs/yd^2 (2.7 kg/m^2)	6.0 lbs/yd^2 (3.2 kg/m^2)
Type I REAS	7.1 lbs/yd^2 (3.8 kg/m^2)	8.4 lbs/yd^2 (4.5 kg/m^2)
Type II REAS	11.3 lbs/yd^2 (6.1 kg/m^2)	15.0 lbs/yd^2 (8.1 kg/m^2)
Type III REAS	22.5 lbs/yd^2 (12.2kg/ m^2)	28.1 lbs/yd^2 (15.2kg/ m^2)

302-4.7.6.3 Corrective Action. When the application rate is less than the minimum shown in the tables above, the Contractor shall apply additional REAS to the nonconforming areas as necessary to conform to the Specifications.

When the application rate exceeds the maximum shown in the tables above, the nonconforming material shall be removed and replaced, or be left in place at no additional cost to the Agency, as determined by the Engineer.

302-4.8 Spreading and Application. Prior to spreading, the Contractor shall clean the existing pavement unless otherwise specified in the Special Provisions. Immediately ahead of the spreader truck, the existing pavement shall be pre-wetted by a pressurized water distribution system equipped with a fog-type spray bar capable of completely covering the surface of the pavement.

Slurry seal mixed at the Work site shall be spread by a spreader box attached to a continuous-flow mixer truck conforming to 302-4.3. REAS produced at a central mixing plant shall be spread by a spreader box attached to an agitator truck conforming 203-5.5.3.2.

The spreader box shall be equipped with flexible material in continuous contact with the existing pavement and shall be capable of controlling the rate of application. The spreader box shall have adjustable width and strike-off height, and be capable of controlling and providing uniform spreading.

The maximum speed of the spreader truck shall not exceed 270 feet per minute (80 meters per minute).

REAS mixed at a central mixing plant shall be continuously agitated during spreading.

Hand squeegees and other equipment shall be provided for spreading and spillage removal in areas inaccessible to the spreader box.

Slurry seal shall be applied in such a manner that no ridges remain. Areas in which there is evidence of solidification of the emulsified asphalt, balling or lumping of the aggregates, or uncoated aggregates shall be removed and replaced to the satisfaction of the Engineer.

The Contractor shall prevent slurry seal from being deposited on other than asphalt concrete surfaces and shall remove it from surfaces not designated to be sealed. The method of removal shall be approved by the Engineer.

Where the completed slurry seal surfacing is not uniform in color, the street shall be treated by a method approved by the Engineer to eliminate the color variation.

302-4.9 Field Sampling and Testing.

302-4.9.1 Field Sampling. During the performance of the Work, the Contractor shall provide the Engineer with at least 2 field samples, from separate loads, of mixed slurry seal per mixer per day. WTAT specimens shall be cast and struck off within 60 seconds of obtaining the sample. WTAT specimens shall not be transported until the slurry seal has set as defined by ASTM D3910.

Field samples shall conform to the requirements shown in Table 302-4.9.1 (A):

TABLE 302-4.9.1 (A)

Tests	ASTM Test Method	Requirements	
		Min.	Max.
Wet Track Abrasion Test, Weight loss, gm/ft^2 (gm/m^2) Type Fine Aggregate	D3910[1]	0	50 (540)
Wet Track Abrasion Test, Weight loss, gm/ft^2 (gm/m^2) Type I Aggregate	D3910[1]	0	50 (540)
Wet Track Abrasion Test, Weight loss, gm/ft^2 (gm/m^2) Type II Aggregate	D3910[1]	0	60 (650)
Wet Track Abrasion Test, Weight loss, gm/ft^2 (gm/m^2) Type III Aggregate[4]	D3910[1]	0	60 (650)
Consistency Test (mm)	D3910[1]	20	40
Extraction Test (Calculated Emulsion Content, %)	D6307[2], CT 382[2]	± 1 % of mix design for EAS. ± 3 % of mix design for REAS	
Water Content (% of Dry Slurry)	See Note 3	Type I, II and III EAS < 25 Type Fine and I REAS < 40 Type II and III REAS < 31	

1. Modified ASTM D3910 to include No. 4 (4.75mm) aggregate or greater and to be performed using field samples. Subsection 6.4.4.7, ASTM D3910 may be modified to use a microwave oven for drying the specimen after the abrasion cycle is complete and the debris washed off.
2. Modified ASTM D6307 and California Test Method 382 to allow a minimum of 500 ±50 gram sample.
3. Weigh a minimum of 500 grams of homogenized mixed slurry into a previously tared quart can with a friction lid. The lid shall be placed on the can to prevent loss of material during transportation. Place the can with the lid off in an oven and dry to constant mass at 220^{0}F± 10^{0}F (110^0 C ± 5^{0}C).
4. The 3/8 inch (9.5mm) template shall be used.

If the test results fail to meet the Specifications, the Contractor shall cease spreading slurry seal produced by the nonconforming mixer until the Contractor demonstrates the mixer is producing slurry seal which conforms to the Specifications.

302-4.10 Measurement.

302-4.10.1 General. The basis of measurement shall be the weight of materials, in tons (Tonnes), used in the Work, as determined by licensed weighmaster's certificates. Upon completion of the Work, the Contractor shall submit to the Engineer licensed weighmaster's certificates for materials delivered to the Work site and for excess materials not incorporated into the Work.

302-4.10.2 Slurry Seal Mixed in Continuous-Flow Mixers. Slurry seal mixed in continuous-flow mixers shall be measured by each ton (Tonne) of emulsified asphalt and each ton (Tonne) of each type of aggregate used in the Work.

302-4.10.3 REAS Mixed at a Central Mixing Plant. REAS mixed in a central mixing plant shall be measured by each ton (Tonne) used in the Work, including aggregate, RPME, additives and water.

302-4.11 Payment.

302-4.11.1 Payment Reduction for Noncompliance.

302-4.11.1.1 General. Payment to the Contractor will be reduced for failure of the field test samples to conform to the WTAT requirements specified in 302-4.9.1.

302-4.11.1.2 Reduction in Payment Based on WTAT. If the average of all WTATs performed per mixer, per day, fails to conform to the requirements specified in 302-4.10.1, the Contractor agrees that payment for the Work represented by the failed tests shall be reduced as shown in Table 302-4. 11.1.2 (A) or (B).

TABLE 302-4.11.1.2 (A)

WTAT Loss gm/ft^2 (gm/m^2)	Payment Reduction (Percent) Type Fine & I Aggregate
0 – 50 (0 – 540)	0
50.1 – 60 (540.1 – 650)	5
60.1 – 70 (650.1 – 750)	15
70.1 – 80 (750.1 – 860)	30
80.1 – 99 (860.1 – 1070)	70
99.1 or greater (1070.1 or greater[1])	100

1. Slurry seal surfacing with WTAT loss greater than 99.1 gm/m^2 (1070.1 gm/ft^2) shall be removed to the satisfaction of the Engineer.

TABLE 302-4.11.1.2 (B)

WTAT Loss gm/ft^2 (gm/m^2)	Payment Reduction (Percent) Type II and III Aggregate
0 – 60 (0 – 540)	0
60.1 – 75 (650.1 – 810)	15
75.1 – 80 (810.1 – 860)	30
80.1 – 99 (860.1 – 1070)	70
99.1 or greater (1070.1 or greater[1])	100

1. Slurry seal surfacing with WTAT loss greater than 99.1 gm/m^2 (1070.1 gm/ft^2) shall be removed to the satisfaction of the Engineer.

302-4.11.2 Slurry Seal Mixed in Continuous-Flow Mixers. Payment for slurry seal surfacing for slurry seal mixed in continuous-flow mixers shall be at the Contract Unit Price per ton (Tonne) for emulsified asphalt and the Contract Unit Price per ton (Tonne) for each type of aggregate. No separate payment will be made for calibration, scheduling, public convenience, or traffic control unless otherwise specified in the Special Provisions.

302-4.11.3 REAS Mixed at a Central Mixing Plant. Payment for slurry seal surfacing for REAS mixed in a central mixing plant shall be at the Contract Unit Price per ton (Tonne). No separate payment will be made for calibration, scheduling, public convenience, or traffic control unless otherwise specified in the Special Provisions.

302-5 ASPHALT CONCRETE PAVEMENT.

302-5.1 General. Asphalt concrete pavement shall consist of one or more courses of a mixture of paving asphalt and graded aggregate as specified in 203-6, placed upon a prepared roadbed or base, or over existing pavement. The courses shall be of the type of mixture and the dimensions shown on the Plans or Specifications.

302-5.2 (Not Used).

302-5.3 Prime Coat. When specified, a prime coat consisting of Grade SC-250 liquid asphalt shall be applied at a rate between 0.10 and 0.25 gallon per square yard (0.45 L/m^2 and 1.15 L/m^2). Grade SC-70 liquid asphalt may be used when approved by the Engineer.

302-5.4 Tack Coat. If the asphalt concrete pavement is being constructed directly upon an existing hard-surfaced pavement, a tack coat of PG 64-10 paving asphalt at an approximate rate of 0.05 gallon per square yard (0.25 L/m^2) or SS-1h emulsified asphalt at an approximate rate of 0.05 to 0.10 gallon per square yard (0.25 L/m^2 to 0.45 L/m^2) shall be uniformly applied upon the existing pavement preceding the placement of the asphalt concrete. The surface shall be free of water, foreign material, or dust when the tack coat is applied. To minimize public inconvenience, no greater area shall be treated in any one day than is planned to be covered by asphalt concrete during the same day, unless otherwise approved by the Engineer.

A similar tack coat shall be applied to the surface of any course, if the surface is such that a satisfactory bond cannot be obtained between it and a succeeding course.

The contact surfaces of all cold pavement joints, curbs, gutters, manholes, and the like shall be painted with either SS-1h emulsified asphalt or PG 64-10 paving asphalt immediately before the adjoining asphalt concrete is placed.

302-5.5 Distribution and Spreading. The Contractor shall provide and install a header upon the line of termination of asphalt pavement where shown on the Plans or required by the Specifications. Such headers shall remain in place upon completion of the improvements.

Headers shall be 2-inch (50mm) nominal size lumber, the vertical dimension of which shall be within 1/2 inch (15mm) of the thickness of the pavement at the header line. The minimum vertical dimension of a header shall be 3-1/2 inches (88mm). The headers shall have a firm bearing on the header subgrade and the top edges shall be set to conform to the grade of the proposed street surface. Side stakes 2 inches by 3 inches (50mm x 75mm) nominal size, 18 inches (450mm) long, or longer, and spaced not over 4 feet (1.2m) apart, shall be driven on the outside of the headers to a depth of 1 inch (25mm) below the top edge and then nailed to the header. The joints between the individual boards being used as headers shall be spliced with a 1 inch (25mm) thick nominal size board of the same height as the header and not less than 24 inches (600mm) long.

The temperature of the mixture directly behind the paving machine, before the breakdown roller, shall not be lower than 270°F (132°C) or higher than 320°F (160°C), the lower limit to be approached in warm weather and the higher in cold weather.

Asphalt concrete shall not be placed unless the atmospheric temperature is at least 50°F (10°C) and rising or during unsuitable weather.

The asphalt concrete shall be evenly spread upon the subgrade or base to such a depth that, after rolling, it will be of the specified cross section and grade of the course being constructed.

The depositing, distributing, and spreading of the asphalt concrete shall be accomplished in a single, continuous operation by means of a self-propelled mechanical spreading and finishing machine designed specially for that purpose. The machine shall be equipped with a suitable full-width compacting screed capable of being accurately regulated and adjusted to distribute a layer of the material to a definite predetermined thickness. When paving is of a size or in a location that use of a self-propelled machine is impractical the Engineer may waive the self-propelled requirements.

The asphalt concrete as delivered shall be deposited directly into the hopper of the spreading and finishing machine. With the approval of the Engineer, the Contractor may deposit the asphalt concrete material from bottom-dump trucks into a uniformly sized windrow, then pick up the material and convey it to the spreading machine with loading equipment provided:

a) The spreading machine shall be of such design that the material will fall into a hopper, having a movable bottom conveyor or screw to feed the screed.

b) The loader (pickup machine) shall be constructed and operated so that substantially all of the material deposited on the roadbed is picked up and deposited in the spreading machine.

c) The windrow may be deposited no more than two truck loads ahead of the pickup machine and the temperature of the material in the windrow shall not fall below 270°F (132°C).

d) Allowable windrow lengths may be increased if approved by the Engineer, but the windrow shall not block intersections prior to the arrival of the paving machine.

e) A skip loader shall be on site to remove any windrowed material not meeting temperature requirement or which will overload the paving machine hopper.

Asphalt concrete of the class indicated in the following table shall be laid in courses not exceeding 4 inches (100mm) in thickness unless otherwise approved by the Engineer.

TABLE 302-5.5 (A)

Specified Total Thickness of Pavement		Minimum Number of Courses	Class of Mixture
Greater Than inches (mm)	But Not More Than inches (mm)		
0	1 (25)	1	D1 or D2
1 (25)	1-1/2 (38)	1	C1, C2, D1 or D2
1-1/2 (38)	3 (75)	1	C1, C2, B or A (as directed)
3 (75)	4 (100)	1	B or A (as directed)
4 (100)	5 (125)	2	C1, C2, B or A (as directed)
5 (125)	—	2 or as directed	C1, C2, B or A (as directed)

Spreading, once commenced, must be continued without interruption. No greater amount of the mixture shall be delivered in any one day than can be properly distributed and rolled during that day.

Successive courses may be laid upon previously laid courses as soon as the previous course has cooled sufficiently to show no displacement under equipment or loaded material delivery trucks.

The asphalt concrete surface of an alley shall be warped up to meet paved driveways which are 6 inches (150mm) or less above grade. Such warping shall not extend more than 18 inches (450mm) into the alley and shall be accomplished by thickening the pavement.

302-5.6 Rolling.

302-5.6.1 General. Asphalt concrete shall be thoroughly compacted by rolling. The number of rollers used with each paving operation shall not be less than specified below. Each roller shall have a separate operator.

TABLE 302-5.6.1 (A)

Rollers Required[1]		
Tons (Tonnes) Placed per Hour	Compacted Thickness	
	1-1/2" (38mm) or less	More than 1-1/2" (38mm)
Less than 100 (90)	1	1
100 to 200 (90 to 180)	2	2
201 to 300 (181 to 270)	3	2
300 (More than 270)	3	3

1. Additional rollers may be necessary to meet the requirements herein.

Self-propelled compacting rollers shall meet the following criteria:

a) Each roller manufactured after 1998 shall have a Manufacturer's identification plate that is readily accessible and readable with the following information:

1) Name of Manufacturer.

2) Model Number.

3) Static pounds per lineal inch (PLI) (newton per millimeter (N/mm)) of each drum.

4) Static PLI (N/mm) of ballasted drum.

5) PLI (N/mm) of each drum in vibratory mode.

Contractors using rollers manufactured prior to 1999 shall have the manufacturer's specifications, providing the information requested above, available to the Engineer upon request. Any roller not having this information shall not be used and shall be removed from the jobsite.

b) Tandem rollers in the static mode used for breakdown or intermediate rolling shall be such that the ballasted or unballasted weight on at least one drum is a minimum 250 PLI (44 N/mm).

c) Vibratory rollers used for breakdown or intermediate rolling shall have a compactive effort of not less than 250 PLI (44 N/mm) of centrifugal force at the setting indicated by the manufacturer's ID plate.

d) Finish rolling shall be performed by static or vibratory steel rollers in static mode

e) Pneumatic-tired rollers used for intermediate rolling shall be the oscillating type having a width of not less than 4 feet (1.2m) and equipped with pneumatic-tires of equal size and diameter, having treads satisfactory to the Engineer. Wobble-wheel rollers will not be permitted. The tires shall be so spaced that the gap between adjacent tires will be covered by the tread of the following tire. The tires shall be inflated to 90 psi (620 kPa) or such lower pressure as designated by the Engineer, and maintained so that the air pressure will vary not more 5 psi (35 kPa) from the designated pressure. Pneumatic-tired rollers shall be so constructed that the total mass of the roller can be varied to produce an operating mass per tire of not less than 2,000 pounds (900 kg). The total operating mass of the roller shall be varied as directed by the Engineer.

For areas to be paved that are not to be subjected to vehicular traffic and when the asphalt is placed in these areas at a rate less than 100 tons (90 tonnes) per hour the roller shall have a minimum compactive force of 150 PLI (26 N/mm).

Other rollers may be used subject to prior approval by the Engineer.

As soon as the layer of asphalt concrete has been placed the breakdown rolling shall commence. The layer of asphalt shall have an internal temperature greater than 260ºF (126ºC) and not displace under the roller. Except when compacting lifts greater than 4 inches (100mm) in compacted thickness, rolling shall be commenced along the lower edge of the area to be rolled and continued until the edge is thoroughly compacted, after which the roller shall be gradually advanced to the crown point, both sides being rolled in like manner. When vibratory rollers are used, they shall be operated at the highest frequency possible without breaking rock. All breakdown and intermediate rolling shall be completed prior to the surface of the mat reaching 180°F (82°C). Rolling shall be continued until the pavement layer has become thoroughly compacted throughout and is true to grade and cross section. The finish rolling shall remove all marks from the surface of the mat.

For lifts greater than 4 inches (100mm) in compacted thickness, rolling shall be commenced in the middle of the mat, after which the roller shall be gradually advanced to both edges. The roller should be advanced to a supported edge first, if applicable. Rolling of an unsupported edge may be delayed provided the required densities are obtained after the completion of the finishing rolling.

All rollers must be maintained in good mechanical condition. Those that cannot be driven along a straight path, operated without jerking, or the amplitude or frequency cannot be adjusted shall not be used and shall be removed from the Work site. No leakage of petroleum products from any roller shall be allowed to come in contact with pavement being constructed, nor shall any roller be permitted to stand motionless on any portion of the work. The surfaces of all roller wheels shall be treated with sufficient water to prevent the pickup of bituminous materials, but under no circumstances shall the quantity of water used be detrimental to the surface of the pavement being rolled.

302-5.6.2 Density and Smoothness. Asphalt concrete pavement shall be true to grade and cross section. When a 10-foot (3m) straightedge is laid on the finished surface parallel to the centerline of the roadway, the finished surface shall not vary from the edge of the straightedge more than 1/8 inch (3mm), except at intersections or at changes of grade. Any areas that are not within this tolerance shall be brought to grade immediately following the initial rolling. If the paving material has cooled below the lower limits of the spreading temperatures prescribed in 302-5.5 or 302-9.4, the surface of the pavement shall be brought to a true grade cross section. The paving material in the area to be repaired shall be removed, by cold milling, to provide a minimum laying depth of 1 inch (25mm), or 2 times the maximum aggregate size, whichever is greater, of the new pavement at the join line. Repairs shall not be made to the pavement surface by tapering the thickness at the join lines.

The compaction after rolling shall be a minimum of 95 percent of the density obtained on samples compacted with the California Kneading Compactor per California Test 304. The density shall be determined in accordance with California Test 308, Method A. Method C may be used if the absorption of the compacted specimen is less than 2 percent.

The field density of compacted asphalt concrete shall be determined by:

a) A properly calibrated nuclear asphalt testing device in the field, or

b) California Test 308, Method A when slabs or cores are taken for laboratory testing.

In case of dispute, method b) (above) shall be used.

Paved areas not to be subject to vehicular traffic shall be compacted to 90 percent of the density determined as specified above.

302-5.7 Joints. Joints between successive runs shall be vertical and at right angles to the line of the improvement. Care shall be exercised in connection with the construction of all joints to ensure that the surface of the pavement is true to grade and cross section. Lapped joints will not be permitted.

When terminating paving operations for the day, the Contractor shall construct temporary hot-mix ramps at all vertical joints which are greater than 1-1/2 inches (38mm) in height and transverse to through traffic. Temporary hot-mix ramp dimensions and compaction shall be approved by the Engineer. Prior to resuming paving operations, the Contractor shall remove temporary hot-mix ramps to provide for a vertical face and a full depth lift joint and apply a tack coat to the faces of the joint in accordance with 302-5.4.

302-5.8 Manholes (and Other Structures). Sewer and storm drain structures extending 2 inches (50mm) or more above the new subgrade shall be removed by the Contractor to the new subgrade before paving. Other structures shall be lowered by the owners. Structures projecting less than 2 inches (50mm) above the subgrade may be paved over and later adjusted to grade. All debris and foreign material shall be removed per 301-1.6. The top of reset manholes and other structures shall meet the smoothness requirement as specified in 302-5.6.2.

All structures from which manhole frames and covers have been removed to facilitate paving shall be temporarily covered with a steel plate by the Contractor. When this procedure is impractical, such as for large vaults, special structures, or where portland cement concrete pavement is to be constructed, all remodeling or reconstruction shall be completed to finish permanent surface prior to paving operations. The Contractor shall notify utility owners, at least 2 working days in advance, of the need to commence work required prior to paving operations and again for work required after paving operations. If it is found to be impractical for the utility owner to complete the final remodeling or adjustment of structures within a reasonable time after paving operations, as evaluated by the Engineer, then the Contractor shall be absolved of further responsibility in connection therewith, and the structure shall be adjusted to grade by the utility owner under permit or ordinance procedure established by the Agency for utility cuts in pavement.

After the pavement has been completed, the necessary portions of the subgrade, base, and pavement shall be neatly removed, the structure built up, and the manhole frame set to be backfilled to within 1-1/2 inches (38mm) of the surface with portland cement concrete conforming to 302-6.1 by the party responsible for adjustment of the frame and cover. The Contractor shall fill the remaining 1-1/2 inches (38mm) with an asphalt concrete wearing surface mixture to match the project surface course. This material shall be placed and compacted in a workmanlike manner to conform to the appearance of the surrounding pavement.

302-5.9 Measurement and Payment. Asphalt concrete pavement will be paid for at the Contract Unit Price per square foot (square meter), or at the Contract Unit Price per ton (tonne) as shown in the Bid. Such price shall constitute full compensation for the preparation of subgrade and applying tack coat if required. Resetting, reconstructing, or adjusting manhole or vault frames and covers to grade will be paid for as provided in 301-1.7.

The asphalt concrete Bid item which utilizes the latex modifier shall be identified as latex modified asphalt concrete or asphalt concrete with latex. The Contract Unit Price paid for latex modified asphalt concrete or asphalt concrete with latex shall include all costs for furnishing, mixing, and placement.

When payment is to be made on a tonnage basis, the Contractor shall furnish to the Engineer at the time of delivery of the material to the Work site a legible copy of a licensed weighmaster's certificate showing gross, tare, and net weights of each truckload of asphalt concrete mixture. When an automatic batching system is used, the licensed weighmaster's certificate may show only the net weight of material in the truck load. Failure of the Contractor to provide a certificate to the Engineer by the end of the day on which the material represented by such certificate is delivered to the Work site may, at the discretion of the Engineer, result in the forfeiture of all payment for such material, including any labor and equipment costs included in the price for furnishing and placing the asphalt concrete.

Payment for installing headers, where required, will be made at the Contract Unit Price per linear foot (meter) for headers.

302-6 PORTLAND CEMENT CONCRETE PAVEMENT.

302-6.1 General. Unless otherwise specified, portland cement concrete pavement shall be constructed of concrete prepared as prescribed in 201-1.

Concrete pavement shall be removed in accordance with 300-1.3.

302-6.2 Forms and Headers.

302-6.2.1 General. Forms and headers shall be either wood or metal. They shall be set plumb and true to line and grade, with the upper edge thereof set to the grade of the pavement to be constructed; and shall be rigidly installed on a true alignment and so maintained for a distance in advance of placing the pavement to provide for at least a 1-day run of concrete. Headers shall rest firmly on the subgrade or base. They shall be oiled immediately prior to the placing of the concrete and shall remain in place for at least 12 hours after concrete has been placed. Forms and headers must be removed before the work will be accepted.

302-6.2.2 Wooden Forms. Wooden forms shall be constructed of 3-inch (75mm) nominal lumber in pieces not less than 16 feet (4.9m) long, except where changes in alignment or grade necessitate the use of material of smaller dimensions. The lumber used shall be free from warp and other imperfections which would impair the strength for the use intended; shall have square edges (which may be slightly beveled) and square ends; shall be surfaced on the upper edge; and shall be not more than 1/2 inch (13mm) less in depth than the specified thickness of the edge of the pavement.

Such forms shall be secured by nailing to side stakes spaced not more than 4 feet (1.2m) apart and driven into the subgrade vertically to a depth not less than 12 inches (300mm), and so that the tops will be below the upper edge of the header. The stakes shall be of sufficient length and cross-sectional area to adequately resist lateral displacement of the headers during the paving operations.

Wooden headers shall be spliced by nailing a board to the outside of the headers. The board shall be at least 4 feet (1.2m) long, 1 inch (25mm) thick, and at least 6 inches (150mm) wide (or the depth of the header, whichever is least), and shall be centered on the joint.

302-6.2.3 Metal Forms. Metal forms shall be free from warp, have sufficient rigidity to resist springing during the paving operations, and shall be not less in depth than the specified thickness of the edge of the pavement being constructed. They shall be secured by means of metal stakes spaced not more than 5 feet (1.5m) apart and driven below the top of the forms. They shall be designed so as to be driven through openings in the forms to lock them in position.

302-6.3 Placing Concrete.

302-6.3.1 General. Concrete shall be placed on a subgrade sufficiently dampened to ensure that no moisture will be absorbed from the fresh concrete.

Immediately after being mixed, the concrete shall be deposited on the subgrade to the required depth over the entire width of the section.

At the end of each day's run, or at any time when operations are stopped for a period of more than 40 minutes, a rigid transverse header shall be placed vertically and at a right angle across the improvement at the location designated by the Engineer and the pavement shall be finished to form a square vertical joint against which the work may be resumed. Hand mixing may be used only if necessary to provide sufficient concrete to complete paving to the expedient header.

302-6.3.2 Slip-Form Construction. At the option of the Contractor, and with the approval of the Engineer, concrete pavement may be constructed by the use of slip-form paving equipment.

Slip-form paving equipment shall be provided with traveling side forms of sufficient dimensions, shape, and strength to support the concrete laterally for a sufficient length of time during placement to produce pavement of the required cross section, and it shall spread, consolidate, screed, and float-finish the freshly placed concrete to provide a dense and homogeneous pavement.

The concrete shall be distributed uniformly into final position by the slip-form paver and the horizontal deviation in alignment of the edges shall not exceed 1-1/4 inches (32mm) from the alignment established by the Engineer.

The concrete, for the full paving width, shall be effectively consolidated by internal vibration, with transverse vibrating units, or with a series of longitudinal vibrating units. Internal vibration shall mean vibration by means of vibrating units loaded within the specified thickness of pavement section and at a minimum distance ahead of the screed equal to the pavement thickness.

When concrete is being placed adjacent to an existing pavement, that part of the equipment which is supported on the existing pavement shall be equipped with protective pads on crawler tracks or rubber-tired wheels, offset to run a sufficient distance from the edge of the pavement to avoid breaking or cracking the pavement edge.

After the concrete has been given a preliminary finish by finishing devices incorporated in the slip-form paving equipment, the surface of the fresh concrete shall be checked by the Contractor with a straightedge to the tolerances and finish required in 302-6.4.

Final finishing for slip-form pavement construction shall be as specified in 302-6.4.4.

302-6.4 Finishing.

302-6.4.1 General. The concrete shall be consolidated, and the surface finished true to grade and cross section. Upon completion, the surface shall be free of any unevenness greater than 1/8 inch (3mm) when checked with a 10-foot (3m) straightedge placed on the surface of the pavement. The 10-foot (3m) straightedge shall be furnished by the Contractor and shall be at the Work site prior to the commencing of the placing of the concrete.

302-6.4.2 Tamping. The concrete shall be distributed uniformly between the side forms as soon as it is placed, after which the concrete shall be struck off and tamped by means of a mechanical tamper. The tamper shall be operated at right angles to the centerline of the pavement, and tamping continued until the concrete is thoroughly consolidated to the specified cross section and sufficient mortar for finishing purposes has been brought to the surface.

Steel-shod hand tampers or vibrating bars may be substituted in those cases where the use of a mechanical spreader and tamper would be obviously impracticable.

Approved concrete vibrating equipment shall be used in conjunction with the mechanical tamper to consolidate the concrete adjacent to the forms or existing pavement.

302-6.4.3 Floating.

a) **General.** After tamping, the surface of the concrete shall be floated by either the finishing-machine method or the transverse-float method described below. Bridge decks may be floated by the longitudinal-float method.

b) **Finishing-Machine Method.** The concrete shall be floated smooth and true to grade with an approved finishing machine.

c) **Transverse-Float Method.** The concrete shall be floated at least twice with a long-handled float at least 5 feet (1.5m) wide, following which the surface of the concrete shall be finished smooth and true to grade, with a wooden float 16 feet (4.9m) long, 2 inches (50mm) thick, and 6 inches (150mm) wide. It shall be rigidly ribbed and with adjustable screws between the rib and float board to ensure a true and flat surface on the under side at all times. The float shall be operated from the side of the pavement, and parallel with the centerline.

The edge of the float shall be used to cut down all high areas, and the material so removed shall be floated into the depression until a true surface is obtained. Each successive pass of the float shall half-lap the previous pass.

The float shall be operated as far behind the tamping machine as the workability of the concrete will permit before its initial set.

d) **Longitudinal-Float Method.** The concrete shall first be floated with a double-handled longitudinal float not less than 16 feet (4.9m) nor more than 20 feet (6.1m) in length, having a troweling surface not less than 8 inches (200mm) nor more than 10 inches (255mm) wide.

The float shall be operated from bridges over the pavement with its length parallel to the centerline of the improvement, and shall be worked back and forth transversely across the slab, planing off high spots and filling depressions.

This operation shall be continued until the surface is reasonably smooth, after which the bridges may be advanced not to exceed two-thirds the length of the surface so floated, and the operation continued.

302-6.4.4 Final Finishing. After being finished by one of the above methods, the outside edges of pavement shall be rounded to a 1/2 inch (13mm) radius; and transverse contact joints, expansion joints, and joints adjacent to an existing pavement shall be rounded to a 1/4 inch (6mm) radius.

A strip of wetted burlap shall be provided, of a length not less than the width of the pavement slab. It shall be attached by one edge to a rigid frame supported over the pavement so that the free edge of the burlap will rest or drag on the surface of the concrete. The burlap shall be dragged back and forth longitudinally along the pavement until the surface of the slab is of uniform texture and appearance throughout its entire length.

302-6.5 Joints.

302-6.5.1 General. Joints in concrete pavement will be designated as longitudinal and transverse construction joints, transverse expansion joints, and longitudinal and transverse weakened-plane joints.

Unless otherwise specified, transverse joints shall be constructed perpendicularly to the centerline of the pavement, longitudinal joints shall be constructed parallel to the centerline of the pavement, and the faces of all joints shall be perpendicular to the finished surface of the pavement.

Joint filler, when required, shall be as designated on the Plans or in the Special Provisions.

302-6.5.2 Construction Joints. Construction joints are those made by placing fresh concrete against hardened concrete at planned locations. They shall be constructed at the locations and in the manner shown on the Plans.

Longitudinal construction joints shall be constructed by one of the following methods:

1) A plain face,
2) Use of tie bars,
3) Construction of keyways.

The bars or keyways shall be as designated on the Plans or in the Special Provisions.

302-6.5.3 Transverse Expansion Joints. Transverse expansion joints shall be installed at locations shown on the Plans. Expansion joint filler material shall have a minimum thickness of 1/2 inch (12mm), a maximum thickness of 3/4 inch (19mm), a depth equal to the thickness of the pavement, and shall be composed of materials as specified or approved by the Engineer. After the concrete has been finished, an edger of 1/4 inch (6mm) radius shall be used on each side of the expansion joint filler. The expansion joint filler shall be cleaned of all concrete mortar.

302-6.5.4 Weakened-Plane Joints. Weakened-plane joints shall be constructed at the locations shown on the Plans and shall be formed by cutting a groove in the pavement with a power-driven saw. The groove for a transverse joint shall be cut to a minimum depth of 1-1/2 inches (38mm) or one-sixth of the pavement thickness, whichever is greater; the groove for a longitudinal joint shall be cut to a minimum depth of 1-1/2 inches (38mm) or one-fourth of the pavement thickness, whichever is greater; and the width shall be the minimum width possible with the saw being used, but in no case shall the width exceed 1/4 inch (6mm). Any portion of the sealing compound which has been disturbed by sawing operations shall be restored by spraying the areas with additional sealing compound.

In the initial lane of concrete, the first transverse weakened-plane joint immediately following a transverse construction joint, and every fourth weakened-plane joint thereafter, shall be sawed within 10 to 24 hours after the concrete has been placed. The time lapse will be subject to the approval of the Engineer. Every second transverse weakened-plane joint shall be sawed within 24 hours after the concrete is placed, and the remaining weakened-plane joints may be sawed at such time as the Contractor may elect; except that in any lane, all weakened-plane joints shall be sawed before concrete is placed in succeeding adjacent lanes and before any traffic whatsoever is permitted to use the pavement.

In succeeding adjacent lanes of concrete pavement transverse weakened-plane joints opposite those which have opened in the initial lane shall be sawed within 10 to 24 hours after the concrete has been placed. The time lapse will be subject to the approval of the Engineer. In all cases, no more than three consecutive, transverse, weakened-plane joints shall be bypassed.

Longitudinal weakened-plane joints may be used at traffic lane lines in multi-lane, monolithic concrete pavement in lieu of longitudinal construction joints. Dowel requirements shall be as shown on the Plans or in the Special Provisions.

302-6.6 Curing. The pavement shall be cured by a concrete curing compound conforming to the requirements of 201-4.1.

Curing shall commence as soon as free water leaves the surface of the concrete, but not later than 3 hours following the deposit of the concrete upon the subgrade.

Spraying equipment shall be of the fully atomizing type, equipped with a tank agitator of an approved type which provides for continual agitation of the compound during application. The use of nonagitating-type, hand-pumped garden sprayers will not be permitted except for small and inaccessible areas as may be permitted by the Engineer.

302-6.7 Traffic and Use Provisions. The concrete pavement shall be immediately barricaded upon its installation, and no vehicular traffic will be permitted thereon until the expiration of at least 7 days.

Pavement constructed of concrete which has been treated in accordance with 201-1 to obtain an early increase in strength may be opened to traffic 3 days after it is placed, if directed by the Engineer.

At least 3 days shall elapse from the time the concrete is placed before any mechanical tamper, spreader, or finisher which will be supported by the edge of the new pavement, may be operated in adjacent lanes.

302-6.8 Measurement and Payment. Payment for concrete pavement will be made on a cubic yard (cubic meter) or square foot (square meter) basis as shown on the Bid.

Payment for reconstructing or adjusting manholes to grade will be made as a separate item as provided in 301-1.7. If no such item is provided, payment will be deemed included in the other items of work.

302-7 PAVEMENT FABRIC.

302-7.1 General. Pavement fabric material shall conform to 213-1.

302-7.2 Placement.

302-7.2.1 General. The existing pavement surface shall be prepared as shown on the Plans or specified in the Special Provisions prior to placement of the tack coat and pavement fabric. The entire surface to be covered shall be free of water, foreign matter, vegetation, and dust before application of the tack coat. The temperature of the underlying asphalt concrete shall not exceed 150°F (66°C) when the fabric is placed.

After placement, pavement fabric material shall be overlaid with asphalt concrete pavement or covered with a chip seal. Pavement fabric shall not be placed in areas where the asphalt concrete overlay is less than 1-1/2 inches (38mm) thick.

302-7.2.2 Tack Coat.

302-7.2.2.1 General. Tack coat shall be PG 64-10 or PG 70-10 paving asphalt as specified in the Special Provisions. Tack coat shall be applied uniformly prior to placing fabric.

Tack coat shall be sprayed with a truck-mounted sprayer at the rate of 0.25 ± 0.02 gallon per square yard (1.13 ± 0.09 L/m^2). On a new asphalt concrete leveling course, the rate shall be 0.2 ± 0.02 gallon per square yard (0.91 ± 0.09 L/m^2). The application rate may be adjusted as directed by the Engineer. Hand spraying shall be kept to a minimum.

The width of the sprayer application shall be no more than 6 inches (150mm) and no less than 2 inches (50mm) wider than the fabric width.

The temperature of the tack coat shall not exceed 325°F (163°C) when the fabric is placed.

302-7.2.2.2 Calibration of Truck-Mounted Sprayer Units. The Contractor shall calibrate each sprayer unit used in the Work prior to the start of spreading the tack coat. Spreading shall not proceed until the calibration has been accepted by the Engineer.

302-7.2.2.3 Testing Apparatus. Testing apparatus shall consist of the following:

a) Portable measuring scale accurate to within ± 2 grams.
b) Test units shall be a 12-inch x 12-inch (300mm x 300mm) square of 1/8 inch (3mm) or 1/4 inch (6mm) thick hardboard or plywood.

302-7.2.2.4 Calibration Procedure. The calibration procedure shall be as follows:

a) Pre-weigh hardboard or plywood square (minimum of 1, maximum of 3) and write the weight on the underside. The number of test units to be used shall be determined by the Engineer.
b) Locate the truck mounted sprayer unit at the starting point of tack coat placement.
c) The operator sets the application to that rate specified in 302-7.2.2.1 or as specified in the Special Provisions.
d) Place test units on the pavement directly in front of the truck mounted sprayer unit. If 1 unit is used, place it in the center of the truck mounted sprayer unit. If 3 units are used, place 1 unit in center of the sprayer unit and 1 unit outside of each wheel path.
e) Retrieve test units after the truck mounted sprayer unit has passed over them and re-weigh.

302-7.2.2.5 Calculation of the Application Rate. The application rate shall be calculated as follows:

a) Subtract the original weight, recorded on the underside of the test unit, from the weight of the test unit including the tack coat.
b) The difference is the weight of the tack coat on the test unit.

c) Multiply by 9 to calculate the number of grams applied per square yard.

d) Divide the number of grams per square yard by 3861.03 grams per gallon at 60°F.

e) The tack coat application rate shall be expressed in gallons per square yard.

f) Retest as necessary until the specified application rate is confirmed.

302-7.2.3 Laydown. If manual laydown methods are used, pavement fabric material shall be unrolled, stretched, aligned, and placed in increments of approximately 15 feet (4m).

Adjacent borders of the fabric material shall be lapped 2 to 4 inches (50mm to 100mm). The preceding roll shall lap 2 to 4 inches (50mm to 100mm) over the following roll in the direction of paving at ends of rolls or at any break. If the lap exceeds 4 inches (100mm), tack coat shall be placed to bond the layers of fabric material together and both the tack coat and pavement fabric material shall lap by the same amount.

Pavement fabric material shall be placed with the treated side up and shall be seated with brooms or pneumatic rolling equipment after placing. Turning of the paving machine and other vehicles shall be gradual and kept to a minimum to avoid damage.

Pavement fabric material shall not be placed more than 800 feet (180m) in advance of paving operations or chip sealing unless otherwise approved by the Engineer. No more pavement fabric material shall be placed than can be covered that day.

If the pavement fabric material is placed within 50 feet (15m) of the tack coat spray unit, the first 15 feet (4m) of each roll shall be placed by hand, if directed by the Engineer, to allow inspection of the tack coat application.

Pavement fabric material shall be placed with no wrinkles that lap. The test for lapping shall be made as follows:

a) By gathering together the pavement fabric material in a wrinkle.

b) The two sides of the wrinkle shall be pressed together from pavement surface to a fold point with equal amounts of pavement fabric material on both sides of the fold point down to the pavement surface.

c) If the height of the doubled portion of extra pavement fabric material exceeds 1/2 inch (13mm), the pavement fabric material shall be cleanly cut to remove the wrinkle.

d) The cut shall be made on the side of the wrinkle away from the paving operation.

e) The opposite or longer side shall be lapped over the shorter and the re-laid wrinkle area shall be pressed and smoothed into place against the pavement surface.

f) Any lap in excess of 2 inches (50mm) shall be cleanly cut away, and then overlapped in the same manner as for smaller laps.

g) A minimum 1/2 inch (13mm) overlap shall be provided in all cases when a wrinkle area is re-laid.

Pavement fabric material shall not be reduced more than 2 inches (50mm) on each side after being placed on the tack coat. If the overall width is reduced more than 4 inches (100mm), then the operation shall be stopped and the temperature of the tack coat immediately prior to placement of the pavement fabric material or the type of pavement fabric material shall be changed.

Care shall be taken to avoid tracking tack coat onto the pavement fabric material or distorting the fabric during seating of the fabric material with rolling equipment. If necessary, exposed tack coat shall be covered lightly with sand.

A small quantity of asphalt concrete, to be determined by the Engineer, may be spread over the fabric material immediately in advance of placing asphalt concrete overlay in order to prevent pavement fabric material from being picked up by construction equipment.

Public traffic shall not be allowed to drive over bare pavement fabric material. However, public traffic may be allowed to cross the fabric material under traffic control. The Contractor may place a small quantity of asphalt concrete over the fabric material to protect it from damage if so directed by the Engineer.

302-7.3 Measurement. Pavement fabric material will be measured by the square yard (square meter) for the actual pavement area covered.

302-7.4 Payment. Payment for pavement fabric shall be made at the Contract Unit Price per square yard (square meter). The Contract Unit Price for pavement fabric shall include cleaning of the existing pavement, tack coat, calibration of the truck mounted sprayer unit, and furnishing and placing pavement fabric material.

Payment for advance spreading of asphalt concrete over the fabric shall be considered as included in the Contract Unit Price for asphalt concrete.

302-8 SEALCOAT FOR MISCELLANEOUS AREAS.

302-8.1 Materials. The materials for sealcoat shall conform to 203-9. Before incorporation in the Work, the Contractor shall submit a 2-quart (2 liter) sample of undiluted sealcoat at no cost to the Agency per Section 4.

302-8.2 Application.

302-8.2.1 General. The work shall consist of spreading sealcoat material on the pavement where shown on the Plans. The sealcoat material shall be applied in 2 applications. Unless otherwise specified, the total quantity applied shall be 50 gallons per 1,000 square feet (2.0 L/m^2).

Sealcoat material shall be diluted as directed by the Engineer using clean, potable water in an amount not to exceed 20 percent of the total volume.

A tack coat, if required by Contract Documents, shall consist of 3 parts clean, potable water and one part SS-1h emulsion and shall be applied at the rate of 0.05 gallon per square yard (0.25 L/m^2).

Sealcoat material shall not be placed over new asphalt pavement until the pavement has cured for 30 days or as required by the Engineer.

302-8.2.2 Spreading. Sealcoat shall only be applied when the atmospheric temperature is greater than 55°F (13°C) and if rain is not forecast for the period of 24 hours after application.

Prior to applying sealcoat material, all cracked and broken pavement shall be removed and patched in accordance with the Plans and Specifications. Cracks wider than 1/8 inch (3mm) shall be cleaned, treated with weed killer, and filled with an asphalt-based crack filler specified by the Engineer. The pavement surface shall be clean and free from dirt, oil, and grease deposits.

When ambient temperatures are over 80°F (27°C) or the pavement is excessively aged or porous, the surface shall be sprayed with a mist of water in an amount that will leave the surface damp, but with no visible puddles of water. This procedure is not required if a tack coat per 302-8.2.1 has been applied.

Sealcoat material shall be applied using a truck-mounted tank or wheeled container in continuous parallel lines and spread by means of brooms or rubber-faced squeegees either by hand or machine and in such a manner as to eliminate all ridges, lap marks, and air pockets. Raised pavement markers, valve box covers, and manhole covers shall be protected and kept free of sealcoat material.

Sealcoat material shall be homogeneous prior to spreading, with no visible separation of solids and liquids.

The second coat of sealcoat material shall not be applied until the first coat has dried to the touch.

The Contractor shall exercise care to prevent sealcoat material from being deposited on other than specified surfaces and shall remove sealcoat material from surfaces not designated to be sealed, at no cost to the Agency.

Traffic control shall be per 302-4.4.

302-8.3 Measurement and Payment. Sealcoat material shall be paid for at the Contract Unit Price per square yard (square meter) or as shown in the Bid.

302-9 ASPHALT RUBBER HOT MIX (ARHM).

302-9.1 General. ARHM shall conform to 203-11.

302-9.2 Tack Coat. Tack coat shall conform to 302-5.4.

302-9.3 Distribution and Spreading. Distribution and spreading shall conform to 302-5.5 except that at the time of delivery to the Work site, the temperature of the ARHM shall be between 300°F (149°C) and 330°F (166°C). The atmospheric temperature shall be 50°F (10°C) and rising. When the atmospheric temperature is above 85°F (29°C), the temperature of the ARHM at the time of delivery to the Work site may be reduced to 290°F (143°C) if so approved by the Engineer.

302-9.4 Rolling. Rolling shall conform to 302-5.6 except that vibratory rollers using the vibratory mode shall be used for initial breakdown rolling. The initial coverage of breakdown rolling shall commence before the ARHM temperature falls below 290°F (143°C). If the atmospheric temperature is above 85°F (29°C), the initial breakdown rolling temperature may be reduced to 280°F (138°C). Pneumatic rollers shall not be used. If the initial breakdown rolling temperature is reduced, the Contractor shall not be relieved of the density requirements.

302-9.5 Joints. Joints shall conform to 302-5.7.

302-9.6 Manholes (and Other Structures). Manhole and other structures shall conform to 302-5.8.

302-9.7 Rock Dust Blotter. At the option of the Engineer, when traffic conditions warrant, a rock dust blotter may be required to avoid tracking. Rock dust blotter shall conform to 200-1.2 and be uniformly applied using a mechanical spreader at a rate of 2 pounds per square foot (1.1 kilograms per square meter) to 4 pounds per square foot (2.2 kilograms per square meter). When the ARHM pavement has cooled to below 150°F (66°C), the rock dust blotter may not be required.

302-9.8 Measurement. ARHM will be measured by the ton (tonne) or by the square foot (square meter).

The Contractor shall furnish to the Engineer at the time of delivery to the Work site, a legible copy of a licensed weighmaster certificate showing gross, tare, and net weights of each truckload of ARHM. When an automatic batching system is used, the certificate may show only the net weight of ARHM in the truckload.

302-9.9 Payment. Payment for ARHM will be made at Contract Unit Price per ton (tonne) or square foot (square meter). No separate payment for tack coat, rock dust blotter, or sweeping will be made.

Failure of the Contractor to provide a licensed weighmaster certificate to the Engineer by the end of the day on which the ARHM represented by such certificate was delivered to the Work site may, at the discretion of the Engineer, result in forfeiture of payment.

302-10 ASPHALT RUBBER AND AGGREGATE MEMBRANE (ARAM).

302-10.1 Application. Asphalt rubber shall be placed upon a clean dry surface. The pavement surface temperature shall be a minimum of 55°F (13°C) including shaded areas; the atmospheric temperature shall be a minimum of 60°F (16°C) ; the wind shall not adversely affect spray distribution; and all necessary equipment shall be in position and ready to commence placement operations prior to starting the work. The Contractor shall take temperature readings with a temperature measuring device approved by the Engineer.

Asphalt rubber shall be applied by distributor equipment meeting the requirements of the following:

a) Distributor trucks shall meet the requirements for distributing equipment of 203-2.5 and be equipped with an internal heating device capable of evenly heating the material to a temperature of 425°F (218°C).

b) The distributor shall have a platform on the rear of the vehicle and an observer shall accompany the distributor.

c) The observer shall ride in a position so that all spray bar tips are in full view and readily accessible for unplugging if a plugged tip should occur.

d) Material shall be applied at a rate between 0.55 to 0.65 gallons per square yard (2.5 to 3.0 L/m^2) as directed by the Engineer.

e) Material spreading shall not be in excess of that which can be covered with aggregate within 15 minutes maximum.

f) At the time of delivery to the Work site, the temperature of the asphalt rubber shall be 300°F (149°C) minimum.

The asphalt rubber mixture may be applied to the roadway immediately following mixing and reacting at a temperature between 375°F (191°C) minimum to 425°F (218°C) maximum. However, if the material is not to be used within 6 hours of mixing, the mixture shall be allowed to cool below 300°F (149°C) for 12 hours maximum, or to ambient temperature for longer periods, and shall be uniformly reheated to a temperature between 375°F (149°C) minimum to 425°F (218°C) maximum at time of placement and conform to the viscosity requirements.

When joining edges against areas with cover aggregate, the joint shall be swept clean of excess aggregate prior to the adjacent application of asphalt rubber material. Transverse joints of this type shall be constructed by placing building paper across and over the end of the previous asphalt rubber application. Once the spraying has progressed beyond the paper, the paper shall be removed immediately.

Joints between areas of asphalt rubber without cover aggregate shall be made by overlapping asphalt rubber distributions. The excess material shall be dispersed by spreading with a squeegee or rake over a larger area of freshly applied asphalt rubber. The longitudinal joint between adjacent applications of screenings shall coincide with the line between designated traffic lines. All longitudinal joints shall be overlapped for complete coverage. The overlap shall not exceed 4 inches (100mm). At longitudinal joints, the edge shall be broomed back and blended so there are no gaps and the elevations are the same, free from ridges and depressions.

The application of asphalt rubber to areas not accessible with the distributor bar on the truck shall be accomplished by using pressurized hand wands or other means approved by the Engineer.

Application of asphalt rubber shall be discontinued sufficiently early in the workday to permit completion of initial sweeping prior to the termination of traffic control.

302-10.2 Screenings. Following the application of asphalt rubber, screenings conforming to 203-12.2 shall be placed over all areas receiving asphalt rubber. Screenings shall be applied (within a maximum of 15 minutes) at a temperature between 260°F (127°C) minimum to 325°F (163°C) maximum at a rate of 28 to 40 pounds per square yard (15 to 22 kg/m^2) as directed by the Engineer. Stockpiling of screenings after preheating and precoating with paving asphalt will not be permitted.

The Contractor shall prevent any vehicle, including construction equipment, from driving on the uncovered asphalt rubber. Screenings shall be placed with a self-propelled aggregate-spreading machine that can be adjusted to accurately spread the specified amounts per square yard (meter). Trucks for hauling screening material shall conform to 302-2.3.

Initial rolling shall commence within 90 seconds following the placement of screenings. Rolling shall be accomplished by 3 self-propelled, pneumatic-tired rollers meeting the requirements of 302-5.6.1 except that the tires shall be inflated to 100 psi (690 kPa). The operating weight of each roller shall be a minimum of 16,000 pounds (7200 kg). If in the opinion of the Engineer, complete coverage may be provided by 2 rollers in one pass, then 2 pneumatic-tired rollers are sufficient. The initial rolling equipment shall maintain a distance of not more than 200 feet (60m) behind the cover-aggregate spreader on the first pass. There shall be at least 4 complete coverages (single pass in one direction) by the pneumatic-tired rollers before final roller coverage. A steel-drum roller weighing 8 tons (7.2 tonnes) minimum to 10 tons (9.1 tonnes) maximum shall complete the final roller coverage.

Sweeping shall be a multi-step operation following final rolling of the aggregate. A power broom shall be used to remove loose material without dislodging aggregate set in the asphalt rubber. The initial sweeping shall be a light brooming on the same day as ARAM placement. The ARAM shall be maintained free of loose screenings for a minimum of 5 working days following placement. During this period, the surface shall be swept as necessary to remove any loose cover material as directed by the Engineer. Final sweeping shall be done and all loose aggregate shall be removed prior to acceptance. The sweeping operations shall be accomplished without the use of gutter brooms or steel-tined brooms.

Immediately upon opening the street to traffic, the Contractor shall start removing loose aggregate from parkways, sidewalks, and intersecting streets. Both operations shall continue until all excess or loose aggregate is removed from the roadway surface and abutting adjacent areas.

At the option of the Engineer, rock dust blotter material shall be applied immediately after the initial pass of the rollers or after sweeping, but prior to opening to traffic, to prevent bleeding and pickup of the asphalt rubber material. Rock dust blotter conforming to 200-1.2 shall be uniformly applied using a mechanical spreader at a rate of 2 lbs/yd^2 (1.1 kg/m^2) minimum to 4 lbs/yd^2 (2.2 kg/m^2) maximum. If the ARAM is to be used as a finished surface, a flush coat shall be used.

The Contractor shall protect all existing manhole, valve, survey monument, and other miscellaneous frames and covers. The Contractor shall cooperate with the owners of any frames and covers and shall cover and completely protect them with heavy roofing paper or other suitable material. Petroleum-based release agents shall not be used for this purpose.

302-10.3 Flush Coat. If required a flush coat shall be applied. This work shall consist of an application of fog seal coat and rock dust blotter material to the surface of ARAM.

Flush coat shall be applied to the ARAM immediately after removal of excess screenings following initial brooming of the ARAM and prior to opening the lane to uncontrolled public traffic, as directed by the Engineer.

Asphaltic emulsion (fog seal coat) shall be applied to the surface of the ARAM and shall be grade CSS1 or CSS1H, per 203-3, unless otherwise ordered by the Engineer.

The application rate of the fog seal coat (asphaltic emulsion and an equal amount of water) shall be such that the diluted asphaltic emulsion will be spread at a rate of 0.06 gal/yd^2 (0.27 L/m^2) minimum to 0.12 gal/yd^2 (0.54 L/m^2) maximum. The exact rate of application will be determined by the Engineer.

During fog seal coat operations the surface upon which an ARAM is being applied shall be closed to public traffic. Care shall be taken to avoid tracking fog seal coat material onto existing pavement surfaces beyond the limits of construction.

302-10.4 Public Convenience and Traffic Control. The Contractor shall prohibit traffic on the street until the initial sweeping is complete. Prior to opening the streets to traffic, "Loose Gravel," C6 signs, and appropriate speed-reduction signs conforming to local, State, and Federal regulations shall be posted and maintained by the Contractor. These signs shall remain in place until there is no further dislodging of the cover aggregate.

302-10.5 Measurement and Payment. ARAM including asphalt rubber and cover aggregate will be paid for at the Contract Unit Price per square yard (square meter). Unless otherwise specified, such price shall include full compensation for pavement preparation, furnishing and placing materials required, including rock dust blotter, sweeping, and incidentals needed to complete the work in place.

The Contractor shall include in the Contract Unit Price for ARAM the full cost of applying rock dust blotter to all areas of ARAM, as directed by the Engineer.

The Contractor shall include in the Contract Unit Price for ARAM the full cost of applying a flush coat (if required) to all areas of ARAM, as directed by the Engineer.

302-11 CRUMB RUBBER MODIFIED ASPHALT CONCRETE-GAP GRADED (CRUMAC-GG).

302-11.1 General. CRUMAC-GG shall conform to 203-13.

302-11.2 Tack Coat. Tack coat shall conform to 302-5.4.

302-11.3 Distribution and Spreading. Distribution and spreading shall conform to 302-9.3, except that at the time of delivery to the Work site the maximum temperature shall be 340 °F (171 °C).

302-11.4 Rolling. Rolling shall conform to 302-9.4.

302-11.5 Joints. Joints shall conform to 302-5.7.

302-11.6 Manholes (and Other Structures). Manholes and other structures shall conform to 302-5.8.

302-11.7 Rock Dust Blotter. Rock dust blotter shall conform to 302-9.7.

302-11.8 Measurement. CRUMAC-GG will be measured by the ton (tonne) or by the square foot (square meter).

The Contractor shall furnish to the Engineer at the time of delivery to the Work site, a legible copy of a licensed weighmaster certificate showing gross, tare, and net weights of each truckload of CRUMAC-GG. When an automatic batching system is used, the certificate may show only the net weight of CRUMAC-GG in the truckload.

302-11.9 Payment. Payment for CRUMAC-GG will be made at the Contract Unit Price per ton (tonne) or at the Contract Unit Price per square foot (square meter). No separate payment for tack coat, rock dust blotter, or sweeping will be made.

Failure of the Contractor to provide a licensed weighmaster certificate to the Engineer by the end of the day on which the ARHM represented by such certificate was delivered to the Work site may, at the discretion of the Engineer, result in forfeiture of payment.

302-12 TIRE RUBBER MODIFIED ASPHALT CONCRETE (TRMAC).

302-12.1 General. TRMAC shall conform to 203-14.

302-12.2 Tack Coat. Tack coat shall conform to 302-5.4.

302-12.3 Distribution and Spreading. Distribution and spreading shall conform to 302-9.3.

302-12.4 Rolling. Rolling shall conform to 302-9.4.

302-12.5 Joints. Joints shall conform to 302-5.7.

302-12.6 Manholes (and Other Structures). Manholes and other structures shall conform to 302-5.8.

302-12.7 Rock Dust Blotter. Rock dust blotter shall conform to 302-9.7.

302-12.8 Measurement. TRMAC will be made at the Contract Unit Price per ton (tonne) or at the Contract Unit Price per square foot (square meter).

The Contractor shall furnish to the Engineer at the time of delivery to the Work site, a legible copy of a licensed weighmaster certificate showing gross, tare, and net weights of each truckload of TRMAC. When an automatic batching system is used, the certificate may show only the net weight of TRMAC in the truckload.

302-12.9 Payment. Payment for TRMAC will be made at the Contract Unit Price per ton (tonne) or at the Contract Unit Price per square foot (square meter). No separate payment for tack coat, rock dust blotter, or sweeping will be made.

Failure of the Contractor to provide a licensed weighmaster certificate to the Engineer by the end of the day on which the TRMAC represented by such certificate was delivered to the Work site may, at the discretion of the engineer, result in forfeiture of payment.

302-13 POROUS ASPHALT CONCRETE PAVEMENT.

302-13.1 General. Porous asphalt concrete shall conform to 203-15.

302-13.2 Distribution and Spreading. Distribution and spreading shall conform to 302-5.5, except each successive course shall be laid upon previously laid courses as soon as the previous course has cooled sufficiently to show no displacement under equipment or loaded material delivery trucks and preclude the necessity of tack coat between courses.

302-13.3 Rolling. Rolling shall be performed using an 8 to 10 ton static steel drum roller conforming to 302-5.6. The surface temperature during rolling shall be between 210°F (99°C) and 260°F (127°C). Rolling shall be limited to a maximum of 2 passes. No vehicular traffic shall be permitted on porous asphalt concrete pavement for 24 hours after rolling.

302-13.4 Infiltration Rate. Porous asphalt concrete pavement shall have an infiltration rate of a minimum of 100 inches per hour (254 cm/hr) when tested in accordance with 211-3 unless otherwise specified in the Special Provisions.

302-13.5 Measurement. Porous asphalt concrete will be measured by the ton (tonne). The Contractor shall furnish to the Engineer at the time of delivery to the Work site a legible copy of a licensed weighmaster's certificate showing gross, tare, and net weights of each truckload of porous asphalt concrete. When an automatic batching system is used, the licensed weighmaster's certificate may show only the net weight of material in the truckload.

302-13.6 Payment. Payment for porous asphalt concrete will be made at the Contract Unit Price per ton (tonne). Failure to provide a certificate to the Engineer by the end of the day on which the material represented by such certificate is delivered to the Work site, may, at the discretion of the Engineer, result in the forfeiture of payment for such material.

SECTION 303 - CONCRETE AND MASONRY CONSTRUCTION

303-1 CONCRETE STRUCTURES.

303-1.1 General. Concrete bridges, culverts, catch basins, retaining walls, abutments, piers, footings, foundations, and similar structures shall be constructed in conformity with the Plans and Specifications. Concrete for use in work constructed under this section shall conform to the requirements of 201-1.

Safe and suitable ladders shall be provided to permit access to all portions of the work.

The compressive strength of the concrete referred to in this section will be based on the results of concrete test cylinders made and tested by the Engineer in accordance with the requirements of 201-1. The cylinders shall be cured in accordance with ASTM C31, unless otherwise approved by the Engineer.

When plastic-lined concrete structures are required by the Plans, the plastic liner materials shall comply with 210-2 and the installation of the liner shall be in accordance with 311-1.

303-1.2 Subgrade for Concrete Structures. Earth subgrade upon which concrete is placed shall be firm and free from standing water. Groundwater shall be kept below subgrade until the concrete has set. When the subgrade is in dry earth, it shall be thoroughly dampened with water to ensure that no moisture will be absorbed from the fresh concrete.

When the design details for the Work provide for the construction of filter or drain material consisting of gravel (or combination of gravel and sand), which material will be subgrade for concrete, the placing of steel reinforcement and placement of concrete shall follow the installation of the filter or drain material as closely as practical. The filter or drain material shall be kept dewatered to the extent necessary to prevent any portion of concrete materials being deposited in water. No payment will be made for dewatering other than as may be included in the prices bid for various items of work or when an item for dewatering is provided.

When the concrete is to be deposited on rock, the rock shall be fully uncovered, cleaned, and its surface shall be removed to a depth sufficient to expose sound rock. Bedrock shall be roughly leveled off or cut to approximately horizontal and vertical steps. Seams in the rock shall be grouted under pressure or otherwise treated as the Engineer may direct. Grouting seams in rock or otherwise treating them will be paid for as provided in the Specifications.

303-1.3 Forms. Forms shall be of suitable material and of a type, size, shape, quality, and strength to ensure construction as designed. The forms shall be true to line and grade, mortar-tight, and sufficiently rigid to resist deflection during placing of the concrete. The responsibility for their adequacy shall rest with the Contractor. All dirt, chips, sawdust, nails, and other foreign matter shall be completely removed from forms before any concrete is deposited therein. The surfaces of forms shall be smooth and free from irregularities, dents, sags, and holes that would deface the finished surfaces. Forms previously used shall be thoroughly cleaned of all dirt, mortar, and foreign matter before being reused. Before concrete is placed in forms, all inside surfaces of the forms shall be thoroughly treated with an approved releasing agent which will leave no objectionable film on the surface of the forms that can be absorbed by the concrete. Care shall be exercised that no releasing agent is deposited on previously placed concrete.

Forms for all surfaces that will not be completely enclosed or hidden below the permanent surface of the ground shall be made of surfaced lumber, or material which will provide a surface at least equal to surfaced lumber or plywood. Any lumber or material which becomes badly checked or warped, prior to placing concrete, shall not be used.

Forms for all exposed surfaces of bridges, viaducts, overcrossings, and similar structures shall be constructed of plywood or an approved equal. Plywood for forms shall be of the grade "Exterior B-B (concrete form)", conforming to the latest Product Standard for Soft Plywood, Construction and Industrial, of the National Institute of Standards and Technology. Plywood shall be furnished and placed in no less than 48-inch (1.2m) widths and in uniform lengths of not less than 96 inches (2.4m) except where the dimension of the member formed is less than the specified panel dimension. Plywood shall be placed with the grain of the outer plies in the direction of the span. Where plywood is attached directly to the studs or joists, the panels shall be not less than 5/8 inch (15mm) thick, and the studs or joists shall be spaced not more than 12 inches (300mm), center to center. Plywood less than 5/8 inch (15mm) thick, otherwise conforming to the requirements specified herein, may be used with a continuous backing of 3/4 inch (19mm) sheeting. All form panels shall be placed in a neat, symmetrical pattern with the horizontal joints level and continuous. All joints shall be filled with an approved quick-setting compound and finished flush with the interior of the form.

Wooden forms for copings and curbs shall have a thickness of not less than 1-1/2 inches (38mm) and a width of not less than the full depth of coping or curb.

Unless otherwise shown on the Plans, all sharp edges shall be chamfered with 3/4 inch x 3/4 inch (20mm x 20mm) triangular fillets. Forms for curved surfaces shall be so constructed and placed that the finished surface will not deviate from the arc of the curve.

Forms shall be so constructed that portions, where finishing is required, may be removed without disturbing portions of forms to remain in place.

Joists and stringers supporting slabs and overhangs shall be considered as falsework and designed in accordance with 303-1.6.

Forms for girders and slabs shall be cambered as may be required by the Engineer.

Forms shall, as far as practicable, be so constructed that the form marks will conform to the general lines of the structure. Concrete bridge railings shall be constructed to present a smooth uniform appearance in their final position, conforming to the lines shown on the Plans. The height of the concrete railings shall be adjusted as directed by the Engineer to compensate for the camber and deadload deflection of the superstructure.

Form clamps or bolts, approved by the Engineer, shall be used to fasten forms. The use of twisted-wire loop ties to hold forms in position will not be permitted, nor shall wooden spreaders be used unless authorized by the Engineer. Clamps or bolts shall be of sufficient strength and number to prevent spreading of the forms. They shall be of such type that they can be entirely removed or cut back 1 inch (25mm) inside the finished surface of the concrete. All forms for outside surfaces shall be constructed with stiff wales at right angles to the studs, and all form clamps or bolts shall extend through and fasten such wales.

Forms for cast-in-place concrete drain conduits or sewer structures will not be required for concrete to be placed directly against the sides of the excavation or sheeting, provided the following conditions are met:

a) If concrete is placed directly against the faces of the excavation, the faces must be firm, compact, able to stand without sloughing, and must be outside the concrete lines shown on the Plans at all points. The entire faces of excavation, against which concrete is to be placed without the use of outside forms, shall be gunited to sufficient thickness to prevent raveling of the exposed earth faces during the placing of reinforcing steel, forms, and concrete.

b) If concrete is placed against sheeting, such sheeting shall be closely fitted and all points shall be outside the concrete lines shown on the Plans. Those surfaces against which the concrete is to be placed shall be faced with building paper. Except as otherwise specified herein, all sheeting shall be removed, but not until at least 7 days after placing concrete, or until the concrete has attained strength in compression of 2,000 psi (14MPa).

   Care shall be used in removing sheeting so as to avoid damaging the concrete. Voids left by the removal of sheeting, piles or similar sheeting components shall be backfilled with material having a sand equivalent of not less than 30 and consolidated by jetting as directed by the Engineer. When field conditions or the type of sheeting or methods of construction used by the Contractor are such as to make the removal of sheeting impracticable, that portion of the sheeting against which concrete has been placed may be left in place.

c) The reinforcing steel shall be set accurately and held firmly in place.

d) The Contractor shall assume all risks of damage to the Work or to existing improvements that may be attributable to this method of construction.

e) Should this method of construction prove unsatisfactory, the Contractor shall discontinue this method and construct the conduit by using outside forms.

f) No direct payment will be made for building paper, sheeting, gunite, for concrete placed outside of concrete lines shown on the Plans, or for cement used in such gunite and concrete. The cost thereof shall be included in the prices in the Bid for the various items of work.

303-1.4 Removal of Forms.

303-1.4.1 General. The periods of time for form removal set forth herein are permissive only and subject to the Contractor assuming all risks that may be involved. The time periods are minimum with no allowance therein for external loads. At times of low temperature, or other adverse conditions, the Engineer may require the forms to be kept in place for longer periods of time.

The time periods are predicted on the use of concrete to which no admixtures have been added for the purpose of obtaining a high early strength, and upon the use of the same type of cement throughout the structure. The Engineer may permit the use of admixtures, additional cement, or different types of cement in accordance with 201-1.2.4. If such permission is granted, the minimum time periods for stripping forms will be established by the Engineer in accordance with the materials, methods to be used, and the stresses to which the structure may be subjected.

When the Contractor elects to use Type IP (MS) cement in accordance with 201-1.2.1 minimum form removal times may be longer than indicated in the following subsections.

303-1.4.2 Bridges. The periods of time set forth herein are based on the use of Type II cement.

Forms and falsework supporting concrete beams, arch ribs, slabs, or other members subject to direct bending stress shall not be removed in less than 21 days after the concrete has been placed, unless concrete test cylinders show a strength of not less than 3,000 psi (21MPa) in compression, when cured under conditions similar to those affecting the structure.

Forms and falsework supporting the bottom slab of the superstructure of box girder structures shall remain in place 14 days after placing of the deck of the superstructure. Forms for the webs of box girders shall be removed before the deck slab is placed. Forms for the upper deck slab which are to remain in place shall be supported by bolts through the girder webs or some equally satisfactory method that will prevent the transfer of any load to the lower deck slab. Forms supporting the concrete deck slab of box girders may be left in place. All interior forms in box girders, except those permitted to remain in place, shall be removed completely and the inside of the box girder cleaned of all loose material.

Side forms for beams, girders, columns, railings, or other members in which the forms do not resist dead load bending, may be removed within a period of 2 to 5 days, as authorized by the Engineer, provided that satisfactory arrangements are made to cure and protect the concrete thus exposed.

Side forms for arch rings, columns, and piers shall be removed before the members of the structure which they support are cast so that the quality of the previously placed concrete may be inspected. Such forms shall be so constructed that they may be removed without disturbing other forms which support direct load or resist bending stress.

303-1.4.3 Miscellaneous Structures. The periods of time set forth herein are based on the use of Type II cement.

Forms for concrete members (except bridges) subject to bending stresses, where the member relies upon forms for vertical support, may be removed 7 days after concrete is placed.

Curb forms shall not be removed until the concrete has set sufficiently to hold its shape but shall be removed in time to permit proper finishing.

Stairway forms shall be removed and the finish of the steps completed on the day the concrete is placed. Metal stairway treads, if required by the Plans, shall be installed immediately after the steps have been poured.

303-1.4.4 Standard Structures.

a) **General.** Except as otherwise stipulated, the periods of time set forth herein for removal of forms are based on the use of Types II, III, IV, or V portland cement.

b) **Standard Catch Basins.**

1) Outside forms and inside wall forms which do not support the top slab forms - 16 hours.

2) Top slab forms - 48 hours if Type II or V cement is used; 24 hours if Type III cement is used.

c) **Standard Transition Structures.**

1) Outside forms and inside wall forms which do not support the top slab form - 16 hours.

2) Top slab forms - as specified for box section slab forms.

303-1.4.5 Channels and Conduits.

a) **General.** Except as otherwise specified, the periods of time set forth herein are based on the use of Types II, III, IV, or V portland cement.

b) **Forms Removal.** Forms for open channels and forms and shoring for box sections and arch sections of sewers and storm drains may be removed as follows:

1) Forms for open channel walls - 16 hours.

2) Outside forms of box sections and inside wall forms of box sections which do not support the slab forms - 16 hours.

3) Arch sections in open cut - 12 hours.

4) Slab forms for box sections -

(a) Type II cement - 48 hours or 6 hours per foot (20 hours per meter) of span between supports, whichever is greater.

(b) Type III cement - 24 hours or 3 hours per foot (10 hours per meter) of span between supports, whichever is greater.

(c) Type V cement - 56 hours or 7 hours per foot (23 hours per meter) of span between supports, whichever is greater.

303-1.5 Removal of Forms for Box Sections. In lieu of form removal as specified in 303-1.4, the forms may be stripped by the following method:

If the walls and top slab of the box structure are placed monolithically, the forms may be removed when the concrete has attained the compressive strength as computed from the following formula:

$C_{US} = 20\ S + 1000$ For U.S. Standard Measures

($C_{SI} = 0.45\ S + 7$ For Metric Units)

Where S = Span length in feet (meters) from center to center of supports (maximum span 20 feet (6m) unless otherwise approved by the Engineer).

Where C_{US} (C_{SI}) = Required compressive strength in pounds per square inch (MPa) of the concrete as determined in accordance with the requirements below:

If the top slab is not placed monolithically with side walls and if the wall forms do not support the top slab forms, the forms for the walls may be removed when the concrete has attained a compressive strength of 1,000 psi (7MPa). The forms for the top slab may be removed when the concrete has attained a compressive strength equal to that computed by the above formula; provided that the concrete in the walls has attained a compressive strength at least equal to that determined for the top slab at the time it is proposed to remove the top slab forms.

The strengths set forth herein at which the Contractor may remove forms in the walls and top slab of box sections are permissive only, and subject to the Contractor assuming all risks that may be involved in such removals. No allowance for external loads is included in the specified strength. At times of low temperature, or other adverse conditions, the Engineer may require the forms to be kept in place until greater concrete strength have been attained.

303-1.6 Falsework.

303-1.6.1 General. The Contractor shall, in accordance with 2-5.3, submit detailed Working Drawings of the falsework proposed to be used. Such plans shall be in sufficient detail to indicate the general layout, sizes of members, anticipated stresses, grade of materials to be used in the falsework, and typical soil conditions.

All falsework shall be designed and constructed to provide the necessary rigidity and to support the loads. Falsework for the support of a superstructure shall be designed to support the loads that would be imposed if the entire superstructure were placed at one time.

All falsework, staging, walkways, forms, ladders, cofferdams, and similar accessories shall equal or exceed the minimum requirements of the State Division of Occupational Safety and Health. Compliance with such requirements shall not relieve the Contractor from full responsibility for the adequacy of safety measures.

Sufficient inspection walkways and access thereto shall be provided under the deck to permit inspection of all forms. The walkways shall be not more than 8 feet (2.4m) below the forms to be inspected.

Falsework shall be placed upon a solid footing safe against undermining and protected from softening. When the falsework is supported on timber piles, the piles shall be driven to a bearing value as determined by the formula specified in 305-1.5, equal to the total calculated pile loading. The maximum calculated pile loading shall not exceed 20 tons (180 kN).

Construction of falsework which is to be supported on a concrete invert may be started 48 hours after the concrete is placed, provided no heavy equipment or concentrated loads are placed on the concrete invert. When heavy equipment must be placed on the concrete invert to erect falsework, the concrete shall be 7 days old or shall have attained a compressive strength of 2,000 psi (14MPa) before falsework erection is started. Unless otherwise directed by the Engineer, concentrated falsework loads shall be so spread over the supporting concrete slab as to reduce the soil pressure under it not over 2,000 pounds per square foot (96 kN/m^2), assuming 45-degree lines of distribution through the slab.

Falsework and forms shall be so constructed as to produce in the finish structure the lines and grades indicated on the Plans. Suitable jacks or wedges shall be used in connection with the falsework to set the forms to the grade or the form work before or during the placing of concrete. Single wedges for this purpose will not be permitted, it being required that all such wedges be in pairs to ensure uniform bearing. Dead load deflection in stringers and joists will be compensated for by varying the depths of the joists or by using varying depth nailing strips.

Arch centering shall be removed uniformly and gradually, beginning at the crown and working toward the haunches to permit the arch to take its load slowly and evenly. Centering for adjacent arch spans shall be struck simultaneously.

Falsework under any continuous unit or rigid frame shall be struck simultaneously; the supporting wedges being released gradually and uniformly, starting at the center and working both ways toward the supports.

303-1.6.2 Falsework Design. Falsework shall be designed to carry all loads and pressures applied to it. The construction loads to be applied are as follows:

a) Tunnel centering - 100 percent of the concrete load where concrete is placed by pneumatic process.

b) All other structures - a live load of 30 pounds per square foot (1.5 kN/m^2) of horizontal area.

c) Transverse and longitudinal bracing - a horizontal force equal to 2 percent of the vertical load.

The unit stresses shown below shall be the maximum allowable for falsework design for used material. The Contractor may elect to furnish new grade-marked material. In such event, the falsework plans submitted shall indicate the grade. The unit stresses allowed shall be those recommended in the latest issue of "Standard Grading Rules for West Coast Lumber" (published by the West Coast Lumber Inspection Bureau), increased 25 percent for short time loading.

MAXIMUM UNIT STRESSES FOR FALSEWORK

Axial tension and bending.. 1,400 psi (9.7MPa)

Compression perpendicular to grain.. 350 psi (2.4MPa)

End grain bearing .... 1,300 psi (9.0MPa)

Shear parallel to grain.. 125 psi (0.9MPa)

Timber Columns:

U.S. Standard Measure............................. $S_{US} = 1300$ (1 - L/60d)

Metric Units... $S_{SI} = 9.0$ (1 - L/60d)

Where:

S_{US} (S_{SI}) = Maximum allowable stress in psi (MPa).

d = Least side dimension of column in inches (mm).

L = Unbraced length of column in inches (mm).

L/d shall not exceed 50 for bracing.

L/d shall not exceed 30 for main members.

Falsework may be bolted or spiked at the option of the Contractor, but the use of bolts and spikes shall not be combined in the same connection. The allowable spacings and connection values of bolts and spikes shall be in accordance with the National Design Specifications for Stress-Grade Lumber and its Fastenings as recommended by National Lumber Manufacturers Association, except that an additional allowance of 25 percent for temporary use shall be added to the connection values for bolts and spikes.

Ends of columns bearing on wedges shall be tied in both directions by girts.

303-1.7 Placing Reinforcement.

303-1.7.1 General. Before placing reinforcing steel, the Contractor shall submit a reinforcing steel placing plan Working Drawing in accordance with 2-5.3.

Reinforcing bars shall be placed in accordance with the size and spacing shown on the Plans. Reinforcing bars shall be firmly and securely held in position in accordance with the "Manual of Standard Practice" of the Concrete Reinforcing Steel Institute, using concrete or metal chairs, spacers, metal hangers, supporting wires, and other approved devices of sufficient strength to resist crushing under full load. Metal chairs which extend to the surface of the concrete (except where shown on the Plans) and wooden supports, shall not be used. Tack welding on reinforcing bars will not be permitted.

Placing bars on layers of fresh concrete as the work progresses and adjusting bars during the placing of concrete will not be permitted. Before placing in the form, all reinforcing steel shall be cleaned thoroughly of mortar, oil, dirt, loose mill scale, loose or thick rust, and coatings of any character that would destroy or reduce the bond. No concrete shall be deposited until the placing of the reinforcing steel has been inspected and approved.

Bar spacing is center to center of bars. Bar cover is clear distance between surface of bar and face of concrete and shall be 2 inches (50mm) unless otherwise noted on the Plans. Reinforcement shall terminate 2 inches (50mm) from concrete surfaces and expansion joints, unless otherwise noted on the Plans.

Reinforcement used in post-tensioned concrete shall be adjusted or relocated during the installation of prestressing ducts or tendons, as required to provide planned clearances to the prestressing tendons, anchorages, jacks, and equipment, as approved by the Engineer.

303-1.7.2 Splicing. Splices of bars shall be made only where shown on the Plans or as approved by the Engineer. Where bars are spliced, they shall be lapped at least 30 diameters, unless otherwise shown on the Plans.

Splicing shall be accomplished by placing the bars in contact with each other and wiring them together.

Welding of reinforcing steel will not be permitted unless specifically authorized by the Engineer.

303-1.7.3 Bending Reinforcement. Bends and hooks in bars shall be made in the manner prescribed in the "Manual of Standard Practice" of the Concrete Reinforcing Steel Institute.

Bars shall not be bent or straightened in a manner which will injure the material. Bars with kinks or unspecified bends shall not be used.

303-1.7.4 Welded Wire Fabric. Welded wire fabric shall be spliced not less than 2 meshes. It shall be lifted carefully into its specified position after the concrete is placed but still plastic.

303-1.8 Placing Concrete.

303-1.8.1 General. Concrete shall be conveyed, deposited, and consolidated by any method which will preclude the segregation or loss of ingredients. Equipment used in conveying and depositing concrete shall not have any aluminum or other material component in direct extended contact with the concrete which results in deleterious or injurious effects to the concrete.

All surfaces against which concrete is to be placed shall be thoroughly moistened with water immediately before placing concrete. All ponded and excess water shall be removed to leave surfaces moist but not flooded.

Chutes used in conveying concrete shall be sloped to permit concrete of the consistency required to flow without segregation. Where necessary to prevent segregation, chutes shall be provided with baffle boards or a reversed section at the outlet.

Where a sequence for placing concrete is shown on the Plans, no deviation will be permitted unless approved in writing by the Engineer.

303-1.8.2 Grouting. Where concrete is to be deposited against hardened concrete at horizontal construction joints, placing operations shall begin by conveying a grout mixture through the placing system and equipment and depositing the mixture on the joint. Unless otherwise approved by the Engineer, the grout mixture shall have a combined aggregate E grading in accordance with 201-1.3.2, and meet all other requirements consistent with the approved mix design. Consolidation techniques shall be utilized to ensure proper bond at the joint.

303-1.8.3 Depositing. To avoid segregation, concrete shall be deposited as near to its final position as is practicable. The use of vibrators for extensive shifting of the mass of concrete will not be permitted. Concrete that has partially hardened, has been retempered, or is contaminated by foreign materials shall not be deposited in the structure.

Concrete shall be placed in horizontal layers insofar as practical. Placing shall start at the low point and proceed upgrade unless otherwise permitted by the Engineer. Concrete shall be placed in a continuous operation between construction joints and shall be terminated with square ends and level tops unless otherwise shown on the Plans.

Concrete shall not be permitted to free fall more than 6 feet (1.8m) without the use of tremies or other suitable conveyance. Tremies shall be at least 6 inches (150mm) in diameter, or the equivalent cross-sectional area for rectangular sections. Concrete shall not be placed in horizontal members or sections until the concrete in the supporting vertical members or sections has been consolidated and a 2-hour period has elapsed to permit shrinkage to occur.

303-1.8.4 Consolidating. Concrete shall be thoroughly consolidated in a manner that will encase the reinforcement and inserts, fill the forms, and produce a surface of uniform texture free of rock pockets and excessive voids.

Structural concrete, except slope paving such as spillway aprons and channel lining, and concrete placed under water, shall be consolidated by means of high frequency internal vibrators of a type, size, and number approved by the Engineer. The location, manner, and duration of the application of the vibrators shall be such as to secure maximum consolidation of the concrete without separation of the mortar and coarse aggregate, and without causing water or cement paste to flush to the surface. Internal vibrators shall not be held against the forms or reinforcing steel.

The number of vibrators employed shall be sufficient to consolidate the concrete within 15 minutes after it has been deposited in the forms. At least 2 vibrators in good operating condition shall be available at the site of the structure in which more than 25 cubic yards ($19m^3$) of concrete is to be placed.

Approved external vibrators for consolidating concrete will be permitted when the concrete is not accessible to internal vibration. Forms and falsework shall be designed and constructed to resist displacement or damage from external vibration.

303-1.8.5 Walkways. Walkways and platforms shall be provided for personnel and equipment at a level convenient for the concrete placement and to permit the performance of all operations necessary for the completion of such work including finishing.

Where bridge decks are to be constructed to final roadway grade, walkways shall be provided outside the deck area along each side and for the full length of the structure. These walkways shall be of sufficient width and so constructed as to provide for the support of the bridges from which the longitudinal floats specified are to be operated.

303-1.8.6 Joints. The work shall be so prosecuted that construction joints will occur at designated places shown on the Plans unless otherwise authorized by the Engineer. The Contractor shall construct, in one continuous concrete placing operation, all work comprised between such joints. Joints shall be kept moist until adjacent concrete is placed.

All construction joints having a keyed, stepped, or roughened surface shall be cleaned by sandblasting prior placement of the adjacent concrete, unless otherwise directed by the Engineer. Any quality of sand may be used which will accomplish the desired results.

The sandblasting operations shall be continued until all unsatisfactory concrete, laitance, coatings, stains, debris, and other foreign materials are removed. The surface of the concrete shall be washed thoroughly to remove all loose material. The method used in disposing of wastewater employed in washing the concrete surfaces shall be such that the wastewater will not stain, discolor, or affect exposed surfaces of the structures, and will be subject to the approval of the Engineer.

Expansion and contraction joints in concrete structures shall be formed where shown on the Plans. No reinforcement shall be extended through the joints, except where specifically noted or detailed on the Plans.

Asphalt paint used in joints shall be as specified in 203-8. Premolded asphalt filler shall be as specified in 201-3.

No direct payment will be made for furnishing and placing asphalt paint, premolded asphalt filler, or other types of joint separators. The cost therefore shall be included in the Bid price for the item of work of which they are a part.

303-1.8.7 Application of Joint Sealants.

a) **General.** All joints sealants shall conform to the general requirements 201-3 and the special requirements shown below. Prior to sealing joints containing waterstops, the expansion joint filler, hardboard, concrete spillage, and all foreign material shall be removed from the deck joint down to a depth of the waterstops. All such material shall be removed from the entire depth of joints in curbs, sidewalks, railings, and the overhanging portion of deck slabs. Immediately before applying the joint sealant, the joint shall be thoroughly cleaned by abrasive blasting or other approved means to remove all mortar, laitance, scale, dirt, dust, oil, curing compounds, and other foreign material. The joints shall be blown out with high pressure compressed air to remove all residue.

If a sealant is shown in the sidewalk, saw-cutting of grooves at concrete railing locations shall be completed prior to constructing the railings. Joint seal material shall be protected during the construction of the railing.

At the time of applying the joint sealant, the joint shall be surface dry, and acceptable to the Engineer. No sealant shall be placed during unsuitable weather, when the atmospheric temperature is below 50°F (10°C), or when weather conditions indicate that the temperature may fall below 32°F (0°C) within 24 hours.

The joint shall be filled from the bottom to the top without formation of voids. The top of the finished joint seal shall be between 1/4 inch (6mm) and 3/8 inch (10mm) below the finished surface.

All adjoining surfaces shall be carefully protected during the joint sealing operations, and any stains, marks or damage thereto, as a result of the Contractor's operations, shall be corrected in a manner satisfactory to the Engineer.

b) **Type "A" Seal.** The top edges of the seal shall remain in continuous contact with the sides of the groove over the entire range of joint movement and such contact shall not break when thumb pressure is applied vertically to the seal at the centerline of the seal. The seal shall satisfactorily resist the intrusion of foreign material and water.

Grooves for joint seals shall be cut to a uniform width and depth and to the alignment shown on the Plans or as ordered by the Engineer. Both sides of the groove shall have saw-cut surfaces for their full depth and shall expose sound concrete.

The concrete saw for cutting the grooves shall be fitted with diamond blades having a core (disk) thickness not less than 3/16 inch (5mm). Double blades, cutting both sides of the groove simultaneously, shall be used for the initial cut. The completed groove measured at the top shall be within 1/8 inch (3mm) of the width shown on the Plans. The groove width measured at the bottom shall not vary from the top width by more than 1/16 inch (1.5mm) for each 2 inches (50mm) of depth.

Immediately following cutting, the lip of the groove shall be beveled by grinding, as shown on the Plans, and the groove shall be thoroughly washed with water under pressure and blown out with high pressure air jets to remove all residue and foreign materials.

At least 48 hours prior to installing the seal, the Contractor shall repair all spalls, fractures, breaks, or voids in the concrete surfaces of the joint groove by methods approved by the Engineer.

If any of the requirements of this subsection are not satisfied, the Contractor at its expense shall remove the seal, repair or reconstruct the groove, and replace the seal or install new seal material.

c) **Type "B" Seal.** All of the requirements of 303-1.8.7 b) shall apply to the application of Type "B" seals. In addition, the following requirements shall apply:

Sawcutting of grooves shall not be started until seal material has been tested, approved, and delivered to the Work site.

The elastomeric joint seal shall be installed with mechanical equipment specifically designed for this purpose.

The equipment shall not elongate the seal longitudinally nor cause structural damage to the seal or the concrete. The equipment shall place the seal to the depth shown on the Plans.

The equipment shall not twist, distort, or cause other malformations in the completed seal. Equipment that does not provide a properly installed seal shall not be used.

A combination lubricant and adhesive shall be applied to the dry sides of the seal and to all vertical surfaces which will be in contact with the seal immediately prior to installation. The rate of application shall be as recommended by the manufacturer of the seal. The contact surfaces of the seal shall be cleaned with normal butyl acetate, using clean rags or mops, just prior to applying the lubricant-adhesive.

If the completed joint seal is out of position or out of shape, if it does not maintain uniform folding, if the top edges pull away from the sides of the groove, or if the seal does not satisfactorily resist the intrusion of foreign material and water, the Contractor at its expense shall remove the seal, repair or reconstruct the groove, and replace the seal.

d) **Type "C" Seal.** At no time shall the emulsion be subjected to a temperature below 40°F (4°C). Prior to application, the joint sealant may be warmed if necessary to permit proper filling of the joints. The heating shall be carefully controlled to avoid overheating of any part of the container or mixture and under no circumstance shall the emulsion be heated to a temperature greater than 130°F (54°C).

Immediately before applying the sealant, the emulsion shall be mixed with the proper amount of pastesetting agent. The components shall be mixed, preferably with a powder mixer, for 5 minutes to produce a homogeneous material.

e) **Type "E" Seal.** The rubber rod shall be compressed into the clean joint to a position such that the top is 1-1/4 inches (31mm) below the level of the finished surface.

The sealing material components shall be maintained at temperatures of not less than 60°F (16°C). To ensure adequate mixing and application, they may be preheated to a maximum of 90°F (32°), by means other than the application of direct flame.

The materials shall be power mixed for at least 8 minutes to produce a homogeneous material. When the amount of component "A" is less than 25 percent by weight of component "B", extreme care must be used to be certain that all of component "A" is incorporated in the mix.

Joint sealant in place shall comply with the following test: within 24 hours of application, a 1-cent coin shall be pressed edgewise one-half of its diameter into the joint sealer. Within 1 minute after release, the coin shall be ejected by the joint sealant, leaving the sealing compound free of abrasions or indentations deeper than 1/16 inch (1.5mm).

Joint sealant, which does not cure to a homogeneous, rubber-like compound, bond to the joint faces, or comply with any other requirements of this section, shall be removed. The joint shall be recleaned and new joint sealant placed by the Contractor at its expense.

303-1.8.8 Placing Concrete Under Adverse Weather Conditions. Concrete for structures shall not be placed on frozen ground nor shall it be mixed or placed while the atmospheric temperature is below 35°F (2°C), unless adequate means are employed to heat the aggregates and water, and satisfactory provisions have been made for protecting the work.

Concrete shall not be placed on frozen ground, nor shall concrete be mixed or placed when the atmospheric temperature is below 35°F (2°C), or when conditions indicate that the temperature may fall to 35°F (2°C) within 24 hours, except with the written permission of the Engineer and only after such precautionary measures for the protection of the concrete have been taken as approved by the Engineer.

Concrete shall be effectively protected from freezing or frost for a period of 5 days after placing.

Concrete for structures shall not be mixed or placed while the atmospheric temperature is above 115°F (46°C) unless adequate means are employed to cool the aggregate and water and satisfactory provisions have been made for protecting the work. In any case, the temperature of the concrete as placed shall not exceed 90°F (32°C).

Concrete placement shall be stopped when rainfall is sufficient to cause damage to the work.

303-1.8.9 Concrete Deposited Under Water. When conditions render it impossible or inadvisable to dewater excavations before placing concrete, the Contractor shall deposit underwater, by means of a tremie or underwater bottom-dump bucket, a layer of concrete of sufficient thickness to thoroughly seal the cofferdam. To prevent segregation, the concrete shall be carefully placed in a compact mass and shall not be disturbed after being deposited. Water shall be maintained in a still condition at the point of deposit.

A tremie shall consist of a watertight tube having a diameter of not less than 10 inches (250mm) with a hopper at the top. The tube shall be equipped with a device that will close the discharge end and prevent water from entering the tube while it is being charged with concrete. The tremie shall be supported to permit free movement of the discharge end over the entire top surface of the work and to permit rapid lowering when necessary to retard or stop the flow of concrete. The discharge end shall be closed at the start of the work to prevent water entering the tube and shall be entirely sealed at all times, except when the concrete is being placed. The tremie tube shall be kept full of concrete. When a batch is dumped into the hopper, the flow of concrete shall be induced by slightly raising the discharge end, always keeping it in the deposited concrete. The flow shall be continuous until the work is completed and the resulting concrete seal shall be monolithic and homogeneous.

The underwater bucket shall have an open top and the bottom doors shall open freely and outwardly when tripped. The bucket shall be completely filled and slowly lowered to avoid backwash and shall not be dumped until it rests on the surface upon which the concrete is to be deposited. After discharge, the bucket shall be raised slowly until well above the concrete.

303-1.9 Surface Finishes.

303-1.9.1 General. The classes of surface finish described herein shall be applied to various parts of concrete structures as specified. Exposed box sections and bridge decks shall be finished in conformance with 302-6.4, except that final finish shall be accomplished with a drag broom in lieu of burlap drags.

The invert of cast-in-place sewers and sewer structures shall be given a steel-trowel finish. The invert in circular conduit is defined as the unlined portion of lined conduit or the bottom 60 degrees of circumference of the inside of unlined conduit. Unless otherwise specified, the invert of cast-in-place storm drains shall be given a wood-float finish.

303-1.9.2 Ordinary Surface Finish. Immediately after the forms have been removed, all exterior form bolts shall be removed to a depth of at least 1 inch (25mm) inside the surface of the concrete and the resulting holes or depressions cleaned and filled with mortar, except on the interior surfaces of box girders, the bolts may be removed flush with the surface of the concrete. Mortar shall be Class “C”. White cement shall be added to the mortar in an amount sufficient to tint the mortar a shade lighter than the concrete to be repaired. Mortar shall be mixed approximately 45 minutes in advance of use. Care shall be exercised to obtain a good bond with the concrete. After the mortar has thoroughly hardened, the surface shall be rubbed with a carborundum stone in order to obtain the same color in the mortar as in the surrounding concrete. All fins caused by form joints, and other projections shall be removed and all pockets cleaned and filled. Mortar for filling pockets shall be treated as specified for bolt holes.

Ordinary surface finish shall be applied to all concrete surfaces either as a final finish or preparatory to a higher-class finish. On surfaces which are to be buried underground or surfaces which are completely enclosed (such as the cells of box girders), the removal of fins and form marks and the rubbing of a mortared surface to a uniform color will not be required. Ordinary surface finish, unless otherwise specified, shall be considered as a final finish on the following surfaces:

a) The undersurfaces of slab spans, box girders, filled-spandrel arch spans, and floor slabs between T-girders of superstructures except for grade separation structures.

b) The exposed surfaces of channel walls and the inside vertical surface of T-girders of superstructures except for grade separation structures.

c) Surfaces which are to be buried underground, covered with fill, or for surfaces of culverts above finish grade which are not visible from the traveled way.

d) Top surfaces which are to be buried underground shall be struck off and given a float finish.

303-1.9.3 Class 1 Surface Finish. Class 1 surface finish shall be applied to the following surfaces, unless otherwise specified:

a) All surfaces of superstructures for grade separation structures.

b) All surfaces of bridge piers, columns, and abutments; culvert headwalls; and retaining walls above finished ground and to at least 1 foot (300mm) below finished ground.

c) The outside vertical surfaces and bottom surface of outside girders, and the outside vertical surfaces and the undersurfaces of cantilever sidewalks, safety curbs, and floor slabs overhanging outside girders.

d) All surfaces of open spandrel arch rings, spandrel columns, and abutment towers.

e) Surfaces inside of culvert barrels having a height of 4 feet (1.2m) or more for a distance inside the barrel at least equal to the height of the culvert.

f) All interior surfaces of pumping plant motor and control rooms and the engine-generator room.

After completion of the ordinary surface finish, the entire surface specified shall be sanded with a power sander or other approved abrasive means as required to obtain a uniform color and texture.

The use of power carborundum stones or discs will be required to remove unsightly bulges or irregularities. The Class 1 surface finish shall not be applied until after the surfaces have been exposed to the elements for a period of 30 days, or until a uniform appearance of the surface can be secured.

The specifications for a Class 1 finish require a smooth, even surface of uniform appearance with unsightly bulges removed and depressions due to form marks and other imperfections repaired. The degree of care in building forms and the character of materials used in formwork are a contributing factor in the amount of such sanding and grinding required, and the Engineer shall determine the extent of such work required to meet the standard of this class of finish.

303-1.9.4 Class 2 Surface Finish. Class 2 surface finish shall be applied to the following surfaces unless otherwise specified:

All surfaces of concrete railings, including barrier railings, rail posts, rail endposts, and rail bases.

When Class 2 surface finish is specified, the ordinary surface finish and Class 1 surface finish shall be completed in succession. The process specified under Class 2 surface finish shall then be deferred until all other work which would in any way affect or mar the final finish is complete. The Contractor shall then apply a brush coat or surface film of Class "A" mortar.

303-1.10 Curing. All concrete shall be cured after the completion of the specified finishing operations and as soon as the condition of the concrete will permit without damaging the concrete. All exposed surfaces shall be cured either by continuous application of water; or by being covered with plastic sheeting, soil, sand, or burlap; or by application of a curing compound. Curing materials shall conform to 201-4.

Concrete that is water cured shall be kept continuously wet for at least 10 days after being placed; preferably being covered with at least two layers of not lighter than 10-ounce per linear yard, 40 inches wide (305 g/m^2) burlap. Handrail, base rail, railing posts, tops of walls, and similar parts of the structure, if water cured, shall be covered with burlap as prescribed above, immediately following the finishing treatment specified therefore, and such covering shall not be removed in less than 10 days.

Roadway areas, floors, slabs, curbs, walks, and the like, that are water cured may be ponded or covered with sand to a depth of at least 2 inches (50mm) in lieu of the burlap as prescribed above, as soon as the condition of the concrete will permit, and such covering shall remain wet and in place for at least 10 days, unless otherwise directed by the Engineer or prescribed by the Special Provisions.

When the surface is covered with plastic sheeting, it shall remain covered for at least 10 consecutive days. The plastic sheeting shall be laid either with edges butted together and sealed with a 2-inch (50mm) wide sealing tape or with edges lapped not less than 3 inches (75mm) and fastened with waterproof adhesive.

When a membrane curing compound is used, it shall be applied in a manner and quantity to entirely cover and seal all exposed surfaces of the concrete with a uniform film. The membrane shall not be applied to any surface until finishing operations have been completed. Such surfaces shall be kept damp until the membrane is applied. All surfaces on which a bond is required, such as construction joints, shear planes, reinforcing steel, and the like, shall be adequately covered and protected before starting the application of the curing compound to prevent any of the compound from being deposited thereon; and any such surface with which the compound may have come in contact shall immediately be cleaned. Care shall be exercised to prevent damage to the membrane seal during the curing period. Should the seal be damaged before the expiration of 10 days after the placing of the concrete, additional impervious membrane shall be immediately applied over the damaged area.

The top surface of highway bridge decks shall be cured by a combination of both the curing compound method and the water method, except that the curing compound shall be Type 2. The curing compound shall be applied continuously and progressively during the deck finishing operations immediately after the finishing operations are completed on each individual portion of the deck. The water cure shall be applied as soon as the curing compound has formed a continuous membrane, but not later than 4 hours after completion of deck finishing or, for portions of the decks on which finishing is completed after normal working hours, the water cure shall be applied as directed by the Engineer.

Should any forms be removed sooner than 10 days after the placing of the concrete, the exposed surface shall either be immediately coated with curing compound or kept continuously wet by the use of burlap or other suitable means until such concrete has cured for at least 10 days.

When tops of walls are cured by the curing compound method, the side forms, except for metal forms, shall be kept continuously wet for the 10 days following the placing of the concrete.

If there is any likelihood of the fresh concrete checking or cracking prior to the commencement of the curing operations (due to weather conditions, materials used, or for any other reason), it shall be kept damp, but not wet, by means of an indirect fine spray of water until it is not likely that checking or cracking will occur, or until the curing operations are started in the area affected.

303-1.11 Payment. Payment for concrete structures will be made in conformance with the terms of the Contract and will be based on Contract Unit Prices or lump sums as set forth in the Bid.

Where concrete is scheduled for payment on the basis of cubic yards (cubic meters), the calculation of the quantity of concrete for payment will be made only to the neat lines of the structures as shown on the Plans and on the basis of the concrete having the specified dimensions. However, all concrete shall be placed to line and grade within such tolerances, as determined by the Engineer, as are reasonable and acceptable for the type of work involved. The quantity of such concrete will be calculated considering the mortar used to cover construction joints as being concrete and no deductions will be made for rounded or beveled edges, space occupied by reinforcing steel, or metal inserts or openings 5 square feet ($0.5m^2$) or less in area. The cost of cement used in mortar for covering construction joints, patching, or other uses in the structure being constructed, in excess of that required for the design mix of the adjacent concrete, shall be included in the item of work of which said mortar is a part.

The quantity of reinforcing steel, when scheduled as a separate item, will be calculated for payment on the basis of the number of each type bar actually placed in accordance with the Plans and approved changes. The weight will be calculated using the actual lengths of bars placed and the unit weights per linear foot (meter) specified in ASTM A615/615M, A616/616M, and A617/617M.

Steel for laps indicated on the Plans, or required by the Engineer, will be paid for at the Contract Unit Price. No payment will be made for reinforcing steel in laps (whether specified or optional) which are not used, and payment will not be made for additional steel in laps which are requested by the Contractor for its convenience, or for steel used in chairs or other devices for supporting the required reinforcement.

Payment for longitudinal steel reinforcement will be made on the basis that the longest standard mill lengths will be placed; and not more than one lap will be paid for between 2 consecutive construction joints, unless otherwise authorized by the Engineer. The standard mill length for bar sizes No. 4 (No. 13M) and larger is 60 feet (18m).

When optional longitudinal construction joints are indicated on the Plans or specified, the Contractor will be permitted to lap the transverse reinforcing steel at said joints and the reinforcing steel used in such laps will be paid for at the Contract Unit Price.

303-2 AIR-PLACED CONCRETE.

303-2.1 Requirements.

303-2.1.1 General. Air-placed concrete construction shall be in accordance with this subsection and the applicable provisions of 303-1.

Only personnel skilled in the techniques of air placement of concrete shall be utilized for air-placed concrete construction.

Unless otherwise specified, air-placed concrete shall be applied by one of the following methods.

303-2.1.2 Method A (Gunite). A proportional combination of portland cement and aggregate pneumatically transported in a dry state through a pipe or hose to a nozzle where water is added immediately prior to discharge.

303-2.1.3 Method B (Shotcrete). A proportioned combination of portland cement, aggregate, and water mixed by mechanical methods, pumped in a plastic state through a pipe or hose to the nozzle where, by the addition of air, the mixture is forcibly propelled to the work.

303-2.2 Equipment. For Method A, the minimum air pressure shall be 45 psi (310 kPa) on the gun tank when 100 feet (30m) or less of hose is used and the pressure shall be increased 5 psi (35 kPa) for each additional 50 feet (15m) of hose. The pressure shall also be increased 5 psi (35 kPa) for each 25 feet (8m) that the nozzle is located above the elevation of the gun tank. The maximum nozzle diameter shall be 1-5/8 inches (42mm) unless otherwise permitted by the Engineer. Water pressure at the nozzle shall be at least 15 psi (100 kPa) above the air pressure at the nozzle.

For Method B, the pump system utilized to convey premixed concrete shall deliver a uniform and uninterrupted flow of material, without segregation or loss of the ingredients. The main run from the pump to the work shall be at least 3-inch (75mm) diameter steel pipe or flexible hose reduced to 2-inch (50mm) diameter at the point of expulsion. Aluminum pipe will not be permitted. The air compressor shall have the capacity to deliver at least 100 cubic feet ($2.8m^3$) per minute for each operating nozzle.

303-2.3 Materials, Proportioning, and Mixing.

303-2.3.1 Method A. Aggregate and portland cement shall comply with 200-1.5.4 and 201-1.2.1, respectively. Unless otherwise specified, the proportions by volume shall be 1 part cement to 4-1/2 parts sand. The sand shall contain not less than 3 percent nor more than 6 percent moisture by weight. The cement and sand shall be mixed thoroughly in a power mixer for at least 1-1/2 minutes. The dry-mixed material shall be used promptly after mixing and any material that has been mixed for more than 45 minutes shall be rejected and removed from the Work site.

303-2.3.2 Method B. The concrete class shall comply with 201-1.1.2.

303-2.4 Tests. The Contractor shall make the work accessible to facilitate the preparation of test specimens.

The strength of air-placed concrete shall be determined from cores cut from the completed work, cores cut from test panels, compression test cylinders or a combination of these methods as directed by the Engineer.

Compression test cylinders shall be prepared by the Contractor in the presence of the Engineer in 6-inch (150mm) diameter x 12-inch (300mm) long containers of 3/4 inch-square (19mm x 19mm) hardware cloth, utilizing the same mix, air pressure, water pressure and nozzle tip as for the material placed in the structure. Cylinders shall be cured in accordance with ASTM C31 and tested in accordance with ASTM C39.

Test panels prepared for core tests shall be constructed by the Contractor, of material that is representative of that used in the structure. The size of the test panel shall be as directed by the Engineer. Four-inch (100mm) minimum-diameter core specimens shall be obtained from the completed work or test panels and tested in accordance with ASTM C42 at the Contractor's expense. Core holes in the completed work shall be repaired with Class C mortar per 207-3.3.1 b) and 201-5.1.

A compressive strength test of air-placed concrete shall consist of 3 specimens. If the test specimens are 6-inch x 12-inch (150mm x 300mm) cylinders, one shall be tested at 7 days. If the test specimens are cores, one shall be tested at 14 days. The remaining 2 specimens shall be tested at 28 days. At least one set of test specimens shall be obtained for each day's work from each nozzleperson employed.

The minimum strength of test specimens shall be:

7-day (cylinders)..2,000 psi (14MPa)
14-day (cores) ....2,300 psi (16MPa)
28-day...3,250 psi (23MPa)

When a test specimen shows deficient strength, 2 cores taken from adjacent areas at the Contractor's expense may be required for each deficient specimen. Should either core prove deficient, the work shall be subject to rejection.

303-2.5 Preparation of Surfaces. Earth subgrade for air-placed concrete shall be neatly trimmed to line and grade and shall be free of all loose material. The subgrade shall be compacted as required by the Plans or Specifications.

Overexcavation shall be backfilled with earth compacted to 90 percent relative density, or air-placed concrete at the Contractor's expense.

Masonry, rock, asphalt, and concrete surfaces to be covered by air-placed concrete shall be free of loose material. Dust, dirt, grease, organic material, or other deleterious substances shall be removed and the surface washed with water.

303-2.6 Placement. All surfaces shall be dampened before application and material shall not be applied to a surface on which free water exists.

The velocity of the material as it leaves the nozzle shall be maintained uniformly at a rate satisfactory for the job conditions. Material that rebounds and does not fall clear of the work, or which collects on the surfaces, shall be removed. Rebound shall not be used in any portion of the work.

The nozzle shall be held at such distance and position that the stream of flowing material will impinge approximately at right angles to the surface being covered. Any portion of the in-place material which sags, is soft, contains sand pockets, or shows other evidence of being defective, shall be removed and replaced with new material. Reinforcement damaged or destroyed by such repairs shall be replaced by properly lapped additional steel.

Mortar blocks, metal chairs, clips, or spacers with wire ties, or other acceptable means shall be used to secure the reinforcement firmly in the position shown on the Plans.

Where material is placed on overhead surfaces, the amount of water in the mix shall be controlled to permit placement of layers of material approximately 3/4 inch (20mm) thick without sag or slough.

303-2.7 Forms and Ground Wires. The forms shall be built in accordance with the applicable provisions of 303-1. All forms shall be constructed so as to permit the escape of air and rebound.

Ground wires shall be installed in such a manner that they accurately outline the finished surface as indicated on the Plans. They shall be located at intervals sufficient to ensure proper thickness throughout. Wires shall be stretched tight and shall not be removed prior to application of the finish coat.

Headers will be required where the Plans indicate a formed edge or joint.

303-2.8 Joints. Construction joints shall be sloped off at an angle of approximately 45 degrees to the surface to which air-placed material is being applied. Before applying air-placed material in the adjacent sections, the sloped portion shall be thoroughly cleaned and wetted by means of air and water blasting.

Control joints shall be formed at the locations designated on the Plans.

303-2.9 Finish. Upon reaching the thickness and shape outlined by forms and ground wires, the surface shall be rodded off to true line and grade. Low spots or depressions shall be brought up to proper grade by placing additional air-placed material. Ground wires shall then be removed and, unless otherwise specified, the surface shall then be broom-finished to secure a uniform surface texture. Rodding and working with a wood float shall be held to a minimum.

Rebound or accumulated loose sand shall be removed an disposed of by the Contractor.

When a nozzle finish is specified on the Plans, the surface upon which the finish is to be applied shall be at the proper grade and prepared by sand and water blasting to remove all laitance prior to application of the concrete.

303-2.10 Curing. Air-placed concrete (gunite or shotcrete) shall be cured in accordance with the provisions of 303-1.10.

The Contractor shall, at all times, protect the finished work from being scarred or damaged.

303-2.11 Measurement and Payment. Quantities of air-placed concrete will be computed from measurements of actual areas in the plane of the work and the dimensions shown on the Plans. No compensation will be allowed for material placed in excess of the dimensions shown on the Plans.

The Contract Unit Price for air-placed concrete shall include full compensation for preparing the foundation, setting all formwork and grounds, furnishing and placing reinforcement, placing the concrete, finishing surfaces, curing, and structure backfill as shown on the Plans or in the Special Provisions.

303-3 PRESTRESSED CONCRETE CONSTRUCTION.

303-3.1 General. This work shall consist of furnishing and placing pretensioned or post-tensioned prestressed concrete members, and shall include the manufacture, transportation, and storage of girders, slabs, piling, and other structural of prestressed concrete, and placing of all prestressed concrete members, except piling which shall be placed as provided in 305-1.

The members shall be furnished complete including all concrete, prestressing steel, reinforcing steel, and incidental materials.

Prestressing shall be performed by either pretensioning or post-tensioning methods. The method of prestressing to be used shall be optional with the Contractor, within the limitations of these specifications.

Prior to casting any members, the Contractor shall, in accordance with 2-5.3, submit complete Working Drawings and details of the method, materials, concrete mix, and equipment to be used in the prestressing operations. Any additions or rearrangement of reinforcing steel from that shown on the Plans shall be specially noted. Such details shall outline the method and sequence of stressing and shall include complete specifications and details of the prestressing steel and anchoring devices, anchoring stresses, type of enclosures, and all other data pertaining to the prestressing operations, including the arrangement of the prestressing steel in the members, pressure grouting materials, and equipment. The Working Drawings shall also include details of the holdups and holddowns if the pretensioned method is used. Friction losses at these locations shall be included in the calculations submitted to the Engineer in accordance with 2-5.3. For any rearrangement of stress pattern, the stress calculations, signed by a Civil or Structural Engineer registered by the State of California, shall be submitted for approval by the Engineer.

303-3.2 Concrete. Concrete construction shall conform to the applicable provisions in 303-1.

The design of the precast prestressed concrete members is based on the use of concrete having an ultimate compressive strength at 28 days of not less than the values shown on the Plans.

The Contractor shall be responsible for furnishing concrete for prestressed members which contains not less than 560 pounds (330 kg) nor more than 750 pounds (445 kg) of cement per cubic yard (cubic meter) of concrete, which is workable, and which conforms to the strength requirements specified. Variation from the above cement content shall have prior approval of the Engineer.

The compressive strength of the concrete will be determined from concrete test cylinders cured under conditions similar to those affecting the member.

The use of admixtures shall be as specified in 201-1.2.4.

Concrete shall not be placed in the forms until the Engineer has inspected the placing of the reinforcement, enclosures, anchorages, and prestressing steel.

The concrete shall be vibrated internally or externally, or both, as required to consolidate the concrete. The vibrating shall be done with care and in such a manner that displacement of reinforcement, enclosures, and prestressing steel will be avoided.

Holes for anchor bars and for diaphragm dowels which pass through the member, openings for connection rods, recesses for grout, and holes for railing bolts shall be provided in the members in accordance with the details shown on the Plans. Where diaphragm dowels do not pass through the member, the dowels may be anchored in the member by embedment in the concrete or by means of an approved threaded insert.

Forms for interior cells or holes in the members shall be constructed of a material that will resist breakage or deformation during the placing of concrete and will not materially increase the weight of the member.

Lifting anchors shall be installed as detailed on the Working Drawings. In members to be placed in bridge decks, all portions of the anchor above the concrete shall be removed after the members are placed.

Side forms for prestressed members may be removed after a period of 24 hours, provided arrangements satisfactory to the Engineer are made for curing and protecting the concrete. If side forms are lift in place until transfer strength has been attained, no further curing will be required.

The steam curing method or other methods approved by the Engineer may be used for curing precast prestressed concrete members in lieu of water curing. Steam curing, if elected by the Contractor, shall conform to the following provisions:

a) After placement of the concrete, members shall be held for a 4-hour-minimum presteaming period. If the ambient air temperature is below 50°F (10°C), steam shall be applied during the presteaming period to hold the air surrounding the members at a temperature between 50°F and 90°F (10°C and 32°C).

b) All exposed surfaces of the members shall be kept wet continuously during the holding and curing period.

c) The steam shall be saturated, low pressure and shall be distributed uniformly over all exposed surfaces of the member and shall not impinge on the exposed concrete surfaces.

d) The steam hood shall be equipped with temperature recording devices that will furnish an accurate, continuous, permanent record of the temperatures under the hood during the curing period. The position of the temperature devices shall be approved by the Engineer.

e) During application of the steam the temperature gradient within the enclosure shall not exceed 40°F (22°C) per hour. The curing temperature shall not exceed 150°F (66°C) and shall be maintained at a constant level for a sufficient time necessary to develop the required compressive strength.

f) The members shall be protected from sudden temperature and moisture loss after completion of steam curing. In discontinuing the steam application, the ambient air temperature shall decrease at a rate not to exceed 40°F (22°C) per hour until a temperature has been reached 20°F (11°C) above the temperature of the air to which the concrete will be exposed.

g) After steam curing is complete, a copy of the steam charts shall be submitted to the Engineer.

303-3.3 Prestressing Steel. Prestressing steel shall be high-tensile wire conforming to ASTM A421/A421M, high-tensile wire strand conforming to ASTM A416/A416M, or uncoated high-strength steel bars conforming to ASTM A722/A722M, including all supplementary requirements.

Bars of different ultimate strength shall not be used interchangeably in the same member, unless otherwise permitted by the Engineer.

In handling and shipping bars, care shall be taken to avoid bending, injury from deflection, scraping, or overstressing of the bars. All damaged bars will be rejected.

When bars are to be extended by the use of couplers, the assembled units shall have tensile strength of not less than the specified minimum ultimate tensile strength. Failure of any one sample to meet this requirement will be cause for rejection of the heat of bars and lot of couplers. The location of couplers in the member shall be subject to approval by the Engineer.

All wires, strands, and bars shall be:

a) Protected from corrosion during shipping by a factory treatment or process.

b) Protected against abrasion during shipment and handling.

c) Installed in members to be post-tensioned after steam curing, when steam curing is used, unless otherwise approved by the Engineer.

d) Grouted in the enclosures of post-tensioned members within 48 hours after the wire or strand has been tensioned, and within 10 calendar days after removal of the prestressing steel from shipping containers, unless adequate provisions are made to inhibit corrosion. In all cases, grouting shall be completed within 96 hours after the wire or strand has been tensioned.

Wires shall be straightened, if necessary, to produce equal stress in all wires of wire groups or parallel-lay cables that are to be stressed simultaneously, or when necessary to ensure proper positioning in the enclosures.

When wires are button-headed, the buttons shall be cold-formed symmetrically about the axes of the wires, and shall develop the full strength of the wire. No cold-forming process shall be used that results in indentations in the wire.

When the button-headed wire assembly is tested as a unit in tension in accordance with California Test 641, at least 90 percent of the failures at or above the minimum guaranteed ultimate strength of the wire shall occur in the wire and not in the buttons.

Until finally encased in concrete or grouted in the member, all prestressing steel shall be protected against corrosion and damage, and shall be free of all dirt, scale, oil, grease, and other deleterious substances. Evidence of mishandling or inadequate protection such as physical damage or development of visible rust or other results of corrosion shall be cause for rejection.

No welds or grounds for welding equipment shall be made on any prestressing steel. If arc welding is utilized on other parts of a prestressed structure, the ground shall be attached directly to the part being welded. All grounding and welding operations performed after the prestressing steel has been installed shall be approved by the Engineer.

303-3.4 Anchorages and Distribution. All post-tensioned prestressing steel shall be secured at the ends by means of approved permanent anchoring devices. The anchors shall be of such design that they will not kink, neck down, or otherwise damage the prestressing steel.

The load from the anchoring device shall be distributed to the concrete by means of approved devices that will effectively distribute the load to the concrete.

All anchorage devices for post-tensioning shall hold the prestressing steel at a load producing stress of not less than 95 percent of the guaranteed minimum tensile strength of the prestressing steel, when tested in accordance with California Test 641.

Where the end of a post-tensioned assembly will not be covered by concrete, the anchoring devices shall be recessed so that the ends of the prestressing steel and all parts on the anchor devices will be at least 2 inches (50mm) inside of the end surface of the members, unless a greater embedment is shown on the Plans. Following post-tensioning, the recesses shall be filled with Class "A" mortar and finished flush.

When headed wires are used, the outside edge of any hole for prestressing wire through a stressing washer, or through an unthreaded bearing ring or plate, shall not be less than 1/4 inch (6mm) from the root of the thread of the washer or from the edge of the ring or plate.

Distribution plates or assemblies shall conform to the following requirements:

a) The final unit compressive stress on the concrete directly underneath the plate or assembly shall not exceed 3,000 psi (21MPa), and a suitable grillage of reinforcing steel shall be used in the stressed area.

b) Bending stresses in the plates or assemblies induced by the pull of the prestressing steel shall not exceed the yield point of the material or cause visible distortion in the anchorage plate when 100 percent of the ultimate load is applied.

Materials and workmanship shall conform to 304.

Should the Contractor elect to furnish anchoring devices of a type which are sufficiently large and which are used in conjunction with a steel grillage embedded in the concrete that effectively distributes the compressive stresses to the concrete, the steel distribution plates, or assemblies may be omitted.

303-3.5 Duct Enclosures. Duct enclosures for prestressing steel shall be rigid, mortar-tight, accurately placed at Plan locations, and free of angle changes, crimping, or flattening.

Ducts shall be rigid, galvanized, ferrous metal tubes with either welded or interlocked seams having sufficient strength to maintain correct alignment during placing of concrete. Galvanizing of the welded seam will not be required. Joints between sections shall be positive metallic connections sealed with waterproof tape. Transition couplings connecting ducts to anchoring devices need not be galvanized.

All duct openings or anchorage assemblies shall be provided with pipes or other suitable connections for the injection of grout after prestressing. After installation in the forms, the ends of ducts shall be covered to prevent the entry of water or debris. The Contractor shall demonstrate by positive means to the satisfaction of the Engineer that the ducts are free of water and debris prior to the installation of the prestressing steel. The inside diameter of ducts shall be at least 3/8 inch (10mm) greater than the outside diameter of the tendon.

Ducts shall be securely fastened in place to prevent movement during the placement of concrete. Vents shall be 1/2 inch (12mm) minimum-diameter standard pipe connected at the high points in the duct profile to ducts with metallic structural fasteners sealed with waterproof tape. Ends of vents shall be removed 2 inches (50mm) below the roadway surface after grouting has been completed.

303-3.6 Prestressing. All prestressing steel shall be tensioned by means of hydraulic jacks. Each jack shall be equipped with an accurate pressure gage with a dial at least 6 inches (150mm) in diameter, and each jack and its gage shall be accompanied by a recent certified calibration chart acceptable to the Engineer, showing the relationship between gage readings and total load applied by the ram. At the option of the Contractor, with the approval of the Engineer, a reverse calibrated load-cell may be used. Except where the compressive strength of concrete at time of initial prestress is specified on the Plans, tension shall not be applied or transferred to any member until the concrete in the member has attained 80 percent of the design compressive strength shown on the Plans.

Subject to prior approval by the Engineer, a portion of the total prestressing force may be applied to a member when the strength of the concrete in the member is less than the value shown on the Plans and the member may then be moved. Approval by the Engineer of such partial prestressing and moving shall in no way relieve the Contractor of full responsibility for successfully constructing the members.

The cutting and release of prestressing steel in pretensioned members shall be performed in such an order that the eccentricity of prestress will be a minimum. The prestressing steel shall be cut off flush with the end of the member and the exposed ends of the prestressing steel shall be heavily coated with roofing asphalt or an approved epoxy.

Post-tensioning will not be permitted until it is demonstrated that the prestressing steel is free and unbonded in the enclosure. In addition, prior to placing forms for the closing of box girder cells, the Contractor shall demonstrate that adjacent ducts are unobstructed.

The tensioning process, as applied to post-tensioned members, shall be so conducted that tension being applied and the elongation of the prestressing steel may be measured and recorded at all times. The record of gage pressures and elongations shall be submitted to the Engineer for approval.

Prestressing steel in post-tensioned members shall be tensioned by simultaneous jacking at both ends of the assembly, except that simple span members may be tensioned by jacking from one end only. Where jacking from one end is permitted, half of the prestressing steel in the member shall be stressed from one end and the other half from the opposite end.

Determination of the jacking stresses shall be supported by calculations, or both calculations and field tests when specified, prepared by the Contractor. The Contractor shall, prior to making field tests, submit to the Engineer for approval, its calculations and details of its proposed gages and load devices for determining the jacking load at each end of the test prestressing unit. The stress at the center will be calculated from the average of the end test loads. Jacking stresses within 2 percent of the specified values will be considered satisfactory.

The following friction coefficients shall be used in calculating friction losses. K represents the wobble of the ducts, and U represents the curvature in draped cables:

TABLE 303-3.6 (A)

Type of Steel	Type of Duct	K	U
Bright metal wire or strand	Galvanized-rigid	0.0002	0.25
Bright metal bars	Galvanized-rigid	0.0002	0.15

The maximum temporary tensile stress (jacking stress) in prestressing steel shall not exceed 75 percent of the ultimate tensile strength of the prestressing steel. The prestressing steel shall be anchored at stresses (initial stress) that will result in the ultimate retention of working forces of not less than those shown on the Plans, but in no case shall the initial stress exceed 70 percent of the ultimate tensile strength of the prestressing steel. The value to be used for ultimate tensile strength of prestressing steel shall be the specified minimum ultimate tensile strength, unless satisfactory evidence is furnished that the actual ultimate tensile strength exceeds the specified minimum ultimate tensile strength. Such evidence shall be furnished to the Engineer at the time the Contractor submits details as specified.

The loss in stress in post-tensioned prestressing steel due to creep and shrinkage of concrete, creep of steel, and sequence of stressing shall be assumed to be 25,000 psi (172MPa).

The loss in stress in pretensioned prestressing steel due to creep and shrinkage of concrete, creep of steel, and elastic compression of concrete shall be assumed to be 35,000 psi (241MPa).

Longitudinal prestressing steel in pretensioned members shall not be cut or released until tests on concrete cylinders indicate that the concrete in the member has attained a compressive strength of not less than the value shown on the Plans or the following values, whichever is greater:

TABLE 303-3.6 (B)

Diameter of Strand, inches (mm)	Compressive Strength psi (MPa)
3/8 (9.53)	3,500 (24)
7/16 (11.11)	4,000 (28)
1/2 (12.70)	4,000 (28)

The working force in the prestressing steel shall be not less than the value shown on the Plans. Unless otherwise specified or shown on the Plans, the average working stress in the prestressing steel shall not exceed 60 percent of the ultimate tensile strength of the prestressing steel.

Working force and working stress will be considered as the force and stress remaining in the prestressing steel after all losses, including creep and shrinkage of concrete, creep of steel, losses in post-tensioned prestressing steel due to sequence of stressing, friction and takeup of anchorages, and all other losses inherent in the method or system of prestressing, have taken place.

The minimum clear spacing of prestressing steel at the end of pretensioned beams shall be three times the diameter of the steel or one and one-half times the maximum size of the concrete aggregate, whichever is greater. In post-tensioned beams, the minimum clear distance between ducts at the ends of the beam shall be 1-1/2 inches (38mm) or 1-1/2 times the maximum size of the concrete aggregate, whichever is the greater.

303-3.7 Bonding and Grouting. Post-tensioned prestressing steel shall be bonded to the concrete by completely filling the entire void space between the duct and the tendon with grout placed under pressure.

Grout shall consist of Type II portland cement, water, and a nonshrinking or expansive admixture in compliance with the requirements of 201-1.

The grout shall be mixed in mechanical mixing equipment of a type that will produce uniform and thoroughly mixed grout. The water content shall be not more than 5 gallons per 100 pounds (42 L per 100 kg) of cement and shall first be added to the mixer followed by cement and admixture. Retempering of grout will not be permitted. Grout shall be continuously agitated until it is pumped.

The pumpability of the grout shall be determined by the Engineer in accordance with the U.S. Army Corps of Engineers Test Method CRD-C 79. The efflux time of a grout sample immediately after mixing shall be not less than 11 seconds at zero-quiescent time. Efflux time is the amount of time that a sample of grout requires to run out of the flow cone after the plug is pulled. Quiescent time is the amount of time that a sample of grout remains undisturbed in the flow cone.

Grouting equipment shall be capable of grouting at a pressure of at least 100 psi (700 kPa) and shall be furnished with a pressure gage having a full-scale reading of not more than 300 psi (2100 kPa).

Standby flushing equipment capable of developing a pumping pressure of 250 psi (1700 kPa) and of sufficient capacity to flush out any partially grouted ducts shall be provided.

All ducts shall be clean and free of deleterious materials that would impair bond of the grout or interfere with grouting procedures. Immediately prior to the grouting, each duct shall be thoroughly flushed with water containing 0.1 pounds (12 grams) of hydrated lime or quicklime per gallon (liter) and then blown out with oil-free air, or cleaned by another method approved by the Engineer.

All grout shall pass through a screen with 0.07 inch (1.8mm) maximum clear openings prior to being introduced into the grout pump.

Grout injection pipes shall be fitted with positive mechanical shutoff valves. Vents and ejecting pipes shall be fitted with valves, caps or other devices capable of withstanding the pumping pressures. Valves and caps shall not be removed or opened until the grout has set. Leakage of grout through anchorage assemblies shall be prevented by mechanical capping or other positive devices capable of withstanding the grouting pressure.

Grout shall be pumped through the duct and continuously wasted at the outlet until 15 seconds after all visible slugs of water or air are ejected. The outlet pipe shall then be closed and the pumping pressure held momentarily. The valve at the inlet shall then be closed while maintaining this pressure.

After post-tensioned prestressing steel has been pressure grouted, the members shall not be moved or otherwise disturbed until at least 24 hours have elapsed.

The anchorage assemblies shall not be encased in concrete until the duct grouting has been completed and the concrete surfaces against which the encasement is to be placed have been cleaned by abrasive blasting to expose the aggregate.

303-3.8 Samples for Testing. Sampling and testing shall conform to the specifications of ASTM A416/A416M, A421/A421M, and as specified in this subsection.

Samples from each size, lot, and heat of prestressing steel wires and bars, from each manufactured reel of prestressing steel strand, and from each lot of anchorage assemblies and bar couplers to be used, shall be furnished for testing. With each sample of prestressing steel wires, bars, and strands furnished for testing, there shall be submitted a certification stating the manufacturer's minimum guaranteed ultimate tensile strength of the sample furnished.

All materials for testing shall be furnished by the Contractor at its expense. The Contractor shall anticipate and furnish far enough in advance of the need in the work to allow reasonable time for testing and shall have no claim for additional compensation in the event its work is delayed awaiting approval.

All wire or bars, of each size from each mill lot, and all strands from each manufactured reel to be shipped to the site, shall be assigned an individual lot number and shall be tagged in such a manner that each such lot can be accurately identified at the jobsite. Each lot of anchorage assemblies and bar couplers to be installed at the site shall be likewise identified. All unidentified prestressing steel, anchorage assemblies, or bar couplers received at the site will be rejected.

Samples of material and tendons, selected by the Engineer from the prestressing steel at the plant or jobsite well in advance of anticipated use, shall be furnished by the Contractor as follows:

a) For wire or strand, two 7-foot (2.1m) long samples shall be furnished for each heat or reel; for bars, two 6 foot (1.8m) long samples shall be furnished for each heat.

b) If the prestressing tendon is to be prefabricated, one completely fabricated prestressing tendon 5 feet (1.5m) in length for each size of tendon shall be furnished, including anchorage assemblies. If the prestressing tendon is to be assembled at the jobsite, sufficient wire or strand and end fittings to make up one complete prestressing tendon 5 feet (1.5m) in length for each size of tendon shall be furnished, including anchorage assemblies.

c) If the prestressing tendon is a bar, one 6-foot (1.8m) length complete with two end anchorages shall be furnished, and, in addition, if couplers are to be used with the bars, two 3-foot (1.0m) lengths of bar equipped with one coupler and fabricated to fit the coupler shall be furnished.

For prefabricated tendons, the Contractor shall give the Engineer at least 10 days notice before commencing the installation of end-fittings or the heading of wires. The Engineer will inspect all end-fitting installations and wire headings while such fabrication is in progress at the plant and will arrange for all testing required.

No prefabricated tendon shall be shipped to the site without first having been released by the Engineer, and each tendon shall be tagged before shipment for identification purposes at the site. All unidentified tendons received at the site will be rejected.

Jobsite or site as referred to herein shall be considered to mean the location where the members are to be manufactured, whether at the Work site or a casting yard elsewhere.

The release of any material by the Engineer shall not preclude subsequent rejection of the material if it is damaged in transit or later found to be defective.

303-3.9 Handling. Extreme care shall be exercised in handling, storing, moving, or erecting precast prestressed concrete members to avoid twisting, racking, or other distortion that would result in cracking or damage to the members. Every precast prestressed member shall be handled, transported, and erected in an upright position and the points of support and directions of the reactions with respect to the member shall be approximately the same during transportation and storage as when the member is in its position.

After erection, the prestressed girders shall be adequately supported and braced until after the concrete of the diaphragms or of other girder bracing members has hardened.

Precast prestressed concrete piling shall be placed in accordance with the provisions for concrete piling as specified in 305-1.

303-3.10 Measurement and Payment. Precast prestressed concrete members, except piling, will be paid for at the Contract Unit Price in the Bid for furnishing and erecting precast prestressed concrete members of the various types and lengths.

Precast prestressed concrete piling will be measured and paid for as provided in 305-1.8.

Full compensation for furnishing and placing transverse connections, anchor rods, expansion joints material, and for grouting spaces and recesses between the members shall be considered as included in the Contract Unit Price for furnishing and erecting the member and no additional compensation will be allowed therefore.

303-4 MASONRY CONSTRUCTION.

303-4.1 Concrete Block Masonry[1].

1. Portions reprinted through courtesy of Concrete Masonry Association of California.

303-4.1.1 General. All materials for concrete block masonry shall conform to the requirements of 202-2.

303-4.1.2 Construction. All work shall be performed in a workmanlike manner and in full compliance with the applicable building ordinances.

All masonry walls shall be laid true, level, and plumb in accordance with the Plans.

Masonry units shall be cured, dry, and surfaces shall be clean when laid in the walls.

During construction, all partially laid walls as well as units in storage shall be protected from moisture. All concrete block units and any partially laid walls which become wet during the construction shall be permitted to dry for at least 1 week or longer, if required by weather conditions, before recommencing work.

Proper masonry units shall be used to provide for all windows, doors, bond beams, lintels, pilasters, etc., with a minimum of unit cutting. Where masonry unit cutting is necessary, all cuts shall be neat and regular and edges exposed in the finished work shall be cut with a power-driven abrasive saw.

Where no bond pattern is shown, the wall shall be laid up in straight uniform courses with regular running bond and alternate header joints in vertical alignment.

Intersecting masonry walls and partitions shall be bonded by the use of 1/4 inch (6mm) minimum-diameter steel ties at 24 inches (600mm) on centers (maximum).

Where stack bond is indicated on the Plans, approved metal ties shall be provided horizontally at 24 inches (600mm) on centers (maximum).

Mortar joints shall be straight, clean, and uniform in thickness. Unless otherwise specified or detailed on the Plans, horizontal and vertical joints shall be approximately 3/8 inch (10mm) thick with full mortar coverage on the face shells and on the webs surrounding cells to be filled. Units shall be laid with "push joints". No slushing or grouting of a joint will be permitted, nor shall a joint be made by working in mortar after the units have been laid.

Exposed walls shall have joints tooled with a round bar (or V-shaped bar) to produce a dense, slightly concave surface well-bonded to the block at the edges. Tooling shall be done when the mortar is partially set but still sufficiently plastic to bond. All tooling shall be done with a tool which compacts the mortar, pressing the excess mortar out of the joint rather than dragging it out.

If it is necessary to move a block so as to open a joint, the block shall be removed from the wall, cleaned, and set in fresh mortar.

303-4.1.3 Placing Reinforcing Steel. Reinforcing steel shall be placed as indicated on the Plans. Splices shall be lapped a minimum of 40 diameters, except that dowels other than column dowels need be lapped only 30 diameters. Column dowels shall lap 50 diameters.

Outside horizontal steel shall lap around corners 40 diameters and be carried through columns unless otherwise shown on the Plans. Inside horizontal steel shall extend as far as possible and bend into corner core. A dowel shall be provided in the foundation for each vertical bar.

Where horizontal courses are to be filled, metal stops shall be used. Use of paper stops will not be permitted. All horizontal reinforcing steel shall be laid in a course of bond beam blocks filled with grout.

Vertical cores containing steel shall be filled solid with grout, and thoroughly rodded.

Where knockout blocks are used, steel shall be erected and wired in place before three courses have been laid. Vertical cores at steel locations shall be filled as construction progresses.

Where knockout blocks are not used, vertical cores at steel locations shall be filled in lifts of not more than 4 feet (1.2m) The maximum height of pour shall be 8 feet (2.4m). Cores shall be cleaned of debris and mortar and shall have reinforcing steel held straight and in place. If ordered by the Engineer, inspection and cleanout holes shall be provided at the bottom of each core to be filled.

Reinforcing steel shall be inspected prior to placing grout.

303-4.1.4 Protection and Curing. During construction operations, all adjoining work shall be protected from mortar droppings. Concrete block masonry shall be protected from the sun and rain. When approved in advance by the Engineer, completed masonry construction may be protected with a curing compound. Except in hot weather when it may be fog sprayed sufficiently to dampen the surface, finished concrete block masonry shall not be wetted.

303-4.1.5 Measurement and Payment. Payment for concrete block masonry will be made as shown in the Bid.

Unless otherwise specified, concrete block masonry walls will be measured parallel to the finished grade, deducting the widths of full-height openings.

303-4.2 Brick Masonry.

303-4.2.1 Materials. Unless otherwise specified, brick masonry shall be constructed of Grade MW brick and cement mortar as described in 202-1.

303-4.2.2 Bricklaying. Brick shall be clean, wetted immediately before laying, and shall be laid on a full mortar bed with "push joints". In no event will slushing or grouting of a joint be permitted, nor shall a joint be made by working in mortar after the brick has been laid. Joints between courses of bricks shall be of a uniform thickness of 3/8 inch (10mm) as nearly as possible. Joints on surfaces which are not to be plastered, or on any surface that will be exposed upon completion of the work, shall be neatly struck and pointed. In all cases, the work shall be well-bonded, and if new work is to be joined to the existing or unfinished work, the contact surfaces of the latter shall first be properly cleaned and moistened.

Brickwork shall not be constructed upon a concrete foundation until at least 24 hours after such foundation has been placed. No brick shall be laid in water nor shall water be permitted to stand or run on any brickwork until the mortar has thoroughly set, except as provided in 303-4.2.3.

303-4.2.3 Protection and Curing. During construction operations, all adjoining work shall be protected from mortar droppings. Brickwork shall be protected from thc sun and rain.

Except in hot weather when it may be fog sprayed sufficient to dampen the surface, finished brick masonry shall not be wetted.

303-4.2.4 Measurement and Payment. Payment for brick masonry will be made as shown in the Bid.

Unless otherwise specified, brick masonry walls will be measured parallel to the finished grade, deducting the width of full-height openings.

303-5 CONCRETE CURBS, WALKS, GUTTERS, CROSS GUTTERS, ALLEY INTERSECTIONS, ACCESS RAMPS, AND DRIVEWAYS.

303-5.1 Requirements.

303-5.1.1 General. Concrete curbs, walks, gutters, cross gutters, alley intersections, access ramps, and driveways shall be constructed of portland cement concrete of the class and other requirements prescribed in 201-1. The finish coat to be applied to curbs shall consist of Class "B" mortar prepared as prescribed in 201-5.1. Subgrade preparation shall conform to the requirements of 301-1.

Unless otherwise specified on the Plans, and except as otherwise prescribed in 303-5.1.3 under the heading "Driveway Entrances", the minimum thickness of walks shall be 3 inches (75mm). The thickness of gutters, cross gutters, alley intersections, access ramps, and driveway aprons shall be as shown on the Plans.

303-5.1.2 Drainage Outlets Through Curb. Where existing building drains occur along the line of work, the new curb shall be suitably sleeved to provide for such drains. Similar sleeves shall be installed to serve low areas on adjacent property where drainage has been affected by the work.

The location and size of the sleeves and construction of connecting sidewalk drains shall be in accordance with 306-7.

303-5.1.3 Driveway Entrances. Driveway entrances shall be provided in new curb at all existing driveways along the line of the work, at locations shown on the Plans, and at such other locations as may be designated by the Engineer.

The fully depressed curb opening at driveway entrances shall be 1 inch (25mm) above gutter flowline at the curb face. The top of the fully depressed portion of the curb shall be finished to a transverse slope toward the gutter of 3/4 inch (19mm).

Where a walk is to be constructed across driveways to commercial establishments, the thickness thereof shall be 6 inches (150mm) unless otherwise specified or indicated on the Plans. At residential driveways, the thickness of the walk will be 4 inches (100mm) unless otherwise specified.

303-5.2 Forms.

303-5.2.1 Standard Forms. Form material shall be free from warp, with smooth and straight upper edges, and if used for the face of curb, shall be surfaced on the side against which the concrete is to be placed. Wooden forms for straight work shall have a net thickness of at least 1-1/2 inches (38mm). Metal forms for such a work shall be of a gage that will provide equivalent rigidity and strength. Curb face forms used on monolithic curb and gutter construction shall be of a single plank width when the curb face is 10 inches (250mm) or less, except for those used on curb returns. All forms used on curb returns shall be not less than 3/4 inch (19mm) in thickness, cut in the length and radius as shown on the Plans, and held rigidly in place by the use of metal stakes and clamps. The curb face form shall be cut to conform exactly with the curb face batter as well as being cut to the required length and radius. Forms shall be of sufficient rigidity and strength, and shall be supported to adequately resist springing or deflection from placing and tamping the concrete.

Form material shall be clean at the time it is used and shall be given a coating of light oil, or other equally suitable material, immediately prior to the placing of the concrete.

All forms except back planks of curb shall be set with the upper edges flush with the specified grade of the finished surface of the improvement to be constructed, and all forms shall be not less than a depth equivalent to the full specified thickness of the concrete to be placed.

Back forms shall be held securely in place by means of stakes driven in pairs at intervals not to exceed 4 feet (1.2m), one at the front form and one at the back. Clamps, spreaders, and braces shall be used to such extent as may be necessary to ensure proper form rigidity. Forms for walk, gutter, and similar work shall be firmly secured by means of stakes driven flush with the upper edge of the form at intervals not to exceed 5 feet (1.5m). Form stakes shall be of sufficient size and be driven so as to adequately resist lateral displacement.

Commercial form clamps for the curb and gutter may be used provided they fulfill the requirements specified herein.

303-5.2.2 Slip-Forms. At the option of the Contractor and with the approval of the Engineer, slip-form equipment may be used for the construction of concrete curb and gutter.

Slip-form equipment shall be provided with traveling side and top forms of suitable dimensions, shapes, and strength to support the concrete for a sufficient length of time during placement to produce curb and gutter of the required cross section. The equipment shall spread, consolidate, and screed the freshly placed concrete to provide a dense and homogeneous product.

The slip-form equipment shall have automatic sensor controls which operate from an offset control line. The line and grade of the slip-form equipment shall be automatically controlled.

303-5.3 Placing Concrete. Concrete shall be placed on a subgrade sufficiently dampened to ensure that no moisture will be absorbed from the fresh concrete.

Concrete shall be placed in curb, gutter, and curb and gutter forms in horizontal layers not exceeding 6 inches (150mm) in thickness, each layer being spaded along the forms and thoroughly tamped. Concrete may be placed in layers of more than 6 inches (150mm) in thickness only when authorized by the Engineer and the spading and tamping is sufficient to consolidate the concrete for its entire depth.

After the concrete for walk has been placed, a strikeoff shall be used to bring the surface to the proper elevation when compacted. It shall be spaded along the form faces and tamped to assure a dense and compact mass, and to force the larger aggregate down while bringing to the surface not less than 3/8 inch (10mm) of the free mortar for finishing purposes.

Concrete shall be placed in cross gutters in horizontal layers of not more than 4 inches (100mm) in thickness, each layer being spaded along the form faces and thoroughly tamped into a dense and compact mass. If internal vibrators are used, the full specified thickness may be placed in one operation.

After the concrete has been placed and tamped, the upper surface shall be struck off to the specified grade.

303-5.4 Joints.

303-5.4.1 General. Joints in concrete curb, gutter, and walk shall be designated as expansion joints and weakened plane joints.

303-5.4.2 Expansion Joints. Expansion joints shall be constructed in curb, walk, and gutter as shown on the Plans or as specified herein. Such joints shall be filled with premolded joint filler conforming with the requirements prescribed in 201-3.2. No such joints shall be constructed in cross gutters, alley intersections, access ramps, or driveways except as may be approved by the Engineer.

Joints 1/4 inch (6mm) thick shall be constructed in curb and gutter at the end of all returns except where cross gutter transitions extend beyond the curb return, in which case they shall be placed at the ends of the cross gutter transition. No joints shall be constructed in returns. Where monolithic curb and gutter is constructed adjacent to concrete pavement, no expansion joints will be required except at the EC and BC of curb returns.

Expansion joint filler 1/4 inch (6mm) thick shall be placed in walk at the EC and BC of all walk returns and around all utility poles which may project into the concrete along the line of the work. Joints 1/4 inch (6mm) thick shall be constructed in walk returns between the walk and the back of curb returns when required by the Engineer. At the EC and BC and around utility poles, the joint filler-strips shall extend the full depth of the concrete being placed. Joint filler-strips between walk and curb shall be the depth of the walk plus 1 inch (25mm) with the top set flush with the specified grade of the top of curb.

All expansion joint filler strips shall be installed vertically, and shall extend to the full depth and width of the work in which they are installed, and be constructed perpendicular to straight curb or radially to the line of the curb constructed on a curve. Expansion joint filler materials shall completely fill these joints to within 1/4 inch (6mm) of any surface of the concrete. Excess filler material shall be trimmed off to the specified dimension in a neat and workmanlike manner. During the placing and tamping of the concrete, the filler strips shall be held rigidly and securely in proper position.

303-5.4.3 Weakened Plane Joints.

a) **General.** Weakened plane joints shall be straight and constructed in accordance with Subsections b) or c) below, unless otherwise shown on the Plans.

   In walks, joints shall be transverse to the line of work and at regular intervals not exceeding 10 feet (3m). At curves and walk returns, the joint shall be radial.

   In gutter, including gutter integral with curb, joints shall be at regular intervals not exceeding 20 feet (6m). Where integral curb and gutter is adjacent to concrete pavement, the joint shall be aligned with the pavement joints where practical.

b) **Control Joint.** After preliminary troweling, the concrete shall be parted to a depth of 2 inches (50mm) with a straightedge to create a division in the coarse aggregate. The concrete shall then be refloated to fill the parted joint with mortar. Headers shall be marked to locate the weakened plane for final joint finishing, which shall be accomplished with a jointer tool having a depth of 1/2 inch (13mm) and a radius of 1/8 inch (5mm). The finished joint opening shall not be wider than 1/8 inch (3mm).

c) **Plastic Control Joint.** The joint material shall be a T-shaped plastic strip at least 1 inch (25mm) deep, having suitable anchorage to prevent vertical movement, and having a removable stiffener with a width of at least 3/4 inch (19mm). After preliminary troweling, the concrete shall be parted to a depth of 2 inches (50mm) with a straightedge. The plastic strip shall be inserted in the impression so that the upper surface of the removable stiffener is flush with the concrete. After floating the concrete to fill all adjacent voids, the removable stiffener shall be stripped. During final troweling, the edges shall be finished to a radius of 1/8 inch (3mm), using a slit jointer tool.

303-5.5 Finishing.

303-5.5.1 General. Finishing shall be completed as specified herein for the type work being performed.

303-5.5.2 Curb. The front forms may be stripped as soon as the concrete has set sufficiently. Class "B" mortar, as prescribed in 201-5.1 and thinned to the consistency of grout, shall be immediately applied to the top and face of the curb. If monolithic curb and gutter is being constructed, this mortar shall be applied to the full exposed curb face; otherwise, it shall extend 2 inches (50mm) below the gutter surface.

The face and top of the curb shall then be carefully troweled to a smooth and even finish; the top being finished to a transverse slope of 1/4 inch (6mm) toward the gutter, with both edges rounded to a radius of 1/2 inch (13mm). The troweled surface shall be finished with a fine-hair broom applied parallel with the line of the work. The edge of the concrete at all expansion joints shall be rounded to a 1/4 inch (6mm) radius. The surface of the work shall be finished as prescribed; after which the name of the Contractor, together with the year in which the improvement is constructed, shall be stamped therein to a depth of 1/4 inch (6mm) in letters not less than 3/4 inch (19mm) high, at the BC and EC of curb returns.

Joints shall conform to 303-5.4.

303-5.5.3 Walk. The forms shall be set to place the finished surface in a plane sloping up from the top of curb 2 percent when measured at right angles to the curb.

Following placing, the concrete shall be screeded to the required grade, tamped to consolidate the concrete and to bring a thin layer of mortar to the surface, and floated to a smooth, flat, uniform surface. The concrete shall then be edged at all headers, given a preliminary troweling and provided with weakened plane joints.

Walk shall be steel troweled to a smooth and even finish. All formed edges shall be rounded to a radius of 1/2 inch (13mm). Edges at expansion joints shall be rounded to a radius of 1/8 inch (3mm). Preliminary troweling may be done with a longhandled trowel or "Fresno", but the finish troweling, shall be done with a hand trowel. After final troweling, walk on grades of less than 6 percent shall be given a fine-hair-broom finish applied transversely to the centerline. On grades exceeding 6 percent, walk shall be finished by hand with a wood float. Walk shall be remarked as necessary after final finish, to assure neat uniform edges, joints, and score lines.

Scoring lines, where required, shall have a minimum depth of 1/4 inch (6mm) and a radius of 1/8 inch (3mm). When longitudinal scoring lines are required, they shall be parallel to, or concentric with, the lines of the work. Walk 20 feet (6m) or more in width shall have a longitudinal center scoring line. In walk returns, one scoring line shall be made radially midway between the BCR and ECR. When directed by the Engineer, longitudinal and transverse scoring lines shall match the adjacent walk. The Contractor shall have sufficient metal bars, straightedges, and joint tools on the Work site.

Headers shall remain in place for at least 16 hours after completion of the walk but must be removed before the Work is accepted.

303-5.5.4 Gutter. After the concrete has been thoroughly tamped to force the larger aggregate into the concrete and bring to the top sufficient free mortar for finishing, the surface shall be worked to a true and even grade by means of a float, troweled with a longhandled trowel or "Fresno", and wood-float finished. The flowline of the gutter shall be troweled smooth for a width of approximately 4 inches (100mm) for integral curb and gutter and 4 inches (100mm) on either side of the flowline on cross gutters and longitudinal gutters. The outer edges of the gutter shall be rounded to a radius of 1/2 inch (13mm).

Side forms shall remain in place for at least 24 hours after completion of the gutter, but must be removed before the work will be accepted.

Joints shall conform to 303-5.4.

303-5.5.5 Alley Intersections, Access Ramps, and Driveways. Alley intersections, access ramps, and driveways shall be constructed as specified for concrete pavement in 302-6, except final finishing for alley intersections, access ramps, and the sloping portion of driveways shall be done by hand with a wood float and the remaining portion of the driveway finished as specified for walks in accordance with 303-5.5.3.

303-5.6 Curing. Immediately after finishing operations are completed, curing compound conforming to 201-4.1 shall be applied.

The curing compound shall be applied in a manner to entirely cover all exposed surfaces of the concrete with a continuous membrane.

No power equipment used for the preparation of subgrade will be permitted adjacent to concrete curb, gutter, or alley intersections until the fourth day following placement of the concrete. The placement of bituminous pavement adjacent to concrete curb, gutter, or alley intersections will not be permitted until the seventh day following the placement of concrete nor will concrete paving operations be permitted until the seventh day where placing or finishing equipment will ride on the previously placed concrete. If admixtures, additional cement or Type III cement is used to obtain high early strength concrete in accordance with 201-1, grading operations will be permitted on the second day following the placement of the concrete and paving operations on the third day.

303-5.7 Repairs and Replacements. Any new work found to be defective or damaged prior to its acceptance shall be repaired or replaced as approved by the Engineer.

303-5.8 Backfilling and Cleanup. Backfilling to the finished surface of the newly constructed improvement must be completed before acceptance of the Work.

Upon completion of the work the surface of the concrete shall be thoroughly cleaned and the site left in a neat and orderly condition.

303-5.9 Measurement and Payment. Payment for concrete curbs, walks, gutters, cross gutters, alley intersections, access ramps, and driveways will be made as shown in the Bid.

303-6 STAMPED CONCRETE.

303-6.1 General. Stamped concrete shall be imprinted with special tools to provide the pattern specified. Colored stamped concrete shall also conform to 303-7.

The Contractor shall install a sample for each pattern included in the Work. The sample shall be a minimum of 10 ft^2 ($1m^2$) which shall be subject to inspection and approval by the Engineer. All other areas to be installed shall match the texture of the approved area.

All coloring and curing compounds used in the Work shall be from the same manufacturer.

303-6.2 Concrete Placement. Placing of concrete shall conform to 302-6 and 303-5. The minimum slab thickness shall be 4 inches (100mm). The maximum size aggregate in the top 2 inches (50mm) shall be 3/8 inch (9.5mm).

303-6.3 Pattern. The pattern of stamped concrete shall be implanted, indented, imprinted, or stamped into the surface by means of forms, molds, or other approved devices. The impressions shall be approximately 3/8 inch (9.5mm) in width, not to exceed 1/2 inch (13mm) in depth and be ungrouted unless otherwise specified. Expansion joints and control joints shall be located so as not to disrupt the pattern.

Joints shall conform to 303-5.4.

303-6.4 Curing. Curing shall conform to 303-1.10. Curing compound used in the Work shall be of a single type and manufacturer.

303-7 COLORED CONCRETE.

303-7.1 General. Colored concrete shall be produced by Method A or B as specified below. The Contractor shall provide a sample in the Work for each color specified of a size satisfactory to the Engineer. The sample shall be inspected and approved by the Engineer before proceeding with the Work.

303-7.2 Method A (Dry Shake). Color hardener shall be applied evenly to the plastic surface by a dry shake method using approved manufacturer's printed instructions, otherwise, it shall be applied in two applications, wood floated after each, and troweled only after the final floating.

303-7.3 Method B (Integral Color).

a) **Color Conditioning Admixture.** Color conditioning admixture shall be added to the concrete in accordance with approved manufacturer's printed instructions. No calcium chloride shall be added to the concrete. Other non-chloride admixtures may be added subject to approval of the Engineer.

b) **Pure Mineral Pigments.** Pure mineral pigments shall be added to the concrete in accordance with approved manufacturer's printed instructions. Other admixtures specified or approved by the Engineer shall be added to the concrete in accordance with 201-1.2.4, except that no calcium chloride, or other admixture containing chloride ions shall be used.

303-7.4 Curing. Colored concrete shall be cured with a liquid curing compound in matching color and complying with the requirements of ASTM C309. The curing compound shall be applied in accordance with approved manufacturer's printed instructions. Curing with clear, white-pigmented or fugitive-dye curing compounds, or with plastic or other waterproof membranes will not be allowed. When approved by the Engineer, colored concrete may be cured by a continuous indirect fine spray of water for a minimum of 10 days.

303-8 Pervious Concrete.

303-8.1 General. Pervious concrete shall conform to 201-1.6. Pervious concrete shall be constructed a minimum of 6 inches (150mm) thick, unless otherwise specified in the Special Provisions or shown on the Plans.

303-8.2 Construction Test Section. The Contractor shall construct a test section using the same equipment, and placing crew as it proposes to use for the Work. The test section shall be a minimum of 225 square feet (7m^2) and may be placed non-contiguously. The test section will be tested by the Agency in accordance with 211-3. The test section(s) may be incorporated into the Work if so approved by the Engineer. The Engineer shall be notified at least 24 hours in advance of construction of the test section.

303-8.3 Subgrade. The subgrade shall be compacted to a relative compaction of between 92 and 96 percent. Chemical stabilization is not be permitted. The subgrade shall have a minimum permeability of 0.5 inch per hour as determined by ASTM D3385.

303-8.4 Forms. Forms shall conform to 302-6.2. Forms shall be placed to the full depth thickness of the pervious concrete. Forms shall be capable of supporting mechanical equipment without deformation during or following spreading, strike-off and compaction operations.

303-8.5 Placement. Pervious concrete shall be deposited over the entire width of the formed area. The concrete shall be placed and finished by a combination of vibration and consolidation using a vibrating truss screed and rollers such that the result is a coherent, stable mass of interconnected voids. Strike-off by hand-rodding may be allowed in small, constrained areas if so approved by the Engineer. During initial placement and prior to strike-off, pervious concrete shall not be walked on. Combined strikeoff/compaction equipment will be allowed if the compaction force exceeds 10 psi (70 kPa). The compactive force shall be calculated by dividing the weight of the compaction device by the contact area. The contact area for a roller shall be one percent of the total surface area of the roller.

Immediately after strike-off, the surface shall be consolidated to the cross-section shown on the Plans. Consolidation shall be performed by use of a roller or plate compactor with a vertically applied pressure of 10 psi (70kPa). Hand tampers shall be used to consolidate the pervious concrete along the slab edges immediately adjacent to the forms. The Contractor shall exercise caution during consolidation to insure that the surface voids do not become sealed. Over-consolidation and accumulation of excess paste at the concrete surface will not be allowed.

When tested in accordance with 302-5.6.2 the finished surface shall not deviate more than 3/8 inch (9.5mm) from a 10 foot (3m) straightedge laid on the surface. Surface depressions shall be corrected immediately after consolidation by placing fresh pervious concrete in the depressions and compacting using a hand tamper.

After consolidation the surface shall be immediately water fogged until curing material is applied.

303.8.6 Weakened-Plane Joints.

303-8.6.1 General. Weakened-plane joints shall be constructed at the locations shown on the Plans or at regular intervals not to exceed two times the width of the placement or 40 feet (12m) on-center, whichever is less. Joints shall be constructed to a depth of 1/4 the thickness of the pervious concrete with a minimum depth of 1 1/2 inches (38mm). Transverse construction joints shall be installed whenever placing is suspended for more than 20 minutes.

303-8.6.2 Construction. Weakened-plane joints may be constructed by one of the following methods as specified in the Special Provisions.

a) Weakened-plane joints shall be constructed in fresh concrete and shall be formed by rolling a small roller to which a beveled fin has been attached around the circumference. Such joints are constructed immediately after compaction and prior to curing.

b) Weakened-plane joints shall be constructed by saw cutting. Saw cut joints shall be constructed as soon as the concrete can be saw cut without raveling the joint edge. Only the area occupied by the saw shall be exposed. Immediately after sawing each joint, the area shall be heavily misted with water and recovered in accordance with 302-8.7.

303-8.7 Curing. Curing procedures shall begin within 20 minutes of concrete placement or 15 minutes when the wind is in excess of 5 miles per hour (8 kilometers per hour). When the wind is in excess of 5 miles per hour (8 kilometers per hour), evaporation retardant and curing compound or water fogging shall be used. The surface is then covered securely with polyethylene sheeting. Evaporation retardants, polyethylene sheeting and curing compound shall conform to 201-4. Polyethylene sheeting shall have a minimum thickness of 6 mils (152μm). The cover shall remain securely in place for a minimum of 7 days.

303-8.8 Acceptance. Pervious concrete shall have an infiltration rate of a minimum of 100 inches per hour (254 cm/hr) or as specified in the Special Provisions, when tested in accordance with 211-3.

303-8.9 Measurement and Payment. Payment for pervious concrete shall be made at the Contract Unit Price per square foot (square meter) or cubic yard (cubic meter) as shown on the Bid.

SECTION 304 - METAL FABRICATION AND CONSTRUCTION

304-1 STRUCTURAL STEEL.

304-1.1 General.

304-1.1.1 Shop Drawings. The Contractor shall, in accordance with 2-5.3, submit shop drawings which show details, dimensions, sizes of material, and all information and data necessary for the fabrication of the metal work, including full details of the match markings. Shop Drawings required to be submitted by the Contractor shall conform to the applicable provisions of 304-1.4.

When required by the Plans or Special Provisions, the Contractor shall furnish to the Engineer, before acceptance of the Work, detailed drawings of the structure to be built. Inasmuch as the drawings will be retained by the Agency as permanent records, they must be in the form of printable transparencies of quality satisfactory to the Engineer.

304-1.1.2 Falsework Plans. The Contractor shall, in accordance with the provisions of 2-5.3 and 303-1.6, submit detailed Working Drawings of falsework to be used. Approval of such Working Drawings will be based upon compliance with the design criteria set forth for falsework for concrete structures in 303-1.6.2. Approval of falsework Working Drawings will not relieve the Contractor of responsibility for the results obtained by use of such Working Drawings. The Contractor shall be fully responsible for providing falsework capable of supporting all loads which are applied.

304-1.2 Methods and Equipment. When requested before starting erection of any structural members, the Contractor shall inform the Engineer fully as to the methods it proposes to follow and the amount and character of equipment proposed for use in such work. The use of such methods and equipment shall be subject to the approval of the Engineer, but this approval shall not be considered as relieving the Contractor of the responsibility for the safety of its methods or equipment, or for carrying out the work in full accordance with the Plans and Specifications.

304-1.3 Inspection. An inspector or other authorized representative of the Engineer will examine the metals and metal items to be fabricated before they are worked in the shop and may exercise constant surveillance over the Work during its progress, with full power to reject all materials or workmanship not conforming to the Plans and Specifications.

The Contractor shall give the Engineer 5 days minimum advance notice before commencement of the fabricating operations to permit ample time for the inspection of the materials.

The Engineer shall be furnished complete copies of all mill reports prior to commencing fabrication. The Contractor shall furnish ample means and assistance for sampling all materials. Arrangements shall be made for the Engineer to have free access at all times to any portion of the workshops where work is being done under these specifications.

No fabricating, machining, cutting, welding, assembling, or painting shall be done except with the knowledge of the Engineer. Any work done otherwise will be subject to rejection.

The acceptance of any material or finished member by the Engineer shall not preclude subsequent rejection if it is later found to be defective. Rejected material and workmanship shall be promptly repaired or replaced by the Contractor.

Samples of materials, except castings, shall be cut from stock designated by the Engineer or will be selected from items furnished. Gray iron, steel, and bronze castings shall be cast with test coupons.

304-1.4 Steel Structures. Fabrication and erection of structures shall conform to "Specifications for the Design, Fabrication and Erection of Structural Steel for Buildings" of the American Institute of Steel Construction (AISC), except for any conflicts with the applicable building code which may exist, and except that the following sections are considered as excluded from the AISC Specifications:

1) Section 1.1 Plans and Drawings
2) Section 1.4 Material
3) Section 1.24 Shop Painting
4) Section 1.26 Inspection

The subject matter excluded from the foregoing AISC Specifications shall be superseded by the applicable provisions of these Specifications.

In addition to complying with AISC Specifications, when work involving the use of the high strength bolts is included in the Work, the design and construction of such work shall conform to ASTM A 325.

The design, fabrication and erection of structural steel and all similar work incidental or appurtenant to steel construction for highway bridges shall be performed and accomplished in accordance with the latest Standard Specifications for Highway Bridges adopted by the American Association of State Highway and Transportation Officials.

The Plans or Specifications will designate the members to be painted or galvanized.

304-1.5 Workmanship.

304-1.5.1 General. Workmanship and finish shall be equal to the best general practice in modern steel fabricating shops.

Before being laid out or worked, rolled material shall be straight. If straightening is necessary, it shall be done by methods approved by the Engineer. Kinks and bends in the material will be cause for rejection. Heat shrinking of low alloy structural steels will not be permitted.

If straightening is necessary in the field, only methods approved by the Engineer shall be used.

Following the straightening of a bend or buckle, the surface of the metal shall be carefully inspected for evidence of fracture.

Portions of the work exposed to view shall be finished neatly. Shearing, flame cutting, and chipping shall be done carefully and accurately. Undercut gusset plates will not be accepted. All sharp corners and edges, and edges that are marred, cut, or roughened in handling or erection, shall be slightly rounded by grinding or other suitable means.

304-1.5.2 Holes for Bolts or Rivets.

General. Holes shall be either punched full size, punched and reamed, or drilled. The finished hole shall be 1/16 inch (1.5mm) larger than the nominal diameter of the rivet.

Holes punched full size shall have all burrs and sharp edges removed. The diameter of the die shall not exceed that of the punch by more than 3/32 inch (2.4mm).

a) **Shop Rivets.** Holes for shop rivets shall be subpunched, or subdrilled at the fabricator's option, 1/4 inch (6mm) less in diameter than that of the finished holes, and shall be reamed to size with the parts assembled, with the following exceptions:

1) Holes in material thicker than 7/8 inch (22mm) shall not be punched; however, at the fabricator's option, they may be subdrilled to the diameter specified for subpunching or may be drilled full size with the parts assembled, provided that the parts are adequately bolted or clamped together.

2) Holes in rolled beams and plate girders, including stiffeners and active fillers at bearing points, may be subpunched 1/8 inch (3mm) less in diameter than that of the finished holes; and reamed to size (after assembly) in material not thicker than the nominal diameter of the rivet less 1/8 inch (3mm).

3) Holes in material not more than 7/8 inch (22mm) thick, for rivets which do not transfer stress caused by external vertical loading, may be punched full size or, at the fabricator's option, may be subpunched 1/8 inch (3mm) less in diameter than the finished holes and reamed to size after assembly. This applies to holes for stitch rivets, lateral, longitudinal, or sway bracing and their connecting material, lacing, stay plates, diaphragms which do not transfer shear or stress, inactive fillers, and stiffeners not at bearing points. However, holes through assembled material shall not pass through both reamed plies and plies punched full size unless the reamed holes have been subpunched for the fabricator's convenience, or the assembled material is not over five plies thick, of which the main material consists of not more than three plies.

b) **Field Rivets.** Holes for field rivets shall be subpunched or subdrilled at the fabricator's option, 1/4 inch (6mm) less in diameter than that of the finished holes, and shall be reamed to size through steel templates with hardened steel bushings, with the following exceptions:

1) Field splices in plate girders and in the chords of trusses shall be reamed with the members assembled. Other field connections may be reamed with the members assembled, at the

fabricator's option. Chord splices in truss members shall, in all cases, be reamed or drilled with at least three abutting sections assembled and with milled ends of compression chords in full bearing.

2) Assemblies, such as floor systems to girders, complete trusses, rolled beam spans connected by diaphragms, and portals to trusses shall be reamed with the members assembled if so indicated on the Plans, and otherwise at the fabricator's option.
3) Field connections of lateral, longitudinal, or sway bracing shall conform to the requirements of holes for shop rivets.
4) Holes in material thicker than 7/8 inch (22mm) shall not be punched, but shall be subdrilled to the diameter specified for subpunching, or drilled full size with parts assembled.

The accuracy of the punching shall be such that for any group of holes when assembled, 75 percent shall admit a rod equal to the diameter of the cold rivet at right angles to the plane of the connection. Otherwise the holes shall be reamed. When the extent of the reaming is such that the holes cannot be properly filled or accurately adjusted after reaming, the faulty member shall be discarded and replaced.

Mispunched members shall not be corrected by welding without the approval of the Engineer.

304-1.5.3 Reamed Work. Reaming shall be done after the pieces forming a member are assembled and so firmly bolted together that the surfaces are in close contact. Burrs and sharp edges of each reamed hole under both rivet heads shall be removed with a countersinking tool making 1/16 inch (1.5mm) fillets. The pieces shall be taken apart before riveting, if necessary, and any shavings removed. If it is necessary to take the members apart for shipping or handling, the pieces reamed together shall be so marked that they may be reassembled in the same position. Reamed parts shall not be interchanged.

304-1.5.4 Drilled Holes. Drilled holes shall be 1/16 inch (1.5mm) larger than the nominal diameter of the rivet. Burrs and sharp edges of each drilled hole under both rivet heads shall be removed with a countersinking tool making a 1/16 inch (1.5mm) fillet. Burrs on the outside surfaces shall be removed. If members are drilled while assembled, the parts shall be held securely together while the drilling is being done.

Drilled holes shall be drilled to finish size while all of the thicknesses of metal are assembled, or subdrilled and reamed as required for punched and reamed holes.

Holes shall be clean-cut, without torn or ragged edges. Holes that must be enlarged to admit rivets shall be reamed. Drilling shall be done accurately.

304-1.5.5 Assembling Steel.

a) **General.** Steel parts shall be assembled in the shop or in the field in accordance with 304-1.5.5 b) and 304-1.5.5 c).

b) **Shop Work.** At the time of assembling and riveting, bolting, or welding, steel surfaces in contact for shop or field connection shall be thoroughly cleaned of rust, loose mill scale, dirt, grease, or other material foreign to the steel. No paint shall be applied to contact surfaces prior to riveting, bolting or welding.

   Riveted or bolted trusses, continuous plate girder and I-beam spans, skew portals, skew connections, rigid frames, bents, and towers shall be completely assembled in the shop and accurately adjusted to line and camber. Holes for field connections shall be drilled or reamed while assembled. Holes for other connections, except those in lateral, longitudinal, and sway bracing shall be drilled or reamed in the shop with the connecting parts assembled; or drilled or reamed to a metal template with hardened bushing, without assembling.

   Long-span truss work shall be assembled in lengths of not less than three abutting panels, the members adjusted for line and camber, and holes for field connections drilled or reamed while assembled.

Field riveted or bolted joints for girders shall be completely assembled, the members adjusted for line and camber, and holes for field connections drilled or reamed while assembled.

Field butt joints for welded girders shall be completely assembled with the members adjusted for line and camber and prepared to fit for welding.

All machinery shall be completely assembled. All bearing shall be fitted to the specified clearances and alignment. Gear reductions and all line gears shall have gear center distances set and the gears properly matchmarked.

c) **Field Work.** The parts shall be accurately assembled as shown on the Plans and all matchmarks shall be followed. The material shall be carefully handled so that no parts will be bent, broken, or otherwise damaged. Hammering which will injure or distort the members will not be permitted. Bearing surfaces and surfaces to be in permanent contact shall be cleaned before the members are assembled. Unless erected by the cantilever method, truss spans shall be erected on blocking so placed as to give the trusses proper camber. The blocking shall be left in place until the tension chord splices are fully riveted or bolted and all other truss connections pinned and bolted. Rivets or bolts in splices of butt joints of compression members and rivets or bolts in railings shall not be driven or torqued until the span has been erected in place, temporarily bolted, and the member is supporting its own weight. Splices and field connections shall have half of the holes filled with bolts and cylindrical erection pins (in approximately equal numbers) before riveting or bolting.

Splices and connections carrying traffic during erection shall have three-fourths of the holes so filled.

Fitting-up bolts shall be of the same nominal diameter as the rivets, and cylindrical erection pins shall be 1/32 inch (0.8mm) larger.

The drifting done during assembling shall be only such as to bring the parts into position, and not sufficient to enlarge the holes or distort the metal.

If any holes must be enlarged to admit the rivets, they shall be reamed.

Connecting parts assembled in the shop for the purpose of reaming holes in field connections shall be matchmarked, and a diagram showing such marks shall be furnished to the Engineer.

304-1.5.6 Riveting.

a) **Shop Work.** Rivets shall be heated uniformly to a light cherry-red color and shall be driven while hot. Rivets, when heated and ready for driving, shall be free from slag, scale, and other adhering matter. When driven, they shall completely fill the holes. The heads shall be of approved shape, full size, neatly formed, concentric with the shank, free from fins, and in full contact with the surface of the member.

Loose, burned, or otherwise defective rivets shall be replaced. In removing rivets, care shall be taken not to injure the adjacent metal. Caulking or recupping will not be permitted.

Rivets shall be driven by direct-acting riveters where practicable. If rivets are driven with a pneumatic hammer, a pneumatic bucker shall be used if practicable.

b) **Field Work.** Pneumatic hammers shall be used for field riveting. Connections shall be accurately and securely fitted up before the rivets are driven.

Drifting shall be only such as to draw the parts into position and not sufficient to enlarge the holes or distort the metal. Unfair holes shall be reamed or drilled. Rivets shall be heated uniformly to a light cherry-red color and shall be driven while hot. They shall not be overheated or burned.

Rivet heads shall be full and symmetrical, concentric with the shank, and shall have full bearing all around. They shall not be smaller than the heads of the shop rivets. Rivets shall be tight and shall grip the connected parts securely together. Cup-faced dollies, fitting the head closely to ensure good bearing, shall be used. Sufficient air compressor capacity shall be maintained to keep the air pressure at 100 psi (700 kPa) at the hammers.

Caulking or recupping will not be permitted. In removing rivets, the surrounding metal shall not be injured. The removal of loose or defective rivets by flame cutting will not be permitted, except upon written permission of the Engineer.

304-1.5.7 Bolted Connections. Bolts shall be unfinished bolts, turned bolts, or high strength steel bolts, as shown on the Plans or in the Special Provisions.

Unfinished or turned bolts shall have hexagonal heads and nuts and shall be of such length that they will extend entirely through the nut but not more than 1/4 inch (6mm) beyond. Bolts in tension shall have two nuts.

Unfinished bolts in shear shall have not more than one thread within the grip. The diameter of the unfinished bolt shall not be more than 1/32 inch (0.8mm) smaller than the diameter of the hole.

The threads of turned bolts shall be entirely outside the grip. The holes for turned bolts shall be reamed and the bolts shall be finished to provide a driving fit. Approved nut locks or flat washers 1/4 inch (6mm) thick shall be furnished, as specified.

Bolted connections using high strength steel bolts shall conform to ASTM A 325.

Holes for bolted connections using high strength steel bolts shall conform to the requirements specified in 304-1.5.2, except as follows: Holes that are required to be subdrilled and reamed may be subdrilled 1/8 inch (3mm) less in diameter than that of the finished hole, provided that the offset of any hole in any ply measured from the outer ply after the hole is finished does not exceed 1/32 inch (0.8mm) and that no more than 20 percent of the holes shall provide as much offset as 1/32 inch (0.8mm).

304-1.6 Joint and Connections.

304-1.6.1 Edge Planing. Sheared edges of plates more than 5/8 inch (16mm) in thickness and carrying calculated stress shall be planed to a depth of 1/4 inch (6mm).

304-1.6.2 Facing of Bearing Surfaces. Surfaces of bearing and base plates and other metal bearing surfaces that are to come in contact with each other, with ground concrete surfaces, or with asbestos sheet packing shall be finish-machined flat to within 1/32 inch (0.8mm) tolerance in 12 inches (300mm) and to within 1/16 inch (1.5mm) tolerance overall. Surfaces of bearing and base plates and other metal bearing surfaces that are to come in contact with preformed fabric pads, elastomeric and elastic bearing pads, or portland cement grout shall be finish-machined flat to within 1/8 inch (3mm) tolerance in 12 inches (300mm) and to within 3/16 inch (5mm) tolerance overall.

Steel slabs, where not in contact with other metal bearing surfaces, may be heat-straightened in lieu of machining, provided the above tolerances are met.

304-1.6.3 Abutting Joints. When shown on the Plans, abutting joints shall be faced and brought to an even bearing. Where joints are not faced, the opening shall not exceed 1/4 inch (6mm).

304-1.6.4 End Connection Angles. Floor beams, stringers, and girders having end connection angles shall be built to exact length back to back of connection angles. If end connections are faced, the finished thickness of the angle shall not be less than that shown on the detail drawings.

304-1.6.5 Web Plates. In girders having no cover plates and which are not to be encased in concrete, the top edge of the web plate shall not extend above the backs of the flange angles and shall be not more than 1/8 inch (3mm) below at any point.

304-1.6.6 Fit of Stiffeners. End stiffener angles of girders and stiffener angles intended as supports for concentrated load shall be milled or ground to secure an even bearing against the flange angles. All fillers under stiffener angles shall fit sufficiently tight to exclude water after being painted.

304-1.6.7 Pin and Bolted Connections. Pilot and driving nuts shall be used in driving pins. Pins shall be so driven that the members will take full bearing on them. In field assembling, the pin nuts on pin connections and the bolts on bolted connections shall be screwed up tight and the threads, except when high strength bolts are used, burred at the face of the nuts with a pointed tool.

304-1.6.8 Pins and Rollers. Pins and rollers shall be accurately turned to the dimensions shown on the drawings and shall be straight, smooth, and free from flaws. The final surface shall be produced by a finishing cut.

Pins and rollers more than 7 inches (175mm) in diameter shall be forged and annealed.

In pins larger than 9 inches (230mm) in diameter, the forging shall be permitted to cool to a temperature below the critical range under suitable conditions to prevent injury by too rapid cooling, and a hole not less than 2 inches (50mm) in diameter shall be bored full length along the axis of the pin before being annealed.

Pin holes in structural members shall be bored true to the specified diameter, smooth and straight, at right angles with the axis of the member and parallel with each other unless otherwise required. The final surface shall be produced by a finishing cut.

The distance outside to outside of holes in tension members and inside to inside of holes in compression members shall not vary from that specified more than 1/32 inch (0.8mm). Holes in built-up members shall be bored after riveting, bolting, or welding is completed.

The diameter of the pin hole shall not exceed that of the pin by more than 1/50 inch (0.5mm) for pins 5 inches (125mm) or less in diameter, or 1/32 inch (0.8mm) for larger pins.

304-1.6.9 Screw Threads. Screw threads shall make close fits in the nuts and shall be Unified Standard Series conforming to ANSI B1.1-1960.

304-1.7 Bearings and Anchorage. Anchor bolts shall be either headed bolts, installed with or without pipe sleeves, or swage bolts installed in drilled holes, as detailed on the Plans. The anchor bolts shall be carefully installed to permit true positioning of the bearing assemblies.

When anchor bolts are installed in pipe sleeves, the pipes shall be completely filled with grout at the time the grout pads are constructed or at the time the bearing assemblies or masonry plates are placed. Swage bolts installed in holes shall be either sulphured in or grouted in as shown on the Plans.

All bearing assemblies shall be set level and to the elevations shown on the Plans. Adjustments in the horizontal positions of bearing assemblies shall be made for temperature as directed by the Engineer.

In conformance with the details shown on the Plans, masonry plates and the bearing plates of bearing assemblies shall be set on ground concrete surfaces, on preformed fabric pads, or on grout pads.

Grout to be placed below masonry plates or bearing plates of the bearing assemblies and in anchor bolt sleeves shall consist of Class "E" mortar. Concrete areas to be in contact with the grout shall be cleaned of all loose or foreign matter that would in any way prevent bond between the mortar and the concrete surfaces and shall be kept thoroughly saturated with water for a period of not less than 24 hours immediately prior to placing the grout. The grout shall contain only sufficient moisture to permit packing and shaping. The grout shall completely fill the anchor bolt sleeves and shall be tightly packed under the masonry or bearing plates to provide full bearing. After placing, all exposed surfaces of the grout pads shall be kept covered with a heavy thickness of burlap saturated with water for a period of 3 days. All improperly cured or otherwise defective grout shall be removed and replaced at the Contractor's expense.

Immediately before setting bearing assemblies or masonry plates directly on ground concrete surfaces, the Contractor shall thoroughly clean the surfaces of the concrete and the metal to be in contact and shall apply a coating of non-sag polysulfide or polyurethane caulking conforming to Federal Specification TT-S-230, Type II, to contact areas to provide full bedding of the metal in the caulking.

Preformed fabric pads shall be furnished and installed at the locations and in accordance with the details shown on the Plans.

The preformed fabric pads shall be composed of multiple layers of 8 ounces per square yard (270 g/m^2) cotton duck impregnated and bound with high-quality natural rubber or of equivalent and equally suitable materials compressed into resilient pads of uniform thickness. The number of plies shall be sufficient to produce the specified thickness, after compression and vulcanizing. The finish pads shall withstand compression loads perpendicular to the plane of the laminations of not less than 10,000 psi (70MPa) without extrusion or detrimental reduction in thickness.

304-1.8 Expansion and Rotation Assemblies. Before leaving the shop or foundry, the rockers or roller nests shall be completely assembled with the bearing plates for checking and approval by the Engineer.

304-1.9 Welding. All welding shall conform to the requirements of the *Structural Welding Code AWS D1.1*, and these Specifications.

Inspection of welding made to control the quality of welds and workmanship will be performed in accordance with the requirements of the AWS. All welding may be subject to radiographic or other nondestructive testing. Such nondestructive testing will be performed without charge to the Contractor except that if a weld is shown to be defective, all costs involved in reinspection shall be borne by the Contractor.

Weld metal shall be sound throughout, except that very small gas pockets and small inclusions of oxide or slag may be permitted if well dispersed and if none exceeds 1/16 inch (1.5mm) in greatest dimension, and provided further that the sum of the greatest dimension of all such defects in any 1 square inch (625mm^2) weld area does not exceed 3/8 inch (9.5mm).

All welding shall be performed in such a manner that the Brinnell hardness of the weld metal and heat-affected zone is within the following limits:

Minimum Brinnell Hardness

$$\text{In US Std. Units} = \frac{\text{Minimum specified tensile strength of parent metal (psi)}}{500}$$

$$\text{In SI Units} = \frac{\text{Minimum specified tensile strength of parent metal (MPa)}}{3.5}$$

Maximum Brinnell Hardness

$$\text{In US Std. Units} = \frac{\text{Maximum specified or tested tensile strength of parent metal (psi)} + 50}{500}$$

$$\text{In SI Units} = \frac{\text{Maximum specified or tested tensile strength of parent metal (MPa)} + 50}{3.5}$$

All welding of structural steel (ASTM A36, A242, and A441) shall be performed by either the submerged or gas-shielded arc process, or with low hydrogen electrodes. Low hydrogen electrodes for welding low alloy steel shall conform to the requirements of the Military Specifications for Electrodes (mineral covered, low hydrogen) for Welding Medium and High Tensile Steels, MIL-E-18038 (Ships). All welding of low alloy structural steel shall be qualified by procedure tests before fabrication is commenced.

Low hydrogen electrodes shall be stored for holding in an approved low hydrogen oven at a temperature of from 300°F to 400°F (150°C to 205°C) to control the moisture in the coating on the electrode.

Low hydrogen electrodes not packaged in a moisture-proof container shall be restored by rebaking for 2 hours or more at a temperature of from 450°F to 500°F (232°C to 260°C) in an approved oven.

Low hydrogen electrodes which have been removed from their moisture-proof containers shall be stored in an approved oven at a temperature of from 300°F to 400°F (150°C to 205°C) after rebaking.

Areas contiguous to welding operations shall be preheated to a minimum temperature of 300°F (150°C) when necessary to prevent distortion of weld cracking. Preheating to a temperature in excess of 400°F (205°C) will not be required.

Unless otherwise shown on the Plans or specified in the Special Provisions, bearing assemblies that are to be machined after welding shall be stress-relieved by heat treatment before machining, in accordance with the "Structural Welding Code" AWS D1.1.

Portions of members in bearing assemblies or in direct bearing shall be straightened, planed, or otherwise corrected after fabrication as necessary to provide fill bearing on bearing assemblies or bearing areas on level bearing plates.

Where the end of a stiffener plate is shown "tight fit" on the Plans, the end of the plate shall be so fitted that it bears on the beam flange with at least point bearing. Local clearances between the end of the plate and the flange shall not exceed 1/16 inch (1.5mm).

Unless otherwise shown on the Plans or specified in the Special Provisions, erection bolts required for welded splices or welded connections may be left in place and the ends of all such erection bolts which project beyond the nut shall be burned off flush with the face of the nut. Where the bolt does not project, the end of the bolt and nut shall be tack welded to prevent loosening of the nut. Burning off projecting bolt ends and tack welding shall be performed prior to painting.

Welders, welding operators, and tackers shall be prequalified in accordance with the specifications of AWS D1.1 and shall produce written evidence of qualification satisfactory to the Engineer.

Electroslag and electrogas welding will not be permitted without the written approval of the Engineer.

304-1.10 Torch Cutting. The use of a cutting torch is permissible if the metal being cut is not carrying stress during the operation. The radius of re-entrant flame-cut fillets shall be as large as possible, but never less than 3/4 inch (19mm). To determine the net area of members so cut, 1/8 inch (3mm) shall be deducted from the flame-cut edges. Stresses shall not be transmitted through a flame-cut surface.

Where cutting with a torch, cuts shall be true to line with a maximum deviation of 1/16 inch (1.5mm). All burned edges shall be finished by grinding.

304-1.11 Bent Plates. Cold-bent, load-carrying, rolled steel plates shall conform to the following:

a) They shall be taken from the stock plates so that the direction of bending will be at right angles to the direction of rolling.

b) The radius of bend, measured to the concave face of the metal, shall not be less and preferably shall be greater than shown in the following table, in which "T" is the thickness of the plate:

c) Before bending, the corners of the plate shall be rounded to a radius of 1/16 inch (1.5mm) throughout that portion of the plate at which the bending is to occur.

If a shorter radius is essential, the plates shall be bent hot, and such plates shall conform to requirement 1) above.

TABLE 304-1.11 (A)

Angle Through Which Plate is Bent	Minimum Radius
61° to 90°	1.0 T
91° to 120°	1.5 T
121° to 150°	2.0 T

304-1.12 Measurement and Payment.

304-1.12.1 General. Steel structures will be paid for at a Contract Unit Price per pound (kg) for structural steel, and at Contract Unit Prices per pound (kg) for cast steel and cast iron or as indicated in the Bid. The pay quantities shall be determined by scale weights or, if permitted by the Engineer, by computed weights obtained as provided in this 304-1.12.3. The Contractor will be paid only for material actually used in the completed structure.

Computed weights shall be used to determine pay quantities of alloy and carbon steel when members contain both alloy and carbon steel.

The weight of erection bolts, paint, boxes, crates, and other containers used for packing and the materials used for supporting members during transportation will not be included in the weights of material for pay purposes.

The weight of structural steel to be paid for shall not exceed the computed weight by more than 1-1/2 percent. The weight of cast steel or cast iron to be paid for shall not exceed the computed weights by more than 7-1/2 percent. If the scale weight of any member is less than 97-1/2 percent of the computed weight of that member, the member will be rejected and will not be included in pay quantities.

If computed weights are used, the weight to be paid for will be the calculated weight as established by the Engineer, and no allowance will be made for weight in excess thereof.

When the estimated quantities of structural steel, cast steel, and cast iron required for the work as described in the Specifications and shown on the Plans as final quantities, said estimated quantities shall be the final quantities for which payment will be made unless the dimensions of the work as shown on the Plans are revised by the Engineer. If such revisions result in an increase or decrease in the quantity of structural steel, cast steel, or cast iron, the final quantities for payment will be revised accordingly. The estimated quantities of structural steel required for the work shall be considered as approximate only. No allowance will be made in the event quantities determined from the details and dimensions as shown on the Plans do not equal the estimated quantities. These provisions concerning basis of payment shall not be construed to waive the specification of ASTM A6/A6M.

304-1.12.2 Scale Weights. Scale weight shall be actual weight of the members as determined on accurate scales. When carload or truck weights are used, a record shall be submitted to the Engineer, which shall contain an itemized statement of the dunnage and the members included in each lot.

For any protective coating, scale weights will be reduced as follows:

0.25 percent for each coat of oil

0.5 percent for each shop coat of paint

3.5 percent for hot-dip galvanizing

Scale weights of members will not be required when the quantities of structural steel are designated on the Plans or in the Specifications as final quantities.

304-1.12.3 Computed Weights. The computed weight shall be obtained by the use of the following rules and assumptions:

a) The density of structural and cast steel shall be assumed at 0.2833 pounds per cubic inch (7849 kg/m^3). The density of cast iron shall be assumed at 0.2604 pounds per cubic inch (7208 kg/m^3).

b) The weights of rolled shapes and of structural plates shall be computed on the basis of their nominal weights and dimensions, as shown on the Shop Drawings, deducting for copes, cuts, and open holes, exclusive of rivet and bolt holes.

c) Since no deduction is made in the computed weight of structural steel members by reason of rivet or bolt holes, the computed weights of the completed members will be obtained by adding to the above weights the weights of the heads of all rivets and bolts in the structure, both shop-driven and field-driven. Full compensation for all rivets and bolts furnished in excess of the actual number in place in the completed structure shall be considered as included in the Contract Unit Price per pound (kg) for structural steel in place and no additional compensation will be allowed therefore.

Should the computed weights be used to determine pay quantities, the weight of rivet heads shall be assumed as follows:

TABLE 304-1.12.3 (A)

Diameter of Rivet inches (mm)	Weight of 100 Heads lbs (kg)
1/2 (13)	5.0 (2.3)
5/8 (16)	9.7 (4.4)
3/4 (19)	16.0 (7.3)
7/8 (22)	24.0 (10.9)
1 (25)	35.0 (15.9)
1-1/8 (29)	49.0 (22.2)
1-1/4 (32)	78.0 (35.4)

Should computed weights be used to determine pay quantities of high strength steel bolts, the weights of portions of bolts outside the grip (including two washers and one nut) shall be assumed as follows:

TABLE 304-1.12.3 (B)

Diameter of Bolt inches (mm)	Weight of 100 Bolts (Each complete with 2 washers and one nut, less grip length) lbs (kg)
5/8 (16)	46 (21)
3/4 (19)	71 (32)
7/8 (22)	105 (48)
1 (25)	145 (66)
1-1/8 (29)	194 (88)
1-1/4 (32)	259 (117)

d) The weight of castings and fillets shall be computed from the dimensions shown on the Shop Drawings, deducting for all openings or cuts in the finished casting.

e) The weight of pins and rollers shall be computed from the dimensions shown on the Shop Drawings, deducting for all holes, openings, pockets, and metal removed by machine finishing.

Pilot nuts and driving nuts for each size of pin shall be furnished for erection work and the weights of such nuts will not be included in the weight of structural steel for which payment is made.

f) The weight of bolts, cap screws, anchor bolts, nuts, washers, except as limited by 3) above, and anchor pipe sleeves remaining in the finished structure shall be computed on the basis of their nominal weights and dimensions as shown on the shop drawings.

g) No allowance will be made for the weight of paint in computing the weights of metal for payment.

h) The weight of shop an field fillet welds shall be assumed as follows:

TABLE 304-1.12.3 (C)

Size of Filled Weld inches (mm)	Weights lbs per ft (kg/m)
3/16 (5)	0.08 (0.098)
1/4 (6)	0.14 (0.141)
5/16 (8)	0.22 (0.251)
3/8 (10)	0.30 (0.392)
1/2 (13)	0.55 (0.663)
5/8 (16)	0.80 (1.005)
3/4 (19)	1.10 (1.417)
7/8 (22)	1.50 (1.899)
1 (25)	2.00 (2.453)

i) If computed weights are used to determine the pay quantities of galvanized metal, the weight to be added to the calculated weight of the base metal for the galvanizing shall be determined from the table of weights of zinc coatings specified by ASTM A 153.

304-1.12.4 Payment. Structural steel, cast steel, cast bronze, or cast iron will be paid for by the pound as shown in the Bid. If the Plans or Specifications require the metal to be galvanized, the weight of the metal shall include the weight of the zinc coating.

Full compensation for furnishing and placing sheet packing, preformed fabric pads, elastomeric or elastic bearing pads, and caulking, and for grouting masonry or bearing plates as shown on the Plans, shall be considered as included in the Contract Unit Price per pound for structural steel.

Cleaning and painting structural steel will be paid for as provided in 310.

304-2 METAL RAILINGS.

304-2.1 Metal Hand Railings.

304-2.1.1 General. The materials for metal hand railings shall conform to the requirements of 206-5.1. Except where a Standard Plan is referred to on the Plans, the Contractor shall in accordance with 2-5.3, submit Shop Drawings showing the details and dimensions of all metal hand railings.

304-2.1.2 Fabrication. Welding shall conform to the requirements of the "Structural Welding Code" AWS D1.1 for steel, and to the requirements of the "Specifications for Aluminum Structures" of the Aluminum Association, for aluminum alloys. All exposed welds shall be ground flush with adjacent surfaces.

Railing panels shall be straight and true to dimensions. Adjacent railing panels shall align with each other with a variation not to exceed 1/16 inch (1.5mm). Joints shall be matchmarked.

For structures on curves, either horizontal or vertical, the railing shall conform closely to the curvature of the structure by means of series of short chords. The lengths of the chords shall be the distance center to center of rail posts.

Steel railing units shall be galvanized after fabrication in accordance with 210-3.

Completed aluminum railing units shall be anodized after fabrication conforming to the Aluminum Association Standard for Anodized Architectural Aluminum, Class I Anodic Coating, AA-C22-A41.

304-2.1.3 Installation. The railing shall be erected in accordance with the Plans on anchor bolts, or in holes formed by inserts provided in the concrete railing base to receive the railing posts. Sheet metal inserts shall be removed before the erection of the railing.

No railing shall be erected on the structure until the sidewalk to which it is to be attached is completed and all falsework supporting the system is released.

The railing shall be carefully erected, true to line and grade. Posts and balusters shall be vertical with the deviation from the vertical for the full height of the panel not exceeding 1/8 inch (3mm).

After erecting the railing, any abrasions or exposed steel shall be repaired in accordance with 210-3.5.

304-2.1.4 Measurement and Payment. The various types of railing will be measured and paid for by the linear foot (meter) from end to end along the face of the railing, including terminal sections.

304-2.2 Flexible Metal Guard Rail.

304-2.2.1 General. This work shall consist of the construction of metal beam guard railing, at the locations and in accordance with the Plans and Specifications. Materials for metal beam guard rail shall conform to 206-5.2.

304-2.2.2 Installation. The posts shall be firmly placed in the ground. The space around posts shall be backfilled with selected earth, free of rock, placed in layers approximately 4 inches (100mm) thick, and each layer shall be moistened and thoroughly compacted.

Posts shall be placed at equal intervals as shown on the Plans, except that the end posts may be spaced closer to adjacent posts if directed by the Engineer.

The bolted connection of the rail element to the post shall withstand a 5,000-pound (22 kN) pull at right angles to the line of the railing.

The metal work shall be fabricated in the shop, and no punching, cutting, or welding will be permitted in the field. Rail elements shall be lapped so that the exposed ends will not face approaching traffic. Terminal sections shall be installed in accordance with the manufacturer's recommendations.

Surplus excavated material remaining after the guard rail has been constructed shall be removed from the Work site.

304-2.2.3 Measurement and Payment. The railing will be measured and paid for by the linear foot from end to end along the face of the railing, including terminal sections.

304-3 CHAIN LINK FENCE.

304-3.1 General. Materials for chain link fence shall conform to 206-6.

304-3.2 Fence Construction. Posts shall be spaced at not more than 10 foot (3m) intervals, measured from center to center of posts and shall be vertical.

Corner posts shall be installed at changes in fence line where the horizontal angle of deflection is greater than 30 degrees. Slope posts shall be installed at changes in surface grade greater than 5 percent.

Footings for fence posts shall be concrete of the class specified in Table 201-1.1.2 (A). Footings shall be crowned at the top to shed water. Line post footings for fabric 5 feet (1.5m) or less in height shall have a minimum depth of 30 inches (750mm) and a minimum diameter of 8 inches (200mm). Line post footings for fabric more than 5 feet (1.5m) in height shall have a minimum depth of 36 inches (900mm) and a minimum diameter of 8 inches (200mm). All other footings, unless otherwise indicated on the Plans or in the Special Provisions, shall have a minimum depth of 36 inches (900mm) and a minimum diameter of 12 inches (300mm).

End and gate posts shall be braced to the nearest line post. Corner and slope posts shall be braced to the nearest line post on each side. The bracing shall consist of a horizontal brace 12 inches (300mm) below the top of the fence fabric and a diagonal tension member. The tension member shall be a 3/8 inch (9.5mm) steel rod with turnbuckle or other approved tightening device. When a top rail is specified, the horizontal brace shall be omitted at intermediate, slope, end, and corner posts.

End and gate posts shall be braced to the nearest line post. Corner and slope posts shall be braced to the nearest line post on each side. The bracing shall consist of a horizontal brace 12 inches (300mm) below the top of the fence fabric and a diagonal tension member. The tension member shall be a 3/8 inch (9.5mm) rod with turnbuckle or other approved tightening device. When a top rail is specified, the horizontal brace shall be omitted at intermediate, slope, end, and corner posts.

Unless otherwise specified, all fences shall be installed with a top rail and a bottom tension wire. When the top rail is omitted, a top and bottom tension wire shall be used.

The fabric shall be placed on the outward facing side of the posts and shall be installed so that the top edge projects over the top rail of the fence and the bottom tension wire by 3 inches (75mm). The fabric shall be stretched taut and securely fastened to the posts, the top rail, and the bottom tension wire. The tension wire shall be installed parallel to the line of the fabric. The bottom of the fabric shall extend to within 2 inches (50mm) of the natural ground or paved surface. High points of ground shall be excavated to clear the bottom of the fabric and depressions shall be filled and compacted to within 1 inch (25mm) of the bottom of fabric.

The fabric shall be fastened to end, corner, slope, and gate posts with 3/16 inch (5mm) thick by 3/4 inch (19mm) wide carbon steel tension bars and the tension bars shall be fastened with steel tension bar bands spaced at 16 inches (400mm) intervals. Steel tension bar bands shall have a minimum thickness of 1/8 inch (3mm) and a minimum width of 1 inch (25mm). The fabric shall be fastened to line posts, top rails, and tension wires with tie wires or metal bands. Tie wires or metal bands shall be placed on line posts at intervals of approximately 16 inches (400mm), and on top rail and tension wire at intervals of approximately 24 inches (600mm).

304-3.3 Installation of Gates. The widths of any gates to be installed will be indicated on the Plans.

Gates with fabric 7 feet (2m) or more in height shall have a horizontal stiffener. Vertical stiffeners shall be installed at a maximum of 8-foot (2.4m) centers. A 3/8 inch (9.5mm) adjustable tension rod shall be installed on all gates over 4 feet (1.2m) in width.

The corners of gate frames shall be fastened together and reinforced with a fitting designed for the purpose or by welding. All welds shall be ground smooth.

Chain link fence fabric shall be attached to the gate frame by the use of tension bars and tie wires as specified for fence construction, and suitable tension connectors spaced at approximately 16-inch (400mm) intervals.

The swing gates shall be hung by at least two steel or malleable iron hinges, designed to securely clamp to the gate post and permit the gate to be swung back 180 degrees.

Gates shall be provided with a combination steel or malleable iron catch and locking attachment of approved design. Stops to hold gates open and a center rest with catch shall be provided where required.

304-3.4 Measurement and Payment. Chain link fence will be measured parallel to the ground slope along the line of the completed fence, deducting the widths of gates and openings.

Gates will be paid for at the Contract Unit Price for each size of gate required by the Plans, which price shall include full compensation for furnishing the gates together with all necessary fittings and hardware, and doing all the work involved in installing the gate complete in place. If double gates are required, each double gate will be paid for at the Contract Unit Price and such unit price shall include furnishing and installing both leaves.

Full compensation for clearing the line of the fence and disposing of the resulting material, excavating high points in the existing ground between posts, excavating and furnishing and placing concrete footings, connecting new fences to structures and existing fence as shown on Plans and any other related work shall be considered as included in the Contract Unit Price per linear foot (meter) of fence and no additional compensation will be made therefore.

304-3.5 Security Fencing. Security fencing shall conform to 206-6.6, the Plans, and the Special Provisions. Security fencing shall completely enclose all open excavations and remain in place until backfill has been placed to the approximate level of the adjacent ground.

304-3.5.1 Payment. Unless otherwise specified in the Special Provisions, payment for security fencing shall be included in the various Bid items that require security fencing.

304-4 STRUCTURAL STEEL PLATE PIPE AND ARCH CONSTRUCTION.

304-4.1 General. The materials for structural steel plate pipe and arches shall be as designated on the Plans or in the Special Provisions and as specified in 207-12.

Damage to galvanized surfaces during construction shall be repaired in accordance with 210-3.5. Damage to other coatings shall be repaired in accordance with the coating requirements for the original material. Asphalt mastic coating in accordance with 207-11.5.4 may be applied after field assembly of the structure.

304-4.2 Structures and Footings. Footings and arch bearings shall be constructed as shown on the Plans. Where shown on the Plans, inlet and outlet structures shall be constructed in connection with structural steel plate installations. When such structures are constructed, the ends of plates shall be placed flush or cut off flush with the structures face, unless otherwise directed by the Engineer.

304-4.3 Plate Assembly. Structural steel plates shall be assembled in accordance with the manufacturer's instructions. A copy of the manufacturer's instructions shall be furnished to the Engineer prior to assembling the plates. Each side of the arch shall rest on a galvanized steel angle or channel anchored to the footings. When plates of 2 dissimilar thicknesses are involved in one cross section of an installation, the thickness of structure steel plates will be identified in the Plans or in the Special Provisions.

Distorted circular plate pipes shall be placed with the major axis vertical. When distortion is accomplished by use of field devices, they shall not be removed until the supporting earthfill is completed, unless otherwise permitted by the Engineer. The devices shall be removed prior to construction of inlet or outlet structures and before acceptance of project.

The method for distorting plates in the field shall conform to details shown on the Plans. The vertical diameter throughout that portion of the pipe between outer shoulder lines of the roadway shall be increased to the approximate percentages listed in the following table:

Pipes using 0.280 in or 0.249 inch (7.11mm or 6.32mm) top and side plates .... 1%.

Pipes using 0.218 in or 0.188 inch (5.54mm or 4.78mm) top and side plates ....2%.

Pipes using 0.168 inch, 0.138 inch, or 0.109 inch (4.27mm, 3.31mm, or 2.77mm) top and side plates ....3%.

Between the outer shoulder lines of the roadway and the outer ends of the pipe, the distortion may be decreased uniformly to zero.

SECTION 305 - PILE DRIVING AND TIMBER CONSTRUCTION

305-1 PILE DRIVING.

305-1.1 General. Piles shall be accurately located, and driven either vertically or to the prescribed batter as indicated on the Plans. No greater variation from the vertical or specified batter line than 1/4 inch per foot (20mm/m) of length will be permitted. Piles driven with greater variation and those seriously damaged in driving shall be removed or cut off, and replaced with new piles. Should any pile be heaved by the subsequent driving of adjacent piles it shall be redriven.

The pile tip elevations shown on the Plans are approximate, and are to be used as a basis for establishing quantities for piling, including exploratory piles, for bidding purpose only.

When required in the Special Provisions, one pile of the type selected or designated for the Work shall be driven in each pier and abutment area as an exploratory pile. The location of these piles will be determined by the Engineer.

Exploratory piles shall be driven to determine the length and penetration that will be required for the balance of the piles. No piles other than exploratory piles shall be driven at each pier or abutment until such determination has been made by the Engineer, and has been reported to the Contractor.

The conditions under which the exploratory piles are to be driven shall be as ordered by the Engineer. These exploratory piles shall be furnished and driven by the Contractor, and under normal circumstances shall be left in place and utilized as one of the specified piles.

Exploratory piles shall be driven with the same size and type hammer operating with the same effective energy and efficiency as that to be used in driving the remainder of the piles.

The Engineer will specify the tip elevation to which the piling shall be driven for each pier or abutment. All piles shall be driven to the tip elevation established by the Engineer, or deeper if necessary, to develop the bearing value as determined by the formula prescribed in 305-1.5.

Excavations required in the areas through which the piles are to be driven shall be made before any pile is driven. No excavation may be made below the bottom of the pile footing elevation, unless approved by the Engineer.

When piles are to be driven through the bridge approach embankment and the depth of the embankment at the pile location is in excess of 5 feet (1.5m), the pile shall be driven in a hole drilled through the embankment. The hole shall have a diameter of not less than the butt diameter of the pile plus 6 inches (150mm). After driving the pile, the annular space around the pile shall be filled to ground surface with dry sand or pea gravel.

No piles shall be driven within 25 feet (7.5m) of any concrete that has not attained a minimum compressive strength of 2,000 psi (14MPa).

To eliminate hazard to life and to preclude dirt or debris from falling or being thrown into them, the tops of driven pile shells or drilled holes shall be securely covered immediately upon withdrawal of the mandrel or drilling equipment.

305-1.2 Driving Equipment. Pile hammers shall be approved types that develop sufficient energy to drive the pile at a penetration rate of not less than 1/8 inch (3mm) per blow at the required bearing value, and shall develop energy per blow at each full stroke of the piston of not less than 1 foot-pound for each pound (3 Joules for each kilogram) of weight driven. Vibratory pile hammers may be used only when approved by the Engineer.

Drop hammers may be used on timber piles only. Drop hammers shall weigh not less than 3,000 pounds (1350 kg) and shall be equipped with proper leads and hoisting equipment to handle the work efficiently. The fall of the hammer shall not exceed 10 feet (3m).

Steam or air hammers shall be furnished with boiler or air capacity at least equal to that specified by the manufacturers of the hammers being used. The boiler or compressor shall be equipped at all times with an accurate pressure gage. The valve mechanism and other parts of steam or air hammers shall be maintained in first-class condition so that the length of stroke and number of blows per minute for which the hammer is designed can be obtained at all times. Steam or air hammers not meeting the Specifications shall be removed from the Work.

When necessary to obtain the specified penetration and with the approval of the Engineer, the Contractor may be required to supply and operate one or more water jets and pumps; or to furnish the necessary drilling apparatus and drill holes and drive the piles therein as specified in 305-1.3.

The use of jets at locations where the stability of embankments or other improvements would be endangered will not be permitted. Jetting normally will not be permitted in cohesive soils. All jetting must be suspended and the pile driven for the last 3 feet (1m) to specified bearing.

The cost of any jetting or drilling that may be required shall be included in the Contract Unit Price for driving piles, or for other applicable items of work.

The use of followers, underwater hammers, or hammers not in leads will not be permitted unless authorized by the Engineer. When a follower or underwater hammer is authorized, the first pile in each bent shall be furnished sufficiently long for it to be driven without a follower or underwater hammer, and the bearing value and penetration shall be determined from this pile.

305-1.3 Drilled Holes.

305-1.3.1 Driven Piles. When approved by the Engineer, piles may be driven in predrilled holes. The holes shall have a diameter not greater than the diameter of the pile at ground surface. The depth of the predrilled hole shall be adjusted by the Contractor (as directed by the Engineer) as the work proceeds in order to maintain adequate bearing. Piles shall be driven sufficiently to secure full bearing. Minimum penetration of the pile below the bottom of the predrilled hole shall be 5 feet (1.5m) unless otherwise authorized by the Engineer.

305-1.3.2 Drilled Holes for Cast-in-Place-Piles. Holes for cast-in-place concrete piles shall be drilled dry to the tip elevations shown on the Plans or determined by the Engineer. All holes shall be inspected for straightness prior to placing concrete therein. When viewed from the top, more than one-half of the entire bottom area must be visible.

305-1.3.3 Drilling Material. All loose material existing at the bottom of the hole after drilling operations have been completed shall be removed before placing concrete in the hole.

Material resulting from drilling holes shall be disposed of as provided in 7-8.1.

305-1.3.4 Water. The use of water for drilling operations, or for any other purpose where it may enter the hole, will not be permitted. Surface water shall not be permitted to enter the hole and all water which may have infiltrated into the hole shall be removed before placing concrete therein.

305-1.3.5 Casings. Suitable casings shall be furnished and placed when required to prevent caving of the hole before concrete is placed therein. Casing used in drilling operations shall be removed from the hole as concrete is placed therein. The bottom of the casing shall be maintained not more than 5 feet (1.5m) nor less than 1 foot (0.3m) below the top of the concrete during withdrawal and placing operations, unless otherwise permitted by the Engineer. Separation of the concrete during withdrawal operations shall be avoided by hammering or otherwise vibrating the casing.

305-1.3.6 Reinforcing Cage. The reinforcing cage shall be placed and secured symmetrically about the axis of the pile and shall be securely blocked to clear the sides of the hole.

305-1.3.7 Concrete. Care shall be exercised to ensure that the concrete in the hole is dense and homogeneous. Vibration of the concrete during placing will not be required. After the hole has been filled with concrete, the top 10 feet (3m) of the concrete, or the length of the reinforcing, whichever is the greater, shall be vibrated.

305-1.4 Driving. During driving operations, the pile heads shall be protected and held in position by the use of a steel driving block or anvil. Timber piles shall be shaped to closely fit the driving head. The heads of the piles may be protected by means of heavy steel or wrought iron rings. The heads of concrete piles or casings shall be protected from direct impact of the hammer by a cushion block which shall be maintained in good condition during the entire driving operation. This cushion block shall be arranged so that any reinforcing bars projecting above the piles will not be displaced or damaged in driving. For driving steel H-beam piles and shells without a mandrel for cast-in-place concrete piles, steel combination driving heads and pilots shall be used. The driving heads shall closely fit the top of the steel pile or shell and shall extend down the sides of the pile at least 4 inches (100mm). Piles materially out of line, as determined by the Engineer, shall be pulled and replaced.

305-1.5 Bearing Value. Piles shall be driven to the penetration and bearing value shown on the Plans as a minimum. Timber piles shall not be driven to a bearing value exceeding 20 tons (180 kN). The bearing value shall be determined from the applicable formula in the following schedule:

For piles driven with a drop hammer

U.S. Standard Measures Equation: $$P = \frac{2WL}{s + 1}$$

Metric Equation: $$P = \frac{WL}{6\,(s + 2.54)}$$

For piles driven with a single acting steam or air hammer

U.S. Standard Measures Equations: $$P = \frac{2WL}{s + 0.1} \quad \text{or} \quad P = \frac{2E}{s + 0.1}$$

Metric Equations: $$P = \frac{WL}{6\,(s + 2.54)} \quad \text{or} \quad P = \frac{E}{6\,(s + 2.54)}$$

For piles driven with a double acting steam or air hammer

U.S. Standard Measures Equations: $$P = \frac{2L\,(W + ap)}{s + 0.1} \quad \text{or} \quad P = \frac{2E}{s + 0.1}$$

Metric Equations: $$P = \frac{L\,(W + ap)}{6\,(s + 2.54)} \quad \text{or} \quad P = \frac{E}{6\,(s + 2.54)}$$

Where:

P = Safe bearing load developed by the pile in pounds (kN).

W = Weight of the hammer in pounds (newtons).

L = Length of stroke or height of fall of the hammer in feet (meters).

s = Penetration of the pile into the ground per blow in inches (millimeters) taken as the average over the last 10 blows. Penetration shall be measured at a time when there is no appreciable rebound of the hammer and the preceding blow was struck upon a sound pile head or driving block.

a = Effective area of the piston in in^2 (m^2).

p = Mean effective steam pressure in the case of steam hammers or means effective air pressure in the case of air hammers, in psi (pascals).

E = Manufacturer's rating of energy developed by the hammer in foot-pounds (joules).

305-1.6 Cutoff and Extension. Timber piles which are to be capped shall be accurately cut off so that true bearing is obtained on every pile without the use of shims. Other timber piles shall be cut off on the square at the elevation designated. Piles inaccurately cut off shall be replaced. Splicing of timber piles will not be permitted, except upon written permission of the Engineer.

Except for piles that are to be capped with concrete, the tops of treated piles, after cutoff, shall be treated as specified in 204-2.3.

Concrete piles shall be cut off at such elevations that they will extend into the cap or footing as indicated on the Plans. Concrete piles may be cast the full length of the reinforcing bars, provided that the concrete is cut off to expose the steel as shown on the Plans after the piles have been driven. When concrete piles are driven or cut off below the elevation of the bottom of the cap, the pile section shall be extended to the elevation of the bottom of the cap by means of a reinforced concrete extension constructed in accordance with the details shown on the Plans. Concrete shall be removed from the end of the pile to expose sufficient reinforcing steel to permit a lap of at least 35 diameters.

Steel shells or concrete casings for cast-in-place concrete piles shall be cut off at the designated elevations. The work of cutting off precast concrete piles or concrete casings shall be performed in such a manner as to avoid spalling or damaging the pile below cutoff. In case of such damage, the pile shall be replaced or repaired as required by the Engineer.

All cut off lengths of piling shall become the property of the Contractor and shall be disposed of outside the Work area.

305-1.7 Load Testing. If load tests are required, they shall be performed on the exploratory piles. The loading shall not be applied until 48 hours after the pile is driven or, in the case of cast-in-place piles, the concrete has attained a minimum compressive strength of 2,000 psi (14MPa).

A loading test shall consist of the continuous application of a load of twice the design load to the pile being tested. The pile shall be considered to have a bearing value equal to the design load if the permanent settlement produced by such test loading is not greater than 1/4 inch (6mm).

Unless otherwise permitted by the Engineer, the loading tests shall be completed before the remaining piles are cast or driven.

When a loading test is required, the Contractor shall provide suitable facilities and equipment by means of which a prescribed test load can be transmitted vertically to each pile to be tested. Provisions for varying the applied load shall also be made, and the loads must be in known and measurable increments, applied axially to the pile.

The marks, gages, dials, or other instruments of any loading equipment required to determine settlement of the pile, shall be arranged so as to provide convenient observation thereof without danger to the observer or the equipment. All test equipment shall be accurately calibrated and shall be approved by the Engineer.

The test loads shall be applied under the direction of the Engineer and at such rate or in such increments as he may specify. When a load test of a pile is commenced, the test shall be continuous, and the Contractor shall furnish all facilities on a 24-hour-day, 7-day-week basis until the test is completed. Forty-eight hours after all deflection and settlement has ceased, or sooner if directed by the Engineer, the test load shall be removed at such rate or in such increments as the Engineer may direct. If the results of the above-described operations indicate that excessive permanent settlement of the test pile has occurred, the pile shall be driven to such additional depths as the Engineer may specify, and the above-described test loading operations repeated. Each complete operation, which shall include loading and unloading as above-prescribed, shall be considered as an individual test.

305-1.8 Payment.

305-1.8.1 General. Timber, steel, and concrete piles will be paid for at the Contract Unit Price per linear foot (meter) for furnishing piling and the Contract Unit Price per pile for driving piles. Load tests will be paid for at the Contract Unit Price per load test. Test piles that become a part of the completed structure will be paid for at the Contract Unit Prices for furnishing piling and for driving piles. No payment will be made for piles rejected prior to driving or for piles which are driven out of place or are damaged in handling or driving.

305-1.8.2 Payment for Furnishing Piles. The length of timber, steel, and concrete piles to be paid for shall be the total length in place of the completed work, measured from the tip of pile to the plane of pile cutoff, except when otherwise specified that the Engineer will determine the length of pile to be furnished.

The Contract Unit Price paid per linear foot (meter) for furnishing timber, concrete, or steel piling shall include full compensation for furnishing the piles at the site for driving, including steel shells and concrete casings and the filling materials for cast-in-place concrete piles, and constructing reinforced concrete extensions as shown on the Plans or in the Special Provisions.

Payment for furnishing piles shall also include full compensation for the attaching and fitting of steel shoes when they are specified for timber piles, and the furnishing and attachment of brackets, lugs, core stoppers, and cap plates necessary, including fins, brackets, plates, or other devices ordered by the Engineer to increase the bearing value of steel piles.

If the Contractor manufactures concrete piles to the full length of the reinforcement bars to facilitate driving, no payment will be made for that portion over Plan length where concrete must be removed in order that the bars may project as shown on the Plans.

305-1.8.3 Payment for Driving Piles. The Contract Unit Price per pile for driving piles shall include full compensation for doing all the work involved in driving timber, concrete, and steel piles; driving steel shells or concrete casings for cast-in-place concrete piles; drilling holes for concrete piles cast in drilled holes; placing concrete for cast-in-place concrete piles; and cutting off piles, all complete in place to the required bearing and penetration as shown on the Plans or in the Special Provisions.

Full compensation for all jetting, drilling, or other work necessary to obtain the specified penetration and bearing of the piles, for drilling holes through embankment and filling the space remaining around the pile with sand or pea gravel, for disposing of material resulting from drilling holes, for splicing steel piles, and for all excavation and backfill involved in construction of concrete extensions as shown on the Plans or in the Special Provisions, shall be considered as included in the Contract Unit Price for driving piles.

305-2 TIMBER STRUCTURES AND TIMBER CONSTRUCTION.

305-2.1 General. Timber structures erected under these specifications shall conform to the dimensions and details of design shown on the Plans.

305-2.2 Materials. The materials to be used shall conform with the requirements or indicated on the Plans or in the Special Provisions.

Timber and lumber that is stored prior to its use shall be neatly piled on skids to raise it from the ground, and shall be protected from the sun when so required. The materials shall be stored or piled in such a manner to permit ready access for the purpose of inspection.

The use of cant hooks, peavies, or other pointed tools and hooks, will not be permitted in the handling of structural timber, lumber, or piles. Precautions shall be exercised in handling treated material to prevent damage to the surface thereof to the extent that untreated wood is exposed. Any piece so damaged will be rejected.

If treated timber or piling is cut after treatment, such cuts shall be treated in accordance with 204-2.3. This requirement shall also apply to any surface that has become abraded to the extent of exposing untreated wood. All borings and holes in such material shall be similarly treated, and holes which are not to be used for rods, bolts, pins, screws, spikes, and the like, or which will not subsequently be otherwise closed, shall be tightly filled with treated plugs.

Timber for floors and decks, and that which is to be used in the construction of split ring or shear plate connected trusses, shall be well seasoned and thoroughly air dried before being placed or incorporated in the work. This requirement shall apply to treated material as well as to that which is untreated.

305-2.3 Workmanship. Workmanship shall be first class throughout. Framing shall be true and exact and none but thoroughly competent workers shall be employed or engaged in connection with the erection of any structure under these specifications. All lumber and timber shall be cut and framed to a close fit and shall have even bearing over the entire contact surfaces. No shimming will be permitted in making joints. All members shall be true to size for the full depth thereof.

Holes for drift pins in untreated lumber shall be bored with a bit 1/16 inch (1.5mm) less in diameter than the pin or dowel. Holes for drift pins and dowels in treated lumber shall be bored with a bit of the same diameter as the pin or dowel. Holes for truss rods or bolts shall be bored with a bit 1/16 inch (1.5mm) larger than the rod or bolt. Holes for lag screws shall be bored with a bit not larger than the base of the thread. In small timbers where the prevention of splitting is necessary, holes shall be bored for spikes with a bit having a diameter not larger than that of the spike.

In the installation of metal timber connectors, care shall be exercised to ensure that the connector is installed concentric with its corresponding bolt; and if more than one connector bolt is installed in any individual joint, all bolts in such joint shall be drawn up to an even and uniform tension. The grooves for split-ring and shear-plate connectors shall be carefully cut to a uniform width and depth for the full perimeter therefore. The dimensions of these grooves, and the manner and means of cutting, shall be as recommended by the manufacturer of the particular connector to be installed, and any special tool or equipment used in cutting the grooves shall be operated in the manner and at the speed similarly recommended. Toothed-ring and spiked-grid connectors shall be installed by means of pressure equipment of a type intended for the purpose. However, split-ring connectors shall not be forced on, but shall be expanded to such an extent as to readily slip over the core formed by the groove without damaging the wood.

Unless otherwise indicated on the Plans, all bolts shall be 3/4 inch (19mm) in diameter or larger and shall be of sufficient length to project beyond the nut when the nut is drawn tight. Bolts shall be fitted at each end with either a malleable iron (ogee) washer or a steel plate at least 3 inch square (75mm x 75mm) and not less than 3/8 inch (9.5mm) thick, or as otherwise shown on the Plans.

305-2.4 Framing. Mudsills shall be firmly and evenly bedded on solid material. Sills and caps shall have a full, even bearing on the pedestals, posts, or piles and shall be secured in place as indicated on the Plans.

Bents shall be accurately aligned before the bracing is placed. Bracing shall be fastened at the ends and at each intersection by bolts. Bracing shall be of such length as will provide a minimum distance of 8 inches (200mm) between the outside bolt and the end of the brace. Treated posts or piles shall not be cut to accommodate the bracing. Treated filler blocks shall be used if necessary to fill any space that may occur between the bracing and the member of the bent.

In placing joists, the best edge shall be placed down. The elevation of the tops of adjacent joists shall not vary more than 1/8 inch (3mm). Outside joists shall have butt joints, Interior joists shall be lapped and shall extend the full width of the cap to obtain full bearing. Bridging between joists shall be solid and fastened to the joists near the top of the block and on each side of the bottom of the block. Bridging shall be accurately cut to fit closely between the joists.

Trusses, when completed, shall show no irregularity of line. Chords shall be straight and true from end to end in horizontal projection, and in vertical projection shall show a smooth curve through panel points conforming to the correct camber. Uneven and rough cuts at the points of bearing will be cause for rejection of the piece containing the defect.

Laminated bridge floors shall be constructed as shown on the Plans. The planks shall be laid with the best edge down.

Spiking of deck planking in roadway areas of bridges and similar structures shall be accomplished by the means of an air hammer equipped with a suitable driving head so designed and constructed as to ensure that the spikes are driven to sufficient depth to draw the planking tightly to the joints without damaging or abrading the surface of the plank.

305-2.5 Painting. The railing of timber bridges, including the posts, the entire outer edge of bridge decks, except treated surfaces, and any other surfaces indicated on the Plans to be painted, shall be painted as prescribed in 210-1.5 and 310.

The surface of wooden guard rails above the ground shall be painted as prescribed in 210-1.5 and 310.

The lumber shall be cut to fit and the entire surface shall be given the specified prime coat. The remaining coats shall be applied after the structure has been erected.

305-2.6 Measurement and Payment. Timber structures will be paid for as provided in the Bid. Where board measure (thousand cubic meters) is used as the basis of payment, the quantity to be paid for will be determined from actual (nominal) widths and thicknesses and the actual lengths of the pieces in the finished structure, except that in the case of laminated timber flooring, the number of laminations to be paid for shall be the required number of the size specified after dressing and the length of each lamination shall be considered as the full width or length of the floor.

The Contract Unit Price per board measure (thousand cubic meters), per linear foot (meter) of structure, or the lump sum, shall include full compensation for furnishing all nails, hardware, paint, and wood preservative.

SECTION 306 - UNDERGROUND CONDUIT CONSTRUCTION

306-1 OPEN TRENCH OPERATIONS.

306-1.1 Trench Excavation.

306-1.1.1 General. For the purpose of shoring or bracing, a trench is defined as an excavation in which the depth is greater than the width of the bottom of the excavation.

Excavations for appurtenant structures, such as but not limited to manholes, transition structures, junction structures, vaults, valve boxes, catch basins, thrust blocks, and boring pits shall, for the purpose of shoring and bracing, be deemed to be in the category of trench excavation.

Excavation shall include the removal of all water and materials of any nature which interfere with the construction work. Removal of groundwater to a level below the structure subgrade will be necessary only when required by the Plans or Special Provisions.

Excavation for conduits shall be by open trench unless otherwise specified or shown on the Plans. However, should the Contractor elect to tunnel or jack any portion not so specified, he shall first obtain approval from the Engineer. Payment for such work will be made as though the specified methods of construction had been used.

306-1.1.2 Maximum Length of Open Trench. Except by permission of the Engineer, the maximum length of open trench where prefabricated pipe is used shall be 500 feet (150m) or the distance necessary to accommodate the amount of pipe installed in a single day, whichever is the greater. The distance is the collective length at any location, including open excavation, pipe laying and appurtenant construction and backfill which has not been temporarily resurfaced.

Except by permission of the Engineer, the maximum length of open trench in any one location where concrete structures are cast in place will be that which is necessary to permit uninterrupted progress. Construction shall be pursued as follows: excavation, setting of reinforcing steel, placing of floor slab, walls, and cover slab or arch. Each shall follow the other without any one operation preceding the next nearest operation by more than 200 feet (60m).

Failure by the Contractor to comply with the limitations specified herein may result in an order to halt the work until such time as compliance has been achieved.

306-1.1.3 Maximum and Minimum Width of Trench. For pipe (except corrugated steel pipe), the minimum and maximum width of trench permitted shall be as indicated on the Plans or Standard Plans.

For corrugated steel pipe, the trench shall be at least 16 inches (400mm) wider than the diameter of the pipe to be installed.

If the maximum trench width is exceeded, the Contractor shall provide additional bedding, another type of bedding, or a higher strength of pipe, as shown on Plans or approved by the Engineer, at no additional cost to the Agency.

Additional payments or deductions from the Contract Unit Price per trench excavation for conduits will be based upon a calculated volume. The width used in calculating the volume of excavation for prefabricated conduit will be the maximum width of trench shown on the Plan and measured at the top of the pipe. In case of sewers or storm drains formed and cast in place, such volume will be based upon the outside width of the structure being constructed plus 3 feet (0.9m).

Additional payment or deductions from Contract Price for trench resurfacing will be based upon an area determined by the maximum width of trench as specified herein.

306-1.1.4 Access to Trenches. Safe and suitable ladders which project 2 feet (0.6m) above the top of the trench shall be provided for all trenches over 4 feet (1.2m) in depth. One ladder shall be provided for each 50 feet (15m) of open trench, or fraction thereof, and be so located that workers in the trench need not move more than 25 feet (7.5m) to a ladder.

306-1.1.5 Removal and Replacement of Surface Improvements. Bituminous pavement, concrete pavement, curbs, sidewalks, or driveways removed in connection with construction shall be removed in accordance with 300-1.3 and reconstructed in accordance with 302 or 303-5.

306-1.1.6 Bracing Excavations. The manner of bracing excavations shall be as set forth in the rules, orders, and regulations of the Division of Occupational Safety and Health of the State of California.

Prior to commencing the excavation of a trench 5 feet (1.5m) in depth or greater and into which a person will be required to descend, the Contractor shall first obtain a permit to do so from the Division of Occupational Safety and Health pursuant to 7-10.4.1.

Should the bracing system utilize steel H-beams or piles or other similar vertical supports, driving of said vertical supports will not be permitted except for the last 4 feet (1.2m). The vertical supports shall be placed in holes drilled to a depth of 4 feet (1.2m) above the proposed bottom of pile, except where this procedure is impracticable. The vertical support may then be driven to the required depth, not to exceed 4 feet (1.2m). During the drilling and driving operations, the Contractor shall take care to avoid damage to utilities.

At locations where the drilling of such holes is impracticable because of the existence of rocks, running sand, or other similar conditions, and provided said impracticability is demonstrated to the satisfaction of the Engineer by actual drilling operations by the Contractor, the Engineer may, upon request of the Contractor, approve the use of means other than drilling for the purpose of placing the vertical support. Such other means, however, must be of a nature which will accomplish, as nearly as possible, the purpose of the drilling, namely, the prevention of damage to existing surface or subsurface improvements, both public and private. All costs for this work shall be included in the prices bid for the items involved.

If sheeting is used to support the excavated trench, the sheeting shall be removed by the Contractor, and no such sheeting will be permitted to remain in the trench. When field conditions, the type of sheeting, or methods of construction used by the Contractor are such as to make the removal of sheeting impracticable, the Engineer may permit portions of the sheeting to be cut off to a specified depth and remain in the trench.

306-1.2 Installation of Pipe.

306-1.2.1 Bedding.

306-1.2.1.1 General. Bedding material and placement shall conform to this subsection except as otherwise specified in 306-1.2.1.2 or 306-1.2.1.3. The bedding zone shall be defined as the area containing the material specified that is supporting, surrounding, and extending to 1 foot (0.3m) above the top of pipe. Where it becomes necessary to remove boulders or other interfering objects at subgrade for bedding, any void below such subgrade shall be filled with the bedding material designated on the Plans. Where concrete is specified to cover the pipe, the top of the concrete shall be considered as the top of the bedding.

If soft, spongy, unstable, or other unsuitable material is encountered upon which the bedding material or pipe is to be placed, this material shall be removed to a depth ordered by the Engineer and replaced with bedding material suitably densified. Additional bedding so ordered, over the amount required by the Plans or Special Provisions, will be paid for as provided in the Bid. If the necessity for such additional bedding material has been caused by an act or failure to act on the part of the Contractor or is required for the control of groundwater, the Contractor shall bear the expense of the additional excavation and bedding.

Bedding material shall first be placed on a firm and unyielding subgrade so that the pipe is supported for the full length of the barrel. There shall be 4 inches (100mm) minimum of bedding below the pipe barrel and 1 inch (25mm) minimum clearance below a projecting bell for sewer, storm drain and water pipe. There shall be a minimum side clearance of 6 inches (150mm) on each side of the pipe barrel. For pipes and cables not requiring bedding by their owners, the trench width shall be appropriate to obtain the required compaction around the installation by mechanical equipment, except where cement slurry or CLSM is used. The material in the bedding zone shall be placed, and densified either by jetting or by mechanical compaction. When jetting is used, the lifts shall not exceed 4 feet (1.2m). Mechanical compaction shall be per 306-1.3.2.

When densifying the bedding material by jetting, the trench wall shall have a minimum sand equivalent of 15. The jet pipe shall meet the requirements of 306-1.3.3 and be of sufficient length to reach within 2 feet (0.6m) of the bottom of the pipe. Jetting shall provide enough water to thoroughly saturate and densify, without voids, the bedding material around the pipe. The jet pipe shall be inserted at intervals of 3 feet (1m) maximum, contiguous along each side of the pipe. Neither flooding, nor free standing water will be permitted. Unless the sheeting or shoring is to be cut off and left in place, densification of bedding for pipe shall be accomplished after the sheeting or shoring has been removed form the bedding zone, and prior to the placement of backfill.

Except where otherwise specified, bedding material shall be sand, gravel, crushed aggregate, or native free-draining granular material having a sand equivalent of not less than 30 or having a coefficient of permeability greater than 1-1/2 inches/hour (35mm per hour), or other material approved by the Engineer.

Concrete used for bedding shall be as specified in 201-1. Concrete shall be cured for the indicated time period prior to backfilling.

In cases where native free-draining granular material is suitable for use as bedding, the trench may be excavated to a point above the invert grade and the trench bottom hand shaped so that the bottom segment of the pipe is firmly supported on undisturbed material.

Unless otherwise specified, special pipe bedding will not be required for steel or iron water pipe, and the trench bottom need not be shaped to the outside diameter of the pipe. However, the trench bottom shall provide firm and uniform bearing.

306-1.2.1.2 Bedding for Narrow Trenches. Narrow trenches are defined as 10 inches (250mm) or less in width. Bedding for narrow trenches shall be placed as specified in 306-1.2.1.1 except as modified herein.

Bedding for narrow trenches meeting the requirements of this subsection shall be specified by the owner of the installation. When bedding is specified by the owner, it shall be placed on firm and unyielding subgrade so as to support the pipe or cable for its full length. Bedding shall not be jetted unless specified by the owner of the installation.

When the Contractor is permitted to place the pipe or cable without bedding, it shall be placed on firm and unyielding subgrade.

These excavations shall be backfilled as specified per 306-1.3.4. The Contractor shall take all necessary precautions to prevent the pipe or substructure from floating when placing trench backfill slurry or CLSM.

306-1.2.1.3 Bedding for Plastic Pipe and Fittings. Plastic pipe and fittings, including but not limited to:

Pipe	Subsection
ABS Solid Wall Pipe	207-15
ABS or PVC Composite Pipe	207-16
PVC Solid Wall Pipe	207-17
Annular High-Density Polyethylene Pipe with Smooth Interior, Corrugated Exterior, with Bell-and-Spigot Joints (Type S)	207-18
PE Solid Wall Pipe	207-19
Centrifugally Cast Fiberglass Reinforced Plastic Mortar Pipe	207-20

shall be bedded as shown on the Plans and specified in the Special Provisions.

No internal or external bracing of the pipe shall be allowed during or after pipe placement, bedding or backfill.

The bedding zone shall begin 4 inches (100mm) below the pipe or bell, whichever is lower, and shall extend to a minimum of 12 inches (300mm) above the pipe or bell, whichever is higher. The bedding zone limits shall extend on each side of the pipe or bell shall be in accordance with Table 306-1.2.1.3 (A).

TABLE 306-1.2.1.3 (A)

Nominal Pipe Size inches (mm)	Side Clearance inches (mm)	
	Min.	Max.
Up to and including 15 (375)	8 (200)	12 (300)
Over 15 (375) to and including 30 (750)	12 (300)	18 (450)
Over 30 (750)	Per the Plans	

Bedding materials shall be placed and densified to the requirements shown on the Plans or specified in the Special Provisions. For ABS and PVC composite pipe, the bedding material shall conform to 306-1.2.1.1 (A). For all other pipe and fittings, the bedding material shall be crushed rock conforming to 200-1.2 and Table 306-1.2.1.3 (B) unless otherwise specified in the Special Provisions or shown on the Plans. No minimum relative compaction requirement shall apply to crushed rock. The crushed rock shall be placed by slicing, shovel-spading, or shovel rodding to insure complete filling of the haunch areas below the pipe.

TABLE 306-1.2.1.3 (B)

Nominal Pipe Size inches (mm)	Maximum Rock Gradation
Up to and including 15 inches (375mm)	1/2 inch (12mm)
Over 15 inches (375mm)	3/4 inch (19mm)

Bedding for fittings shall be portland cement concrete, Class 330-C-23(560-C-3250), conforming to 201-1.1.2.

306-1.2.2 Pipe Laying. Pipe will be inspected in the field before and after laying. If any cause for rejection is discovered in a pipe after it has been laid, it shall be subject to rejection. Any corrective work shall be approved by the Engineer and shall be at no cost to the Agency.

When connections are to be made to any existing pipe, conduit, or other appurtenances, the actual elevation or position of which cannot be determined without excavation, the Contractor shall excavate for, and expose, the existing improvement before laying any pipe or conduit. The Engineer shall be given the opportunity to inspect the existing pipe or conduit before connection is made. Any adjustments in line or grade which may be necessary to accomplish the intent of the Plans will be made, and the Contractor will be paid for any additional work resulting from such change in line or grade in the manner provided in 3-2.

Pipe shall be laid up-grade with the socket or collar ends of the pipe up-grade unless otherwise authorized by the Engineer.

Corrugated metal pipe shall be laid with external laps of the circumferential seams toward the inlet end. Corrugated pipe shall be shipped and handled in such a manner as to prevent damage to protective coatings.

When specified, circular corrugated steel pipe shall be elongated in the shop or in the field before backfilling. The pipe shall be vertically elongated from a true circle to provide an increase in the diameter of approximately 5 percent for the full length.

Installation of slotted corrugated steel pipe shall not be started until after paving of the traffic lanes adjacent to the pipe has been completed at the locations where the pipe is to be placed. The slot shall be covered with roofing paper or other approved covering during backfilling operations to prevent infiltration of material into the pipe.

Pipe shall be laid to Plan line and grade, with uniform bearing under the full length of the barrel of the pipe. Suitable excavation shall be made to receive the socket or collar, which shall not bear upon the subgrade or bedding. Any pipe which is not in true alignment or shows any undue settlement after laying shall be taken up and re-laid at the Contractor's expense.

Pipe sections shall be laid and jointed in such a manner that the offset of the inside of the pipe at any joint will be held to a minimum at the invert. The maximum offset at the invert of pipe shall be 1 percent of the inside diameter of the pipe or 3/8 inch (10mm), whichever is smaller.

In joining socket-and-spigot pipe, the spigot of each pipe shall be so seated in the socket of the adjacent pipe as to give a minimum of 3/8 inch (10mm) annular space all around the pipe in the socket. Unavailable offsets shall be distributed around the circumference of the pipe in such a manner that the minimum offset occurs at the invert.

When pipe is laid in a sheeted trench, all sheeting against which concrete cradle is to be placed shall be faced with at least one thickness of building paper and the sheeting shall be withdrawn without displacing or damaging the cradle, except as otherwise provided in 306-1.1.6.

After the joints have been made, the pipe shall not be disturbed in any manner.

At the close of work each day, or whenever the work ceases for any reason, the end of the pipe shall be securely closed unless otherwise permitted by the Engineer.

306-1.2.3 Field Jointing of Clay Pipe. Unless otherwise indicated on the Plans, any of the following joints may be used for sewers constructed of clay pipe:

a) **Type "D" Joints (Rubber-Sleeve Coupling with Shear Ring for Plain-End Clay Pipe).** Pipe joints shall be made with the couplings described in 208-2.2. Unless otherwise specified, pipe shall be delivered to the jobsite with the rubber sleeve and shear ring installed on one end of the pipe or fitting. Before installing compression bands, the surface of the rubber sleeve shall be thoroughly wetted with a silicone base lubricant. This lubricant shall not be injurious to the sleeve, stainless steel bands, or plastic shear ring. Joints installed on to pipe in the plant shall have compression bands torqued to 70 inch-pounds (8 N·m), minimum. When the joint is installed in the field, the plain end of the pipe to be joined shall be inserted into the sleeve and the compression bands torqued to 70 inch-pounds (8 N·m), minimum, and shall provide uniform tension. Type "D" Joints may be used on pipe on curves in accordance with Item c) below.

b) **Type "G" Joints (Polyurethane Compression).** Type "G" joints shall be made with pipe prepared as specified in 208-2.3.

   Prior to jointing, the matting surfaces shall be clean, and lubricated with a lubricant recommended by the pipe supplier. The pipe shall be joined spigot into socket and when jointing is completed shall be within the following joint space tolerance:

TABLE 306-1.2.3 (A)

Pipe Size inches (mm)	Joint Space inches (mm)
15 - 18 (375 - 450)	5/8 (16)
21 - 42 (525 - 1050)	7/8 (22)

This joint space shall not be increased because of deflected joints on curve. Straight pipe with Type "G" joints is permitted for pipelines on curves in accordance with Item c) below.

c) **Straight Nonbeveled Pipe On Curves.** Straight nonbeveled pipe with Type "D" or "G" joints is permitted for pipelines on curves, provided the radius of curvature is not less than that shown in the following table. For radius of curvature less than that shown, beveled pipe or shorter lengths shall be provided.

TABLE 306-1.2.3 (B)

D Pipe Size inches (mm)	For Pipe Length ft (m)	Min. Radius of Curvature ft (m)	Max. Deflection Per Joint Degrees	Max. Deflection Per Length inches (mm)
6 to 12 (150 to 300)	5 (1.5)	120 (37)	2.4	2-1/2 (63)
	5-1/2 (1.7)	132 (40)	2.4	2-3/4 (70)
	6 (1.8)	144 (44)	2.4	3 (76)
15 to 24 (375 to 600)	5 (1.5)	160 (49)	1.8	1-7/8 (47)
	5-1/2 (1.7)	176 (54)	1.8	2-1/16 (52)
	6 (1.8)	192 (59)	1.8	2-1/4 (57)
	7-1/2 (2.3)	240 (73)	1.8	2-13/16 (71)
27 to 36 (675 to 900)	5 (1.5)	240 (73)	1.2	1-1/4 (31)
	5-1/2 (1.7)	264 (80)	1.2	1-3/8 (34)
	6 (1.8)	288 (88)	1.2	1-1/2 (38)
	7-1/2 (2.3)	360 (110)	1.2	1-7/8 (47)
39 to 42 (975 to 1050)	5 (1.5)	320 (97)	0.9	15/16 (23)
	5-1/2 (1.7)	352 (107)	0.9	1-1/16 (27)
	6 (1.8)	384 (117)	0.9	1-1/8 (28)

For pipe lengths not included above, use the following:

TABLE 306-1.2.3 (C)

D Pipe Size inches (mm)	Maximum Allowable Deflection Δd inches per ft (mm per m) of pipe	Equation for Minimum Radius of Curvature (L = Pipe Length)
6 to 12 (150 to 300)	1/2 (42)	r = 24L
15 to 24 (375 to 600)	3/8 (31)	r = 32L
27 to 36 (675 to 900)	1/4 (21)	r = 48L
39 to 42 (975 to 1050)	3/16 (16)	r = 64L

306-1.2.4 Installation, Field Jointing, and Inspection of Reinforced Concrete Pipe.

a) **General.** Circular concrete pipe with elliptical reinforcement shall be laid with the minor axis of the reinforcement cage in the vertical position. The minor axis shall be marked by the manufacturer with a 4 inch (100mm) high "T". Normally lay pipe with socket end up hill starting at the bottom of the line. Plastic lined reinforced concrete pipe shall be jointed and otherwise treated in accordance with this subsection and the applicable requirements of 311-1.

b) **Tongue-and-Groove (T&G) Self-Centering Joints.** All joints shall be cleaned with a wire brush and wetted before mortaring. All mortar shall conform to the applicable provisions of 201-5.1 for Class C mortar.

No mortaring of outside joints will be required except where concrete pipe is used on curves, unless otherwise specified.

Pipes used on curves shall have one or both ends beveled, or shall be pulled to provide a smooth curve. Pipes used on curves shall have one or both ends beveled, or shall be pulled to provide a smooth curve. If the extreme ends of the pipe do not overlap and the resulting clear space between the extreme ends does not exceed 1 inch (25mm) the space shall be filled with Class C mortar for the full thickness of the pipe wall. If the clear space between the extreme ends of the pipes is more than 1 inch (25mm) but less than 3 inches (75mm), a concrete cover is required for the joint using Class 520-C-2500 (310-C-17) concrete to a minimum depth of 6 inches (150mm) for a width of 15 inches (380mm) centered about the joint. Such concrete cover shall be placed from the bottom of the pipe to a point where the extreme ends of the pipe overlap. Sandbags or dirt sacks will be acceptable side forms. The inside of the joint shall be mortared as above. If the clear space between the extreme ends of the pipe is 3 inches (75mm) or greater but less than 6 inches (150mm) a concrete collar is required. If the clear space is 6 inches (150mm) or greater, a transition structure is required.

When pipe is under 24 inches (600mm) in diameter, the joints shall be made by filling the outer joint space with mortar.

When the pipe is 24 inches (600mm) or greater in diameter, the entire interior annular space of the joint shall be filled to its full depth to ensure a strong, level, and tight joint. The mortar shall be forced tightly into the joint, completing the joint with a level, smooth, troweled finish. The joint area shall then be wiped clean of excess mortar. The jointing procedure shall be as follows:

1) When the entire trench is to be jetted or flooded, no joints shall be mortared before the next two joints in advance are laid. However, the mortaring of joints shall be completed as specified herein before jetting is started.

2) When the entire trench is to be compacted mechanically, no interior joints shall be mortared until the compaction has been completed. The joint shall then be completed as specified herein.

3) Where the lower portion of the trench is to be jetted and the remainder mechanically compacted, the joints shall be mortared in two operations. Before jetting is started, the inside joints shall be filled to within approximately 1 inch (25mm) of the inside surface, the mortar being pressed into place to make a firm and tight joint.

   After jetting and compaction have both been completed, the inside joints shall be cleaned and completed as specified herein.

4) For gravity sewer pipe, the top half of the outside joint shall be filled with mortar by means of troweling or wiping prior to placement of backfill, and the inside joints shall be completed as specified herein.

   In all cases, the entire depth of the finished inside joint shall be filled with mortar in such a manner as to ensure a strong, watertight joint.

c) **Collar Joints.** Pipe with collar joints shall be laid with the collar end up-grade. The pipes shall be tightly butted together and uniform caulking space left between the pipe and the collar. When the entering pipe has been placed and checked for line and grade, the body of the pipe shall be backfilled with earth on both sides to hold the pipe firmly in place. The caulking space shall then be completely filled with stiff mortar tamped firmly in small increments by means of a caulking tool and hammer.

d) **Gasket-Type Joints for Reinforced Concrete Pipe.** The ends of the pipe shall be so formed that, when the pipes are laid together and joined, they shall make a continuous and uniform line of pipe with a smooth and regular surface.

Joints shall be watertight and flexible. Each joint shall contain a gasket per 208-3 or other material approved by the Engineer, which shall be the sole element responsible for water tightness of the joint. This gasket shall be a circular or profile-type cross section unless otherwise approved by the Engineer. The length and cross-sectional dimension of the gasket, the annular space provided for the gasket, and all other joint details shall be such as to produce a watertight joint. The slope of the longitudinal gasket contact surfaces of the joint with respect to the longitudinal axis of the pipe shall not exceed 2 degrees.

Under ordinary laying conditions, the work shall be scheduled so that the socket end of the pipe faces in the direction of laying.

For O-ring type gaskets, prior to placing the spigot into the socket of the pipe previously laid, the spigot groove, the gasket, and the first 2 inches (50mm) of the inside surface of the socket shall be thoroughly cleaned, then lubricated with a soft vegetable soap compound.

The gasket after lubrication shall be uniformly stretched when placing it in the spigot groove so that the gasket is distributed evenly around the circumference.

For pipe in which the inside joints are to be pointed, suitable spacers shall be placed against the inside shoulder of the socket to provide the proper space between abutting ends of the pipe.

For profile type gaskets, follow the requirements of the manufacturer for lubrication and assembly to provide a watertight joint.

Where steel joint rings are used, a suitable cloth, plastic, or paper band shall be placed around the outside of the pipe and centered over the joint to prevent dirt from entering the joint recess.

The joint band shall be bound to the pipe by the use of steel box strapping or by an equivalent method, and shall completely and snugly encase the outside joint except for an opening near the top where grout is to be poured into the joint recess. Grout shall be poured and allowed to set before densification of bedding and backfill materials by jetting or flooding methods. In any case, joints shall be grouted before backfill is placed over the top of the pipe. With the jointing band properly secured, the joint recess shall be moistened with water and then filled with Class "C" mortar. The mortar grout shall completely fill the outside annular space between the ends of the pipe and around the complete circumference. After the recess has been filled, the jointing band shall be replaced over the opening left for pouring and the mortar allowed to set. After the bedding and backfill have been densified, the inside joint recess shall first be moistened, then filled with stiff Class "C" mortar. The finished joint shall be smooth and flush with the adjacent pipe surfaces.

After the joint is assembled, a thin metal feeler gage shall be inserted between the socket and the spigot and the position of the gasket checked around the complete circumference of the pipe. If the gasket is not in the proper position, the pipe shall be withdrawn, the gasket checked to see that it is not cut or damaged, the pipe relayed, and the gasket position again checked.

e) **Field Inspection of Reinforced Concrete Pipe.** Installed pipe shall be inspected after completion of the backfill operation for:

1. Joint offset tolerances per 306-1.2.2.
2. Joints installed per this subsection.
3. Compliance with 207-2.8 or 207-3.2 for the respective pipe material except as modified below.

Unless otherwise specified, installed and backfilled pipe shall not exhibit cracks with widths in excess of those shown in Table 207-2.9.2 (A).

306-1.2.5 Field Jointing of Nonreinforced Concrete Pipe.

a) **Tongue-and-Groove Joints.** The groove end of pipe shall be buttered with a stiff mixture of Class "C" mortar prior to joining pipe. The pipe joint shall then be carefully wiped on the inside.

b) **Socket-and-Spigot Mortar Joints.** In making the joints, the entire annular space shall be completely and compactly filled with Class "C" mortar.

Mortar placed in the joint to assist in the assembling and centering of the pipe shall not be considered as filling that portion of the joint in which it is placed. The mortar shall be beveled on a 1-to-1 slope from the outer edge of the socket, and the interior of the pipe cleaned of surplus mortar or other foreign material.

When approved by the Engineer, a narrow gasket of oakum or lead may be caulked into each joint in wet trenches, after which the mortar shall be placed therein.

Interior joints in pipe shall be neatly wiped on the inside.

c) **Socket-and-Spigot Gasket Joints.** The outside of the spigot and the inside of the socket of the pipe shall be thoroughly cleaned prior to laying. The gasket and the socket interior shall be lubricated with a soft vegetable soap compound before the pipes are joined.

306-1.2.6 Field Jointing of Iron Pipe.

a) **General.** The type of joint to be used will be indicated on the Plans or in the Specifications. If not designated, the type of joint may be any of those listed below:

b) **Cement Joints.** A gasket of untarred jute or oakum twisted into a rope of about the same diameter as the joint space and thoroughly saturated with neat cement grout shall be well driven against the base of the socket. After placing this gasket, caulking cement shall be pushed into the socket with a steel caulking tool until the interior of the socket is completely filled and then it shall be thoroughly tamped with a caulking tool.

The joint shall then be beveled off from the outer edge of the socket to the sides of the pipe, special care being taken to obtain good work on the lower part of the joint.

The joints shall be protected from the sun immediately after they are caulked.

c) **Lead Joints.** A gasket of twisted or braided jute or oakum shall be driven tightly into the socket so that the lead, after having been poured and caulked, shall have a depth of at least 2 inches (50mm) in the socket. The socket shall be free from dirt, grease, and water, and the runner shall be firmly held in place before the lead is poured. The melting pot shall be kept near the joint to be poured and each joint shall be made at one pouring. Dross and slag shall not be allowed to accumulate in the melting pot ladles. After the joints have cooled sufficiently, they shall be properly caulked by hand or mechanical methods so as to secure a tight joint.

d) **Flanged Joints.** Flanged joints shall be firmly and fully bolted with machine bolts of proper size. Approved gaskets shall be used at all flanged joints.

e) **Mechanical Joints.** The outside of the spigot and the inside of the socket shall be thoroughly cleaned of foreign matter. The gland and gasket shall then be slipped on to the spigot end of the pipe. The gasket shall be pressed evenly into the socket only after the spigot is seated in the socket.

The gland shall be brought up evenly by tightening alternately the nuts spaced 180 degrees apart.

f) **Slip-On Joint.** The gasket and gasket seal inside the socket shall be wiped clean before the gasket is inserted. A thin film of soft vegetable soap compound shall be applied to the gasket and the outside of the spigot end of the pipe. The spigot shall then be positioned inside the socket and shoved home. Lubricant other than that furnished with the pipe shall not be used unless approved by the Engineer.

306-1.2.7 Field Jointing of Corrugated Metal Pipe. Where metal pipe and/or couplings and metal pipe and/or couplings with metallic coatings are joined with dissimilar metals, the contact points shall be coated with asphalt mastic per 207-11.5.2. Pipe sections shall be laid in the trench with a maximum spacing between sections of 1-1/2 inches (38mm). Annular corrugated pipe shall be laid in the trench with outside laps or circumferential joints upgrade. The pipe coupling corrugations or projections shall properly engage the pipe sections before bolts are tightened. Care shall be taken to ensure that dirt or other particles do not get between the outside of pipe and the pipe coupling.

Aluminum pipe and aluminized steel pipe shall not be in contact with reinforcing steel or structural steel members. Aluminum pipe and aluminized steel pipe shall be coated with asphalt mastic coating per 207-11.5.2 where concrete or slurry backfill is required or where the pipe is to be embedded in concrete.

Paved inverts shall be placed and centered on the bottom of the trench. Any damage to the protective lining and coating shall be repaired prior to the backfilling around the pipe.

Watertight joints, when required by the Plans or the Special Provisions, shall conform to 207-11.2.2.

306-1.2.8 (Deleted).

306-1.2.9 Field Jointing of Solvent-Welded ABS and PVC Pipe. Solvent-welded jointing of ABS and PVC pipe shall be in accordance with the approved manufacturer's printed instructions which shall be furnished to the Engineer. Solvent cement shall be in accordance with 207-15.1 for ABS pipe and 207-17.3.3 for PVC pipe.

The spigot end shall be inserted to the proper depth of the socket as indicated by the home mark.

306-1.2.10 Jointing of Gasket-Type ABS, CHDPE and PVC Pipe. Jointing shall conform to the manufacturer's specifications which shall be submitted to the Engineer per 2-5.3. Gaskets shall conform to 208-4.

The spigot end shall be inserted to the proper depth of the socket as indicated by the "home mark." The home mark shall be shown as a circumferential line or by the words "home mark" on the outside of the pipe. The home mark shall be clearly and permanently indicated on the spigot end of the pipe at the factory.

306-1.2.11 Jointing of Injection-Sealed PVC Pipe. Jointing shall conform to the manufacturer's specifications which shall be submitted to the Engineer per 2-5.3.

The spigot end shall be inserted to the full depth of the socket as indicated by the "home mark" and driven into the locking taper as recommended by the manufacturer.

The ports in the socket end shall be positioned to allow observance of flow of the adhesive compound from the exhaust port. The adhesive compound shall be injected until air is no longer observed to bubble from the exhaust port. Escape of adhesive compound beyond the retainer ring shall be cause for rejection of the joint.

306-1.2.12 Maximum Allowable Deflection for Plastic Pipe and Fittings.

306-1.2.12.1 General. Pipe and fittings shall be tested to ensure that vertical deflections and measured diameter do not exceed the maximum allowable deflection. The maximum allowable deflection shall be:

a) Three percent of the maximum average ID for ABS or PVC Composite Pipe and fittings.

b) For pipe and fittings other than ABS or PVC Composite, the following percentage of the maximum average ID shall conform to Table 306-1.2.12.1 (A):

TABLE 306-1.2.12.1 (A)[2, 3]

Nominal Pipe Size		Percentage Deflection Allowed[1]
Inches	Millimeters	
Up to and including 12	Up to and including 300	5.0
Over 12-to and including 30	Over 300 -to and including 750	4.0
Over 30-to and including 60	Over 750 -to and including 1500	3.0
Over 60-to and including 90	Over 1500-to and including 2250	2.5
Over 90-to and including 120	Over 2250-to and including 3000	2.0
Over 120	Over 3000	1.5

1. 30 days after Installation.
2. Deflections of up to 6.5 percent of the in-field measured diameter are acceptable for storm drain applications.
3. Deflection tests shall not be performed sooner than 30 days after completion of placement and compaction of backfill. The pipe and fittings shall be cleaned and inspected for offsets and obstructions prior to testing.

For solid wall flexible pipe and fittings, the maximum average ID shall be equal to the average OD minus 2 minimum wall thicknesses in accordance with the applicable ASTM(s). Manufacturing and other tolerances shall not be considered when determining maximum allowable deflections.

For CHDPE pipe and fittings, the ID for each lot (as defined in 207-18.5.1) shall be calculated by averaging the diameter at 8 equally spaced points around the circumference of a section of unloaded pipe. For storm drain applications, the mandrel size shall be 93.5 percent of the average ID calculated at the Work site. The Contractor shall be responsible for measuring, recording and providing the ID's to the Engineer prior to installation.

306-1.2.12.2 ID's 24 inches (600mm) or Smaller. A mandrel shall be pulled through the pipe by hand to ensure the maximum allowable deflection has not been exceeded. Fittings shall be visually inspected to ensure the maximum allowable deflection has not been exceeded.

Any over-deflected pipe shall be uncovered and, if not damaged, reinstalled. Damaged pipe shall not be reinstalled, but shall be removed from the Work site. Any pipe subjected to any method or process other than removal, which attempts, even successfully, to reduce or cure any over deflection, shall be uncovered, removed from the worksite, and replaced with new pipe. The mandrel shall:

a) Be a rigid, nonadjustable, odd-numbered-leg (9 legs minimum) mandrel having an effective length not less than its nominal diameter.

b) Have a minimum diameter at any point along the full length conforming to Table 306-1.2.12.2 (A).

c) Be fabricated from steel and be fitted with pulling rings at each end.

d) Be stamped or engraved on a segment other than the runner, with the pipe material specification, nominal size, and mandrel OD (e.g. PVC D3034-200mm-187.10mm; ABS Composite D2680-250mm-243.43mm; PVC D3034-8"-7.366"; ABS Composite D2680-10"-9.584").

TABLE 306-1.2.12.2 (A)

	Nominal Size		Minimum Mandrel Diameter[1]	
Pipe Material	Inches	mm	Inches	mm
PVC-ASTM D3034 (SDR 26)	6	150	5.503	139.78
	8	200	7.366	187.10
	10	250	9.207	233.86
	12	300	10.961	278.41
	15	375	13.559	344.40
PVC-ASTM D3034 (SDR 35)	6	150	5.619	142.72
	8	200	7.524	191.11
	10	250	9.405	238.89
	12	300	11.191	284.25
	15	375	13.849	351.76
PVC-ASTM F679 (T-1 Wall)	18	450	16.924	429.87
	21	525	19.952	506.78
	24	600	22.446	570.13
	27	675	25.297	642.54
	30	750	28.502	723.95
ABS or PVC Composite Pipe ASTM D2680	6	150	5.636	143.15
	8	200	7.663	194.64
	10	250	9.584	243.43
	12	300	11.475	291.47
	15	375	14.356	364.64
CCFRPM ASTM D3262, 46 psi (318KPa)	12	300	11.820	300.23
	18	450	17.729	450.32
	20	500	19.661	499.39
	24	600	23.527	597.59
	30	750	29.229	742.42
	36	900	35.403	899.24

1. Metric mandrel diameters are direct conversions of mandrel diameters in U.S. Standard Measures. If the above types of pipe are available and specified by the appropriate ASTM in metric dimensions, as the primary measure, the Engineer will determine the appropriate mandrel diameter which conforms to this subsection.

Prior to use, the mandrel shall be certified by the Engineer or by another entity approved by the Engineer. Use of an uncertified mandrel or a mandrel altered or modified after certification will invalidate the test.

If the mandrel fails to pass through the pipe, the pipe will be deemed to be over-deflected.

306-1.2.12.3 ID's Greater Than 24 inches (600mm) to Less Than or Equal to 36 inches (900mm). Deflections shall be determined by a method submitted to the Engineer per 2-5.3. Fittings shall be visually inspected to ensure the maximum allowable deflection has not been exceeded.

306-1.2.12.4 ID's Greater Than 36 inches (900mm). Deflections shall be determined by using a 1 inch (25mm) diameter, nonadjustable metal bar; a minimum-radius rigid template; or by a method approved by the Engineer. Fittings shall be visually inspected to ensure the maximum allowable deflection has not been exceeded.

306-1.2.12.5 Acceptance. Any over-deflected pipe shall be uncovered and, if not damaged, reinstalled. Damaged pipe shall not be reinstalled, but shall be removed from the Work site. Any pipe subjected to any method or process other than removal, which attempts, even successfully, to reduce or cure any over-deflection, shall be uncovered, removed from the Work site, and replaced with new pipe.

306-1.2.13 Installation of Plastic Pipe and Fittings. Plastic pipe and fittings shall be placed as shown on the Plans and specified in the Specifications. Plastic pipe bedding shall be bedded conformance with 306-1.2.1.3.

Connections of pipe and fittings to a manhole shall be watertight. The use of manhole water stops shall be approved by the Engineer. Water stops shall be installed in conformance with the manufacturer's recommendations. The manufacturer's recommendations shall be submitted to the Engineer per 2-5.3. Junctions connecting pipe or fitting to a pipe shall utilize a "wye" fitting. "Tee" connections will not be permitted. Pipe may be used on curves only if deflection fittings or couplings are used, or if solid wall pipe is bent without any application of heat. If deflection fittings or couplings are proposed for use on curves, the proposed alignment and method of joining shall be submitted to the Engineer per 2-5.3. Solid wall pipe proposed for use on curves shall be bent per the manufacturer's recommendations. The manufacturer's recommendations shall be submitted to the Engineer per 2-5.3. The minimum centerline radius to be achieved by bending solid wall pipe shall be based on the pipe diameter, dimension ratio (ratio of the OD to the minimum wall thickness) and the manufacturer's recommendations.

306-1.3 Backfill and Densification.

306-1.3.1 General. Backfill shall be placed as follows except as modified in 306-1.3.4. Backfill shall be considered as starting at the top of the bedding zone. For concrete encasement the backfill shall be considered as starting at the top of the concrete encasement.

Backfill or fill shall start at the subgrade for cast-in-place structures such as, but not limited to, manholes, transition structures, junction structures, vaults, valve boxes, and reinforced concrete box conduits

Backfill, except that within State Highway, shall be placed as specified in 306-1.3.2 and 306-1.3.3. Compaction requirements shall be in accordance with 306-1.3.5.

When the depth of cover of the top pipe or cable is less than 30 inches (760mm), the top 24 inches (600mm) of backfill, measured from the surface, shall be compacted to 90 percent relative compaction.

Except where the pipe must remain exposed for force main leakage test and subject to the provisions herein, the Contractor shall proceed with backfilling operations as soon as possible. Care shall be exercised so that the pipe will not be damaged or displaced. If the pipe is supported by concrete bedding that does not cover the pipe, the remainder of any bedding material shall be placed to 1 foot (0.3m) over the top of the pipe. The backfill above the concrete bedding shall not be placed nor sheeting pulled until concrete bedding has been cured per 201-1.

The Contractor may place fill or backfill against or over the top of any cast-in-place structure in accordance with Table 306-1.3.1 (A), unless otherwise specified or approved by the Engineer.

TABLE 306-1.3.1 (A)

Operation	Location	
	Against Sides of Structures (Days)	Over Top of Structure (Days)
Placement of Loose Backfill	5	21
Densification of Backfill	7	28[1]

1. Or 100 percent of the specified compressive strength.

Rocks greater than 6 inches (150mm) in any dimension will not be permitted in backfill placed between 1 foot (0.3m) above the top of any pipe or box and 1 foot (0.3m) below pavement subgrade.

When the trench is wider than 3 feet (0.9m), rocks not exceeding 12 inches (300mm) in greatest dimension, which originate from the trench, will be permitted in the backfill from 1 foot (0.3m) above the top of any pipe or box to 5 feet (1.5m) below the finished surface.

Rocks greater than 2-1/2 inches (60mm) in any dimension will not be permitted in backfill placed within 1 foot (0.3m) of pavement subgrade.

Where rocks are included in the backfill, they shall be mixed with suitable excavated materials to eliminate voids.

Subject to the provisions specified herein, the material obtained from project excavations may be used as backfill provided that all organic material, rubbish, debris, and other objectionable materials are first removed. However, broken portland cement concrete and bituminous-type pavement obtained form the project excavations will be permitted in the backfill subject to the same limitations as rocks.

Where it becomes necessary to excavate beyond the limits of normal excavation lines in order to remove boulders or other interfering objects, the voids remaining after the removal of the boulders shall be backfilled with suitable material and densified as approved by the Engineer.

The removal of all boulders or other interfering objects and the backfilling of voids left by such removals shall be at the expense of the Contractor and no direct payment for the cost of such work will be made. The cost of such work shall be included in the prices in the Bid for the various items of work.

Voids left by the removal of sheeting, piles and similar sheeting supports shall be immediately backfilled with clean sand which shall be jetted or vibrated into place to ensure dense and complete filling of the voids.

Densification shall proceed as soon as possible.

If the Engineer determines that it is not practical to attain the required compaction by mechanical methods, or jetting, such as in areas around utilities, vaults, or other structures, trench backfill slurry per Table 201-1.1.2 (A) will be required.

306-1.3.2 Mechanically Compacted Backfill. Backfill shall be mechanically compacted by means of tamping, sheepsfoot, pneumatic tire, or vibrating rollers, or other mechanical tampers. All such equipment shall be of size and type approved by the Engineer. Impact-type pavement breakers (stompers) shall not be permitted over or adjacent to pipe, duct, or cable, unless permitted by the Engineer.

Permission to use specific compaction equipment shall not relieve the contractor from responsibility to ensure that the use of such equipment will not result in damage to adjacent ground, existing improvements, or improvements installed under the Contract. The Contractor shall make its own determination in this regard.

Mechanically compacted backfill shall be placed in horizontal layers of thickness compatible to the material being placed and the type of equipment being used. Each layer shall be evenly spread, moistened (or dried, if necessary), and then tamped or rolled until the specified relative compaction has been attained.

Unless otherwise approved by the Engineer, material for mechanically compacted backfill shall be placed in lifts which, prior to compaction, shall not exceed the thickness specified below for the various types of equipment:

a) Impact, free fall, or “stomping” equipment- maximum lift thickness of 24 inches (0.6m)
b) Vibratory equipment, including vibratory plates on backhoe dipsticks, vibratory smooth-wheel rollers, and vibratory pneumatic-tired rollers - maximum lift thickness of 18 inches (0.5m).

c) Rolling equipment, including sheepsfoot (both vibratory and nonvibratory), grid, smooth-wheel (nonvibratory), grid, smooth wheel (nonvibratory), and segmented wheels - maximum lift thickness of 8 inches (0.2m).

d) Hand-directed mechanical compactors such as vibratory plates or tamper - maximum lift thickness of 4 inches (100mm).

306-1.3.3 Jetted Backfill. Backfill to be densified by water shall be jetted. Jetting will be permitted only if the soils of the trench walls have a minimum sand equivalent of 15. Jetting shall be accomplished by the use of a jet pipe to which a hose is attached, carrying a continuous supply of water under pressure.

Backfill shall be jetted in accordance with the following requirements:

a) The jet pipe shall consist of a minimum 1-1/2-inch (38mm) diameter pipe to which a minimum 2-inch (50mm) diameter hose is attached at the upper end. The jet shall be of sufficient length to project to within 2 feet (0.6m) of the bottom of the lift being densified.

b) The Contractor shall jet to within 2 feet (0.6m) of the bottom of the lift and apply water in a manner, quantity and at a rate sufficient to thoroughly saturate the thickness of the lift being densified. The jet pipe shall not be moved until the backfill has collapsed and the water has been forced to the surface.

c) The lift of backfill shall not exceed that which can be readily densified by jetting, but in no case shall the undensified lift exceed 15 feet (4.5m).

d) Suitable backfill material to be jetted shall have a sand equivalent of 15 or greater.

e) Where the nature of the material excavated from the trench is generally unsuitable for densification with water, the Contractor may, at no cost to the Agency, import from an approved source suitable material for jetting or densify the excavated material by other methods as approved by the Engineer. The backfill shall be allowed to thoroughly drain until the surface of the backfill is in a firm and unyielding condition prior to commencement of any subsequent improvements. The Engineer may require the Contractor, at the Contractor's expense, to dig a sump and provide a pump to remove any accumulated water.

f) The Contractor shall make its own determination that jetting will not result in damage to adjacent structures or facilities. Any resulting damage shall be repaired at the Contractor's expense.

g) The Contractor shall have available a continuous supply of water at a minimum pressure of 40 psi (275 kPa) gage. If a water truck is used to supply water, it shall have a pump capable of supplying water at 40 psi (275 kPa) gage and shall have the capacity to jet the trench without refill.

h) After jetting trench backfill, the Contractor shall prepare the top of the backfill to comply with 306-1.3.5 and to provide a firm and unyielding subgrade conforming to 301-1. Jetting maybe supplemented with mechanical methods.

306-1.3.4 Backfilling Narrow Trenches. Narrow trenches are defined as 10 inches (250mm) or less in width. Backfill for narrow trenches shall be placed as specified in 306-1.3.1 except as modified herein. Narrow trenches shall be backfilled by the use of trench backfill slurry per 201-1 or CLSM per 201-6, unless otherwise approved by the Engineer.

When narrow trenches are backfilled using trench backfill slurry or CLSM, the Contractor may place the material in a single lift using vibrators for consolidation. The Contractor shall take all necessary precautions to prevent the pipe or substructure from floating or becoming displaced. The top of the trench backfill slurry or CLSM shall be placed flush with top of the pavement when steel plates are not placed over narrow trenches. The trench backfill slurry or CLSM shall be cut back to a minimum of 1 inch (25mm) but no greater than 8 inches (200mm) below the existing pavement prior to placing permanent paving. For trenches 6 inches (150mm) or less in width, the compacted thickness of asphalt concrete shall be 3 inches (75mm).

Backfill to be mechanically compacted in narrow trenches shall be placed per 306-1.3.1 and 306-1.3.2, except as modified herein. Backfill shall not have any rocks greater in any dimension, than 1/4 the width of the trench. Mechanically compacted backfill shall meet the relative compaction requirements of 306-1.3.6. In place density for narrow trenches shall be determined in accordance with ASTM D2937 or by a method approved by Engineer.

306-1.3.5 Jetted Bedding and Backfill Compaction Requirements. Except as specified otherwise, trench bedding and backfill densified thru jetting shall be densified to the following minimum relative compaction:

a) 85 percent relative compaction:
   1) From the bottom of the trench to the beginning of the upper 3 feet (0.9m), measured from the pavement surface (or finish grade where there is no pavement) within native material or unengineered embankments.
   2) Outside the traveled way, shoulders, and under sidewalks, in the upper 3 feet (0.9m), measured from the pavement surface (or finish grade where there is no pavement).
   3) Under sidewalks.

b) 90 percent relative compaction:
   1) In the upper 3 feet (0.9m), measured from the pavement surface (or finish grade where there is no pavement), within the existing or future traveled way, shoulders, and other paved areas (or areas to receive pavement).
   2) Within engineered embankments.
   3) Where lateral support is required for existing or proposed structures.
   4) 95 percent relative compaction where required by 301-1.3.

306-1.3.6 Mechanical Compaction Requirements. Except as specified otherwise, mechanically compacted trench backfill shall be densified to the following minimum relative compaction:

a) 85 percent relative compaction:
   1) In the bedding zone.
   2) Outside the traveled way and other paved areas (or areas to receive pavement).
   3) Under Sidewalks.

b) 90 percent relative compaction:
   1) In the upper 3 feet (0.9m) measured from the pavement surface (or finish grade where there is no pavement), within the existing or future traveled way, shoulders, and other paved areas (or areas to receive pavement).
   2) Within engineered embankments.
   3) Where lateral support is required for existing or proposed structures.

c) 95 percent Relative Compaction where required by 301-1.3.

306-1.3.7 Imported Backfill. If the Contractor elects to import material from a source outside the project limits for use as backfill, said materials shall be clean soil, free from organic material, trash, debris, rubbish, broken portland cement concrete, bituminous materials, or other objectionable substances.

Whenever the Contractor elects to use imported material for backfill, it shall deliver, not less than 10 days prior to intended use, a sample of the material to the Engineer. The sample shall have a minimum dry weight of 100 pounds (45 kg) and shall be clearly identified as to source, including street address and community of origin. The Engineer will determine the suitability, the minimum relative compaction to be attained, and the placement method.

Should the imported material not be substantially the same as the approved sample, it shall not be used for backfill and shall be removed from the Work site at the Contractor's expense.

The densification method for imported material authorized by the Engineer will be dependent upon its composition, the composition of the in-place soil at the point of placement, and the relative compaction to be obtained.

306-1.3.8 Transported Backfill. The Contractor may transport or back-haul material to be used as backfill material from any portion or line of the Work to any other portion or line of the Work. Such transported material shall be clean soil, free from organic material, trash, debris, rubbish, or other objectionable substances except that broken portland cement concrete or bituminous type paving as specified in 306-1.3.1 may be included.

306-1.4 Testing Pipelines.

306-1.4.1 General. All leakage tests and all post-installation closed circuit television (CCTV) inspections shall be completed and approved prior to placing of permanent resurfacing.

When leakage or infiltration exceeds the amount allowed by the Specifications, the Contractor at its expense shall locate the leaks and make the necessary repairs or replacements in accordance with the Specifications to reduce the leakage or infiltration to the specified limits. Any individually detectable leaks shall be repaired, regardless of the results of the tests. Leakage tests shall be made on completed pipelines as follows:

a) Storm Drains - Not required unless called for on the Plans or in the Special Provisions.

b) Gravity Sanitary Sewers 24 inches (600mm) or less in diameter where difference in elevation between inverts of adjacent manholes is 10 feet (3m) or less - Water exfiltration test or water infiltration test as directed. The Engineer may allow substitution of an air pressure test for the water exfiltration test.

c) Gravity Sanitary Sewers 24 inches (600mm) or less in diameter where difference in elevation between inverts of adjacent manholes is greater than 10 feet (3m) - Air pressure test or water infiltration test as directed.

d) Gravity Sanitary Sewers 24 inches (600mm) or greater in diameter - Air pressure test or water infiltration test as directed.

e) Gravity Sanitary Sewers which are in service and a bypass system is not available - the Contractor shall perform post-installation CCTV inspection per 500-1.1.5 and payment shall be per 500-1.1.9.

f) Pressure Sanitary Sewers (force mains) - Water pressure test at 120 percent of maximum operating pressure.

g) Water Pipelines - Water pressure test: Pipe specified by pressure classification, 50 psi (350 kPa) over pressure classification. Other type of pipe, 120 percent of maximum operating pressure.

306-1.4.2 Water Exfiltration Test. Each section of sewer shall be tested between successive manholes by closing the lower end of the sewer to be tested and the inlet sewer of the upper manhole with stoppers. The pipe and manhole shall be filled with water to a point 4 feet (1.2m) above the invert of the sewer at the center of the upper manhole; or if groundwater is present, 4 feet (1.2m) above the average adjacent groundwater level.

The allowable leakage will be computed by the formulae:

$E_{US} = 0.0001\ LD\ \sqrt{H}$ for mortared joints.

($E_{SI} = 0.00009\ LD\ \sqrt{H}$ for mortared joints)

$E_{US} = 0.00002\ LD\ \sqrt{H}$ for all other joints.

($E_{SI} = 0.000018\ LD\ \sqrt{H}$ for all other joints.)

Where:

L = length of sewer and house connections tested, in feet (meters).

E_{US} (E_{SI}) = the allowable leakage in gallons (liters) per minute of sewer tested.

D = the internal diameter of the pipe in inches (millimeters).

H = is the difference in elevation feet (meters) between the water surface in the upper manhole and the invert of the pipe at the lower manhole; or if groundwater is present above the invert of the pipe in the lower manhole, the difference in elevation between the water surface in the upper manhole and the groundwater at the lower manhole.

The Contractor shall, at its expense, furnish all water, materials and labor for making the required test. All tests shall be made in the presence of the Engineer.

306-1.4.3 Water Infiltration Test. If, in the opinion of the Engineer, excessive groundwater is encountered in the construction of a section of the sewer, the exfiltration test for leakage shall not be used.

The end of the sewer at the upper structure shall be closed sufficiently to prevent the entrance of water, and pumping of groundwater shall be discontinued for at least 3 days, after which the section shall be tested for infiltration.

The infiltration into each individual reach of sewer between adjoining manholes shall not exceed that allowed by the formula in 306-1.4.2 where H is the difference in the elevation in feet (meters) between the groundwater surface and the invert of the sewer at the downstream manhole.

Unless otherwise specified, infiltration will be measured by the Engineer using measuring devices furnished by the Agency.

306-1.4.4 Air Pressure Test. The Contractor shall, at its expense, furnish all materials, equipment, and labor for making an air test. Air test equipment shall be approved by the Engineer unless otherwise provided on the Plans or in the Special Provisions.

The Contractor may conduct an initial air test of the sewer mainline after densification of the backfill, but prior to installation of the house connection sewers. Such tests will be considered to be for the Contractor's convenience and need not be performed in the presence of the Engineer.

Each section of sewer shall be tested between successive manholes by plugging and bracing all openings in the sewer mainline and the upper ends of all house connection sewers. Prior to any air pressure testing, all pipe plugs shall be checked with a soap solution to detect any air leakage. If any leaks are found, the air pressure shall be released, the leaks eliminated, and the test procedure started over again. The Contractor has the option of wetting the interior of the pipe prior to the test.

The final leakage test of the sewer mainline and branching house connection sewers, shall be conducted in the presence of the Engineer in the following manner:

Air shall be introduced into the pipeline until 3.0 psi (20 kPa) gage pressure has been reached, at which time the flow of air shall be reduced and the internal air pressure shall be maintained between 2.5 and 3.5 psi (17 kPa and 24 kPa) gage pressure for at least 2 minutes to allow the air temperature to come to equilibrium with the temperature of the pipe walls. Pressure in the pipeline shall be constantly monitored by a gage and hose arrangement separate from hose used to introduce air into the line. Pressure in the pipeline shall not be allowed to exceed 5 psi (34 kPa) gage pressure.

After the temperature has stabilized and no air leaks at the plugs have been found, the air pressure shall be permitted to drop and, when the internal pressure has reached 2.5 psi (17 kPa) gage pressure, a stopwatch or sweep-second-hand watch shall be used to determine the time lapse required for the air pressure to drop to 1.5 psi (10 kPa) gage pressure.

If the time lapse (in seconds) required for the air pressure to decrease from 2.5 to 1.5 psi (17kPa to 10kPa) gage pressure exceeds that shown in Table 306-1.4.4 (A), the pipe shall be presumed to be within acceptance limits for leakage.

If the time lapse is less than that shown in the table, the Contractor shall make the necessary corrections to reduce the leakage to acceptance limits.

TABLE 306-1.4.4 (A)

Time in Seconds for Pressure to Drop from 2.5 to 1.5 psi (17kPa to 10 kPa) Gage Pressure

Main Line — 4 inch (100mm) House Connection

Nominal Diameter inches (mm)	Length ft (m)	House Connection Length 0 ft (0m)	100 ft (30m)	200 ft (60m)	300 ft (90m)	400 ft (120m)
8 (200)	0 (0)	0	20	40	50	70
	50 (15)	40	50	70	90	80
	100 (30)	70	90	100	100	90
	150 (45)	110	120	110	100	100
	200 (60)	140	120	110	110	100
	300 (90)	140	130	120	110	110
	400 (120)	140	130	120	120	110
250 (10)	15 (50)	50	70	90	100	90
	30 (100)	110	130	120	110	110
	60 (200)	170	150	140	130	120
	90 (300)	170	160	150	140	130
	120 (400)	170	160	150	150	140
300 (12)	15 (50)	80	100	110	110	110
	30 (100)	160	170	150	140	130
	60 (200)	200	180	170	160	150
	90 (300)	200	190	180	170	160
	120 (400)	200	190	180	180	170

Main Line — 6 inch (150mm) House Connection

Nominal Diameter inches (mm)	Length ft (m)	House Connection Length 0 ft (0m)	100 ft (30m)	200 ft (60m)	300 ft (90m)	400 ft (120m)
8 (200)	0 (0)	0	40	80	100	100
	50 (15)	40	70	110	110	110
	100 (30)	70	110	120	110	110
	150 (45)	110	120	120	120	110
	200 (60)	140	130	120	120	120
	300 (90)	140	130	120	120	120
	400 (120)	140	130	130	120	120
250 (10)	15 (50)	50	90	120	120	110
	30 (100)	110	140	130	130	120
	60 (200)	170	150	140	140	130
	90 (300)	170	160	150	140	140
	120 (400)	170	160	150	150	140
300 (12)	15 (50)	80	120	140	130	120
	30 (100)	160	170	150	140	140
	60 (200)	200	180	170	160	150
	90 (300)	200	190	180	170	160
	120 (400)	200	190	180	180	170

TABLE 306-1.4.4 (A) *(Continued)*

Main Line		4 inch (100mm) House Connection				
Nominal Diameter inches (mm)	Length ft (m)	House Connection Length				
		0 ft (0m)	100 ft (30m)	200 ft (60m)	300 ft (90m)	400 ft (120m)
375 (15)	15 (50)	120	140	160	140	130
	30 (100)	250	220	190	170	160
	60 (200)	260	230	220	200	190
	90 (300)	260	240	230	220	210
	120 (400)	260	240	230	220	220
450 (18)	15 (50)	180	200	190	170	150
	30 (100)	310	260	230	210	190
	60 (200)	310	280	260	250	230
	90 (300)	310	290	280	260	250
	120 (400)	310	290	280	270	260
525 (21)	15 (50)	240	260	230	200	180
	30 (100)	360	310	280	250	230
	60 (200)	360	330	310	290	280
	90 (300)	360	340	320	310	300
	120 (400)	360	340	330	320	310
600 (24)	15 (50)	320	320	270	240	210
	30 (100)	410	360	320	290	270
	60 (200)	410	380	360	340	320
	90 (300)	410	390	370	360	350
	120 (400)	410	390	380	370	360
675 (27)	15 (50)	400	370	310	280	250
	30 (100)	460	410	370	340	310
	60 (200)	460	430	410	390	370
	90 (300)	460	440	420	410	390
	120 (400)	460	450	430	420	410
750 (30)	15 (50)	490	420	360	310	280
	30 (100)	510	460	420	380	360
	60 (200)	510	480	460	440	420
	90 (300)	510	490	470	460	440
	120 (400)	510	500	480	470	460
825 (33)	15 (50)	560	460	400	350	320
	30 (100)	560	510	460	430	400
	60 (200)	560	530	510	490	460
	90 (300)	560	540	520	510	490
	120 (400)	560	550	530	520	510
900 (36)	15 (50)	610	510	440	390	360
	30 (100)	610	560	510	480	440
	60 (200)	610	580	560	530	510
	90 (300)	610	600	580	560	540
	120 (400)	610	600	580	570	560

Main Line		6 inch (150mm) House Connection				
Nominal Diameter inches (mm)	Length ft (m)	House Connection Length				
		0 ft (0m)	100 ft (30m)	200 ft (60m)	300 ft (90m)	400 ft (120m)
375 (15)	15 (50)	120	160	160	150	140
	30 (100)	250	210	190	170	160
	60 (200)	260	230	210	200	190
	90 (300)	260	240	220	210	200
	120 (400)	260	240	230	220	210
450 (18)	15 (50)	180	220	190	170	160
	30 (100)	310	260	220	200	190
	60 (200)	310	280	260	240	220
	90 (300)	310	290	270	260	240
	120 (400)	310	290	280	270	260
525 (21)	15 (50)	240	260	220	200	180
	30 (100)	360	300	260	240	200
	60 (200)	360	330	300	280	260
	90 (300)	360	330	320	300	290
	120 (400)	360	340	330	310	300
600 (24)	15 (50)	320	310	260	220	200
	30 (100)	410	350	310	280	260
	60 (200)	410	370	350	320	310
	90 (300)	410	380	360	350	330
	120 (400)	410	390	370	360	350
675 (27)	15 (50)	400	350	290	260	230
	30 (100)	460	390	350	320	290
	60 (200)	460	420	390	370	350
	90 (300)	460	430	410	390	380
	120 (400)	460	440	420	410	390
750 (30)	15 (50)	480	490	330	290	260
	30 (100)	510	440	390	360	330
	60 (200)	510	470	440	420	390
	90 (300)	510	480	460	440	420
	120 (400)	510	490	470	460	440
825 (33)	15 (50)	560	440	370	320	290
	30 (100)	560	490	440	400	370
	60 (200)	560	520	490	460	440
	90 (300)	560	530	510	490	470
	120 (400)	560	540	520	510	490
900 (36)	15 (50)	610	480	410	360	320
	30 (100)	610	540	480	440	410
	60 (200)	610	570	540	510	480
	90 (300)	610	590	560	540	520
	120 (400)	610	590	570	560	540

TABLE 306-1.4.4 (A) *(Continued)*

Main Line		4 inch (100mm) House Connection					Main Line		6 inch (150mm) House Connection				
Nominal Diameter inches (mm)	Length ft (m)	House Connection Length					Nominal Diameter inches (mm)	Length ft (m)	House Connection Length				
		0 ft (0m)	100 ft (30m)	200 ft (60m)	300 ft (90m)	400 ft (120m)			0 ft (0m)	100 ft (30m)	200 ft (60m)	300 ft (90m)	400 ft (120m)
975 (39)	15 (50)	660	560	490	440	390	975 (39)	15 (50)	660	530	450	390	350
	30 (100)	660	610	560	520	490		30 (100)	660	590	530	480	450
	60 (200)	660	630	610	580	560		60 (200)	660	620	590	560	530
	90 (300)	660	640	620	610	590		90 (300)	660	640	610	590	570
	120 (400)	660	650	630	620	610		120 (400)	660	640	620	610	590
1050 (42)	15 (50)	710	610	540	480	430	1050 (42)	15 (50)	710	580	490	430	390
	30 (100)	710	660	610	570	540		30 (100)	710	640	580	530	490
	60 (200)	710	680	660	630	610		60 (200)	710	670	640	610	580
	90 (300)	710	690	680	660	640		90 (300)	710	690	660	640	620
	120 (400)	710	700	680	670	660		120 (400)	710	690	670	650	640

306-1.4.5 Water Pressure Test. Preparatory to testing, the section of the pipeline to be tested shall be filled with water and placed under a slight pressure for at least 48 hours. The pipeline shall then be brought up to the test pressure specified and maintained on the section under test for a period of not less than 4 hours.

Accurate means shall be provided for measuring the quantity of water required to maintain full pressure on the line for the test period, which volume shall not exceed:

For U.S. Standard Measures: $L_{US} = \dfrac{CND\sqrt{P}}{1{,}850}$

For SI Units: $L_{SI} = \dfrac{CND\sqrt{P}}{32{,}600}$

Where:

L_{US} (L_{SI}) = Maximum allowable leakage in gallons (liters) per hour for section of pipeline tested.
N = Number of joints in length tested.
D = Diameter of pipe in inches (mm).
P = Test pressure in psi (kPa).
C = 1.0 for reinforced concrete pressure pipe with rubber joints, cylinder type.
C = 3.0 for reinforced concrete pressure pipe with rubber joints, non-cylinder type.
C = 0.50 for cast iron pipe with mechanical or rubber gasket joints.
C = 1.0 for other type of cast iron joints (caulked and other types of pipe.)

No leakage is allowed for welded steel pipe with welded joints.

306-1.4.6 Leakage Test for Corrugated Metal Pipelines. After the pipe has been laid and assembled, and when required, the pipeline shall be filled with water to a hydrostatic pressure head of 10 feet (3m) above the point in the line to be tested.

A hydrostatic test shall be conducted for a period of not less than 24 hours, during which time an accurate measure of the water required to maintain the test pressure shall be made. Any leakage developed by the test shall not exceed 0.60 gallon per inch (90mL per mm) of inside diameter per 100 feet (30m) of pipe per hour. Any leakage in excess of this amount shall be stopped in a manner satisfactory to the Engineer, and the test repeated until the total leakage does not exceed the amount specified. All obvious leaks shall be stopped in a manner satisfactory to the Engineer, whether or not the leakage from the line exceeds that permitted herein.

306-1.5 Trench Resurfacing.

306-1.5.1 Temporary Resurfacing. Unless permanent pavement is placed immediately, temporary bituminous resurfacing 2 inches (50mm) thick shall be placed and maintained at locations determined by the Engineer wherever excavation is made through pavement, sidewalk or driveways. In sidewalk areas the temporary bituminous resurfacing shall be at least 1 inch (25mm) thick; in all other areas it shall be at least 2 inches (50mm) thick. At major intersections and other critical locations, a greater thickness may be ordered. Temporary resurfacing shall be placed as soon as the condition of the backfill is suitable to receive it and shall remain in place until the condition of the backfill is suitable for permanent resurfacing.

The bituminous mixture used for temporary trench resurfacing shall conform to Class D2 asphalt concrete mixture in 203-6.4.3; and bitumen conforming to grade SC-800 liquid asphalt in the Slow Curing Product table, 203-2.4.

The mixture may be furnished from stockpiles or directly from the plant and may be laid cold, at the option of the Contractor. Prior to placing temporary resurfacing, the Contractor shall level and compact the backfill on which the surfacing is to be placed. The grade of the backfill on which the resurfacing is to be placed shall be such as to provide the full thickness of temporary resurfacing specified. The temporary resurfacing shall be placed, rolled, maintained, and removed and disposed of by the Contractor.

On improvements being constructed under contract with the Agency, the Proposal will contain a Bid item for an estimated number of tons (tonnes) of temporary bituminous resurfacing materials. The Contract Unit Price per ton (tonne) shall include full compensation for furnishing, placing, maintaining, removing, and disposing of such temporary resurfacing materials.

Payment will be limited to that quantity of material ordered placed by the Engineer and shall include material used to maintain the temporary resurfacing until the permanent resurfacing is placed. Material which is placed by the Contractor for its convenience shall be at no cost to the Agency.

306-1.5.2 Permanent Resurfacing. Unless otherwise shown on the Plans, Permit or in the Special Provisions, all surface improvements damaged or removed as a result of the Contractor's operations shall be reconstructed by the Contractor to the same dimensions, except for the pavement thickness, and with the same type of materials used in the original work. Trench and excavation resurfacing shall be 1 inch (25mm) greater in thickness than existing pavement.

Subgrade for trench resurfacing shall conform to 301 and the pavement reconstruction shall comply with the applicable provisions of 302. Asphalt concrete pavement shall also comply with 306-1.5.3. Aggregate base, when encountered within the structural section area, shall be compacted to a minimum density of 95 percent and compacted in lifts per 301-2.2. The thickness of aggregate base shall be equal to that existing adjacent to the excavation.

306-1.5.3 Placement of Permanent Repair Hot Mixed Asphalt Concrete. The asphalt concrete shall be placed in compacted lifts per Table 306-1.5.3.

TABLE 306-1.5.3

Compaction Equipment	Maximum Compacted Thickness inches (mm)
Vibratory Plate	1-1/2 (38)
Pneumatic Plate	2 (50)
Vibratory Rammers	2 (50)
Steel Wheel Roller*	2-1/2 (63)
Vibratory Roller*	3 (75)
Pneumatic Tired Rollers	Not Permitted

* Rollers must fit entirely within the trench.

After placement of the backfill and/or aggregate base, the sides of the excavation shall be cleaned prior to the application of an asphalt tack coat. The tack coat may be an emulsified asphalt conforming to 203-3 or a paving asphalt conforming to 203-1. This tack coat when cured or cooled shall be of sufficient thickness to uniformly cover all vertical surfaces of the existing asphalt concrete. An extra heavy application of the tack to the vertical edges will not be cause for rejection. Excess tack on the horizontal surface of the aggregate base or subgrade shall be spread uniformly over the surface and may require the application of a blotting sand to prevent bleed through. Areas that are not sufficiently coated shall have the tack reapplied. Care must be exercised by the Contractor to insure that the tack coat is not damaged during the placement of the asphalt concrete.

306-1.5.4 Base Course for Asphalt Concrete Placement The base course shall be a B or C gradation and shall be placed by either a spreader box, paving machine or "shoe" attachment.

For trenches less than 3 feet (1m) wide and individual excavations or bore holes having an area of less than 50 square feet ($5m^2$), the base course pavement shall be placed in such a manner as to obtain the specified density and smoothness.

The compacted surface shall not deviate from the planned base course elevation by more than 1/4 inch (6mm).

306-1.5.5 Finish Course for Asphalt Concrete Placement. The finish course shall be a C or D gradation. For trenches 8 feet (2.5m) or greater in width, the final lift of asphalt concrete shall be placed with a paving machine or a full width spreader box. When the total tonnage required for the final lift of asphalt concrete on the Work is greater than 110 tons (100 tonnes), a paving machine shall be used.

For trench widths 3 feet (1m) or greater and less than 8 feet (2.5m), the final lift shall be placed with a narrow paving machine or a spreader box when the total tonnage required for the final lift of asphalt concrete on the Work is greater than 17 tons (15 tonnes).

For trenches less than 3 feet (1m) wide and individual excavations or bore holes having an area of less than 50 ft^2 ($5m^2$), the final lift shall be placed in such a manner as to obtain the specified density and smoothness.

306-1.5.6 Density and Smoothness. For trench width 3 feet (1m) or greater, the Contractor shall compact all lifts with a self propelled steel wheeled roller meeting the PLI (N/mm) requirement specified in 302-5.6.

For trench widths less than 3 feet (1m) wide, the Contractor shall compact all lifts by steel wheel rollers, vibratory plates, or rammers of such width to fit within the sides of the excavation. The PLI (N/mm) requirements of 302-5.6 shall not apply except for the final lift. The final lift shall be compacted using a steel wheel roller meeting the N/mm (PLI) requirements of 302-5.6.

For individual excavations or bore holes having an area of less than 50 square feet ($5m^2$), the Contractor shall compact all lifts by steel wheel rollers, vibratory plates, or rammers of such width to fit within the sides of the excavation. The PLI (N/mm) requirements of 302-5.6 shall not apply.

Pneumatic tire rollers or truck tires shall not be allowed on any of the lifts.

Trenches of any width backfilled with CLSM or trench backfill slurry will not require aggregate base. Asphalt concrete shall be replaced to the full depth of existing asphalt concrete plus 1 inch (25mm), except for trenches specified in 306-1.3.4.

For trench widths 3 feet (1m) or greater, the compaction temperatures of the asphalt concrete mat shall be per 302-5.6. For trench width less than 3 feet (1m), the compaction of the asphalt concrete mat shall be initiated before the material cools to less than 200°F (94°C).

The minimum compaction after rolling shall be 95 percent of the density obtained in accordance with the methods specified in 302-5.6.2. When the density is determined by a core sample, it shall be based on a full depth sample, as specified in 302-5.6.2.

The final pavement surface for trenches wider than 3 feet (1m) and parallel to the centerline of the street shall meet the smoothness requirements of 302-5.6.2. Trenches less than 3 feet (1m) wide, individual excavations or bore holes having an area less than 50 ft^2 ($5m^2$), and trenches of any width not parallel to the centerline of the street shall match the smoothness of the existing pavement, except final pavement surface tolerances are minus 0 to plus 1/8 inch (3mm) based on the existing pavement on either side of the excavation. Final pavement below the existing surface is not acceptable.

Finish courses with deviations exceeding the above requirements shall be removed and replaced. Such pavement shall be removed to a minimum depth of 1-1/2 inches (38mm) for the full width of the trench. The minimum length of removal along the trench shall extend for 4 feet (1.2m) beyond the ends of the deviations, but in no case exceed the limit of the original excavation.

306-1.5.7 Concrete Resurfacing. Replacement of PCC pavement for trench or individual excavations or bore holes shall be 1 inch (25mm) greater in thickness than existing the pavement. The existing concrete pavement shall be saw cut per 300-1.3. The concrete shall conform to and be placed per 302-6.

306-1.6 Basis of Payment for Open Trench Installations. Pipe and conduit shall be measured along the longitudinal axis between the ends as laid and shall include the actual pipe in place and shall not include the inside dimensions of structures. House connection sewers shall be measured from the center of the main sewer to the upper end of the house connection sewer. Catch basin connections shall be measured from the inside face of the catch basin to the inside face of conduit or structure to which connection is being made. Chimney pipe shall be measured vertically from the upper end of the chimney to the invert of the sewer.

The Contract Unit Price per linear foot (meter) for pipe and conduit in place shall be considered full compensation for all wyes, tees, bends, monolithic catch basin connections, and specials shown on the Plans; the removal of interfering portions of existing sewers, storm drains, and improvements; the closing or removing of abandoned conduit and structures; the excavations of the trench; the control of ground and surface waters; the preparation of subgrade; placing and joining pipe; backfilling the trench; permanent resurfacing; and all other work (excluding temporary resurfacing) necessary to install the pipe or conduit, complete in place.

Payment for structures such as manholes, junction structures, lamp holes, and catch basins shall be made at the price in the Bid for each structure and shall be full payment for each structure complete in place, including excavation, backfill, constructing inverts, furnishing and installing castings, restoration of the street surface, and all other work, excluding temporary resurfacing, necessary to complete the Work.

306-1.7 Precast Reinforced Concrete Box (PRCB).

306-1.7.1 General. These specifications cover the construction of single-cell PRCB sections intended to be used for the conveyance of storm water. PRCB sections shall conform to 216.

306-1.7.2 Repairs. PRCB sections damaged due to imperfections in fabrication or handling shall be repaired by a method approved by the Engineer.

306-1.7.3 Subgrade. Subgrade material shall be densified to 90 percent relative compaction. Unsuitable subgrade material shall be removed to the depth shown on the Plans or determined by the Engineer and replaced with leveling bed material. Voids below subgrade shall be filled with leveling bed material prior to densification.

306-1.7.4 Leveling Bed Material. Leveling bed material shall conform to 216-2.4 and be densified to 90 percent relative compaction.

306-1.7.5 Installation. PRCB sections shall be laid up-grade with the groove ends up-grade unless otherwise approved by the Engineer. Connections shall be constructed as shown on the Plans.

At the close of work each day, or whenever the work ceases for any reason, each end shall be securely closed as approved by the Engineer.

306-1.7.5.1 Tongue-and-Groove Joints. Tongue and groove joints shall be constructed in accordance with 306-1.2.4 (b) modified as follows:

a) Only one end shall be beveled for PRCB sections placed on curves.

b) Concrete used to fill clear spaces more than 1 inch (25mm) and less than 3 inches (75mm) shall be 560-C-3250 (330-C-23) or Class C mortar conforming to 201-5 unless otherwise specified in the Special Provisions.

Preformed flexible joint sealant conforming to ASTM C990 or AASHTO M198 may be used. Preformed flexible joint sealant shall be installed in accordance with the manufacturer's specifications on the tongue and groove, in order to fill the joint annular space on the inside of the PRCB section. Flexible plastic gaskets shall not be used on PRCB pulled to provide a curve.

Preformed flexible joint sealant bands conforming to ASTM C877 may be used, in conjunction with mastic or mortar, when installed in accordance with the manufacturer's specifications.

306-1.7.5.2 Structure Backfill. Structure backfill shall conform to 300-3.5.1. Structure backfill material shall be placed 12 inches (300mm) from the top and 24 inches (600mm) from each side.

306-1.7.6 Measurement. PRCB will be measured for payment along the longitudinal axis between the ends laid for each size. The length shall include the actual length of the PRCB in place but it shall not include the inside dimensions of structures.

306-1.7.7 Payment. Payment shall conform to 306-1.6. Payment for additional leveling bed material shall conform to 3-3 unless otherwise specified in the Special Provisions.

306-2 JACKING OPERATIONS.

306-2.1 General. Before starting excavation, the Contractor shall, in accordance with 2-5.3, submit Working Drawings of jacking pit bracing, casing (or conduit), and jacking head proposed to be used.

Unless otherwise specified, the methods and equipment used in jacking casing or conduit shall be optional with the Contractor, provided that the proposed method is approved by the Engineer. Such approval, however, shall in no way relieve the Contractor of the responsibility for making a satisfactory installation meeting the criteria set forth herein. Only workers experienced in jacking operations shall be used in performing the work.

The leading section of conduit shall be equipped with a jacking head securely anchored thereto to prevent any wobble or variation in alignment during the jacking operation.

The driving ends of the conduit shall be properly protected against spalling and other damage, and intermediate joints shall be similarly protected by the installation of sufficient bearing shims to properly distribute the jacking stresses. Any section of conduit showing signs of failure shall be removed and replaced with a new section of precast conduit, or with a cast-in-place section, which is adequate to carry the loads imposed upon it.

Excavation shall not be made in excess of the outer dimensions of the conduit being jacked unless approved by the Engineer. Every effort shall be made to avoid any loss of earth outside the jacking head. Excavated material shall be removed from the conduit as excavation progresses, and no accumulation of such material within the conduit will be permitted.

Once the jacking operation has commenced, it shall be continued uninterrupted around the clock until the conduit has been jacked between the specified limits. This requirement may be modified if the Contractor submits to the Engineer for prior approval methods and details that shall prevent the "freezing" of the conduit and ensure that the heading is stable at all times.

Upon completion of the jacking operations, all voids around the outside face of the conduit shall be filled by grouting.

Grouting equipment and material shall be on the Work site before jacking operations and drilling of grout holes are completed in order that grouting around the jacked conduit may be started immediately after the jacking operations have finished.

Should appreciable loss of ground occur during the jacking operation, the voids shall be backpacked promptly to the extent practicable with soil-cement consisting of a slightly moistened mixture of 1 part cement to 5 parts granular material. Where the soil is not suitable for this purpose, the Contractor shall import suitable material at its expense. The soil-cement shall be thoroughly mixed and rammed into place as soon as possible after the loss of ground.

306-2.2 Jacking Reinforced Concrete Pipe. When pipe is specified to be jacked into place, the design of such pipe is based upon the superimposed loads and not upon the loads which may be placed upon the pipe as a result of the jacking operations. Any increase in pipe strength in order to withstand jacking loads shall be the responsibility of the Contractor.

Where pipe 60 inches (1500mm) or greater in nominal inside diameter is to be jacked for a distance greater than 32 feet (10m), a pilot tunnel shall be constructed first to ensure accuracy of grade and alignment. The dimension and support of the pilot tunnel will be optional with the Contractor subject to the approval of the Engineer. Such approval shall in no way relieve the Contractor of the responsibility for damage of any nature which might occur as a result of the method used.

Supports for pilot tunnels shall be removed as jacking progresses.

Unless the Contractor submits an alternate proposal to the Engineer for approval, the following method shall be used for supporting and guiding the pipe:

After the pilot tunnel has been constructed, a concrete cradle shall be placed true to line and grade and conforming to the outside radius of the pipe. The cradle shall be of such dimensions as to adequately and uniformly support the pipe under the lower 60-degree sector measured on the outside of the pipe. The curved surface shall be formed or accurately screeded to the proper dimensions. It shall be reinforced with not less than 0.3 percent of longitudinal steel and not less than 0.5 percent of transverse steel with respect to the cross-sectional area of the cradle. The transverse steel shall be bent on a radius equal to the radius of the outside of the pipe plus 2 inches (50mm) and shall extend to within 1 inch (25mm) of the edge of the cradle.

In lieu of the concrete cradle specified above, the Contractor may, subject to the approval of details by the Engineer, set steel rails in the concrete base slab to true line and grade.

The Contractor shall place grout holes, pipe, and fittings in the pipe invert on centers not greater than 5 feet (1.5m) and shall perform such pressure grouting as is necessary to fill voids and to secure uniform bearing between the cradle and the pipe. The grout shall be neat cement grout. Grouting pressures shall be as determined in the field by the Engineer.

All costs involved in the performance of the work of constructing pilot tunnels and cradles shall be included in the price in the Bid for jacking pipe.

306-2.3 Jacking Steel Casing. Unless otherwise specified on the Plans, the size and wall thickness of the casing to be jacked to accommodate the contract pipeline shall be at the Contractor's option, except that the casing thickness shall be not less than 3/8 inch (9.5mm), and the Contractor shall be fully responsible for the sufficiency of the casing provided.

The joints of sections of casing to be jacked shall be welded with a continuous circumferential weld. It shall be the Contractor's responsibility to provide stress transfer across the joints which is capable of resisting the jacking forces involved.

All clay pipe installed in a jacked casing shall have mechanical compression joints. The pipe shall be braced or filled to prevent shifting or flotation during backfilling operations.

Backfill shall be gunite sand, gunite concrete, or pressure concrete, except where specified otherwise in the Plans or in the Specifications. Pressure concrete shall not be placed until the mix design, placement method, and equipment have been approved by the Engineer.

If the pressure concrete mix cannot be readily pumped or placed by the placing equipment, additional water may be added, provided the water-cement ratio of the approved mix design is not exceeded.

Gunite sand backfill shall conform to 306-3.7. Where gunite sand backfill is used, the pipe shall be laid on a concrete subbase or on gravel bedding where shown on the Plans or approved by the Engineer.

Where gunite concrete or pressure concrete backfill is to be used, the pipe shall be laid on a subbase of pipe bedding concrete as specified in 201-1 at least 5 inches (125mm) thick at the centerline.

The pipe barrels shall rest upon concrete support blocks with the pipe sockets clearing the concrete subbase by at least 1/2 inch (13mm).

In addition to submitting details of the jacking pit bracing, casing, and jacking head required in 306-2, the Contractor shall submit to the Engineer for approval details of the following in advance of the proposed jacking operation: concrete support blocks, bracing to prevent pipe shifting or flotation, and pressure concrete mix design, placement method, and equipment.

306-2.4 Jacking Corrugated Steel Pipe. Corrugated steel pipe to be jacked in place between the limits shown on the Plans shall conform to provisions of these specifications and the following: The thickness of the pipe designated in the Contract item will be the minimum thickness permitted. Any heavier thickness of pipe or other facilities required to withstand jacking pressure shall be determined and furnished by the Contractor at its expense.

Corrugated pipe lengths may be joined by field riveting. Variation from theoretical alignment and grade at the time of completion of placing shall not exceed 1 inch per 100 feet (8mm per 10m).

The diameter of the excavated hole shall not be more than 0.1 foot (30mm) greater than the outside diameter of the pipe. Sluicing or jetting with water will not be permitted. When material tends to cave in from outside these limits, a shield shall be used ahead of the first section of pipe or the face of the excavation shall not extend beyond the end of the pipe greater than 1-1/2 feet (0.5m), unless permitted by the Engineer.

306-2.5 Tolerances. Concrete conduit shall be jacked true to line and grade and the Contractor shall modify the jacking operation to correct any deviation. Unless otherwise shown on the Plans or in the Specifications, when a pilot tunnel is required to be constructed in connection with jacking reinforced concrete pipe or box sections, the Contractor will be permitted a tolerance from exact grade or alignment of 1 inch per 100 feet (8mm per 10m).

306-2.6 Payment. The Contract Unit Price per foot (meter) of jacked conduit shall include full compensation for excavation; constructing, supporting, and removing pilot tunnels; constructing reinforced concrete cradles where required; providing grout holes, grout, and grouting where necessary; and doing whatever else is appurtenant to jacking conduit within the limits shown on the Plans and as specified herein.

Except when a Bid item is provided for jacked casing, the cost of furnishing and jacking casing in place shall be included in the Contract Unit Price per linear foot (meter) for that portion of the pipeline or conduit to be installed within the casing.

When a section of reinforced concrete pipe conduit is specified to be constructed by jacking methods, the specified limits for jacking may be increased by the Contractor with the approval of the Engineer. Such increased limits may require an increase in the strength of the pipe to be jacked. When reinforced concrete pipe conduit is specified to be constructed by open trench method, the Contractor may construct said conduit by jacking methods, with the approval of the Engineer. Such methods may require an increase in strength of the pipe.

When a change in construction method or an increase in jacking limits as specified herein is requested by the Contractor and authorized by the Engineer, payment for the work will be based on the Contract Unit Prices as though the specified method had been used.

306-3 TUNNELING OPERATIONS.

306-3.1 General. Required pipe tunnel locations and lengths are shown on the Plans. However, tunnels may be constructed at the Contractor's option in lieu of trench construction. The Contractor shall, in accordance with 2-5.3, submit a proposed plan of tunnel operation which shall include working drawings showing details of the following:

a) Tunnel shaft bracing and dimensions.

b) Tunnel supports (see 306-3.4).

c) Method of backpacking tunnel supports.

d) Method of transporting pipe in tunnel.

e) Bracing to prevent pipe shifting and flotation.

f) Pressure concrete mix design, placement method and equipment.

Isolated tunnels or undercrossings less than 20 feet (6m) in length shall be adequately supported, subject to inspection and approval by the Engineer in the field. Submission of Working Drawings in accordance with 2-5.3 will not be required for this case.

All provisions regarding backpacking and backfilling contained in 306-3.6 and 306-3.7 shall apply to such tunnels, except that if the roof of any such tunnel, or portion thereof, is sloped upward toward the ends of the tunnel for the full width of the excavation at an angle of 45 degrees or greater with the horizontal, the backfill within the sloped portion of the tunnel may be made with material removed from the excavation and densified by flooding or jetting, or mechanically compacted to a minimum relative density of 85 percent.

If the supporting base of any substructure is disturbed or any sewer or storm drain is exposed or partially exposed, it shall be supported with a concrete wall.

306-3.2 Excavations. Access shafts or portals shall be located where shown on the Plans or designated in the Special Provisions. Where no such locations are given, the Contractor shall have the option of determining such locations subject to approval by the Engineer. In general, access shafts or portals will not be permitted within street intersections.

The Contractor shall excavate all materials encountered in the tunnel within the width and height necessary to install tunnel supports, place pipe, make joints, properly place backfill to fill all void space around the pipe and do whatever else is necessary to complete the pipe installation in the tunnel.

Clearances shown on the tunnel details on the Plans are minimum and no encroachment within the dimensions shown will be permitted. The spring line clearances shown shall be increased by 3 inches (75mm) for any tunnel to be constructed on a curve with a centerline radius of less than 300 feet (90m).

All drilling and blasting shall be performed in such a manner to avoid undue shattering or loosening of material. The Contractor shall remove all material which is likely to fall or appears dangerous to workers or the Work. The fact that such removal may enlarge the excavation beyond the required limits shall not relieve the Contractor from the necessity of performing such work, and the Contractor shall not be entitled to any additional compensation by reason of such tunnel enlargement.

Loose material in the invert shall be removed to a reasonably clean rock surface or undisturbed foundation prior to placing pipe bedding and installing pipe. Deep depressions may be filled with suitable material approved by the Engineer. The work of removing loosened invert material and filling the resulting depressions or enlargement of the tunnel from overshooting or overexcavating shall be considered a part of tunnel excavation and no additional compensation will be allowed therefore.

306-3.3 Dewatering. All water encountered in constructing the tunnel shall be disposed of by the Contractor in such manner as will not damage public or private property or create a nuisance or health menace. The Contractor shall furnish, install, and operate pumps, pipes, appliances, and equipment of sufficient capacity to keep all tunnel excavations and accesses free from water until the tunnel is backfilled, unless otherwise authorized by the Engineer. The Contractor shall provide all means or facilities necessary to conduct water to the pumps. Water, if odorless and stable, may be discharged into an existing storm drain, channel, or street gutter in a manner approved by the Engineer. When required by the Engineer, a means shall be provided for desilting the water before discharging it.

306-3.4 Tunnel Supports. Unless otherwise shown on the Plans, the materials used for tunnel supports may be timber, metal, concrete, or a combination thereof at the option of the Contractor. Steel liner plates, if used, shall be provided with grout connections sufficient in number to permit backpacking by means of grout, should such action prove necessary. All tunnel supports shall conform to the requirements set forth in the Tunnel Safety Orders of the State of California. The Contractor shall, in accordance with 2-5.3, submit Working Drawings of tunnel supports proposed to be used. Such drawings shall include full details of the proposed tunnel supports (including connections), longitudinal and transverse bracing and foot blocks, the proposed method of pipe installations, the proposed method of backpacking tunnel supports, and other pertinent details.

The tops of foot blocks shall be installed below the pipe barrel a distance of 1/16 the pipe diameter or a minimum of 4 inches (100mm), whichever is greater. Transverse timber struts, spreaders, and footings will be permitted only where necessary to support horizontal thrust from the tunnel sides. Timber bracing, where necessary, may be left in place provided it lies entirely below the bottom of the pipe the distance specified herein for foot blocks and does not occupy more than 15 percent of the bottom area of the tunnel.

Vertical and horizontal clearance dimensions between pipe sockets and the inside face of continuous tunnel supports, lagging, splining, or steel liner plates as specified herein or as shown on the Plans, will be considered minimum dimensions. The clearance dimensions between pipe sockets and such intermittent timber and steel members as timber sets or steel rib sets are also minimum dimensions and no encroachment within the dimensions specified will be permitted. It shall be the responsibility of the Contractor to increase tunnel dimensions where necessary in order to provide adequate room for workers and equipment and such space shall be at no increase in cost to the Agency.

Unless otherwise specified or shown on the Plans, the minimum clearances shall be as follows:

For tunnels to be backfilled with pressure concrete, the minimum side clearance at the spring line of pipe sockets to continuous steel or timber shall be 12 inches (300mm), and to intermittent sets or ribs shall be 10 inches (250mm). The minimum overhead clearance from pipe sockets to nearest inside face of any steel or timber member shall be 10 inches (250mm).

For tunnels to be backfilled with gunite concrete or gunite sand, the minimum side clearance at the springline of pipe sockets shall be as for pressure concrete backfill specified above, but the minimum overhead clearance shall be increased to 18 inches (450mm).

The minimum side and top clearances prescribed herein shall be increased by 3 inches (75mm) for pipe without projecting sockets or collars and shall apply to the barrel of the pipe.

The minimum side and top clearances prescribed herein shall be increased by 3 inches (75mm) for pipe without projecting sockets or collars and shall apply to the barrel of the pipe.

No exterior work will be required on the following types of joints:

a) Socket and spigot pipe with rubber gasket or mechanical compression joints.

b) Pipe 24 inches (600mm) or larger in internal diameter.

c) Steel-ring-and-gasket-type reinforced concrete pipe for sewers, if the tunnel backfill and bedding under the pipe are concrete; or where the tunnel backfill is concrete and the bedding material under the pipe is granular and the Contractor beds the pipe for 4 inches (100mm) on each side of the joint in fresh mortar at least 3 inches (75mm) thick and extending 2 inches (50mm) above the top of the granular bedding material.

The Contractor will be required to do such reconstruction of tunnel supports at its expense as may be necessary to meet the foregoing requirements. The Agency may make minor revisions in the horizontal tunnel alignment where possible in sections at least 50 feet (15m) long to minimize the extent of such reconstruction. Similar changes in vertical alignment will generally not be approved.

All timber collar braces and, to the extent practicable, timber supports, lagging and blocking shall be removed prior to backfilling tunnels, except where such removals would be hazardous to persons or the structure. Material to remain in place shall be cleaned of adhering tunnel muck or other material not suitable for backfill.

306-3.5 Subgrade and Bedding. Pipe shall be placed and bedded as shown on the Plans or Working Drawings specifying the methods of laying and bedding pipe in trenches.

If an invert slab is required or otherwise placed separately, it shall be 5 inches (125mm) minimum thickness, the full width of the tunnel, and the concrete shall be tunnel backfill concrete as specified in 201-1. Concrete shall not be placed until placement method and equipment have been approved. The slab, when placed separately, shall be cured for at least 5 days prior to the application of heavy loading.

Payment for rock or other base material required to the extent shown on the Plans or Working Drawings for pipe bedding in trenches, shall be considered as included in the Contract Unit Price for pipe.

All rock required to fill voids caused by overexcavation or to maintain the tunnel bottom to support construction equipment and tunnel supporting members, or to control water throughout the period of tunnel excavation, shall be furnished and placed by the Contractor and the cost thereof shall be included in the Contract Unit Price for pipe complete in place and no additional compensation will be made therefore.

When ordered by the Engineer, rock, in addition to that required by the Plans for bedding, shall be placed by the Contractor and payment therefore will be made as provided in 3-2.

306-3.6 Backpacking Tunnel Supports. Voids behind all temporary or permanent tunnel support systems, including overbreak, cave-ins, and chimneys, shall be backpacked as specified herein. Backpacking shall be placed progressively as soon as practicable after placement of tunnel supports. When ordered by the Engineer, the Contractor shall place backpacking immediately. The non-backpacked length of tunnel shall be held to the minimum practicable for the method of backpacking utilized by the Contractor.

Tunnels in rock supported by timber lagging, steel liner plate, or bolted steel plate tunnel lining shall be backpacked either with pressure grout or soil-cement except that tunnel spoil may be used to the mid-height of the tunnel. When voids 1 cubic foot ($0.03m^3$) in size or larger exist behind lagging or sheeting in tunnels so supported in soil, the Contractor shall backpack behind such supports with either pressure grout or tunnel spoil when ordered by the Engineer.

Tunnels in rock or soil and supported by timber or steel sets with partial timber or metal lagging may be backpacked to the mid-height of the tunnel with tunnel spoil.

All spaces not filled with such backpacking shall be filled at the time of, and with material selected for, tunnel backfilling around the pipe.

Tunnel spoil used for backpacking shall be selected from the better spoil material available, and shall contain sufficient fines to fill all voids. Such material shall be rammed into place. Soft or wet clay may be used only if satisfactory compaction can be obtained. Otherwise the Contractor will be required to import granular material for backpacking at no additional cost to the Agency.

Soil-cement for backpacking lagged or fully lined tunnels shall consist of a slightly moistened mixture of 1 part cement to 5 parts of granular material selected from the tunnel spoil when such material is suitable. Otherwise, granular material shall be imported at the Contractor's expense. The soil-cement shall be thoroughly mixed and rammed into place immediately following placement of tunnel supports. The placement interval shall not exceed three rings of liner plate or the distance between tunnel sets. Mechanically or pneumatically operated tampers shall be used to ram the soil-cement into place unless another placing method is approved by the Engineer.

306-3.7 Tunnel Backfill. Pipe laying operations in tunnels shall not precede tunnel backfill by more than 150 feet (45m) without the approval of the Engineer. Longer reaches may be approved if tunnel clearances are increased from the minimums shown in order to obtain additional working space around the pipe.

The space between the tunnel supports and the pipe shall be completely backfilled with the materials and methods specified herein. The backfill material shall be forced or packed into all the crevices and around all timber sets or steel ribs from the tunnel invert to its crown. The Contractor shall provide whatever wedging or bracing is needed to ensure against pipe movement during placement of backfill.

Backfill for tunnels in rock shall be limited to pressure concrete or gunite concrete.

The approval of the use of gunite concrete for backfill is contingent upon the prior backpacking of tunnel supports with acceptable materials other than gunite concrete.

Unless the Plans for tunnels to be constructed in soil require the use of pressure concrete or gunite concrete for backfill, the Contractor may use gunite sand for backfill.

Gunite sand shall be placed with a pneumatic gun in accordance with the requirements for placing gunite concrete except that no portland cement need be added. The Contractor may add up to 100 pounds of cement per cubic yard (60 kg of cement per cubic meter) to improve placement stability at its option and expense. In either case, water sufficient to saturate the material and ensure proper packing and minimize rebound shall be added to the mixture. The nozzleperson shall operate in the immediate vicinity of the backfill face to ensure compaction and complete filling of voids.

The Contractor shall submit to the Engineer for approval, at least 30 days prior to backfill operations, a proposed mix design and method of placing concrete, including placing equipment. No pressure concrete backfill shall be placed until mix design, placement method, and equipment have been approved. If the approved mix cannot be readily pumped or placed by the Contractor's placing equipment, additional water may be added, provided the water-cement ratio of the approved mix design is not exceeded.

The pressure concrete shall be placed by methods capable of forcing it into crevices and filling all void spaces in the tunnel. Unless otherwise provided on the Plans, the concrete backfill shall be placed under pressure by means of a "slick" line and pneumatic or positive displacement pumps.

The combined length of the slick line and delivery line shall not exceed the recommendation of the manufacturer of the concrete pump or, if no manufacturer's performance data is available, 150 feet (45m). The discharge end of the slick line shall be rigid conduit with a minimum length of 10 feet (3m). It shall be kept buried in at least 5 feet (1.5m) of fresh concrete during concrete placement. Concrete shall be pumped continuously during withdrawal of the slick line to eliminate voids.

306-3.8 Pressure Grouting of Voids. Where the Engineer has reasonable doubt that the tunnel void spaces are completely filled, the Contractor shall pressure grout such locations as ordered through grout pipes to be installed either from the ground surface or from within the conduit. At least two grout holes will be required at each location to permit escape of air. The location of surface grout pipes may be adjusted as may be required, dependent upon traffic requirements on overhead streets.

Grout for filling voids shall be low pressure grout (less than 10 psi (70 kPa)). Neat cement grout shall be used except that large voids shall be filled with pressure concrete or grout containing sand.

Grout shall be placed by means of pumps of positive displacement or pneumatic type and capable of placing grout at pressures up to 100 psi (700 kPa) unless otherwise approved by the Engineer. Grout shall be placed at pressures which are requisite for the conditions encountered, and will ordinarily be less than 10 psi (70 kPa) except in cases where large cave-ins or other adverse conditions may require higher pressures.

Regardless of the materials or methods of backfilling or filling voids used, the Engineer shall reserve the right to require filling of void spaces known to remain by additional grouting. Such work will be deemed to have resulted from the Contractor's operations and shall be done at its expense.

306-3.9 Payment. Unless the Special Provisions or the Bid provide for Contract Unit Prices for individual work items included in tunnel work, the lump sum or Contract Unit Price per linear foot (meter) for tunnel as set forth in the Bid shall include full compensation for dewatering, backpacking, maintaining tunnel supports, placing tunnel backfill, low pressure grouting, providing access shafts of portals including excavation, backfill and replacement of surface or other improvements, furnishing and installing pipe, and doing whatever else is appurtenant to tunnel construction within the limits shown on the Plans or in the Special Provisions. Unless otherwise specified, payment for tunnel excavation shall include the excavating of any type of material encountered. High-pressure grouting required by the Engineer, and not resulting from an act or failure to act on the part of the Contractor, will be paid for as extra work as provided in 3-2.

For pipe laid through tunnel access shafts, payment shall be made as provided in 306-1.6.

306-4 CAST-IN-PLACE NONREINFORCED CONCRETE PIPE (CIPCP).

306-4.1 General. These specifications are for cast-in-place nonreinforced concrete pipe intended to be used for gravity and low head drains and irrigation systems.

a) Trench construction shall conform to 306-1.1.2, and

b) Where soils encountered are not capable of standing unsupported from the bottom of trench to the top of the pipe without sloughing and where soils are saturated or contain water quantities or other conditions harmful to the concrete, the Contractor shall install an alternate pipe as directed by the Engineer. The substitution of alternate pipe shall be at no additional expense to the Agency.

306-4.2 Materials. Concrete, unless otherwise specified, shall conform to 201-1, Class 560-C-3250 (Class 330-C-23), except that:

a) The slump shall be 1 inch (25mm) minimum and 3 inches (75mm) maximum and measured only after all water has been added. No water shall be added after the slump test material has been sampled, and

b) Concrete shall not be placed when temperature of the concrete exceeds 90°F (32°C) or is less than 50°F (10°C). The soil adjacent to the trench shall be at a temperature above freezing, and

c) Batch proportions shall be designed by the Contractor and submitted to the Engineer for written approval 7 days in advance of any work.

306-4.3 Excavation. Trenches shall be excavated to ensure the pipe is constructed on the alignment and to the grades shown on the Plans. The subgrade shall be fine graded to the tolerances specified in 306-4.6.5. No concrete shall be placed unless the trench is within the specified grade and alignment tolerances.

306-4.3.1 Trench Width. Except for curves and structures, the trench shall not exceed the width of the pipe OD plus 2 inches (50mm) for a height of 1 foot (0.3m) above the top of the pipe. The bottom of the trench shall be shaped to serve as the outside form of the pipe. The trench shall provide full, firm, and uniform support over the bottom 210 degrees of the pipe, which is referred to as the "trench form". See Plans for curves and structures.

306-4.3.2 Isolated Rock. Where isolated rock is encountered within the trench form, it shall be removed. If the rock is too large to be removed by hand, all portions of the rock within 6 inches (150mm) of the lower 90 degrees of the trench form shall be removed. The void shall be refilled with the monolithically placed concrete, if approved by the Engineer, prior to construction of the pipe placed or backfilled with soil compacted to a minimum relative compaction of 90 percent.

306-4.3.3 Extensive Rock. Where extensive rock is encountered, the bottom 90 degrees of soil and rock shall be overexcavated to a depth of 6 inches (150mm) below the trench form and 12 inches (300mm) on the remaining portions of the trench forms. The void shall be refilled with the class of concrete used for the CIPCP or with soil compacted to a minimum relative compaction of 90 percent when approved by the Engineer.

306-4.3.4 Soil Moisture. At the time of concrete placement, all soils to be in contact with cast-in-place pipe shall be moistened, but shall not contain standing, seeping, or flowing water. Provisions shall be made to dewater the trench so that flowing or standing water is eliminated. The Contractor may, at its sole cost, place a layer of 1 inch (25mm) maximum size rock 6 inches (150mm) thick below the trench invert to assist in water control.

306-4.4 Placement.

306-4.4.1 General. Concrete placement shall be in accordance with 303-1.8 except as provided herein. The flowline grade and alignment of the finished pipe shall conform to the tolerances stated in 306-4.6.5.

306-4.4.2 Concrete Forms. The concrete shall be placed around the full circumference of the pipe in one operation by means of fixed forms and traveling forms. The internal fixed forms shall be of sufficient strength to withstand the vibrating or tamping of concrete. Inflatable internal forms shall not be used. The internal fixed forms shall be of sufficient strength to prevent deformation during construction of the placed concrete. The concrete shall be vibrated, tamped, or worked with suitable devices until the concrete has been consolidated and completely fills the forms.

306-4.4.3 Pipe Junction. Where junction structures are to be constructed, the pipe shall be continuous through the structure locations. The pipe shall be cut away to the specified opening prior to the final set of concrete. Alternate methods may be used as approved in writing by the Engineer.

306-4.4.4 Construction Stoppage. When placement is stopped for a period of time that initial set is likely to occur or 20 minutes, whichever is less, a construction joint shall be made by sloping the end of the pipe at approximately 45 degrees and inserting 24 inches (600mm) long No. 3 (No. 10M) dowels 12 inches (300mm) into the center of the pipe wall at approximately 18 inches (450mm) intervals around the pipe circumference. The total exposed face shall be left in a roughened condition.

Before placing operations may resume, the concrete placed at the construction stoppage joint shall attain sufficient strength to permit an excavation to be made on each side of the joint to form a concrete collar. This collar shall be centered on the joint and have a minimum thickness of one and one-half times the pipe wall thickness and a length of 24 inches (600mm). The joint shall be cleaned of laitance, foreign, and loose materials before resuming concrete placement.

Before placing operations may resume, the concrete placed at the construction stoppage joint shall attain sufficient strength to permit an excavation to be made on each side of the joint to form a concrete collar. This collar shall be centered on the joint and have a minimum thickness of one and one-half times the pipe wall thickness and a length of 600mm (24 inches). The joint shall be cleaned of laitance, foreign, and loose materials before resuming concrete placement.

306-4.4.5 Form Removal. Internal fixed forms shall remain in place until the concrete is self-supporting, after which they may be loosened but shall not be removed for at least 6 hours after placement. As soon as practical thereafter, the forms shall be removed to facilitate inspection and prompt repair. At times of low temperature or other adverse conditions the forms may be kept in place for longer periods of time.

306-4.4.6 Finishing. The interior of the pipe shall be at least as smooth as a steel trowel finish except for the form lap ridges permitted in 306-4.6.4.

306-4.4.7 Curing. Immediately after concrete placement, the exposed top portion of the pipe shall be cured by placing a polyethylene film at least 0.0015 inch (0.038mm) thick so as to completely cover the top surface. All openings in the pipe shall be covered with 0.0015 inch (0.038mm) polyethylene and securely fastened for at least 7 days immediately after placement. At locations where work on the pipe is required, and only during the period that such work is actually in progress, shall necessary openings be uncovered.

306-4.4.8 Repairing. After the internal fixed forms have been removed, the inside of the pipe shall be inspected by the Engineer. All rock pockets, blisters, voids, or similar defects not extending through the wall and less than 2 square feet ($0.18m^2$) in area, shall be repaired immediately by removing the defective concrete and replacing it with properly bonded and cured mortar or other patching material approved by the Engineer.

All rock pockets, blisters, voids, or other defects greater than 2 square feet ($0.18m^2$) or which extend through the pipe wall shall be repaired by removing the entire pipe for 1 foot (300mm) on each side beyond the limits of the defect.

Cracks shall not be repaired until the entire backfill is in place. However, the Contractor may remove and replace cracked pipe prior to placement of the entire backfill.

Subsequent to placement of the entire backfill, the Contractor shall notify the Engineer when the pipe is ready for reinspection. Cracks less than 0.01 inch (0.255mm) in width or cracks greater than 0.01 inch (0.255mm) but less than 12 inches (300mm) long shall be painted with a cement paste. Longitudinal cracks exceeding 0.01 inch (0.255mm) in width and 12 inches (300mm) in length must be repaired by epoxy pressure grouting provided the total length of cracks for any reach is less than 25 percent. If the total length of cracks exceeds 25 percent, the entire reach shall be removed and replaced. A reach is any length between two structures.

Circumferential cracks exceeding 0.01 inch (0.255mm) in width and 12 inches (300mm) in length shall be repaired by removing at least 1 inch (25mm) of concrete in width for a depth of at least 1/2 the wall thickness. After cleaning this area, it shall be filled with properly bonded and cured mortar.

Alternate repair methods shall be submitted in writing not less than 7 days prior to use for approval by the Engineer. Any repairs performed shall ensure the specified structural strength is not compromised and by techniques which have been approved by the Engineer.

306-4.4.9 Rejection. Pipe will be rejected for any of the following reasons:

a) Longitudinal cracks exceeding 0.01 inch (0.255mm) in width and 12 inches (300mm) or greater in length unless repaired per 306-4.4.8. If longitudinal cracks occur intermittently in 25 percent or more of a reach of pipe, the pipe shall not be repaired and shall be removed and replaced.

b) Circumferential cracks exceeding 0.01 inch (0.255mm) in width and 12 inches (300mm) or greater in length unless repaired per 306-4.4.8.

c) Longitudinal cracks exceeding 1/1000 the internal diameter or a maximum 1/16 inch (1.5mm) in width.

d) Rock pockets, honeycombing, blisters, voids, or other defects that extend through the pipe wall.

e) A wall thickness less than the minimum as listed in 306-4.6.3.

f) A diameter that does not meet the requirements of 306-4.6.2.

g) Application of any wash coat of cement, grout, or other material prior to reinspection after the entire backfill has been placed.

h) Air bubble voids (bugholes) on the interior surface of the pipe exceeding 1/4 inch (6mm) in depth unless pointed with mortar or other approved material.

i) Unpaired offsets or indentations, including transverse and longitudinal form offsets exceeding those allowed in 306-4.6.4

j) Deviation or departure from true grade or alignment exceeding that allowed in 306-4.6.5.

k) Concrete used that has a slump of less than 1 inch (25mm) or more than 3 inches (75mm) per 306-4.2. Concrete that has had water added after slump and/or cylinder samples have been taken or that does not meet the proportioning requirements of 201-1.

l) Concrete that has core strengths less than that required per 306-4.2.

m) The pipe does not pass the load test per 306-4.7.4.

n) The pipe has been damaged in any manner including but not limited to placing or compacting the backfill.

o) Concrete that was placed when the concrete temperature exceeded 90°F (32°C) or was less than 50°F (10°C), or when the soil adjacent to the trench was below freezing.

p) The trench does not provide full, firm, and uniform support over the bottom 210 degrees of the pipe or the trench width exceeds the OD per 306-4.6.3 by more than 2 inches (50mm), except when meeting the requirements of 306-4.3.1, 4.3.2, and 4.3.3.

q) The interior of the pipe is not at least as smooth as a steel trowel finish except for the form lap ridges.

306-4.5 Backfill. Backfill for cast-in-place pipe shall be considered as starting at the top of the trench form and shall conform to 306-1.3. The method of backfilling shall be subject to the approval of the Engineer. The equipment used in placing the backfill shall not cause damage to the pipe or cause loads to be placed on the pipe which are in excess of design loads.

Backfilling will not be permitted over cast-in-place concrete pipe until the concrete attains the strength specified. The Contractor may place backfill prior to 28 days upon written approval by the Engineer provided the required 28-day strength has been attained and verified by a laboratory.

306-4.6 Dimensions and Tolerances.

306-4.6.1 General. The minimum nominal size of cast-in-place pipe shall be 24-inch (600mm) ID.

306-4.6.2 Diameter. The internal diameter of the pipe at any point shall not be less than 99 percent nor more than 105 percent of the nominal diameter, and the average of any four measurements of the internal diameter made at 45-degree intervals shall not be less than the nominal diameter.

306-4.6.3 Wall Thickness. The wall thickness at any point shall not be less than specified on the Plans and in no case less than in the following table. Any length which fails to meet the thickness requirements will be rejected. The grade and alignment shall be controlled so that the wall thickness of the pipe is symmetrical.

TABLE 306-4.6.3 (A)

Internal Diameter inches (mm)	Minimum Wall Thickness inches (mm)
24 and 30 (600 and 750)	3 (76)
36 (900)	3-1/2 (89)
42 (1050)	4 (102)
48 (1200)	5 (127)
54 (1350)	5-1/2 (140)
60 (1500)	6 (153)
66 (1650)	6-1/2 (165)
72 (1800)	7 (178)
78 (1950)	7-1/2 (191)
84 (2100)	8 (203)
90 (2250)	8-1/2 (216)
96 (2400)	9 (229)
108 (2700)	10 (254)
120 (3000)	12 (305)
132 (3300)	14 (356)
144 (3600)	15 (381)

For any ID not indicated above, the minimum wall thickness shall be equal to the next size larger pipe. Wall thickness will be determined in accordance with 306-4.7.4, or when applicable, with 306-4.7.1.

306-4.6.4 Offsets and Indentations. Any offset or indentation, including transverse and longitudinal form offsets and construction stoppage joints, shall not exceed 1/4 inch (6mm) for pipe with specified ID of 42 inches (1050mm) or less, and 3/8 inch (10mm) for pipe with specified ID over 42 inches (1050mm) and less than 72 inches (1800mm), and 1/2 inch (13mm) for all pipe diameters equal to or greater than 72 inches (1800mm).

Reaches having offsets or indentations in excess of these limits shall be repaired as approved by the Engineer.

306-4.6.5 Grade and Alignment. Departure from and return to established grade shall not exceed 3/8 inch per foot (10mm/m) and maximum departure shall not exceed 1.0 inch (25mm). Maximum departure from established alignment shall not exceed 2 inches (50mm) on tangents and 4 inches

(100mm) on curves. Departure from and return to established alignment shall not exceed 1/4 inch per foot (20mm/m). A laser grade control shall be used for all trenches.

If the departure exceeds the maximum allowed, the work shall be stopped and the necessary adjustments made. The affected portions of the conduit exceeding the above departure shall be removed and replaced at the proper grade and alignment.

306-4.7 Test Requirements.

306-4.7.1 Sequence of Sampling and Testing. Sampling and testing shall be performed in the sequence described herein. Tests for portland cement concrete will be taken during concrete placing operations in accordance with 201-1.1.4. If the concrete cylinders do not meet the required 28-day strength, cores shall be obtained from the completed pipe. Cores shall be used to determine thickness and compressive strength. Strength test results shall be verified by a laboratory approved by the Engineer. If the strength of these cores fails to meet the design requirements, the pipe will be rejected.

The Engineer will determine the number and location of the samples and tests. The location shall be identified by station, and where applicable, the angle from vertical measured clockwise facing up-station.

306-4.7.2 Thickness. The Engineer will determine the wall thickness of the pipe as follows:

a) The thickness at the invert and crown of the pipe will be measured by probing at approximately 25 foot (7.5m) intervals during placement of the concrete. The probe shall be forced through the concrete to make firm contact with the form at the crown and shall be held in a position normal to the surface when the measurement is taken. The invert shall be inspected by removing a small portion and measuring the thickness. The probe shall be 3/8 inch (9.5mm) round bar, at least 2 inches (50mm) longer than the wall thickness to be measured, rounded on one end with a tee handle on the other.

b) Thickness at the invert and springline will be measured through holes drilled by the Contractor. The holes shall be at least 3/4 inch (19mm) in diameter and shall be drilled after the removal of the forms and within 72 hours of concrete placement.

Three holes shall be drilled every 50 feet (15m) at the invert and both springlines and shall be located as determined by the Engineer. The Engineer may require additional holes on curves to determine the extent of thin sections.

After measurement, the Contractor shall fill all holes using Class C mortar per 201-5. All costs of probing, drilling, removing, and repairing shall be borne by the Contractor.

306-4.7.3 Concrete Cores. Cores, where required, shall be obtained from pipe and tested in accordance ASTM C42. The cores shall have a length-to-diameter ratio of not less than one. The diameter of cores shall be at least three times the maximum size of the aggregate used in the concrete, except where the wall thickness if such that the length-to-diameter will be less than one, in which case the core diameter may be reduced to two and one-half times the maximum aggregate size used.

There shall be at least four cores taken for each 200 linear feet (60 linear meters), or fraction thereof, of pipe. Cores shall be taken at the following points at stations selected by the Engineer: one through the crown, one through the invert, and two in the lower half of the pipe 45 degrees from the vertical. The Engineer may require additional cores at any location. The Contractor shall patch all core holes in such a manner that the patch will be permanent, will not leak, and will have a smooth finish flush with the interior surface of the conduit. All costs of coring, testing, and patching core holes shall be borne by the Contractor.

306-4.7.4 Load Bearing. Load bearing tests shall be required for every 1,000 feet (300m) of pipe having the same size and wall thickness, with a minimum of one per size and two per project. The test locations will be specified by the Engineer. The test shall be performed in the presence of the Engineer, and the Contractor shall be responsible for all costs and risks involved. Failure of the test section will be cause for rejection of the conduit represented by the test.

The method and apparatus requirements for load bearing tests are as follows:

a) The test shall be performed with only the trench form providing bottom support. If the pipe has been constructed so that more than 210 degrees is in contact with the natural soil, the trench wall shall be re-excavated to provide 210 degrees of trench form without altering the existing bedded condition of the trench form.

b) The test length shall be at least 4 feet (1.2m) and not more than 5 feet (1.5m). At the option of the Contractor, the test section may be isolated from the completed pipe.

c) The test load shall be applied by use of a "sand box," consisting of a frame and bearing plate, in such a manner that sand carefully placed in the sand box forms a bearing symmetrically about the centerline and over the entire length of the test section. The width of the bedding shall be 0.7 times the specified ID of the pipe. The minimum thickness of the sand shall be 0.25 times the specified ID.

d) The frame and bearing plate shall be sufficiently rigid so that they will distribute the load uniformly and will not deform under the loaded condition. The interior surfaces of the frame shall be smooth. The lower surface of the bearing plate shall be a true plane. Cloth or plastic film shall be attached to the inside of the frame along the lower edges to prevent the loss of sand through the gap between the pipe and the frame. This type of apparatus is described in ACI Specification 346.

e) The frame shall be properly located on the pipe test section and filled with sand. The sand shall be clean and graded so that it will pass a No. 4 (4.75mm) sieve. The sand shall be struck off level and covered with the bearing plate. During the test, the bearing plate shall not contact the frame.

f) The load shall be applied symmetrically on the bearing plate until the total required has been attained. The pipe shall remain loaded until the interior of the pipe has been inspected by the Engineer and results have been observed and recorded.

g) The applied load, in pounds (newtons), shall equal the test load multiplied by the length of the test section, in feet (meters). The test load shall be calculated as follows:

U.S. Standard Measures:

$$\text{Test Load} = (127.5H + 1.5LL + 5.56T)\,OD + 34.0(ID)^2$$

SI Units:

$$\text{Test Load} = (20030H + 1.5LL + 10.48T)\,OD + 5340\,(ID)^2$$

Where:

ID = Specified inside diameter of the pipe in feet (meters).

T = Specified wall thickness of the pipe in inches (millimeters).

OD = ID + 2T/12 = Outside diameter of pipe in feet.

(SI Units: OD = ID + 2T/1000 = Outside diameter of pipe in meters.)

H = Depth of cover on pipe in feet (meters).

LL = Live load on pipe in pounds per square foot (kPa).

TABLE 306-4.7.4 (A)

Depth of Cover ft (m)	Live Load (LL) lbs/ft^2 (kPa)
3 (0.9)	489 (23.4)
4 (1.2)	314 (15.0)
5 (1.5)	234 (11.2)
6 (1.8)	182 (8.71)
7 (2.0)	145 (6.94)
8 (2.4)	119 (5.70)
9 (2.7)	120 (5.75)
10 (3.0)	90 (4.31)
Over 10 (3.0)	N/A

h) The total test load shall be supported by the test section without the development of any additional cracking.

i) After the satisfactory completion of the test, the Contractor shall repair the pipe, resulting from isolating the test section, in a manner satisfactory to the Engineer.

In lieu of using a "sand box" as described above, the Contractor may conduct a wheel load test on a 4-foot (1.2m) section of pipe when approved in writing by the Engineer. The load applied shall be determined by the equation in item 7 above applied to a section of pipe. The total test load shall be supported by the test section without the development of any additional cracking.

306-4.8 Payment. The Contract Unit Price per linear foot for CIPCP shall include full compensation for doing all work involved in constructing the pipeline, including excavation, backfill, testing, repair, and replacement as specified or as directed by the Engineer.

306-5 ABANDONMENT OF CONDUITS AND STRUCTURES. When sanitary sewer or storm drain conduits have been or are to be abandoned and are found to interfere with construction, the interfering portion shall be removed and the remaining open portion securely sealed. Where the greater internal dimension of the conduit is 4 feet (1.2m) or less, the seal shall consist of a wall of concrete not less than 6 inches (150mm) thick or a 8 inch (200mm) wall of brick and mortar. For larger openings, details of the seal will be shown on the Plans. In the case of catch basin connector pipes, the inlet opening to the mainline pipe shall also be sealed.

When a sanitary sewer or storm drain is to be abandoned within specified limits, all structures and appurtenances within said limits shall also be abandoned.

When catch basins or manholes are to be abandoned, the upper portion shall be removed to a depth of at least 1 foot (0.3m) below street subgrade and the conduits connecting to the structure shall be sealed as provided herein. The bottom of such structures shall be perforated or broken to prevent the entrapment of water.

Structures designated on the plans to be removed shall be removed to the full depth of the structure, including its foundation. Voids resulting from abandoned or removed structures shall be filled with suitable material compacted to a relative compaction of 90 percent.

Cover sets, gratings, and other steel components (except reinforcing bars) of removed or abandoned structures shall be salvaged. The Contractor shall contact the owners and, if required, shall deliver to and load such material in the owner's truck at the Work site. Otherwise, such material shall become the property of the Contractor and shall be disposed of by it away from the Work site.

306-6 REMODELING EXISTING SEWER FACILITIES. Where the Plans indicate construction involving existing sewer facilities, the Contractor shall provide temporary seals, enclosures, forced ventilation, or other devices as may be necessary to prevent odor nuisance during construction. Sewers shall be open to the atmosphere only for a reasonable time necessary for construction.

All existing sewer facilities shall be considered potential permit-required confined spaces in accordance with 7-10.4.4. Hazards to which workers may be exposed, include, but are not limited to engulfment, hydrogen sulfide gas, explosive/flammable gases, and/or oxygen deficiency. When required, the Contractor shall implement a permit space program in accordance with 7-10.4.4.

Where a manhole bottom is to be remodeled on an existing sewer, the portion to be remodeled shall be removed to a minimum depth of 3 inches (75mm) to permit construction of new channels and shelves. Sewage in new and remodeled manholes shall be controlled across the manhole in such a manner that sewage does not flow over concrete channels until they have cured for 24 hours. The controls shall prevent backup of sewage upstream from the manhole unless otherwise approved by the Engineer.

Where required by the Plans or Special Provisions, the Contractor shall submit Working Drawings for control operations in accordance with 2-5.3.

306-7 CURB DRAINS. Drains shall be constructed beneath the sidewalk to connect building drains to curb outlets and to serve low areas on adjacent property as shown on the Plans or as directed by the Engineer.

The drain shall be a 3-inch (75mm) diameter pipe for a 6-inch (150mm) curb face, and a 4-inch (100mm) diameter pipe for an 8-inch (200mm) curb face or greater. The invert of the drain shall be located 1/2 inch (13mm) above the gutter flowline. The drain pipe shall have a minimum 2-inch (50mm) clearance from top of curb and be laid on a straight grade with a minimum slope of 3/8 inch per foot (30mm/m) and terminate 1 inch (25mm) back of the curb face.

Curb drains may be constructed using pipe materials specified in 207 or other pipe materials approved by the Engineer. The pipe shall be suitably joined in accordance with the manufacturer's standard jointing system.

306-8 MICROTUNNELING.

306-8.1 General.

306-8.1.1 Description. Microtunneling is an unmanned entry method that uses a remotely operated microtunnel boring machine (MTBM) to install pipes underground with minimal surface disruption. Microtunneling continuously installs pipe behind a remotely controlled, steerable, laser-guided, full-face controlled, articulated MTBM. The pipe to be installed is connected to and follows the MTBM.

The length of drive which is possible to achieve is dependent upon the jacking force, pipe material, and pipe size.

The jacking force required is a function of many variables including the soil conditions, depth of the pipeline, annular space between the pipe and the soil, lubrication of the pipe, water table location, overburden loads, installation time, pipe material, out-of-roundness, diameter, and strength.

306-8.1.2 Minimum Soil Cover. The minimum depth of cover to the top of the installed pipe using this process shall be one and one-half to three times the outside diameter of the pipe being installed, or 6 feet (2m), whichever is greater depending on the soil conditions. With prior approval of the Engineer, the minimum depth of cover may be reduced.

306-8.1.3 Surface Description. Unless otherwise noted in the Contract Documents, settlement of heave at the ground surface during and after construction shall not exceed 1/2 inch (13mm) or unless specified in the Contract Documents as measured along the centerline of the conduit being installed. Zero settlement or heave may be required when specified in the Contract Documents or required by applicable permits.

306-8.1.4 Definitions.

Annular Space - The void created between the outside diameter of pipe being installed and extreme outer limits created by MTBM bore process.

Earth Pressure Balance - MTBM pressure applied to the cutting face equals the pressure of earth against the cutting face.

Full Face Control - Complete mechanical support of the excavated face at all times.

Lubricant- A substance applied between the pipe and soil to minimize friction and to fill the annular space.

Microtunneling Boring Machine (MTBM) - A remotely controlled, steerable, laser guided microtunnel boring machine consisting of an articulated boring machine shield and a rotating cutting head.

Pipe String - The succession of joined individual pipes being used to advance the excavation equipment.

Shaft or Pit - A vertical excavation to insert or receive microtunneling equipment and pipe.

Slurry - Water mixture, which may contain additives, that is used to transport spoils and counterbalance any ground water pressure.

306-8.1.5 Submittals. The Contractor shall submit the following items for review and approval by the Engineer in accordance with 2-5.3. Approval of the submittal by the Engineer shall be obtained prior to ordering pipe materials and/or start of the microtunneling operations.

a) Manufacturers' data sheets and specifications describing in detail the microtunneling system to be used.

b) Description of similar projects with references on which the proposed system had been successfully used by contractor/operator.

c) Description of method to remove and dispose of spoil.

d) Maximum anticipated jacking loads and supporting calculations.

e) Description of methods to control and dispose of groundwater, spoil, temporary shoring, and other materials encountered in the maintenance and construction of pits and shafts.

f) Shaft dimensions, locations, surface construction profile, depth, method of excavation, shoring, bracing, and thrust block design.

g) Pipe design data and specifications.

h) A description of the grade and alignment control system.

i) Intermediate jacking station locations and design.

j) Description of lubrication and/or grouting system.

k) Layout plans and descriptions of operational sequence.

l) A detailed plan for monitoring ground surface movement (settlement or heave) due to the microtunneling operation. The plan shall address the method and frequency of survey measurement. At minimum, the plan shall measure the ground movement of all structures, roadways, parking lots, and any other areas of concern within 25 feet (8m) on both sides of all microtunneling pipelines at a maximum spacing of 100 feet (30m) along the pipeline route, or as required by the Engineer.

m) Contingency plans for approval for the following potential conditions: damage to pipeline structural integrity and repair; loss and return to line and grade; and loss of ground.

n) Procedures to meet all applicable OSHA requirements. These procedures shall be submitted for a record purpose only and will not be subject to approval by the Engineer. At a minimum, the Contractor shall provide the following:

1) Protection against soil instability and ground-water inflow.
2) Safety for shaft access and exit, including ladders, stairs, walkways, and hoists.
3) Protection against mechanical and hydraulic equipment operations, and for lifting and hoisting equipment and material.
4) Ventilation and lighting.
5) Monitoring for hazardous gases.
6) Protection against flooding and means for emergency evacuation.
7) Protection of shaft, including traffic barriers, accidental or unauthorized entry, and falling objects.
8) Emergency protection equipment.
9) Safety supervising responsibilities.

o) Annular space grouting plan if required by the Contract Documents.

306-8.1.6 Subsurface Conditions.

a) **Microtunneling Specified by the Agency.** The Agency will make accessible to the Contractor all available subsurface information, if any, which is listed in 2-7 and 306-8.1.6 c). All subsurface investigations deemed necessary by the Contractor to complete the work shall be included at no additional cost to the Agency. Copies of all reports and information obtained by the Contractor shall be provided to the Agency.

b) **Microtunneling Requested by the Contractor.** When microtunneling is proposed by the Contractor as an alternative to the specified methods of conduit installation, the Contractor shall obtain and provide the Agency with the copies of the information and reports listed in 2-7 and 306-8.1.6 c) at no additional cost to the Agency. Microtunneling operations must be approved by the Agency prior to commencement of work.

c) **Subsurface Data.** The following subsurface information will affect equipment selection and the progress and practicality of microtunneling. The actual test data required will vary depending upon the scope of the Work and soil conditions encountered and may include but not be limited to the following:

Particle-size analysis ASTM D422

Soil Classification ASTM D2487

Plastic limit ASTM D4318

Liquid limits ASTM D4318

Plasticity index ASTM D4318

Expansion index ASTM D4829

Density ASTM D1556, D2037, D5195, D4564.

Water (Moisture) content ASTM D4959, D2216, D5220, D3017, D4643, D4944.

Shear strength:

Direct ASTM D3080

Triaxial, C.U. ASTM D4767

Unconfined compressive strength ASTM D2166

Permeability ASTM D2434

Apparent or unconfined soil cohesion.

Standard penetration test (SPT) ASTM D1586.

Water table depth

Nature of fill material

Nature of pollutants

Rock type and color

Fracture index

Rock quality designation (RQD)

Core recovery, TCR.

All reasonable attempts will be made to collect subsurface test samples within 20 feet (6m) horizontally of the centerline of the proposed conduit location. Subsurface test samples are to be collected to a minimum depth of one pipe diameter below conduit invert. The test samples should typically be collected at 200-foot (60m) intervals or at manhole locations.

306-8.1.7 Site Cleanup. Unless otherwise specified, all existing surface improvements damaged or removed as a result of microtunneling operations shall be restored to their original condition.

306-8.2 Major Components of a Microtunneling System.

306-8.2.1 MTBM. The MTBM selected shall be capable of installing the pipe while being compatible with the anticipated soil and geotechnical conditions. The MTBM cutter face shall at all times be capable of supporting the full excavated area without the use of ground stabilization and have the capability of measuring the earth pressure at the face and setting a calculated earth balancing pressure. The maximum radial annular space shall not exceed 1 inch (25mm), unless otherwise specified in the Contract Documents.

The MTBM shall be capable of controlling shield rotation by means of a bi-directional drive on the cutter head or by use of mechanical fins or grippers. The MTBM shall be mechanically articulated to enable remotely controlled steering of the shield. The MTBM shall control groundwater during excavation without the use of external dewatering equipment. The measuring and balancing of earth and groundwater pressure shall be achieved by use of a slurry or cased auger system. The system shall be capable of incremental adjustments to maintain face stability for the soil conditions encountered.

306-8.2.2 Jacking Equipment. The main jacks shall be mounted in a jacking frame and located in the jacking shaft The MTBM shall be moved forward by the jacks advancing a successive string of connected pipes toward a receiving shaft

A pipe lubrication system may be used to lower the friction developed on the surface of the pipe during jacking with approval of the Engineer. An approved lubricant, typically betonite or polymers, may be injected at the rear of the MTBM or through lubrication ports. The pipe lubrication system pressure shall be continuously monitored, recorded, and controlled to prevent pipe buckling and/or ground heave.

A thrust block is required to transfer jacking loads into the soil. The thrust block shall be perpendicular to the proposed pipe alignment. The thrust block shall be designed to support the maximum jacking pressure developed by the main jacking system. Special care shall be taken when securing the pipe guide rails and/or jacking frame in the jacking shaft to ensure correctness of the alignment, grade, and stability of the pipe. If a concrete thrust block or treated soil zone is utilized to transfer jacking loads into the soil, the MTBM shall not be jacked until the concrete or other materials have attained the required strength.

When intermediate jacking stations are utilized, the maximum jacking force shall not exceed the maximum allowable jacking load of the pipe.

306-8.2.3 Excavation Controls. The control equipment shall integrate the method of excavation and removal of soil and its simultaneous replacement by a pipe. As each pipe section is jacked forward, the control system shall synchronize spoils removal, excavation, and jacking speeds.

Operations shall be stopped when they result in pipe damage or any surface disruption. The Contractor shall propose immediate action for review and approval by the Engineer to remedy the problem at no additional cost to the Agency.

306-8.2.4 Automated Spoils Transportation. The MTBM shall include one of the following:

a) **Slurry System.** The system shall be capable of measuring earth and groundwater pressure and making the adjustments required to counter-balance the earth and groundwater pressure to prevent loss of slurry or uncontrolled soil and groundwater inflow.

   1) The slurry pressure at the excavation face shall be controlled by use of slurry pumps.

   2) A slurry bypass method shall be included to allow for a change in direction of flow to be made and/or isolated.

   3) Provide a separation process, properly sized for the tunnel being constructed, the soil type being excavated, and the workspace available at each area. Separate the spoil from the slurry so that slurry may be returned to the cutting face for reuse.

   4) Monitor the composition of the slurry to maintain the slurry density and viscosity limits as approved in the submittals.

b) **Cased Auger System.** The system shall monitor and continuously balance the soil and groundwater pressure. The system shall be capable of adjustments required to maintain face stability for the particular soil condition to be encountered to prevent loss of soil or uncontrolled groundwater inflow.

   1) Maintain the pressure at the excavation face by controlling the volume of spoil removal with respect to the advance rate. Monitor the speed of the rotation of the auger and the amount of water added.

   2) Submit an evaluation of equipment's ability to balance earth and water pressure at the face, stability of the soils, and the significance of the groundwater present for the Engineer's review.

306-8.2.5 Active Steering Controls. A remotely controlled steering mechanism shall be provided that allows for the operation of the system without the need for personnel to enter the microtunnel.

The steering information shall be monitored and transmitted to the operation console. the minimum steering information available to the operator on the control console shall include the position of the shield relative to the design reference, roll, inclination, attitude, rate of advance, installed length, thrust force, and cutter head torque.

306-8.2.6 Guidance/Monitoring Equipment. The MTBM display equipment shall continuously show and automatically record the position of the shield with respect to the project design line and grade. The automated recording system shall include real time information such as earth and ground pressure, roll, pitch, attitude, rate of advance, installed length, cutter head torque, jacking loads, slurry pressure, slurry flow, and slurry valve positions.

Line and grade shall be controlled by a guidance system that relates the actual position of the MTBM to a design reference (e.g., by a laser beam transmitted from the jacking shaft along the line of the pipe to a target mounted in the shield). The line and grade tolerances of pipe installed shall be ± 1 inch (25mm) on grade and 1-1/2 inches (38mm) in line between shafts, unless otherwise stated in the Contract Documents or approved by the Engineer.

The rate of return to line and grade shall not exceed 1 inch in 25 feet (1:300), unless otherwise specified.

306-8.3 Methods. Prior to pipe installation, the Contractor shall implement the approved plan submittals to monitor ground movement.

306-8.3.1 Intermediate Shafts. If an intermediate shaft is requested, the Contractor shall obtain a written approval from the Engineer. The intermediate shaft shall not be located in areas prohibited the Contract Documents. The Contractor's request shall include all necessary permits and approvals, minimize public inconvenience and minimize impacting existing facilities.

306-8.3.2 Annular Space Grouting. The annular space created by the overcut of the MTBM in excess of 3/4 inch (19mm) shall be filled with an approved material, unless otherwise specified in the Contract Documents.

When grouting is specified, pressure-injected grout shall fill voids outside the limits of the excavation created by caving or collapse of earth cover over the excavation.

The Contractor shall furnish and operate suitable equipment for any required grouting operations depending on the condition of the application. The grouting operation shall not damage adjacent utilities or other properties. Grout shall be injected at a pressure that will not distort or imperil any portion of the work or existing installations or structures.

306-8.3.3 Work Hours. Work hours are not restricted unless stated otherwise in the Contract Documents. Multiple shifts may be used where permitted by the Contact Documents and if noise levels do not exceed local ordinances. Continuous microtunneling will be permitted where:

a) Expansive soils are encountered; or

b) The actual jacking forces required approach either the capacity of the jacking system or the designed jacking capacity of the pipe.

306-8.3.4 Construction Zone. Any microtunneling construction zone in the public right-of-way shall be limited to one lane traffic, or the Contractor shall maintain a minimum of one lane of traffic in each direction. The Engineer may specify a larger or smaller zone if circumstances warrant.

306-8.3.5 Shafts. Shafts shall be of a size commensurate with safe working practices and located as described in the Contract Documents. With the written approval of the Engineer, the Contractor may relocate shafts to better suit the capabilities of the microtunneling equipment proposed.

Shaft locations shall, where possible, be kept clear of road intersections and within a single traffic lane, in order to minimize disruption to the flow of traffic.

The design of the shafts shall ensure safe MTBM exit from the driving shaft and entry into the receiving shaft The Contractor shall furnish and install equipment to keep the jacking shaft free of excess water. The Contractor shall also provide surface protection during the period of construction to ensure that surface runoff does not enter shafts.

All shafts shall be backfilled per 306-1.3. All shoring materials, bracing, temporary supports, rubbish, and construction materials shall be removed from the jobsite and disposed of.

306-8.3.6 Installation and Field Testing. The pipe installation shall be tested in accordance with 306-1.4.

306-8.4 Payment. Unless the Specifications provide for Contract Unit Prices for individual work items included in microtunneling work, the lump sum or Contract Unit Prices per linear foot for microtunneling as set forth in the Bid shall include full compensation for grouting and lubricants; providing jacking/receiving/recovery shafts including excavation, disposal, dewatering, backfill and replacement of surface or other improvements; furnishing and installing pipe, excavating, and disposal of materials encountered by installation of the pipe; and all other work appurtenant to microtunneling within the limits described in the Contract Documents.

306-8.5 Pipe Specification. In general, pipe used for this subsection shall be specifically designed for microtunneling by the pipe manufacturer. The pipe shall be round, smooth, and with flush-jointed outer surfaces. The ends of the pipe shall be perpendicular to the longitudinal axis of the pipe with a maximum deviation of no more than 1/16 inch per foot (5mm/m) of pipe diameter, with a maximum of 1/4 inch (6mm), measured with a square and a straight edge across the end of the pipe. Pipe ends shall be square and smooth so that jacking loads are evenly distributed against the pipe end faces without point loads when the pipe is jacked. Pipe used for microtunneling shall be capable of withstanding the jacking forces imposed by the process of installation, as well as the final in-place loading conditions. The driving ends of the pipe and intermediate joints shall be protected against damage. The detailed method proposed to cushion and distribute the jacking forces shall be submitted to the Engineer for approval.

Damaged pipe shall be jacked through to the reception shaft and be removed. Other methods of repairing the damaged conduit may be used, as recommended by the manufacturer and approved by the Engineer.

The pipe manufacturer's design jacking loads shall not be exceeded during the installation process. The pipe shall be designed to take full account of all temporary installation loads. The pipe materials acceptable for microtunneling will be specified in the Contract Documents.

The maximum jacking capacity used shall not exceed the allowable jacking capacity of the pipe that has a minimum factor of safety of 2.5.

SECTION 307 - STREET LIGHTING AND TRAFFIC SIGNAL SYSTEMS

307-1 GENERAL. The work shall consist of furnishing and installing, modifying, or removing one or more street lighting and/or traffic signal systems, as shown on the Plans and specified in the Specifications. The work shall conform to the following codes:

a) California Administrative Code, Title 8, Chapter 4, Subchapter 5, and

b) NEC.

Incidental parts that are not shown on the Plans or specified in the Specifications and are necessary to complete the Work, shall be furnished and installed as though such parts were shown on the Plans or specified in the Specifications.

Unless otherwise approved by the Engineer, the Contractor shall submit a list of equipment and materials to be installed in accordance with 2-5.3. The list shall include the name of the manufacturer, size, and identifying number of each item. In addition, the Contractor shall, in accordance with 2-5.3, submit detailed Shop Drawings and wiring diagrams for the electrical equipment to be used.

If requested by the Engineer, the Contractor shall submit for review samples of the material proposed for use. After review, the samples will be returned to the Contractor.

Upon completion of the Work, the Contractor shall submit one complete set of "as-built" Plans showing in detail all construction changes.

Payment for furnishing the list of equipment and materials, material samples and corrected Plans shall be considered as included in the Contract Unit Price or lump sum price in the Bid for the equipment, labor, and material required to make the street lighting or traffic signal system fully operational.

307-2 MAINTENANCE OF EXISTING AND TEMPORARY SYSTEMS. The Contractor, during the progress of the Work, shall maintain existing or temporary street lighting and traffic signal systems.

307-3 COORDINATION WITH SERVING UTILITY.

307-3.1 General. Safety clearances between new or existing Standards (including luminaires, arms or signal heads) and overhead power and communication lines shall conform to the State of California Industrial Safety Orders, General Order 95. Construction in close proximity to high-voltage overhead lines shall be performed in accordance with the California Code of Regulations, Title 8 and the latest revision of Article 86, State of California High Voltage Electric Safety Orders. The Contractor shall inspect the location of each Standard for safety clearance requirements and notify the Engineer and the serving utility, in writing, of the locations where safety clearances are required. The Contractor shall be responsible for the necessary coordination with the serving utility.

The Contractor shall take the serving utility's schedule into account in preparing the construction schedule. Requests for safety clearances shall be made at least 21 days in advance of the date the Contractor will be working at each location requiring a safety clearance by the serving utility. If required, the Contractor shall make the necessary arrangements with the serving utility to raise their overhead facilities in order to provide for required clearances.

Coordination with the serving utility shall be shown as individual activities on the construction schedule specified in 6-1.

307-3.2 Payment. Payment for coordinating with the serving utility and temporary relocation of power and communication lines shall be considered as included in the Contract Unit Price or lump sum price in the Bid for the equipment and material required to make the systems fully operational.

307-4 DAMAGE TO EXISTING SYSTEMS. Should any damage to an existing system occur, which is not part of the work shown on the Plans, the Contractor shall immediately notify the Engineer and arrange for the immediate repair and restoration of service. The Contractor shall commence repairs or replacements within 24 hours of damaging the system or receiving approval of the equipment and materials by the Engineer, whichever takes longer. Electrical safety clearance shall be obtained from the serving utility before performing any work on existing energized circuits. Equipment and materials used for repairs or replacements shall conform to 209 and be approved by the Engineer.

307-5 TEMPORARY STREET LIGHTING AND TRAFFIC SIGNAL SYSTEMS.

307-5.1 General. The Contractor shall provide temporary street lighting and/or traffic signal systems whenever the existing systems are removed or relocated. Temporary systems shall be in operation prior to the removal or relocation of the existing systems. Temporary street lighting systems shall provide an average illumination and uniformity ratio (avg./min.) that matches or exceeds the existing lighting levels utilizing standard roadway lighting optics. The Contractor shall submit working drawings in accordance with 2-5.3. Construction of the temporary systems shall be shown as an individual activity on the construction schedule specified in 6-1.

Should any damage occur to a temporary system, the Contractor shall immediately notify and arrange for repair and restoration of service in accordance with 307-4.

307-5.2 Temporary Wiring. Temporary wiring specified herein shall not apply to circuits exceeding 150 volts to ground. Temporary wiring for circuits exceeding 150 volts to ground shall be as specified in the Special Provisions.

Temporary wiring may utilize either overhead or underground conductors. Temporary overhead traffic circuit runs shall be 28-conductor cable (except service wire).

Temporary overhead conductors shall be slack-spanned with 20-foot (6m) minimum overhead clearance above thoroughfares and 12-foot (4m) minimum clearance above sidewalk areas. Temporary conductors may not run on top of the ground or across any sidewalk area unless adequately protected in an electrical conduit. Temporary conductors less than 10 feet (3m) above ground level must be protected in an electrical conduit. Temporary overhead conductors shall be multi-conductor cable or single conductors, securely tied or taped at intervals not to exceed 5 feet (2m). No spare conductors are required.

Splices within 10 feet (3m) above ground level shall be enclosed in metal junction boxes. Splices made at ground level shall also be enclosed in pull boxes. All splices shall be made in accordance with 307-13.3. All circuits shall be grounded in accordance with 307-13.4.

307-5.3 Temporary Standards, Signals, Luminaires and Lamps. Temporary Standards shall be installed adjacent to existing Standards. Combination poles that allow mounting of both street lighting luminaires and traffic signal heads shall not be utilized for temporary Standards.

For street lighting systems, the mounting height shall be a minimum of 30 feet (9m). The lamp size shall be 200 watts. Luminaires shall be cutoff with a type 3 distribution pattern.

Temporary traffic signal heads shall provide a minimum of 2 clearly visible signal faces, for each phase, for traffic from each direction. One signal face shall be adjacent to the left side of the traveled way and the other signal face shall be adjacent to the right side of the traveled way. Temporary mast arms shall be installed as necessary. Additional signal faces and their location will be determined by the Engineer and paid for in accordance with 3-3.

Temporary traffic signals shall be securely mounted approximately 10 feet (3m) high, on wood poles, platform standards, or semi-permanent structures. Mast arms, where required, shall provide a minimum clearance of 17 feet (5m) from the traveled way to the bottom of the traffic signal head. Primary and mast arm traffic signals shall have backplates. Mast arm traffic signals and arrow indications shall be 12- inch (300mm) in size and other signals shall be 8-inch (200mm) in size.

Traffic signal shutdown periods shall be as specified in the Special Provisions. Preliminary work associated with the shutdown shall be performed prior to the actual shutdown in order to minimize the amount of time necessary for the completion of the Work.

Shutdowns, flashing operations, and turn-ons shall be requested at least 24 hours in advance and will be overseen or performed by the Engineer.

307-5.4 Payment. Payment for furnishing and installing temporary systems complete and in place shall be considered as included in the Contract Unit Price or lump sum price in the Bid for the work that required a temporary system.

307-6 ORDERING MATERIALS. The Contractor shall submit a letter to the Engineer stating that the Contractor-furnished street lighting and/or traffic signal materials have been ordered. The letter shall contain the names and addresses of the suppliers and the estimated delivery dates. The letter shall be submitted to the Engineer within 5 working days after acceptance of the submittal for materials specified in 2-5.3.

307-7 EXCAVATION AND BACKFILL.

307-7.1 General. Excavation required for the installation of conduit, foundations, and other equipment shall be performed in such a manner as to cause the least possible damage to the streets, sidewalks, and other existing improvements. Excavation shall not be performed until just prior to installation of equipment. Material generated from excavations shall be placed in a location that will cause the least obstruction to surface drainage and vehicular and pedestrian traffic.

Excavations that are required in concrete sidewalk, cuts and joints shall conform to 300-1.3.

At the end of each working day and at other times when construction operations are suspended, equipment, material, and debris shall be removed from that portion of the right of way open for vehicular and pedestrian traffic. Barricades shall be erected at excavations not backfilled or finished to final grade.

Excavations that are required shall be kept backfilled and maintained in a smooth and well-drained condition until permanent repairs are completed. Excavations, including those resulting from removal of existing equipment as specified on the Plans, shall be backfilled, compacted, and the surface restored to match existing improvements in conformance with 7-9, 306-1.3.4, and 306-1.5.

307-7.2 Trenches. Trenches shall not be excavated wider than necessary for the proper installation of the conduit. Trenches for conduit shall be 4 inches (100mm) wide. Trenching is not permitted through portland cement concrete improvements, such as bus pads, spandrels and cross gutters. When encountered, galvanized rigid steel conduit shall be jacked, or rigid non-metallic conduit shall be bored beneath the improvements.

307-7.3 Payment. Payment for excavation and backfill of trenches shall be considered as included in the Contract Unit Price or lump sum price in the Bid for the work that required excavation, and backfill of trenches.

307-8 FOUNDATIONS, FOUNDATION CAPS AND SLABS.

307-8.1 General. Work shall conform to the lines, elevations and grades as shown on the Plans or established by the Engineer. Construction of concrete foundations, caps and slabs construction shall conform to 303-1, 303-5, and 305-1.3. Forms shall be constructed in accordance with 303-1.3, and removed in accordance with 303-1.4.

307-8.2 Foundations. Foundations shall be constructed in a single placement of portland cement concrete of the class specified in 201-1.1.2. The bottom of the foundations shall rest securely on firm, unyielding soil. When firm, unyielding soil cannot be obtained at the depth shown on the Plans, or Standard Plans, or where the foundation cannot be constructed to the dimensions shown on the Plans or Standard Plans due to obstructions, the foundation shall be constructed as directed by the Engineer and paid for in accordance with 3-3.

Foundations shall cure for 24 hours before erecting Standards and 72 hours before erecting mast arms. Pile foundations shall cure for 48 hours before erecting Standards and 7 days before erecting mast arms.

Foundations constructed within the sidewalk or parkway shall pose no hazard to pedestrian traffic. The above-ground portion of a foundation, if any, and/or anchor bolts, conduits etc. shall be protected with protection devices (i.e. cones, delineators, barricades etc.) approved by the Engineer. The Contractor shall connect the protection devices to the foundation. Protection devices shall protect pedestrians from the above ground portion of the foundation, and/or exposed anchor bolts, conduit, etc. Access meeting ADA requirements shall be provided adjacent to the foundations. Protection devices shall remain and be maintained in place until the related equipment is installed on the foundation.

307-8.3 Foundation Caps. Foundation caps shall be the same color, finish, and material as the adjacent sidewalk, and be a minimum of 3 inches (75mm) thick or unless otherwise specified in the Special Provisions. Foundation caps shall be placed after the Standard is set in its final position. The longitudinal grade shall be the same as the grade for the top of the existing curb. If there is no curb, the longitudinal grade will be established by the Engineer.

The transverse grade shall be established as follows:

a) Existing curb and no sidewalk - by sloping upward from the top of the back face of curb at the rate of 1/4 inch/foot (20mm/m).

b) Existing curb and sidewalk – by straight grade from the top of the back face of curb to the top of the near edge of sidewalk, and shall join all around in full-width sidewalk or sidewalk constructed adjacent to the curb.

c) Service road parkways – by a straight line between the top of the back face of one curb to the top of the back face of the other curb.

d) If the lateral grade of the existing parkway exceeds a slope of ± 1 inch/foot (80mm/m), the Contractor shall construct retaining curbs and sidewalk as directed by Engineer. Retaining curbs not shown on the Plans will be considered as Extra Work per 3-3.

307-8.4 Concrete Slabs. Wherever the edge of a concrete foundation extends within 18 inches (450mm) of any existing concrete improvement, a concrete slab with a minimum thickness of 3 inches (75mm) or as specified in the Special Provisions shall be extended to meet the existing improvement.

307-8.5 Measurement. Foundations shall be measured by the lump sum or by the cubic yard (m^3) for each type and size.

307-8.6 Payment. Payment for foundations, including protection devices, bolt circles, reinforcement and foundation cap, shall be at the Contract Unit Price or lump sum price in the Bid for each size and type. Payment for concrete slabs shall conform to 303-5.9.

307-9 ANCHOR BOLTS, NUTS, AND WASHERS.

307-9.1 General. Anchor bolts, nuts, and washers, including those required for the relocation of existing standards , shall be furnished by the Contractor. Anchor bolts, nuts and washers shall be of the type and size shown on the Plans or Standard Plans and shall conform to 209-3.2.

307-9.2 Payment. Payment for furnishing and installing anchor bolts, nuts and washers shall be considered as included in the Contract Unit Price or lump sum price in the Bid for foundations or for the work that required the anchor bolts, nuts or washers.

307-10 STANDARDS, PEDESTALS AND MAST ARMS.

307-10.1 General. Street lighting, and traffic signal Standards, pedestals for cabinets, mast arms and other similar equipment furnished shall be as shown on the Plans and/or Standard Plans and conform to 209.

Plumbing of standards shall be accomplished by adjusting the nuts on the anchor bolts before the foundation cap is placed. Shims or other similar devices for plumbing or raking shall not be used.

If base covers or foundation caps are not used, anchor bolts shall be cut 1/4 inch (6mm) above the nuts. If anchor bolts are cut, the cut surfaces shall be repaired in accordance with 210-3.5.

Holes left in the shafts of existing standards, due to removal of equipment, shall be repaired as follows:

a) Steel shafts – by welding a suitable disc, grinding smooth, and painting as specified in 210-3.

b) Concrete shafts – by grouting to match the existing texture and color.

307-10.2 Mast Arms. Mast arms for Standards shall conform to the dimensions shown on the Plans or Standard Plans and 209-3.4. The joint between a Standard and a mast arm shall be rain-tight.

307-10.3 Measurement. Standards and/or mast arms or pedestals shall be measured by each type and size.

307-10.4 Payment. Payment for Standards and/or mast arms or pedestals shall be at the Contract Unit Price or lump sum price in the Bid for each size and type.

307-11 PULL BOXES.

307-11.1 General. Pull boxes, covers, and extensions shall be of the type, size, and details shown on the Plans or Standard Plans, or as specified in the Special Provisions, and shall conform to 209-3.7.

Covers for pull boxes shall be marked as shown on the Plans, or Standard Plans or as specified in the Special Provisions. Where the surrounding sidewalk surface is composed of a special material (terrazzo, pavers etc.), pull box covers shall be made of matching material unless otherwise shown on the Plans or specified in the Special Provisions.

Pull boxes shall be installed at the locations shown on the Plans. If not shown, they shall be approximately equally spaced, but not further apart than:

a) 300 feet (90m) for traffic signal interconnect,

b) 200 feet (60m) for street lighting systems, and

c) 600 feet (180m) apart for fiber optic cable systems.

Pull boxes shall be placed:

a) Adjacent to standards with a 3 feet (900mm) clearance from the side of the foundation unless otherwise directed by the Engineer.

b) A minimum of 5-feet (1500mm) from the top of the "X" dimension of driveways and access ramps. Pull boxes shall not be installed in any part of a driveway, curb ramp area or other traveled way unless otherwise specified in the Special Provisions.

c) At least 6 inches (150mm) from any substructure or back of curb unless otherwise approved by the Engineer. Pull boxes shall be installed with the long side parallel to the curb unless otherwise approved by the Engineer.

The bottom of the pull box shall rest firmly on a 12-inch (300mm)-thick bed of 1 inch (25mm) crushed rock conforming to 200-1.2, extending 6 inches (150mm) beyond the outside edges of the pull box. The grade for the top of pull box shall be established as specified in 307- 8.3.

Where ballasts or transformers are installed in a pull box, a pull box extension shall be used.

The Contractor may, at its own expense and subject to the approval of the Engineer, install additional pull boxes to facilitate the Work.

307-11.2 Measurement. Pull boxes shall be measured by each type.

307-11.3 Payment. Payment for pull box shall be at the Contract Unit Price or lump sum price in the Bid for each type.

307-12 CONDUIT.

307-12.1 General. Conduit shall conform to 209-3.5 and shall be of the type and size indicated on the Plans, or as specified in the Special Provisions. If so specified in the Special Provisions, galvanized pipe or rigid non-metallic conduit may be substituted for galvanized electrical conduit. Conduit shall be installed by trenching, jacking or directional boring methods as specified in the Special Provisions.

Conduit shall be placed, bored or jacked, to a depth of not less than 30 inches (750mm) nor more than 60 inches (1500mm) below the flowline grade. Conduit placed behind a curb shall not be less than

14 inches (350mm) nor more than 36 inches (900mm) below the top of curb. Conduit placed under railroad tracks shall not be less than 36 inches (900mm) nor more than 60 inches (1500mm) below bottom of the ties, unless otherwise required by the jurisdictional railroad. Conduit shall be placed directly behind the curb if approved by the Engineer. If there are obstructions such as foundations, pull boxes, water meter vaults, etc., the conduit may be placed further behind the curb. In no case shall the conduit be placed more than 36 inches (900mm) behind the curb unless otherwise approved by the Engineer. Conduit may be laid on top of the existing pavement within curbed medians being constructed on top of said pavement.

Pavement shall not be disturbed without permission from the Engineer, except at potholes to expose utility lines in accordance with 5-1. Jacking or drilling pits shall be kept 2 feet (600mm) clear of the edge of any type of pavement unless otherwise approved by the Engineer. Excessive use of water, such that the pavement may become undermined, or softened, will not be permitted.

Jacking pits adjacent to railroad tracks shall be constructed not less than 25 feet (7.6m) from the centerline of track. When a jacking pit or open trench is left overnight, it shall be covered with steel plates as specified in the Special Provisions.

Conduit laid in an open trench shall not be covered nor shall any trench or inspection hole be backfilled until the Engineer has approved the installation.

Detector, telephone interconnect, or street lighting conduit shall be 1 inch (25mm) nominal size unless otherwise specified in the Special Provisions. Direct interconnect, utility service, inductive loop detector or traffic signal conduit shall be 2-inch (50mm) nominal size unless otherwise specified in the Special Provisions. The Contractor may use conduit of a larger size than that shown or specified, provided the larger size is used for the entire length of the run. Reducing couplings shall not be used. If the Contractor chooses to use a larger size conduit it shall be at no additional cost to the Agency.

Conduit entering a pull box shall enter not less than:

a) 5 inches (125mm) below the top of the box,

b) 2 inches (50mm) above the bottom of the box.

Conduit shall terminate within 2 to 4 inches (50mm to 100mm) inside the box wall. The prolongation of the conduit shall pass through the top of the pull box. Conduits shall enter from the direction of the run.

Spare conduit stubs from foundations shall extend a minimum of 6 inches (150mm) from the face and a minimum of 14 inches (360mm) below the top of foundation and shall be capped on each end.

The ends of the conduit, whether shop or field cut, shall be reamed to remove burrs and rough edges. Cuts shall be made so that the ends will come together for the full circumference thereof. Slip joints or running threads shall not be used for coupling conduit. Threads shall be treated with the approved joint compound specified in the Special Provisions before fittings are placed thereon.

Conduit installed on the surface of poles or structures or other exposed locations or in concrete structures and foundations shall be unpainted, except that exposed conduit installed on a painted structure shall be painted the same color as the structure. Conduit runs on the surface of structures shall be secured with galvanized malleable iron clamps spaced not more than 5 feet (1500mm) apart.

A No.12 AWG pull wire, or a tape or rope conforming to 209-3.6 shall be installed in all complete conduit runs that are to receive future conductors. A minimum of 2 feet (600mm) of pull wire, tape or rope shall be secured and extended beyond each end of the conduit run.

Conduit ends shall be capped until the pulling of conductors is started. The conduit shall be blown clean with compressed air prior to installing conductors. In the presence of the Engineer, the Contractor may be required to pass a proper size-testing mandrel through the conduit. Existing conduit that is being incorporated into a new or modified system shall be cleaned with a mandrel or cylindrical wire brush and blown clean with compressed air.

A minimum of 1 foot (300mm) length of continuous oakum or other sealant as specified in the Special Provisions shall be securely packed into all uncapped conduit ends. Oakum, if used, shall be string oakum that is tar or oil saturated to be water resistant. The oakum or sealant shall be packed into the uncapped conduit immediately after the installation of conductors in the conduit. The oakum or sealant shall be packed around the conductors in such a manner that no visible gaps remain between the conductors and conduit, and with sufficient care to avoid damage to conductors, bushings or conduit.

Conduit terminated without a pull box shall be capped and identified by chipping the standard "‡" at a minimum of 3 inches (75mm) in height on the curb or as specified in the Special Provisions.

Conduit abandoned in place shall be terminated a minimum of 1 foot (300mm) below finished grade.

307-12.2 Rigid Metallic Conduit. Rigid metallic conduit shall conform to 209-3.5.2.

Expansion fittings, as detailed on the Plans and/or specified in the Special Provisions, shall be installed where the conduit crosses an expansion joint in a structure. Each expansion fitting shall be provided with a No. 8 AWG copper bonding jumper.

Rigid metallic conduit shall be bent without crimping or flattening and shall have a radius of not less than 6 times the inside diameter of the conduit.

Rigid metallic conduit fittings shall be galvanized steel unless otherwise specified in the Special Provisions. Couplings shall be securely tightened to provide an electrical and mechanical connection throughout the entire length of the conduit run. When a standard coupling cannot be used, a UL or ETL listed threaded union coupling approved by the Engineer shall be used.

When rigid metallic conduit is capped and the caps are removed, the ends of the conduit shall be provided with conduit bushings. Caps and bushings may be manufactured of metal or plastic, threaded or push-on as specified in the Special Provisions.

307-12.3 Galvanized Pipe. Galvanized pipe used as conduit shall conform to 209-3.5.3. Should "necking" occur when any length of pipe is installed, the length of pipe where the "necking" occurred shall be removed.

Couplings shall have a uniform thread which ensures that the ends of the pipe to be joined are in full surface contact when the couplings are in place. Couplings with tapered threads shall not be used.

307-12.4 Rigid Non-Metallic Conduit. Rigid non-metallic conduit shall conform to 209-3.5.4. Rigid non-metallic conduit shall only be installed in underground locations. A separate No. 8 AWG, solid bare ground wire and a pull rope or tape conforming to 209-3.7 shall be inside at the time of installation. A pull box shall be installed whenever an underground conduit changes from a metallic type to a non-metallic type. The Contractor shall pull a mandrel, approved by the Engineer, through all rigid non-metallic conduit.

307-12.5 Measurement. Measurement of conduit shall be the horizontal plane distance plus the minimum depth to grade by the foot (meter).

307-12.6 Payment. Payment for conduit shall be at the Contract Unit Price or lump sum price in the Bid for each size and type of conduit.

307-12.7 Abandonment of Conduit. Conduit that is to be abandoned in place shall have all wires or cable removed. The conduit shall then be removed to a minimum depth of 1 foot (300mm) below the surface and have both ends crimped or capped.

307-12.7.1 Payment. Payment for abandoning conduit in place shall be considered as included in the Contract Unit Price or lump sum price in the Bid for the work that required the conduit to be abandoned.

307-13 WIRES, CONDUCTORS AND CABLES.

307-13.1 General. Wires, conductors and cables shall conform to 209-4.2 for street lighting systems and 209-5.3 for traffic signal systems. Conductors shall be run in conduit except when they are inside Standards, pedestals or for overhead and temporary installations unless otherwise specified in the Special Provisions. Temporary wiring shall conform to 307-5.2.

Where low-voltage conductors are run in Standards containing high-voltage conductors, either the low-voltage or the high-voltage conductors shall be encased in flexible or rigid metallic conduit to a point where the two types of conductors are no longer in the same raceway.

Conductors/cables shall be pulled by hand. Winches or other power-actuated pulling equipment shall not be used. Only lubricants specified in the Special Provisions or approved by the Engineer shall be used in pulling conductors/cables in conduit.

A total of 2 feet (600mm) of slack shall be left at each Standard, and within each pull box sufficient slack shall be left to extend 18 inches (450mm) above the top of the pull box.

Small, permanent identification bands shall be marked as specified in 209-4.2, 209-5.3 or herein. The bands shall be securely attached to conductors in pull boxes and near the termination of each conductor. Where circuit and phase are clearly indicated by conductor insulation, bands need not be used. Permanent identification bands shall be embossed, 6 mil (150μm), oil-resistant polyvinyl chloride tape with pressure-sensitive backing. Tape shall be of a type such that symbols contrast with the background color.

307-13.2 Splices. Splices shall conform to the details shown on the Plans unless otherwise specified in the Special Provisions. Splices shall only be made in pull boxes and Standard bases. Splices shall be capable of operation under continuous submersion in water. Multi-conductor cables shall be spliced and insulated to provide a watertight joint and to prevent absorption of moisture by the cable. Where more than 1 conductor enters the sleeve of a ballast installed in a pull box, the insulation and taping shall be applied between the conductors in such a manner as to provide a watertight joint.

Splice insulation shall conform to 209-3.8 unless otherwise specified in the Special Provisions and shall be applied to a thickness equal to, and well-lapped over, the original insulation.

Wire/conductor insulation shall be well-penciled, trimmed to a conical shape, and roughened before applying the splice insulation. Tape shall be applied to a thickness equal to, and well-lapped over, the original insulation.

When 3 or more conductors are to be enclosed within a single splice using heat-shrink tubing, mastic shall be placed around each conductor prior to being placed inside the heat-shrink tubing. The mastic shall be of the type recommended by the manufacturer of the heat-shrink tubing.

After contraction, the ends and seams of heat-shrink tubing shall be painted with an electrical insulating coating recommended by the manufacturer. Heat-shrink tubing shall not be heated with an open flame. A heating device designed for the purpose is required.

The Contractor may, at the Contractor's option, use either of the following splice insulation methods:

a) A minimum of 2 thicknesses of electrical insulating pad shall be used. Pads shall be applied to the splice in conformance with the manufacturer's recommendations,

b) Heat-shrink tubing conforming to 209-3.8.3.

Conductors shall be joined together by connectors conforming to 209-4.3, unless otherwise specified in the Special Provisions or approved by the Engineer. Connectors and terminals used with aluminum utility power service conductors shall be aluminum and shall be greased with an inhibitor as specified in the Special Provisions or approved by the Engineer.

307-13.3 Payment. Payment for wire, conductor or cable and splicing wire, conductor or cable shall be considered as included in the Contract Unit Price or lump sum price in the Bid for conduit.

307-13.4 Bonding and Grounding. Metallic cable sheaths, metal pull box covers, metallic conduit, equipment grounding conductors, ballast and transformer cases, service equipment, sign switches, anchor bolts, and metal standards that form a continuous system shall be effectively grounded with materials conforming to 209-3.9.

At service points, grounding of metallic conduit, service equipment, and neutral conductors shall conform to the applicable code and the requirements of the serving utility, except that grounding electrode conductors shall be No.6 AWG solid copper wire or larger.

Each multiple service disconnect location shall be grounded. Ground electrodes shall be installed in accordance with the provisions of the Code and the materials utilized shall conform to 209-3.9. The service equipment shall be bonded to the ground electrode by use of a ground clamp or exothermic weld and No. 6 AWG or larger solid copper wire enclosed in a size 16 or larger diameter conduit.

For equipment grounding purposes in rigid non-metallic conduit, a No. 6 AWG solid copper wire shall be run continuously in circuits used for series street lighting, and a No. 8 AWG minimum solid copper wire shall be run continuously in all other street lighting and traffic signal circuits. The bonding wire size shall be increased to match the circuit breaker size, unless otherwise shown on the Plans or as specified in the Special Provisions. Whenever rigid non-metallic conduit is to be installed for future conductors, the solid copper wire may be omitted. Equipment bonding and grounding conductors are not required in conduit which contain only loop lead-in cable or signal interconnect cable or both.

Bonding of Standards shall be accomplished by means of a No. 8 AWG bonding wire attached from a grounding bushing to a foundation bolt or a 3/16 inch (4mm), or larger, brass or bronze bolt installed in the lower portion of the Standard. The bonding jumper in Standards with handholes and traffic signal pull box lid covers shall be attached by a 3/16 inch (4mm) or larger brass bolt and shall be run to the conduit or bonding wire in the adjacent pull box. Standards without handholes shall be bonded by a jumper attached to each anchor bolt, and shall be run to the conduit or bonding wire in the adjacent pull box. The grounding jumper shall be visible after the cap has been placed on the foundation.

Bonding of metallic conduit in metal pull boxes shall be by means of locknuts, 1 inside and 1 outside of the pull box.

Bonding of metallic conduit in nonmetallic pull boxes shall be by means of copper strap or galvanized grounding bushing and bonding jumpers conforming to 209-3.9.

For series circuits, the metallic conduit or bonding conductor system shall be securely grounded at intervals not to exceed 500 feet (150m) by one of the following:

a) galvanized pipe conforming to 209-3.9 driven to a depth of 8 feet (2m) having its upper end not more than 3 inches (75mm) above the conduit; or,

b) copper-coated steel rod conforming to 209-3.9, driven to a depth of 8 feet (2m); or,

c) metal water service pipe on the street side of the meter, with the approval of the owner. The water pipe shall be thoroughly scraped and cleaned prior to connection.

For multiple circuits, the metallic conduit or bonding conductor system shall be grounded at the service points.

On wood poles, all equipment mounted less than 8 feet (2m) above the ground surface shall be grounded.

307-13.4.1 Payment. Payment for furnishing and installing equipment for bonding and grounding circuits shall be considered as included in the Contract Unit Price or lump sum price in the Bid for conduit.

307-14 SERVICES.

307-14.1 General. Before any work is begun, the Contractor shall obtain daily circuit clearance from the serving utility.

The Contractor shall furnish and install all material and equipment necessary to complete the electrical connection between the terminating point of the serving utility and the electrical system, as shown on the Plans. Electrical service installation and materials shall conform to the requirements of the serving utility and 209-3.10. Installation of service equipment shall be scheduled so as to enable the serving utility to complete its work in advance of the completion of the Work.

Upon the Contractor's written request, the Engineer will arrange with the serving utility to complete service connections for permanent installations. The Agency will pay all costs and fees required by the serving utility for the service connections for permanent installations unless otherwise specified in the Special Provisions.

307-14.2 Services on Utility-Owned Poles. When service equipment is to be installed on a utility-owned pole, the Contractor shall furnish and install conduit, conductors and other necessary material to complete the installation of the service. The position of the riser and equipment will be determined by the serving utility unless otherwise shown on the Plans or specified in the Special Provisions.

Each service shall be provided with an in-line fuse holder on each non-grounded service conductor, except the neutral conductor. The in-line fuse holder shall conform to 209-4.3.2.

307-14.3 Services in Vaults. The Contractor shall contact the serving utility to determine the location of stub outs for vault service points.

Thirty feet (9m) of slack in each conductor entering a vault shall be provided for the serving utility to "rack" the conductor on the walls of the vault.

In the service pull box, the Contractor shall furnish and install 2 disconnect splices for each circuit. Disconnect splices shall conform to 209-4.3.1. Disconnect splices shall be installed so that the 2 service wires may be connected together.

307-14.4 Measurement. Services shall be measured by each unit.

307-14.5 Payment. Payment for connection between the "service point" and the nearest pull box will be paid at the Contract Unit Price or lump sum price in the Bid for each service point.

Where the "service point" is indeterminate or shown on the Plans as an "approximate location" or "service point not yet established", payment for the connection between the "service point", when established, and the nearest pull box shown on the Plans will be paid for as Extra Work in accordance with 3-3.

307-15 CIRCUIT BREAKERS.

307-15.1 General. Circuit breakers conforming to 209-3.11 shall be constructed at the locations shown on the Plans. Circuit breakers used as service disconnect equipment shall be enclosed in a NEMA raintight enclosure with a dead-front panel and hasp with a 7/16 inch (11mm) hole for a padlock. The padlock will be furnished by the Agency.

307-15.2 Payment. Payment for circuit breakers shall be considered as included the Contract Unit Price or lump sum price in the Bid for the equipment requiring circuit breakers.

307-16 STREET LIGHTING CONSTRUCTION.

307-16.1 General. Street lighting materials shall conform to 209-3 and 209-4.

307-16.2 Pull Box Covers. Pull box covers shall conform to 209-3.7 and be inscribed "STREET LIGHTING HIGH VOLTAGE."

307-16.3 Conduit. Conduit shall conform to 307-12. Conduit terminating in street lighting Standards shall not be transposed. Conduit shall terminate as near the door of the Standard as possible with the end of the conduit below, but within 1 inch (25mm) of the lower edge of the door. The prolongation of the conduit shall pass through the door opening.

307-16.4 Wiring/Conductors. Single conductors shall be used for all circuits. Conductors shall be copper of the gauge and color specified in 209-4.2 unless otherwise specified in the Special Provisions.

For series-circuit conductor splices, sufficient synthetic oil-resistant rubber tape, conforming to the requirements of ASTM D119, shall be applied over the conductor to fill all voids before placing the vinyl chloride tape specified in 209-3.8.2. The splice shall then be well covered with a coating of insulating material specified in the Special Provisions.

For multiple circuits, a fused disconnect splice connectors conforming to 209-4.3 unless otherwise specified in the Special Provisions shall be installed in each ungrounded conductor between the line and the ballast. The connector shall be installed in the base of the luminaire standard or in an adjacent pull box and be readily accessible.

307-16.5 Fused Splice Connectors. Fused splice connectors shall conform to 209-4.3.2.

In the pull box adjacent to each Standard, a fused disconnect splice connector shall be installed on each ungrounded conductor between the line and the ballast. The connector shall be readily accessible in the pull box regardless of whether the ballast is remote or integral with the luminaire.

Terminals shall be rigidly crimped, using a tool of the type recommended by the manufacturer of the fused splice connector, onto the line conductors and the conductors to the ballasts. Terminals shall be insulated, waterproof and conform to the splice connector manufacturer's recommendations.

Fused splice connectors shall not be used in series circuits.

307-16.6 Terminal Blocks. Terminal blocks shall conform to 209-4.3.3 unless otherwise specified in the Special Provisions. Barrier type terminal blocks rated for 40A minimum shall be provided in each service equipment enclosure. Field conductors shall be terminated using crimped, insulated loop connectors.

307-16.7 Services. Services shall conform to 307-14. For series street lighting systems served from overhead circuits, a switch of 5,000-volt rating shall be connected to control each circuit. The switch shall be enclosed in a NEMA Type 3R, 18-inch x 24-inch x 6-inch (450mm x 600mm x 150mm) terminal box. The terminal box shall be fitted with a cover permanently inscribed "DANGER – HIGH VOLTAGE." The cover shall be attached to the box to form a raintight plate and shall require tools for removal. The terminal box shall be installed not less than 10 feet (3m) above the ground.

307-16.8 Luminaires.

307-16.8.1 General. Luminaires shall conform to 209-4.4 unless otherwise specified in the Special Provisions.

307-16.8.2 Lamp Size and Identification. Lamps shall conform to 209-4.6 unless otherwise specified in the Special Provisions.

Each lamp shall be identified with a tape target conforming to 209-4.6.1. Targets shall be affixed to the underside of the horizontal luminaire or on the body facing oncoming traffic for upright luminaires.

Existing targets shall be removed and disposed of or covered as specified in the Special Provisions.

307-16.8.3 Lamp Ballasts. A Certificate of Compliance conforming to 4-1.5 shall be submitted by the Contractor with each lot of integral ballast luminaires and with each lot of ballasts designed for use outside of luminaires. The certificate shall state that the lot of ballasts conform, in every respect, to 209-4.5, the lamp-ballast specifications of the lamp manufacturer or the Special Provisions.

307-16.8.4 Photoelectric Units. Photoelectric electric units shall be installed as shown on the Plans or as specified in the Special Provisions. Photoelectric electric units shall conform to 209-4.7 unless otherwise specified in the Special Provisions.

307-16.8.5 Measurement. Luminaires shall be measured by each type and size.

307-16.8.6 Payment. Payment for luminaires, including ballasts and photoelectric units, shall be at the Contract Unit Price or lump sum price in the Bid for each size and type of luminaire.

307-17 TRAFFIC SIGNAL CONSTRUCTION.

307-17.1 General. Traffic signal materials shall conform to 209-3 and 209-5.

307-17.2 Controllers

307-17.2.1 General. Controllers, cabinets and auxiliary equipment shall conform to 209-5.4 and the Standard Plans. The Contractor shall supply the controller, cabinet and all auxiliary equipment required to provide a complete functioning controller unless otherwise specified in the Special Provisions.

307-17.2.2 Controller Cabinet Wiring Diagrams. A wiring diagram for field modifications to the existing controller cabinet including, but not limited to, the installation of sensor units, switch packs, etc. shall be submitted to the Engineer at least 10 working days prior to performing work in the controller cabinet.

Controller cabinet documentation, labeling, and placards shall be corrected/updated to reflect any changes made within the controller cabinet that are the result of all current and past modifications.

Prior to acceptance of the Work, the Contractor shall furnish 5 sets of traffic signal controller cabinet schematic wiring diagrams conforming to the phase designations shown on the Plans for the intersection. The diagrams shall show the location of the installation and shall list all equipment installed in each cabinet. In addition, for each signal installation, the Contractor shall furnish an intersection sketch showing Standards, detectors, and phasing. One copy of the controller cabinet diagram and the intersection and phase diagram, as reviewed by the Engineer, shall be placed in a heavy-duty plastic envelope and attached to the inside of the door of each controller cabinet.

307-17.2.3 Controller Cabinet Wiring. Wiring shall conform to 209-5.4.2 unless otherwise specified in the Special Provisions.

307-17.2.4 Service Checks. Service checks shall be performed by the Contractor. The Contractor shall measure the service voltage at the live AC terminal before the main circuit breaker in the controller assembly. If the voltage is less than 110 volts, the Contractor shall notify the Engineer. A resistance measurement shall also be made between the service neutral terminal and the chassis ground terminal. If the resistance is more than 4 ohms, the Contractor shall notify the Engineer. The Engineer will notify the serving utility to rectify the problem.

307-17.2.5 Measurement. Controllers shall be measured by each type and size.

307-17.2.6 Payment. Payment for controllers shall be at the Contract Unit Price or lump sum price in the Bid for each size and type of controller.

307-17.3 Pull Box Covers. Pull box covers to be installed in traffic signal systems, or combined traffic signal and multiple circuit street lighting systems conforming to 209-4.2.3, shall conform to 209-3.7 and be inscribed "TRAFFIC SIGNALS." Pull box covers for underground service points, where both traffic signals and street lighting jointly occupy the same box, shall be inscribed "TS-SL."

307-17.4 Conduit. Conduit shall conform to 209-3.5 and 307-12. Conduit terminating in traffic signal standards shall extend vertically 1 inch (25mm) above the foundation cap and be centered within the bolt circle. Traffic signal conduits shall be separated from street lighting conduits.

Rigid non-metallic conduit shall be installed in open trenches, except in the vicinity of pullboxes where it may be bored in, pre-drilled, augured or through air blown holes.

307-17.5 Wiring, Conductors and Cable.

307-17.5.1 General. Multi-conductor cable shall be used for all circuits in lieu of individual conductors. Multi-conductor cable shall be as shown on the Plans and conform to 209-5.3 and 307-13 unless otherwise specified in the Special Provisions. Wires shall be solid copper of the gauge shown in Table 209-5.3.1 (A), on the Plans, or as otherwise specified in the Special Provisions.

The neutral for pedestrian push-button circuits shall be separate from the signal light circuit neutral.

Interconnect conductors shall be continuous from controller to controller unless splices are approved by the Engineer. Splices shall be made only in pull boxes.

Where telephone circuits are installed adjacent to other electrical circuits, the telephone conductors shall be encased in UL approved metallic conduit conforming to 209-3.5.2.

Multi-conductor cable entering controller cabinets shall be neatly arranged, secured, and tagged. Each cable shall be identified in all controller cabinets by a plastic tag 1/2 inch x 2 inches (13mm x 50mm) in size, stamped with the cable run identification characters in 1/4 inch (6mm) letters and secured to the conductor with 2 nylon tie-wrap devices.

Each multi-conductor cable shall be identified in all communication cabinets by an aluminum tag, 1 inch x 4 inches (25mm x 100mm) in size, stamped with the cable run identification characters in 1/2 inch (13mm) letters and secured to the cable with 2 nylon tie-wrap devices.

Whenever new conductors are to be installed in a conduit with existing individual conductors (service wire excepted), all individual conductors shall be removed and replaced with multi-conductor cable. A 28-conductor cable conforming to 209- 5.3.2.6 shall be installed in new cross street runs unless otherwise shown on the Plans.

Stranded conductors shall be terminated with approved terminal lugs.

307-17.5.2 Conductor Splicing and Termination. Field spliced solid wires shall be twisted together and secured using vinyl, water-tight, spring-tensioned, silicone-filled, direct-burial wire connectors as specified in the Special Provisions. The Contractor shall provide a minimum of 3 feet (1m) of slack. The slack shall be neatly coiled in a clockwise direction within each pullbox. Splice insulation shall conform to 209-3.8.

Stranded conductors shall be terminated with a terminal connector as specified in the Special Provisions. Stranded conductors shall be properly compressed for minimum resistance at the attachment.

Where optimum operation of circuits requires minimum resistance, the connections and/or terminals shall be soldered.

307-17.5.3 Ground Rods. Ground rods shall conform to 209-3.9. Copper ground rods shall be installed in controller foundations and service pull boxes unless otherwise directed by the Engineer.

307-17.5.4 Interconnect.

307-17.5.4.1 Direct Wire/Conductor. Conductors shall conform to 209-5.3.3.1 unless otherwise specified in the Special Provisions. Conductors shall be continuous from the controller, unless splices are specifically approved by the Engineer. Splices, where approved, shall be soldered and shall be secured using vinyl, water-tight, spring tensioned, silicone filled, direct burial wire connectors, unless otherwise specified in the Special Provisions. A minimum of 3 feet (1m) of slack shall be provided at each splice and 7 feet (2m) at each controller cabinet. Spliced conductors shall be insulated with heat-shrink tubing of the appropriate size and overlap the conductor insulation at least 19/32 inch (15mm). The overall cable splice shall be covered with hear-shrink tubing with at least 1- 9/16 inches (40mm) of overlap of the cable jacket. Heat-shrink tubing shall conform to 209-3.8.3.

307-17.5.4.2 Filled Telephone Cable. Filled telephone cable shall conform to 209-5.3.3.2 unless otherwise specified in the Special Provisions. Cable splices shall be made at communication or controller cabinets or other splice points specified in the Special Provisions. Splices shall conform to 209-3.8.

307-17.5.4.3 Identification. Each cable shall be identified in the communication cabinet by an aluminum or plastic tag 1 inch x 4 inches (25mm x 100mm) in size, stamped with the cable run identification characters in 1/2 inch (13mm) letters and secured to the cable with 2 nylon tie-wrap devices.

Each cable shall be identified in all controller cabinets or other specified splice points by a plastic tag 1/2 inch x 2 inches (13mm x 50mm) in size, stamped with the cable run identification characters in 1/4 inch (6mm) letters and secured to the cable with 2 nylon tie-wrap devices.

307-17.5.5 Fiber Optic Cable.

307-17.5.5.1 General. Fiber optic cable shall conform to 209-5.3.4 unless otherwise specified in the Special Provisions.

307-17.5.5.2 Installation. Fiber optic cable installation and handing procedures shall conform to the manufacturer's recommendations.

In pullboxes with outside dimensions of 22 inches x 34 inches (550mm x 850mm), there shall be 10 feet (3m) minimum of additional looped cable. In splice cabinets and vaults, there shall be 100 feet (30m) minimum of additional looped cable for each cable entering or leaving the splice cabinet or vault.

307-17.5.5.3 Splicing of Fiber Optic Cable. Splicing shall be performed by the fusion technique. Cables shall be prepared and spliced in accordance with the cable manufacturer's recommendations. Completed splices shall be protected by either heat shrinkable tubing conforming to 209-3.8.3, or metal protective sleeves unless otherwise specified in the Special Provisions.

Completed splices shall be enclosed in re-enterable splice enclosures that seal to form a moisture resistant enclosure. The splice case or enclosure shall contain a removable splice organizer or crib that shall secure the individual fibers and protect the splices. The splice organizer or crib shall be attached to the steel strength members in the fiber optic cable and shall be bonded to a ground stud on the exterior of the splice case or enclosure. There shall be adequate space inside the enclosure to hold at least 3 feet (1m) of cable. Splice enclosures shall be as specified in the Special Provisions. Fiber optic cable shall only be spliced at the locations shown on the Plans or specified in the Special Provisions.

307-17.6 Signal Heads.

307-17.6.1 General. Signal heads shall conform to 209-5.5.

New vehicle signal heads installed at any individual intersection shall be of the same style and from the same manufacturer, except for programmed visibility heads.

Mast arm and arrow indications shall be a minimum of 12 inches (300mm). Other signal indications shall be 8 inches (200mm).

Backplates, conforming to 209-5.5.10, shall be installed unless otherwise shown the Plans or specified in the Special Provisions. Pedestrian heads shall be equipped with a glare reduction device conforming to 209-5.6.2.6 unless otherwise specified in the Special Provisions.

Signal heads shall not be installed at any intersection until all other signal equipment, including the controller, is in place and ready for operation , except that signal heads may be mounted if the faces are turned away from traffic or are covered as specified in the Special Provisions.

Signal heads shall be located and aimed as shown on the Plans or as directed by the Engineer. Mounting and location on Standards shall be as shown on the Plans or as specified in the Special Provisions.

307-17.6.2 Signal Head Mountings. Fittings for signal head mountings shall conform to 209-5.7. Assemblies shall be installed plumb or level, as applicable, be symmetrically arranged, and securely tightened. Top horizontal members shall be approximately 12 inches (300mm) in length. Construction shall be such that all conductors are concealed within standards or pipe assemblies.

Unless otherwise shown on the Plans or specified in the Special Provisions, heads shall be installed with terminal compartment mountings. For top mounting of a one-way head and mast arm mountings, a slip-fitter without a terminal compartment shall be used. Clamp-type mounting may be used for installation of heads on existing concrete standards if inserts for terminal compartment mountings have not been provided. The terminal compartment shall be mounted on the Standard on the side away from traffic and parallel with the prolongation of the nearest curb face.

307-17.6.3 Visors. Visors shall conform to 209-5.5.8. Vehicle signal indications at signalized intersections shall be provided with removable beveled or long visors if shown on the Plans, and shall conform to the Standard Plans.

307-17.6.4 Directional Louvers. Directional louvers shall conform to 209-5.5.9. Where shown on the Plans or Standard Plans, louvers shall be furnished and installed in the visors of the signal head sections.

307-17.6.5 Measurement. Signal heads shall be measured by each type and size.

307-17.6.6 Payment. Payment for each signal head shall be the Contract Unit Price or lump sum price in the Bid for each type and size.

307-17.7 Vehicle Detectors.

307-17.7.1 General. Vehicle detectors shall conform to 209-5.8. Vehicle detectors shall be of the type or types shown on the Plans or specified in the Special Provisions. The location of each detector shall be as shown on the Plans or as specified in the Special Provisions.

307-17.7.2 Sensor Units. Sensor units shall conform to 209-5.8 unless otherwise specified in the Special Provisions. A minimum of 1 sensor unit shall be provided for each approach for each separately controlled phase of operation unless otherwise specified in the Special Provisions. Sensor units shall be housed in the controller cabinet unless otherwise specified in the Special Provisions.

307-17.7.3 Inductive Loop Detectors.

307-17.7.3.1 General. Inductive loop detectors shall conform to 209-5.8.2 and the Standard Plans unless otherwise specified in the Special Provisions. Inductive loop detectors shall be installed at the locations shown on the Plans. Unless otherwise specified, each loop shall consist of 3 turns of conductor.

The Contractor shall install and activate inductive loop detectors within 14 days of the activation of the controller.

307-17.7.3.2 Installation Details. Each loop shall be installed entirely within one continuous type of paving material. Slots 1/4 inch (6mm) minimum in width shall be cut in the pavement, washed or blown clean, and thoroughly dried before installation. Triangular corner cutoffs, 18 inches (450mm) long on 2 adjacent sides shall be provided on all loops. Residue resulting from slot cutting operations shall not be permitted to flow across lanes occupied by traffic. Residue shall be removed from the pavement surface by vacuuming or other methods approved by the Engineer.

Installation shall conform to the following:

a) Two leads for each loop shall be installed as a pair, twisted at a rate of 360^0 per foot (300mm) in a common saw slot. Inductive loop detector leads may share a common saw slot with leads from other loops. However, inductive loop detector leads shall not cross any loops and shall not be installed within 20 inches (500mm) of any other loop.

b) Lead-in cable shall consist of No.12 AWG-THW 45 mil (1mm) twisted pair and be continuous from the pull box where connections are made to the loops to the cabinet containing the sensor units for inductive loop detectors.

c) Continuity and insulation resistance tests shall be performed after installation on both inductive detector loops and lead-in cables, as specified in 307-13. Measurements shall be made using the conduit system as ground and with the shield (if any) of the lead-in grounded.

Unless otherwise specified in the Special Provisions, loops shall be wound in a clockwise direction. The input (or start) wire shall be tagged with an odd number, the output (or finish) wire with the next higher even number. A plastic tag 1/2 inch (13mm) by 2 inches (50mm) shall be wrapped around each loop pair to identify each pair by timing phase and number of individual conductors.

Where circuits are to be spliced, each splice shall be crimped and soldered with resin core solder. Splices shall be taped and sealed with an Agency-approved electrical waterproofing compound conforming to 209-5.8.2.4 unless otherwise specified in the Special Provisions. Acid core solder or acid paste shall not be used.

Loop conductors shall be installed without splices and shall terminate in the pull box shown on the Plans. In addition to the requirements for splices in detector circuits, the open end of the cable jackets or tubing shall be sealed in a manner similar to the splicing requirements. Loops shall be joined in the pull box in a combination of series and parallel so that optimum sensitivity is obtained at the sensor unit. Final splices between loops and lead-in cable shall not be made until the operation of the loops under actual traffic conditions is approved by the Engineer. Loop conductors for each direction of travel, for the same phase of a traffic signal system, and in the same pull box, shall be spliced to a detector lead-in cable.

Loop conductors for inductive loop detector traffic signal and traffic monitoring installations shall be identified and banded, in pairs, by lane, in the pull box adjacent to the loops. The pull box shall be near the termination of the conductor in the controller or traffic monitoring station cabinet. Bands shall conform to 209-5.3.

Loop conductors for traffic monitoring shall terminate in a pull box or terminal strip in the traffic monitor station cabinet when a cabinet of that type is installed.

Inductive loop detector lead-in cable shall extend from the pull box adjacent to the loop detector to a sensor unit mounted in the controller cabinet. Inductive loop detector lead-in cable from the pullbox to the controller cabinet shall have 4 conductors (consisting of 2 twisted pairs). Detector lead-in cable connections and/or terminations shall be soldered.

After the loops have been installed in the slots cut in the pavement, the slots shall be filled with sealant to within 1/8 inch (3mm) of the pavement surface. The sealant shall be a minimum of 1 inch (25mm) above the top conductor in the saw cut. Before setting, surplus sealant shall be removed from the adjacent road surface without the use of solvents.

If specified in the Special Provisions, the loops shall be installed prior to placing the asphalt concrete surface course. The loops shall be installed, as shown on the Plans, in the compacted layer of asphalt concrete pavement immediately below the asphalt concrete surface course. Installation details shall be as shown on the Plans, except the sealant shall fill the slot flush to the surface.

307-17.7.3.3 Sealants. Sealants shall conform to 209-5.8.2.4. Hot-melt rubberized asphalt sealant shall be melted in a jacketed, double boiler type melting unit. Temperature of the heat transfer medium shall not exceed the flash point of the sealant or 475° F (240° C), whichever is less.

Sealants shall be applied with a pressure-feed applicator or pour pot. Sealants shall be applied when the pavement surface temperature is greater that 40° F (4° C).

307-17.7.4 Magnetometer Detectors. Magnetometer detectors shall be as shown on the Plans or specified in the Special Provisions. Magnetometer detectors that are to be installed in a particular intersection shall be of the same make and type.

Where an existing system is being modified and existing magnetometer detectors are to remain in use, new detection equipment shall be of the same make and type as the existing, or an Agency-approved alternate.

A separate channel shall be provided for each lane. Separate control units shall be provided for each approach.

A separate cable shall be provided for each magnetometer sensing element placed in the pavement. The cable shall be run continuously (without splices) to the lead-in cable.

Slots, 1/4 inch (6mm) minimum in width by 1 inch (25mm) minimum in depth, shall be cut in the pavement between the sensing element holes and the nearest pull box for lead-in conductor cables. Slots and holes cut in the pavement shall be blown clean and dried before installing lead-in conductor cables and sensing elements.

Sensing elements shall be placed in vertically cut holes in the roadway at locations shown on the Plans. Each hole shall be of a diameter large enough to accept the type of sensing element used, with adequate space for the lead-in conductor cable. Holes shall be cut to a depth sufficient to provide mounting depth of the sensing element recommended by the manufacturer. Sensing elements shall be placed in the bottom of the holes in a vertical position. Holes shall be filled with clean, dry sand to approximately 3 inches (75mm) below the pavement surface.

Epoxy sealant for the slots and holes shall conform to 307-17.7.3.2, except the top 3 inches (75mm) of the holes shall be filled with epoxy sealant conforming to 209-5.8.2.4.

307-17.7.5 Magnetic Detectors. Magnetic detectors shall conform to 209-5.8.3 and be located as shown on the Plans or specified in the Special Provisions. Magnetic detectors shall consist of:

a) a sensing element or group of sensing elements installed in the roadway as shown on the Plans or specified in the Special Provisions,

b) a lead-in cable, and

c) an amplifier with integral power supply installed in a controller cabinet.

Magnetic detector sensing elements and amplifiers to be installed at any single intersection or any traffic count station shall be of the same manufacturer and type.

Magnetic detector sensing elements shall be installed within conduit conforming to 307-7.12 unless otherwise specified in the Special Provisions. The conduit shall extend across the full-width of the traveled way (pull box to pull box) as shown on the Plans. The bottom of the conduit shall be placed 12 inches (300mm) below the top of the pavement.

307-17.7.6 Measurement. Detectors complete and in-place shall be measured by each type.

307-17.7.7 Payment. Payment for constructing each detector shall be the Contract Unit Price or lump sum price in the Bid for each type.

307-17.7.8 Removing or Abandoning Existing Pressure-Sensitive Detectors. When a foundation for a pressure-sensitive vehicle detector is to be removed, the trench left by removing the detector frame and foundation shall be backfilled and resurfaced in accordance with 306-1.2.1 unless otherwise specified in the Special Provisions.

When a foundation for a pressure-sensitive vehicle detector is to be abandoned in place, the top of the foundation, anchor bolts, frame and conduits shall be removed to a depth of not less than 6 inches (150mm) below the surface of existing pavement. The resulting trench shall be backfilled and resurfaced in accordance with 306-1.2.1 and 306-1.5 respectively, unless otherwise specified in the Special Provisions.

307-17.8 Pedestrian Signals.

307-17.8.1 General. The Contractor shall provide one visible operating pedestrian head at all times for each direction of each signalized crosswalk while modifying the pedestrian head. Non-functioning pedestrian heads at a signalized crosswalk shall be covered as specified in the Special Provisions. Pedestrian signal heads shall conform to 209-5.6 or be as specified in the Special Provisions and be installed as shown on the Plans.

307-17.8.2 Pedestrian Push Buttons. Where shown on the Plans, pedestrian push button assemblies conforming to 209-5.9 and signs shown on the Plans or specified in the Special Provisions shall be furnished and installed. Pedestrian push button assemblies shall be equipped as shown on the Plans or specified in the Special Provisions. Pedestrian push button assemblies and signs shall be installed on the crosswalk side of the Standard or pedestal unless otherwise shown on the Plans. Arrows on push button signs shall point in the same direction as the corresponding crosswalk.

Where a pedestrian push button is attached to a pedestal, the housing shall be shaped to fit the curvature of the pedestal and secured to provide a rigid installation. Saddles shall be provided to make a neat fit when required. Where a pedestrian push button is mounted on the top of a pedestal, the housing shall be provided with a slip-fitter fitting and the screws required to secure it rigidly to the pedestal.

307-17.8.3 Measurement. Pedestrian signals and push buttons shall be measured by each type.

307-17.8.4 Payment. Payment for pedestrian signals and push buttons shall be at the Contract Unit Price or lump sum price in the Bid for each type.

307-17.9 Controller Slabs. In unpaved areas, a 4-inch (100mm)- thick portland cement concrete slab shall be constructed in front of each controller cabinet. The slab shall extend the full width of the cabinet foundation and extend out 3 feet (1m) from the face of the foundation.

307-18 TESTING.

307-18.1 General. Prior to acceptance of the Work, the Contractor shall perform the following tests on all electrical circuits, in the presence of the Engineer:

a) **Continuity.** Each circuit shall be tested for continuity.

b) **Ground.** Each circuit shall be tested for unintentional ground.

c) **Megger.** A megger test at 500 volts DC shall be performed on each circuit between the circuit and a ground. The insulation resistance shall be not less than 10 megohms on all circuits, except for inductive loop detector circuits which shall have an insulation resistance value of not less than 100 megohms.

d) **Functional.** A functional test shall be performed which shows that the system functions as specified The test shall not commence until approval is received from the Engineer.

The functional test for each new or modified electrical system shall consist of not less than 5 days of continuous, satisfactory operation. If unsatisfactory performance of the system develops, the condition shall be corrected and the test shall be repeated until the 5 days of continuous satisfactory operation are obtained. Functional tests shall not start nor turn-ons be made on a Friday, or on the day preceding a legal holiday. Shutdown caused by factors beyond the Contractor's control shall not constitute discontinuity of the functional test.

Any material revealed by these tests to be faulty in any part of the installation shall be replaced or corrected by the Contractor at its expense in a manner approved by the Engineer, and the same test shall be repeated.

307-18.2 Payment. No separate payment will be made for testing.

307-19 PAINTING AND GALVANIZING. New equipment to be painted or galvanized shall be as shown on the Plans or specified in the Special Provisions. Painted or galvanized equipment that has been relocated shall be repainted or galvanized as shown on the Plans or specified in the Special Provisions. Paint or galvanizing material shall conform to 210. Painting shall conform to 310.

Paint shall be applied evenly and smoothly by either hand brushing or spray methods as specified in the Special Provisions. Spraying shall not be done at the Work site. The use of brushes for the application of paint shall be required when paint spraying does not result in a finish conforming to the Specifications.

307-19.1 Payment. Full compensation for painting or galvanizing shall be considered as included in the Contract Unit Price or lump sum price in the Bid for each item that requires painting.

307-20 SALVAGE. Unless otherwise specified in the Special Provisions, wires/conductors, Standards, pedestals, , electrical equipment, and foundations that are not identified to be salvaged shall become the property of the Contractor and shall be removed from the Work site. Any salvage value shall be reflected in the Bid.

Care shall be exercised in removing equipment to be salvaged so that it will remain in the condition existing prior to its removal.

SECTION 308 - LANDSCAPE AND IRRIGATION INSTALLATION

308-1 GENERAL. This section shall govern the preparation, planting, and irrigation system construction for landscape areas required by the Plans or Specifications.

Existing utilities and improvements not designated for removal shall be protected in place. Removals shall be performed in accordance with applicable provisions of 300-1.3.

Unless otherwise provided, walls, curbs, planter boxes, walks, irrigation systems, and similar improvements required by the Plans or Specifications shall be constructed following rough grading and before landscaping.

All work on the irrigation system, including hydrostatic and coverage tests, preliminary operational tests of the automatic control system, and the backfill and densification of trenches, and other excavations shall be performed after topsoil work and before planting.

308-2 EARTHWORK AND TOPSOIL PLACEMENT.

308-2.1 General. Earthwork and topsoil placement shall include excavation and backfill for the irrigation system and the preparation for and the spreading, densification, cultivation, and raking of topsoil, including fertilization and conditioning.

Planting holes and backfill shall be in accordance with 308-4.

Preliminary rough grading and related work to prepare areas for landscaping work to within 0.1 foot (30mm) of finish grade, or to subgrade for Class A or Class B topsoil, shall be completed in accordance with Section 300.

308-2.2 Trench Excavation and Backfill. Trenches and other excavations shall be sized to accommodate the irrigation system components, conduit, and other required facilities. Additional space shall be provided to assure proper installation and access for inspection.

Unless otherwise specified, the minimum depth of cover over pipelines and conduit shall be as follows:

a) Electrical conduit – 24 inches (600mm) (36 inches (900mm) under roadways).

b) Waterlines continuously pressurized – 24 inches (600mm) (36 inches (900mm) under roadways).

c) Lateral sprinkler lines – 12 inches (300mm).

The bottom of trenches shall be true to grade and free of protruding stones, roots, or other matter which would prevent proper bedding of pipe or other facilities. Bedding of the pipe shall conform to 308-5.2.

Trenches and excavations shall be backfilled so that the specified thickness of topsoil is restored to the upper part of the trench. Backfill shall be flooded in accordance with 306-1.3.3.

308-2.3 Topsoil Preparation and Conditioning.

308-2.3.1 General. The type and thickness of topsoil shall be shown on the Plans, or if not shown, shall be Class A, 6 inches (150mm) thick. Planting areas shall be free of weeds and other extraneous materials to a depth of 12 inches (300mm) below finish grade before topsoil work.

Soil shall not be worked when it is so wet or so dry as to cause excessive compaction or the forming of hard clods or dust.

The existing soil below subgrade for Class A topsoil shall be scarified to a depth of 6 inches (150mm) prior to spreading topsoil.

Class C topsoil shall be scarified and cultivated to a finely divided condition to a depth of 8 inches (200mm) minimum below finish grade. During this operation, all stones over 1 inch (25mm) in greatest dimension shall be removed.

308-2.3.2 Fertilizing and Conditioning Procedures. The planting area shall be brought to finish grade before spreading the fertilizers or conditioning materials specified.

Fertilizing and conditioning materials shall be mechanically spread at a uniform rate. The quantities of materials necessary for the planting area shall be at the site and shall be verified by delivery tickets furnished to the Engineer before spreading.

After spreading, the fertilizing and conditioning materials shall be uniformly cultivated into the upper 6 inches (150mm) of soil by suitable equipment operated in at least two directions approximately at right angles. The resulting soil shall be in a friable condition.

308-2.4 Finish Grading. The finish grade shall be smooth, uniform, and free of abrupt grade changes and depressions to ensure surface drainage.

The finish grade below adjacent paving, curbs or headers shall be 1 inch (25mm) in lawn areas and 2 inches (50mm) in shrub or groundcover areas.

The finish grade below adjacent paving, curbs or headers shall be 1 inch (25mm) in lawn areas and 2 inches (50mm) in shrub or groundcover areas.

After fertilizing and conditioning, the soil shall be watered and allowed to settle to provide a stable surface, not overly densified to the extent that it will prevent aeration and water infiltration. After the soil has dried out to a workable condition, the planting areas shall be regraded, raked, and smoothed to the required grades and contours. Finish surfaces shall be clean and suitable for planting.

308-3 HEADER INSTALLATION. Headers shall be installed at the locations and grades shown on the Plans prior to planting.

Stakes shall be located at splices, corners, and at intervals not to exceed 5 feet (1.5m) and driven slightly below the top of the header. Splice plates shall be used at butt joints. Headers shall be nailed to stakes with 2 nails, clinched 1/2 inch (13mm). Splice plates shall be centered on the joint and nailed to each header with four 10d box nails.

308-4 PLANTING.

308-4.1 General. The types, sizes, and quantities of plant materials shall be as called for in the Special Provisions or on the Plans.

All plants will be inspected prior to planting, including plants previously approved at the nursery. The Contractor shall be responsible for the condition of all plants, planted or otherwise, until acceptance.

Planting shall be performed with materials, equipment, and procedures favorable to the optimum growth of the plants and in compliance with these procedures.

Except as noted for specimen planting, all planting shall follow the completion of the irrigation system.

308-4.2 Protection and Storage. The Contractor shall keep all plant material delivered to the site in a healthy condition for planting. Plants shall not be allowed to dry out. Bare root stock shall be separated and "heeled in" moist earth or other suitable material. Balled and burlapped plants shall have the root ball covered with moist sawdust, wood chips, or other approved material.

308-4.3 Layout and Plant Location. Planting areas will be staked by the Engineer. Detailed layout within the planting areas shall be performed by the Contractor and approved by the Engineer prior to planting. Parkway trees will be located in the field by the Engineer before planting.

The first row of plants in areas designated for center-to-center spacing of plants shall be located at one-half of designated spacing from the edge of the area.

308-4.4 Specimen Planting. Plants in boxes 24 inches (600mm) and larger shall be planted before installation of lateral irrigation lines.

Irrigation lines conflicting with specimen plant locations shall be rerouted to clear the root ball.

308-4.5 Tree and Shrub Planting. Planting holes shall be approximately square with vertical sides twice the depth and width of the plant container or ball, and shall be larger if necessary to permit handling and planting without injury or breakage of the root ball or root system. Any plant with a broken or cracked root ball before or during planting shall not be planted.

Containers shall be opened and removed in such a manner that the root is not injured. Balled plant wrappings shall be loosened or cut back after plant is positioned in the planting hole.

The native soil at the bottom of planting holes shall be scarified to a depth of 6 inches (150mm).

All planting holes shall be backfilled with a prepared soil mix. Soil shall consist of 50 percent of the specified topsoil and 50 percent Type 1, 2, or 3 organic soil amendment.

After planting, the plant shall be plumb, with the root crown at its natural growing depth with respect to finish grade. Planting shall be governed by the following requirements:

1) A layer of prepared soil mix shall be deposited in the planting hole.
2) The plant shall be set approximately at the center of the hole.
3) Prepared soil mix shall be deposited in the remainder of the hole to finish grade.
4) The backfill shall be thoroughly water-settled and additional prepared soil mix added to fill any remaining void below finish grade.
5) A circular watering basin slightly larger than the planting hole, 4 inches (100mm) high for trees and 2 inches (50mm) high for shrubs, shall be left around the plant. The bottom of the basin shall be at approximate finish grade or slightly lower. Type 1, 2, 3, or 4 mulch shall be spread at least 2 inches (50mm) thick in the basin.
6) The plant shall be guyed and staked as specified in 308-4.6.
7) The area around plant shall be regraded to finish grade. The excess soil shall be disposed of by the Contractor.

308-4.6 Plant Staking and Guying.

308-4.6.1 Method "A" Tree Staking. The tree shall be staked with a 1-1/2 inches (38mm) nominal diameter x 10 foot (3m) long galvanized steel pipe, per 212-2.1.2, installed vertically, positioned at least 6 inches (150mm) from the trunk at ground level, and 30 inches (750mm) into the soil in a manner to avoid injury to the roots or breaking the root ball. The trunk shall be secured to the pipe with one tie just below the head of the tree. The tie shall be No. 12 BWG galvanized iron wire covered with garden hose, or other approved tie. The loop shall be 1 inch (25mm) greater in diameter than the trunk. The tie shall be attached to the pipe through a hole drilled at the tie location described above. The wire ties shall be secured by twisting the ends.

308-4.6.2 Method "B" Tree Staking. The tree shall be staked with 2 No. 6 (No. 19M) steel reinforcing bars 6 feet (1.8m) long driven 24 inches (600mm) into the ground. The stakes shall be 18 inches (450mm) from each side of the tree trunk, and stakes and tree shall be in a plane parallel to the street centerline. Ties shall be made of 1 inch (25mm) or wider flexible plastic ribbon material having a minimum tensile strength of 500 pounds (2.2 kN). Four ties shall be used; 2 at 2 inches (50mm) from the top of each stake and 2 at 2 feet (0.6m) above the ground. Ties shall be loops secured to the stake on one end and shall be long enough to provide for 3 inches (75mm) of slack to permit the tree trunk limited movement in any direction.

308-4.6.3 Guying. Trees and other plants, except specimen plants, to be guyed shall be designated in the Contract Documents.

Guying shall be done immediately after planting. Three guys per plant shall be installed in accordance with the following:

1) Each guy shall be secured to the appropriate main branch by a twisted loop of No. 12 BWG galvanized iron wire housed in garden hose.
2) Each guy shall be anchored to a driven stake located at a horizontal distance from the tree equal to the vertical distance from ground to the connection of guy wire on the tree branch.
3) Each guy shall be covered with highly visible garden hose or plastic tubing to a height of 6 feet (1.8m) above grade.
4) Slack in each guy shall be removed by hand so as not to bend or twist the plant.

308-4.7 Ground Cover and Vine Planting. Soil preparation and fine grading shall be completed prior to ground cover planting.

Ground cover and vines shall be planted in moist soil and spaced as indicated on the Plans.

Each plant shall be planted with its proportionate amount of flat soil to minimize root disturbance. Soil moisture shall be such that the soil does not crumble when removing plants.

Following planting, ground cover and vine areas shall be regraded to restore smooth finish grade and to ensure proper surface drainage. A 1 inch (25mm) layer of Type 1, 2, 3, or 4 mulch shall be spread over the planted areas. Watering shall begin immediately following mulching.

When necessary to prevent plant damage from pedestrian traffic during the initial growing stage, the Contractor shall erect temporary protective fencing to be removed at the end of the plant establishment period.

Vines shall be tied to walls, fences, etc., in the manner prescribed on the Plans. Temporary staking shall be removed at the end of the plant establishment period.

308-4.8 Lawn Planting.

308-4.8.1 General. Before planting lawn, all specified soil preparation and fine grading shall be completed.

308-4.8.2 Seed. Seed lawn planting may be accomplished by Method A (dry method) or Method B (hydraulic method). Seeding shall not be performed when the wind velocity exceeds 5 miles per hour (8 km/h), or is detrimental to the uniform distribution of the seed.

a) **Method A.** The area to be seeded shall be lightly raked to provide a seed bed.

   The required seed mixture shall be sown uniformly at the specified rate. Seeding shall be done in two operations with the spreader set to sow one-half the specified amount in each operation. The second sowing shall be at right angles to the first.

   After sowing, the area shall be evenly covered to a depth of 1/4 to 1/2 inch (6mm to 13mm) with an approved mulch.

   The lawn area shall be watered in a manner so as not to cause surface erosion.

   Newly seeded surfaces shall be kept moist continuously throughout the germination period.

b) **Method B.** The seed, fertilizer, fiber, and other materials in the slurry mixture shall be as specified. All materials shall be of such character that they will disperse into a uniform slurry when mixed with water. The mixture shall be such that an absorbent, porous mat will be formed.

   All materials must be available for inspection prior to application. Weights and contents of containers shall be clearly identified. A green coloring additive shall be used in the slurry for visual inspection purposes.

   The slurry shall be applied under pressure at the specified rates.

   Areas to be planted by this method shall be moistened to a depth of 6 inches (150mm) but shall not be surface wet at the time of application.

   The slurry planted areas shall be kept moist during the germination period, but puddling shall be avoided.

308-4.8.3 Sod. The type and thickness of sod and the areas to be sodded shall be in accordance with the Contract Documents.

Subgrade for sod shall be the specified thickness of the sod below finish grades. Soil conditioning and fine grading shall be completed before sodding. No heavy equipment shall operate over the subgrade after grading is completed.

The subgrade shall be moist but not wet when sod is laid. Sod shall be laid with closely fitted joints, and the ends of the strips shall be staggered. Openings shall be plugged with sod or topsoil.

Within 2 hours after installing sod and before rolling, the sod shall be lightly irrigated. All seams and joints shall then be rolled until the sod is well bonded to the subgrade.

The area shall then be watered thoroughly to penetrate the subsoil at least 8 inches (200mm). Watering shall be repeated as necessary to keep the sod moist until rooted into the subgrade. Sodded areas shall be protected against foot traffic until the sod is well established.

308-4.8.4 Stolon. Topsoil preparation and conditioning and finish grading shall be completed in accordance with 308-2.3 and 308-2.4 before stolon planting.

The area to be planted in stolons shall be thoroughly irrigated to a depth of at least 8 inches (200mm). As soon as the soil can be worked, the specified commercial fertilizer shall be worked into the top 1 inch (25mm) of soil.

At the time of planting, the top 2 inches (50mm) of soil shall be friable and contain enough moisture to prevent stolons from drying out during the planting operation. The stolons shall be worked into the soil to a depth of 1/2 to 1-1/2 inches (15mm to 40mm) by a mechanical or hand planter, or broadcast by hand and covered with 1/4 inch (6mm) of mulch.

When the area to be planted exceeds 10,000 ft^2 ($1000m^2$), a mechanical spreader shall be used; when less than 10,000 ft^2 ($1000m^2$) and more than 2,000 ft^2 ($200m^2$), the use of a hand planter or mechanical planter is optional; and when less than 2,000 ft^2 ($200m^2$), handplanting or broadcasting with mulch is optional.

The planted stolons shall not be allowed to dry out. Watering shall begin immediately after planting and the stolons kept moist at all times until the plants are well established.

When overseeding is required, the seed shall be spread in accordance with 308-4.8.2, Method A, immediately after planting stolons.

308-4.9 Erosion Control Planting.

308-4.9.1 General. Erosion control planting shall be for slope protection. Topsoil preparation and conditioning shall be in accordance with 308-2.3.

308-4.9.2 Straw Stabilization. When straw stabilization is specified, Type 6 mulch shall be incorporated into the slope topsoil either by discing or with a steel plate studded roller. The steel plate studs shall be at least 6 inches (150mm) long, not more than 6 inches (150mm) wide, and approximately 1 inch (25mm) thick, with rounded edges. The roller shall be capable of forcing the straw into the soil a sufficient depth to tie down the surface soils.

308-4.9.3 Seeding and Mulching. Seed, fertilizer, mulch, and other specified materials may be applied on slopes by Method A or Method B described in 308-4.8.2.

308-4.9.4 Sprigging. Sprigging shall consist of planting turf grasses, cut stems of plants, and plants with attached root system, but without adhering soil.

Sprigs shall normally be harvested and planted within a 24-hour period. Ice plant sprigs shall be harvested between 48 and 96 hours before planting so that a thin callus is formed over the cut surface of each sprig. Sprigs shall be shaded during callusing, but shall not be moistened.

Turf grasses shall be planted in accordance with 308-4.8.4.

Ice plant sprigs shall be planted in moist soil in holes or furrows 4 inches (100mm) deep and the hole or furrow refilled with soil and made firm around the plant in such a manner that the plant is not damaged.

Sprigs shall be planted individually at specified spacing. When row sprigging is specified, planting shall be in furrows cut along the contour of the slope.

If mulching of sprigged areas is required, it shall immediately follow planting.

308-4.9.5 Watering. All seeded and planted areas shall be kept moist during the establishment period.

Areas containing ice plants shall be maintained in a barely moist condition to a depth of 1 inch (25mm) below the planted root depth.

When a permanent irrigation system is not available, the Contractor shall provide whatever temporary system is necessary to provide adequate watering during the establishment period without erosion detrimental to the planting.

308-5 IRRIGATION SYSTEM INSTALLATION.

308-5.1 General. The Contractor shall furnish all necessary materials, labor, and equipment required to complete the work of installing the irrigation system in accordance with the Specifications.

Large specimen plants shall be planted before installing the irrigation system, as required by 308-4.4.

Unless otherwise provided, the irrigation system layout shown on the Plans shall be considered schematic. With the Engineer's approval, the Contractor may make adjustments where necessary to conform to actual field conditions. The irrigation system shall be operational, with uniform and adequate coverage of the areas to be irrigated, prior to planting.

Service connections shall be as shown on the Plan or designated by the utility company and will be installed by others at no cost to the Contractor. The Contractor shall notify the Engineer at least 3 weeks prior to the time electrical and water services are required. The Contractor shall be responsible for furnishing the labor and materials to connect to the service connection.

Trenches through paved areas shall be resurfaced in accordance with 306-1.5.

After completing the irrigation system, the Contractor shall submit as-built drawings showing the location of pipe, valves, tubing, wiring, controllers, and electrical services as constructed.

308-5.2 Irrigation Pipeline Installation.

308-5.2.1 General. Trench excavation and backfill including the depth of cover over the pipeline shall be in accordance with requirements of 308-2.2.

Pipe fittings shall be installed in accordance with the manufacturer's recommendations and these specifications. When requested by the Engineer, the Contractor shall furnish the manufacturer's printed installation instructions before pipe installation.

Pipe shall be bedded in at least 2 inches (50mm) of finely divided material to provide a firm, uniform bearing. After laying, the pipe shall be surrounded with additional finely divided material to at least 2 inches (50mm) over the top of the pipe. Trench backfill, sufficient to anchor the pipe, may be deposited before the pipeline pressure testing, except that joints shall remain exposed until satisfactory completion of testing.

When 2 or more pipelines are installed in the same trench, they shall be separated by a minimum horizontal clear distance of 6 inches (150mm) and they shall be installed so that each pipeline, valve, or other pipeline component may be serviced or replaced without disturbing the other.

All assemblies shall be assembled as specified and in accordance with the manufacturer's directions.

During installation of pipe, fittings, valves, and other pipeline components, foreign matter shall be prevented from entering the system. All open ends shall be temporarily capped or plugged during cessation of installation operations.

Changes in pipeline size shall be accomplished with reducer fittings.

308-5.2.2 Steel Pipeline. Ends of pipe shall be cut square and reamed to full size with a long taper reamer.

Threads shall be cut with clean, sharp dies and shall conform to American Standards Association Specification B2.

Joints shall be made with a nontoxic, nonhardening joint compound applied to the male threads only.

When wrapped pipe is specified, joints and any remaining unwrapped portion of the pipeline shall be similarly wrapped after pressure testing.

308-5.2.3 Plastic Pipeline. Plastic pipe shall be jointed by socket-type solvent-welded fittings, threaded fittings, rubber-ring fittings, or by other means specified. When plastic pipe is joined to steel pipe, the steel pipe shall be installed first.

Plastic pipe shall be cut square, externally chamfered approximately 10 to 15 degrees, and all burrs and fins removed.

Solvent welded joints shall be made in accordance with ASTM D2855. The solvent recommended by the manufacturer shall be used.

Plastic pipe installation shall be in accordance ASTM D2774 and the requirements herein.

Care shall be exercised in assembling a pipeline with solvent welded joints so that stress on previously made joints is avoided. Handling of the pipe following jointing, such as lowering the assembled pipeline into the trench, shall not occur prior to the set times specified in ASTM D2855.

Solvent shall be applied to pipe ends in such a manner that no material is deposited on the interior surface of the pipe or extruded into the interior of the pipe during jointing. Excess cement on the exterior of the joint shall be wiped clean immediately after assembly.

Threads for plastic pipe shall be as specified in 308-5.2.2. A plug shall be installed in the bore of the pipe to prevent distortion prior to threading.

Threaded pipe joints shall be made using teflon tape or other approved jointing material. Solvent shall not be used with threaded joints.

Pipe shall be protected from tool damage during assembly. Vises shall have pleated jaws and strap wrenches shall be used for installation of fittings and nipples.

Plastic pipe which has been nicked, scarred, or otherwise damaged shall be removed and replaced. Plastic pipe shall be snaked from side to side in the trench to allow 1 foot (1 meter) of expansion and contraction per 100 feet (30m) of straight run.

The pipeline shall not be exposed to water for 24 hours after the last solvent welded joint is made.

308-5.2.4 Copper Pipeline. Copper pipeline shall be made with sweated solder joints.

Before jointing, the end of the pipe for the depth of the fitting, and the interior of the fitting shall be buffed to a bright finish and coated with solder flux. The assembled joint shall be made with a 50-50 tin-lead solder. A continuous solder bead shall show around the joint circumference after soldering.

Copper pipe shall be joined to steel or cast iron pipe with a dielectric union.

308-5.3 Installation of Valves, Valve Boxes, and Special Equipment. Valves, backflow preventers, pressure regulators, and related accessories shall be furnished and installed as specified.

All valves and other equipment shall be installed in a normal upright position unless otherwise recommended by the manufacturer, and shall be readily accessible for operation, maintenance, and replacement. Sectional control valves shall not be located within range of sprinklers they control.

Valves shall be same size as the pipeline in which they are installed.

Gate valves and sectional control valves shall be installed below ground. Gate valves shall be housed in a covered concrete box that will permit access for servicing. Sectional control valves shall be equipped with a sleeve and cap centered on the valve stem.

Quick-coupler valves and garden valves projecting above grade shall be installed 3 feet (0.9m) from curbs, pavement, and walks. In lawn areas, such equipment shall be installed in a covered concrete box set to finish grade. In ground cover and shrubbery areas, quick-coupler valves shall be set 6 inches (150mm) above finish grade, and garden valves shall be set between 12 and 15 inches (300mm and 375mm) above finish grade. Quick-coupler valves and garden valves shall be installed on a double-swing-joint riser assembly as described in 308-5.4.3 and secured to a driven No. 4 (No. 13M) reinforcing steel rod as described in 308-5.4.3.

All valve boxes, pipe sleeves, and caps shall be set to finish grade, and valves shall be set at sufficient depth to provide clearance between the cover and the cap, valve handle, or key when the valve is in the fully open position.

Backflow preventers shall be provided with pipe supports and the accessories necessary to properly secure the assembly. All backflow preventers shall be assembled with pipe and fittings of galvanized steel.

308-5.4 Sprinkler Head Installation and Adjustment.

308-5.4.1 General. In accordance with the requirements of 308-5.6, all mains and laterals, including risers, shall be flushed and pressure tested before installing sprinkler heads, after which a water coverage test shall be performed.

308-5.4.2 Location, Elevation, and Spacing. Sprinkler head spacing shall not exceed the maximum shown on the Plans or recommended by the manufacturer.

In new lawn areas, sprinkler heads shall be installed 3 inches (75mm) above grade and then reset flush with the finish surface just prior to the first mowing. Lawn sprinklers shall be installed 2 inches (50mm) clear of adjacent walks, curbs, paving, headers, and similar improvements.

Sprinkler heads shall be installed 4 inches (100mm) from adjacent vertical elements projecting above grade such as walks, planter boxes, curbs, and fences.

Shrub heads, bubbler heads, and oscillating sprinklers shall be installed 6 inches (150mm) above finish grade.

Nozzle lines shall be installed at least 12 inches (300mm) above finish grade. Sprinkler heads projecting above finish grade shall be at least 12 inches (300mm) from adjacent curbs, walks, paving, and similar improvements.

308-5.4.3 Riser and Nozzle Line Installation. To obtain optimum coverage of the area, risers shall be oriented perpendicular to finish grade.

Risers for oscillating sprinklers and nozzle lines shall be galvanized steel pipe. All other risers may be galvanized steel of Schedule 80 PVC. All pipe between the connection to the lateral or main and the sprinkler head shall be threaded.

Sprinkler head riser assemblies shall be top outlet, single-swing joint, or double-swing joint as specified herein.

Sprinkler head risers and nozzle risers installed above grade within 24 inches (600mm) of roadway paving, curbs, walks, and similar improvements shall be of the double-swing joint type.

A top outlet riser assembly shall consist of a pipe riser threaded into a top outlet ell or tee installed in the lateral supply line.

Double-swing joint and single-swing joint riser assemblies shall utilize a horizontal 6 inch (150mm) pipe nipple threaded into a side outlet ell or tee installed in the lateral supply line. For a double-swing joint, three ells shall be used in the remaining assembly ahead of the vertical riser pipe. For a single-swing joint, one ell shall be used.

Risers for nozzle lines, oscillating sprinklers, and other sprinkler heads installed above grade within 24 inches (600mm) of curbs, walks, roadways, and similar improvements shall be supported by a No. 4 (No. 13M) reinforcing steel rod. The upper end of the rod shall be at finish grade and be of such length that it extends 24 inches (600mm) below the lateral supply line.

Where nozzle lines cannot be supported on adjacent fences, guard rails and the like, they shall be supported on driven 1/2 inch (13mm) pipe stakes, 4 feet (1.2m) long at 8 foot (2.4m) centers. The nozzle line shall be secured to the top of the stake with 3/8 inch (9.5mm) anchor rings, 12 inches (300mm) long.

308-5.4.4 Sprinkler Head Adjustment. When all sprinkler heads are installed and the irrigation system is operating, each section or unit shall be adjusted and balanced, with all section control valves fully open to obtain uniform and adequate coverage.

Sprinkler heads having adjustable pin nozzles or orifices shall have the pins adjusted to provide adequate distribution of water over the coverage pattern. The Contractor shall substitute larger or smaller nozzle cores in nonadjustable sprinkler heads as necessary.

If additional work other than the prescribed above is necessary to correct a deficiency in the system installed as specified, such work will be paid for in accordance with 3-2.

308-5.5 Automatic Control System Installation. The Contractor shall install a complete automatic irrigation control system including the automatic controller, remote control valves and wiring, and all necessary accessories and utility service connection.

The automatic controller shall be installed outside of the coverage pattern of the irrigation system at the location designated in the contract documents. The foundation for the controller shall be walk concrete, as specified in 201-1, of the size shown on the Plan or recommended by the manufacturer. The control components in the controller shall be fused and the chassis shall be grounded.

Remote control valves shall be compatible with the automatic controller. When the valve is to be housed in a concrete box, it shall be installed with at least a 6-inch (150mm) clearance below the concrete cover. The box shall be set to finish grade on a 12-inch (300mm) layer of 1 inch (25mm) crushed rock.

All service wiring shall be installed at the minimum depth specified in 308-2.2 in galvanized steel conduit from the service point to the controller. The applicable provisions of 307-2.8 shall govern the locations of service points. A separate disconnect switch or combination meter socket, as required, shall be installed between the source of power and the controller. The minimum service wire shall be No.12 AWG copper, 600-volt, Type TW, THW, THWN, or larger as required by the Specifications or controller manufacturer. Wire splices shall be located only in specified pull boxes and shall be made with a packaged kit approved for underground use, or as specified in 307-2.6. Pull boxes shall be concrete, set to grade on a 12 inch (300mm) layer of 1 inch (25mm) crushed rock.

Control wiring or hydraulic control tubing shall be housed in conduit between the controller and a concrete pull box installed at least 12 inches (300mm) outside the limits of the controller foundation, or the structure foundation where the controller is housed. All other wiring and hydraulic control tubing issuing from the pull box shall be direct burial installed in main or lateral waterline trenches wherever practicable. The wiring or tubing shall be bundled and secured to the lower quadrant of the irrigation pipeline at 10 foot (3m) intervals with plastic electrical tape. Sufficient slack shall be left in the wiring or tubing to provide for expansion and contraction. When the control wiring or tubing cannot be installed in a pipe trench, it shall be installed a minimum of 18 inches (450mm) below finish grade and a bright colored plastic ribbon with suitable markings shall be installed in the trench 6 inches (150mm) below grade directly over the wire or tubing.

Control wiring shall be suitably color coded as necessary for identification. All common wire shall be the same color. Unless otherwise required, all control wiring shall be direct burial, Type UF, No. 14 AWG copper. Splices in control wire shall be made in accordance with the requirements for service wire. At least 2 feet (0.6m) of slack shall be left at each splice and point of connection in pull boxes and valve boxes.

All wiring shall be tested for continuity, open circuits, and unintentional grounds prior to connecting to equipment. When tested for a period of 4 hours, the hydraulic control system shall maintain a constant test pattern of 125 psi (860 kPa). Upon completion of the Work, the control system shall be in operating condition with an operational chart mounted within the controller cabinet.

Upon completion of the Work, the control system shall be in operating condition with an operational chart mounted within the controller cabinet.

308-5.6 Flushing and Testing.

308-5.6.1 General. After completion and prior to the installation of any terminal fittings, the entire pipeline system shall be thoroughly flushed to remove dirt, scale, or other material. After flushing, the following tests shall be conducted in the sequence listed below. All equipment, materials, and labor necessary to perform the tests shall be furnished by the Contractor and all tests shall be conducted in the presence of the Engineer.

308-5.6.2 Pipeline Pressure Test. A water pressure test shall be performed on all pressure mains and laterals before any couplings, fittings, valves, and the like are concealed. All open ends shall be capped after the water is turned into the line in such a manner that all air will be expelled. Pressure mains shall be tested with all control valves to lateral lines closed. After the pressure main test, all valves shall be opened to test lateral lines. The constant test pressure and the duration of the test are as follows:

Mains 0.6 hours at 125 psi (860 kPa)

Laterals 0.2 hours at 100 psi (690 kPa)

308-5.6.3 Sprinkler Coverage Test. The coverage test shall be performed after sprinkler heads have been installed and shall demonstrate that each section or unit in the irrigation system is balanced to provide uniform and adequate coverage of the areas serviced. The Contractor shall correct any deficiencies in the system in accordance with the requirements of 308-5.4.4.

308-5.6.4 Operational Test. The performance of all components of the automatic control system shall be evaluated for manual and automatic operation.

During the maintenance period and at least 15 days prior to final inspection, the Contractor shall set the controller on automatic operation and the system shall operate satisfactorily during such period. All necessary repairs, replacements, and adjustments shall be made until all equipment, electrical work, controls, and instrumentation are functioning in accordance with the Contract Documents.

308-6 MAINTENANCE AND PLANT ESTABLISHMENT. The Contractor shall maintain all planted areas on a continuous basis as they are completed during the progress of the work and during the establishment period, and shall continue to maintain them until final acceptance.

All planted areas shall be kept free of debris and shall be weeded and cultivated at intervals not to exceed 10 days. The first mowing of lawn areas shall be performed when the grass is 2-1/2 inches (60mm) high and shall be repeated as often as necessary to maintain the lawn at a height of 2 inches (50mm). In no case shall the lawn be cut lower than 1-1/2 inches (40mm) in height.

Any required pruning of plants will be designated by the Engineer at the start of the plant establishment period and the Contractor shall perform the pruning as part of the plant establishment work.

The Contractor shall request a final inspection to begin the plant establishment period after all planting and related work have been completed in accordance with the Contract Documents.

After planting is completed, a field notification will be issued to the Contractor to establish the effective beginning date of the plant establishment period. The plant establishment period shall be for a period of 30 calendar days and shall be extended by the Engineer if the planted areas are improperly maintained, appreciable plant replacement is required, or other corrective work becomes necessary.

Upon completion of the plant establishment period, a final inspection for acceptance will be performed by the Engineer. If the plant establishment period is satisfactorily completed ahead of other work included in the Contract, the maintenance of planted areas shall be continued by the Contractor until all other work has been completed and accepted.

308-7 PAYMENT. The lump sum or Contract Unit Prices shown in the Bid shall include full compensation to complete and maintain the landscape and irrigation work shown on the Plans or in the Specifications.

SECTION 309 - MONUMENTS

309-1 DESCRIPTION. This work shall consist of furnishing and installing portland cement concrete right-of-way monuments and cast-in-place survey monuments at the locations shown on the Plans, in the Specifications, or as directed by the Engineer.

309-2 MATERIALS. The concrete portion of monuments shall be constructed in accordance with the provisions in 201-1 and Section 303.

Marker plates for survey monuments will be furnished by the Agency at the Work site.

309-3 CONSTRUCTION. Survey monuments shall be cast-in-place in neat holes without the use of forms. The exposed surface of the finished monuments shall be uniform, of even texture, and shall be free of holes, cracks, and chipped edges.

Marker plates or copper bars shall be placed in survey monuments before the concrete has acquired its initial set and shall be firmly bedded in the concrete. The concrete shall be so located that when the plate is inserted, the reference point will fall within a 1 inch (25mm) circle in the center of the plate.

Monuments shall be set firmly and vertically in the ground to a depth of at least 3 feet (0.9m).

The tops of survey monument covers shall be set flush with the groundline or pavement surface, whichever applies.

309-4 PAYMENT. The Contract Unit Price paid for any survey monuments shall include full compensation for doing all the work involved in constructing the survey monuments, including necessary excavation and backfill as shown on the Plans or directed by the Engineer.

SECTION 310 - PAINTING

310-1 GENERAL.

310-1.1 Weather Conditions. Paint shall be applied only on thoroughly dry surfaces and during periods of favorable weather. Except as provided herein, painting will not be permitted when weather conditions are such that the atmospheric temperature is at or below 35°F (2°C), or when freshly painted surfaces may become damaged by rain, fog, or condensation, or when it can be anticipated that the atmospheric temperature will drop below 35°F (2°C) during the drying period. If fresh paint is damaged by the elements, it shall be replaced by the Contractor at its expense.

Subject to the approval of the Engineer, the Contractor may provide suitable enclosures to permit painting during inclement weather. Provisions must be made to artificially control atmospheric conditions within limits suitable for painting inside the enclosure throughout the painting operation. The cost of providing and maintaining such enclosures shall be considered as included in the prices paid for the various items of work and no additional payment will be made therefore.

310-1.2 Application. Painting shall be done in a neat and workmanlike manner. Unless otherwise specified, paint shall be applied by brush, roller, or spray methods.

If brushes are used, they shall have sufficient body and length of bristle to spread the paint in a uniform coat. In general, the primary movement of the brush shall be such as to fill thoroughly all irregularities in the surface, after which the coating shall be smoothed by a series of parallel strokes. Paint shall be evenly spread and thoroughly brushed out. The paint will be considered to have been improperly applied if an inordinate amount of residual brush marks remain. If rollers are used, they shall be of a type that do not leave a stippled texture in the paint film.

On all surfaces which are inaccessible for brushing, the paint shall be applied by spray or by sheepskin daubers especially constructed for the purpose, or by other means approved by the Engineer.

If spray methods are used, the operator shall be thoroughly experienced. Runs, sags, thin areas in the paint coat, or skips and holidays shall be considered as evidence the work is unsatisfactory and the Contractor may be required to apply the remainder of the paint by brush.

A water trap acceptable to the Engineer shall be furnished and installed on all equipment used in spray painting.

Mechanical mixers shall be used to mix the paint a sufficient length of time prior to use to thoroughly mix the pigment and vehicle. To keep the pigment in suspension, paint shall be kept thoroughly mixed while being applied.

310-1.3 Thinning Paint. Paints specified are formulated ready for application and no thinning will be permitted. If the paint becomes thick in cool weather, it shall be heated by immersing the container in hot water.

310-1.4 Protection of Work. The Contractor shall protect all parts of the structure against disfigurement as a result of its painting operations. The Contractor shall be responsible for any damage caused by its operations to vehicles, persons, or property, and shall provide at its expense protective means to guard against such damage.

Paint stains on adjacent improvements which result in an unsightly appearance shall be removed by the Contractor at its expense.

When ordered by the Engineer to abate a dust nuisance and to protect the wet paint film, the Contractor shall dampen the adjacent roadbed and shoulders with water at its expense on each side of the location where painting is in progress. The Contractor shall furnish and post at its expense DRIVE SLOWLY signs and take other precautions necessary to prevent dust and dirt from accumulating on freshly painted surfaces.

310-2 SURFACE PREPARATION FOR PAINTING STEEL STRUCTURES*.

* Portions reproduced courtesy of Steel Structures Painting Council.

310-2.1 General. The following methods of surface preparation apply to steel surfaces unless another method is specified.

310-2.2 Hand Cleaning. Hand cleaning is a method of preparing metal surfaces for painting by removing loose mill scale, loose rust, dirt, and loose paint by hand brushing, hand sanding, hand scraping, hand chipping, with other hand impact tools, or by a combination of these methods. It is not intended that all mill scale, rust, and paint be removed by this process, but loose mill scale, rust paint, and other detrimental foreign matter present shall be removed.

Oil, grease, or salts shall first be removed by the methods prescribed in 310-2.3. Other detrimental foreign matter shall be removed by the following operations:

a) Stratified rust (rust scale) shall be removed by hand hammering, hand chipping, other hand impact tools, or a combination thereof.

b) All loose mill scale and all loose or nonadherent rust shall be removed by hand wire brushing, hand sanding, hand scraping, or by a combination of these methods. Rust and mill scale are classified as "loose mill scale" and "loose or nonadherent rust" if they can be removed from a steel surface by vigorous hand brushing with a new, commercially acceptable wire brush, of suitable type, at a rate of 2 square feet ($0.2m^2$) per minute. This test shall be conducted on an area not previously brushed, scraped, or sanded, but from which all detrimental stratified rust

(rust scale), oil, and grease (if present) have been removed. This test establishes a standard for surface preparation and shall not be considered as establishing the production rate by cleaning. Regardless of the methods used for cleaning under this specification, the surface shall be cleaned at least as well as the surface resulting from this test.

In preparing surfaces for repainting, all loose or nonadherent paint shall be removed. Edges of remaining old paint shall be feathered so that the repainted surface will have a smooth appearance. The remaining old paint shall have sufficient adhesion so that it cannot be lifted as a layer by inserting the blade of a dull putty knife under it. All accessible weld flux and spatter shall be removed by hand scraping or by hand impact tools followed by wire brushing. The accessible portions of all partially enclosed steel members shall be cleaned. On new work, areas which will be inaccessible after assembly shall be cleaned before assembly.

All rivets, welds, corners, joints, and openings shall be properly cleaned. The steel wire of the wire brushes shall have sufficient rigidity to clean the surface, shall be kept free of excess foreign matter, and shall be discarded when they are no longer effective. Hand scrapers shall be made of suitable material and shall be kept sharp enough to be effective. The tools shall be operated in such a manner that no burrs or sharp ridges are left on the surface and no sharp cuts made into the steel.

After hand cleaning is completed, dust and other loose matter shall be removed from the surface. If detrimental amounts of grease or oil are still present, these areas shall be spot cleaned with solvent. The pretreatment (if any is specified) or prime coat of paint shall be applied as soon as possible after cleaning and before further deterioration of the surface occurs.

310-2.3 Solvent Cleaning. Solvent cleaning is a procedure for removing detrimental foreign matter such as oil, grease, soil, drawing and cutting compounds, and other contaminants from steel surfaces by the use of solvents, emulsions, cleaning compounds, steam cleaning, or other materials and methods which may not involve a solvent action. It is intended that solvent cleaning, if specified, shall be used prior to the application of paint and with other specified surface preparations for the removal of rust, mill scale, or paint.

Soil, cement spatter, drawing compounds, salts, or other foreign matter (other than grease or oil) shall be removed by brushing with stiff-fiber or wire brushes, by scraping, or by cleaning with solutions of alkaline cleaners provided such cleaners are followed by a fresh water rinse, or by a combination of these methods. When specified, the fresh water rinse shall be followed with a passivating dichromate or dilute chromic acid wash.

Oil or grease shall be removed by any of the following methods:

a) Wiping or scrubbing the surface with rags or brushes wetted with solvent. The final wiping shall be done with clean solvent and clean rags or brushes.

b) Spraying of the surface with solvent. The final spraying shall be done with clean solvent.

c) Complete immersion in a tank or tanks of solvent. Solvent for the last immersion shall not contain detrimental amounts of contaminant.

Emulsion cleaners may be used in lieu of the methods in this subsection provided that after treatment the surface shall be washed to remove detrimental residue.

Steam cleaning, using detergents or cleaners if specified, may be used in place of the methods in this subsection provided that the surface shall finally be steamed or washed to remove detrimental residues.

If chemical paint strippers are used for the removal of paint, all wax from the stripper remaining on the surface shall be removed by the use of suitable solvents. All alkaline residues from the paint strippers shall be removed by washing the surface with fresh water. All detrimental paint residue or stripping agent residue shall be removed.

Regardless of the method used to clean oil, grease, or contaminants from a surface, there shall be no detrimental residue left on the surface.

Solvent-cleaned surfaces shall be primed or prepared as specified before any detrimental corrosion or recontamination occurs.

310-2.4 Power Tool Cleaning. Power tool cleaning is a method of preparing metal surfaces for painting by removing loose mill scale, loose rust, and loose paint with power wire brushes, power impact tools, power grinders, power sanders, or by a combination of these methods. It is not intended that all mill scale, rust, and paint be removed by this process; but loose mill scale, rust, paint, and other detrimental foreign matter present shall be removed.

Oil, grease, and salts shall first be removed by the methods prescribed in 310-2.3. Other detrimental foreign matter will be removed as described below.

Stratified rust (rust scale) shall be removed by power impact tools. If minor quantities of stratified rust are present, they may be removed as outlined in 310-2.2.

Large areas of tight, well-adhered paint, even though they may be removable, shall be removed only if specified. All loose mill scale and all loose or nonadherent rust and all loose paint, as defined below, shall be removed by one or more of the following methods:

a) Power wire brushing using rotary, radial, or cup brushes of suitable size, entering all accessible openings, angles, joints, and corners. The steel wire of such brushes have sufficient rigidity to clean the surface, shall be kept free of excess foreign matter, and shall be discarded when they are no longer effective. The surface shall be cleaned, but not burnished to a detrimental degree.

b) Power impact tool cleaning using power driven chipping or scaling hammers, rotary scalers, single- or multiple-piston scalers, or other similar impact cleaning tools. Cutting edges of such tools shall be kept in effective condition.

c) Power grinding using abrasive wheels or power sanding using abrasive materials. Sanding or abrasive materials shall be discarded when they become ineffective.

Mill scale, rust, and paint are classified as "loose mill scale," "loose and nonadherent rust," and "loose" or "removable paint" if they can be removed from a steel surface by power wire brushing using a commercial electric or air wire brushing machine operated at a speed under load of 3,450 rpm and equipped with a 6-inch (150mm) diameter cup brush of double-row, knotted construction, made of No. 20 gage music wire. The brush shall be held against the steel surface with a force of 16 pounds (71 N), and the rate of cleaning shall be 2 square feet ($0.2m^2$) of surface per minute. This test must be conducted on an area not previously brushed, scraped, or sanded, but from which all detrimental stratified rust (rust scale), oil, and grease, if present, have been removed. This test establishes a standard for surface preparation and shall not be considered as establishing the production rate of cleaning. Regardless of the method used for cleaning under this specification, the surface shall be cleaned at least as well as the surface resulting from this test.

In preparing surfaces for repainting, all loose paint shall be removed. Thick edges of remaining old paint shall be feathered so that the repainted surface will have a smooth appearance. The remaining old paint shall have sufficient adhesion so that it cannot be lifted as a layer by inserting the blade of a dull putty knife under it. All accessible weld flux and spatter shall be removed by power tools. The accessible portions of all partially enclosed steel members shall be cleaned. On new work, areas which will be inaccessible after assembly shall be cleaned before assembly.

Rivet heads, cracks, crevices, lap joints, fillet welds, and re-entrant angles shall be cleaned by the use of power wire brushing, sharp chisels used in chipping, scaling hammers, rotary grinders, sanders, or by a combination of such tools. All tools shall be operated in such a manner that no burrs or sharp ridges are left on the surface and no sharp cuts are made into the steel. Areas inaccessible for cleaning by power tools but accessible for hand cleaning shall be cleaned by methods outlined in 310-2.2.

After these cleaning operations are completed, dust and other loose matter shall be removed from the surface. If detrimental amounts of grease or oil are still present, these areas shall be spot cleaned with solvent. The pretreatment (if any), or the prime coat of paint shall be applied as soon as possible after cleaning and before further deterioration of the surface occurs.

310-2.5 Blast Cleaning.

310-2.5.1 General.

a) **Definition.** Blast cleaning is a method of preparing metal surfaces for painting by removing mill scale, rust, rust scale, paint, or foreign matter by the use of abrasives propelled through nozzles or by centrifugal wheels, to obtain one of the degrees of surface cleanliness described below.

b) **White metal.** A white metal blast-cleaned surface finish is defined as a surface with a gray-white, uniform metallic color, slightly roughened to form a suitable anchor pattern for coatings. The surface, when viewed without magnification, shall be free of all oil, grease, dirt, visible mill scale, rust, corrosion products, oxides, paint, or any other foreign matter. The color of the clean surface may be affected by the particular abrasive medium used.

c) **Near-White.** A near-white blast-cleaned surface finish is defined as one from which all oil, grease, dirt, mill scale, rust, corrosion products, oxides, paint, or other foreign matter have been completely removed from the surface except for very light shadows, very slight streaks, or slight discolorations caused by rust stain, mill scale oxides, or slight, tight residues of paint or coating that may remain.

   At least 95 percent of each square inch of surface area shall be free of all visible residues, and the remainder shall be limited to light discoloration mentioned above.

d) **Commercial.** A commercial blast-cleaned surface finish is defined as one from which all oil, grease, dirt, rust scale, and foreign matter have been completely removed from the surface and all rust, mill scale, and old paint have been completely removed except for slight shadows, streaks, or discolorations caused by rust stain, mill scale oxides, or slight, tight residues of paint or coating that may remain; if the surface is pitted, slight residues of rust or paint may be found in the bottom of pits; at least two-thirds of each square inch ($645 mm^2$) of surface area shall be free of all visible residues and the remainder shall be limited to the light discoloration, slight staining or light residues mentioned above.

e) **Brush-off.** A brush-off blast-cleaned surface finish is defined as one from which all oil, grease, dirt, rust-scale, loose mill scale, loose rust, and loose paint or coatings are removed completely, but tight mill scale and tight-adhered rust, paint, and coating are permitted to remain provided that all mill scale and rust have been exposed to the abrasive blast pattern sufficient to expose numerous flecks of the underlying metal fairly uniformly distributed over the entire surface.

   Heavy deposits of oil or grease shall be removed by the methods prescribed in 310-2.3.

   Small quantities of oil or grease may be removed by the blast-cleaning operation.

   Excessive rust scale shall preferably be removed by impact tools, as prescribed in 310-2.2 and 310-2.4.

310-2.5.2 Methods. The surface of the metal may be blast-cleaned by one of the following methods:

a) Dry sandblasting using compressed air blast nozzles and dry sand of a maximum particle size no larger than that passing through a No. 16 (1.18mm) mesh screen, U.S. sieve series.

b) Wet or water-vapor sandblasting using compressed air blast nozzles, water, and sand of a maximum particle size no larger than that passing through a No. 16 (1.18mm) mesh screen, U.S. sieve series.

c) Grit blasting using compressed air blast nozzles and crushed grit made of cast iron, malleable iron, steel, or synthetic grits other than sand. The largest commercial grade of metal grit permitted by this specification shall be SAE Grit No. G25 abrasive material.

d) Shot blasting using compressed air nozzles and cast iron, malleable iron, steel, or synthetic shot. The largest commercial grade shot permitted by this specification shall be SAE Shot No. S330.

e) Closed, re-circulating nozzle blasting using compressed air, vacuum, and any of the preceding abrasives.

f) Grit blasting using centrifugal wheels and crushed grit made of cast iron, malleable iron, steel, or synthetic grits. The largest commercial grade permitted by this specification shall be SAE Shot No. S330.

The surface, if dry blasted, shall be brushed with clean brushes made of hair, bristle, or fiber, blown off with compressed air (from which detrimental oil and water have been removed), or cleaned by vacuum, for the purpose of removing any traces of blast products from the surface, and also for the removal of abrasive from pockets and corners.

The surface, if wet sandblasted, shall be cleaned by rinsing with fresh water to which sufficient corrosion inhibitor has been added to prevent rusting, or with fresh water followed immediately by an inhibitive treatment. This cleaning shall be supplemented by brushing, if necessary, to remove any residue.

The compressed air used for nozzle blasting shall be free of detrimental amounts of condensed water or oil. Adequate separators and traps shall be provided. Blast cleaning operations shall be done in such a manner that no damage is done to partially or entirely completed portions of the work.

The blast-cleaned surface shall be further treated, or primed, as specified, within 8 hours after blasting when practicable, but in any event not later than 24 hours after blasting and also before any visible or detrimental rusting occurs. Surfaces which rust before painting is accomplished shall be recleaned by the Contractor at its expense.

310-3 SURFACE PREPARATION FOR PAINTING GALVANIZED SURFACES.

310-3.1 Hand Cleaning. Concrete spatter, heavy grease, and other foreign matter shall be removed from galvanized surfaces by hand scraping or wire brushing.

310-3.2 Solvent Cleaning. After hand cleaning, all galvanized surfaces shall be cleaned by the solvent cleaning procedures prescribed in 310-2.3 herein to remove oil, grease, and other detrimental foreign matter. After washing, all areas shall be roughened by abrasive blasting using an abrasive that is no larger than No. 30 mesh (600μm). Galvanizing shall not be removed by this operation.

310-4 SURFACE PREPARATION FOR PAINTING WOOD SURFACES. Wood surfaces shall be prepared for painting by removing all cracked or peeled paint, loose chalky paint, dirt, and other foreign matter by wire brushing, scraping, sanding, or other approved means immediately prior to painting. All surfaces shall be wiped or dry brushed to remove any dust or chalky residue that may result from cleaning operations. All wood designated to be painted shall be thoroughly dry before paint is applied.

310-5 PAINTING VARIOUS SURFACES.

310-5.1 Painting Structural Steel.

310-5.1.1 Paint. Unless otherwise specified, paints shall consist of a primer (applied in not less than two coats), a pre-treatment, and two finish coats. The total dry film thickness of the primer shall not be less than 3 mils (80μm), and the total dry film thickness of two finish coats shall be not less than 2 mils (50μm). The dry film thickness of the paint will be measured in place with a calibrated magnetic film thickness gage. Pre-treatment thickness shall be sufficient to completely coat the underlying primer.

Excessively thick coats of paint will not be permitted. The thickness of each coat shall be limited to that which will result in uniform drying throughout the paint film.

Paint shall conform to 210-1. Succeeding coats of paint, not otherwise materially different in color, shall have carbon black mixed into the paint 0.25 lbs. per gallon (30 g/L) lamp black, Federal Specification TT-P-00350, or slightly varying pigments to produce a shade contrasting with the paint being covered. Such changes shall be in undercoats, and the final finish coat shall be the specified finish color.

310-5.1.2 Cleaning. Unless otherwise specified, after erection and riveting or welding, all surfaces of structural steel which will be exposed to air in the completed structure, shall be commercially blast-cleaned as prescribed by 310-2.5 prior to painting.

In repainting existing steel structures where partial cleaning is required, the method of cleaning will be as specified.

Any damage to sound paint on areas not designated for treatment, which results from the Contractor's operations, shall be repaired as directed by the Engineer.

310-5.1.3 Application of Paint. Painting of finish coats of structural steel, except for sections which will be inaccessible after erection as described herein, shall be done after erection unless otherwise specified. Requests to do any painting other than undercoats prior to erection shall be submitted by the Contractor and approved by the Engineer in writing before such work is started. Any deficiencies in the first coat of paint shall be corrected prior to the application of succeeding coats of paint.

Surfaces exposed to the atmosphere which would be inaccessible for painting after erection shall be painted the full number of coats prior to erection.

All previous coats of paint shall be dry and fully cured and the surface of the paint coat being covered shall be free from moisture, dust, grease, or any other deleterious material which would prevent the bond of the succeeding paint coats. In spot painting, any old paint which lifts after application of the first spot coat, shall be removed by scraping and the area repainted before application of the next coat.

The application of finish coats will not be permitted until the repaired total film thickness of the undercoats of paint, as described in 310-5.1.1, is obtained.

Open seams at contact surfaces of stiffeners and built-up members which would retain moisture shall be caulked with non-sag polysulphide material conforming to Federal Specification TT-S-230, Type 2, or other approved material before applying the second coat of primer.

Except for anchor bolt assemblies, steel surfaces embedded in concrete need not be painted. Ungalvanized anchor bolt assemblies shall be painted or dipped with one coat of zinc-rich primer prior to installation.

The bottom surfaces of masonry plates and surfaces of structural steel to be in contact with elastomeric bearing pad or preformed fabric pads shall be cleaned and painted with the full number of specified coats prior to erection.

With the exception of abutting chord and columns splices and column and truss shoe bases, machine finished surfaces shall be coated with a rust inhibitor which can be easily removed. Surfaces of iron and steel castings which have been machine finished shall be painted with a coat of shop paint.

Zinc-rich primer shall be applied by spray methods. On areas inaccessible to spray application, the paint may be applied by brush or daubers. Mechanical mixers shall be used in mixing the primer. After mixing, the primer shall be strained through a metal No. 30 to 60 (600μm to 250μm) mesh screen or a double layer of cheesecloth immediately prior to or during pouring into the spray pot. An agitating spray pot shall be used in all spray application of primer. The agitator or stirring rod shall reach within 2 inches (50mm) of the bottom of the spray pot and shall be in motion at all times during primer application. Such motion shall be sufficient to keep the primer well mixed. Whenever painting operations are interrupted, the primer remaining in the fluid hose shall be expelled from the hose. Primer shall be free from dust, dirt, salt, or other deleterious deposits and thoroughly dry before applying pre-treatment vinyl wash primer.

The wash primer shall not be applied more than 72 hours before application of finish coats. The vinyl wash primer wash shall be applied by spraying to produce a uniform wet film completely coating the underlying surface.

310-5.1.4 Payment. Full compensation for preparing surfaces and for painting shall be considered as included in the prices for the various contract items of work involving structural steel and no separate payment for such work will be made.

310-5.2 Painting Machinery. Prior to installation, all surfaces of machinery exposed to the atmosphere, which are subject to corrosion and are normally painted, shall be painted with two coats of paint. Unless otherwise specified, after installation of the machinery, such surfaces shall be painted with a finish coat. All coats shall be as specified for structural steel. Full compensation for painting machinery shall be considered as included in the price paid for the machinery or in the item of which the machinery is part.

310-5.3 Painting Galvanized Surfaces. Unless otherwise specified, galvanized surfaces shall be left unpainted. When required to be painted, the surfaces shall be prepared as specified in 310-3 and then painted in accordance with 210-1, one coat of zinc dust-zinc oxide primer. The primer shall be applied by spraying to produce a complete covering of the galvanized surface. After the primer is applied, one coat of pre-treatment vinyl wash primer shall be applied. One finish coat shall be applied the same day as the wash primer is applied.

Full compensation for painting such surfaces shall be included in the price paid for the various contract items involving galvanized metal objects and no separate payment for such painting will be made.

310-5.4 Painting Metal Guard Rails. Metal guard rails, when required to be painted, shall be painted with three coats of paint of the type specified for metal guard rails in 210-1.5. Full compensation for painting guard rails shall be considered as included in the price paid for the guardrails and no separate payment for such painting will be made.

310-5.5 Painting Lumber.

310-5.5.1 Paint. Unless otherwise specified, all new lumber requiring painting shall consist of a primer and two finish coats as prescribed in 210-1.5 for wood structures, or as specified by the Engineer.

On all lumber previously painted the number of coats and types of paint will be as specified.

310-5.5.2 Preparation of Surfaces. Wood surfaces designated to be painted shall be cleaned in accordance with 310-4.

310-5.5.3 Application of Paint. When permitted by the Engineer, the first coat of paint may be applied prior to erection.

After the first coat has dried and the lumber is in place, all cracks, checks, nail holes, etc., shall be puttied flush with the surface and allowed to dry before the second coat is applied.

Skips, holidays, thin areas, or other deficiencies in any one coat of paint shall be corrected before the succeeding coat is applied.

The surface of any paint coat being covered shall be free of deleterious material before additional paint is applied.

310-5.5.4 Payment. Full compensation for preparing surfaces and for painting lumber shall be considered as included in the Bid prices for the various Contract items of work involving lumber and no separate payment for such work will be made.

SECTION 311 - SPECIAL PROTECTIVE MATERIALS

311-1 PLASTIC LINER INSTALLATION.

311-1.1 General. The installation of all plastic liner shall be done in accordance with Plans and Specifications.

Liner shall be applied and secured to the forms, inspected, and approved by the Engineer prior to the placement of reinforcing steel. Forms in contact with plastic liner shall not be oiled. Forms in contact with rigid plastic liner may be lubricated with a biodegradable lubricating material approved by the Engineer.

Liner sheet, weld strip, other liner accessory items, adhesive products, and cleansers to be used in conjunction with the installation of the liner shall conform to 210-2.

311-1.2 Installer Qualifications.

311-1.2.1 Applicators. The application of plastic liner to forms and other surfaces is considered to be specialized work. Personnel performing such work shall be adequately trained in the methods of liner installation and shall demonstrate their ability to the Engineer prior to commencing work.

311-1.2.2 Welders. Each welder shall pass a qualification welding test before doing any welding. Requalification may be required at any time it is deemed necessary by the Engineer. All test welds shall be made in the presence of the Agency's representative and shall consist of the following:

a) Two pieces of liner at least 15 inches (380mm) long and 9 inches (230mm) wide, shall be lapped 1-1/2 inches (38mm) and held in a vertical position.

b) A welding strip shall be positioned over the edge of the lap and welded to both pieces of liner. Each end of the welding strip shall extend at least 2 inches (50mm) beyond the liner to provide tabs.

The weld specimen will be tested by the Engineer as follows:

a) Each welding strip tab, tested separately, shall be subjected to a 10-pound (45 N) pull normal to the face of the liner with the liner secured firmly in place. There shall be no separation between the welding strip and liner.

b) Three test specimens shall be cut from the welded sample and tested in tension across the welds.

   1) If none of these specimens fails when tested as indicated in 210-2.3.5, the weld will be considered as satisfactory.

   2) If one specimen fails to pass the tension test, a retest will be permitted. The retest shall consist of testing three additional specimens cut from the original welded sample. If all three of the retest specimens pass the test, the weld will be considered satisfactory.

   3) If two of three specimens fail, the welder will be considered to be an unqualified welder and shall be disqualified.

A disqualified welder may submit a new welding sample when he has had sufficient off-the-job training or experience to warrant re-examination.

311-1.3 Placing Liner.

311-1.3.1 Coverage. The circumferential coverage shown on the Plan for the liner is the minimum limit of coverage permitted.

After pipe is installed, the offset of each longitudinal terminal edge of sheet on adjoining pipe shall not be greater than 1-1/2 inches (38mm). In cast-in-place structures, no such offset of the lower terminal edge shall be permitted.

At any location as shown on the Plans, where is a difference in, and the longitudinal terminal edges of liner downstream from said location are lower than those upstream, the terminal edges of the liner installed in the section of pipe or structure immediately upstream from the station shall be sloped uniformly for the entire length of the section of pipe or structure from the limits of the smaller coverage to those of the greater coverage. Wherever the longitudinal terminal edges of liner downstream from the station are higher than those upstream, the slope shall be accomplished uniformly throughout the length of the section of pipe or structure immediately downstream from the station. An approved locking extension shall be provided along all sloping lower terminal edges of liner plate.

311-1.3.2 Positioning Liner. All liner installed in pipe shall be positioned so that the locking extensions are parallel to the axis of the pipe.

Liner shall be centered within the form with respect to the "T" of the pipe when the inner form is in position. Liner shall be set flush with the inner edge of the socket end of a pipe section and shall extend either to the spigot end or beyond the spigot end, as required for the type of liner joint to be made with the adjoining pipe.

Rigid liner shall be set flush with the inner edge of the socket-and-spigot end of pipe or structure. Rigid liner may be set back from the edge of the pipe up to 1/2 inch (13mm) to facilitate manufacturing.

Liner installed in cast-in-place structure shall normally be positioned with locking extensions placed in the vertical direction. Horizontal placement may be utilized with the approval of the Engineer.

Liner shall be closely fitted to inner forms. Sheets shall be cut to fit curved and warped surfaces using a minimum number of separate pieces.

Prior to installation, the Contractor shall indicate to the Engineer the proposed layout of liner sheets for cast-in-place structures, including the location and type of all field welds.

The Engineer may require field sketches or the use of patterns or the marking of sheet layouts directly on the forms where complicated or warped surfaces are involved.

At transverse joints between sheets of liner used in cast-in-place structures and at all pipe joints, the space between ends of locking extensions, measured longitudinally, shall not exceed 4 inches (100mm). Where sheets are cut and joined for the purpose of fitting irregular surfaces, this space shall not exceed 2 inches (50mm).

311-1.3.3 Securing Liner in Place. Liner shall be held snugly in place against inner forms. For pipes and similar circular sections, light steel banding straps, prefabricated tubes, or other approved means shall be used. If used, banding straps shall be placed in strap channels. Any method of banding other than in strap channels shall require prior approval by the Engineer. Where form ties or form stabilizing rods pass through the liner, provisions shall be made to maintain the liner in close contact with the forms during concrete placement.

Concrete shall be prevented from flowing around the edges of sheets at joints by welding a weld strip or applying a waterproof tape over the back of the joint.

311-1.3.4 Flexible Liner Weep Channels. At each pipe joint and at transverse joints in cast-in-place structures, a gap not less than 2 inches (50mm) nor greater than 4 inches (100mm) shall be left in all locking extension to provide a transverse weep channel. If locking extensions are removed to provide a weep channel at joints, the base of the extension left on the sheet shall not exceed 32 mils (800μm).

Intermediate weep channels shall be provided as required by the Plans or Special Provisions. Intermediate weep channels shall not be less than 2.5 inches (60mm) nor greater than 4 inches (100mm) in width. If locking extensions are removed to provide intermediate weep channels, the base of the extension left on the sheet shall not exceed 63 mils (1.60mm). Weep channels are not required on pipe containing a steel cylinder or pipe having 360 degree liner coverage.

Any area behind liner, which is not properly served by regular weep channels, shall have additional weep channels 2 inches (50mm) wide provided by cutting away locking extensions. Provisions shall be made to permit any water accumulated behind the liner of concrete manhole shafts to drain into the weep channels of the lined structure.

An additional transverse weep channel shall be provided approximately 12 inches (300mm) away from each liner return where surfaces lined with plastic liner join surfaces which are not lined, and at the terminal edge of the weep channel for reinforcement.

As a part of the work of installing liner, outlets of all weep channels shall be cleared of obstructions which would interfere with their proper function.

Where required by the Plans or Special Provisions, a 12 inch (300mm) long weld strip, 1 inch (25mm) wide, shall be centered over each terminal edge of the weep channel for reinforcement.

311-1.3.5 Liner Returns. A liner return with integral weep channels shall be installed where shown on the Plans and wherever surfaces lined with plastic liner join surfaces which are not so lined (such as brick, unlined concrete pipe, clay pipe, cast iron pipe, manhole frames, and metal or clay tile gate guides).

Unless otherwise indicated by the Plans, the Special Provisions, or the Standard Plans showing liner installation methods, returns shall be made as follows:

a) Each liner return shall be a separate strip of liner at least 3 inches (75mm) wide joined at right angles to the main liner by means of approved corner strips.

b) Flexible corner strips shall continuously heat-welded to the return and to the main liner. Rigid corner strips shall be nonflammable solvent welded.

Each liner return shall be sealed to the adjacent construction with which it is in contact by means of a compound approved by the Engineer. If the joint space is too wide or the joint surface too rough to permit the use of the compound, the joint space shall be filled with 2 inches (50mm) of densely caulked cement mortar, lead wool, or other caulking material approved by the Engineer, and finish coated with a minimum of 1 inch (25mm) of an approved corrosion-resistant material.

311-1.4 Concrete Casting Operations.

311-1.4.1 Concrete Placement. During concrete placement, shall be continuously vibrated to avoid damage to the liner and produce dense concrete securely anchoring the locking extensions into the concrete.

311-1.4.2 Removing Forms. When removing forms, care shall be taken to protect liner from damage. Sharp instruments shall not be used to pry forms from lined surfaces. When forms are removed, any nails that remain in the liner plate shall be pulled and the resulting holes clearly marked. Form tie holes shall be marked before ties are broken off and all areas of abrasion of the liner shall be marked.

Following completion of form removal, liner in pipe and structures shall be cleaned for inspection at the direction of the Engineer. Repairs to the liner in pipe and structures shall be completed and approved by the Engineer prior to shipment of the pipe.

Banding straps used in securing liner to forms shall be removed within the limits of the unlined invert. Voids left in the invert at the edge of the liner shall be filled with cement mortar or other material approved by the Engineer.

311-1.5 Field Jointing of Liner.

311-1.5.1 General. No field joint shall be made in liner until the lined pipe or structure has been backfilled and 7 days have elapsed after the flooding, jetting, or other means of compaction has been completed. Where groundwater is encountered, the joint shall not be made until pumping of groundwater has been discontinued for at least 7 days and no visible leakage is evident at the joint. The liner at the joints shall be free of all mortar and other foreign material and shall be clean and dry before joints are made.

Heated joint compound shall not be brought in contact with liner.

No coating of any kind shall be applied over any joint, corner, or welding strip, except where nonskid coating is applied to liner surfaces.

311-1.5.2 Field Joints in Pipe Installation. Field joints in liner plate at pipe joints shall be performed by utilizing one of the following types:

a) Flexible liner Type P-1 joint shall consist of a 4-inch (100mm) joint strip, centered over the mortared pipe joint and secured along each edge to adjacent liner by means of a welding strip.

b) Flexible liner Type P-2 joint shall be made with an integral joint flap with locking extensions removed per 210-2.4.6, extending a minimum of 3 inches (75mm) beyond the spigot end of the pipe. The flap shall overlap and be welded to the adjacent lined pipe section using a weld strip. Care shall be taken to protect the flap from damage. Excessive tension and distortion while bending the flap back to facilitate laying and joint mortaring shall be avoided.

Any flap which has been bent back and held shall be allowed to return to its original shape and flatness well in advance of making the liner joint.

c) Rigid liner Type CJ-1 shall consist of a co-extruded 7-1/4 inch (185mm) wide joint strip, centered over the mortared pipe joint, and secured along each edge to the adjacent liner by means of solvent welding. The nonflammable solvent shall be approved by the Engineer prior to its use.

d) Field joints in rigid liner at manhole joints shall use and shall consist of a 1-1/2 inch (38mm) by 1-1/2 inch (38mm) factory-installed L-angle. When manholes are installed in the field, polyurethane compound, approved by the Engineer, shall be applied between L-angles for sealing purposes.

Field joints in liner at pipe joints shall not be made until in the mortar in the pipe joint has been allowed to cure for at least 48 hours.

All joints between flexible-lined pipe and flexible-lined cast-in-place structures shall be either Type C-1 or Type C-2 specified herein.

311-1.5.3 Field Joints in Cast-in-Place Structures. Field joints in liner on cast-in-place structures shall be one of the following types:

a) **Flexible Liner.**

1) Type C-1 joint shall be made in the same manner as a Type P-1 joint. The width of space between adjacent sheets of liner in a Type C-1 joint shall not exceed 1/2 inch (13mm). This is the only type of joint permitted at transverse expansion and contraction joints in concrete. Its only other use is for joints between pipes and cast-in-place structures.

2) Type C-2 joints shall be made by overlapping sheets not less than 1-1/2 inches (38mm) and securing the overlap to the adjacent liner by means of a welding strip. The upstream sheet shall overlap the downstream sheet. The length of that part of the overlapping sheet not having locking extensions shall not exceed 4 inches (100mm).

3) Type C-3 joint shall be made by butting sheets of liner together and applying a welding strip over the back of the joint before concrete is placed. After the concrete is in place, apply a welding strip over the front of the joint. A Type C-3 joint is not permitted at a transverse joint which extends to a lower terminal edge of liner or at any joint where the gap between adjoining sheets of liner exceeds 1/8 inch (3mm).

b) **Rigid Liner.**

1) Type CJ-1 shall be used where flexible joints are required. The solvent-welded joint shall be held in place by applying continuous pressure over the entire width and circumferential length of the solvent-welded joint for a minimum of 6 hours.

2) Type CJ-2 shall be used as a liner return as described in 311-2.3.5.

3) Type CJ-3 shall be used for internal and external 90-degree corners.

4) Type CJ-4 shall be used for internal external 135-degree corners.

5) Type CJ-5 shall be used for applications where custom panel widths are required. It may also be used where custom angles are required.

6) Type CJ-6 shall be used for custom or compound angle requirements.

311-1.5.4 Installation of Welding Strips for Flexible Liner. Welding strips shall be fusion welded to joint strips and liner by welders approved by the Engineer using only approved methods and techniques. The welding operation of any joint shall be continuous until that joint has been completed.

Adequate ventilation shall be maintained during all welding operations.

Hot air welding tools shall provide clean effluent air at constant pressure to the surfaces to be joined within a temperature range between 500°F and 600°F (260°C and 315°C).

For lap welds, the welding strip shall be offset so that approximately 1/3 of the width is placed on the high side of the lap and properly fused. A small gap in fusion, not to exceed 1/8 inch (3mm) in width at the lap, is acceptable.

For butt welds, the welding strip shall be centered over the cleaned surfaces to be joined and fused across its entire width. Incomplete fusion, charred, or blistered welds will be rejected.

After repairs have been made, repaired welds shall be reinspected, tested, and approved by the Engineer.

311-1.5.5 Joint Reinforcement. A 12-inch (300mm) long welding strip shall be applied as a reinforcement across each transverse joint, weep channel, or return which extends to the lower terminal edge of liner. These reinforcement strips shall be centered over the joint being reinforced and located as close to the lower edge of liner as practicable. They shall be welded in place after the transverse welding strips have been installed.

311-1.6 Application of Liner to Concrete Surfaces. Application of liner plate to concrete surfaces by means of approved adhesive shall be accomplished by the following steps:

a) The concrete surface shall be etched by sandblasting to develop a slightly granular surface. When permitted by the Engineer, the concrete surface may be acid etched and neutralized in lieu of being sandblasting.

b) After sandblasting, the concrete surface shall be thoroughly cleaned of dust. Surfaces etched with acid shall be thoroughly washed with clean water and completely dried before applying primer. Application of primer, adhesive, and liner shall be in accordance with manufacturer's recommendations as approved by the Engineer.

311-1.7 Non-Skid Surfaces. All surfaces of the liner, shown on the Plans to be non-skid, shall be treated as follows prior to installation:

a) The liner shall be cleaned, dried, and sprayed with an adhesive coating recommended by the manufacturer of the liner plate.

b) Immediately after the adhesive is applied to the liner, the surface shall then be liberally sprinkled with clean, dry, well-graded sand, passing a No. 30 (600µm) sieve but be retained on a No. 70 (212µm) sieve.

After the sanded surface has thoroughly dried, all excess sand shall be brushed away and a seal coat of the adhesive coating shall be sprayed over the sand in sufficient quantity to coat and bond the sand to the liner plate.

c) The coated sand surface shall be allowed to dry thoroughly before handling.

311-1.8 Application of Liner to Steel. All fabrication and welding of steel to be lined with plastic liner shall be completed before the liner is installed except for field welding.

All steel surfaces to which plastic liner is to be applied shall be sandblasted, leaving surfaces free of all mill scale, rust, grease, moisture, and other deleterious substances. All interior weld metal shall be ground smooth and all weld spatter removed. After welds are ground, weld metal shall not project more than 1/16 inch (1.5mm) above the pipe surface. In the event that field welding is required, the plastic liner shall not be installed closer than 12 inches (300mm) to the weld. Plastic liner shall be installed in the weld area after welding on the steel has been completed.

The application of primer, adhesive, activator, and liner to steel surfaces shall conform to the requirements set forth herein for bonding of liner plate to concrete surfaces with adhesive. All field joints shall be tight-fitting butt joints. After the liner has been applied to steel surfaces, corner strips or welding strips shall be applied over all joints and welded in place.

311-1.9 Protection and Repair of Liner. All necessary measures and precautions shall be taken to prevent damage to liner from equipment and materials used in, or taken through the work. Any damage to installed liner plate shall be repaired by the Contractor in accordance with the requirements set forth herein for the repair of liner at no additional cost to the Agency.

For flexible liner, all nail and tie holes and all cut, torn, and seriously abraded areas in the liner plate shall be patched. Patches made entirely with welding strip shall be fused to the liner over the entire patch. The use of this method is limited to patches which can be made with a single welding strip. The use of parallel, overlapping, or adjoining welding strips will not be permitted. Large patches may consist of smooth liner over the damaged area, with edges covered with welding strips fused to the patch and to the liner adjoining the damaged area. The size of a single patch of the latter type shall be limited only as to its width, which shall not exceed 4 inches (100mm).

For rigid liner, all nail and tie holes, and all cut, torn, and seriously abraded areas in the liner plate shall be patched. Patches are available in three standard sizes: 2 inch (50mm), 3 inch (75mm), and 4 inch (100mm) diameter disks with a thickness of 78 mils (1.98mm). Patches shall be solvent welded to the liner plate.

Whenever liner is not properly anchored to concrete, or wherever patches larger than those permitted above are necessary, the repair of liner and the restoration of anchorage shall be as directed by the Engineer.

311-1.10 Field Tests. Upon completion of the installation, the surface of liner shall be cleaned to permit visual inspection and spark testing by the Engineer, using a spark-type detector complying with the requirements of 210-2.3.7. All areas of liner plate failing to meet the field test shall be properly repaired and retested. In addition to the visual inspection and prior to spark testing, all welds shall be tested for adhesion by probing with an instrument, such as a putty knife, to assure proper fusion of the weld strip and liner plate without damage to the weld strip or liner.

The Contractor shall assist in the inspection and spark testing by providing adequate ventilation, ladders for access, barricades, or other traffic control devices, and shall be responsible for opening and closing entrances and exits.

Any spark testing of liner by the Contractor for its purposes shall be done with a detector complying with 210-2.3.7.

311-1.11 Payment. Payment for plastic liner materials and their installation shall be included in the price bid for the pipe or structure to which they are applied.

SECTION 314 - TRAFFIC STRIPING, CURB AND PAVEMENT MARKINGS, AND PAVEMENT MARKERS

314-1 GENERAL. The work shall consist of the removal and application of traffic striping (striping), curb and pavement markings (markings), and pavement markers at the locations shown on the Plans.

For the purposes of these Specifications, traffic striping shall be defined as longitudinal centerlines and lane lines that separate traffic lanes in the same or opposing direction of travel, and longitudinal edge lines that mark the edge of the traveled way or the edge of the lanes. Curb markings shall be defined as colored markings on the curb that denote parking restrictions. Pavement markings shall be defined as transverse markings which include, but are not limited to, word and symbol markings, limit lines (stop lines), crosswalks, shoulder markings, parking stall markings and railroad crossing markings.

The type of material, paint or thermoplastic, to be applied shall be as specified in the Special Provisions or as shown on the Plans.

314-2 REMOVAL OF TRAFFIC STRIPING AND CURB AND PAVEMENT MARKINGS.

314-2.1 General. The Contractor shall remove existing traffic striping and pavement markings by wet or dry sandblasting, high velocity water jet, grinding, or other methods as specified in the Special Provisions. Obliteration with black paint or emulsified asphalt will not be allowed.

Curb markings shall be painted over as specified in the Special Provisions or shown on the Plans.

Conflicting striping and pavement markings shall be removed before the application of new temporary or permanent traffic striping, and curb or pavement markings.

314-2.2 Measurement. Removal of traffic striping and curb and pavement markings will be measured as follows:

a) Traffic striping and curb markings will be measured by the linear foot (meter) for each width and type of traffic stripe or curb marking.
b) Pavement markings will be measured by the square foot (square meter). The space between the stripes or letters will be included in the overall measurement.

314-2.3 Payment. Payment for removal of traffic striping and curb markings will be made at the Contract Unit Price per linear foot for each width and type of traffic stripe or curb marking. Payment for removal of pavement markings will be made at the Contract Unit Price per square foot.

314-3 REMOVAL OF PAVEMENT MARKERS.

314-3.1 General. Removal of existing pavement markers shall be performed in such a manner as to leave the existing pavement undamaged.

Asphalt concrete pavement shall be considered damaged when a depression of more than 1/4 inch (6mm) results. Damaged asphalt concrete pavement shall be patched with E-PG 64-10 asphalt concrete pavement conforming to 203-6 and 302-5.

314-3.2 Measurement. Removal of pavement markers will be measured by each pavement marker removed.

314-3.3 Payment. Payment for removal of pavement markers will be made at the Contract Unit Price for each pavement marker to be removed.

314-4 APPLICATION OF TRAFFIC STRIPING AND CURB AND PAVEMENT MARKINGS.

314-4.1 General. Traffic striping and curb and pavement markings shall conform to the dimensions and details shown on the Plans.

314-4.2 Control of Alignment and Layout.

314-4.2.1 General. When necessary, the Engineer will furnish the necessary control points. The Contractor shall establish traffic striping between these points by string line or other method approved by the Engineer.

Traffic lines shall be spotted in advance of application by using a rope as a guide for marking spots every 5 feet (1.5m), by using a marking wheel mounted on a vehicle, or by another method approved by the Engineer.

314-4.3 Painted Traffic Striping and Curb and Pavement Markings.

314-4.3.1 General. The work shall consist of applying painted (including glass beads) traffic stripes and curb and pavement markings at the locations shown on the Plans. Glass beads shall conform to 214-3. Paint shall conform to 214-4.

Paint shall be applied only when:

a) the pavement surface is dry and clean,
b) the atmospheric temperature is above 40^0F (1^0C) when using acetone-based paint and 50^0 F (10^0C) when using waterborne paint,
c) the weather is not windy, foggy, or humid,
d) the forecast atmospheric temperature will not drop below the aforementioned 40^0F (1^0C) or 50^0 F (10^0C) temperatures during the drying period, and
e) the forecast temperature is within the paint manufacturer's recommended range.

314-4.3.2 Surface Preparation. Before applying paint, the existing pavement surface shall be cleaned. Areas which cannot be satisfactorily cleaned shall be scrubbed with a water solution of tri-sodium phosphate (10 percent Na_3PO_4) or other cleaning solution approved by the Engineer. After cleaning, the surface shall be rinsed with water and dried before painting. The cleaning solution and rinse water shall not be allowed to enter any storm drain or natural water course.

314-4.3.3 Mixing. Mechanical mixers shall be used to mix paint. Prior to application, paint shall be mixed a sufficient length of time to thoroughly mix the pigment and vehicle together, and shall be kept thoroughly agitated during application.

314-4.3.4 Application Equipment.

314-4.3.4.1 General. Application equipment shall be approved by the Engineer prior to use. Application equipment shall include brushes, brooms, compressors, air blowers, mechanical marking machines, heating devices, bead dispensing devices, auxiliary hand spray painting equipment, stencils, paint rollers, and other equipment necessary to complete the work.

314-4.3.4.2 Striping Machines. Striping machines shall:

a) be a spray-type, rubber tired vehicle suitable for applying traffic paint;
b) have sufficient paint capacity for each color with adequate air pressure to perform the Work satisfactorily without excessive stopping;
c) be capable of producing a uniform film thickness without running or spattering;
d) be capable of being guided within the straightness tolerances set forth in these Specifications;
e) have suitable adjustments for painting the line width specified and when required, be equipped with an automatic cycling device to produce intermittent (skip) lines in accordance with 314-4.2.1;
f) be equipped to produce a variable skip pattern, including simultaneous painting of a broken line on one side and a solid line on the other side of a multiple stripe;
g) be capable of accurately superimposing succeeding coats of traffic paint upon the first coat and upon existing stripes at a minimum speed of 5 mph (8km/h);

h) be capable of 2-gun, or 3-gun (1 black and 2 yellow spray guns operating simultaneously or individually) applications;
i) have a wheel base of sufficient length to produce a straight line to meet the straightness tolerance specified in 314-4.2.4; and
j) be capable of producing curved lines without abrupt breaks.

Striping machines shall be equipped with the following:

a) a pointer or sighting device not less than 5 feet (1.5m) long extending from the front of the machine;
b) a pointer or sighting device extending from the side of the machine to gauge the distance from the centerline for painting shoulder stripes;
c) shields or an adjustable air curtain for line control;
d) pressure regulators and gages (if pneumatically operated) that are in within full view of the operator;
e) a paint strainer in the paint supply line;
f) a paint storage tank with a mechanical agitator that operates continuously during painting operations;
g) a glass bead dispenser located behind the paint applicator nozzle which is controlled simultaneously with the paint applicator nozzle; and
h) calibrated rods for measuring the volumes of paint and glass beads in the paint and glass bead tanks.

Where the configuration or location of a traffic stripe is such that the use of striping machines is unsuitable, traffic paint and glass beads may be applied by other methods and equipment approved by the Engineer. The Engineer will determine if the striping machine is unsuitable for a particular use.

Stencils and hand spray equipment shall be used to paint curb and pavement markings. Stencils shall conform to the dimensions shown on the Standard Plans.

314-4.3.5 Application. Each coat of paint for any traffic stripe, including glass beads, shall be applied in one pass of the striping machine, regardless of the number, widths and patterns of individual stripes involved.

On 2-lane streets, when the first coat of the centerline stripe is applied in the same direction as the stationing increases, the right-hand spray gun of the 3 spray guns used to apply a double yellow stripe shall be used to apply a single yellow stripe. When the first coat of the centerline stripe is applied in the direction as the stationing decreases, the left-hand gun of the 3 spray gun used to apply double yellow stripe shall be used to apply a single yellow stripe. The second coat of centerline striping shall be applied in the opposite direction that the first coat was applied.

A 1-coat, 3-inch (75mm) wide black stripe shall be painted between the 2, 4-inch (100mm) wide yellow stripes of a double traffic stripe. If the 2, 4-inch (100mm) wide yellow stripes are to be applied in 2 coats, the black stripe shall be applied concurrently with the second yellow stripe.

On existing pavement surfaces, traffic striping and pavement markings shall be applied in 1 coat. Paint to be applied in 1 coat shall be applied at an initial rate of 1 gallon per 107 square feet (1L per 2.6m^2) unless otherwise specified in the Special Provisions. The final rate shall be approved by the Engineer.

On new pavement surfaces, traffic striping (except the black stripe between the yellow stripes of a double traffic stripe) and pavement markings shall be applied in 2 coats unless otherwise shown on the Plans or specified in the Special Provisions. If the paint is applied in 2 coats, the first coat shall be thoroughly dry before the second coat is applied.

Paint to be applied in 2 coats shall be applied at the initial rates shown in Table 314-4.3.5 (A) unless otherwise specified in the Special Provisions.

TABLE 314-4.3.5 (A)

Paint Type	Square Foot Coverage Per Gallon	
	First Coat	Second Coat
Waterborne Paint	215	215
Acetone-Based Paint	360	150

The final application rates shall be approved by the Engineer. The volume of paint applied shall be measured by inserting a calibrated rod into the paint tank. At the option of the Engineer, if the striping machine is equipped with paint gauges, the volume of paint may be determined by using the gauges.

Glass beads shall be mechanically applied to the surface of each coat of paint, except black paint, prior to the paint setting. Glass beads shall be embedded to a depth of one-half of their diameter. Glass beads shall be applied at a rate of 5 pounds per gallon (2 kilograms per liter) of paint unless otherwise specified in the Special Provisions or directed by the Engineer. The amount of glass beads applied shall be measured by inserting a calibrated rod into the glass bead tank of the striping machine.

314-4.3.6 Measurement. Painted traffic stripes will be measured by the linear foot along the line of the traffic stripes, without deductions for gaps in broken traffic stripes. A double traffic stripe, consisting of 2, 4-inch (100mm) wide yellow stripes separated by a 3-inch (75mm) wide black stripe, will be measured as 1 traffic stripe.

Painted curb markings will be measured by the linear foot (meter).

Painted pavement markings will be measured by the square foot (square meter) for the actual area painted or by lump sum.

314-4.3.7 Payment. Payment for painted traffic stripes and painted curb markings will be made at the Contract Unit Price per linear foot (meter) for the number of coats of paint required, as specified in the Proposal, regardless of the number, widths, and pattern of individual stripes composing each traffic stripe.

Payment for painted pavement markings will be made at the Contract Unit Price per square foot (square meter) or lump sum Bid price for the number of coats of paint required, as specified in the Proposal.

No separate payment will be made for establishing alignment for stripes and layout work.

314-4.4 Thermoplastic Traffic Striping.

314-4.4.1 General. The work shall consist of applying thermoplastic traffic striping and pavement markings at the locations shown on the Plans and Standard Plans. Glass beads shall conform to 214-3. Thermoplastic material shall conform to 214-5.

Thermoplastic traffic striping and pavement markings shall only be applied to dry pavement surfaces with a surface temperature above 50^0 F (10^0 C) unless otherwise specified in the Special Provisions. The wind shall be calm enough that no blowing dirt is deposited on the pavement surface.

314-4.4.2 Surface Preparation. Before applying thermoplastic material, the existing pavement surface shall be mechanically wire brushed to remove all dirt and contaminants. New Portland cement concrete pavement shall be mechanically wire brushed or abrasive blast-cleaned to remove all laitance and curing compound.

314-4.4.3 Application Equipment. Application equipment shall be approved by the Engineer prior to use. Pre-heaters with mixers having a 360-degree rotation shall be used to preheat thermoplastic material before it is applied.

Application equipment shall be capable of:

a) utilizing either the spray or extrusion methods;

b) applications that dry to "no pick up" in accordance with ASTM D711; and

c) producing smooth, continuous lines having sharp dimensions.

Thermoplastic pavement markings shall be applied with equipment and stencils specifically designed and constructed for that purpose.

314-4.4.4 Application. A primer, of the type recommended by the manufacturer of the thermoplastic material, shall be applied to asphalt concrete surfaces over 6 months old and to all Portland cement concrete surfaces. The primer shall be applied immediately in advance of, but concurrent with, the application of thermoplastic material. The primer shall be applied at the rate recommended by the manufacturer and shall not be thinned.

Thermoplastic material shall be applied at a temperature between 400^0 F and 425^0 F (200^0 C and 218^0 C), unless a different temperature is recommended by the manufacturer. Thermoplastic material shall be applied in a single layer. The pavement surface to which the thermoplastic material is applied shall be completely coated by the material and the voids in the pavement surface shall be filled.

Unless otherwise specified in the Special Provisions, thermoplastic material for traffic striping shall be applied at a minimum thickness of 60 mils (1,500μm). Thermoplastic material for pavement markings shall be applied at a thickness of 100 to 150 mils (2,500μm to 3,800μm).

Glass beads conforming to 214-3 shall be mechanically applied to the surface of the molten thermoplastic material at a rate of not less than 8 pounds per 100 square feet (0.39 kilograms per square meter). The amount of glass beads applied shall be measured by inserting a calibrated rod into the glass bead tank.

314-4.4.5 Measurement. Thermoplastic traffic striping will be measured by the linear foot (meter) along the line of the traffic stripes, without deductions for gaps in broken traffic stripes. A double traffic stripe, consisting of 2, 4-inch (100mm) wide, yellow stripes, will be measured as 2 traffic stripes.

Thermoplastic pavement markings will be measured by the square foot (square meter) for the actual area covered or by the lump sum.

314-4.4.6 Payment. Payment for thermoplastic traffic stripes will be made at the Contract Unit Price per linear foot (meter) for each width and pattern, as specified in the Proposal.

Payment for thermoplastic pavement markings will be made at the Contract Unit Price per square foot (square meter).

No separate payment will be made for establishing alignment for stripes and layout work.

314-4.5 Tolerances and Appearance. Traffic striping shall not vary more than 1/2 inch in 50 feet (88mm per 100m) from the specified alignment.

Straight stripes deviating more than 1/2 inch in 50 feet (80mm per 100m) shall be obliterated by sandblasting and the markings corrected. When existing traffic striping and pavement markings are to be repainted, they shall be repainted to completely cover the old markings within 1/4 inch (6mm). Stripe repainting shall be retraced within a longitudinal tolerance of 6 inches (150mm) at the end of each stripe. Abrupt breaks in striping alignment will not be allowed. The striping shall be a continuous operation except where crossovers are required to complete painted medians.

Drips, overspray, improper markings, paint and thermoplastic material tracked by traffic shall be immediately removed, at the Contractor's expense, from the pavement surface by methods approved by the Engineer.

Completed traffic stripes shall have clean and well-defined edges without running or deformation, be uniform, be straight on tangent alignments and be a true arc on curved alignments. The widths of completed traffic stripes shall not deviate more than 1/4 inch (6mm) on tangent nor more than 1/2 inch (13mm) on curves from the width shown on the Standard Plans. Broken traffic stripes shall also conform to the following:

a) The lengths of the gaps and individual stripes that form broken traffic stripes shall not deviate more than 2 inches (50smm) from the lengths shown on the Standard Plans.
b) The lengths of the gaps and individual stripes shall be of such uniformity throughout the entire length of the broken traffic stripes that a striping machine will be able to repeat the pattern and superimpose additional stripes upon the traffic stripe being applied.

Completed curb and pavement markings shall have clean and well-defined edges without running or deformation and conform to the dimensions shown on the Standard Plans, except that minor variations may be accepted by the Engineer.

314-4.6 Protection from Damage. Newly placed traffic stripes and curb and pavement markings shall be protected from damage by public traffic or other causes until the paint is thoroughly dry or the thermoplastic material has sufficiently hardened. All adjacent surfaces shall be protected from disfiguration by spatter, splashes, spillage, and dripping of paint or other material.

314-5 PAVEMENT MARKERS.

314-5.1 General. The work shall consist of applying pavement markers conforming to 214-7 to the existing pavement surface at the locations shown on the Plans.

314-5.2 Adhesives.

314-5.2.1 General. Pavement markers shall be cemented to the pavement with rapid-set epoxy adhesive conforming to 214-7.2.2 or hot-melt bituminous adhesive conforming to 214-7.3 unless otherwise specified in the Special Provisions.

In areas of new construction where the pavement markers are protected from traffic, including the Contractor's vehicles, standard-set epoxy adhesive conforming to 215-5.2.5 may be used. Pavement markers applied with standard-set epoxy adhesive shall be protected from traffic for at least 3 hours after placement when the pavement surface temperature is 55^0F (13^0 C) or above, for at least 24 hours when the pavement surface temperature is between 40^0F and 55^0F (4^0C and 13^0 C), and for at least 48 hours when the pavement surface temperature is 40^0F (4^0C) or below. The Engineer will determine when the adhesive has set sufficiently to bear traffic.

Regardless of the type of adhesive used, pavement markers shall not be placed when the pavement is not surface dry.

314-5.2.2 Epoxy Adhesives.

314-5.2.2.1 General. Pavement markers shall not be placed using epoxy adhesives when either the pavement or the atmospheric temperature is 32^0F (0^0C) or less. Epoxy adhesives shall not be used to apply non-reflective plastic pavement markers.

314- 5.2.2.2 Mixing.

a) General. Automatic mixing equipment shall use positive displacement pumps capable of metering the 2 components in the specified ratio to within ± 5 percent by volume of either component. At the beginning of each working day when directed by the Engineer, the ratio of the 2 components shall be checked by the Contractor in the presence of the Engineer. This check shall be made by disconnecting the mixing heads, or using suitable bypass valves, and filling 2 suitable containers with the unmixed components. The mixing equipment shall mix the 2 components until there is no trace of black or white streaks in the mixed epoxy adhesive.

Voids in a cured, undisturbed sample of the mixed epoxy adhesive obtained from the extrusion nozzle shall not exceed 4 percent.

b) Rapid-Set Epoxy Adhesives. Rapid-set type epoxy adhesives shall not be mixed by hand. Rapid-set epoxy adhesives shall be mixed by a 2-component-type automatic mixing and extrusion apparatus. When machine mixing standard-set epoxy adhesive or rapid-set epoxy adhesive, pavement markers shall be placed within 60 seconds after the epoxy adhesive has been mixed and extruded and no further movement of the marker will be allowed. In addition, the time between when the components are pumped into the mixing head and when the mixed epoxy adhesive is in place on the pavement, and not subject to further movement, shall not exceed 90 seconds. Mixed epoxy adhesives shall not remain in the mixing head for more than 45 seconds. Epoxy adhesive remaining in the mixing head longer than this period shall be disposed of.

c) **Standard-Set Epoxy Adhesives.** When hand mixing standard-set epoxy adhesives, not more than 1 quart shall be mixed at a time, and the pavement markers shall be aligned and pressed into place within 5 minutes after beginning mixing operations. A mixed batch which becomes so viscous that the adhesive cannot be readily extruded from under the marker upon application of slight pressure shall not be used.

314-5.2.3 Hot-Melt Bituminous Adhesives. Pavement markers shall not be placed using hot-melt bituminous adhesives when the pavement surface or atmospheric temperature is 50^0F (10^0C) or less.

Hot-melt bituminous adhesives shall be heated indirectly in an applicator with continuous agitation or recirculation. Hot-melt bituminous adhesives shall not be heated above the maximum safe heating temperature recommended by the manufacturer and shall not be applied at temperatures greater than 425^0F (210^0C) nor less than 375^0F (190^0C).

314-5.3 Surface Preparation. The existing pavement surface shall be free from dirt, curing compound, grease, oil, moisture, loose or unsound layers, paint and any other material that would adversely affect the bond of the adhesive. Cleaning shall be done by blast cleaning on all surfaces regardless of age or type, except that blast cleaning of clean, new asphalt concrete and clean, new seal coat surfaces will not be required when hot-melt bituminous adhesives are used.

314-5.4 Placement. Reflective markers shall be placed in such a manner that the reflective face of the marker is perpendicular to a line parallel to the roadway centerline. Pavement markers shall not be placed over longitudinal or transverse joints of the pavement surface. Pavement markers shall not be placed on new asphalt concrete pavement or seal coats until they have been open to public traffic for a period of not less than 7 days when hot-melt bituminous adhesive is used, or not less than 14 days when epoxy adhesive is used.

Adhesives shall be placed uniformly on the pavement surface or on the bottom of the marker in a quantity sufficient to result in complete coverage of the area of contact with the pavement marker with no voids present and with a slight excess after the marker has been pressed in place. Pavement markers shall be placed in position and pressure applied until firm contact is made with the pavement. When hot-melt bituminous adhesive is used, the markers shall be placed immediately after application of the adhesive. Excess adhesive around the edge of the marker, excess adhesive on the pavement, and adhesive on the exposed surfaces of the markers shall be immediately removed. Markers shall be protected against impact until the adhesive has hardened to the degree designated by the Engineer.

314-5.4 Pavement Recesses. Pavement recesses shall be located along the line or lines of new or existing stripes.

Pavement recesses shall be constructed in new or existing pavement. The method of recess construction shall be selected by the Contractor. Equipment for recess construction shall be power-operated, mechanical and capable of removing the pavement to the dimensions shown on the Plans or Standard Plans without damaging the underlying pavement.

Residue shall be removed from the roadbed by use of vacuum equipment. Residue from the removal operations shall neither flow across the pavement nor into gutters or other drainage facilities. Residue shall be removed from the pavement surface before the residue is blown by action of traffic or wind.

Pavement recesses shall not be constructed on existing structures.

314-5.5 Measurement. Retroreflective and non-reflective pavement markers will be measured by the number of each type of pavement marker placed on the pavement surface. Retroreflective pavement markers placed in a pavement recess will be measured by the number of each type of pavement marker.

314-5.6 Payment. Payment for retroreflective and non-reflective pavement markers placed on the pavement surface will be made at the Contract Unit Price for each type of pavement marker placed on the pavement surface.

Payment for retroreflective pavement markers placed in pavement recesses will be made at the Contract Unit Price for each pavement marker placed in a pavement recess.

PART 4

SECTION 400 - ALTERNATE ROCK PRODUCTS, ASPHALT CONCRETE, PORTLAND CEMENT CONCRETE AND UNTREATED BASE MATERIAL

(This subsection shall apply only when Alternate Rock Material-Type S is specified)

400-1 ROCK PRODUCTS.

400-1.1 Requirements.

400-1.1.1 General. The following specifications set forth the requirements for aggregates for asphalt concrete, portland cement concrete, and untreated base material.

All rock products shall be clean, hard, sound, durable, uniform in quality, and free of any detrimental quantity of soft, friable, thin, elongated or laminated pieces, disintegrated material, organic matter, oil, alkali, or other deleterious substance.

The weight loss, as determined by ASTM C131, shall not exceed 15 percent during 100 revolutions nor 52 percent during 500 revolutions.

Specified gradations represent the limits which determine the suitability of aggregate for use. Actual gradations shall be uniformly graded from coarse through fine, remaining proportionately distant from these limits.

Coarse aggregate is material retained on the No. 4 (4.75mm) sieve and fine aggregate is material passing the No. 4 (4.75mm) sieve.

The Contractor, at its expense, shall provide safe and satisfactory facilities for obtaining representative samples.

Materials may be sampled at any time until final formal acceptance of the Work.

400-1.1.2 Source. Before beginning portland cement concrete and asphalt concrete work, the Contractor shall submit the name of the supplier to the Engineer as specified in 2-5.3. The supplier shall have on file with the Agency mix designs for portland cement concrete conforming to 201-1.1.1, and asphalt concrete conforming to 203-6.2, when required by the Specifications.

The Contractor or supplier shall resubmit required information when any change is made.

400-1.1.3 Statistical Testing. Statistical testing shall conform to the following:

Whenever both individual test results and moving average requirements are specified, materials shall meet both requirements.

Individual samples tested prior to the first use of aggregates from each source, or prior to the first use of aggregates after any changes have been made in aggregate processing procedures, shall conform to the limits specified for the moving average.

Whenever the results of an individual test for any property of a material, other than concrete compressive strength, does not comply with the limits specified for an individual test and if the moving average would not comply with the limit specified for the moving average should the next test be of the same value as that of the test being considered, the production of that material shall be suspended until corrective changes have been made by the Contractor and tests indicate that the quality of the next material to be used in the work complies with that specified for the moving average.

Moving average shall be computed in accordance with 211-5.

400-2 UNTREATED BASE MATERIALS.

400-2.1 General.

400-2.1.1 Requirements. All requirements of 200-2 shall apply except as hereafter provided. When base material without further qualification is specified, the Contractor shall supply processed miscellaneous base.

400-2.2 Disintegrated Granite. Disintegrated granite shall conform to 200-2.7.

400-3 PORTLAND CEMENT CONCRETE. The provisions of 201-1 shall apply except that gradings A, B, and C may be used interchangeably. When no class is specified, Class 560-C-3250 (330-C-23) shall be used.

The concrete ball penetration test, California Test 533, may be used at the discretion of the Engineer if it is calibrated to the slump test for a given mix.

400-3.1 Coarse Aggregate. Coarse aggregate shall meet the requirements of 200-1.4 with the following modifications:

a) The soundness loss, when tested by California Test 214 shall not exceed 10 percent.
b) The minimum specific gravity of the No. 4 (4.75mm) material shall be 2.56.
c) The specific gravity requirement may be waived if the material is stockpiled and approved by the Engineer before use.

The grading requirements for course aggregate are shown in Table 400-3.1 (A) for each aggregate size.

TABLE 400-3.1 (A): PERCENTAGE PASSING

Sieve Sizes	Number 2		Number 3		Number 4
	Individual Test Results	Moving Average	Individual Test Results	Moving Average	Combined Average
2" (50mm)	100	100	-	-	-
1-1/2" (37.5mm)	89-100	91-100	100	100	-
1" (25mm)	X ± 18	X ± 12	89-100	91-100	-
3/4" (19mm)	0-16	0-14	X ± 18	X ± 12	100
3/8" (9.5mm)	0-5	0-4	X ± 18	X ± 12	90-100
No. 4 (4.73mm)	-	-	0-16	0-14	0-30
No. 8 (2.36mm)	-	-	0-5	0-4	0-10

In the above table the symbol "X" is the gradation that the Contractor proposes to furnish for the specific sieve size. The X value shall meet the gradation limits of Table 200-1.4 (B).

400-3.2 Fine Aggregate. The fine aggregate shall meet the requirements in 200-1.5. The relative mortar strength shall not be less than 95 percent. Fine aggregate for mortar and plaster shall conform to the gradation for mortar sand in 200-1.5.5.

Fine aggregate shall be graded within the following limits:

TABLE 400-3.2 (A): PERCENTAGE PASSING

Sieve Sizes	Individual Test Results	Moving Average
3/8" (9.5mm)	100	100
No. 4 (4.75mm)	95 - 100	96 - 100
No. 8 (2.36mm)	61 - 99	66 - 94
No. 16 (1.18mm)	X ± 11	X ± 8
No. 30 (600µm)	X ± 8	X ± 7
No.50 (300µm)	X ± 6	X ± 4
No. 100 (150µm)	1 - 11	3 - 9
No. 200 (75µm)	0 - 5	0 - 4

In the above table, the symbol "X" is the gradation that the Contractor proposes to furnish for the specific sieve size. The X value shall meet the gradation limits of Table 200-1.5.5 (A).

400-4 ASPHALT CONCRETE.

400-4.1 General. Asphalt concrete shall be a mixture of mineral aggregate and up to 15 percent reclaimed asphalt pavement (RAP) with paving asphalt conforming to 203-6.1 at a central mixing plant. RAP shall conform to 203-7.2.2, except the viscosity of RAP asphalt recovered in accordance with ASTM D1856 (Abson Recovery Method) will not be required. RAP asphalt content may be determined in accordance with California Test 382 except the aggregate correction factor shall not be applied.

This material will be designated by the type of asphalt concrete, class and grade, i.e., "III-B2-PG 64-10". Unless otherwise specified on the Plans or in the Special Provisions, III-B3-PG 64-10 shall be used. Asphalt concrete containing up to 15 percent RAP shall be identified by adding the suffix "RAP" to the class and grade, i.e., "III-B2-PG 64-10 RAP".

Acceptance of asphalt concrete shall conform to 203-6.4.1.

400-4.2 Materials.

400-4.2.1 Asphalt. The asphalt to be mixed with the mineral aggregate shall be asphalt binder conforming to 203-1 or 203-2.

400-4.2.2 Aggregate. Coarse and fine aggregate shall consist of any one or a mixture of the following materials:

a) Broken or crushed stone, or crushed gravel.

b) Natural material having sufficient roughness to meet the specified stabilometer requirements when confined within the limits specified for grading in 400-4.3.

Each type of aggregate shall be fed to the dryer at a rate within 10 percent of the amount set. Adequate means shall be provided for controlling and checking the accuracy of the feeder. The aggregate shall conform to the requirements of 400-1.1.

The percentage of crushed particles will be determined by California Test 205.

400-4.2.3 Coarse Aggregate. Coarse aggregate shall consist of material of which at least 25 percent by weight shall be crushed particles.

400-4.2.4 Fine Aggregate. Fine aggregate shall consist of material containing not less than 20 percent by weight of crushed particles in the portion passing the No. 4 (4.75mm) sieve and retained on the No. 8 (2.36mm) sieve. The remainder may consist of natural fine aggregate.

If the fine aggregate is deficient in material passing the No. 200 (75µm) sieve, a mineral filler conforming to the requirements of 203-6.3.3 shall be added to meet the combined grading.

400-4.3 Combined Aggregates. The combined aggregate for asphalt concrete sampled after completion of all processing except adding of asphalt and mineral filler shall conform to Table 400-4.3 (A).

TABLE 400-4.3 (A)

Tests	California Test	Requirements
Both Kc and Kf-Factors (obtained from Centrifuge Kerosene Equivalent Test)	303	1.7 Max.
Sand Equivalent	217	45

The combined aggregate for asphalt concrete shall also conform to Table 400-4.3 (B) when mixed with the amount of asphalt determined to be optimum by California Test 367, which shall in no case be less than the minimum asphalt percentage specified in Table 400-4.3 (C) for the class specified.

TABLE 400-4.3 (B)

Tests	California Test	Requirements
Stabilometer Value*	366	35 Min.
Moisture Vapor Susceptibility (Stabilometer Value)	307	25 Min.
Swell	305	30 mil (760μm) Max.

* S-values shall be as set forth in Section 203-6.2, Section 203-6.4.1, and Table 203-6.4.3 (A)

The grading of the combined aggregates shall conform to one of the gradings shown in Table 400-4.3 (C).

TABLE 400-4.3 (C): TYPE III ASPHALT CONCRETE

CLASS	B2		B3	
Sieve Size	Individual Test Result	Moving Average	Individual Test Result	Moving Average
1" (25.0mm)	100	100	100	100
3/4" (19.0mm)	87 - 100	90 - 100	90 - 100	95 - 100
3/8" (9.5mm)	50 - 80	60 - 75	60 - 84	65 - 80
No. 4 (4.75mm)	30 - 60	40 - 55	40 - 60	45 - 60
No. 8 (2.36mm)	22 - 44	27 - 40	24 - 50	30 - 45
No. 30 (600μm)	8 - 26	12 - 22	11 - 29	15 - 25
No. 200 (75μm)	1 - 8	3 - 6	1 - 9	3 - 7
Asphalt %	4.6 - 6.0		4.6 - 6.0	

CLASS	C2		C3	
Sieve Size	Individual Test Result	Moving Average	Individual Test Result	Moving Average
3/4" (19.0mm)	100	100	100	100
1/2" (12.5mm)	89 - 100	95 - 100	89 - 100	95 - 100
3/8" (9.5mm)	70 - 94	75 - 90	74 - 100	80 - 95
No. 4 (4.75mm)	44 - 72	50 - 67	50 - 78	55 - 72
No. 8 (2.36mm)	30 - 54	35 - 50	32 - 60	38 - 55
No. 30 (600μm)	10 - 34	15 - 30	14 - 38	18 - 33
No. 200 (75μm)	2 - 10	4 - 7	2 - 10	4 - 8
Asphalt %	4.8 - 6.5		4.6 - 7.0	

TABLE 400-4.3 (C): TYPE III ASPHALT CONCRETE *(Continued)*

CLASS	D	F
Sieve Sizes	**Combined Average**	**Combined Average**
1/2" (12.5mm)	100	–
3/8" (9.5mm)	95 - 100	100
No. 4 (4.75mm)	65 - 85	95 - 100
No. 8 (2.36mm)	50 - 70	70 - 80
No. 30 (600µm)	28 - 40	35 - 50
No. 200 (75µm)	5 - 14	7 - 16
Asphalt %	6.0 - 8.0	8.0 - 10.0

The statistics, the basis of which will be the long term record of the plant, will be kept for each batch plant. Evaluation of gradation test results shall conform to the provisions of 400-1.1.3.

400-4.4 Storing, Drying, and Screening Aggregates. Storing, drying, and screening aggregates shall conform to 203-6.5.

400-4.5 Proportioning, Mixing, Asphalt Concrete Storage, and Miscellaneous Requirements. Proportioning shall conform to the provisions of 203-6.6. Mixing shall conform to the provisions of 203-6.7. Asphalt concrete storage shall conform to 203-6.8. Miscellaneous requirements shall conform to 203-6.9.

PART 5

SYSTEM REHABILITATION

SECTION 500 - PIPELINE

500-1 PIPELINE REHABILITATION.

500-1.1 Requirements.

500-1.1.1 General. This subsection provides specifications for various pipeline system rehabilitation.

Section 500 does not address the structural capacity of any of the rehabilitation systems described herein nor their structural requirements. The method shall be capable of bridging cracks, holes, and joint displacements that have been determined not to require point repair. The type of rehabilitation materials and methods for a given Contract will be designated on the Plans and in the Specifications. Unless otherwise provided for in the Special Provisions, proof of meeting the chemical resistance and physical testing shall be submitted to the Engineer for approval in accordance with 2-5.3 as a submittal. The Agency may require testing of the materials and methods prior to commencement of the Work to verify manufacturing compliance with required quality control standards and that no damage occurred to the materials during shipment to the Work site. At the time of installation, materials shall not be more than 6 months old from the date of manufacture. Material safety data sheets (MSDS) shall be available at the Work site.

500-1.1.2 Submittals. Prior to rehabilitation, the Contractor shall submit working drawings of construction details and all other submittals per 2-5.3. The working drawings shall include the location, method of rehabilitation and, when applicable, any bypass locations with sufficient detail to assure that the work can be accomplished without sewage spill. All submittals required by these specifications shall meet the requirements as shown on the Plans and in the Specifications.

500-1.1.3 Storage and Handling. Liner pipes and rehabilitation materials shall be properly stored and handled to prevent damage in accordance with the manufacturer's recommendations and as approved by the Engineer. Damage is described as, but is not limited to, gouging, abrasion, flattening, cutting, puncturing, or ultra-violet (UV) degradation. Thorough inspection of the liner pipes and rehabilitation materials shall be performed prior to installation. Criteria for acceptance/rejection shall be per 500-1.1.8.

500-1.1.4 Cleaning and Preliminary Inspection. Pipeline cleaning shall be performed prior to closed circuit television (CCTV) inspection and rehabilitation. The Contractor shall protect the manholes to withstand forces generated by equipment, water, and air pressure. After cleaning, the Contractor shall also confirm the inside minimum and maximum size (diameter and/or configuration) of the pipeline. The Contractor shall be responsible for the removal of debris from the pipeline and restore the pipeline to a minimum of 95 percent of the original diameter or area, as shown on the Plans or in the Special Provisions. Pipeline debris is described as, but is not limited to, sludge, dirt, sand, rocks, grease, roots, and other solid or semisolid materials.

Some pipeline cleaning methods available are listed herein. When utilizing high-velocity hydraulic cleaning equipment independently or in combination with other cleaning methods, it is recommended that a minimum of 2 passes with the hydraulic nozzle shall be done unless otherwise approved by the Engineer. Root cutters and porcupines shall be attached to the winches if so specified in the Special Provisions or directed by the Engineer. The Contractor shall be responsible for conducting a site inspection of each pipeline prior to rehabilitation to determine which cleaning methods are to be used. These methods shall be submitted to the Engineer for approval in accordance with 2-5.3.

a) **Hydraulically Propelled Equipment.** The equipment shall be a movable-dam type and be constructed in such a way that a portion of the dam may be collapsed at any time during the cleaning operation to prevent flooding of the sewer. The movable dam shall be equal in size to the pipeline being cleaned and provide a flexible scraper around the outer periphery to ensure removal of grease and other debris. If sewer cleaning balls or other equipment which cannot be collapsed are used, special precautions to prevent flooding of the sewers and public or private property shall be taken.

b) **High-Velocity Hydraulic (Hydro-Cleaning) Equipment.** All high-velocity hydraulic cleaning equipment shall carry a water tank, auxiliary engines, pumps, and a hydraulically driven hose reel. The equipment shall have a selection of two or more high velocity nozzles capable of producing a scouring action from 15 to 45 degrees in all size lines designated to be cleaned. The cleaning units shall have high-velocity nozzles for washing and scouring manhole walls and floors. The nozzles shall be capable of producing flows from a fine spray to a solid stream.

c) **Mechanically Powered Equipment.** Bucket machines shall be used in pairs with sufficient power to perform the work in an efficient manner. Machines shall be belt operated or have an overload shutoff device. Machines with a direct drive that could cause damage to the pipe will not be allowed. Bucket machines shall not be used on any host or rehabilitated pipeline that is lined with a plastic pipe or material. A power rodding machine shall be either a sectional or continuous-rod type capable of holding a minimum of 750 feet (230m) of rod. The machine shall be fully enclosed and have an automatic safety clutch or relief valve.

For segmented liner systems 27 inches (675mm) and larger, a standard test section of liner pipe or mandrel shall be inserted prior to sliplining. The mandrel shall have a segment length equal to that of the liner pipe. The outside diameter of the mandrel shall be a minimum of one percent greater than the outside diameter of the liner pipe but shall not exceed 1/2 inch (13mm) without prior approval of the Engineer. The equipment used by the Contractor to insert the test section or mandrel shall conform to Table 500-1.1.4 (A). A baffle plate shall be attached to the test section with adequate height to guarantee removal of any debris which could be present.

If cleaning cannot be completed from one manhole, the equipment shall be moved and set up on the other manhole and cleaning shall be re-attempted. If successful cleaning still cannot be performed or the equipment fails to traverse the entire pipeline section, it shall be assumed that a major blockage exists. Efforts to clean the lines shall be temporarily suspended and the Contractor shall notify the Engineer. Upon removal of the obstruction, the Contractor shall complete the cleaning operation.

The Contractor shall dispose of all debris removed from the pipeline, in accordance with current applicable regulations. Any hazardous waste material encountered during the Contract shall be considered as a changed condition.

TABLE 500-1.1.4 (A)

Nominal ID of Liner Pipe, inches (mm)	Minimum Equipment Insertion Force, Tons (kN) [1]
27 (675) to and including 60 (1500)	25 (220)
Over 60 (1500) to and including 84 (2100)	32.5 (290)
Over 84 (2100)	50 (440)

1. The equipment at the insertion pit shall be capable of withdrawing the test section or mandrel, if necessary.

500-1.1.5 Television Inspection. Closed circuit television (CCTV) inspection will be required prior to rehabilitation to document the condition of the host pipeline and to verify that it was cleaned per 500-1.1.4. A post-installation CCTV inspection shall be performed to determine if the work was completed per the Contract Documents and that all service connections have been re-instated, as required. All video inspections shall be recorded on a four-head VCR in VHS format, standard play mode. All original videotapes, log sheets, and reports shall be submitted to the Engineer and will become property of the Agency.

CCTV equipment shall include television cameras, a television monitor, cables, power sources, and other equipment. Focal distance shall be adjustable through a range from 6 inches (150mm) to infinity. The remote-reading footage counter shall be accurate to less than 1 percent error over the length of the particular section of pipeline being inspected. This distance is measured from the centerline of the manhole to the centerline of the next manhole. The camera and television monitor shall produce a minimum 350 lines per inch (14 lines per mm) resolution. Telephones, radios, or other suitable means of communication shall be set up to ensure that adequate communication exists between members of the crew. The CCTV inspection system to be utilized on the project shall be approved by the Engineer prior to the work being performed. CCTV inspection for re-instating service connections shall be performed utilizing system b) or c).

CCTV inspection shall be performed utilizing one of the following video camera systems:

a) Remote-focus stationary lens cameras;

b) Rotating-lens cameras; or

c) Pan-and-tilt cameras.

The video camera shall be mounted on a skid, floatable raft system, or transporter based on the conditions of the pipeline to be televised.

The Contractor shall televise the pipeline during optimum low-flow level conditions, as pre-approved by the Engineer. The television camera utilized shall be specifically designed and constructed for sewer inspection. The camera shall be operative in 100 percent humidity conditions. Lighting for the camera shall minimize reflective glare. Lighting and picture quality shall be suitable to provide a clear, in-focus picture of the entire periphery of the pipeline for all conditions encountered during the work.

The camera shall be moved through the pipeline in a downstream direction at a uniform rate, stopping when necessary to ensure proper documentation of the sewer's condition, but in no case shall the television camera be pulled at a speed greater than 30 feet per minute (9m/min.). A clear picture shall be provided looking into each service connection for both the pre-lining and post-lining tapes. During the pre-rehabilitation video inspection, if the television camera will not pass through the entire pipeline section, the Contractor shall reset the equipment at the downstream manhole and attempt to inspect the section of pipe from the opposite direction. If the camera fails to pass through the entire section, it shall be assumed that an obstruction exists. Efforts to televise that section of pipe shall be temporarily suspended and the Contractor shall notify the Engineer. Upon removal of the obstruction, the Contractor shall complete the CCTV inspection.

If an obstruction is encountered during the post-rehabilitation or post-installation video inspection, the Contractor shall remove the obstruction by excavation, repair, or other means approved by the Engineer at the Contractor's expense, in order that television inspection may continue.

Documentation shall consist of a color, VHS-format videotape, log sheets, and a written report detailing the post-rehabilitation or post-installation condition of the pipeline and lateral connections/openings. The report shall note the time and date of video inspection, street name, upstream and downstream manhole, direction of view, direction of flow, surface material, pipeline length, pipe section length, pipe size, pipe material, lateral connections, video tape number, counter number, and a detailed logging of defects encountered. Any rejected work shall be repaired, then re-televised. If the quality of the video tape is deemed to be unacceptable by the Engineer, the pipeline shall be re-televised at no additional cost to the Agency. Additional Agency requirements for performing CCTV inspection will be noted on the Plans or in the Special Provisions.

500-1.1.6 Sampling, Testing, and Installation. All materials shall be sampled and tested in accordance with 4-1. Rehabilitation materials shall be tested in accordance with the requirements of 211-2 and conform to the requirements for the specified material, unless otherwise specified in the following subsection. Test methods, specifications, standards, and the required quality control procedures for testing and installation are listed in each subsection. The Contractor shall install only those pipeline rehabilitation system materials that specifically meet the criteria in the following subsections.

500-1.1.7 Miscellaneous.

a) **Service Connections.** The Contractor shall be responsible for locating all service connections and cleanouts. The Contractor shall provide written notification of work activities to all local users and provide interim sewer service, as specified by the Agency.

Service connections shall be re-established as quickly as possible, not to exceed 24 hours, after completion of each liner pipe installation. Services requiring bypasses to be provided by the Contractor will be identified in the Special Provisions. When the service connection is re-established, the invert of the service connection shall match the bottom of the reinstated service opening. The service opening shall be reinstated from a minimum of 95 percent to a maximum of 100 percent of the original service connection. The new edge shall be smooth and crack free with no loose or abraded material.

If the service connection is to be re-established by a remote control device, the Contractor shall have a fully operational backup device on site. If for any reason the Contractor is unable to remotely re-establish the service connections, the Contractor shall immediately re-establish each service connection by open excavation at no additional cost to the Agency.

b) **Segmented Liner Pipes.** The rehabilitation processes specified in 500-1.8, 500-1.11 and 500-1.12 may be accomplished while flow exists in the host pipeline, without diverting the flow or bypass pumping. The Contractor shall consider the effects of varying floor levels on the buoyancy calculations in accordance with 2.5.3. Obstructions, including, but not limited to, roots, large offset joints, rocks, or other debris that could prevent passage or cause damage to the liner sections shall be removed. For variable exterior profile wall liner pipe, the Contractor shall consider the inside surface smoothness of the host pipe to minimize damage to the liner. The existing pipe joints shall be repaired prior to installing the liner pipes. Liner pipes shall be inserted one section at a time through an access/insertion pit constructed above the existing sewer. When segmented liner pipe sections are inserted from 2 locations to a common point, a coupling device shall be provided that is pre-approved by the Engineer. The top of the existing host pipe exposed in the pit shall be evenly removed down to the springline level. Liner pipe sections shall be inserted spigot end first with the bell end trailing and a pushing force shall be applied to the pipe wall and to the inside of the bell, unless otherwise approved by the Engineer.

Before insertion of segmented liner pipe 27 inches (675mm) and larger, a standard test section of liner pipe or an approved mandrel shall be pushed/pulled through the section of pipe being rehabilitated in accordance with 500-1.1.4 to ensure that it has been properly cleaned and all obstructions removed. A jacking or pulling ring shall be used to distribute the push/pull forces uniformly against the bell end perimeter of the liner pipe. The calculated forces shall include the frictional forces of the liner pipe against the invert of the host pipe or the soffit of the host pipe when buoyant forces cause the liner pipe to float. A load-measuring device, approved by the Engineer, shall be used to measure the loads exerted on the liner pipe, so that the manufacturer's approved maximum loads specified in 500-1.8, 500-1.11 and 500-1.12 will not be exceeded.

c) **Access/Insertion Pits.** Upon completion of the rehabilitation process, and/or as directed by the Engineer, the access/insertion pit area shall be restored in accordance with working drawings approved by the Engineer in accordance with 2-5.3 and 7-9.

d) **Manhole Protection.** During the rehabilitation process, the Contractor shall protect the manholes to withstand the forces generated by equipment, water, and air pressures used while completing the rehabilitation installation.

e) **End Seals.** The beginning and end of the new pipe liner shall be sealed to the host pipe with an epoxy or other material. The epoxy or other material shall conform to 211-2 and be submitted for approval to the Engineer in conformance with 2-5.3. The approved epoxy or other material shall be compatible with the lining material and host pipe and shall provide a watertight seal. The finished liner shall protrude a minimum of 1 inch (25mm) and a maximum of 2 inches (50mm) into the manhole, unless otherwise shown on the Plans or specified in the Special Provisions. Liner material shall be cut smooth and parallel with the manhole wall. The interface between the host pipe and the pipe liner shall be sealed 360 degrees. When the pipe liner extends through the manhole it shall be sealed as shown on the Plans and as specified in the Special Provisions.

500-1.1.8 Rejection. If the Contractor has used any material or method that has not been approved by the Engineer, the Contractor shall, at its sole expense, remove the entire rehabilitated pipe and replace it with new pipe as directed by the Engineer. All damaged rehabilitation materials and pipe rejected by the Agency shall be promptly removed from the Work site at the Contractor's expense and disposed of in accordance with current applicable regulations.

500-1.1.9 Measurement and Payment. Pipeline cleaning and inspection, including CCTV inspection, will be paid for at the Contract Unit Price per linear foot (meter). If a separate Bid item is not included, then full compensation shall be considered to be included in the Bid price for the liner pipe and/or the pipeline point repair/replacement pipe.

Pipeline point repair/replacement and rehabilitation shall be measured along the longitudinal axis between the ends of the pipeline, as shown on the Plans, and shall not include the inside dimensions of structures.

The Contract Unit Price per linear foot (meter) or lump sum for pipeline point repair/replacement and rehabilitation shall be considered full compensation for furnishing and installing all fittings, connections, seals, and special work shown on the Plans and in the Specifications. Additionally, the Contract Unit Price shall include all labor, materials, and equipment required; removal of interfering portions of existing sewers, storm drains, and other improvements; closing or removing of abandoned pipelines and structures, if required CCTV inspection and/or leak testing; excavation of the trench and/or access/insertion pits; control of ground and surface waters; preparation of the subgrade; placing and joining of pipe, including any necessary annular space grouting; backfilling of the trench and/or access/insertion pits; temporary and/or permanent resurfacing; and all other work necessary for pipeline point repair/replacement and rehabilitation, complete and in place.

500-1.2 Pipeline Point Repair/Replacement.

500-1.2.1 General. This subsection specifies the point repair and/or replacement of host pipelines. The Contractor shall be responsible for repairing the pipeline where point repairs are identified on the Plans or in the Special Provisions prior to any rehabilitation. If this is not shown, it will constitute Extra Work per 3-3 when approved by the Engineer.

The work shall include verifying the location of the point repair and/or replacement through CCTV or person-entry inspection of the pipeline, locating all interfering utilities, excavation, dewatering, pipe repairs or replacement, backfilling, surface restoration, temporary flow bypassing, sewer dewatering, and traffic control.

500-1.2.2 Materials. The pipe and repair materials shall be the same as the host pipeline unless otherwise indicated, and shall comply with 207 for the type and class required.

500-1.2.3 Excavation. Trenching and excavation shall conform to 306.

500-1.2.4 Sewer Bypassing and Dewatering. When required by the Contract Documents or the process, the Contractor shall bypass the sewer flow around the work and dewater the work area, in accordance with 7-8.5 and 306-3.3.

500-1.2.5 Notification of Work. The Contractor shall notify the Engineer a minimum of 48 hours in advance of the planned time to begin pipeline point repair work/replacement at a particular location.

500-1.2.6 Installation and Field Inspection. The installation of the replacement pipe and/or repair work shall conform to 306. Prior to the demobilization of the Contractor performing open-excavation repairs, the post-cleaning video tape(s) shall be submitted to the Engineer. The results of the post-cleaning video tape(s) may indicate the need for additional excavation and repair prior to lining. The Contractor shall review all the post-cleaning tape(s) and identify any additional point repair, which impacts the placement of the liner and the reinstatement of the service connection(s) and shall provide these locations in writing to the Engineer. All pipeline point repairs/replacement shall be inspected and measured by the Engineer prior to any backfilling and compaction and leak testing/CCTV inspection prior to placing of permanent resurfacing.

500-1.3 HDPE Solid-Wall Pipe Liner.

500-1.3.1 General. HDPE solid-wall liner pipe for use in sanitary sewers, storm drains, and house connection sewers shall comply with ASTM D3350 and ASTM F714. Fittings shall comply with ASTM D2683 or D3261. Fittings fabricated by mitered, butt fusions are also permitted.

500-1.3.2 Material Composition. Pipe and fittings shall be made from HDPE compounds conforming with ASTM D3350, Cell Classification 345434C, D, and E and shall also meet the requirements of 207-19.2.

500-1.3.3 Liner Pipe Acceptance. Liner pipe acceptance shall conform to 207-19.3.

500-1.3.4 Marking. Liner pipe marking shall conform to 207-19.4.

500-1.3.5 Chemical Resistance and Physical Testing. The HDPE liner pipe shall conform to 207-19.5

500-1.3.6 Installation and Field Inspection. The HDPE liner pipe shall conform to 500-1.1 for the cleaning and inspection of the host pipeline; preparation of entry points as needed; and the storage, handling, and joining of HDPE pipe. A proofing pig shall be pulled through the host pipeline prior to liner insertion to verify adequate clearances.

500-1.3.7 Annular Space Grouting. The entire annular space shall be fully grouted. The maximum safe annular grouting pressure in psig for single-stage or multi-stage grouting shall not exceed the values shown in Table 500-1.3.7 (A).

TABLE 500-1.3.7 (A)
DIFFERENTIAL-PRESSURE (VACUUM OR EXTERNAL FLUID) CAPABILITY FOR UNSUPPORTED PIPE AT 73.4°F (23°C) [1]

SDR	kPa (psi)
32.5	4 (28)
26	8 (55)
21	16 (110)
19	21 (145)
17	28 (193)
15.5	36 (248)

1. Safety factor not included.

500-1.3.8 Service Connections and End Seals. The Contractor shall be responsible for locating all service laterals and cleanouts. Service connections shall not be made until the liner pipe has stabilized, which is normally accomplished after a 24-hour waiting period. Service laterals shall be connected to the liner pipe by use of a heat-fused saddle or mechanical saddle as approved by the Engineer.

500-1.3.9 Repair and Rejection. Liner pipe may be repaired for minor superficial pipe damage. Damaged liner pipe which has been penetrated over 10 percent of the wall thickness at either the inner or outer wall surface, shall be repaired by cutting out the damaged section and replacing it with new pipe. All repair methods shall be submitted to the Engineer for prior approval in accordance with 2-5.3. The remaining liner pipe sections shall be a minimum of 8 feet (2.4m) in length. Liner pipes shall be inspected for damage immediately prior to installation. If liner pipe is found to be superficially damaged, the Engineer may allow the pipe to be repaired or may reject it. Rejected liner pipe shall be replaced with a new section of liner pipe.

500-1.4 Cured-In-Place Pipe Liner.

500-1.4.1 General. CIPP liner for the rehabilitation of pipelines shall be either the Type A - inversion process in compliance with ASTM F1216 or the Type B - pull-in-place process in compliance with ASTM F1743 for installation using heated-water cure. The CIPP liner shall use an approved epoxy or epoxy-vinyl ester-resin-impregnated flexible fabric tube. The tube is installed by an inversion method using a hydrostatic head or by pulling it through an existing pipe and inflating by inverting a membrane using a hydrostatic head.

500-1.4.2 Material Composition and Testing. The fabric tube shall consist of one or more layers of flexible, needled felt or an equivalent nonwoven material and have plastic coating(s). The material shall be compatible with and capable of carrying epoxy or epoxy-vinyl-ester resin, be able to withstand installation pressures and curing temperatures, and be compatible with the approved resins used. The approved epoxy or epoxy-vinyl-ester resin shall be compatible with the application and pipeline environment and be able to cure in the presence of water. The initiation temperature for cure shall be as recommended by the resin manufacturer and approved by the Engineer. The CIPP liner shall comply with ASTM D5813 and shall have, as a minimum, the initial structural properties per Table 500-1.4.2 (A).

TABLE 500-1.4.2 (A)

Epoxy Resin Properties	ASTM Test Method[1]	Initial Values psi (MPa)
Flexural Strength	D 790	5,000 (34.5)
Flexural Modulus	D 790	300,000 (2068)
Tensile Strength	D 638	4,000 (27.6)
Tensile Modulus	D 638	250,000 (1724)
Epoxy-Vinyl-Ester Resin Properties	**ASTM Test Method**	**Initial Values psi (MPa)**
Flexural Strength	D 790	4,500 (31.0)
Flexural Modulus	D 790	250,000 (1724)
Tensile Strength	D 638	3,000 (21.0)
Tensile Modulus	D 638	250,000 (1724)

1. The initial values are determined by ASTM D638 and D790.

The Contractor shall provide field-cured samples as directed by the Engineer and as specified in the Special Provisions. The physical properties of the finished CIPP shall be verified through a field-sampling procedure in accordance with ASTM F1216 or ASTM F1743 and in accordance with ASTM D5813.

500-1.4.3 Resin and Tube Acceptance. At the time of resin impregnation, the entire fabric tube shall be inspected for defects. The resin shall not contain fillers, except those required for viscosity control, fire retardance, or extension of pot life. Thixotropic agents that do not interfere with visual inspection may be added for viscosity control. Also, the opacity of the plastic coating shall not interfere with visual inspection. Resins may contain pigments, dyes, or colors that do not interfere with visual inspection of the CIPP liner or its required properties. Additives may be incorporated that enhance the physical and/or chemical resistance.

500-1.4.4 Chemical Resistance. The CIPP liner system shall conform to 211-2 and to the weight change requirement of Table 210-2.4.1(A).

500-1.4.5 Installation. The host pipeline shall be cleaned and televised in accordance with 500-1.1.4 and 500-1.1.5. The OD of the tube being installed shall be properly sized to allow for expansion so that the CIPP can fit tightly against the existing pipe.

The CIPP shall be installed in accordance with ASTM F1216 or ASTM F1743 and the Contractor's recommendations as approved by the Engineer. Immediately prior to installation, the CIPP liner tube shall be saturated with resin (on or off the Work site) and stored/transported at a cool temperature as recommended by the resin manufacturer.

500-1.4.6 Curing. After tube placement is completed, a suitable heat source and distribution equipment shall be provided by the Contractor to distribute or recirculate hot water throughout the installed CIPP liner tube. Temperature shall be maintained during the curing period as recommended by the resin manufacturer and approved by the Engineer. After the tube is cured, a cool-down period shall be used prior to opening the downstream end, reconnection of services, and returning normal flow back into the system. Heat curing of the resin shall occur within the manufacturer's approved recommended time frame (pot life). The water in the CIPP shall be cooled to below 100°F (38°C) before discharge.

500-1.4.7 Service Connections and End Seals. After the curing is complete, existing service connections shall be re-established. This may be done without excavation by means of a remote-control cutting device operating within small diameter pipe. A CCTV camera shall be attached to the cutting device for precise location of service connections and inspection of the CIPP liner.

500-1.4.8 Repair and Rejection. Internal and external repairs may be made to CIPP liner pipe in accordance with the manufacturer's recommendations and approval by the Engineer. Internal repairs may be made with approved fabric and epoxy or epoxy-vinyl-ester resins to restore strength and integrity. External repairs may be made by using standard plastic pipe repair techniques, including replacement of the damaged section using PVC pipe coupled to the CIPP liner, as approved by the Engineer.

500-1.5 PVC Pipe Lining System.

500-1.5.1 General. PVC profile extrusions with annular space grouting shall be installed for use in sanitary sewers and storm drains. This applies to the rehabilitation of small-diameter pipe and person-entry pipe (36 inches (900mm) and larger) or conduits in terms of materials and installations.

500-1.5.2 Material Composition. The material shall be made from unplasticized PVC compounds conforming to 207-17, having a cell classification of 12334, 12454, or 13354 as defined in ASTM D1784.

500-1.5.3 Material and Equipment Acceptance. At the time of manufacture, each lot of plastic strips shall be inspected for defects and the physical properties certified in accordance with the ASTM Standards listed in this subsection, or as indicated in the Special Provisions. There are 2 strips of PVC used in this process. The former strip is a ribbed panel which varies in width and height as a function of pipe diameter. The joiner strip is a "U"-shaped strip of PVC which is used to lock together the former strip edges as the PVC strips or panels are being spirally wound upon themselves. The minimum thickness of the strips and panels shall be per Table 500-1.5.3 (A).

TABLE 500-1.5.3 (A)

Nominal ID of Original Pipe inches (mm)	Minimum Thickness		Minimum Profile Height mils (mm)
	Former Strip mils (mm)	Joiner Strip mils (mm)	
8 to 12 (200 to 300)	25 (0.64)	25 (0.64)	192 (4.88)
15 to 18 (375 to 400)	30 (0.75)	31 (0.79)	242 (6.15)
24 to 36 (600 to 900)	45 (1.15)	58 (1.48)	480 (12.20)
30 to 72 (750 to 1800) [1]	60 (1.53)	–	488 (12.40)

1. In some lining applications for pipes and conduits 30 to 36 inches (750 to 900mm) in diameter, it may be determined to use person-entry techniques.

The initial stiffness factor shall conform to Table 500-1.5.3 (B).

TABLE 500-1.5.3 (B)

Nominal ID of Original Pipe[1] inches (mm)	Stiffness Factor (EI)[(2)] in^3 - lbf/in^2 ($Pa \cdot m^3$)
8 (200)	120 (14)
10 (250)	120 (14)
10 (250)	240 (27)
12 (300)	240 (27)
15 (375)	240 (27)
15 (375)	600 (68)
18 (450)	600 (68)
24 (600)	600 (68)
24 (600)	1600 (181)
30 (750)	1600 (181)
36 (900)	1600 (181)

1. For ID's larger than 36 inches (900mm), see Plans or Special Provisions.
2. Stiffness factors shall be determined in accordance with ASTM D2412. $EI = 0.149R^3 (PS)$.

At the time of delivery, the strips shall be homogeneous throughout, uniform in color, and free of cracks, holes, foreign materials, blisters, or other deleterious faults. For testing purposes, a lot is defined as production during an 8-hour shift, while a batch is defined as each 200 linear feet (60 linear meters) of PVC product. Testing shall be performed every 2 hours and records kept on file in conformance with 500-1.5.4. The Contractor shall furnish and maintain in good condition all necessary equipment required for the proper execution and inspection of the work.

500-1.5.4 Marking. Each PVC continuous strip on each reel shall be distinctively marked on its inside surface end with a coded number which identifies the manufacturer, strip thickness, minimum profile height, size, material, machine, date, and shift on which the material was extruded. These markings shall also appear on the PVC strips with a maximum distance between markings of 5 feet (1.5m), and shall be visible from inside the completed liner.

500-1.5.5 Chemical Resistance. PVC and cured sealant/adhesive shall be tested in accordance with the requirements of 211-2 and conform to the weight change requirements of Table 207-17.5(A).

500-1.5.6 Installation and Field Inspection. The host pipeline shall be cleaned of any obstructions and televised per 500-1.1.4 and 500-1.1.5. The condition shall be approved by the Engineer prior to insertion of the liner. The plastic strips or panels shall be handled with care to ensure that the plastic is not kinked, gouged, or otherwise damaged.

The former and joiner strips shall be engaged and an approved sealant/adhesive shall be injected onto the engaged locks. The Contractor shall ensure that the joiner strip is continuously engaged.

For person-entry pipe, the PVC panels shall be cut and trimmed to fit as near as practical to the internal perimeter of the existing conduit. A bead of approved sealant adhesive shall be applied to the female locking edge of the former strip. End joins shall be made with the plasticized end section, which shall overlap the joint by not less than 4 inches (100mm). End joins shall be staggered and shall remain below the normal flowline of the sewer.

500-1.5.7 Annular Space Grouting. For small-diameter pipe, the annular space between the outside of the liner and the inside of the existing pipe shall be grouted. Grouting of the annular space shall be performed in such a manner to prevent damage or collapse of the liner. Grout shall be pumped into the annular space at manholes, service connections, and wherever the liner is exposed.

Grout shall conform to 500-1.3.7.

For person-entry pipe, the grout shall be injected behind the liner by tubes placed on top of the liner or holes drilled through the liner. Any holes in the plastic shall be covered with a patch of similar material as approved by the Engineer.

500-1.5.8 Service Connections and End Seals. Service lateral connections shall be re-established with the liner in accordance with manufacturer's recommendations as approved by the Engineer.

500-1.5.9 Repair and Rejection. Prior to installation, the PVC shall be inspected for flaws such as cracks, blisters, scratches, blemishes, and other faults. Material rejected for any reason shall be replaced prior to installation. If after installation, flaws of a deleterious nature are detected, they shall be corrected in a manner suitable to the supplier and approved by the Engineer. If flaws are correctable by approved splicing or patching methods, that work shall be completed promptly. If approved corrections cannot be made, lining shall be removed from the pipe utilizing new PVC pipe.

500-1.6 (Not Used).

500-1.7 Deformed/Re-formed HDPE Pipe Liner.

500-1.7.1 General. Deformed HDPE extrusions for rehabilitating sanitary sewers and storm drains without excavation shall comply with ASTM D3350 and ASTM F714. This method applies to the rehabilitation of 4 through 18 inches (100 through 450mm) diameter pipe. Unless otherwise specified, liner for pipe shall have a minimum SDR of 32.5. This rehabilitation system may be capable of expanding up to 10 percent. Pipe stiffness shall conform to Table 500-1.7.1 (A).

TABLE 500-1.7.1 (A)[1]

Pipe Size Inches (mm)	Pipe Stiffness[2]		
	SDR 21	SDR 26	SDR 32.5
4 (100)	61	31	16
6 (150)	61	31	16
8 (200)	61	31	16
10 (250)	61	31	16
12 (300)	61	31	16
15 (375)	61	31	16
18 (400)	61	31	16

1. Minimum pipe stiffness (PS) when tested in accordance with ASTM D2412.
2. PS values are from ASTM F174 Table X1.1

500-1.7.2 Material Composition. Pipe shall be made from HDPE compound complying with ASTM D3350, cell classification 345434C, D, or E and shall also meet the requirements of 207-19.2, except that titanium dioxide pigment may be substituted for the 2 percent carbon black.

The Contractor shall provide the manufacturer's certified test results to the Engineer for approval, stating that the material conforms with the applicable requirements, including crystallization temperatures.

500-1.7.3 Material and Equipment Acceptance. Material and equipment acceptance shall conform to 207-19.3.

500-1.7.4 Marking. Marking shall conform to 207-19.4, except that the material shall be designated by HDPE cell classification.

500-1.7.5 Chemical Resistance and Physical Testing. HDPE pipe specimens shall be tested in accordance with the requirements of 211-2 and conform to 207-19.5, except the requirements shall be met with samples from pipes that have been subjected to the deformation and reforming process.

500-1.7.6 Installation and Field Inspection. HDPE pipe shall be installed as follows:

a) The existing pipeline shall be cleaned and televised per 500-1.1.4 and 500-1.1.5, and the condition approved by the Engineer prior to the insertion of the deformed pipe.

b) A cable shall be strung through the host pipe to be rehabilitated and attached to the deformed pipe through an existing manhole or access point. The pipe shall be pulled through the existing conduit by this cable. Pulling forces shall not exceed the axial strain limits of the deformed pipe. The measured pulling operation limits the pulling force to an allowable tensile stress (1,500 psi (10.3MPa) or 50 percent of the yield) times the pipe wall cross-sectional area. Care shall be taken not to damage the deformed pipe during installation. Appropriate sleeves and rollers shall be used to protect the pipe. Calculations for pulling force limits shall be submitted to the Engineer in accordance with 2-5.3.

c) When the deformed pipe is in place, the pipe shall be cut and the processing manifolds (pipe end-closing assembly used for heat and pressure control within liner) shall be inserted and secured at both pipe ends. The temperature and pressure measuring instruments shall be attached to the deformed pipe at both ends.

d) After the deformed HDPE liner is outfitted with temperature and pressure instruments, steam shall be introduced into the system until a minimum temperature of 226°F (108°C) to a maximum temperature of 244°F (118°C) is reached and shall not exceed the melting temperature of 260°F (127°C). The minimum outside pipe temperature at the terminating end of the pipe shall be 185°F (85°C). This temperature shall be held for a minimum of 20 minutes. The deformed pipe shall be pressurized up to 14.5 psig (100 kPa gage), maximum, while the termination point valves are kept open to provide heat flow. The pressure shall then be increased in increments up to a maximum of 26 psig (179 kPa gage).

e) The Contractor shall cool the re-formed pipe according to the approved manufacturer recommendations. When the temperature reduces to 100°F (38°C), the Contractor shall then slowly raise the pressure to approximately 33 psig (228 kPa gage), while applying air or water for continued cooling. The equipment shall be disconnected after ambient temperature is attained.

f) Temperatures and pressures shall be monitored and recorded throughout the installation process to ensure that each phase of the process is achieved at the approved manufacturer's recommended temperature and pressure levels.

g) If the testing of the installed HDPE liner pipe is required in the Special Provisions, the physical properties of the installed HDPE liner pipe shall be verified through field sampling and laboratory testing, all as approved by the Engineer. Unless the Special Provisions call for more than one sample, a sample shall be cut from a section of reformed/rerounded HDPE liner pipe at the upstream, downstream, or an intermediate manhole/access pit that has been inserted through a same diameter pipe acting as a mold. HDPE liner pipe samples shall be submitted to a certified laboratory which has been pre-approved by the Engineer and tested in accordance with ASTM D638 and ASTM D790 to confirm that the liner pipe conforms to the minimum tensile and elongation requirements per 500-1.7.2. All costs incurred for this testing shall be borne by the Contractor.

500-1.7.7 Service Connections and End Seals. Existing service connections shall be reinstated through the use of a remote control unit or excavation. Service connections and end seals shall conform to 500-1.1.7.

The beginning and end of the new HDPE pipe liner shall be sealed to the rehabilitated host pipeline. If sealing material is required, it shall be compatible with the HDPE pipe and shall provide a watertight seal.

500-1.7.8 Repair and Rejection. The Contractor shall provide an evaluation and repair specification to the Engineer for approval for liner pipe found to be damaged during or after installation. Any liner pipe damaged in transit or on the Work site prior to installation will be rejected and shall be immediately removed from the Work site.

500-1.8 CCFRPM Liner Pipe.

500-1.8.1 General. CCFRPM liner for use in lining sanitary sewers shall comply with ASTM D3262. Unless otherwise indicated, the minimum pipe stiffness shall be 18 psi (124 kPa), or greater, as tested in accordance with ASTM D2412.

500-1.8.2 Material Composition. The amount, location, and orientation of the chopped glass-fiber reinforcement shall be specifically designed for each application. The glass shall be a commercial grade of E-Type glass fibers with a finish compatible with the resin used. The sand shall be a minimum 98 percent silica kiln-dried and graded. The polyester wall resin shall be an isophthalic, orthophthalic or other approved resin with a minimum tensile elongation of 2 percent. A vinyl ester liner resin shall be used to meet the chemical resistance requirements of 211-2 and conform to 207-20.5. Designation per ASTM D3262 shall be Type 1, Liner 2, Grade 3, and a minimum pipe stiffness of 18 psi (124 kPa), unless a higher value is indicated on the Plans or in the Special Provisions. Elastomeric sealing gaskets shall conform to the requirements of ASTM F477.

500-1.8.3 Liner Pipe Acceptance. The liner pipe shall be free of cracks, holes, delaminations, foreign inclusions, blisters or other defects that would, due to their nature, degree, or extent, have a deleterious effect on the pipe performance as determined by the Engineer. Prior to installation, damaged pipe shall be either repaired or field cut to remove the damaged portion as approved by the Engineer.

For testing purposes, a production lot shall consist of all liner pipes having the same lot marking number, but shall not exceed a total of 50 pipes. Pipe length, wall thickness, joint dimensions, pipe stiffness, and deflection characteristics shall be verified by testing for each lot in accordance with ASTM D3262.

500-1.8.4 Marking. Each pipe section shall be marked on the inside and every 5 feet (1.5m) on the outside, to show the manufacturer's name, manufacturing number (identifies factory location, date, shift, and sequence), nominal diameter, pipe stiffness, ASTM D3262 and designation, and lot number.

500-1.8.5 Chemical Resistance and Physical Testing. Pipe liners and gaskets shall be tested in accordance with 211-2 and conform to 208-4 and 207-20.5 respectively. Verification shall be provided that physical testing of the product conforms to ASTM D3262 (qualification test only) and ASTM D2412.

500-1.8.6 Installation and Field Inspection. The existing sewer shall be maintained in operation during the relining process. The existing host pipeline shall be cleaned of any obstructions and televised per 500-1.1.4 and 500-1.1.5. Liner pipes shall be inserted one section at a time through an access pit constructed above the existing sewer. The top of the existing sewer exposed in the pit should be removed down to springline level (halfway). Liner pipes shall be inserted in accordance with 500-1.1.7 b). The pushing force shall be applied to the pipe wall end inside of the bell. Maximum jacking load shall not exceed Table 500-1.8.6 (A).

The pipe installation shall conform to 500-1.1.7 b). Maximum jacking loads shall not exceed the values shown in Table 500-1.8.6 (A).

TABLE 500-1.8.6 (A)

Nominal Diameter inches (mm)	SAFE AXIAL COMPRESSIVE LOAD Tons (kN) [1]	
	18 psi (124 kPa) Pipe Stiffness Tons (kN)	36 psi (249 kPa) Pipe Stiffness Tons (kN)
18 (450)	—	17 (151)
20 (500)	—	19 (169)
24 (600)	23 (205)	31 (276)
30 (750)	36 (320)	52 (463)
36 (900)	52 (463)	66 (587)
42 (1050)	71 (632)	95 (845)
48 (1200)	93 (827)	130 (1157)
54 (1350)	123 (1094)	171 (1521)
60 (1500)	155 (1379)	212 (1886)
66 (1650)	175 (1557)	240 (2135)
72 (1800)	209 (1859)	292 (2598)
78 (1950)	252 (2242)	351 (3123)
84 (2100)	307 (2731)	422 (3754)
90 (2250)	357 (3176)	485 (4315)
96 (2400)	411 (3656)	560 (4982)
102 (2550)	470 (4181)	640 (5694)

1. Factor of safety of 2:1 is included for longitudinal compressive load.

500-1.8.7 Annular Space Grouting. The entire annular space between the outside of the liner and the inside of the host pipe shall be grouted, in accordance with 500-3. The minimum radial annular space shall not be less than 1 inch (25mm) unless approved by the Engineer. In accordance with 2-5.3, the grout mix and placement procedure shall be submitted to the Engineer for approval. Grouting of the annular space shall be done in such a manner to prevent damage or collapse of the liner. Maximum safe grouting pressure is equal to the pipe stiffness divided by 3.

500-1.8.8 Service Connections and End Seals. Service connections and end seals shall conform to 500-1.1.7.

500-1.8.9 Repair and Rejection. Prior to installation, damaged pipe shall be field cut to remove the damaged portion and rejoined by approved methods. Superficial damage may be repaired without field cutting. Liner pipe gouges deeper than 50 percent of the vinyl ester lining shall be field cut and removed per 500-1.1.8.

500-1.9 External In-Place Wrap.

500-1.9.1 General. Existing sewer pipes experiencing crown corrosion may be rehabilitated utilizing a wrap of plastic liner with integral locking extensions followed by a cap of reinforcing steel and concrete.

Plastic liner sheet, weld strip, adhesive products and cleaners shall conform to 210-2. Prior to the plastic liner installation, the existing line shall be uncovered and the pipe exposed to accommodate the coverage shown on the Plans or stated in the Special Provisions. Liner shall be applied and secured to the host pipe and inspected and approved by the Engineer prior to placement of reinforcing steel and concrete.

500-1.9.2 Installer Qualifications. Applicators and welders shall be qualified in accordance with 311-1.2.

500-1.9.3 Preparation of Existing Pipe for Installation of Plastic Liner. The concrete surface shall be etched by sandblasting to develop a slightly granular surface. When permitted by the Engineer, the concrete surface may be acid etched in lieu of sandblasting. After sandblasting, the concrete surface shall be thoroughly cleaned of dust. Surfaces etched with acid shall be neutralized with clean water.

a) **Coverage.** The circumferential coverage shall be the upper 270 degrees unless otherwise indicated on the Plans or in the Special Provisions.

b) **Positioning Liner.** Liner installed on the existing pipe shall be positioned with the locking extensions outward and aligned with or perpendicular to the longitudinal axis of the pipe.

Liner shall be centered with respect to the field top centerline of the pipe. Liner shall be set to fit over the existing pipe joints with the field welded seams located away from the joint portion of the pipes. Liner shall be closely fitted to the existing pipe. Sheets shall be cut to fit curved and warped surfaces using the minimum number of separate pieces.

Prior to installation, the Contractor shall indicate to the Engineer the proposed layout of liner sheets, including the location and type of all field welds.

The Engineer may require working drawings per 2-5.3, the use of patterns, or the marking of sheet layouts directly on the existing pipe where complex or warped surfaces are involved.

At transverse joints between sheets of liner used along the pipeline, the space between ends or edges of locking extensions, measured longitudinally, shall not exceed 4 inches (100mm).

c) **Securing Liner in Place.** Liner shall be held snugly in place against the existing pipe by use of adhesive materials, in accordance with the liner manufacturer's written recommendations as approved by the Engineer.

Liner shall be bonded to the existing pipe a minimum of 6 inches (150mm) along both longitudinal bottom edges. This shall be accomplished by the application of an approved adhesive system.

500-1.9.4 Field Joining of Liner.

a) **General.** Liner joints shall be free of all foreign material and shall be clean and dry before joints are made. All field joints are to be made and tested prior to placement of reinforcing steel and concrete.

b) **Field Joints in Pipe Rehabilitation Installation.** Field joints in the liner plate shall be Type R-2 unless Type R-1 or R-3 is approved by the Engineer.

1) Type R-1 joint shall consist of a 2-inch (50mm) wide weld strip, centered over the 1 inch (25mm) maximum gap between sheets and securely welded along each edge of adjacent liner.

2) Type R-2 joints shall be made with an integral joint flap with locking extensions removed per 210-2.4.6, extending 1-1/2 inches ± 1/4 inch (38mm ± 6mm) beyond the end of the sheet. The sheet shall be overlapped not less than 1/2 inch (13mm) and the overlap secured to the adjacent liner by means of a 1 inch (25mm) welding strip. The downstream sheet shall overlap the upstream sheet.

3) Type R-3 joint shall consist of a 1 inch (25mm) wide weld strip centered over a 1/4 inch (6mm) maximum gap between sheets and secured along each edge of adjacent liner by means of a 1 inch (25mm) welding strip.

c) **Installation of Welding Strips.** Installation of welding strips shall be in accordance with 311-1.5.4.

500-1.9.5 Protection and Repair of Liner. Protection and repair of liner shall be in accordance with 311-1.9.5.

500-1.9.6 Field Testing. Field testing shall be in accordance with 311-1.10.

500-1.9.7 Steel Reinforcement. Before placing reinforcing steel, the Contractor shall submit a reinforcing steel placing plan in accordance with 2-5.3.

Reinforcing bars shall conform to 210-2 and be placed in accordance with 303-1.7. They shall be held in position by the use of concrete or plastic chairs. Metal chairs will not be allowed.

Caution shall be taken when installing reinforcing steel to ensure against puncturing or damaging the liner.

500-1.9.8 Concreting Operations.

a) **General Placement.** Concrete placed against the liner shall be carefully conveyed, deposited, and consolidated to avoid damage to the liner and to produce dense concrete, securely anchoring the locking extensions into the concrete. Vibrators shall be used to consolidate concrete with particular attention along the bottom edge of the liner.

b) **Forms.** The trench walls may serve as the outer form for the new concrete encasement. When outer forms are required, they shall be in accordance with 303-1.3.

500-1.10 Folded and Re-formed PVC Pipe Liner.

500-1.10.1 General. Folded and re-formed PVC liner pipe shall be inserted into sanitary sewers, force mains, and storm drains in order to rehabilitate the existing pipeline system without excavation.

500-1.10.2 Type A Folded and Re-formed PVC Pipe Liner.

a) **Scope.** This method applies to the rehabilitation of 4 through 15 inches (100 through 375mm) diameter pipe. The standard dimension ratio shall be SDR 35, 41, or 50 as specified in the Special Provisions. This rehabilitation system may be capable of expanding up to 10 percent. The initial pipe stiffness factor shall conform to Table 500-1.10.2 (A).

TABLE 500-1.10.2 (A)[2]

Nominal ID of Original Pipe inches (mm)	SDR	Stiffness Factor (EI)[1] in^3- lbf/in^2 ($Pa \cdot m^3$)
4 (100)	35	40 (4.5)
6 (150)	35	133 (15)
8 (200)	50	109 (12)
	41	198 (22)
	35	320 (36)
10 (250)	50	213 (24)
	41	388 (44)
	35	624 (70)
12 (300)	50	369 (42)
	41	671 (76)
	35	1076 (122)
15 (375)	50	720 (81)
	41	1307 (148)
	35	2105 (238)

1. Pipe Stiffness (PS) shall be determined in accordance with ASTM D2412. Stiffness factor is $EI = 0.149r^3$ (PS). The stiffness factors listed in the table are typical values for gravity flow conditions. For pressure applications the stiffness factors are usually higher.
2. Effects of ovality and safety factor are not included.

b) **Material Composition.** The folded pipe shall be made from unplasticized PVC compounds having a cell classification of 13223, as defined in ASTM D1784.

c) **Material and Equipment Acceptance.** At the time of manufacture, the extruded materials shall be inspected for defects and physical properties in accordance with ASTM F1504, to show compliance with 500-1.10.2 b), or as specified in the Special Provisions. Testing shall be performed once per shift, change in material batch or coil. A Certificate of Compliance shall be supplied per 4-1.5.

At the time of installation, the material shall be homogeneous and free of defects, cracks, holes, blisters, foreign materials, or other deleterious faults.

The Contractor shall furnish and maintain in good condition all equipment necessary for proper execution and inspection of the work.

d) **Marking.** Marking shall conform to 207-17.2.1, except that under Item 3, there is no ASTM standard for this product.

e) **Chemical Resistance and Physical Testing.** The PVC material shall be tested in accordance with 211-2 and conform to Table 500-1.10.2 (B). The various requirements shall be met with samples taken from pipe that has experienced the folding and re-forming process.

TABLE 500-1.10.2 (B)

Property	ASTM Test Method	Initial Values	Values After 112 Days Exposure
Tensile Strength psi (MPa)	D638	5,000 (34.5) min.	5,000 (34.5) min.
Impact Strength ft-lbs/inch (J/m)	D256 Method A Size 1/2 x 1/8 x 2-1/2 inches (12.7 x 3.175 x 63.5mm)	1.5 (80) of notch, min.	1.5 (80) of notch, min.
Weight Change %	D543		
Unconditioned			+ 1.5% max.
Conditioned			+ 1.0% max.

f) Installation and Field Inspection.

1) The existing pipeline shall be cleaned of any obstacles and televised per 500-1.1.4 and 500-1.1.5, and the host pipe condition shall be satisfactory to the Engineer prior to the insertion of the folded pipe.

2) If necessary, a flexible heat containment tube shall be permanently placed inside the existing pipe for retention of heat necessary to soften the folded pipe. A cable shall be strung through the heat containment tube.

3) Steam shall be applied to the folded pipe until pliable for a minimum of 15 minutes prior to insertion into the existing pipe. Once the material has become pliable, the cable shall be attached to the folded pipe. Using a winch at the termination point, the folded pipe shall then be inserted into the existing pipe through a manhole or access point. Pulling force shall not exceed 2,000 pounds (8.9kN).

4) After the folded PVC pipe is inserted into the existing pipe, it shall be cut off at the starting point and restrained at the terminating point. Thermocouples shall be placed on the exterior of the liner pipe at both the upstream and downstream ends for monitoring of the re-forming and cool-down process. Steam shall be introduced at the insertion end of the folded pipe until a minimum temperature of 150°F (66°C) is attained at the terminating end. This temperature shall be held for a minimum of 5 minutes and shall not exceed 240°F (115°C).

5) After the material has reached the required temperature, a specifically designed pressure driven rounding device shall be used to progressively round the folded PVC at a maximum rate of 5 feet per second (1.5 meters per second) using steam at 5 to 8 psig (34 to 55 kPa gage). The rounding process shall not cause any scraping, tearing, abrasion, movement, or other damage to the liner.

6) When the rounding process is complete, the steam shall be converted to air, maintaining an internal pressure of 5 to 12 psig (34 to 83 kPa gage). After the conversion to air pressure, water may be introduced until the system is completely filled. A minimum of 8 psig (55 kPa gage) air or water pressure shall be maintained until the system is cooled to at least 120°F (49°C) at both ends. At this point, the pressure shall be relieved and both ends shall be cut off in the manholes.

7) If testing of the installed PVC liner pipe is required in the Special Provisions, the physical properties of the installed PVC liner shall be verified through field sampling and laboratory testing as approved by the Engineer. Unless the Special Provisions call for more than one sample, a sample shall be cut from a section of reformed/re-rounded PVC liner pipe at the upstream, downstream, or an intermediate manhole/access pit that has been inserted through a same diameter pipe acting as a mold. PVC liner pipe samples shall be submitted to a certified laboratory, which has been pre-approved by the Engineer. The samples shall be tested in accordance with ASTM D638 for tensile strength, ASTM D790 for flexural modulus, and ASTM D2444 for impact resistance to confirm that the liner pipe conforms to the minimum requirements per 500-1.10.2 b). All costs incurred for the testing shall be borne by the Contractor.

500-1.10.3 Type B Folded and Re-formed PVC Pipe Liner.

a) **Scope.** This method applies to rehabilitation of 4 through 18 inches (100 through 450mm) diameter pipe. The standard dimension ratio may be SDR 26, 32.5, or 41 as specified in the Special Provisions. This rehabilitation system may be capable of expanding up to 10 percent. The initial pipe stiffness factor shall conform to Table 500-1.10.3 (A):

TABLE 500-1.10.3 (A)

Nominal ID of Original Pipe inch (mm)	CELL CLASS 12111	
	SDR	Stiffness Factor (EI)[1, 2] Pa - m^3 (in^3 - lbf/in^2)
4 (100)	32.5	23 (2.7)
	26	47 (5.3)
6 (150)	32.5	79 (9.0)
	26	153 (17)
8 (200)	32.5	187 (21)
	26	365 (42)
10 (250)	32.5	364 (42)
	26	711 (81)
12 (300)	32.5	630 (72)
	26	1228 (140)
15 (375)	32.5	1226 (140)
	26	2397 (274)
18 (450)	41	1057 (121)

1. Pipe Stiffness (PS) shall be determined in accordance with ASTM D2412. Stiffness factor is $EI = 0.149r^3$ (PS). The stiffness factors listed in the table are typical values for gravity flow conditions. For pressure applications, the stiffness factors are usually higher.
2. Effects of ovality and safety factor are not included.

b) **Material Composition.** The folded pipe shall be made from PVC compounds having a cell classification of 12111, as defined in ASTM D1784.

c) **Material and Equipment Acceptance.** At the time of manufacture, the extruded material shall be inspected for defects and physical properties in accordance with ASTM D7901, D2122, D2152, D2412, and F1057 to show compliance with 500-1.10.3 b) or as indicated in the Special Provisions. Testing shall be performed once per shift, change in material batch, or coil. A Certificate of Compliance shall be supplied per 4-1.5.

At the time of installation, the material shall be homogeneous and free of defects, cracks, holes, blisters, foreign materials, or other deleterious faults.

The Contractor shall furnish and maintain in good condition all equipment necessary for proper execution and inspection of the work.

d) **Marking.** Marking shall conform to 207-17.2.1, except that under Item 3 there is no ASTM standard for this product.

e) **Chemical Resistance and Physical Testing.** The PVC material shall be tested in accordance with 211-2 and conform to Table 500-1.10.2 (B), as modified in Table 500-1.10.3(B) and 210-2.3. The various requirements shall be met with samples taken from pipe that has experienced the folding and re-forming process.

TABLE 500-1.10.3 (B)

CELL CLASS 12111			
Property	**ASTM Test Method**	**Initial Values**	**Values After 112 Days Exposure**
Tensile Strength Yield, psi (MPa) min.	D638	3,500 (24.1) min.	3,500 (24.1) min.
Impact Strength Foot-lbs/inch (J/m) min.	D256 Method A Size 1/2 x 1/8 x 2-1/2 inches (12.7 x 3.75 x 63.5mm)	1.2 (64) of notch, min.	1.2 (64) of notch, min.
Weight Change (%) Unconditioned Conditioned	D 543		 ± 1.5% max. ± 1.5% max.

f) **Installation and Field Inspection.**

1) The existing pipeline shall be cleaned of any obstacles and televised per 500-1.1.4 and 500-1.1.5, and the host pipe condition shall be satisfactory to the Engineer prior to the insertion of the folded pipe.
2) Prior to insertion into the host pipe, heat may be applied to the folded pipe (while on the spool) until pliable. Once the material has become pliable, the pulling cable shall be attached to the insertion end of the folded pipe. Using a winch at the termination point, the folded pipe shall then be inserted through the existing pipe via a manhole or access point. Pulling force shall not exceed 2,000 pounds (8.9kN).
3) After the folded PVC pipe is pulled through the host pipe, it shall be cut off at the starting point and restrained at the terminating point. Thermocouples shall be placed on the exterior of the liner pipe at the downstream end for monitoring of the re-forming and cool-down process. Steam shall be introduced at the insertion end inside the folded pipe until a minimum temperature of 200°F (93°C) is attained at the manifold with an instrument designed to monitor the temperature and pressure during the expanding and cool-down process at the terminating end. This temperature shall be held for a minimum of 5 minutes and shall not exceed 240°F (115°C).
4) After the material has reached the required temperature, a specifically designed pressure driven rounding/expanding device shall be used to progressively round/expand the folded PVC at a maximum rate of 5 feet (1.5m) per second using steam at a maximum of 10 psig (69kPa) gage. The rounding/expanding process shall not cause any scraping, tearing, abrasion, movement, or other damage to the liner.
5) When the rounding/expanding process is complete, the steam shall be transitioned to cooling air, maintaining an internal pressure of up to 10 psig (69kPa) gage. After the conversion to air pressure, the air will be exhausted at the downstream manifold to cool the liner. As the downstream manifold exhaust temperature approaches 100°F (38°C), the water valve on the upstream manifold shall be gradually opened to allow incoming air/water mixture to reach 80°F (27°C). At this point, the air/water pressure shall be relieved and both ends of the rounded/expanded pipe shall be cut off in the manholes.
6) If testing of the installed PVC liner pipe is required in the Special Provisions, the physical properties of the installed PVC liner shall be verified through field sampling and laboratory testing as approved

by the Engineer. Unless the Special Provisions call for more than one sample, a sample shall be cut from a section of rounded/expanded PVC liner pipe at the upstream, downstream, or an intermediate manhole/access pit that has been inserted through a same diameter pipe acting as a mold. PVC liner pipe samples shall be submitted to a certified laboratory which has been pre-approved by the Engineer. The samples shall be tested in accordance with ASTM D638 for tensile strength, ASTM D790 for flexural modulus, ASTM D2444 for impact resistance to confirm that the liner pipe conforms to the minimum requirements per 500-1.10.3 b). All costs incurred for the testing shall be borne by the Contractor.

500-1.10.4 Service Connections and End Seals. Service connections and end seals shall conform to 500-1.1.7.

500-1.10.5 Repair and Rejection. The Contractor shall provide to the Engineer for approval an evaluation and repair specification for liner pipe damaged during or after installation. Any liner pipe damaged shall be removed from the Work site.

500-1.11 HDPE Spirally-Wound Profile Wall Liner Pipe.

500-1.11.1 General. High density polyethylene (HDPE) profile liner pipe for use in lining sanitary sewers shall conform to the requirements of ASTM F894.

Unless otherwise indicated the minimum pipe stiffness shall be 22.5 psi (155 kPa). Pipe stiffness and its respective ring stiffness constant (RSC) are detailed in Table 500-1.11.1. These shall be tested in accordance with ASTM D2412 and ASTM F894. The profile configuration shall be either external or internal unless otherwise specified on the Plans or in the Special Provisions.

500-1.11.2 Material Composition. The material shall conform to 207-19.2. Rubber gaskets shall conform to the requirements of 208-3.

500-1.11.3 Liner Pipe Acceptance. At the time of manufacture, all lot components of the liner pipe and fittings shall be inspected for defects. At the time of delivery, the liner pipe shall be homogeneous throughout, uniform in color, free of cracks, abrasions, holes, foreign materials, blisters, or deleterious faults. For testing purposes, a production lot shall consist of all liner pipe having the same lot marking number, but shall not exceed a total of 50 lengths per day. Pipe length, wall thickness, and joint dimension shall be verified by testing each lot per ASTM F894 or more frequently as required by the Engineer.

TABLE 500-1.11.1 (A)

Nominal Diameter inches (mm)	RSC (Min.)	Pipe Stiffness (PS) (Min.) psi (kPa)
18 (450)	64	22.5 (155)
21 (525)	74	22.5 (155)
24 (600)	84	22.5 (155)
27 (675)	95	22.5 (155)
30 (750)	106	22.5 (155)
33 (825)	116	22.5 (155)
36 (900)	127	22.5 (155)
42 (1050)	148	22.5 (155)
48 (1200)	169	22.5 (155)
54 (1350)	191	22.5 (155)
60 (1500)	212	22.5 (155)
66 (1650)	233	22.5 (155)
72 (1800)	255	22.5 (155)

Note: Higher RSC values with respectively higher pipe stiffness values may be available. RSC values are dimensionless.

500-1.11.4 Marking. Pipe sections having an exterior profile shall be marked at both ends on the inside of each pipe. Each pipe section having an interior profile shall be marked at both ends on the inside and outside of the pipe to show the manufacturer's name, manufacturer's number (identifies factory location, date, shift, and sequence), nominal inside diameter, minimum RSC, pipe stiffness, ASTM F894 designation, and lot number.

500-1.11.5 Chemical Resistance and Physical Testing. Liner pipes and gaskets shall be tested in accordance 211-2 and conform to the requirements of 207-19.5 and 208-3 respectively.

500-1.11.6 Installation and Field Inspection. This pipe installation shall conform to 500-1.1. Maximum jacking loads shall not exceed the values noted in Tables 500-1.11.6 (A) and (B).

TABLE 500-1.11.6 (A) – EXTERIOR PROFILE

Nominal Inside Diameter inches (mm)	Safe Axial Compressive Loads*	
	RSC 100 Tons (kN)	RSC 160 Tons (kN)
18 (450)	2.3 (20.4)	3.3 (29.3)
21 (525)	2.3 (20.4)	3.8 (33.8)
24 (600)	3.3 (29.3)	3.8 (33.8)
27 (675)	3.9 (34.6)	3.8 (33.8)
30 (750)	3.9 (34.6)	4.0 (35.5)
33 (825)	N/A.	5.8 (51.6)
36 (900)	N/A.	5.8 (51.6)
42 (1050)	N/A.	9.4 (83.6)
48 (1200)	N/A.	9.8 (87.1)
54 (1350)	N/A.	17.9 (159.2)

* Includes a Safety Factor of 2:1.

TABLE 500-1.11.6 (B) – INTERIOR PROFILE

Nominal Inside Diameter inches (mm)	Safe Axial Compressive Loads*	
	RSC	Tons (kN)
18 (450)	64	9.5 (84.5)
21 (525)	74	11.0 (97.8)
24 (600)	84	12.5 (111.2)
27 (675)	95	14.0 (124.5)
30 (750)	106	15.5 (137.8)
33 (825)	116	18.5 (164.5)
36 (900)	127	20.0 (177.9)
42 (1050)	148	25.0 (222.4)
48 (1200)	169	30.0 (266.9)
54 (1350)	191	45.0 (400.3)
60 (1500)	212	58.0 (515.0)
66 (1650)	233	72.0 (640.5)
72 (1800)	255	117.0 (1040)

* Includes a Safety Factor of 2:1.

500-1.11.7 Repair and Rejection. Liner pipe may be repaired for minor superficial pipe damage. Major damage which penetrates over 25 percent of the inner or outer wall thickness shall be repaired by cutting out the damaged section and replacing the damaged section with a new pipe. All repair methods shall be submitted to the Engineer for prior approval in accordance with 2-5.3. The liner pipe sections shall be a minimum of 8 feet (2.4m) in length unless shorter sections are authorized by the Engineer. Liner pipe shall be inspected immediately prior to installation for damage. If liner pipe is found to be superficially damaged, the Engineer may allow the pipe to be repaired or may reject it. Major liner pipe damage shall be rejected and replaced with new section of liner pipe.

500-1.11.8 Annular Space Grouting. The entire annular space between the outside of the liner pipe and the inside of the existing host pipe shall be grouted. The grout mix and placement procedure shall conform to 500-3. Grouting of the annular space shall be performed in such a manner as to prevent damage or collapse of the liner pipe. Maximum safe annular space grouting pressure for single-stage or multi-stage grouting shall not exceed the pipe stiffness divided by 4.5.

500-1.11.9 Service Connections and End Seals. Service connections shall be exposed and connected to the liner pipe by use of a saddle approved by the Engineer. Service connections and end seals shall conform to 500-1.1.7.

500-1.12 PVC Closed Profile Liner Pipe.

500-1.12 Polyvinyl Chloride (PVC) Closed Profile Liner Pipe.

500-1.12.1 General. Polyvinyl chloride (PVC) closed profile segmented liner pipe for use in lining sanitary sewers shall conform to ASTM F1803. Unless otherwise specified, the minimum pipe stiffness shall be 46 psi (318 kPa), as tested in accordance with ASM D2412.

500-1.12.2 Material Composition. The material shall be made from unplasticized PVC compounds having a cell classification of 12364 as defined in ASTM D1784. Elastomeric sealing gaskets shall conform to the requirements of 208-4.

500-1.12.3 Liner Pipe Acceptance. The liner pipe shall be free from cracks, holes, blisters, foreign inclusions or other defects that would, due to their nature, degree, or extent, have a deleterious effect on the pipe performance as determined by the Engineer.

For testing purposes, a production lot shall consist of all liner pipe having the same lot marking number, but shall not exceed one shift of production for sizes 21 through 30 inches (525 through 750mm) or 2 shifts of production for sizes 36 through 48 inches (925 through 1200mm). Pipe length, wall thickness and joint dimensions shall be verified by testing for each lot in accordance with ASTM F794. Records of this testing shall be made available if so requested by the Engineer.

500-1.12.4 Marking. Each pipe section shall be marked at one end on the inside and every 5 feet (1.5m) on the outside showing the manufacturers name, manufacturing number (identifies production plant, date, shift), cell classification, lot number, nominal diameter, pipe stiffness and ASTM F794. Internally the pipe shall have a numbered air testing certificate (sticker) that can be correlated through plant records to each piece of pipe. A key of the manufacturer's production and lot codes shall be submitted to the Engineer prior to delivery.

500-1.12.5 Chemical Resistance and Physical Testing. Liner pipe and gaskets shall be tested in accordance with 211-2 and conform to the requirements of Table 500-1.12.5 (A) and 208-4 respectively.

TABLE 500-1.12.5 (A)

Property	ASTM Test Method	Value (Initial and After 112-Day Exposure) Cell Class 12364
Tensile Strength (Yield), psi (MPa), min.	D 638	6000 (41.4)
Impact Strength Ft-lbs/inch (J/m) of notch, min.	D 256 Method A Size 1/2 x 1/8 x 2-1/2 inches (12.7 x 3.17 x 63.5mm)	0.65 (34.7)
Weight Change %	D 543	
Unconditioned		± 1.5 max.
Conditioned		± 1.0 max.

Verification shall be provided that physical testing of the product confirms conformance to ASTM F794 (qualification test only) and ASM D2412.

500-1.12.6 Installation and Field Inspection. The existing sewer may be maintained in operation during the relining process. The host pipeline shall be cleaned of any obstructions and televised per 500-1.1.4 and 500-1.1.5. Liner pipe installation shall conform to 500-1.1.7b). The pushing force shall be applied to the grooved end of the pipe. Maximum pushing loads shall not exceed the values noted in Table 500-1.12.6 (A).

TABLE 500-1.12.6 (A)

Nominal Pipe Diameter in (mm)	46 psi (318 kPa) Pipe Stiffness Maximum Pushing Load – Tons (kN) [1]
21 (525)	12.5 (111)
24 (600)	12.5 (111)
27 (700)	12.5 (111)
30 (750)	12.5 (111)
36 (925)	12.5 (111)
42 (1050)	12.5 (111)
48 (1200)	12.5 (111)

1. A factor of safety of 2:1 is included for the maximum pushing load.

500-1.12.7 Annular Space Grouting. The entire annular space between the outside of the liner pipe and inside of the host pipe shall be grouted in accordance with 500-3. Grouting of the annular space shall be done in such a manner as to prevent damage or collapse of the liner pipe. Maximum safe grouting pressure is 10 psi (69kPa).

500-1.12.8 Service Connections and End Seals. Service connections and end seals shall conform to 500-1.1.7.

500-1.12.9 Repair and Rejection. Prior to installation, liner pipe shall be inspected for damage. Liner pipe with superficial damage may be repaired without field cutting. Major damage which penetrates 50 percent or more of the inner or outer wall shall be rejected. All repair methods shall be submitted to the Engineer for prior approval in accordance with 2-5.3. Rejected liner pipe shall be replaced with a new section of liner pipe.

500-1.13 Spiral Wound Polyvinyl Chloride (PVC) Pipe Liner.

500-1.13.1 General. Spiral wound PVC pipe liner for use in the rehabilitation of circular and non-circular pipelines shall be a PVC profiled strip with a continuously sealed spiral joint. The profile may include steel reinforcing if so specified on the Plans or in the Special Provisions. The profiled strip is wound into the liner shape shown on the Plans to a size ranging from 6 inches through 15 feet (150mm through 4.6m). The profile may be designated as "Type 1" or "Type 2" on the Plans or in the Special Provisions. Type 1 is expandable to fit against the host pipe wall. Type 2 is installed with a fixed dimension, requiring annular space grouting between the liner and the existing host pipe. An end seal shall be provided at each manhole. Installation shall be in accordance with ASTM F1741 as modified herein.

500-1.13.2 Material Composition. The profiled strip shall be made from PVC compounds conforming to ASTM F1697, Section 5. The gasket and/or sealing material shall be as recommended by the manufacturer and shall be submitted in accordance with 2-5.3. When so specified in the Special Provisions, the steel reinforcing strip shall be fabricated from steel conforming to ASTM F1697, Section 5.2. When so specified in the Special Provisions, annular space grout shall be self leveling and consolidating. Structural grout shall be of sufficient strength to support all required loads.

500-1.13.3 Material Acceptance. The material shall consist of a ribbed PVC profiled strip with interlocking, sealed edges, gasket material and steel reinforcing strip, if required. The edges lock together as the strip is wound into a pipe. The profiled strip shall have shaped ribs which vary in height and width as specified in ASTM F1697. The Contractor shall submit, in accordance with 2-5.3, a Certificate of Compliance that states that the PVC profiled strip material, gasket material, and the steel reinforcing strip conforms to ASTM F1697 and the requirements shown on the Plans and Special Provisions. The grout shall conform to ASTM F1741, and the requirements shown on the Plans and in the Special Provisions.

500-1.13.4 Marking. Each PVC profiled strip shall be distinctively marked on its inside surface at intervals not to exceed 30 feet (9m) measured longitudinally along the profiled strip with a coded number which identifies the manufacturer, plant, date of manufacture and shift, cell classification and profile type. This information shall also appear on each reel.

500-1.13.5 Chemical Resistance. The PVC profiled strip, gasket, end seals, sealants, and other material exposed to the sewer environment, as determined by the Engineer, shall be tested in accordance with 211-2 and, conform to the weight change requirements of Table 207-17.5(A).

500-1.13.6 Installation and Field Inspection. The existing pipeline shall be cleaned and televised in accordance with 500-1.1.4 and 500-1.1.5. When so specified in the Special Provisions, the existing pipeline may be inspected and televised by the man-entry method. The condition of the cleaned pipeline shall be approved by the Engineer prior to the installation of the liner pipe.

During this phase of operation all service openings shall be precisely located longitudinally and radially, and logged for subsequent reconnection after the installation of the liner pipe.

At the time of installation, the profiled strip material shall be homogeneous and free of defects, cracks, holes, blisters, or foreign materials.

The installed spiral wound PVC pipe liner shall be inspected and televised in accordance with 500-1.1.4 and 500-1.1.5 or by the man-entry method if so specified in the Special Provisions.

Spiral wound PVC pipe liner shall be of uniform appearance, undamaged, free of cracks, holes, unsealed joints, and shall be installed according to the manufacturer's recommendations and in accordance with ASTM F1741. End seals shall conform to 500-1.1.7, subparagraph "e".

500-1.13.7 Connections. Connections for Type 1 liners shall be re-established in accordance with 500-1.1.7, subparagraph "a".

The Contractor shall submit the data listed in 500-3.1.10, subparagraphs "a" through "i", for the structural grout in accordance with 2-5.3.

The procedure for re-establishing service lateral connections for Type 2 liners shall be submitted to the Engineer in accordance with 2-5.3. This procedure shall include the method of blocking the service connections during grouting and the sleeving system to be used between the liner and the host pipe. The sleeving system shall conform to 500-3 and be submitted to the Engineer in accordance with 2-5.3.

500-1.13.8 Annular Space Grouting. Annular space grouting shall conform to ASTM F1741, Section 6.5. The utilization of structural or non-structural grout shall be as specified in the Special Provisions.

A structural grout mix design shall be submitted to the Engineer in accordance with 2-5.3 and shall have a minimum compressive strength of 5,000 psi in 28 days when tested in accordance with ASTM C39. The submittal shall include the data listed in 500-3.1.10, subparagraphs "a" through "j", "l", and "o", for structural grout.

Non-structural grout material shall conform to 500-3.

The entire annular space shall be grouted. Grout penetration shall be verified by the Contractor. The method of verifying the penetration of the grout shall be submitted to the Engineer in accordance with 2-5.3.

500-1.13.9 Repair. The Contractor shall submit a repair method to the Engineer in accordance with 2-5.3 for any profile strips or liner pipe found to be damaged during or after installation, or if grouting deficiencies are encountered.

500-1.13.10 Measurement and Payment.

500-1.13.10.1 Measurement. Spiral wound PVC pipe liner shall be measured by the linear foot of pipe lined, or by lump sum. Re-establishment of service connections shall be measured by each or by lump sum.

500-1.13.10.2 Payment. Payment for spiral wound PVC liner pipe shall be made at the Contract Unit Price, or lump sum, as specified in the Bid.

Payment for re-establishing service connections shall be made at the Contract Unit Price, or lump sum, as specified in the Bid.

500-2 MANHOLE AND STRUCTURE REHABILITATION.

500-2.1 General. This subsection specifies various lining systems for manholes and structures. The types of rehabilitation materials and methods shall be as shown on the Plans and specified in the Special Provisions. Flow control, if required shall also be as shown on the Plans or as specified in the Special Provisions. Unless otherwise specified in the Special Provisions, proof of meeting the Chemical Resistance test per 211-2 shall be submitted to the Engineer per 2-5.3.

As used in this subsection "holiday" shall be defined as any discontinuity, bare or thin section in a lined or coated area.

500-2.2 Requirements.

500-2.2.1 Installer Qualifications. The installer, whether the Contractor or a subcontractor, shall be a certified installer of the lining system. The installer's personnel shall be adequately trained in maintenance and operation of the required installation equipment, as certified by the lining manufacturer. A letter from the manufacturer of the lining system, verifying the certification of the installer required to be on-site during installation, shall be submitted per 2-5.3.

500-2.2.2 Cleaning, Inspection, and Surface Preparation. Inspection and Surface Preparation. Prior to the installation of the lining system, the host structure shall be prepared to produce a concrete or masonry surface suitable for application and adhesion of the specified lining system. Cleaning and surface preparation shall include the inspection of the host structure for any damage or leaks, and the removal of any protrusions on the surface of the host structure that could interfere with the installation of the lining system. Any damage or leaks shall be reported to the Engineer. Cleaning methods may include high pressure water cleaning at a minimum of 5,000 psi (34.5MPa), abrasive blast, or a method recommended by the manufacturer of the lining system, or another cleaning method submitted to the Engineer for approval per 2-5.3. The Contractor shall protect the host structure from damage by the cleaning equipment, water and air pressure. Flow bypassing, if required by the lining system, shall conform to 7-8.4 and 306-3.3.

Debris from the cleaning operation shall not be allowed to enter the sewer system. The Contractor shall furnish, install and remove any necessary debris containment devices while maintaining sewer flow. The Contractor shall remove and dispose of all debris collected from the cleaning operation. If reinforcing steel is exposed, either before or after removing deteriorated concrete, it shall be thoroughly cleaned to remove all contamination and rust particles. Immediately after the cleaned reinforcing steel is inspected and accepted by the Engineer, the Contractor shall place a protective coating on the exposed reinforcing steel. The protective coating shall be approved by the Engineer in accordance with the manufacturers' specifications.

Manhole steps, pull rings and lifting eyes shall be installed or removed as shown on the Plans or specified in the Special Provisions.

500-2.3 Repair, Resurfacing and Active Infiltration.

500-2.3.1 General. Repair, resurfacing and active infiltration elimination materials shall be compatible with the lining system. Proof of compatibility shall be submitted to the Engineer per 2-5.3.

500-2.3.2 Repair. Prior to installation, patching or localized repairs shall be performed using rapid setting repair mortars compatible with the lining system. Repair mortars shall be used to fill surface irregularities, voids, and deteriorated surfaces and to repair the host structure to a uniform surface. Manufacturer's specifications shall be followed when performing repairs, material handling, mixing, installation and curing. A copy of the manufacturer's specifications confirming the compatibility of the materials used in the repair shall be submitted to the Engineer per 2-5.3.

500-2.3.3 Resurfacing. Air-placed concrete (APC) materials, if required by the Special Provisions, shall conform to 303-2.3.1. Prior to the application of APC, the structure shall be prepared to produce a concrete or masonry surface that is suitable for application and adhesion of the specified APC. The APC shall be applied in continuous lifts of 1/2 inch (12mm) minimum thickness or as shown on the Plans and/or as specified in the Special Provisions. Containment devices, approved by the Engineer, shall be used to prevent rebound (non-adhering excess APC) from entering the sewer system. Immediately following the APC placement operation, the containment device shall be removed and the structure's cover reinstalled to provide a moist curing environment. Where moist conditions within the structure do not exist, the Contractor shall water cure the APC for a minimum of 24 hours prior to installation of the lining system. All defects in the APC shall be repaired per 303-2 prior to the installation of the liner system.

500-2.3.4 Active Infiltration. Active leaks within the host structure shall be eliminated prior to installation of the liner system. Leaks shall be eliminated by pressure grouting with chemical grout as specified in the Special Provisions and/or the application of hydraulic cement conforming to 201-1.2. Chemical grouts and hydraulic cements shall be compatible with the lining system used. A copy of the manufacturer's specifications confirming the compatibility of the materials used in the repair shall be submitted to the Engineer per 2-5.3. The host structures shall be visibly dry with no active infiltration prior to lining.

500-2.4 Inspection, Testing and Repair of Installed Liner Systems.

500-2.4.1 General. The party performing the following tests shall be as specified in the Special Provisions. If the testing party is to be selected by the Contractor, the name of the testing party and information on the testing instruments to be used for adhesion testing and its calibration shall be submitted to the Engineer per 2-5.3.

500-2.4.2 Spark Test. The cured lining system shall be spark tested for holidays with the high voltage holiday detector instrument specified by the coating manufacturer or as specified in the Special Provisions. The voltage shall be set at a minimum of 15,000 volts. For thicknesses greater than 150 mils (4mm), the voltage shall be set at 100 volts per 1 mil (25μm) of thickness of the applied lining material. Identified holidays shall be marked without contaminating the lining surface and repaired in accordance with 500-2.4.5.

500-2.4.3 Mil Gauge Test. During installation, a mil gauge shall be used to verify that the minimum thickness of the lining meets and/or exceeds the minimum thickness specified herein or specified in the Special Provisions.

500-2.4.4 Adhesion Testing. Adhesion testing shall be performed on a minimum of 1 structure or 15 percent of all rehabilitated structures, whichever is greater, or as shown on the Plans and/or specified in the Special Provisions. Adhesion testing shall be conducted after the liner system has cured in accordance with the manufacturer's specifications. Adhesion testing shall be in accordance with ASTM D4541 as modified herein.

A minimum of one 3/4 inch (19mm) dolly shall be affixed to the lined surface of the host structure at the upper section or cone area, the midsection, and at the bottom, unless otherwise specified in the Special Provisions. Each testing location shall be identified by the Engineer. The adhesive used to attach the dollies to the liner shall be rapid setting with tensile strength in excess of the liner material and permitted to cure in accordance with the manufacturer's specifications. The lining material and dollies shall be prepared to receive the adhesive in accordance with the manufacturer's specifications. Prior to the pull test, the tester shall utilize a scoring device to cut through the coating until the substrate is reached. Failure due to improper dolly adhesive or scoring will require retesting. The pull tests in each area shall meet or exceed 200 psi (1,400kPa) and shall include substrate adhered to the back of the dolly or no visual signs of coating material in the test hole. Pull tests with results between a minimum 150 psi (1,000kPa) and 200 psi (1,400kPa) may be acceptable if more than 50 percent of the substrate adhered to the back of the dolly. A test result may be disregarded, as determined by the Engineer, if there is a valid nonstatistical reason as specified in Sections 8.4 and 8.5 of ASTM D4541. If any test fails, a minimum of 3 additional locations in the section of the failure shall be tested, as directed by the Engineer. If any of the retests fail, all loosely adhered or unadhered liner in the failed area, as determined by the Engineer, shall be removed and replaced at the Contractor's expense. If a host structure fails the adhesion test, one additional host structure or 10 percent of the initial number of host structures selected for testing shall be tested as directed by the Engineer or as specified in the Special Provisions.

500-2.4.5 Liner Repairs. Holidays, uncured lining material, blisters, surface imperfections and damage to the liner resulting from the adhesion test shall be repaired to a point 1 inch (25mm) minimum beyond the limits of the damaged area. The repair shall be 125 mils (3mm) thick or the minimum thickness specified in the Special Provisions. Holidays shall be primed and recoated with the same lining system to a minimum additional thickness of 30 mils (750μm) unless otherwise specified by the liner manufacturer or approved by the Engineer. Blisters, uncured lining and surface imperfections shall be completely removed and the areas recoated with appropriate lining material to 1 inch (25mm) minimum beyond the repair areas at a minimum thickness of 100 mils (500μm). Additional spark testing shall be performed after repairs are completed.

500-2.5 Integral Locking PVC Manhole and Structure Lining System.

500-2.5.1 General. This subsection specifies an integral locking PVC lining system. This system consists of temporarily erecting a form inside an existing structure, installing PVC lining material with integral locking devices and filling the annular space between the erected form and PVC lining material and the walls of the host structure with portland cement concrete resulting in a new PVC-lined, monolithic structure within the host structure.

500-2.5.2 Materials.

500-2.5.2.1 Portland Cement Concrete. Portland cement concrete shall be Class 330-C-23 (560-C-3250) conforming to 201-1.1.2, or as specified in Special Provisions.

500-2.5.2.2 Integral Locking PVC Liner. The liner shall conform to the specifications in 210-2 and be as specified in the Special Provisions. A Certificate of Compliance conforming to 4-1.5 shall be submitted per 2-5.3.

500-2.5.3 Installation and Field Inspection. If existing, steps shall be removed flush with the inside wall of the host structure. Formwork for the lining system shall be installed in a manner that fits the existing walls and creates an equal and approximate 3-inch (75mm) annular space. Portland cement concrete shall be used to fill the annular space. The installation of the liner shall conform to 311-1. Replacement of steps, if required, shall conform to the requirements shown on the Plans and/or specified in the Special Provisions.

Prior to placement of the portland cement concrete, the installer shall connect the existing mainline pipe to the host structure as shown on the Plans and/or as specified in the Special Provisions. Exposed portland cement concrete surfaces within the host structure shall be protected as specified in the Special Provisions.

Field testing shall conform to 210-2.3.4 and 311-1.10 unless otherwise specified in the Special Provisions.

500-2.6 Segmented PVC Lining System.

500-2.6.1 General. This subsection specifies a lining system consisting of PVC liner strips placed so that an annular space is created between the liner and the walls of the host structure. This annular space is then filled with cementitious grout. A Certificate of Compliance conforming to 4-1.5 shall be submitted to the Engineer per 2-5.3 for each material. Surface preparation of the host structure and flow bypass shall conform to 500-2.1.3 unless otherwise specified in the Special Provisions.

500-2.6.2 Materials.

500-2.6.2.1 PVC Liner. PVC liner shall conform to 500-1.5.2 through 500-1.5.5 and Table 207-17.5(A).

500-2.6.2.2 Cementitious Grout. Cementitious grout shall conform to ASTM F1741, Section 6.4 and be submitted to the Engineer for approval per 2-5.3 or be as specified in the Special Provisions.

500-2.6.2.3 Sealant/Adhesive. The sealant/adhesive shall be compatible with the PVC lining material and cementitious grout. A copy of the manufacturer's specifications confirming the compatibility of the sealant/adhesive materials shall be submitted to the Engineer per 2-5.3.

500-2.6.3 Installation and Inspection. Installation shall be performed by either manually spirally winding PVC strips or placing PVC panels and engaging the complementary locks (male/female) at the edges of the strips/panels in a manner that creates an annular space, of 1 inch (25mm) minimum thickness, unless otherwise specified in the Special Provisions.

A bead of approved sealant/adhesive shall be applied to the female locking edge of the strip/panel prior to engaging the locking edges. Grouting of the annular space shall be performed in such a manner as to prevent damage or collapse of the liner and completely fill the annular space. Installation shall conform to 500-1.5.6 through 500-1.5.9. Testing shall be performed in accordance with 500-2.4, except 500-2.4.4. Holes in the liner shall be covered with a patch of sealant materials as approved by the Engineer.

500-2.7 Polyurethane and Epoxy Protective Lining System.

500-2.7.1 General. This subsection specifies a polyurethane and epoxy primer protective lining system.

500-2.7.2 Lining Material. Lining material shall consist of 100 percent solid polyurethane material and moisture tolerant epoxy. Polyurethane lining material shall be capable of spray application to 125 mils (3mm) minimum thickness in one continuous coat. Epoxy shall be capable of spray application to 5 mils (127μm) thickness in one continuous coat.

500-2.7.3 Installation and Curing. Lining material shall be applied to all prepared surfaces from 1 inch (25mm) below the low-flow water level to the base of the ring and cover unless otherwise specified in Special Provisions. All termination points of the lining material to the existing subsurface shall be keyed into the subsurface by mechanically scoring a minimum 1/4 inch x 1/4 inch (6mm x 6mm) keyway. Prior to application of the polyurethane, the subsurface shall be primed with the epoxy primer to a thickness of 3 mils (76μm) minimum to 5 mils (127μm) maximum. Polyurethane shall be applied to a thickness of 125 mils (3mm) immediately prior to the epoxy primer becoming tack-free. Lining material shall be uniform in color, fully cured, free of holidays, surface imperfections, blisters and sags and adequately adhered to the subsurface.

500-2.7.4 Inspection and Testing. The set or cured lining materials shall be tested in accordance with 500-2.4 unless otherwise specified in the Special Provisions.

500-2.7.5 Performance Requirements. The lining system shall meet or exceed the requirements specified in Table 500-2.7.5(A).

TABLE 500-2.7.5 (A)

	Polyurethane	Epoxy Primer
Tensile Strength ASTM D638, Type IV, psi (MPa) (min)	2,000 (14)	6,000 (41)
Elongation at Break, % ASTM D638, Type IV	40	5
Wear Resistance, mg. wt. Loss Taber abrasion, ASTM D4060	60[1]	100[1]
Hardness, Shore D, Durometer ASTM D2240	55	75
Tear Resistance, ppi (kg/mm) ASTM D624	150 (2.7)	N/A
Peel Strength, Concrete, pli (g/mm) ASTM D903	7[2] (125)	7[2] (125)
Weight Change[3]	± 1.5%	± 1.5%

1. Abrasive wheel No. CS-17, maximum value.
2. Tested as a system. Test results shall be verified on a per job basis or as specified in the Special Provisions.
3. Tested in conformance with 211-2.

500 - 2.8 Epoxy Lining System.

500-2.8.1 General. This subsection specifies an epoxy lining system.

500-2.8.2 Lining Material. Lining material shall consist of solvent free, high-build epoxy resin capable of spray application to 125mils (3mm) minimum thickness in one continuous coat.

500-2.8.3 Installation and Curing. Lining material shall be applied to all prepared surfaces from 1 inch (25mm) below the low-flow water level to the base of the ring and cover unless otherwise specified in the Special Provisions. Termination points of the lining to the existing subsurface shall be keyed into the subsurface by mechanically scoring a minimum 1/4 inch x 1/4 inch (6mm x 6mm) keyway. Epoxy shall be applied to a thickness of 125 mils (3mm). Lining material shall be uniform in color, fully cured, free of holidays, surface imperfections, blisters and sags and adequately adhered to the subsurface.

500-2.8.4 Inspection and Testing. The set or cured lining materials shall be tested in accordance with 500-2.4 unless otherwise specified in the Special Provisions.

500-2.8.5 Performance Requirements. The lining system shall meet or exceed the specifications in Table 500-2.8.5 (A)

TABLE 500-2.8.5 (A)

	Epoxy Liner
Tensile Strength ASTM D638, Type IV, psi (MPa) (min)	3,000 (21)
Elongation at Break, % ASTM D638, Type IV	0.9
Wear Resistance, mg. wt. Loss Taber abrasion, ASTM D4060	115[1]
Hardness, Shore D, Durometer ASTM D2240	80
Weight Change[2]	± 1.5%

1. Abrasive wheel No. CS-17, maximum value.
2. Tested in conformance with 211-2.

500-2.9 Epoxy Mastic and Flexible PVC Liner System.

500-2.9.1 General. This subsection describes the installation of the mastic primer and epoxy mastic that bonds to the cleaned, repaired and prepared interior concrete substrate, then follows with the mechanical locking of the flexible PVC liner into the epoxy mastic. The integral locking extensions in the flexible PVC liner are embedded to their full depth into the epoxy mastic.

500-2.9.2 Materials.

500-2.9.2.1 Plastic Liner. The liner shall be manufactured from a PVC compound in accordance with 210-2. The plastic liner shall be a flexible PVC liner with a minimum thickness of 1/16 inch (1.65mm) and conform to 210-2.4.

500-2.9.2.2 Mastic Primer. The epoxy mastic primer shall be a two-part coating that is applied to the prepared concrete substrate.

500-2.9.2.3 Epoxy Mastic. The epoxy mastic shall be a 2-part 100 percent epoxy coating for bonding and filling voids in properly prepared concrete substrate.

500-2.9.3 Locking Extensions. All liner to be embedded in the epoxy mastic shall have integral locking extensions and meet the requirements of 210-2.2.4 except for the dimensions requirement. The locking extensions shall have a shape, height, web thickness, and spacing that will allow the liner to be held permanently in place and be able to meet the pull-out requirements of 500-2.4.4.

500-2.9.4 Chemical Resistance and Physical Property Testing. The plastic liner sheet and accessories shall conform to 211-2 and 210-2.4.

500-2.9.5 Preparation and Repair of Substrate. Prior to applying the mastic primer, the structure shall be cleaned and prepared in accordance with 311-1.6 and repaired in accordance with 311-1.9.

500-2.9.6 Installer Qualifications. Applicators and welders of the plastic liner shall be qualified in accordance with 311-1.2.

500-2.9.7 Installation.

500-2.9.7.1 Priming. The mastic primer shall be applied to 3 mils (76μm) minimum to 5 mils (127μm) maximum thickness. The primer shall be allowed to dry before applying the epoxy mastic.

500-2.9.7.2 Epoxy Mastic Application. A finishing trowel or other suitable tool shall be used to apply the epoxy mastic to a uniform minimum thickness of 1/4 inch (6mm).

500-2.9.7.3 Plastic Liner Application. The plastic liner shall be placed while the wetting ability of the epoxy mastic is at its optimum, be pressed into the mastic, and rolled to remove any trapped air. The lining system shall be allowed to cure for the amount of time recommended by the lining manufacturer. The average dry film thickness of the cured lining system, including the liner sheets and the applied epoxy mastic, when completed shall not be less than 315 mils (8mm).

500-2.9.8 Field Jointing of Liner.

500-2.9.8.1 General. The Contractor shall utilize the maximum size plastic liner sheet that is practical and will provide the minimum number of seams. Vertical and horizontal seams shall overlap a minimum of 1/2 inch (13mm) and shall be welded with 1 inch (25mm) weld strips. Corner strips may be used at inside and outside corners, or liner may be wrapped around corners. The Contractor shall be allowed to heat the liner to facilitate turning corners. Excessive heating of the liner material to facilitate turning corners shall be avoided.

500-2.9.8.2 Field Joints in Manhole and Structure Rehabilitation. Field joints in the liner shall be Type AL-2 unless AL-1 or AL-3 is approved by the Engineer.

a) Type AL-1 joint shall consist of a 4-inch (100mm) wide joint strip centered over a 1 inch (25mm) maximum gap between sheets and securely welded along each edge of adjacent liner with a 1 inch (25mm) welding strip.

b) Type AL-2 joints shall be made with integral joint flaps with locking extensions removed 1 inch (25mm) from one side per 210-2.2.6. The flap shall be overlapped a minimum of 1/2 inch (13mm) and the overlap secured to the adjacent liner by means of a 1 inch (25mm) welding strip.

c) Type AL-3 joints shall consist of a 1 inch (25mm) wide weld strip centered over a 1/4 inch (6mm) maximum gap between sheets and securely welded along each edge of adjacent liner.

500-2.9.8.3 Installation of Welding Strips. Installation of welding strips shall conform to 311-1.5.4.

500-2.9.9 Pull Test for Locking Extensions. Liner locking extensions embedded in epoxy mastic shall withstand a test pull of at least 20 pounds per linear inch (3.5N per linear mm), applied perpendicularly to the concrete surface for a period of 1 minute without rupture of the locking extensions or withdrawal from the epoxy mastic or delaminating of the mastic from the concrete substrate. This test shall be made at a temperature between 70°F to 80°F (21°C to 27°C).

500-2.9.10 Inspection. After installation of the protective lining system, the surface of the liner shall be cleaned and prepared by the Contractor and then inspected by the Engineer. Field testing shall be in accordance with 311-1.10.

500-2.9.11 Repair of Defects and Holidays. The Contractor shall repair all defects and damage in the plastic liner in accordance with 311-1.9.

500-2.10 Measurement and Payment.

500-2.10.1 Measurement. Manhole and structure rehabilitation shall be measured by each.

500-2.10.2 Payment. Payment for manhole and structure rehabilitation shall be made at the Contract Unit Price or lump sum price in the Bid for each structure. The Bid price shall include the installation of the lining system as well as surface preparation and repairs, and performance of testing, unless otherwise specified in Special Provisions.

500-3 ANNULAR SPACE GROUTING.

500-3.1 Requirements.

500-3.1.1 General. This subsection covers various requirements of continuous annular space grouting of sliplining systems. The annular space (void between the host and liner pipes) shall be completely grouted to support the liner and provide long-term stability. The Contractor shall engage the services of an Agency approved testing laboratory to certify that the proposed materials and methods comply with these requirements. All proposals shall be submitted to the Engineer per 2-5.3, 201-1, and 500-3.1.10.

500-3.1.2 Preparation. Upon completion of sliplining but prior to grouting, bulkheading of the ends and appropriate venting will be required. This is to seal the annular space from sewer flow to permit the grout to set and withstand the loads imposed by the grout and groundwater. The Contractor shall test the integrity of the installed liner pipe and constructed bulkheads for any leaks by performing the following:

a) Dewater and inject dye water into the annular space (this alternative will not be permitted if the crown or any portion of the host pipe is severely deteriorated to the point where water may escape through the host pipe).

b) Pressurize the annular space to the maximum permissible grouting pressure per the manufacturer's recommendation with approval by the Engineer.

c) The Contractor shall submit a detailed plan to the Engineer that will hold the liner pipe on the invert for a period of time long enough to allow the grout to set where buoyant uplift is a factor.

500-3.1.3 Planned Vents. The Contractor shall submit working drawings, including the proposed number and location of vents relative to pipe diameter and stiffness and the depth of sewer flow in the pipeline for the grouting operation.

500-3.1.4 Materials. The grout materials shall consist of portland cement, portland cement and fly ash, and/or additives, providing materials are not biodegradable.

a) **Compressive Strength.** The grout shall have a minimum penetration resistance of 100 psi (690 kPa) in 24 hours when tested in accordance with ASTM C403 and a minimum compressive strength of 300 psi (2070 kPa) in 28 days when tested in accordance with ASTM C495 or C109.

b) **Performance Requirements.** The Contractor shall submit the proposed grout mixes, methods, plans, and criteria of the grouting operations. The grouting system shall have sufficient gauges, monitoring devices, and tests to determine the effectiveness of the grouting operation and to ensure compliance with the liner pipe specifications and design parameters.

c) **Mix Design.** One or more mixes shall be developed to completely fill the annular space based on the following requirements:

1) Size of the annular void
2) Void (size) of the surrounding soil
3) Absence or presence of groundwater
4) Sufficient strength and durability to prevent movement of the pipe
5) Provide adequate retardation, and
6) Provide less than 1 percent shrinkage by volume

d) **Density and Viscosity.** The Contractor shall design a grout mix with a density to meet the requirements of 500-3.1.6 and to prevent floating of the liner pipe. The apparent viscosity shall not exceed 20 seconds in accordance with ASTM C939 unless otherwise approved by the Engineer.

500-3.1.5 Qualifications. The Contractor shall demonstrate to the Engineer its worker's capabilities of filling the annular space and performing their work in conformance with the Plans and the Specifications.

500-3.1.6 Grouting Equipment. The materials shall be mixed in equipment of sufficient size and capacity to provide the desired amount of grout material for each stage in a single operation. The equipment shall be capable of mixing the grout at densities required for the approved procedure and shall also be capable of changing density as dictated by field conditions any time during the grouting operation.

500-3.1.7 Injection Procedure and Pressure. The gauged pumping pressure shall not exceed the liner pipe manufacturer's approved recommendations as stated in 500-1.3.7, 500-1.8.7, and 500-1.11.8. Pumping equipment shall be of a size sufficient to inject grout at a volume, velocity and pressure compatible with the size of the annular space and degree of host pipe corrosion. Once grouting operations begin, grouting shall proceed uninterrupted from bulkhead to bulkhead. Grout placement shall not be terminated until the following conditions are met, unless otherwise approved by the Engineer:

a) the estimated annular volume of grout has been injected;
b) the exhausted grout at each vent is not less than 85 percent of the density of freshly injected grout;
c) the exhausted grout at each vent is not less than 85 percent of the original viscosity of the freshly injected grout; and
d) when recommended by the grout installer.

A grout pressure gauge and recorder shall be installed immediately adjacent to each injection port. During operations, the recorder shall continuously record the actual grouting pressure versus the time on paper with ink. The gauge shall conform to an accuracy of ± 0.5 psi (± 3.5 kPa). The range of the gauge shall not be more than 100 percent greater than the design and attached to a saddle-type diaphragm seal (gauge saver) to prevent clogging with grout. All gauges shall be certified and calibrated in accordance with ANSI B40, Grade 2A. The grout pressure recordings shall be identified, as a minimum, with the date, batch, and time of day grouting was performed and shall be submitted to the Engineer at the end of the work day that grouting was performed.

500-3.1.8 Onsite Test. For each batch, the Contractor shall provide all equipment and personnel necessary to perform the following tests in the presence of the Engineer:

a) Density per ASTM C138 or by other methods as approved by the Engineer.
b) Viscosity per ASTM C939.

Grout that exceeds ± 3 pounds per cubic foot (48 kg/m^3) of the design density will be rejected.

500-3.1.9 Test Section. The Contractor will be required to perform a test on each type of grout and grout system proposed to be used. The test section to be grouted and the size of the annular space considered for each type of grout system shall be determined by the Contractor and approved by the Engineer.

500-3.1.10 Submittals and Required Calculations. The Contractor shall submit the following to the Engineer in accordance with 2-5.3:

a) The proposed grouting mix.
b) The proposed grout densities and viscosity.
c) Initial set time of the grout.
d) The 24-hour and 28-day minimum grout compressive strengths.
e) The grout working time before a 15 percent change in density or viscosity occurs.
f) The proposed grouting method and procedures.
g) The maximum injection pressures.
h) Proposed grout stage volumes (e.g., Stage 1, to springline; Stage 2, fully grouted).
i) Bulkhead designs and locations.
j) Buoyant force calculations during grouting.
k) Flow control.
l) Provisions for re-establishment of service connections.

m) Pressure gauge, recorder, and field equipment certifications (e.g., calibration by an approved certified lab).

n) Vent location plans.

o) Written confirmation that the Contractor has coordinated grouting procedures with the grout installer and the liner pipe manufacturer.

Data for a) through e) shall be derived from trial grout batches by an approved, independent testing laboratory.

For each different type of grout or variation in procedure or installation, a complete package shall be submitted. The submittal shall include each of the above items and the sewer locations or conditions to which it applies. The Contractor shall obtain approval from the Engineer for any changes to be made in grout mix, grouting procedure, or installation prior to commencement of grouting operations.

500-4 SERVICE LATERAL CONNECTION SEALING.

500-4.1 General. This subsection specifies various SLC sealing systems, and methods of installation. The type of sealing systems and methods to be used shall be as shown on the Plans or specified in the Special Provisions. Unless otherwise specified in the Special Provisions, proof of compliance with the Chemical Resistance Test specified in 211-2 using the weight change specified in Table 500-4.1 (A) shall be submitted to the Engineer in accordance with 2-5.3.

SLC sealing systems shall consist of either a cured-in-place resin saturated felt or fiberglass lining material and tube installed in an existing mainline and HC. Dry or unsaturated areas are not acceptable. The lining material and tube shall be sized such that when installed they are properly aligned, tight fitting and without wrinkles. SLC sealing systems shall be manufactured so as to provide smooth tapered edges after curing. The curing method and schedule shall be submitted to the Engineer in accordance with 2-5.3. The cured SLC sealing system shall meet or exceed the specifications in Table 500-4.1 (A).

TABLE 500-4.1 (A)

Property	ASTM Test Method	Initial. PSI (MPa)
Flexural Strength	D790	3,000 (20.7)
Flexural Modulus	D790	250,000 (1,724)
Tensile Strength	D638	3,000 (20.7)
Tensile Modulus	D638	250,000 (1,724)
Weight Change[1]	-	±1.5%

1. Tested in Conformance with 211-2

Bonding materials used with SLC sealing systems shall be compatible with the existing mainline and HC or with the lining system used in the mainline and HC, and shall be submitted in accordance with 2-5.3.

500-4.2 Requirements.

500-4.2.1 Installer Qualifications. The installer shall be certified by the manufacturer of the SLC sealing system. Personnel installing the SLC sealing system shall be adequately trained in maintenance and operation of the required installation equipment. A letter from the manufacturer of the SLC sealing system, verifying the certification of the installer, shall be submitted in accordance with 2-5.3. The installer shall be on site at all times during installation.

Prior to installation, the Contractor shall submit a detailed installation plan to the Engineer in accordance with 2-5.3. This installation plan shall provide verification of compliance with the physical properties specified in Table 500-4.1 (A), the manufacturer's specified curing time, chemical composition and a detailed description of the SLC sealing system.

500-4.2.2 Cleaning, Inspection, and Surface Preparation. Prior to the installation of the SLC sealing system, the HC, SLC and mainline shall be prepared, as specified by the manufacturer, to produce a surface that is suitable for application of the specified sealing system. Cleaning methods shall be as specified by the manufacturer of the SLC sealing system and submitted to the Engineer in accordance with 2-5.3. Cleaning and surface preparation shall include CCTV inspection of the mainline and HC for locating any damage or leaks. CCTV inspection shall conform to 500-1.1.5. The HC's shall be inspected 16 inches (400mm) beyond the end of the proposed lining unless otherwise specified in the Special Provisions. Any protrusions on the surface of the mainline and HC that could interfere with the installation of the SLC sealing system shall be removed. All roots larger than 1/8 inch (3mm) shall be removed during the cleaning operation, and any damage or leaks shall be reported to the Engineer. Flow bypassing, if required, shall conform to 7-8.5.2, 7-8.5.3, and 306-3.3.

Debris from the cleaning operation shall not be allowed to enter the sewer system. The Contractor shall furnish, install, and remove any necessary debris containment devices while maintaining sewer flow. The Contractor shall remove and dispose of all debris collected from the cleaning operation in accordance with 500-1.1.4.

500-4.3 Repair and Active Infiltration.

500-4.3.1 General. Material used to repair active infiltration shall be compatible with the SLC sealing system and mainline lining material. Proof of compatibility shall be submitted to the Engineer in accordance with 2-5.3.

500-4.3.2 Active Infiltration. Active infiltration shall be eliminated by pressure grouting with chemical grout as specified by the SLC system manufacturer or specified in the Special Provisions. Upon the completion of pressure grouting, if required, the area to be sealed shall be visibly clean with no excess grout prior to lining.

500-4.4 Field Inspection Testing and Repair of Installed SLC Sealing Systems.

500-4.4.1 General. The Contractor shall submit information on the equipment to be used in testing installed SLC sealing systems to the Engineer in accordance with 2-5.3.

500-4.4.2 CCTV Inspection. After the installation is complete, the Contractor shall perform CCTV inspection in accordance with 500-1.1.5.

500-4.4.3 Adhesion Testing. Adhesion testing shall be performed on each SLC seal installed. The Contractor shall notify the Engineer 24 hours prior to performing adhesion testing, if the Engineer cannot be on site during testing, the Contractor shall videotape the test at each seal and shall submit the videotape to the Engineer in accordance with 2-5.3.

Adhesion testing shall be conducted after the SLC seal has cured in accordance with manufacturer's specifications and before the final video inspection is performed. Adhesion testing shall consist of inserting a high velocity, hydraulic cleaning type, 360 degree spinning nozzle and CCTV camera in the mainline pipe and positioning the nozzle at the SLC seal. The water from the nozzle shall be directed downstream for a minimum of 1 minute, at each edge of the SLC seal in the mainline at the minimum pressure of 1500 psi (10MPa) and minimum flow rate of 65 gpm (246 liters per minute).

500-4.4.4 Air Pressure Test. Unless otherwise specified in the Special Provisions, cured SLC sealing systems shall be air pressure tested. A test plug shall be placed adjacent to the upstream and downstream end of the SLC sealing system in the mainline and adjacent to the upstream end in the HC Sewer. The test pressure shall be 4 psi (0.3MPa) for a 3 minute test time during which the pressure shall not drop below 3.5 psi (0.2MPa). If the SLC sealing system fails this test, the test plug in the HC Sewer may be moved onto the SLC seal and the test conducted again. If the second test passes, the SLC sealing system will be deemed to have passed the test.

500-4.4.5 Repairs of Sealing Systems. If the SLC sealing system fails either the adhesion or the second air pressure test, the Contractor shall remove and replace or repair the SLC sealing system as recommended by the manufacturer and approved by the Engineer.

500-4.5 Full Wrap "T-Style" SLC Sealing System.

500-4.5.1 General. This subsection specifies a full wrap "T-style" SLC sealing system.

500-4.5.2 Sealing Material. The sealing material shall consist of a cured-in-place resin saturated felt tube that provides a full-wrap tube in the mainline centered at the HC, with a tube section extended into the HC. The length of the tube section in the mainline and in the HC shall be as specified in the Special Provisions.

500-4.5.3 Installation and Curing. Installation shall conform to Section 7.0 of ASTM F2561. The installation method shall provide an air-tight seal of the SLC sealing system to the mainline pipe and HC Sewer as specified by the manufacturer and shall be submitted to the Engineer in accordance with 2-5.3. Prior to installation, the felt tube shall be saturated with resin at the Work site and stored at the temperature specified by the resin manufacturer. After installation, the felt tube shall be cured as specified by the resin manufacturer. The method of curing shall be submitted to the Engineer in accordance with 2-5.3.

500-4.6 "Brim Style" SLC Sealing System.

500.4.6.1 General. This subsection specifies a "brim style" SLC sealing system.

500.4.6.2 Sealing Material. The sealing material shall consist of a cured-in-place resin saturated fiberglass or felt material that provides a brim section in the mainline with the brim centered around the HC with a tube section extended in the HC. The length of the tube section in the HC and the size of the brim section in the mainline shall be as specified in the Special Provisions.

500.4.6.3 Installation and Curing. The fiberglass or felt material and tube shall be saturated with resin at the Work site or at the factory and stored at the temperature specified by the resin manufacturer. The resin saturated SLC sealing system shall be loaded on an applicator apparatus, attached to a robotic device and positioned in the mainline at the HC to be sealed. The robotic device shall be equipped with a CCTV camera which shall be used to align and center the SLC sealing system within the HC opening. The applicator apparatus shall include a bladder or an approved mechanical device of sufficient length in the mainline and HC such that the inflated bladder or approved mechanical device extends beyond the end of the SLC seal. The insertion pressure shall be adjusted to fully deploy the SLC sealing system in the HC and to hold the ends of the SLC seal against the pipe walls. The SLC sealing system shall produce a smooth transition between the SLC seal and the pipe walls without a ridge or gap between the SLC seal and the inner diameter of the mainline and HC. The insertion pressure shall be maintained for the duration of the curing process. Curing shall be as specified by the resin manufacturer. The method of curing shall be submitted to the Engineer in accordance with 2-5.3.

500-4.7 Measurement and Payment.

500-4.7.1 Measurement. SLC sealing systems shall be measured by "each".

500-4.7.2 Payment. Payment for SLC sealing systems will be made at the Contract Unit Price or lump sum price in the Bid for each SLC. The Contract Unit Price or lump sum price in the Bid shall include the installation of the SLC sealing system, surface preparation and repairs, preparation and tape submittal of all pre- and post-construction CCTV inspection, bypassing if required, and testing, unless otherwise specified in the Special Provisions.

APPENDIX A

SAMPLE ENABLING ORDINANCE

(For use by municipalities in adopting Standard Specifications For Public Works Construction by reference)

Ordinance No _________

An ordinance regulating materials of construction and their use in the erection, installation, alteration, repair, removal, conversion, demolition, and construction of public works improvements in the City (County) of ....; providing for the administration of contracts as well as permits issued in connection with such improvements; and repealing all ordinances and parts of ordinances in conflict therewith.

The People of the City (County) of..........................do ordain as follows:

Except as may otherwise be provided herein, the provisions of the 2009 edition of the "Standard Specifications for Public Works Construction," prepared and promulgated by Public Works Standards, Inc., a California Nonprofit Mutual Benefit Corporation, whose members are the American Public works Association, the Associated General Contractors of California, the Engineering Contractors' Association, the Southern California Contractors Association and BNI Publications, Inc., are adopted and applicable to all public works construction undertaken after the effective date of this ordinance.

REFERENCE DOCUMENTS

Following is a list of documents referred to in this book, arranged alphabetically.

AMERICAN FOREST AND PAPER ASSOCIATION. 1111 19th St., N.W. Suite 800, Washington, D.C. 20036.

AMERICAN NATIONAL STANDARDS INSTITUTE. (Formally the United States of American Standards Institute, also the American Standards Association). Published by the American National Standards Institute, 25 West 43rd St., 4th Floor, New York, New York 10036.

AMERICAN PETROLEUM INSTITUTE STANDARDS. Published by the American Petroleum Institute, 1220 L St., N.W. Washington, D.C. 20005.

AMERICAN SOCIETY FOR TESTING AND MATERIALS. Standards and Metric Practice Guide E380. Published by the American Society for Testing and Materials, 100 Barr Harbor Dr., Conshohocken, PA 19428-2959.

AMERICAN WATER WORKS ASSOCIATION STANDARDS (AWWA). Published by the American Water Works Association, 6666 W. Quincy Avenue, Denver, CO 80235.

AMERICAN WELDING SOCIETY (AWS) STANDARDS AND SPECIFICATIONS. Published by the American Welding Society, 550 N.W. LeJeune Road, Miami, FL 33126.

AMERICAN WOOD PRESERVERS ASSOCIATION (AWPA) BOOK OF STANDARDS. Published by the American Wood Preservers Association, 2345 Grand Blvd Ste 500, Kansas City, MO 64108-2625.

ASPHALT PROTECTIVE COATING FOR PIPELINES. Published by the Asphalt Institute, Research Park Dr., P.O. Box 14052, Lexington, KY 40512-4052.

FEDERAL SPECIFICATIONS. Available through the Federal Supply Service Bureau, Specification Section, 470 East L'Enfant Plaza S.W., Suite 8100, Washington, D.C. 20407.

MANUAL OF STANDARD PRACTICE. Published by the Concrete Reinforcing Steel Institute, 933 N. Plum Grove Rd., Schaumburg, IL 60173-4758.

MANUAL OF STEEL CONSTRUCTION (AISC). Published by the American Institute of Steel Construction, Incorporated, One East Wacker Drive, Suite 3100, Chicago, IL 60601-2001.

MANUAL OF WARNING SIGNS LIGHTS, AND DEVICES.

MATERIALS MANUAL, VOLUMES I, II, & III TESTING AND CONTROL PROCEDURES.

MILITARY SPECIFICATIONS. Available through the Superintendent of Documents, US Government Printing Office, Washington, DC 20402.

NATIONAL ELECTRIC CODE. Published by the National Fire Protection Association, One Batterymarch Park, Quincy, MA 02269.

NATIONAL ELECTRICAL MANUFACTURERS ASSOCIATION STANDARDS. Published by the National Electrical Manufacturers Association, 1300 N 17th St. Ste 1847, Rosslyn, VA 22209.

PLACING REINFORCING BARS. (Formerly Recommended Practice for Placing Reinforcing Bars). Published by the Concrete Reinforcing Steel Institute, 933 N. Plum Grove Road, Schaumburg, IL 60173-4758.

RULES FOR OVERHEAD ELECTRIC LINE CONSTRUCTION- GENERAL ORDER NO. 95. Published by the State of California, Public Utilities Commission, 505 Van Ness Ave, San Francisco, CA 94102.

SAFETY ORDERS-BASIC ELECTRICAL REGULATIONS, TITLE 24, CONSTRUCTION, GENERAL INDUSTRY, TUNNEL. Published by the Office of Procurement, Publications Section[1], P.O. Box 1015, North Highlands, California 95660.

SOFTWOOD PLYWOOD, CONSTRUCTION & INDUSTRIAL. (Formerly Douglas Fir Plywood, Construction & Industry). Published by the U.S. Department of Commerce, Business and Defense Services Administration, Office of Technical Services, Commodity Standards Division; Superintendent of Documents, U.S. Government Printing Office, Washington, D.C. 20402.

SPECIFICATIONS FOR THE DESIGN, FABRICATION AND ERECTION OF STRUCTURAL STEEL FOR BUILDINGS. Published by the American Institute of Steel Construction, One East Wacker Dr., Suite 3100, Chicago, IL 60601-2001.

SPECIFICATION FOR WELDED HIGHWAY AND RAILWAY BRIDGES. Published by the American Welding Society, 550 N.W. LeJeune Rd., Miami, FL 33126.

STANDARD GRADING RULES FOR WEST COAST LUMBER. (Formerly Standard Grading and Dressing Rules for West Coast Lumber). Published by the West Coast Lumber Inspection Bureau, P.O. Box 23145, Portland, OR 97281-3145.

STANDARD SPECIFICATIONS.

STANDARD SPECIFICATIONS FOR GRADES OF CALIFORNIA REDWOOD LUMBER. Published by the California Redwood Association, 405 Enfrente Dr., Suite 200, Novato, CA 94949.

STANDARD SPECIFICATIONS FOR HIGHWAYS MATERIALS AND METHODS OF SAMPLING AND TESTING. Published by the American Association of State Highway and Transportation Officials, PO Box 96716, Washington DC, 20090-6716.

STANDARDS FOR INSULATED POWER CABLE. Published by the Insulated Cable Engineers Association, P.O. Box 1568, Carrollton, GA 30112.

STANDARDS OF OPERATIONS OF TRUCK MIXERS AND AGITATORS. (National Ready Mixed Concrete Association). Published by the National Ready Mixed Concrete Association, 900 Spring Street, Silver Spring, MD 20910.

TRUCK MIXER & AGITATORS STANDARDS. (Truck Mixer Manufacturers Bureau). Published by the Truck Mixer Manufacturers Bureau, 900 Spring Street, Silver Spring, MD 20910.

UNDERWRITERS LABORATORY, INC. LISTINGS AND STANDARDS. Published by the Underwriters Laboratory, Inc., 333 Pfingsten Rd., Northbrook, IL 60062-2096.

U.S. CORPS OF ENGINEERS TEST METHODS. Available through the Corps of Engineers, U.S. Army, Waterways Experimental Stations, 3909 Halls Ferry Rd., Vicksburg, MS 39180-6199.

B

C

D

E

G

J

L

M

N

O

P

Index

Q

R

S

T

U

V